*Case Studies in
Atomic Collision Physics II*

Case Studies in Atomic Collision Physics II

edited by

E. W. McDANIEL
Georgia Institute of Technology, Atlanta, Georgia, U.S.A.

M. R. C. McDOWELL
Royal Holloway College, Englefield Green, Surrey, England

1972

NORTH-HOLLAND PUBLISHING COMPANY – AMSTERDAM · LONDON

© NORTH-HOLLAND PUBLISHING COMPANY – 1972

All rights reserved. No part of this publication may be reproduced, stored in a retrieval system or transmitted, in any form or by any means, electronic, mechanical, photocopying, recording, or otherwise without the prior permission of the copyright owner

Library of Congress Catalog Card Number 79-91451
North-Holland I.S.B.N. 0 7204 0225 5
American Elsevier I.S.B.N. 0 444 10118 7

Publishers:
NORTH-HOLLAND PUBLISHING COMPANY – AMSTERDAM
NORTH-HOLLAND PUBLISHING COMPANY, LTD. – LONDON

Sole distributors for the U.S.A. and Canada:
AMERICAN ELSEVIER PUBLISHING COMPANY, INC.
52 VANDERBILT AVENUE
NEW YORK, N.Y. 10017

PRINTED IN THE NETHERLANDS

PREFACE

This volume contains eight studies of specific areas of Atomic Collision Physics which we believe were in need of detailed treatment at this time. We have attempted to maintain a balance between experimental and theoretical viewpoints.

In particular we have asked colleagues whose work lies in Astrophysics to prepare studies of the role of atomic collision processes in astrophysical plasmas, and these studies form Chapters 4 and 5 of this volume. The remaining studies in this volume cover a wide range, but in each case give an authoritative view by leading workers of the present position and future possibilities of their field of interest.

The success of the first volume, and the interest shown in the preparation of the second have persuaded us of the need for a continuing effort in providing such studies. It has been decided in collaboration with Drs. W. H. Wimmers of North-Holland to continue publication in Journal form. A number of well known atomic physicists have agreed to act as advisory editors, and the first number of the new journal (Case Studies in Atomic Physics) will appear early in 1972. Each bimonthly issue will consist of a single study; the field has been widened to include all of atomic (and molecular) physics.

M. R. C. McDowell

E. W. McDaniel

CONTENTS

Chapter 1. THREE-BODY RECOMBINATION OF POSITIVE AND NEGATIVE IONS
M. R. Flannery

1-1. Introduction to Ionic Recombination	3
1-2. Historical Survey	4
A. The classical theory of Thomson, and related theories	6
B. Quasi-equilibrium and diffusion theories	9
C. Experimental methods	11
1-3. Quasi-Equilibrium Statistical Theory at Low Densities	13
A. Formula for the three-body recombination coefficient	14
B. Quasi-equilibrium distribution	16
1-4. Ions Recombining in Their Parent Gas	19
A. Energy-change rate coefficient	20
B. Calculation of the recombination coefficient	25
C. The non-thermal effect	28
1-5. General Third Body	29
A. General aspects of a collision	29
B. Formula for the rate coefficient	34
C. Reduced parameters and detailed balance	37
D. The rate coefficient for special cases	39
1-6. Calculation of the Recombination Coefficient	41
A. Differential scattering cross sections	42
B. Computations	48
C. Partial recombination coefficient	50
D. Quasi-equilibrium distribution function	53
E. Temperature and interaction	55
F. Mass effect	57
1-7. Theoretical and Experimental Three-Body Ionic Recombination Coefficient	62
1-8. Simple Treatment of Three-Body Ionic Recombination	65
A. Recombination as a Markov process	66
1. The Fokker-Planck equation	67
2. Integral equation for the steady-state distribution	67
3. Relationship with the diffusion theory	69

B. Recombination coefficient deduced from the diffusion, effective-gradient and modified effective-gradient methods 70
   1. The diffusion method 71
   2. The effective-gradient method 73
   3. Accuracy of diffusion and effective-gradient methods 75

1-9. THREE-BODY RECOMBINATION AT MODERATE AND HIGH GAS-DENSITIES . . 77
  A. Natanson's theory 78
  B. Modification to Natanson's theory 80
  C. Comparison with experiment 83

Appendix. THE THOMSON THEORY 85
Acknowledgments 88
References 88

CHAPTER 2. PRECISION MEASUREMENTS OF ELECTRON TRANSPORT COEFFICIENTS

M. T. ELFORD

2-1. INTRODUCTION 94

2-2. THEORY OF ELECTRON SWARMS 95
  A. The electron energy distribution function 95
    1. When only elastic scattering occurs 97
    2. When both elastic and inelastic scattering occurs 97
  B. Transport coefficients 98
    1. Electron drift velocity 98
    2. The diffusion coefficient 99
    3. The characteristic energy 100
    4. The magnetic deflection coefficient 101
  C. The derivation of cross sections from transport coefficients . . . 103
    1. When only elastic scattering occurs 103
    2. When both elastic and inelastic scattering occur 106

2-3. THE ELECTRON DRIFT VELOCITY 107
  A. Review of previous methods of measurement 107
    1. Oscillographic methods 107
    2. Methods using counters 108
    3. Electrical shutter methods 109
  B. The Bradbury-Nielson method 111
    1. Electrical shutter 111
    2. Uniform electric field and geometrical accuracy 112
    3. Electron source 114
    4. Gas temperature measurement and control 114
    5. Gas purity and vacuum requirements 116
    6. Pressure measurement 119
    7. The determination of the current maxima 119
    8. Sources of systematic error 121
    9. Correction for the effect of diffusion 124
    10. The linear variation of the drift velocity with gas number density 125
  C. Experimental data 126
    1. Presentation of data 126

	2. Analysis of error	129
	3. Results and discussion	129
2-4.	The Measurement of D_T/μ by the Townsend-Huxley Method	137
	A. Introduction	137
	1. Historical survey	137
	B. The solution of the diffusion equation and the current ratio formula	139
	1. The current ratio formula	139
	2. Effect of the spatial dependence of the electron energy distribution function	141
	C. Experimental method	141
	1. Diffusion chamber geometry	141
	2. Production of a uniform electric field	142
	3. The collector	145
	4. The current ratio	146
	5. Temperature measurement and control	147
	6. Electron source	147
	7. Gaseous impurities	147
	D. Sources of error	148
	1. Nonuniformity of the electric field	148
	2. Finite size source hole	150
	3. Incorrect axial alignment	150
	4. Presence of negative ions	150
	E. Experimental results	151
2-5.	Conclusion	155
Acknowledgements		156
References		156

Chapter 3. DIFFERENTIAL CROSS SECTIONS IN ELECTRON IMPACT IONIZATION

H. Ehrhardt, K. H. Hesselbacher, K. Jung, K. Willmann

3-1.	Introduction	161
3-2.	Kinematics and Notations	162
3-3.	Different Types of Cross Sections	163
3-4.	Choice of Target Gas and Collision Variables	166
3-5.	The Apparatus	167
	A. Description of the electron impact spectrometer	167
	B. Detection electronics and measurement procedures	172
	C. Intensities	176
	D. Background problems	178
	E. Stray fields	179
	F. Modes of operation	180
	G. Normalization problems	182
3-6.	Experimental Results	182
	A. Double differential cross sections	183
	B. The triple differential cross section	188
3-7.	Comparison with Theory	196

3-8. MEASUREMENTS CLOSE TO THE IONIZATION THRESHOLD 202
Acknowledgement 206
References 206

CHAPTER 4. INTERPRETATION OF SPECTRAL INTENSITIES FROM LABORATORY AND ASTROPHYSICAL PLASMAS
A. H. GABRIEL, CAROLE JORDAN

4-1. INTRODUCTION 211
4-2. THEORETICAL METHODS 212
   A. Statistical equilibrium for excited states 212
   B. Effects of metastable levels 214
   C. Collisional excitation 215
   D. Levels above the first ionization limit 219
   E. Line ratio measurement of electron temperature 222
   F. Line ratio measurement of electron density 223
4-3. COLLISION RATE EXPERIMENTS 224
   A. General methods 224
   B. Electron beam excitation 225
   C. Laboratory plasma experiments 225
   D. Théta-pinches 227
4-4. LITHIUM-LIKE IONS 229
   A. General description 229
   B. Energy levels 231
   C. Transition probabilities 231
   D. Collisional excitation rates 232
   E. Statistical equilibrium 234
   F. Line intensities in laboratory sources 235
   G. Solar line intensities 237
   H. Dielectronic satellite lines 241
4-5. BERYLLIUM-LIKE IONS 242
   A. General description 242
   B. Energy levels 243
   C. Transition probabilities 244
   D. Collisional excitation rates 244
   E. Equations of statistical equilibrium 249
   F. Laboratory plasmas 252
   G. Solar line intensities 255
   H. Low density objects, quasars, and planetary nebulae 264
   I. Recent developments 265
4-6. HELIUM-LIKE IONS 265
   A. General description 265
   B. Energy levels 267
   C. Transition probabilities 268
   D. Collisional excitation rates 270
   E. Low density plasmas 274
   F. High density plasmas 277
   G. Intermediate densities 281

H. Line ratios, N_e and T_e	283
I. Dielectronic satellite lines	286
J. Absolute intensities	286
K. Recent developments	287
Acknowledgements	288
References	288

Chapter 5. ATOMIC PROCESSES IN ASTROPHYSICAL PLASMAS

Valerie P. Myerscough, G. Peach

5-1. Introduction	295
5-2. Stellar Atmospheres	296
A. General theory of radiative transfer	296
B. The construction of model atmospheres	300
C. Line radiation	303
D. Departures from local thermodynamic equilibrium	305
5-3. The Continuous Spectrum	312
A. General formulae	312
B. Absorption by neutral atoms and positive ions	316
C. Photodetachment of negative ions	322
D. Coherent scattering	328
5-4. The Line Spectrum	329
A. Transition probabilities	329
B. The line profiles – general formulae	333
C. Pressure broadening of spectral lines	336
5-5. Collisional Excitation and Ionization	354
A. The quantum mechanical theory – general formulae and approximations	354
B. Classical and semi-classical approximations	361
5-6. Interpretation of Observations	366
A. Curve of growth analysis	366
B. The solar atmosphere	368
C. The early-type stars and planetary nebulae	373
D. The late-type stars; molecular absorption	379
E. Stars of non-solar composition	381
1. Weak helium line stars	381
2. Hydrogen deficient stars	383
3. Carbon stars	384
5-7. Problems Requiring Further Study	387
Acknowledgments	389
References	389

Chapter 6. POLARIZED ORBITAL APPROXIMATIONS

R. J. Drachman, A. Temkin

6-1. Introduction	401
6-2. The Basic Method and Notation	402

6-3. THE POLARIZED TARGET WAVE FUNCTION 404
    A. The adiabatic approximation 404
    B. Perturbation theory 405
    C. The one-electron target 406
    D. Variational-perturbation methods 414
    E. The exact static solution for the one-electron target 417
    F. The many-electron target; Sternheimer's approximation 421
6-4. THE TOTAL POLARIZED ORBITAL WAVE FUNCTION AND THE SCATTERING
    PROBLEM 423
    A. Introductory remarks 423
    B. The total wave function and non-variational methods 424
    C. Variational and variationally motivated methods 426
6-5. APPLICATIONS TO ELECTRON SCATTERING AND REACTIONS 433
    A. Scattering from one-electron targets 433
        1. Elastic scattering 433
        2. Inelastic scattering 440
    B. Scattering from helium 445
    C. Scattering from other (non-highly polarizable) atoms 448
    D. Scattering from highly polarizable atomic systems 451
    E. Electron–molecule scattering 454
    F. Photoionization, autoionization and ionization 456
6-6. APPLICATION TO POSITRON SCATTERING AND ANNIHILATION 462
    A. Positron–hydrogen scattering 463
    B. Positron–helium scattering and annihilation 469

References 477

CHAPTER 7. PHOTODETACHMENT: CROSS SECTIONS AND ELECTRON AFFINITIES

B. STEINER

7-1. INTRODUCTION 485
7-2. PHOTODETACHMENT CROSS SECTIONS 487
    A. Introduction 487
    B. Theory 490
    C. Experiment 491
    D. Double photon detachment 492
7-3. PHOTODETACHMENT THRESHOLDS 493
    A. Experimental energy determination 493
    B. Threshold behavior theory 494
        1. Atoms 494
        2. Molecules 495
    C. Comparison of experiment with theory 496
7-4. EXPERIMENTAL APPROACHES 497
    A. Plasma 497
    B. Beams 498
        1. Introduction 498

2. Classical experimental approach		498
a. Negative ion sources		498
b. Initial ion focusing		502
c. Mass analysis		502
d. Final ion focusing		504
e. Ion monitoring		504
f. Electron monitoring		505
g. Photon optics		511
h. Beam overlap		512
i. Instrument alignment		514
3. Detachment using tunable dye laser		514
4. Detachment using fixed laser and electron energy analysis		515
5. Double quantum detachment		517
7-5. ELECTRON AFFINITIES		517
A. Techniques		517
1. Theory		517
2. Empirical fitting		518
a. Ionization energy		519
b. First excitation energy		519
c. Binding energy per p electron		519
d. Configuration energy		519
e. Electron affinity horizontal analysis		520
f. Radius–vertical analysis		520
g. Correlation energy		520
3. Experiment		520
a. Photodetachment and radiative attachment		521
b. Photoionization		521
c. Dissociative electron attachment		522
d. Thermochemistry		523
e. Born–Haber cycle		525
f. Charge exchange		525
g. Solid spectroscopy		525
h. Field detachment		526
i. Electron scattering		526
B. Case study: S^-		526
C. A selected list of electron affinities		528
Acknowledgment		539
References		539

CHAPTER 8. THE ROLE OF METASTABLE PARTICLES IN COLLISION PROCESSES

R. D. RUNDEL, R. F. STEBBINGS

8-1. INTRODUCTION	549
8-2. PRODUCTION OF METASTABLES	551
A. Thermal metastables	551
1. Recoil effects	552
2. Production cross sections	555
B. Fast metastables	559

8-3. Detection and Identification . . . 561
  A. Collisions with surfaces . . . 561
  B. Collisions with gas atoms or molecules . . . 568
  C. Collisions with photons . . . 569
  D. Collisions with electrons . . . 570
  E. Interaction with d.c. electric and magnetic fields . . . 571
  F. Magnetic resonance techniques . . . 572

8-4. Chemiionization . . . 574
  A. General principles of thermal energy metastable collision experiments 576
  B. Absolute measurements . . . 577
    1. The afterglow technique . . . 578
    2. The beam technique . . . 583
    3. Results . . . 588
  C. Measurements of singlet–triplet cross section ratios . . . 593
  D. Models of chemiionization reactions . . . 596
  E. Theoretical treatments of chemiionization . . . 608
  F. Chemiionization involving highly excited atoms . . . 612

8-5. Ion Beam Studies . . . 613

References . . . 624

Author Index . . . 631

Subject Index . . . 647

CHAPTER 1

THREE-BODY RECOMBINATION OF POSITIVE AND NEGATIVE IONS

BY

M. R. FLANNERY*

Harvard College Observatory and
Smithsonian Astrophysical Observatory,
Cambridge, Massachusetts,
U.S.A.

---

* Now at School of Physics, Georgia Institute of Technology, Atlanta, Georgia, U.S.A.

Contents

	Page
1-1. Introduction to ionic recombination	3
1-2. Historical survey	4
A. The classical theory of Thomson and related theories	6
B. Quasi-equilibrium and diffusion theories	9
C. Experimental methods	11
1-3. Quasi-equilibrium statistical theory at low densities	13
A. Formula for the three-body recombination coefficient	14
B. Quasi-equilibrium distribution	16
1-4. Ions recombining in their parent gas	19
A. Energy-change rate coefficient	20
B. Calculation of the recombination coefficient	25
C. The non-thermal effect	28
1-5. General third body	29
A. General aspects of a collision	29
B. Formula for the rate coefficient	34
C. Reduced parameters and detailed balance	37
D. The rate coefficient for special cases	39
1-6. Calculation of the recombination coefficient	41
A. Differential scattering cross sections	42
B. Computations	48
C. Partial recombination coefficients	50
D. Quasi-equilibrium distribution function	53
E. Temperature and interaction	55
F. Mass effect	57
1-7. Theoretical and experimental three-body ionic recombination coefficients	62
1-8. Simple treatments of three-body ionic recombination	65
A. Recombination as a Markov process	66
1. The Fokker-Planck equation	67
2. Integral equation for the steady state distribution	67
3. Relationship with the diffusion theory	69
B. Recombination coefficient deduced from the diffusion, effective-gradient, and modified effective-gradient methods	70
1. The diffusion method	71
2. The effective-gradient method	73
3. Accuracy of diffusion and effective-gradient methods	75
1-9. Three-body recombination at moderate and high gas-densities	77
A. Natanson's theory	78
B. Modification of Natanson's theory	80
C. Comparison with experiment	83
Appendix: The Thomson theory	85
Acknowledgments	88
References	88

§ 1-1. Introduction to ionic recombination

Ionic recombination is a general term used to denote the various processes, other than diffusion, that contribute to the overall loss of ionization in a plasma containing positive and negative ions in a background neutral gas. The efficiency of such reactions naturally depends on the ability of each ion-pair to dispose of sufficient internal energy for charge-neutralization to proceed; the mechanism causing this energy-reduction characterizes the particular type of recombination process. *This excess energy of recombination*, in the case involving atomic ions, for example, is at least the difference between the ionization potential of the positive ion and the electron affinity of the negative ion. It can be removed (a) by photon-emission or (b) by causing excitation of the neutralized products or else (c) by being transferred to a third body. The *probability* of recombination by any of these processes is conveniently measured by means of a recombination coefficient α (cm^3 sec^{-1}) defined in terms of the rate of ion loss R by the relationship

$$R = \frac{dN_1}{dt} = \frac{dN_2}{dt} = -\alpha N_1 N_2 \text{ cm}^{-3} \text{ sec}^{-1} \qquad (1\text{-}1\text{-}1)$$

where N_1 and N_2 denote the positive and negative ion concentrations (cm^{-3}) respectively at time t such that $\alpha N_1 N_2$ is the number of recombination events occurring in unit volume per second.

In the *radiative recombination process* (a)

$$X^+ + Y^- \rightarrow XY + h\nu \qquad (1\text{-}1\text{-}2)$$

the two ions X^+ and Y^- approach each other along a potential curve of an excited molecular state (with dissociation products X^+ and Y^-) which spontaneously emits a photon of energy $h\nu$ to one of the vibrational levels of some lower electronic (or ground state) of the molecule XY, that can combine with the initial molecular ionic state. Since the kinetic energy of the colliding ions usually covers an extensive range a continuous spectrum (*the recombination continuum*) results. This process, which is the direct inverse of photodissociation into ionic products, is described by a recombination coefficient of the order of 10^{-14} cm^3 sec^{-1} which decreases with increase of ionic temperature and which is negligible compared with that ($\sim 10^{-7}$ cm^3 sec^{-1}) for the two-body mutual neutralization process (b)

$$X^+ + Y^- \rightarrow X^* + Y^* + \Delta E \qquad (1\text{-}1\text{-}3)$$

which occurs through the crossing of the molecular potential-energy curves describing the initial and final quasi-molecular states. Here, the required energy-release is distributed among the various internal modes of excitation (electronic, vibrational, rotational, dissociative) of the neutral atomic or molecular species, X and Y, formed when the positive ion captures an electron from the negative ion. Any further excess energy ΔE consistent with energy-conservation and excitation-probability serves to increase the kinetic energy of relative motion of the neutralized products. With the introduction of a background neutral gas Z, the binary-recombination (1-1-3) becomes overshadowed with increasing gas pressure by the following *three-body ionic recombination process* (c)

$$X^+ + Y^- + Z \to [XY] + Z \tag{1-1-4}$$

where the third body Z absorbs the excess energy of recombination either as increased kinetic energy or as internal excitation and where the brackets denote that the product molecule may become unbound by dissociating into neutrals. The resulting effective recombination coefficient depends, of course, on the number of density N_3 of the third bodies Z and can have magnitudes 10^{-6} cm^3 sec^{-1} even at densities corresponding to low pressures of a few torr at room temperature. Thus, except in the limit of vanishing pressure when (1-1-3) dominates, charge-neutrality of a weakly ionized plasma with heavy-particle constituents is primarily controlled by three-body ionic recombination which will therefore be the subject of this review.

§ 1-2. Historical survey

The problem of ionic recombination was first attacked in 1903 by Langevin [1]. He assumed that the ions denoted by 1 and 2 with charges $\pm e$ separated by r drift towards one another under the influence of the mutually attractive field of intensity e/r^2 in a medium composed of the neutral gas particles 3. The relative speed of approach v was therefore determined by the mobility K_{i3} of each ion in the gas to give

$$v = (K_{13} + K_{23})e/r^2. \tag{1-2-1}$$

Recombination between the ions was assumed to occur once one ion crossed an imaginary sphere centered about the other ion. The recombination coefficient is thus

$$\alpha_L = 4\pi e(K_{13} + K_{23}) \tag{1-2-2}$$

which is independent of the sphere's characteristics.

The mobility, and hence α_L, varies inversely with the neutral-gas density and is relatively insensitive at constant density to temperature change, predictions that are qualitatively consistent with Langevin's measurements, although (1-2-2) yielded considerably higher magnitudes. With decrease in density, however, his and other experimental data became so inaccurate and contradictory that no firm conclusion was possible.

The Langevin treatment above, with the subsequent and arbitrary introduction to the right-hand side of (1-2-2) of a fractional multiplier designed to express the probability that a given encounter between the positive and negative ions indeed results in neutralization, was therefore accepted as valid for all densities until 1924. Then Thomson [2] pointed out that, at low densities the basic mechanism responsible for energy-reduction of ion-pairs was provided through individual collisions of each ion with the neutral-gas atoms. On this basis, he developed a theory which predicted that the recombination coefficient increases linearly with density in the low-density region and approaches a saturation value at densities corresponding to atmospheric pressure thus apparently contradicting the fundamental postulates of the Langevin theory. The saturation effect was, however, unrealistic and in 1932, Harper [3] reasoned that, over a large pressure range above one atmosphere, diffusion dominates the recombination. For pressures greater than 100 atmospheres the motion of the ions was inhibited by the dense neutral gas and the purely attractive-drift assumption of Langevin was valid. From standard diffusion theory, Harper however reproduced eq.(1-2-2) for α, a coincidence which is remarkable since the mechanisms advocated by Langevin and Harper are, in principle, different and are valid over mutually exclusive regions.

The Thomson treatment received corroboration in 1938 from the low-density measurements of Gardner [4] and from the independent data of Sayers [5]. Sayers examined a wider density range and noted that α increased according to the Thomson model, reached a maximum and thereafter followed a decrease consistent with the Langevin-Harper description. The prescription of adopting the theories of Thomson for low-densities and of Langevin and Harper for high densities was invoked up until 1959 when Natanson [6] successfully bridged the gap with a theory that reproduced both density limits. Since then, however, serious questions were raised regarding the validity of the basic assumptions of Thomson and so other approaches were sought.

Up to the present, theoretical treatments of the three-body ionic recombination process (1-1-4) have essentially followed one of three independent descriptions, designed for the limit of low neutral densities, all of which are conceptually and formally quite different. These methods are

(1) the classical theory initially proposed by Thomson [2] in 1924,

(2) the diffusion method originally devised by Pitaevskii [7] in 1962 in connection with electronic recombination, and applied to ionic recombination in 1968 by Landon and Keck [8] and by Mahan [9], and

(3) the effectively exact statistical treatments advanced by Bates and and Moffett [10] in 1966 for ions recombining in their parent gas and by Bates and Flannery [11] in 1968 for recombination between arbitrary ionic species in the presence of a general third body.

The connection between these three low-density approaches, the first two of which are simple and approximate, is far from obvious. Comparison between them will be the subject of future discussion. In this section, we will briefly outline these treatments, making particular reference to the basic assumptions involved.

For intermediate and high gas densities, Bates and Flannery [12] in 1969 modified the expression of Natanson [6] so as to bring it into accord with their low-density predictions. This procedure yielded satisfactory agreement with experiment.

Prior to 1966, effort was directed mainly towards elaborating the simple theory of Thomson by refining the underlying mathematical analysis and by removing certain simplifying assumptions, while retaining, however, the basic artifice of a trapping radius introduced by Thomson, inside which recombination proceeds. We will limit ourselves only to a brief discussion of the important work of Thomson which also serves as a useful basis for comparison with the more detailed treatments. We will not dwell on the individual aspects of the theory since a detailed and clear account, together with the modifications proposed by Natanson [6] in 1959, may be found in the book by McDaniel [13].

A. The Classical Theory of Thomson, and Related Theories

Thomson [2] originally pointed out that recombination proceeds when collisions with the neutral gas atoms reduce the kinetic energy of relative motion of the ion-pairs to such an extent that their total internal energy becomes negative so that their freedom to separate into the ionic constituents is lost. Eventual electron-transfer between the positive and negative ions trapped within the resulting closed orbits completes the recombination. Thomson assumed that *all* the kinetic energy gained from the attractive Coulombic field produced as the ions of masses M_1 and M_2, and charges

$\pm e$, approach one another is removed by the third body of mass M_3 so that the ion-pairs are reduced to their initial state of thermal equilibrium with the background gas at temperature θ. The artificial concept of a fictitious trapping distance r_T was then introduced so that *all* ion-neutral collisions occurring within the critical sphere of radius r_T centered about either ion leads to recombination, r_T being determined by equating the magnitude of the potential energy with the mean energy to yield

$$r_T = e^2/\beta k\theta, \quad \beta = \tfrac{3}{2} \tag{1-2-3}$$

where k is Boltzmann's constant. The total probability of such collisional events can be written as [14]

$$w = w_1 + w_2 \tag{1-2-4}$$

where

$$w_i = \tfrac{4}{3} r_T/\lambda_i, \quad i = 1, 2 \tag{1-2-5}$$

are the respective probabilities that ions 1 and 2 experience a collision with the neutral particle 3 while inside the critical sphere centered at the other ion, λ_i being the mean free paths of these ions. The deduced recombination coefficient in the low-density region is then

$$\alpha_T = \pi r_T^2 w (v_1^2 + v_2^2)^{\frac{1}{2}} \tag{1-2-6}$$

with each ion-neutral relative velocity taken as the root-mean-square speed v_i corresponding to the temperature θ,

$$v_i = (3k\theta/M_i)^{\frac{1}{2}}, \quad i = 1, 2 \tag{1-2-7}$$

to give

$$\alpha_T = \frac{4\pi e^6}{\beta^3 (k\theta)^{\frac{5}{2}}} \left(\frac{1}{3M_{12}}\right)^{\frac{1}{2}} \left(\frac{1}{\lambda_1} + \frac{1}{\lambda_2}\right) \tag{1-2-8}$$

where M_{ij} denotes the reduced mass of particles i and j.

Each ionic mean free path λ_i may be conveniently determined from the knowledge of the diffusion cross section Q_{i3}, or equivalently, from experimental mobility-measurements K_{i3} (in units of cm^2/statvolt-sec) for ions moving in a neutral gas with number density N_3. Since

$$\lambda_i = \frac{1}{N_3 Q_{i3}} = \frac{16}{3e} \left(\frac{M_{i3} k\theta}{2\pi}\right)^{\frac{1}{2}} K_{i3} \tag{1-2-9}$$

we observe that α_T increases linearly with gas-density, and varies with temperature as θ^{-3} and has mass dependence $M_{12}^{-\frac{1}{2}}(M_{13}^{-\frac{1}{2}} + M_{23}^{-\frac{1}{2}})$. The formula (1-2-8) yields results correct to better than an order of magnitude when compared with the earlier experiments of recombination in air, for which it is assumed that O_2^+ and O_2^- are the main ionic species recombining in O_2 (see, however, § 1-7).

However, this simple model of Thomson is clearly inadequate in several respects: (1) the assumed net rate coefficient for collisions leading to free-bound transitions in the ion-pair is probably too high since the electron-transfer time for neutralization is likely to be much longer than the ion-neutral collision time, thus providing the possibility that subsequent bound-free "dissociative" collisions occur; (2) the role of bound-bound transitions is ignored and is presumably important; (3) simple averages (appropriate at best to equal-mass reactants) were taken over the collision parameters and (4) recombination could well take place between ions with separation larger than r_T.

The error resulting from these omissions is very difficult to assess without resort to a more detailed treatment. See the appendix where the approximations inherent in Thomson's treatment are discussed in the light of modern terminology.

In an effort to improve the kinematics of the ion-neutral encounter within the classical framework of Thomson, several investigators have proposed modifications which essentially assign various values to the parameter β assumed to be temperature independent. For example, values of 6, $\frac{12}{5}$ and 4 for β have been suggested respectively by Loeb and Marshall [15], by Natanson [6], and by Brueckner [16] who also makes some allowance for the effect of dissociative collisions. In addition, the Coulombic deflection of the ions has been acknowledged by multiplying the right-hand side of (1-2-8) by

$$\mathscr{C} = 1 + \beta/c \qquad (1\text{-}2\text{-}10)$$

in which Natanson and Brueckner take c to be 1 and $\frac{3}{2}$ respectively. Parks [17] has recently improved the mathematical analysis of the Thomson theory and generalized it to the case involving unequal mass particles. All the above refinements cause considerable variation in the recombination coefficient. However, the reality of modifications which leave the basic artifice of a trapping distance unaltered is very uncertain.

A quite different mechanism involving the formation of positive ion-neutral complexes which subsequently charge-exchange with the negative ions to form neutralized products has been proposed by Fueno, Eyring and

Ree [18] but Mahan and Person [19] have put forward compelling arguments against its acceptance.

In 1965, Feibelman [20] avoided the introduction of a capture radius by using the Monte Carlo technique to follow the lives of each ion-pair and considered the requirement of stability of the recombined ion-pair against further collisions with the neutral gas. He assumed that the ions and neutral interact as hard spheres and that recombination ensues when further collisions with the neutral body result in a more tightly bound ion-pair. If these collisions tend to weaken the binding, then recombination will not occur. The injection of this simple but arbitrary stability-criterion provides, however, only a partial account of the effects of subsequent collisions. Nonetheless, Feibelman's interesting study of ion-recombination verified previous doubt and speculation by demonstrating the lack of validity of the capture radius concept; a significant contribution to the net recombination rate results from ions having separation greater than r_T and not all collisions between neutrals and ion-pairs with separation less than r_T lead to deactivation. Moreover, once deactivated, an ion-pair may frequently be redissociated by subsequent collisions, and the numerical calculations of Feibelman indeed showed that the fate of an ion-pair was impossible to determine until it has survived at least in the order of ten collisions. His investigation strongly suggests that the recombination process should not be described in terms of a single deactivating collision. Rather, it tends to substantiate the approach adopted by Bates, Flannery and Moffett who reasoned that a quasi-equilibrium of bound ion-pairs was first established and that the subsequent course of the recombination was determined by the net downflow of ion-pairs past some negative energy level. Feibelman's study also suggested that Pitaevskii's concept that recombination be regarded as collisionally induced "diffusion" of the ion-pairs in energy-space was feasible and merited investigation.

B. QUASI-EQUILIBRIUM AND DIFFUSION THEORIES

Because of the inadequacy of the Thomson model, a formally quite different quasi-equilibrium statistical theory was developed by Bates and Moffett [10], who treated the special case of ions in their parent gas (with symmetrical resonance charge-transfer collisions controlling the course of events), and by Bates and Flannery [11], who treated the more interesting general case in which the ion-neutral interaction is of the Langevin type (hard-sphere repulsion, polarization attraction). This theory involves the binary rate coefficient describing how collisions with the neutral atoms change the internal energy of the ion-pairs. The subsequent analysis exploits the fact that a quasi-equilibrium of ion-pairs is quickly established in which the rate at which bound ion-pairs are formed and destroyed is very much greater than the rate at which their number density changes (compare with Bates, Kingston and McWhirter [21] for electronic recombination). This quasi-equilibrium distribution of ion-pairs amongst their bound states is then determined by assuming that all elements of phase space, which are accessible at a particular energy, have equal populations. This assumption requires the mean time between collisions λ_i/v_i (determined, for example, from (1-2-7) and (1-2-9)) to be long compared with the orbiting times ($\sim 2\pi r_T/v_i$) of the ion-pairs, and is, of course, valid at low densities since λ_i is then very large and $\gg r_T$ ($\sim 10^{-6}$ cm at room temperature). The net rate of downflow of ion-pairs past any fixed negative energy level is the difference between the number per second which are de-activated to below this fixed energy level and the number per second that are excited past this given level. Hence with this knowledge the rate of recombination can be calculated.

The implementation of the above prescription, however, necessitated extremely lengthy computational work, the results of which are fortunately now available for a wide range of masses, interaction-strengths and temperatures. The predictions show satisfactory agreement with the rather scarce experimental data. The satisfactory agreement is to be expected since the only significant physical approximation invoked is the approximation of taking the motion of the center of mass of the ion-pairs to be thermal. Bates, Hays and Sprevak [22] have recently examined this approximation and have verified that it introduces little error as anticipated originally.

Other treatments requiring less computational work have been suggested by Pitaevskii [7], by Landon and Keck [8], and by Mahan [9], but have met with limited success [23]. These approaches regarded the recombination process at sufficiently low temperatures as diffusion in energy space with the free ion-pairs *gradually* losing internal energy to the neutral particles, eventually proceeding through an imaginary permeable membrane located at the dissociation limit, and then down the negative energy ladder to some energy level at which the recombination becomes stabilized by, for example, charge-transfer. A simple formula for the recombination coefficient can thus be derived from the Fokker-Planck equation [7, 8]. The main defects of this model are the assumptions that the negative energy levels of each bound ion-pair form a continuum, and that the energy-transferred in the collisions is small ($\ll k\theta$).

In the following sections, we will develop the quasi-equilibrium model in detail and by comparison, examine critically the approximate diffusion-methods. Finally, we will consider the implications involved when the neutral density is raised from the low-density limit, and seek a theory valid for all densities. However, a few remarks on the current status of experimental measurements are essential.

C. EXPERIMENTAL METHODS

Early measurements of the recombination of ions in air have already been reviewed by Thomson and Thomson [24] in 1928, by Loeb [25] in 1939, by McDaniel [13] in 1964 and by Sayers [26] in 1965. Recombination coefficients of about 1.6×10^{-6} cm^3 sec^{-1} for air at STP were observed. Dependence of the measured α on the ion age was suggested by Rümelin [27] in 1908 and 1914, by Plimpton [28] in 1913, and was confirmed by Marshall [29] in 1929 and others (cf. Loeb [25]). High vacuum techniques for the control of the composition of the background gas were later introduced by Gardner [4] and by Sayers [5]. Because of this refinement, Gardner's value of 2×10^{-6} cm^3 sec^{-1} for oxygen and Sayer's value of 2.3×10^{-6} cm^3 sec^{-1} for air, at one atmosphere, have received some acceptance as the correct recombination coefficients to be employed in dosimetry measurements.

In most applications of dosimetry ion chambers use ordinary laboratory air as the filling gas. Air which has not been carefully dried and filtered, and has not had any organic contaminants removed, does contain very different species of ions from purified air. Moreover, the formation of ion-neutral complexes which change as the gas-density is varied, inhibits the direct measurement of recombination between definite ionic species.

Therefore, in 1967, McGowan [30] attempted, in a most careful series of experiments, to examine the extent of variations of α in laboratory air. He obtained recombination coefficients for ions formed in oxygen, air, nitrogen with 1 % oxygen, and nitrogen with 0.001 % oxygen, at pressures between 50 and 800 torr. At one atmosphere and 25 °C, the observed values of α for oxygen (2×10^{-6} cm^3 sec^{-1}) and for air (2.2×10^{-6} cm^3 sec^{-1}) agree with those of Gardner and Sayers. McGowan also demonstrated that air not passed through cold traps yields considerably lower recombination coefficients which arise because the attachment of water molecules and organic vapors to the ions produce heavier ionic mass which consequently reduces α. This possibly explains the low rate measured by Ebert, Booz and Koepp [31] in 1964.

Apart from the introduction of devices designed to correct for the ion-distribution and diffusion losses, and of filters to obtain pure gas samples, present-day methods are similar to those employed in the pioneering days of recombination measurements. These methods are based on the following general relation for the rate of increase of ion-density in a radiation field when recombination is the sole ion-loss mechanism

$$dN_1/dt = dN_2/dt = Q - \alpha N_1 N_2, \tag{1-2-11}$$

N_1 and N_2 being the positive and negative ion-concentrations assumed to be uniform over the volume under observation and Q being the rate of ion-pair production per unit volume.

(a) In the *Pulsed-irradiation method*, the ionizing source is removed so that with N_1 and N_2 each equal to N,

$$\alpha = d(N^{-1})/dt. \tag{1-2-12}$$

Short pulses of radiation produce ions in the gas. The time during which recombination occurs (the delay time) was varied by changing the time interval between the ion-source pulse and the application of an electric field designed to collect the ions. From the delay time, and the charge due to the collected ions and to the ionized volume, the N^{-1} versus t plot yields α directly in accordance with (1-2-12).

(b) In the *Continuous-irradiation method*, equilibrium is established between the rate of ion production and the rate of recombination so that, with N_1 and N_2 each equal to N_E, the equilibrium ion density, (1-2-11) becomes

$$\alpha = Q/N_E^2. \tag{1-2-13}$$

A voltage large enough to produce saturation current is applied periodically to the ion chamber. The time between voltage pulses is kept long enough for attainment of the equilibrium ion density, while the length of the pulses are varied. As the electric field is applied, a current due to the equilibrium ions first reaches the collector electrode followed by a smaller ion current determined by the rate of production and the ionized volume. By plotting the collected number of ions against the voltage pulse length, the equilibrium ions can be separated from those being produced continuously and N_E and Q can be calculated to yield α.

McGowan has shown that measurements using the above two techniques were in good agreement. His reports [30] provide further details of the experiment and of the methods used to correct for ion distribution and diffusion losses in plane parallel ion chambers. Mahan and his associates [32] have carried out a series of experiments for various ions recombining in different neutral gases. The situation of all experiments performed to the present day is complicated by the tendency of the ions to form different complexes as the neutral-number density is raised so that reasonable guesses have to be made as to their identity. Data taken from ion-mobility measurements [33] based on the use of drift tube mass spectrometers are suggestive but they do not extend to pressures beyond 10 torr. At higher pressures, particularly in static systems, the ions may all be clustered to fairly high-order, there may be several different species of ions present of each sign, and impurity ions may be dominant (see, for example, Snuggs et al. [33]).

Thus, the lack of knowledge of the ionic species at given densities and the build-up of impurity-ions appear to be the major sources of difficulty encountered in present-day experiments on ionic recombination. A possible solution to these obstacles is for careful measurements to be performed with ultra-high vacuum systems, through which ultra-pure gas is circulated to prevent the build-up of impurities and for careful mass and time analysis of the ions arriving at the walls, a procedure similar to that adopted by McDaniel [33] and others in mobility determinations. Some method of volumetric sampling of the ions arriving at the walls would be even better. There is less known with any great certainty about the experimental side of ion-ion recombination than is known about almost any other subject in atomic collisions.

§ 1-3. Quasi-equilibrium statistical theory at low densities

The reduction in the internal energy of the ion-pairs in the recombination process

$$X^+ + Y^- + Z \rightarrow [XY] + Z \tag{1-3-1}$$

is accomplished through collisions with the neutral atoms Z. The collisional aspects of this energy-loss and the formulation of the subsequent course of the deactivation with time provide the two ingredients essential for a realistic theoretical description of the recombination coefficient.

A full quantum-mechanical treatment of this collision would demand, in principle, a detailed account of the molecular energy surfaces formed by the three particles, and even with this knowledge, the ensuing calculation would then be prohibitively difficult. Resort must therefore be made to semi-quantal techniques which involve a basic framework of classical mechanics together with a realistic injection of quantum ideas, a procedure that represents a powerful and very adequate approach for heavy-particle collisions.

Since the ion-ion Coulombic interaction is stronger and has longer range than the ion-atom interaction, the three-body collision can be viewed as two binary encounters taking place between the atom and either ion. Each binary impact changes the internal energy of the ion-pair. Assuming a knowledge of this energy-change rate coefficient $K(E_i, E_f)\, dE_f$ describing the rate of encounters between an atom and an ion-pair in which the internal energy of the ion-pair is changed from a given E_i to between E_f and $E_f + dE_f$, we will now develop an expression for the recombination coefficient α, and later describe an elegant theory from which $K(E_i, E_f)$ can be calculated.

A. Formula for the Three-Body Recombination Coefficient

Let the number densities of free positive and negative ions, and third bodies be $N(X^+)$, $N(Y^-)$ and $N(Z)$ respectively. The collisional rate coefficient is such that of the total number of ion pairs $n(E)\, dE$, initially possessing internal energy between E and $E + dE$, a number $n(E)\, dE\, N(Z)\, K(E, E_f) \times dE_f$ become formed by collisions with the gas atoms in unit volume per second in the energy interval dE_f about E_f. It is convenient, for subsequent calculations to normalize the internal energies to the mean thermal energy $k\theta$ of the gas at temperature θ by putting

$$\lambda = -E_i/k\theta, \quad \mu = -E_f/k\theta \tag{1-3-2}$$

such that λ is positive for ion-pairs in various stages of binding ($E_i < 0$) and negative for free ion-pairs ($E_i > 0$). Write

$$K(E_i, E_f)|dE_f| = \mathcal{K}(\lambda, \mu)\, d\mu. \tag{1-3-3}$$

Normalize also the number densities $n(E_i)\, dE_i$ of ion-pairs with energy E_i when recombination is proceeding to the number densities $n_T(E_i)\, dE_i$ where there is thermodynamic equilibrium between the bound and free ion-pairs by defining

$$n(E_i)\, dE_i = \eta(\lambda)\, d\lambda, \quad n_T(E_i)\, dE_i = \eta_T(\lambda)\, d\lambda \tag{1-3-4}$$

by introducing the distribution function

$$\rho(\lambda) = \eta(\lambda)/\eta_T(\lambda) \leq 1$$
$$= \begin{cases} 1, & \lambda < 0 \\ 0, & \lambda > \varepsilon \end{cases} \tag{1-3-5}$$

where

$$\varepsilon = -E_s/k\theta \tag{1-3-6}$$

with $|E_s|$ being the binding energy of the ion-pair at which the recombination may be regarded as being stabilized by, for example, the charge-neutralization process

$$[X^+ + Y^-] \to X + Y \tag{1-3-7}$$

where the brackets indicate that the ion-pair is bound. The probability of such ion-loss by charge-transfer is an increasing function of the binding energy of the ion-pair and its effect on the recombination coefficient may be assessed by supposing that the rate of loss of ion-pairs with binding energy greater than $|E_s|$ is so fast that collisions involving them may be neglected. The presence of this sink is necessary for the occurrence of recombination and is therefore acknowledged by assigning the value of zero to $\rho(\lambda)$ in (1-3-5). For negative λ, corresponding to free ions, $\rho(\lambda)$ is of course unity.

The recombination rate $\alpha\, N(X^+)\, N(Y^-)$ is represented by the difference of the upflow of ion-pairs past some arbitrary (negative) energy level $E(= -\nu k\theta)$, located between zero and E_s, from the downflow past this level. The upflow is given by (see Fig. 1-3-1 for diagrammatic energy levels)

$$R\!\uparrow = N(Z) \int_{E_c}^{E} n(E_i)\,dE_i \int_{E}^{\infty} K(E_i, E_f)\,dE_f \tag{1-3-8}$$

where $-E_c (= \omega k\theta)$ is the greatest binding energy an ion-pair may have (even transitorily) while the downflow is

$$R\!\downarrow = N(Z) \int_{E}^{\infty} n(E_f)\,dE_f \int_{E_c}^{E} K(E_f, E_i)\,dE_i \tag{1-3-9}$$

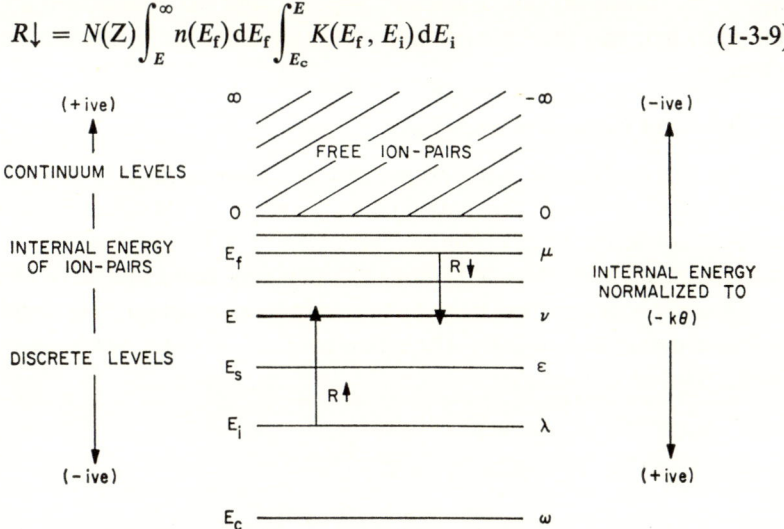

Fig. 1-3-1. Energy-level diagram of ion-pairs.

such that, with the aid of (1-3-2)–(1-3-4), the recombination rate is simply

$$\alpha N(X^+) N(Y^-) = R\downarrow - R\uparrow$$

$$= N(Z) \int_v^\omega d\lambda \int_{-\infty}^v \{\eta(\mu)\mathcal{K}(\mu, \lambda) - \eta(\lambda)\mathcal{K}(\lambda, \mu)\} d\mu. \quad (1\text{-}3\text{-}10)$$

In thermodynamic (Saha-Boltzmann) equilibrium the principle of detailed balance is satisfied between the forward and reverse rates of the transition between energy levels E_i and E_f in the form

$$[n_T(E_i) dE_i][K(E_i, E_f) dE_f] = [n_T(E_f) dE_f][K(E_f, E_i) dE_i] \quad (1\text{-}3\text{-}11)$$

or alternatively, in normalized energy-space,

$$\eta_T(\lambda) \mathcal{K}(\lambda, \mu) = \eta_T(\mu) \mathcal{K}(\mu, \lambda) \quad (1\text{-}3\text{-}12)$$

such that (1-3-10) can be rewritten in terms of the distribution function $\rho(\lambda)$ as

$$\alpha N(X^+) N(Y^-) = N(Z) \int_v^\omega \eta_T(\lambda) d\lambda \int_{-\infty}^v [\rho(\mu) - \rho(\lambda)] \mathcal{K}(\lambda, \mu) d\mu. \quad (1\text{-}3\text{-}13)$$

Thus, from a knowledge of the thermodynamic equilibrium populations $\eta_T(\lambda)$, the collisional rate coefficient $\mathcal{K}(\lambda, \mu)$, and the distribution $\rho(\lambda)$ of ion-pairs amongst their bound states, the recombination coefficient can be obtained.

B. Quasi-equilibrium distribution

In thermodynamic equilibrium, the distribution (in number-density) of ion-pairs amongst the bound-states is given by Maxwell-Boltzmann statistics while the ratio of the numbers $N(X^+) N(Y^-)$ of free ions to the total population N of bound ion-pairs (including those that have charge-exchanged) is determined from Saha's ionization equation [34], valid for collision-dominated regions. The elimination of N yields the Saha-Boltzmann relation in the form

$$\frac{n_T(E_i) dE_i}{N(X^+) N(Y^-)} = \frac{h^3 \exp(-E_i/k\theta)}{2(2\pi M_{12} k\theta)^{3/2}} g(E_i) dE_i \quad (1\text{-}3\text{-}14)$$

where h is Planck's constant and $g(E_i) dE_i$ denotes the statistical weight $2p^2 dp$ of the discrete levels p (assumed hydrogenic) in the energy-range dE_i

around E_i. The energies of these levels, appropriate for a Coulomb field, are of course

$$E_i = -(2\pi^2 M_{12} e^4/h^2 p^2) \qquad (1\text{-}3\text{-}15)$$

and hence

$$\frac{n_T(E_i) dE_i}{N(X^+) N(Y^-)} = \frac{\pi^{\frac{3}{2}} e^6 \exp(-E_i/k\theta)}{2(k\theta)^{\frac{3}{2}} |E_i|^{\frac{5}{2}}} dE_i \qquad (1\text{-}3\text{-}16)$$

so that, by using (1-3-2) and (1-3-4)

$$\frac{\eta_T(\lambda) d\lambda}{N(X^+) N(Y^-)} = \frac{\pi^{\frac{3}{2}} e^6}{2(k\theta)^3} \lambda^{-\frac{5}{2}} \exp(\lambda) d\lambda$$

$$= 1.3 \times 10^{-8} \theta^{-3} \lambda^{-\frac{5}{2}} \exp(\lambda) d\lambda, \quad \lambda \text{ positive} \qquad (1\text{-}3\text{-}17)$$

which is to be inserted in (1-3-13) and which determines the populations of the bound ion-pairs when in thermodynamic equilibrium with the free ions, under the assumption that there is a pure Coulomb interaction between the ions.

When recombination is proceeding, however, we do not have thermodynamic equilibrium because of the charge-transfer sink and the net rate of growth of the ion-pairs $\eta(\lambda)$ in level λ at time t is described by

$$\frac{\partial \eta(\lambda)}{\partial t} = [A(\lambda) - B(\lambda)\eta(\lambda)] N(Z) \qquad (1\text{-}3\text{-}18)$$

where

$$A(\lambda) = \int_{-\infty}^{\varepsilon} \eta(\mu) \mathscr{K}(\mu, \lambda) d\mu \qquad (1\text{-}3\text{-}19)$$

represents all collisional events that tend to increase the number density of level λ and

$$B(\lambda) = \int_{-\infty}^{\omega} \mathscr{K}(\lambda, \mu) d\mu \qquad (1\text{-}3\text{-}20)$$

corresponds to those collisions depleting that level.

Suppose a steady state is attained for all levels (when $\partial\eta(\lambda)/\partial t$ equals zero) and that at time t equals zero, we cause a small perturbation which affects *only* the equilibrium population $\eta_T(\lambda_i)$ of a given level λ_i by increasing it by $\eta_D(\lambda_i)$. This disturbance decays exponentially according to

$$\eta(\lambda_i) = \eta_T(\lambda_i) + \eta_D(\lambda_i) \exp[-B(\lambda_i) N(Z) t] \qquad (1\text{-}3\text{-}21)$$

and the relaxation time for the number density $\eta(\lambda_i)$ to return to its previous equilibrium value $\eta_T(\lambda_i)$ is thus of the order

$$\tau(\lambda_i) = \left[N(Z) \int_{-\infty}^{\omega} \mathscr{K}(\lambda, \mu) \, d\mu \right]^{-1}. \tag{1-3-22}$$

Two important conclusions regarding the course of the recombination follow directly from equations (1-3-17) and (1-3-22):

(a) Because the collisional rate coefficients $\mathscr{K}(\lambda, \mu)$ for the transitions between highly excited levels are so much greater than those involving the lower levels, on account of the relatively smaller energy separations, the highly excited states assume equilibrium more readily than the lowest states. Moreover, since the excited states can relax by both excitation and de-excitation processes (while only excitations contribute to the ground-state relaxation), the attainment of equilibrium for the excited levels is considerably hastened.

(b) It is apparent from (1-3-17) that unless the number densities of free ions are unusually high

$$N(X^+), N(Y^-) \text{ and } N(X^+ - Y^-|0) \gg N(X^+ - Y^-|*) \tag{1-3-23}$$

where $N(X^+ - Y^-|0)$ is the number-density of normal (ground-state) ion-pairs, including those that have dissociated into free atoms by charge-transfer and where

$$N(X^+ - Y^-|*) = \int_{[\lambda]} \eta(\lambda) \, d\lambda \tag{1-3-24}$$

is the total number density of ion-pairs in the range $[\lambda]$ of excited states of importance in the recombination process. We note that it is essentially the charge-transfer mechanism that prevents the attainment of Saha-Boltzmann equilibrium, and which provides the channel leading to recombination.

The arguments above demonstrate that, once recombination begins, the system is disturbed and a quasi-equilibrium of excited states is established almost instantaneously (see (a)) without the number density of the positive and negative ions being appreciably altered (since from (1-3-23) $N(X^+ - Y^-|*)$ is very small by comparison); and that, thereafter, according to (a), the rates per unit volume at which the excited system of interest in $[\lambda]$ are produced and destroyed are very much greater than the rates at which the number densities of these rare systems change as the ionization changes.

The integral equation for this quasi-equilibrium distribution of excited levels is simply obtained by setting the left-hand side of (1-3-18) equal to zero and by using detailed balance to give

$$\rho(\lambda)\int_{-\infty}^{\omega}\mathscr{K}(\lambda,\mu)\,d\mu = \int_{-\infty}^{\varepsilon}\rho(\mu)\mathscr{K}(\lambda,\mu)\,d\mu. \qquad (1\text{-}3\text{-}25)$$

While the recombination is proceeding, the ion-pairs with internal energy in a given interval are in quasi-equilibrium in the sense that the rates of flow of ion-pairs into and out of this interval do balance. They are not, however, in thermodynamic equilibrium. The internal energy of a particular ion-pair tends to decrease with time, principally as a result of collisions with the third bodies, until it reaches a critical energy E_s at which the one-way mutual neutralization process (1-3-7) removes the ion-pair and recombination is considered complete. Some of the energy of the ion-pair is also converted into kinetic energy associated with the center-of-mass of the ion-pair. This conversion gives rise to the non-thermal effect (see § 1-4C), and then the centers-of-mass of the ion-pairs are not in thermodynamic equilibrium with the surrounding neutral gas.

We observe that, without the introduction of a sink provided by the charge-transfer process (1-3-7), the solution that follows from (1-3-25) is simply unity (as it should be), and hence the occurrence of recombination is theoretically ignored. Thus (1-3-25) must be solved subject to the boundary conditions (1-3-5) for the fulfillment of recombination.

The situation above is analogous to that encountered by Bates, Kingston and McWhirter [21] in their treatment of collisional-radiative electronic-recombination between electrons and bare nuclei, in which the attainment of Saha-Boltzmann equilibrium was prevented by the radiative terms describing the spontaneous and stimulated emission and absorption of photons. They therefore set all the $\partial \eta/\partial t$ to zero except that for the ground-state (since the mean thermal energy was assumed to be much less than the first excitation energy) and hence, the quasi-equilibrium was referred to a particular set of number densities of atoms in the ground-state, free electrons and bare nuclei. In the present instance, the lowest bound states of the ion-pair are depopulated almost immediately by the almost irreversible charge-transfer process (1-3-7) in such a manner that we have the equation (1-3-25) for the quasi-equilibrium restricted by (1-3-5). The next stage in the analysis is the formulation of an expression for the collisional rate coefficient $\mathscr{K}(\lambda, \mu)$.

§ 1-4. Ions recombining in their parent gas

For the special case of ions recombining in their parent gas [10]

$$X^+ + X^- + X \rightarrow [X_2] + X \qquad (1\text{-}4\text{-}1)*$$

the reduction in the internal energy of the ion-pair is brought about by the extremely rapid symmetrical resonance charge-transfer collisions

$$X^+ + X \rightarrow X + X^+ \qquad (1\text{-}4\text{-}2)$$

and

$$X^- + X \rightarrow X + X^-. \qquad (1\text{-}4\text{-}3)$$

Here, it is a sufficient approximation to regard the velocity vectors of the ion and the atom as being simply interchanged by such an encounter. We require the change-of-energy rate coefficient $K(E_i, E_f)\,dE_f$ that describes how efficiently collisions between X atoms and $[X^+ - X^-]$ pairs change the mutual internal energy of the ions from E_i to between E_f and $E_f + dE_f$.

The formulation of this rate coefficient proves much simpler than that met in the case involving general ions and atoms for which the mechanisms (1-4-2) and (1-4-3) together with their resulting simplifications cannot be invoked.

It may be noted here that the Thomson-value of the recombination coefficient is

$$\alpha_T = \frac{64\pi e^6}{27(k\theta)^{\frac{5}{2}}} \frac{N(X)Q_D}{(6M)^{\frac{1}{2}}} \qquad (1\text{-}4\text{-}4)$$

where M is the mass of either ion, $N(X)$ is the number density of gaseous atoms and Q_D is the total momentum-transfer cross section for both ions in the gas.

A. ENERGY-CHANGE RATE COEFFICIENT

Let u and v be the initial velocities of a colliding ion and atom, both relative to the centre of mass of the ion-pair, $-u$ being the velocity vector of the other ion assumed to be unaffected by the collision. In the case of symmetrical resonance charge-transfer, u and v are, to a close approximation, simply interchanged by the ion-atom collision. Hence, the encounter alters the kinetic energy of relative motion of the ion-pair from

$$T_i = \tfrac{1}{2}M_{12}(2u)^2 = Mu^2 \qquad (1\text{-}4\text{-}5)$$

to

$$T_f = \tfrac{1}{2}M_{12}(u+v)^2 = \tfrac{1}{4}M(u+v)^2. \qquad (1\text{-}4\text{-}6)$$

* The molecule may not remain bound but may dissociate into neutral products.

1-4 IONS RECOMBINING IN THEIR PARENT GAS

If the interaction distance for processes (1-4-2) and (1-4-3), which are described by cross sections q^+ and q^- respectively, is very much less than some typical ion-ion separation, r_T for example, we can assume that each binary collision occurs instantaneously without affecting the other ion or the potential energy of the ion pair. This assumption is valid since, at room temperature

$$4\pi a^{\pm 2} = q^{\pm} \sim 10^{-16} \text{ cm}^2 \ll \pi r_T^2 \sim 5 \times 10^{-11} \text{ cm}^2 \qquad (1\text{-}4\text{-}7)$$

and so the range of the ion-neutral interactions, determined from the scattering lengths $a^{\pm}$, are much shorter than r_T.

Therefore, we have

$$E_f - E_i = T_f - T_i = \tfrac{1}{4}M(v^2 + 2\gamma uv - 3u^2), \quad \gamma = \hat{\boldsymbol{u}} \cdot \hat{\boldsymbol{v}} = \cos\phi \qquad (1\text{-}4\text{-}8)$$

so that, with E_i, u and v given, we can reach any E_f in the range

$$E_i + \tfrac{1}{4}M(v^2 - 2uv - 3v^2) \leq E_f \leq E_i + \tfrac{1}{4}(v^2 + 2uv - 3u^2) \qquad (1\text{-}4\text{-}9)$$

by varying the angle ϕ between the directions of $\boldsymbol{u}$ and $\boldsymbol{v}$ from zero to 2π, in which range a given E_f is passed through twice. Alternatively, we readily deduce that a specific E_f can be reached only from a definite E_i and v only when the initial kinetic energy of relative motion is in the range

$$T^- < T_i < T^+ \qquad (1\text{-}4\text{-}10)$$

where

$$9T^{\pm} = 5Mv^2 + 12(E_i - E_f) \pm 4M^{\frac{1}{2}}v(Mv^2 + 3E_i - 3E_f)^{\frac{1}{2}} \qquad (1\text{-}4\text{-}11)$$

which is real only for

$$v_1 < v < \infty \qquad (1\text{-}4\text{-}12)$$

where, for the respective cases of excitation and de-excitation

$$v_1^2 = \begin{cases} \dfrac{3}{M}(E_f - E_i), & E_f > E_i \\ 0, & E_f < E_i. \end{cases} \qquad (1\text{-}4\text{-}13)$$

Therefore, the occurrence of excitation of the ion-pair requires that the incoming atom must possess sufficient kinetic energy (at least $\tfrac{1}{3}Mv_1^2$) of motion relative to C, the center of mass of the ion-pair system and the third body; while, from (1-4-11), in order for an extremely slow atom (with $v \approx 0$) to cause de-excitation, the kinetic energy of motion of the ion-pair relative to its center of mass must at least be $\frac{4}{3}(E_i-E_f)$, of which (E_i-E_f) is that relative to C, and the remaining $\frac{1}{3}(E_i-E_f)$ represents the energy of motion of C.

The mutual interaction between the ions determines the probability $f(p)\,dp$ that the momentum of one ion relative to the other is initially in an interval dp around p and it provides hydrogenic internal energies E_i: that is, the orbital kinetic energy has a distribution about the mean appropriate to classical motion in a central Coulomb field. Thus, assuming that all elements of phase space (p, r) associated with a given internal energy E_i are equally populated

$$f(\mathbf{p},\mathbf{r}) = A\delta(E_i - p^2/M + e^2/r) \tag{1-4-14}$$

is the distribution function for an ion-pair with reduced mass $\frac{1}{2}M$ and, internal energy E_i and vector separation r. The coefficient A is determined from the normalization condition for a single ion-pair

$$\int f(\mathbf{p},\mathbf{r})\,d\mathbf{p}\,d\mathbf{r} = 1 \tag{1-4-15}$$

with the result that

$$A = -\frac{2|E_i|^{\frac{5}{2}}}{\pi^3 e^6 M^{\frac{3}{2}}}. \tag{1-4-16}$$

Hence, the normalized number of ion-pairs with kinetic energy (defined by (1-4-5)) between T_i and $T_i + dT_i$ is

$$\mathscr{F}(T_i)\,dT_i = \frac{16 T_i^{\frac{3}{2}} |E_i|^{\frac{5}{2}}}{\pi (T_i - E_i)^4}\,dT_i. \tag{1-4-17}$$

We note that the above microcanonical distribution is identical with the quantum-mechanical distribution for a hydrogenic system

$$n^{-2}\sum_{l,m}|\psi_{nlm}(\mathbf{p})|^2 = \frac{8|E_n|^{\frac{5}{2}}}{\pi^2 (2M_{12})^{\frac{3}{2}}(p^2/2M_{12} - E_n)^4} \equiv \int f(\mathbf{p},\mathbf{r})\,d\mathbf{r} \tag{1-4-18}$$

averaged over angular momentum l and its projection m of an ion-pair of reduced mass M_{12} with internal motion described by the momentum wavefunction $\psi_{nlm}(\mathbf{p})$ with energies E_n.

Further assuming that the center-of-mass of the ion-pair has thermal motion corresponding to the temperature θ of the gas (see [22]), we have that the normalized distribution function $\mathscr{G}(v)\,dv$ is given by

$$\mathcal{G}(v)\,dv = 4\pi v^2(\mathcal{M}/2\pi k\theta)^{\frac{3}{2}} \exp\left(-\tfrac{1}{2}\mathcal{M}v^2/k\theta\right)dv \tag{1-4-19}$$

where k is Boltzmann's constant, and the reduced mass of the third body and the ion-pair system is

$$\mathcal{M} = \tfrac{2}{3}M. \tag{1-4-20}$$

The interaction between the ions and the parent gas is acknowledged by the use of the total charge-transfer cross section,

$$Q = q^+ + q^-. \tag{1-4-21}$$

The rate of encounters with γ (defined by (1-4-8)) varying between γ and $\gamma + d\gamma$ is

$$dK = dK^+ + dK^- = \tfrac{1}{2}|v - u|Q\,d\gamma \tag{1-4-22}$$

where dK^+ is the partial rate arising from only positive ion-neutral encounters, and dK^- is the rate that would arise if only negative ion-neutral collisions occurred. Therefore, since

$$dE_{\mathrm{f}} = \tfrac{1}{2}Muv\,d\gamma \tag{1-4-23}$$

the rate coefficient describing encounters in which the internal energy is changed from E_{i} to between E_{f} and $E_{\mathrm{f}} + dE_{\mathrm{f}}$ is, on integrating over the above distribution functions for the ion-pair and the incoming atom,

$$K(E_{\mathrm{i}}, E_{\mathrm{f}})\,dE_{\mathrm{f}} = \tag{1-4-24}$$
$$= \frac{2^{\frac{1}{2}}}{M}\,dE_{\mathrm{f}} \int_{v_1}^{\infty} v^{-1}\mathcal{G}(v)\,dv \int_{T^-}^{T^+} T_{\mathrm{i}}^{-\frac{1}{2}} \{Mv^2 + 2(E_{\mathrm{i}} - E_{\mathrm{f}}) - T_{\mathrm{i}}\}^{\frac{1}{2}} Q\mathcal{F}(T_{\mathrm{i}})\,dT_{\mathrm{i}}.$$

It is convenient to write, as before

$$\begin{aligned}\lambda &= -E_{\mathrm{i}}/k\theta, & \mu &= -E_{\mathrm{f}}/k\theta, \\ x &= T_{\mathrm{i}}/k\theta, & y &= Mv^2/k\theta\end{aligned} \tag{1-4-25}$$

and

$$K(E_{\mathrm{i}}, E_{\mathrm{f}})|dE_{\mathrm{f}}| = \mathcal{K}(\lambda, \mu)\,d\mu. \tag{1-4-26}$$

Substitution from (1-4-17) and (1-4-19) then gives

$$\mathcal{K}(\lambda, \mu) = \frac{32(6k\theta)^{\frac{1}{2}}|\lambda|^{\frac{3}{2}}Q}{9\pi^{\frac{3}{2}}M^{\frac{1}{2}}} \int_{y_1}^{\infty} \exp(-\tfrac{1}{3}y)\,dy \int_{x_1}^{x_2} \frac{(y + 2\mu - 2\lambda - x)^{\frac{1}{2}}}{(x+\lambda)^4}\,dx \tag{1-4-27}$$

in which

$$9x_{2,1} = 5y + 12\mu - 12\lambda \pm 4y^{\frac{1}{2}}(y + 3\mu - 3\lambda)^{\frac{1}{2}} \quad (1\text{-}4\text{-}28)$$

and

$$y_1 = \begin{cases} 3(\lambda - \mu), & \lambda > \mu \\ 0, & \lambda < \mu. \end{cases} \quad (1\text{-}4\text{-}29)$$

Because Q is a slowly varying function of the relative velocity, it has been taken outside the integral. On carrying out the integration analytically over x we obtain

$$\mathcal{K}(\lambda, \mu) = \frac{32(6k\theta)^{\frac{1}{2}}|\lambda|^{\frac{3}{2}}Q}{9\pi^{\frac{3}{2}}M^{\frac{1}{2}}} \int_{y_1}^{\infty} \mathcal{I}(y; \lambda, \mu) \exp\left(-\tfrac{1}{3}y\right) dy \quad (1\text{-}4\text{-}30)$$

where we write

$$\mathcal{I}(y; \lambda, \mu) = \frac{1}{24q^2} \sum_{j=1}^{2} (-1)^j \left[\frac{p_j^{\frac{1}{2}}}{r_j^3}(8q^2 - 2qr_j - 3r_j^2) + \frac{3}{2q^{\frac{1}{2}}} \ln \left| \frac{p_j^{\frac{1}{2}} - q^{\frac{1}{2}}}{p_j^{\frac{1}{2}} + q^{\frac{1}{2}}} \right| \right] \quad (1\text{-}4\text{-}31)$$

TABLE 1-4-1

Quantities $F(\lambda, \mu)$ and $\mathscr{F}^{-}(\lambda)$ defined in (1-4-36) and (1-4-37), respectively

μ \ λ	0.5	1.0	1.5	2.0	3.0	4.0	5.0
				$F(\lambda, \mu)$			
−5.0	1.30^{-10}	6.86^{-11}	3.75^{-11}	2.06^{-11}	6.53^{-12}	2.13^{-12}	7.11^{-13}
−4.0	3.77^{-10}	1.96^{-10}	1.05^{-10}	5.78^{-11}	1.81^{-11}	5.89^{-12}	1.96^{-12}
−3.0	1.05^{-9}	5.41^{-10}	2.90^{-10}	1.59^{-10}	4.97^{-11}	1.61^{-11}	5.34^{-12}
−2.0	2.73^{-9}	1.42^{-9}	7.69^{-10}	4.23^{-10}	1.33^{-10}	4.34^{-11}	1.44^{-11}
−1.5	4.22^{-9}	2.25^{-9}	1.23^{-9}	6.81^{-10}	2.16^{-10}	7.08^{-11}	2.36^{-11}
−1.0	6.22^{-9}	3.45^{-9}	1.92^{-9}	1.08^{-9}	3.48^{-10}	1.15^{-10}	3.85^{-11}
−0.5	8.49^{-9}	5.09^{-9}	2.94^{-9}	1.68^{-9}	5.55^{-10}	1.85^{-10}	6.23^{-11}
0	9.96^{-9}	7.04^{-9}	4.33^{-9}	2.56^{-9}	8.73^{-10}	2.95^{-10}	1.00^{-10}
+0.5	7.42^{-9}	8.67^{-9}	6.03^{-9}	3.77^{-9}	1.35^{-9}	4.67^{-10}	1.61^{-10}
+1.0	2.53^{-9}	8.01^{-9}	7.61^{-9}	5.28^{-9}	2.04^{-9}	7.31^{-10}	2.56^{-10}
+1.5	1.05^{-9}	4.55^{-9}	7.71^{-9}	6.76^{-9}	3.00^{-9}	1.12^{-9}	4.03^{-10}
+2.0	5.29^{-10}	2.54^{-9}	5.43^{-9}	7.22^{-9}	4.22^{-9}	1.70^{-9}	6.27^{-10}
+3.0	1.87^{-10}	9.69^{-10}	2.38^{-9}	4.16^{-9}	6.22^{-9}	3.50^{-9}	1.45^{-9}
+4.0	8.55^{-11}	4.59^{-10}	1.18^{-9}	2.22^{-9}	4.63^{-9}	5.39^{-9}	2.98^{-9}
+5.0	4.58^{-11}	2.50^{-10}	6.58^{-10}	1.27^{-9}	2.98^{-9}	4.64^{-9}	4.72^{-9}
$\mathscr{F}^{-}(\lambda)$	1.55^{-8}	8.91^{-9}	5.05^{-9}	2.87^{-9}	9.33^{-10}	3.09^{-10}	1.04^{-10}

The index gives the power of 10 by which the entry must be multiplied.

in which

$$9p_{2,1} = 2(y^{\frac{1}{2}} \pm s^{\frac{1}{2}})^2 \tag{1-4-32}$$

$$q = y + 2\mu - \lambda \tag{1-4-33}$$

$$9r_{2,1} = 5y + 12\mu - 3\lambda \mp 4y^{\frac{1}{2}}s^{\frac{1}{2}} \tag{1-4-34}$$

with

$$s = y + 3\mu - 3\lambda. \tag{1-4-35}$$

The final integration over y is performed numerically. Some values of the quantity

$$F(\lambda, \mu) = \frac{M^{\frac{1}{2}}}{\theta^{\frac{1}{2}} Q} \mathcal{K}(\lambda, \mu) \tag{1-4-36}$$

were calculated by Bates and Moffett [10], and are presented in Table 1-4-1 together with some values of the related quantity

$$\mathcal{F}^-(\lambda) = \int_{-\infty}^{0} F(\lambda, \mu) \, d\mu \tag{1-4-37}$$

which is of interest in connection with the rate of disruption of bound ion-pairs. M is measured in gram, θ in degrees Kelvin, Q in cm^2 and $\mathcal{K}(\lambda, \mu)$ in cm^3/sec.

By writing down $\mathcal{K}(\mu, \lambda) \, d\lambda$ for the reverse rate and by substitution to the dummy variables

$$x_1 = x - (\lambda - \mu), \quad y_1 = y + 3(\lambda - \mu) \tag{1-4-38}$$

we note that for Saha-Boltzmann equilibrium, the principle of detailed balance

$$\eta_T(\lambda) \mathcal{K}(\lambda, \mu) = \eta_T(\mu) \mathcal{K}(\mu, \lambda) \tag{1-4-39}$$

is satisfied theoretically in the form

$$\frac{\exp(\lambda) \mathcal{K}(\lambda, \mu)}{|\lambda|^{\frac{3}{2}}} = \frac{\exp(\mu) \mathcal{K}(\mu, \lambda)}{|\mu|^{\frac{3}{2}}}. \tag{1-4-40}$$

B. CALCULATION OF THE RECOMBINATION COEFFICIENT

On introducing $F(\lambda, \mu)$ of (1-4-36), a function which is independent of M, θ and Q, we can rewrite the integral equation (1-3-25) for the quasi-equilibrium distribution $\rho(\lambda)$ as the following

$$\rho(\lambda)\{\mathscr{F}^+(\lambda)+\mathscr{F}^-(\lambda)\} = \mathscr{F}^-(\lambda) + \int_0^\omega \rho(\mu) F(\lambda, \mu)\,d\mu \qquad (1\text{-}4\text{-}41)$$

with $\mathscr{F}^-(\lambda)$ as in (1-4-37) and, correspondingly,

$$\mathscr{F}^+(\lambda) = \int_0^\omega F(\lambda, \mu)\,d\mu. \qquad (1\text{-}4\text{-}42)$$

The integral on the right of (1-4-41) may be evaluated by use of the quadrature formula

$$\int_0^\omega \rho(\mu) F(\lambda, \mu)\,d\mu = \sum_{j=1}^m \phi_j \rho(\mu_j) F(\lambda, \mu_j), \quad \mu_m = \omega \qquad (1\text{-}4\text{-}43)$$

ϕ_j being weighting factors, which can simply be those appropriate to Simpson's rule:

$$\phi_j = \begin{cases} \tfrac{4}{3}h, & j \text{ even} \\ \tfrac{2}{3}h, & j \text{ odd} \\ \tfrac{1}{3}h, & j = 1, m \text{ (odd)} \end{cases} \qquad (1\text{-}4\text{-}44)$$

where h is a constant interval that determines the pivots of the μ-integration. This procedure reduces the integral equation (1-4-41) to the following set of linear equations

$$\rho(\mu_i)\{\mathscr{F}^+(\mu_i)+\mathscr{F}^-(\mu_i)-\phi_i F(\mu_i,\mu_i)\} - \sum_{\substack{j=1 \\ j\neq i}}^m \phi_j \rho(\mu_j) F(\mu_i,\mu_j) = \mathscr{F}^-(\mu_i),$$
$$i = 1, 2, \ldots, m, \qquad (1\text{-}4\text{-}45)$$

where μ_i covers the complete range (≥ 0) and

$$h = \mu_{j+1} - \mu_j \ll 1. \qquad (1\text{-}4\text{-}46)$$

Thus these linear equations can be solved, by standard means, subject to the boundary conditions (1-3-5),

$$\rho(\mu_i) = \begin{cases} 1 & \mu_i \leq 0 \\ 0 & \mu_i > \varepsilon \end{cases} \qquad (1\text{-}4\text{-}47)$$

to yield the quasi-equilibrium distribution. The width h is simply a computational parameter, its value to be decreased until convergence is obtained for $\rho(\mu_i)$. The energy denoted by ε at which the recombination is regarded as being stabilized, and the greatest binding energy ω an ion-pair may have (even transitorily), are ill-determined physical parameters. The binding energy is expected to be at least 5 eV while the calculations of Bates and Boyd [35] on curve crossing indicate that ε is unlikely to correspond to values lower than 0.5 eV. Thus, with $\rho(\lambda)$ determined, the recombination coefficient for ions recombining in their parent gas is calculated from (1-3-13) to give

$$\alpha = \frac{\pi^{\frac{3}{2}} e^6 Q N(X)}{2k^3 \theta^{\frac{3}{2}} M^{\frac{1}{2}}} \int_{\lambda=v}^{\omega} \lambda^{-\frac{5}{2}} \exp(\lambda) \, d\lambda \int_{\mu=-\infty}^{v} \{\rho(\mu) - \rho(\lambda)\} F(\lambda, \mu) \, d\mu \quad (1\text{-}4\text{-}48)$$

where the first term on the right represents the downflow of ion-pairs past some arbitrary (negative) energy level v and the second term corresponds to the upflow past this level. The value assigned to v naturally does not affect the difference between these flows.

With h taken as $\frac{1}{16}$, ε and ω to be 10 and 30, respectively (to which choice the results converged), Bates and Moffett solved equation (1-4-43) for the quasi-equilibrium distribution (reproduced in Fig. 1-4-1) which they

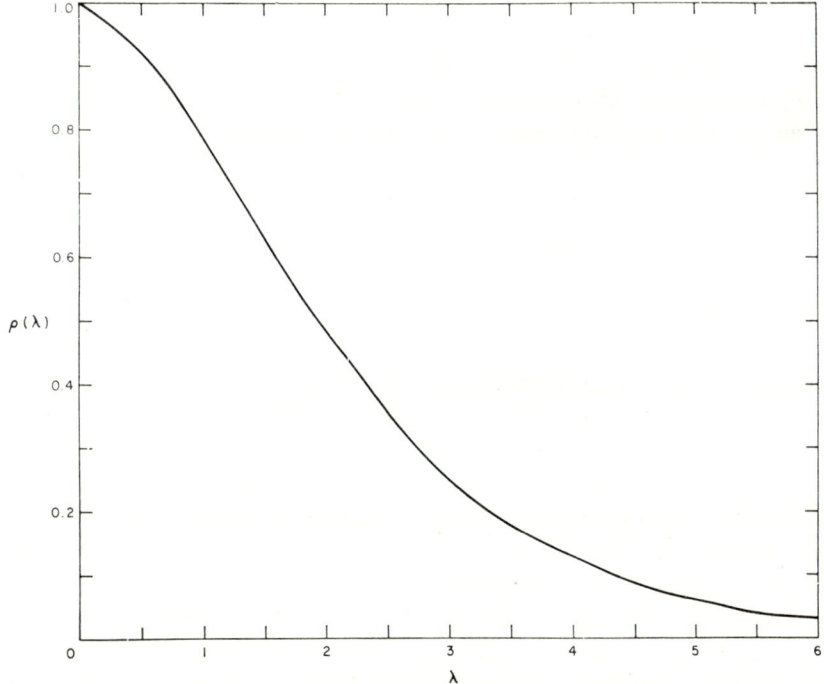

Fig. 1-4-1. Population distribution function $\rho(\lambda)$ for ions recombining in their parent gas.

then used in (1-4-48) to yield α. By comparison with the Thomson recombination coefficient α_T for which the momentum-transfer (or diffusion) cross section Q_D is taken to be *twice* the symmetrical resonance charge-transfer cross section Q [36] we find that the ratio

$$\mathscr{R} = \alpha/\alpha_T = 0.485. \qquad (1\text{-}4\text{-}49)$$

The Thomson theory thus overestimates the absolute magnitude of the recombination coefficient by only a factor slightly greater than 2, which is not large in view of the fact that α_T is proportional to the cube of the trapping distance, r_T, the value for which has been the subject of much controversy. For example, Loeb and Marshall [15], Brueckner [16] and Natanson [6] have suggested that r_T be multiplied by $\frac{1}{4}$, $\frac{3}{8}$ and $\frac{5}{8}$, respectively, leading to α being decreased by factors of 64, 19 and 4. To allow for Coulomb deflexion of the ions, however, Brueckner and Natanson multiplied their values of α by $\frac{11}{3}$ and $\frac{17}{5}$, respectively, thus bringing Natanson's value ($0.83\alpha_T$) within closest agreement with Thomson's result.

It is apparent that accord with the calculations of Bates and Moffett may be obtained by arbitrarily reducing r_T only by about 20%. Thus, the overall accuracy of the Thomson theory is much higher than the accuracies of the various parts of the physical model on which it is based. Further insight regarding this agreement is furnished from the theory of Bates and Flannery [11] to be discussed in the following section.

C. THE NON-THERMAL EFFECT

Before the above theory was regarded as being complete, Bates et al. [22] examined the validity of the physical assumption that the centers-of-mass of the ion-pairs have approximately thermal motion corresponding to the temperature of the ambient gases. In general, while the bulk of the internal energy of an ion-pair is transferred to the third bodies, some of it is converted into kinetic energy of its center-of-mass. This additional energy-loss mechanism gives rise to the non-thermal effect in which the center-of-mass of the ion-pair is not in thermal equilibrium with the surrounding gas. This effect would become most marked for ions recombining in their parent gas, because the subsequent course of events is controlled by symmetrical-resonance charge-transfer collisions that make the mean change in the internal energy per collision relatively high. However, Bates et al. [22] found that, while the mean kinetic energy associated with the centers-of-mass is rather greater than $3/2k\theta$ and is dependent on the internal energy of the ion-pair, the effect on the recombination coefficient is insignificant.

§ 1-5. General third body

In contrast to the special and relatively simple case discussed in the preceding section, the main difficulty envisaged in the more general case

$$X^+ + Y^- + Z \rightarrow [XY] + Z \qquad (1\text{-}5\text{-}1)^*$$

is the theoretical formulation and resulting calculation of the change-in-energy rate coefficient $K(E_i, E_f) \, dE_f$ describing the rate of collisions between atoms and ion-pairs for which the internal energy of the pair is changed from E_i to between E_f and $E_f + dE_f$. As in the previous process (1-4-1), the range of the ion-ion interaction is so much larger than that of the ion-atom interaction that it is permissible to view these collisions as binary encounters between the atom and either ion. Therefore,

$$K(E_i, E_f) = K_{13}(E_i, E_f) + K_{23}(E_i, E_f) \qquad (1\text{-}5\text{-}2)$$

where K_{13} is the rate arising from positive ion-neutral collisions and K_{23} is that calculated assuming negative ion-neutral collisions.

We will now develop the expression of Bates and Flannery [11] for these energy-change collisions between ion-pairs and third bodies in a sufficiently general fashion that the method employed, as well as being particularly suited to the present instance in (1-5-1), can moreover, as shown by Flannery [37], be adapted to the examination of electronic excitation, ionization and de-excitation arising from the impact between two heavy particles in the process

$$A + B \rightarrow A^* + B^* \qquad (1\text{-}5\text{-}3)$$

in which the asterisk denotes internal energy change in the target and projectile atoms.

A. GENERAL ASPECTS OF A COLLISION

We shall call the positive and negative ions 1 and 2, respectively, and the neutral atom 3, and shall affix these numbers as subscripts to the symbols for the various physical quantities.

For thermal energies of impact, we can assume in the calculation of K_{13}, that the positive ion of mass M_1 and velocity v_1, in an orbit (hyperbolic or closed) about the negative ion of mass M_2 and velocity v_2, collides elastically with the gaseous atom or molecule of mass M_3 and velocity v_3 such that the effect of this collision is to cause an increase (excitation or

---

* The square brackets indicate that the molecule may not remain bound.

disruption), a decrease (de-activation to an orbit more tightly bound) or no change at all (elastic process) in the internal energy of the (1, 2) ion-pair.

We shall adopt, as the simplest realistic model of the ion-atom interaction, the Langevin-potential [38] in the form

$$V_{i3}(R) = \begin{cases} -\dfrac{pe^2}{2R^4}, & R > S_{i3} \\ \infty, & R \leq S_{i3} \end{cases} \qquad (1\text{-}5\text{-}4)$$

where R is the distance between the nuclei, $\pm e$ are the charges of the positive and negative ions, p is the polarizability of the neutral atom and S_{i3} is the hard-sphere core parameter. In § 1-7 we will show how to account for the effects arising from the departure of the physical potential from (1-5-4). For the potential-combination (1-5-4), the projectile either follows a hyperbolic orbit that corresponds to large angular momentum or else, for small angular momentum, has an inward spiraling motion towards the scattering center until the repulsive force reverses the trend. Also, closed orbiting collisions can occur but, as will be evident from a later section, do not contribute to energy-change of the ion-pair.

Each ion is assumed to suffer a binary impact with the gas atoms, the encounter being so brief that it does not affect the ion-ion separation (the mutual potential energy) or the velocity vector of the other ion. The *binary-encounter* assumption is certainly valid since, as previously noted, the range of (1-5-4) is much shorter than the Coulombic interaction between the ions which can therefore be regarded as being independent scattering centers. Also, for each collision to occur *instantaneously*, the collision time τ_c during which the ion-neutral interaction is effective, must be very much less than a typical time $\tau_{\text{orb}}(\sim 2\pi r_T/v_i)$ for the ion-pair to complete one orbit; which is exactly true in the hard-sphere limit when

$$S_{i3} \gg r_0 = \left(\dfrac{e^2 p}{M_{i3} v_i^2}\right)^{\frac{1}{4}} \qquad (1\text{-}5\text{-}5)$$

where r_0 is the orbiting radius appropriate to (1-5-4). Even in a pure-polarization field or small-ion limit when $S_{i3} \ll r_0$, then τ_c is of the order $2\pi r_0/v_i$ which is much smaller than τ_{orb} when $p \ll 10^{-19}$ cm^3, which, of course, is normally true. Thus, this three-body collision furnishes a perfect example for which the assumption of a brief binary impact is remarkably well justified.

We denote the velocities of the three bodies after the collision by v'_1, v'_2 and v'_3 and the reduced mass and the center-of-mass velocity of the (i, j) system by M_{ij} and V_{ij} respectively. The collision is therefore so brief that the velocity of the negative ion is unaffected, i.e.,

$$v_2 = v_2' \tag{1-5-6}$$

and with the aid of

$$v_1 = V_{13} + (M_{13}/M_1)g_{13} \tag{1-5-7}$$

the change in the kinetic energy of particle 1 is

$$\tfrac{1}{2}M_1(v_1'^2 - v_1^2) = M_{13}V_{13} \cdot (g_{13}' - g_{13}) \tag{1-5-8}$$

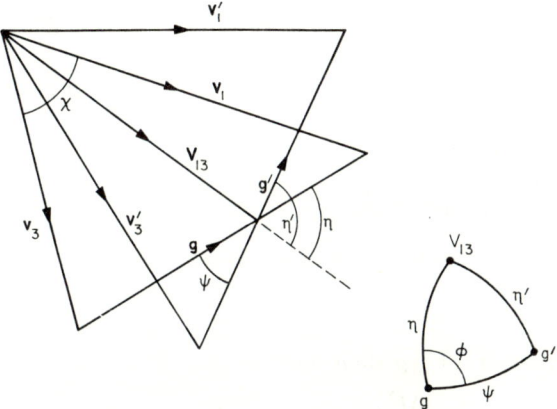

Fig. 1-5-1. Geometry of the 1-3 collision.

where g_{13} and g_{13}' are the initial and final relative velocities of the (1, 3) collision partners; and since the collision is elastic the center-of-mass velocity V_{13} is unchanged (cf. Fig. 1-5-1). Equation (1-5-8) above shows that the kinetic energy change depends on the coordinate system from which the collision is observed. In particular, if the observer should be moving with the center-of-mass of the ion-neutral system, then V_{13} equals zero and no energy change would be apparent.

Choose the reference frame in which the (1, 2) center-of-mass is initially at rest such that

$$v_2' = -(M_1/M_2)v_1 \tag{1-5-9}$$

with the result that the initial kinetic energy of relative motion of the ion-pair is

$$T_i = \tfrac{1}{2}M_{12}(v_1 - v_2)^2 = \tfrac{1}{2}(M_1^2/M_{12})v_1^2 \equiv Cv_1^2 \tag{1-5-10}$$

in which we have defined C for later use. The kinetic energy change in this relative motion caused by the elastic collision with the third body is then, by analogy with (1-5-8),

$$T_f - T_i = \tfrac{1}{2} M_{12}[(\mathbf{v}_1' - \mathbf{v}_2)^2 - (\mathbf{v}_1 - \mathbf{v}_2)^2]$$
$$= M_{13} \mathbf{V}_{13} \cdot (\mathbf{g}_{13}' - \mathbf{g}_{13}) - \frac{M_{13}^2}{(M_1 + M_2)} g^2 (1 - \cos \psi) \qquad (1\text{-}5\text{-}11)$$

in which the (1, 3) relative velocity vector, of constant magnitude g, is simply rotated through the scattering angle ψ.

Assuming that the potential energy is not affected appreciably during the encounter, the change in the internal energy of the (1, 2) pair is, thus,

$$(E_f - E_i) = M_{13} V_{13} g (\cos \eta' - \cos \eta) - \frac{M_{13}^2}{(M_1 + M_2)} g^2 (1 - \cos \psi),$$
$$\cos \eta^{(\prime)} = \hat{\mathbf{V}}_{13} \cdot \hat{\mathbf{g}}_{13}^{(\prime)}. \qquad (1\text{-}5\text{-}12)$$

From Fig. 1-5-1 we note that,

$$\cos \eta' = \cos \eta \cos \psi + \sin \eta \sin \psi \cos \phi \qquad (1\text{-}5\text{-}13)$$

from which

$$(E_f - E_i) = M_{13} V_{13} g \sin \eta \sin \psi \cos \phi$$
$$- \left\{ \frac{M_{13}^2}{(M_1 + M_2)} g^2 + M_{13} V_{13} g \cos \eta \right\} (1 - \cos \psi). \qquad (1\text{-}5\text{-}14)$$

When ϕ, which is the angle the plane defined by $\mathbf{V}_{13}$ and $\mathbf{g}_{13}$ makes with that containing $\mathbf{g}_{13}$ and $\mathbf{g}_{13}'$, i.e.,

$$\phi = \widehat{(\mathbf{V}_{13} \mathbf{g}_{13})(\mathbf{g}_{13} \mathbf{g}_{13}')}, \qquad (1\text{-}5\text{-}15)$$

is varied between 0 and 2π with ψ, η, g, V_{13} and E_i held constant, each value of E_f in the range defined by

$$E_i + M_{13} g V_{13} \{\cos(\psi + \eta) - \cos \eta\} - \frac{M_{13}^2}{(M_1 + M_2)} g^2 (1 - \cos \psi) \leq E_f$$
$$\leq E_i + M_{13} g V_{13} \{\cos(\psi - \eta) - \cos \eta\} - \frac{M_{13}^2}{(M_1 + M_2)} g^2 (1 - \cos \psi) \qquad (1\text{-}5\text{-}16)$$

is passed through twice. Alternatively, if the value of E_f is given and ψ is treated as a variable, then for this specific change in internal energy, the scattering is confined within the limiting angles $\psi^{\pm}$, which satisfy

$$(E_f - E_i) - M_{13} g V_{13} \{\cos(\psi^{\pm} \pm \eta) - \cos \eta\} + \frac{M_{13}^2}{(M_1 + M_2)} g^2 (1 - \cos \psi^{\pm}) = 0 \tag{1-5-17}$$

with solutions

$$\cos \psi^{\pm} = \gamma^{-1} \{\alpha(\alpha + E_f - E_i) \pm \beta[\gamma^2 - (\alpha + E_f - E_i)^2]^{\frac{1}{2}}\} \tag{1-5-18}$$

in which the parameters determined by the velocities before the collision reduce to

$$\alpha \equiv \frac{M_{13}^2}{(M_1 + M_2)} g^2 + M_{13} V_{13} \cos \eta$$

$$= \tfrac{1}{2} M_{13} \left(v_1^2 - v_3^2 + \frac{1-a}{1+a} g^2 \right) \tag{1-5-19}$$

$$\beta \equiv M_{13} V_{13} g \sin \eta$$

$$= \tfrac{1}{2} M_{13} [g^2 (2v_1^2 + 2v_3^2 - g^2) - (v_1^2 - v_3^2)^2]^{\frac{1}{2}} \tag{1-5-20}$$

and

$$\gamma^2 \equiv \alpha^2 + \beta^2 = \left(\frac{M_{13} g}{1+a}\right)^2 [(1+a)(v_1^2 + a v_3^2) - a g^2] \tag{1-5-21}$$

where the convenient mass-ratio parameter a is given by

$$a = \frac{M_2 M_3}{M_1(M_1 + M_2 + M_3)}. \tag{1-5-22}$$

Note that, with these initial mass and velocity parameters α and β preset, we have

$$E_f - E_i = \beta \sin \psi \cos \phi - \alpha(1 - \cos \psi) \tag{1-5-23}$$

which demonstrates that a given change in internal energy can be accomplished through suitable deflection of the (1, 3) relative velocity vector by the spherical angles (ψ, ϕ) in a coordinate frame with the initial g_{13} along the Z-axis and V_{13} in the XZ plane (see Fig. 1-5-1). The range of the variables ψ and ϕ are however mutually dependent in that as ϕ ranges from zero to 2π, the scattering angle is confined to within $\psi^{\pm}$ for a given energy-change. The relative velocity g obeys the relation

$$g_1 \leq (v_1 - v_3) \leq g \leq (v_1 + v_3) = g_2 \tag{1-5-24}$$

but is further restricted by the requirement that $\psi^{\pm}$ must be real. This observation entails that the following conditions must be satisfied by the velocities,

$$X \equiv \tfrac{1}{2}(M_1^2/M_{12})v_1^2 - (E_i - E_f) \geq 0 \qquad (1\text{-}5\text{-}25)$$

$$Y \equiv \tfrac{1}{2}(M_1^2/M_{12})v_3^2 + (E_i - E_f)/a \geq 0 \qquad (1\text{-}5\text{-}26)$$

which can be, equivalently, written as

$$\tfrac{1}{2}\mathcal{M}v_3^2 \geq (E_f - E_i) \qquad (1\text{-}5\text{-}27)$$

where $\mathcal{M}$, the reduced mass of the third body and the ion-pair, is

$$\mathcal{M} = \frac{(M_1 + M_2)M_3}{M_1 + M_2 + M_3}. \qquad (1\text{-}5\text{-}28)$$

Also, g must vary between $G_{1,2}$ defined by

$$\tfrac{1}{2}(M_1^2/M_{12})G_{1,2}^2 = [X^{\frac{1}{2}} \mp Y^{\frac{1}{2}}]^2. \qquad (1\text{-}5\text{-}29)$$

Equations (1-5-25) and (1-5-27) simply state that for de-excitation of the ion-pair to occur, the ion-pair must have at least the energy $(E_i - E_f)$ of de-activation, while the neutral body must possess sufficient kinetic energy of relative motion to cause excitation. The physical limits of g must simultaneously satisfy both (1-5-24) and (1-5-29). Hence, for a known change in internal energy of the ion-pair, only certain regions of the multidimensional space (v_1, v_3, g, ψ) defined by the above equations are accessible. The variables v_1, v_3 and g correspond to conditions prior to the collision, while the angles ψ and ϕ are determined by the amount of internal energy to be gained or lost as a result of the collision. The recognition that, from all the variables describing the collisional aspects, the above choice is the most appropriate, provides the key for the straightforward solution of the problem.

B. Formula for the Rate Coefficient

Let $\sigma(g, \psi)$ be the elastic differential cross section for ion-neutral (1, 3) collisions with scattering angle ψ and relative velocity g given by

$$g^2 = v_1^2 + v_3^2 - 2\mu v_1 v_3, \quad \mu = \cos\chi \qquad (1\text{-}5\text{-}30)$$

where χ is the angle between v_1 and v_3. Then, for a given v_1 and v_3, the rate coefficient, for collisions in which μ is between μ and $\mu + d\mu$ and the scattering is into unit solid angle $\sin\psi \, d\psi \, d\phi$ about the direction (ψ, ϕ), is

$$dK = \tfrac{1}{2}g\,\{\sigma(g,\psi)\sin\psi\,d\psi\,d\phi\}\,d\mu. \qquad (1\text{-}5\text{-}31)$$

Differentiating (1-5-23) subject to the initial conditions and with ψ given yields

$$dE_\text{f} = -\beta\sin\psi\sin\phi\,d\phi \qquad (1\text{-}5\text{-}32)$$

and since, from (1-5-23),

$$\cos\phi = \{(E_\text{f}-E_\text{i})+\alpha(1-\cos\psi)\}/\beta\sin\psi \qquad (1\text{-}5\text{-}33)$$

we obtain, after some straightforward algebraic reduction that

$$d\phi = -dE_\text{f}/R(v_1,v_3,g,\psi|E_\text{f}-E_\text{i}) \qquad (1\text{-}5\text{-}34)$$

where

$$R^2(v_1,v_3,g,\psi|\varDelta) = \beta^2-(\alpha+\varDelta)^2+2\alpha(\alpha+\varDelta)\cos\psi-(\alpha^2+\beta^2)\cos^2\psi \qquad (1\text{-}5\text{-}35)$$

which, with the aid of (1-5-18) for the angular limits $\psi^\pm$ to the scattering for a given g becomes

$$R(v_1,v_3,g,\psi|E_\text{f}-E_\text{i}) = \gamma[(\cos\psi-\cos\psi^-)(\cos\psi^+-\cos\psi)]^{\tfrac{1}{2}}. \qquad (1\text{-}5\text{-}36)$$

Noting that

$$d\mu = -g\,dg/v_1 v_3 \qquad (1\text{-}5\text{-}37)$$

and integrating over all permissible ψ and g, and remembering that each element $d\phi$ contains *two* elements dE_f, we find that the rate coefficient for $(1, 3)$ collisions that leave the $(1, 2)$ internal energy in the internal dE_f about E_f is, for a given v_1 and v_3, $\Gamma\,dE_\text{f}$, where

$$\Gamma(E_\text{i},E_\text{f};v_1,v_3) =$$
$$= \frac{1}{v_1 v_3}\int_{g^-}^{g^+}\frac{g^2\,dg}{(\alpha^2+\beta^2)^{\tfrac{1}{2}}}\int_{\psi^-}^{\psi^+}\frac{\sigma_{13}(g,\psi)\,d(\cos\psi)}{[(\cos\psi^+-\cos\psi)(\cos\psi-\cos\psi^-)]^{\tfrac{1}{2}}} \qquad (1\text{-}5\text{-}38)$$

in which g^- is the larger of g_1 and G_1, where g^+ is the smaller of g_2 and G_2 defined by (1-5-24) and (1-5-29), such that $g^+ > g^-$ for non-zero Γ.

Taking the interaction between the ions to be Coulombic and assuming that all elements of phase space associated with (negative) energy E_i are equally populated, we have (as before) that

$$\mathscr{F}(T_\text{i})\,dT_\text{i} = \frac{16T_\text{i}^{\tfrac{1}{2}}|E_\text{i}|^{\tfrac{5}{2}}}{\pi(T_\text{i}-E_\text{i})^4}\,dT_\text{i} \qquad (1\text{-}5\text{-}39)$$

describes the normalized distribution functions for ion pairs, with internal energy E_i, and with kinetic energy T_i between T_i and $T_i+\mathrm{d}T_i$. If it may be assumed that the center-of-mass of the ion-pair has thermal motion corresponding to the temperature θ of the gas we have also

$$\mathscr{G}(v_3)\,\mathrm{d}v_3 = 4\pi v_3^2 (\mathscr{M}/2\pi k\theta)^{\frac{3}{2}} \exp\left(-\tfrac{1}{2}\mathscr{M}v_3^2/k\theta\right)\mathrm{d}v_3 \qquad (1\text{-}5\text{-}40)$$

as being the distribution function for the incident velocities v_3, with the reduced mass of the third body and the ion-pair given by

$$\mathscr{M} = (M_1+M_2)M_3/(M_1+M_2+M_3). \qquad (1\text{-}5\text{-}41)$$

In this general case, the non-thermal effect (cf. § 1-4C) would be expected to assume most importance for light ions recombining in a heavy ambient gas. In this instance, there would be rapid transfer between the internal energy and the kinetic energy of the center-of-mass of the ion-pair; but there would be only slow transfer of energy to the third bodies, and indeed, in the limit of infinite M_3, there would be no such transfer. Falsely attributing thermal motion to the centers-of-mass of the ion-pairs would reduce the back-transfer of their kinetic energy into internal energy and would therefore lead to an overestimate of the recombination coefficient. However, the non-thermal effect is much less for this general case than for ions recombining in their parent gas, an instance in which Bates et al. [22] found the effect to be unimportant as far as the recombination coefficient was concerned.

Thus, the overall rate coefficient which of course is independent of the reference frame chosen is

$$K_{13}(E_i, E_f)\,\mathrm{d}E_f = \mathrm{d}E_f \int_{v_0}^{\infty} \mathscr{G}(v_3)\,\mathrm{d}v_3 \int_{T_0}^{\infty} \Gamma(E_i, E_f; T_i, v_3)\mathscr{F}(T_i)\,\mathrm{d}T_i \qquad (1\text{-}5\text{-}42)$$

where the integration-limits are, for excitation ($E_f \geqq E_i$)

$$T_0 = 0, \quad v_0^2 = 2(E_f-E_i)/\mathscr{M} \qquad (1\text{-}5\text{-}43)$$

and, for de-excitation ($E_f \leqq E_i$)

$$T_0 = (E_i-E_f), \quad v_0^2 = 0. \qquad (1\text{-}5\text{-}44)$$

However, in the laboratory-frame, the corresponding cross sections (which *do* depend on the reference system) for collisions between an incoming atom 3 of speed v_3 relative to the center-of-mass of the target-ion pair whose internal energy is changed from E_i to between E_f and $E_f+\Delta E_f$ is

$$q_{13}(E_i, E_f; v_3) =$$
$$= v_3^{-2} \int_{T_0}^{\infty} v_1^{-1} \mathscr{F}(T_i) \, dT_i \int_{g^-}^{g^+} g^2 \, dg \int_{\psi^-}^{\psi^+} R^{-1} \sigma(g, \psi) \, d(\cos \psi) \, \Delta E_f.$$
(1-5-45)

C. Reduced Parameters and Detailed Balance

When we adopt the normalized set of dimensionless energy variables

$$\lambda = -E_i/k\theta, \quad \mu = -E_f/k\theta \qquad (1\text{-}5\text{-}46)$$

$$x = T_i/k\theta = \tfrac{1}{2}(M_1^2/M_{12})v_1^2/k\theta \equiv Cv_1^2/k\theta$$

$$y = Cv_3^2/k\theta, \quad z = Cg^2/k\theta \qquad (1\text{-}5\text{-}47)$$

and define the corresponding rate

$$\mathscr{K}_{13}(\lambda, \mu) \, d\mu = K_{13}(E_i, E_f) |dE_f| \qquad (1\text{-}5\text{-}48)$$

we have from (1-5-38) and (1-5-42), following some algebraic reduction that

$$\mathscr{K}_{13}(\lambda, \mu) = (16\sqrt{2}/M_{13}^{\frac{1}{2}} \pi^{\frac{3}{2}}) |\lambda|^{\frac{3}{2}} (k\theta)^{\frac{1}{2}} a(1+a)^{\frac{3}{2}} \mathscr{I}(\lambda, \mu) \qquad (1\text{-}5\text{-}49)$$

with the four-dimensional integral

$$\mathscr{I}(\lambda, \mu) = \int_{x_c}^{\infty} (x+\lambda)^{-4} \, dx \int_{y_c}^{\infty} \exp(-ay) \, dy \int_{z^-(x, y; \lambda, \mu)}^{z^+(x, y; \lambda, \mu)} v^{-1} z^{\frac{1}{2}} \, dz$$
$$\times \int_{\omega^-(x, y, z; \lambda, \mu)}^{\omega^+(x, y, z; \lambda, \mu)} [(\omega^+ - \omega)(\omega - \omega^-)]^{-\frac{1}{2}} \sigma(g, \psi) \, d\omega, \quad \omega = \cos \psi$$
(1-5-50)

where the limits are, for activation ($\lambda \geqq \mu$)

$$x_c = 0, \quad y_c = (\lambda - \mu)/a \qquad (1\text{-}5\text{-}51)$$

and for de-activation ($\lambda \leqq \mu$)

$$x_c = (\mu - \lambda), \quad y_c = 0. \qquad (1\text{-}5\text{-}52)$$

In keeping with (1-5-24) and (1-5-29), the smaller of Z_2 and z_2 is z^+, and z^- is the larger of Z_1 and z_1, where

$$z_{2,1}(x, y) = [x^{\frac{1}{2}} \pm y^{\frac{1}{2}}]^2, \qquad (1\text{-}5\text{-}53)$$

and

$$Z_{2,1}(x, y; \lambda, \mu) = [(x+\lambda-\mu)^{\frac{1}{2}} \pm \{y-(\lambda-\mu)/a\}^{\frac{1}{2}}]^2. \qquad (1\text{-}5\text{-}54)$$

The angular limits reduce to

$$\omega^{\pm}(x, y, z; \lambda, \mu) = \cos\psi^{\pm}$$
$$= v^{-2}[\xi\{\xi+(1+1/a)(\lambda-\mu)\} \pm \tau[v^2-\{\xi+(1+1/a)(\lambda-\mu)\}^2]^{\frac{1}{2}}] \quad (1\text{-}5\text{-}55)$$

with the functions

$$\xi(x, y, z) = x-y+(1-a)z/(1+a) \tag{1-5-56}$$

$$\tau^2(x, y, z) = 2z(x+y)-z^2-(x-y)^2 \equiv (z_2-z)(z-z_1) \tag{1-5-57}$$

and

$$v^2(x, y, z) = 4z\{x+ay-az/(1+a)\}/(1+a) = \xi^2+\tau^2 \tag{1-5-58}$$

such that

$$v^2(x, y, z) - \{\xi(x, y, z)+(1+1/a)(\lambda-\mu)\}^2 = \tau^2(x+\lambda-\mu, y-(\lambda-\mu)/a, z)$$
$$\equiv (Z_2-z)(z-Z_1). \tag{1-5-59}$$

Excitation, for example, requires that the lower limits to x and y be zero and $(\lambda-\mu)$, respectively, while the reverse rate $\mathcal{K}_{13}(\mu, \lambda)$ is given by (1-5-9) with λ and μ interchanged and, in accordance with (1-5-51) and (1-5-52)

$$\mathcal{I}_{13}(\mu, \lambda) = \int_{\lambda-\mu}^{\infty}(x+\mu)^{-4}dx \int_0^{\infty}\exp(-ay)dy \int_{g^-}^{g^+}v^{-1}z^{\frac{1}{2}}dz$$
$$\times \int_{\omega^-(x,y,z;\mu,\lambda)}^{\omega^+(x,y,z;\mu,\lambda)}[(\omega^+-\omega)(\omega-\omega^-)]^{-\frac{1}{2}}\sigma(g, \psi)d\omega \quad (1\text{-}5\text{-}60)$$

where now

$$g^- = \max[z_1, Z_1(x, y; \mu, \lambda)]; \quad g^+ = \min[z_2, Z_2(x, y; \mu, \lambda)]. \tag{1-5-61}$$

Changing the integration variables x and y by the transformation

$$x_1 = x+(\mu-\lambda); \quad y_1 = y-(\mu-\lambda)/a \tag{1-5-62}$$

we have that

$$Z_{2,1}(x_1, y_1; \lambda, \mu) = z_{2,1}(x, y) \tag{1-5-63}$$

and hence

$$g^- \equiv \max[Z_1(x_1, y_1; \lambda, \mu), z_1(x_1, y_1)] = z^-(x_1, y_1)$$
$$g^+ = \min[Z_2(x_1, y_1; \lambda, \mu), z_2(x_1, y_1)] = z^+(x_1, y_1). \tag{1-5-64}$$

Also

$$\xi(x, y, z) + (1+1/a)(\mu - \lambda) = \xi(x_1, y_1, z) \tag{1-5-65}$$
$$\tau^2(x, y, z) = v^2(x_1, y_1, z) - \{\xi(x_1, y_1, z) + (1+1/a)(\lambda - \mu)\}^2 \tag{1-5-66}$$
$$v^2(x, y, z) - \{\xi(x, y, z) + (1+1/a)(\mu - \lambda)\}^2 = \tau^2(x_1, y_1, z) \tag{1-5-67}$$

and

$$v^2(x, y, z) = v^2(x_1, y_1, z) \tag{1-5-68}$$

all of which, when taken in conjunction with (1-5-55) and (1-5-60) demonstrate that

$$\omega^{\pm}(x, y, z; \mu, \lambda) = \omega^{\pm}(x_1, y_1, z; \lambda, \mu) \tag{1-5-69}$$

and all the (x_1, y_1, z) limits and integrands are identical with those for the forward rate $\mathscr{K}_{13}(\lambda, \mu)$, except for the additional factor of $\exp(\lambda - \mu)$. Therefore, we have

$$\frac{\mathscr{K}_{13}(\lambda, \mu) \exp \lambda}{|\lambda|^{\frac{5}{2}}} = \frac{\mathscr{K}_{13}(\mu, \lambda) \exp \mu}{|\mu|^{\frac{5}{2}}} \tag{1-5-70}$$

which is a statement of the principle of detailed balance between the forward and reverse rates for the transition ($\lambda \rightleftharpoons \mu$), the equilibrium number density of ion-pairs being taken according to the Saha-Boltzmann distribution (1-3-17).

D. The Rate Coefficient for Special Cases

The key expression for the (1, 2) energy-change brought about by ion-neutral (1–3) collisions is eq. (1-5-38) to which great simplification occurs for the following instances:

(i) For isotropic scattering, the elastic cross section $\sigma(g, \psi)$ for (1–3) impacts is independent of ψ and hence (1-5-38) reduces to

$$\Gamma(E_i, E_f; v_1, v_3) = \frac{\pi}{v_1 v_3} \int_{g^-}^{g^+} \frac{g^2 \sigma(g) \, dg}{(\alpha^2 + \beta^2)^{\frac{1}{2}}} \tag{1-5-71}$$

which is valid in the limit of small polarizability p of the neutral and large hard-sphere radius S_{13}. Moreover, as this limit is approached, $\sigma(g)$ becomes increasingly independent of the relative velocity g (cf. eq. (1-6-24) and Table 1-6-1 of the following section).

(ii) When $\sigma(g, \psi)$ can be expressed as $\sigma'(P)$, a function only of P ($\equiv 2M_{13} g \sin \frac{1}{2}\psi$), the momentum exchanged during the collision, then it is more convenient to proceed as before to eq. (1-5-14) and then to consider the limits to g for a fixed P and energy change $\varepsilon(\equiv E_f - E_i)$, instead of the limits to ψ for constant g and ε. Thus, it is a simple exercise to show that eq. (1-5-34) can be rewritten alternatively as,

$$d\phi = -\frac{dE_f}{R(v_1, v_3, g, 2\sin^{-1}(P/2M_{13}g)|\varepsilon)}$$

$$= -\frac{2g}{P}\frac{dE_f}{(-g^4 + bg^2 - c)^{\frac{1}{2}}} \tag{1-5-72}$$

where

$$b = \frac{a}{(1+a)^2}\frac{P^2}{M_{13}^2} + (V_1^2 + v_1^2 + V_3^2 + v_3^2) - 4\varepsilon^2/P^2 \tag{1-5-73}$$

and

$$c = \frac{v_1^2 + av_3^2}{1+a}\frac{P^2}{M_{13}^2} + (v_1^2 - v_3^2)(V_1^2 - V_3^2) \tag{1-5-74}$$

in which the *final* velocities of particles 1 and 3 are respectively given by

$$V_1^2 = v_1^2 + \frac{2a\varepsilon}{(1+a)M_{13}} \equiv v_1^2 + \frac{2\varepsilon}{M}, \quad M = M_1\left(1 + \frac{M_1}{M_2}\right) \tag{1-5-75}$$

and

$$V_3^2 = v_3^2 - \frac{2\varepsilon}{(1+a)M_{13}} \equiv v_3^2 - 2\varepsilon/\mathcal{M}, \tag{1-5-76}$$

$\mathcal{M}$ being the reduced mass (1-5-41) of the ion-pair neutral system. With v_1, v_3, P and g held fixed, a given ε in (1-5-23) is passed through twice as ϕ is varied between 0 and 2π, and is accessible, for constant v_1, v_3 and P, when g^2 is within the limits

$$g_{\pm}^2 = \tfrac{1}{2}b \pm \{(\tfrac{1}{2}b)^2 - c\}^{\frac{1}{2}}. \tag{1-5-77}$$

Thus, by substituting

$$d(\cos\psi) = P\,dP/M_{13}^2 g^2 \tag{1-5-78}$$

and

$$d\phi = -\frac{2g\,dE_f}{P[(g_+^2 - g^2)(g^2 - g_-^2)]^{\frac{1}{2}}} \tag{1-5-79}$$

in (1-5-31), by proceeding as before and by performing the g-integration analytically, we now have in lieu of (1-5-38) the single integral

$$\Gamma(E_i, E_f; v_1, v_3) = \frac{\pi}{v_1 v_3 M_{13}^2} \int_{P^-}^{P^+} \sigma'(P) \, dP. \tag{1-5-80}$$

The limits $P^\pm$ ensure real $g_\pm^2$ and can be written as

$$P^+ = \min\left[M(V_1+v_1), \mathcal{M}(V_3+v_3)\right] \tag{1-5-81}$$

and

$$P^- = \max\left[M|V_1-v_1|, \mathcal{M}|V_3-v_3|\right] \tag{1-5-82}$$

where $M = M_1(1+M_1/M_2)$ tends to M_1 for infinite M_2. The significance of (1-5-81) and (1-5-82) becomes apparent when, for $\mathcal{M} \geq M$ and $\varepsilon > 0$ (say), P^+ is either $M(V_1+v_1)$ or $\mathcal{M}(V_3+v_3)$ whenever

$$\varepsilon \lessgtr \varepsilon^+ = \frac{4M\mathcal{M}}{(M+\mathcal{M})^2}\left[\tfrac{1}{2}\mathcal{M}v_3^2 - \tfrac{1}{2}Mv_1^2 + \tfrac{1}{2}(\mathcal{M}-M_1)v_1 v_3\right], \tag{1-5-83}$$

respectively, and P^- is either $M(V_1-v_1)$ or $\mathcal{M}(v_3-V_3)$ whenever

$$\varepsilon \lessgtr \varepsilon^- = \frac{4M\mathcal{M}}{(M+\mathcal{M})^2}\left[\tfrac{1}{2}\mathcal{M}v_3^2 - \tfrac{1}{2}Mv_1^2 - \tfrac{1}{2}(\mathcal{M}-M_1)v_1 v_3\right], \tag{1-5-84}$$

respectively. We note that in an elastic collision $\varepsilon^\pm$ represent the kinetic energies transferred from a point-particle of mass $\mathcal{M}$ initially with speed v_3 to another point-particle of mass M moving with velocity v_1 either directed towards or away from the projectile.

Finally, the cross section describing a process in which an incident particle 3 with speed v_3 collides with a particle 1 with speed v_1 attached to particle 2 with speed v_2 in such a manner that the kinetic energy of relative motion of particles 1 and 2 is changed from E_i to between E_f and $E_f + dE_f$ is

$$dQ(v_1, v_3; E_i, E_f) = v_3^{-1}\Gamma(E_i, E_f; v_1, v_3) dE_f \tag{1-5-85}$$

in the laboratory frame with the (1–2) center-of-mass initially at rest. The function Γ is given by one of (1-5-38), (1-5-71) or (1-5-80) depending on the properties of the (1–3) elastic scattering differential cross section.

§ 1-6. Calculation of the recombination coefficient

The energy-change rate coefficient $K_{13}(E_i, E_f)$ given by (1-5-38) and (1-5-42) contains the differential cross section $\sigma_{13}(g, \psi)$ for elastic scattering of the ion 1 by the neutral body 3 as a variable within a four-dimensional integral. In order to render the resulting calculation of $K(E_i, E_f)$ feasible, it is therefore necessary to devise a relatively simple, yet accurate, technique for the evaluation of σ_{13}. Mason [39] has emphasized that numerical differentiation of the impact parameter with respect to the angle of scatter can lead to errors as large as several percent. This difficulty was obviated by Flannery [40] who, for the case of scattering by any potential field, derived a general expression for σ_{13} which is capable of simple computation to any desired degree of accuracy. In this section this procedure will be briefly described and applied to scattering by the Langevin field (1-5-4) with the eventual aim of calculating $K(E_i, E_f)$, the quasi-equilibrium distribution $\rho(\lambda)$ with the aid of (1-3-25), and hence, the recombination coefficient α given by (1-3-13).

A. DIFFERENTIAL SCATTERING CROSS SECTIONS

For an incident particle of mass M_{13} kinetic energy E, moving with speed g in the potential field $V(r)$ of a fixed target, the deflection angle ψ is expressed in terms of the impact parameter b by the following collision integral [41]

$$\psi = \pi - 2b \int_{r_c}^{\infty} r^{-2} \{1 - V(r)/E - b^2/r^2\}^{-\frac{1}{2}} dr \qquad (1\text{-}6\text{-}1)$$

where r_c is the classical distance of closest approach, given by the outermost zero of the function

$$F(r) = 1 - V(r)/E - b^2/r^2. \qquad (1\text{-}6\text{-}2)$$

In general, b is *not* a single-valued function of ψ and so the differential cross section is defined by

$$\sigma_{13}(g, \psi) = \sum_{\psi} \left| \frac{b\,db}{d(\cos \psi)} \right| \qquad (1\text{-}6\text{-}3)$$

in which the summation is taken over all impact parameters that yield the angles $\pm\psi$, $-2\pi\pm\psi$, $-4\pi\pm\psi$ etc. (with $0 \leq \psi \leq \pi$). The pole in the integrand of (1-6-1) can be separated out by changing the integration variable to

$$x = r_c/r \qquad (1\text{-}6\text{-}4)$$

with the result

$$\psi = \pi - 2(b/r_c) \int_0^1 (1-x^2)^{-\frac{1}{2}} f(x)\,dx \qquad (1\text{-}6\text{-}5)$$

in which

$$f(x) = (1-x^2)^{\frac{1}{2}} g(x) \tag{1-6-6}$$

with

$$g(x) = \{1 - V(r_c/x)/E - b^2 x^2/r_c^2\}^{-\frac{1}{2}} \tag{1-6-7}$$

is well behaved over the range of integration. Therefore, the integral can be expanded by using the Gauss-Mehler quadrature formula

$$\int_{-1}^{+1} (1-x^2)^{-\frac{1}{2}} f(x) \, dx = \frac{\pi}{n} \sum_{j=1}^{n} f\left(\cos \frac{2j-1}{2n} \pi\right) \tag{1-6-8}$$

where n is an even integer [42] to yield

$$\psi = \pi - 2(b\pi/nr_c) \sum_{j=1}^{\frac{1}{2}n} a_k g(a_j) \tag{1-6-9}$$

with

$$a_j = \cos\left(\frac{2j-1}{2n}\pi\right), \quad k = \tfrac{1}{2}n + 1 - j \tag{1-6-10}$$

since $g(x)$ is an even function of x. Smith [43] has noted that (1-6-9) converges very rapidly with n, and is in general more accurate and simpler to use than other integration schemes (see, however, O'Hara and Smith [43]). The customary procedure for evaluating σ_{13} has been to tabulate ψ as a function of b and then to differentiate numerically in accordance with (1-6-3). However, this method is subject to unpredictable error, and an exact expression for (1-6-3) may be obtained by differentiating each term of the summation (1-6-9) analytically according to the formula

$$\frac{d\psi}{db} = \frac{\partial \psi}{\partial b} + \frac{\partial \psi}{\partial r_c} \frac{dr_c}{db} \tag{1-6-11}$$

to obtain the equation

$$\frac{d\psi}{db} = \left(\frac{dr_c}{db}\right) \frac{\pi}{nE} \left[\frac{r_c}{b} \sum_{j=1}^{\frac{1}{2}n} a_k f(a_j) V'(r_c) + \frac{b}{r_c} \sum_{j=1}^{\frac{1}{2}n} a_k f^3(a_j) \{a_j^2 V'(r_c) - V'(r_c/a_j)\}\right] \tag{1-6-12}$$

with

$$\frac{dr_c}{db} = \left[\frac{b}{r_c} - \frac{r_c^2}{2bE} V'(r_c)\right]^{-1}, \quad V'(r_c/a_j) = \left[\frac{dV(r/a_j)}{dr}\right]_{r=r_c}. \tag{1-6-13}$$

The foregoing procedure will now be applied to the following (Langevin) interaction between an ion of charge e and a neutral of polarizability p

$$V(r) = \begin{cases} -\dfrac{pe^2}{2r^4}, & r > S \\ \infty, & r \leq S \end{cases} \tag{1-6-14}$$

where r is the interparticle distance and S is the hard-sphere core parameter. Hassé [38] inserted the above potential into (1-6-1) which he then expressed in terms of elliptic functions (see the book by McDaniel [13]). It is convenient to change to the variables defined by

$$Y = \left(\frac{M_{13}}{pe^2}\right)^{\frac{1}{2}} b^2 g, \quad Z = \frac{S^2}{b^2} Y \tag{1-6-15}$$

such that (1-6-1) is transformed to the set

$$\psi_{AR} = \pi - (Y/Z)^{\frac{1}{2}} \int_{-1}^{+1} (1 - YZ^{-1}\rho^2 + Z^{-2}\rho^4)^{-\frac{1}{2}} d\rho, \quad Y \leq Z + Z^{-1} \tag{1-6-16}$$

and

$$\psi_A = \pi - \rho_0 \int_{-1}^{+1} \{1 - (\rho_0\rho)^2 + Y^{-2}(\rho_0\rho)^4\}^{-\frac{1}{2}} d\rho, \quad Y \geq Z + Z^{-1} \tag{1-6-17}$$

with

$$2\rho_0^2 = Y\{Y - (Y^2 - 4)^{\frac{1}{2}}\}. \tag{1-6-18}$$

The value of ψ at $Y = Z + Z^{-1}$ defines the collision glancing angle ψ_c which separates the scattering $\psi_A (\leq \psi_c)$ arising from the attraction of the induced dipole alone from $\psi_{AR} (\geq \psi_c)$ which is caused by the combined effects of the hard-sphere and dipole potentials. By using the Gauss-Mehler integration formula (1-6-8) we can present ψ as a function of Y for an extensive range of Z. Typical results are plotted in Fig. 1-6-1 from which we note that the curves for all finite $Z(\geq 1)$ eventually connect at $Y = Z + Z^{-1}$ with the lower curve which describes the scattering (through negative angles) by the polarization potential. Y equals 2 represents the condition for orbiting collisions with the cross section obtained from (1-6-15) to give

$$\pi b_0^2 = \frac{2\pi}{g}\left(\frac{pe^2}{M_{13}}\right)^{\frac{1}{2}}, \quad \sigma_{\text{orb}}(g, \psi) = \tfrac{1}{4} b_0^2 \tag{1-6-19}$$

in agreement with Gioumousis and Stevenson [44].

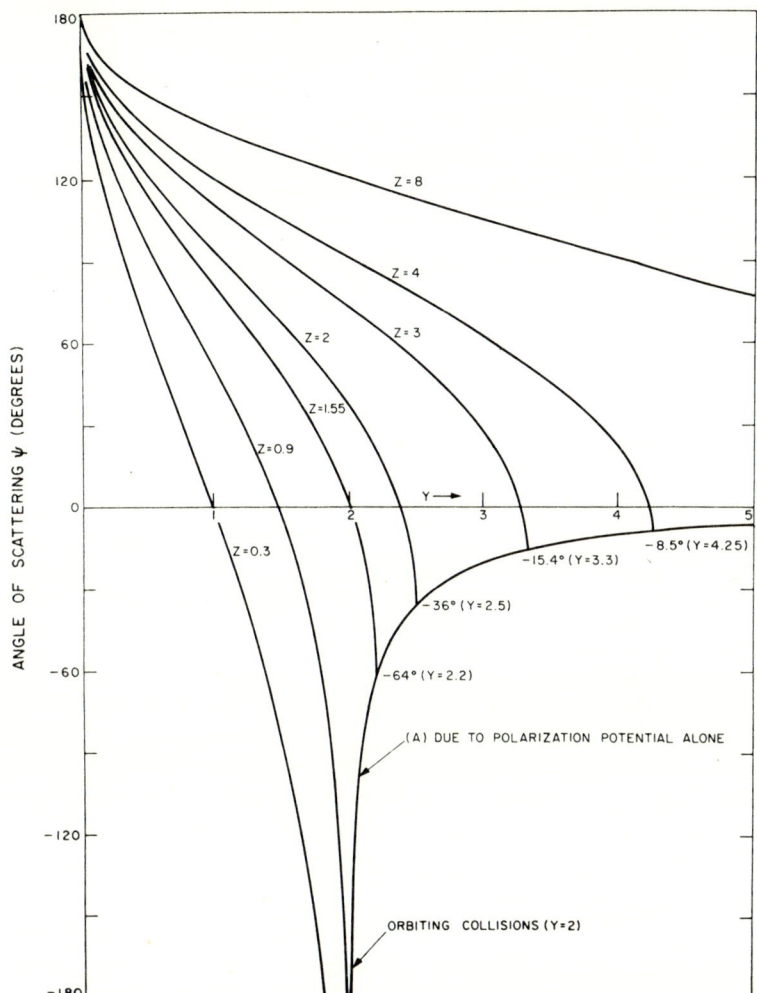

Fig. 1-6-1. Deflection angle for scattering by the Langevin potential as a function of Y and Z. (A) denotes scattering by the purely polarization potential. Curves for $Z \geq 1$ eventually connect with (A) at $Y = Z + Z^{-1}$, which determine collision glancing angles indicated on each curve. Orbiting occurs at $Y = 2$, which is accessible for $Z \leq 1$.

Since the distance of closest approach for hard-sphere repulsion is S, independent of b, we have from (1-6-12) that

$$\frac{d\psi_{AR}}{dY} = -\frac{\pi}{n}\left(\frac{Y}{Z}\right)^{\frac{1}{2}\frac{1}{2}n} \sum_{j=1} a_k g_1(a_j)\{Y^{-1} + Z^{-1}a_j^2 g_1^2(a_j)\} \tag{1-6-20}$$

and

$$\frac{d\psi_A}{dY} = \frac{2\pi}{n}\left(\frac{d\rho_0}{dY}\right) \sum_{j=1}^{\frac{1}{2}n} a_k g_2(a_j)\{\rho_0^2 a_j^2(a_j^2-1)g_2^2(a_j)-1\} \quad (1\text{-}6\text{-}21)$$

TABLE 1-6-1

Values of $\Delta(\psi, Z)$ for scattering by the Langevin potential

Z	0.3	0.9	1.55	2	3	4	8
ψ_c*	—	—	−64°	−36°	−15.4°	−8.5°	−2.1°
ψ (degrees)							
2	3344	3346	3341	3334	3335	3333	3332
10	61.11	61.38	60.41	60.07	59.52	3.202	4.634
20	11.85	12.00	11.54	11.43	2.135	2.508	4.269
30	4.917	5.034	4.758	4.722	1.866	2.290	4.151
40	2.804	2.911	2.752	1.323	1.741	2.189	4.096
50	1.898	2.001	1.906	1.233	1.670	2.132	4.066
60	1.435	1.539	1.477	1.176	1.627	2.098	4.048
80	1.002	1.103	0.911	1.113	1.580	2.061	4.029
100	0.832	0.926	0.874	1.080	1.556	2.042	4.019
120	0.780	0.864	0.852	1.063	1.542	2.032	4.014
140	0.888	0.909	0.841	1.053	1.535	2.026	4.010
160	1.352	1.195	0.834	1.046	1.532	2.023	4.009

* ψ_c is the glancing angle (in degrees) for a collision with a given Z. Δ is a measure of the departure of the elastic cross section from the cross section (1-6-19) for orbiting collisions.

where $g_1(\rho)$ and $g_2(\rho)$ denote the integrands of (1-6-16) and (1-6-17), respectively. The above expressions converge very rapidly for n of the order 10 to 20. Table 1-6-1 contains values of the departure coefficient

$$\Delta(\psi, Z) = \sum_\psi \left|\frac{dY}{d(\cos\psi)}\right| \quad (1\text{-}6\text{-}22)$$

such that the rate of $(1, 3)$ collisions into unit solid angle about the (ψ, ϕ) direction can be written as

$$g\sigma_{13}(g,\psi) = \frac{1}{2}\left(\frac{pe^2}{M_{13}}\right)^{\frac{1}{2}} \Delta(\psi, Z) \equiv \frac{1}{2}\frac{b^2}{Y}g\Delta \equiv \frac{1}{2}\frac{S^2}{Z}g\Delta \equiv \tfrac{1}{4}b_0^2 g\Delta. \quad (1\text{-}6\text{-}23)$$

The entries in Table 1-6-1 represent the analytical differentiation (1-6-22) of the (ψ, Y) curves of Fig. 1-6-1 for given ψ and Z. The large values of the differential cross section obtained for small ψ simply reflect the failure of classical mechanics to properly describe the small-angle scattering that arises from large Y (or large impact-parameters). The pole in $\sigma(g,\psi)$ essentially arises from the asymptotic behaviour of curve A in Fig.

1-6-1. However, the angular limits $\psi^{\pm}$ in eq. (1-5-18) serve as exact cut-offs to the scattering and so prevent the occurrence of this divergence.

We also note from the table that for $Z > 1$, the departure coefficient decreases monotonically with angle towards a limiting value which, for large Z, tends to $\tfrac{1}{2}Z$. Moreover, for large Z (which result from an increase in either the hard-sphere radius S and/or the speed g) the scattering becomes increasingly isotropic (independent of ψ) in accordance with the relations

$$Y = \tfrac{1}{2}Z(1+\cos\psi), \quad \varDelta(\psi, Z) = \tfrac{1}{2}Z \qquad (1\text{-}6\text{-}24)$$

which represent hard-sphere repulsive scattering, valid when the influence of the attractive potential is negligible.

For small Z (<1), the cross section decreases when ψ reaches a minimum at about 130° and then increases. This behaviour can be directly attributed to the fact that, in this region, $\sin\psi$ decreases more rapidly than $dY/d\psi$ such that $dY/d(\cos\psi)$ increases. On the other hand, for larger Z, $dY/d\psi$ decreases faster than $\sin\psi$ and for sufficiently large Z, the rates of decrease of both functions are equal and exactly compensate each other so that their ratio is a constant in keeping with (1-6-24).

With the knowledge of $\varDelta(\psi, Z)$ we can therefore proceed with the evaluation of $\mathscr{K}_{13}(\lambda, \mu)$ in (1-5-49) which now reduces to

$$\mathscr{K}_{13}(\lambda, \mu) = 8e(1+a)\left[\frac{ap|\lambda|^5}{\pi^3 M_{13}}\right]^{\tfrac{1}{2}} \mathscr{I}(\lambda, \mu) \qquad (1\text{-}6\text{-}25)$$

with

$$\mathscr{I}(\lambda, \mu) = \int_{x_c}^{\infty} (x+\lambda)^{-4}\,dx \int_{y_c}^{\infty} \mathscr{J}(x, y; \lambda, \mu)\exp(-ay)\,dy \qquad (1\text{-}6\text{-}26)$$

$$\mathscr{J}(x, y; \lambda, \mu) = \int_{z^{-}(\lambda, \mu)}^{z^{+}(\lambda, \mu)} \mathscr{H}(x, y, z; \lambda, \mu)v^{-1}(x, y, z)\,dz \qquad (1\text{-}6\text{-}27)$$

$$\mathscr{H}(x, y, z; \lambda, \mu) = \int_{\omega^{-}(\lambda, \mu)}^{\omega^{+}(\lambda, \mu)} \frac{\varDelta(\psi, \{az/(1+a)\}^{\tfrac{1}{2}}A)}{[(\omega^{+}-\omega)(\omega-\omega^{-})]^{\tfrac{1}{2}}}\,d\omega; \quad \omega = \cos\psi$$
$$(1\text{-}6\text{-}28)$$

in which a is the *mass ratio parameter* given by

$$a = M_2 M_3 / M_1(M_1 + M_2 + M_3) \qquad (1\text{-}6\text{-}29)$$

and A the *interaction parameter* defined as

$$A = (2k\theta)^{\tfrac{1}{2}} S_{13}^2 / p^{\tfrac{1}{2}} e. \qquad (1\text{-}6\text{-}30)$$

All the remaining quantities, i.e., x, y, z, ω, λ and μ have their usual significance.

It is seen that $\mathscr{K}_{13}(\lambda, \mu)$ depends on the polarizability p, on the reduced mass M_{13} and on the mass ratio and interaction parameters a and A. Reminders of the dependence are omitted for the sake of compactness.

The poles $\omega^{\pm}$ in (1-6-28) can be removed by transforming the integration variable to

$$\tilde{\omega} = [\omega - \tfrac{1}{2}(\omega^{+} + \omega^{-})]/(\omega^{+} - \omega^{-}) \qquad (1\text{-}6\text{-}31)$$

and adoption of the Gauss-Mehler quadrature yields

$$\mathscr{H}(x, y, z; \lambda, \mu) = (\pi/n) \sum_{j=1}^{n} \Delta(\psi_j, \{az/(1+a)\}^{\frac{1}{2}}A) \qquad (1\text{-}6\text{-}32)$$

where the pivots are

$$\omega_j \equiv \cos \psi_j = \tfrac{1}{2}(\omega^{+} + \omega^{-}) + \tfrac{1}{2}(\omega^{+} - \omega^{-}) \cos \frac{2j-1}{2n} \pi. \qquad (1\text{-}6\text{-}33)$$

B. Computations

Since the integrals of (1-6-25) involve only the mass ratio parameter a and the interaction parameter A, Bates and Flannery were able to carry out computations which cover any particular case within a certain range. The range of most interest is

$$0.1 < a < 0.9, \quad 0.3 < A < 2.0 \qquad (1\text{-}6\text{-}34)$$

The classical deflection angle ψ was calculated by Gauss-Mehler quadrature, each term of which being differentiated analytically in accordance with (1-6-20) and (1-6-21) to give $\Delta(\psi, Z)$, and hence the further use of the Gauss-Mehler formula yielded $\mathscr{H}$ given by (1-6-32). Straightforward Gaussian quadrature was used both for the integral $\mathscr{J}$ and for the integral over y in (1-6-26), the decreasing exponential tail being integrated analytically after being curve-fitted. The final integration over x to give $\mathscr{I}$ was carried out similarly except in the region of the sharp peak where Simpson's rule, with a built-in error check, was applied. Though the complete integral is, in effect, five-dimensional, an accuracy of well within 1 % was expected. Evidence that this accuracy was achieved was provided by the closeness with which the results satisfy the principle of detailed balance (1-5-70).

In order to calculate the recombination coefficient (1-3-13), it is necessary to have a large number of values of $\mathscr{K}_{13}(\lambda, \mu)$ readily available.

With this in mind, the semi-infinite region, $\lambda > \mu$, was divided into two and the constants of the following representations were determined by the method of least squares:

$$\ln \mathscr{K}_{13}(\lambda, \mu) = a_1(\lambda-\mu)^3 + b_1\lambda^2 + c_1\mu^2 + d_1\lambda\mu + e_1\lambda + f_1\mu$$
$$+ g_1 + h_1/(\lambda-\mu), \quad 0 < (\lambda-\mu) \leq 2 \quad (1\text{-}6\text{-}35)$$

$$\ln \mathscr{K}_{13}(\lambda, \mu) = b_2\lambda^2 + c_2\mu^2 + d_2\lambda\mu + e_2\lambda + f_2\mu + g_2$$
$$+ h_2/(\lambda-\mu), \quad 2 < (\lambda-\mu). \quad (1\text{-}6\text{-}36)$$

Flannery [45] has tabulated the coefficients which give the values of $\mathscr{K}_{13}(\lambda, \mu)$ correct to within 1 %. Because of detailed balance they also cover the other semi-infinite region, $\lambda < \mu$.

TABLE 1-6-2

Energy change rate coefficient

μ \ λ	Function $I(\lambda, \mu\|0.3, 0.3)$ of eq. (1-6-37)				
	1.0	2.0	3.0	4.0	5.0
−5.0	1.89^{-3}	6.21^{-4}	2.08^{-4}	6.94^{-5}	2.37^{-5}
−3.0	2.63^{-2}	7.51^{-3}	2.29^{-3}	7.21^{-4}	2.34^{-4}
−2.0	9.79^{-2}	2.65^{-2}	7.70^{-3}	2.36^{-3}	7.54^{-4}
−1.0	4.04^{-1}	9.75^{-2}	2.68^{-2}	7.80^{-3}	2.40^{-3}
−0.6	7.17^{-1}	1.66^{-1}	4.43^{-2}	1.27^{-2}	3.86^{-3}
−0.3	1.15^{0}	2.51^{-1}	6.50^{-2}	1.84^{-2}	5.50^{-3}
0.0	1.89^{0}	3.82^{-1}	9.45^{-2}	2.65^{-2}	7.82^{-3}
0.3	3.28^{0}	5.86^{-1}	1.42^{-1}	3.85^{-2}	1.21^{-2}
0.5	5.09^{0}	7.86^{-1}	1.86^{-1}	4.94^{-2}	1.42^{-2}
0.7	9.12^{0}	1.07^{0}	2.43^{-1}	6.35^{-2}	1.81^{-2}
0.9	2.60^{1}	1.47^{0}	3.21^{-1}	8.18^{-2}	2.30^{-2}
1.0	—	1.74^{0}	3.60^{-1}	9.19^{-2}	2.62^{-2}
2.0	—	—	1.63^{0}	3.45^{-1}	8.97^{-2}
3.0	—	—	—	1.56^{0}	3.33^{-1}
4.0	—	—	—	—	1.50^{0}

The index gives the power of 10 by which the entry must be multiplied.

The form of $\mathscr{K}_{13}(\lambda, \mu)$ may be seen from Table 1-6-2 which gives some values of $I(\lambda, \mu|a, A)$ which is dimensionless and such that

$$\mathscr{K}_{13}(\lambda, \mu) = ep^{\frac{1}{2}}a^{\frac{1}{2}}(1+a)M_{13}^{-\frac{1}{2}}I(\lambda, \mu|a, A). \quad (1\text{-}6\text{-}37)$$

As would be expected $\mathscr{K}_{13}(\lambda, \mu)$ falls off very rapidly as $|\lambda-\mu|$ is increased from zero. In the case where symmetrical resonance charge-transfer collisions are responsible for the energy change, the corresponding function has no pole and its variation with $|\lambda-\mu|$ is relatively slow (cf. Table 1-4-1).

C. Partial recombination coefficient

The function analogous to $\mathscr{K}_{13}(\lambda, \mu)$ which describes the energy-change brought about by (2–3) collisions is clearly

$$\mathscr{K}_{23}(\lambda, \mu) = ep^{\frac{1}{2}}b^{\frac{1}{2}}(1+b)M_{23}^{-\frac{1}{2}}I(\lambda, \mu|b, B) \tag{1-6-38}$$

where

$$b = M_1 M_3/M_2(M_1+M_2+M_3), \quad B = (2k\theta)^{\frac{1}{2}}S_{23}^2/p^{\frac{1}{2}}e. \tag{1-6-39}$$

Thus, the rate $\mathscr{K}(\lambda, \mu)$ to be used in the integral equation (1-3-25) for the quasi-equilibrium distribution $\rho(\lambda)$ is given by

$$\mathscr{K}(\lambda, \mu) = \mathscr{K}_{13}(\lambda, \mu) + \mathscr{K}_{23}(\lambda, \mu). \tag{1-6-40}$$

The low-density limit to the recombination coefficient for process (1-5-1) is the following expression given by (1-3-13) and (1-3-17)

$$\alpha = \mathscr{C}\int_{\mu=-\infty}^{v} d\mu \int_{\lambda=v}^{\omega} \lambda^{-\frac{3}{2}} \exp(\lambda)\{\rho(\mu)-\rho(\lambda)\}\mathscr{K}(\lambda, \mu) d\lambda \quad \text{cm}^3 \text{ sec}^{-1} \tag{1-6-41}$$

with

$$\mathscr{C} = \frac{\pi^{\frac{3}{2}}e^6 N(Z)}{2(k\theta)^3} = 1.3 \times 10^{-8} N(Z)\theta^{-3}$$

where the parameters have their customary significance and where the quasi-equilibrium distribution $\rho(\lambda)$ can be determined from the reduced form (1-4-45) of the integral equation (1-3-25). Thus, by using (1-6-38)

$$\alpha = ep^{\frac{1}{2}}\mathscr{C}[M_{13}^{-\frac{1}{2}}\gamma(a, A) + M_{23}^{-\frac{1}{2}}\gamma(b, B)] \tag{1-6-42}$$

with

$$\gamma(c, C) = c^{\frac{1}{2}}(1+c)\int_{\mu=-\infty}^{v} d\mu \int_{\lambda=v}^{\omega} \lambda^{-\frac{3}{2}} \exp(\lambda)\{\rho(\mu)-\rho(\lambda)\} I(\lambda, \mu|c, C). \tag{1-6-43}$$

However, at this stage, it is convenient to introduce the following *partial* recombination coefficients together with the corresponding distribution functions: α_{13} and $\rho_{13}(\lambda)$ the recombination coefficient and distribution function that would arise if *only* 1–3 collisions occurred; and α_{23} and $\rho_{23}(\lambda)$

the recombination coefficient and distribution function that would arise if *only* 2–3 collisions occurred. If $\rho_{13}(\lambda)$ and $\rho_{23}(\lambda)$ are the same as each other it is apparent from (1-3-25) that they are also the same as $\rho(\lambda)$ and it follows from (1-6-41) that in this circumstance

$$\alpha = \alpha_{13} + \alpha_{23} \qquad (1\text{-}6\text{-}44)$$

which is clearly satisfied if the masses M_1 and M_2 and the hard-sphere radii S_{13} and S_{23} are equal. According to the theory of Thomson, the low-density limit to the recombination coefficient is (cf. eq. (1-2-8)),

$$\alpha_T = \frac{32\pi e^6 N(Z)}{27(k\theta)^{\frac{5}{2}}(3M_{12})^{\frac{1}{2}}}[Q_{13} + Q_{23}] \qquad (1\text{-}6\text{-}45)$$

where Q_{13} and Q_{23} are the effective cross sections for 1–3 and 2–3 collisions. Thus on this theory, (1-6-44) is of general validity.

TABLE 1-6-3
Combination rule (1-6-44) ($a = 0.3$)

		$A = 0.9, B = 0.9$		$A = 1.5, B = 0.9$		$A = 2.0, B = 0.9$	
b	M_2/M_1	$M_3^{\frac{1}{3}}\beta$	$M_3^{\frac{1}{3}}(\beta_{13}+\beta_{23})$	$M_3^{\frac{1}{3}}\beta$	$M_3^{\frac{1}{3}}(\beta_{13}+\beta_{23})$	$M_3^{\frac{1}{3}}\beta$	$M_3^{\frac{1}{3}}(\beta_{13}+\beta_{23})$
0.1	1.732	2.066	2.022	2.478	2.425	2.966	2.910
0.2	1.225	2.740	2.730	3.167	3.152	3.678	3.659
0.3	1.000	3.323	3.323	3.762	3.761	4.291	4.288
0.4	0.866	3.768	3.767	4.222	4.221	4.768	4.766
0.5	0.775	4.154	4.154	4.624	4.622	5.188	5.186
0.9	0.577	5.430	5.435	5.973	5.969	6.599	6.600
1.2	0.500	6.265	6.275	6.849	6.855	7.547	7.552
1.6	0.433	7.476	7.495	8.145	8.156	8.940	8.952
2.0	0.387	8.864	8.887	9.648	9.659	10.578	10.588

Note: M_3 is measured in grammes; $M_3^{\frac{1}{3}}\beta$ and $M_3^{\frac{1}{3}}(\beta_{13}+\beta_{23})$ are dimensionless.

In order to explore the position further, Bates and Flannery used (1-6-25), (1-6-40), (1-3-25) and (1-6-41) to determine $\rho(\lambda)$ and α in a number of representative cases, and also, replacing $\mathscr{K}(\lambda, \mu)$ by $\mathscr{K}_{13}(\lambda, \mu)$ and then by $\mathscr{K}_{23}(\lambda, \mu)$ to determine the corresponding $\rho_{13}(\lambda)$ and α_{13} and the corresponding $\rho_{23}(\lambda)$ and α_{23}, respectively. Table 1-6-3 compares $M_3^{\frac{1}{3}}$ times the values of

$$\beta \equiv \alpha/(ep^{\frac{1}{2}}\mathscr{C}) \qquad (1\text{-}6\text{-}46)$$

with $M_3^{\frac{1}{2}}$ times the values of

$$\beta_{13}+\beta_{23} \equiv (\alpha_{13}+\alpha_{23})/(ep^{\frac{1}{2}}\mathscr{C}) \tag{1-6-47}$$

for a number of mass ratio and interaction parameters. It may be seen that (1-6-44) is indeed satisfied to a close approximation. The existence of this accurate combination rule facilitates the task of presenting the results. Thus the difficulty arising from the multiplicity of parameters on which the recombination coefficient depends may be reduced by concentrating attenion on the partial recombination coefficients or the related dimensionless coefficients

$$\gamma_{13} \equiv \alpha_{13} M_{13}^{\frac{1}{2}}/(ep^{\frac{1}{2}}\mathscr{C}), \quad \gamma_{23} \equiv \alpha_{23} M_{23}^{\frac{1}{2}}/(ep^{\frac{1}{2}}\mathscr{C}). \tag{1-6-48}$$

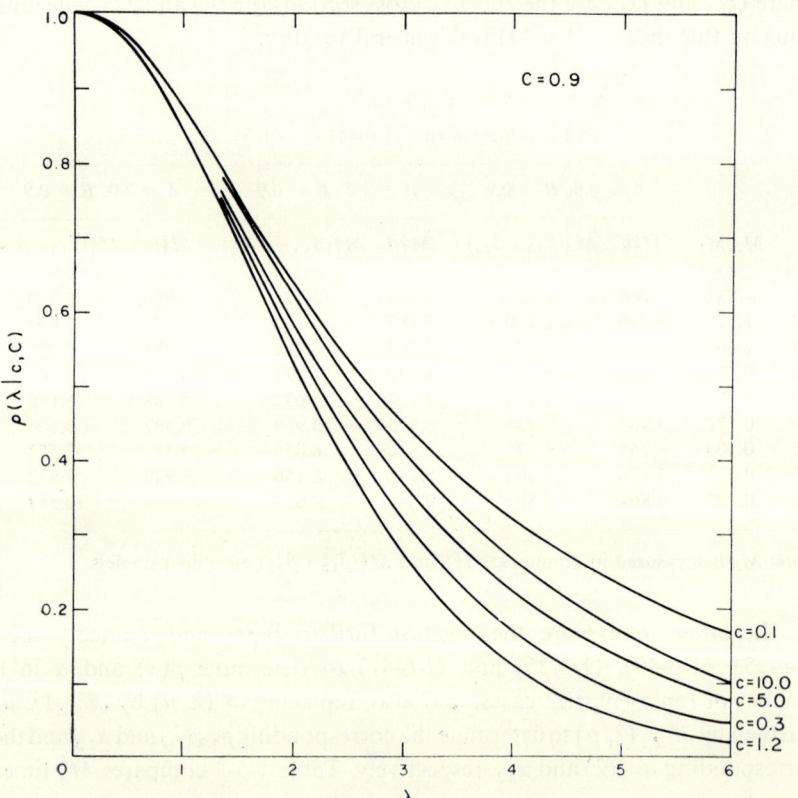

Fig. 1-6-2. Distribution function $\rho(\lambda|c, C)$ for selected values of the mass-ratio parameter c (indicated on right).

Moreover, it is apparent that there is only one basic distribution function, $\rho(\lambda|c, C)$, and only one basic coefficient, $\lambda(c, C)$, both dependent on two parameters and such that

$$\rho_{13}(\lambda) = \rho(\lambda|a, A), \quad \rho_{23}(\lambda) = \rho(\lambda|b, B) \tag{1-6-49}$$

$$\gamma_{13} = \gamma(a, A), \quad \gamma_{23} = \gamma(b, B). \tag{1-6-50}$$

D. Quasi-equilibrium Distribution Function

Some of the distribution functions $\rho(\lambda|c, C)$ obtained from the theory of the previous sections with h, ε and ω taken as $\frac{1}{6}$, 10 and 30, respectively, are shown in Figs. 1-6-2 and 1-6-3. It may be observed from Fig. 1-6-2 that the

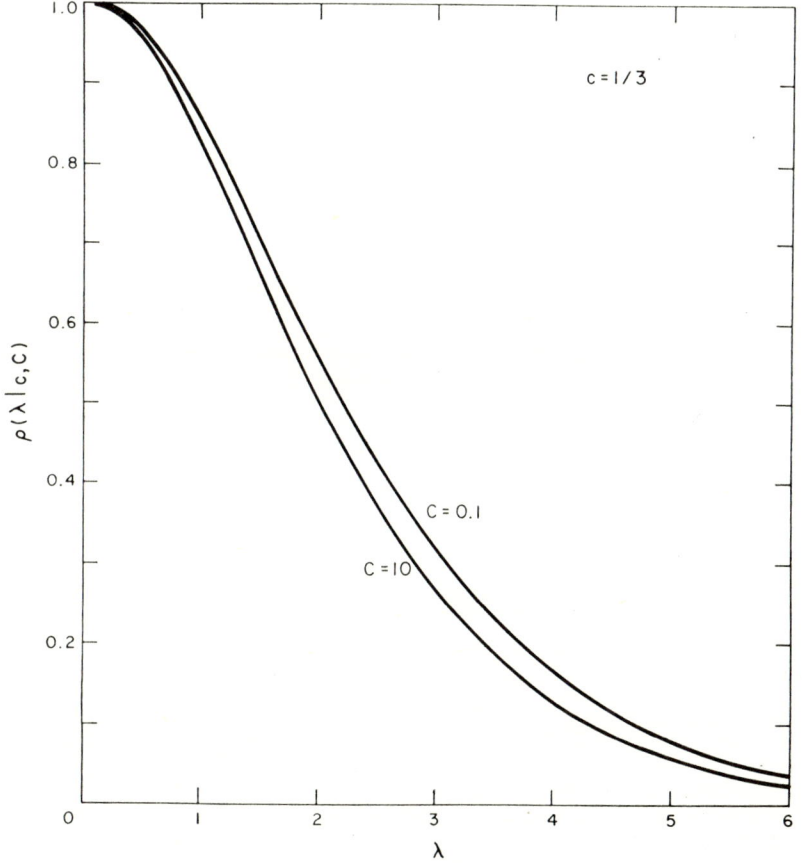

Fig. 1-6-3. Distribution function $\rho(\lambda|c, C)$ for selected values of the interaction parameter C (indicated on curves).

rate of fall-off as λ is increased is relatively slow if c is either small or large compared with unity (a necessary condition for which is that one of the three masses concerned is very different from the others). The effect of C on the distribution functions (cf. Fig. 1-6-3), is slight: small C gives a distribution function characteristic of a polarization interaction (small-ion limit) while large C gives one characteristic of a repulsive hard-sphere core interaction. The rate of fall-off just below the continuum $0 < \lambda < 0.25$ is much less than the corresponding rate in the symmetrical resonance charge-transfer case (cf. Fig. 1-4-1). This effect is due to the marked difference between the forms of the energy-change rate coefficients when $|\lambda-\mu|$ is small.

The computed values of $\gamma(c, C)$ are presented in Table 1-6-4. Substitution in (1-6-42) gives the recombination coefficients. Use of the table is facilitated by noting that

$$\frac{M_1}{M_3} = \frac{1-(ab)^{\frac{1}{2}}}{a^{\frac{1}{2}}(a^{\frac{1}{2}}+b^{\frac{1}{2}})} \tag{1-6-51}$$

and

$$\frac{M_2}{M_3} = \frac{1-(ab)^{\frac{1}{2}}}{b^{\frac{1}{2}}(a^{\frac{1}{2}}+b^{\frac{1}{2}})}. \tag{1-6-52}$$

TABLE 1-6-4

Dimensionless coefficient $\gamma(c, C)$ which appears in (1-6-50) and is related to recombination coefficient by (1-6-48)

		$\gamma(c, C)$					
C	c	0.2	0.3	0.4	0.5	0.9	
0.3		0.947	1.271	1.447	1.575	1.765	
0.9		0.897	1.219	1.399	1.519	1.724	
1.5		1.092	1.540	1.784	1.947	2.206	
2.0		1.352	1.927	2.231	2.435	2.754	
c	0.1	0.6	1.2	1.6	2.0	5.0	10.0
$\gamma(c, 0.9)$	0.428	1.598	1.739	1.714	1.633	1.151	0.746
C		0.01	0.1	0.2	10		
$\gamma(\frac{1}{3}, C)$		1.418	1.364	1.343	(9.667)*		

* In evaluating $\gamma(\frac{1}{3}, 10)$ the approximation that, see eq. (1-6-24), $Y = \frac{1}{2}Z(1+\cos\psi)$ was made in the region $Z > 10$ to save computing time (which would otherwise have been extremely lengthy in this particular case). The error introduced is difficult to estimate but is certainly less than 1%. Our only purpose in taking C to be as high as 10 was to verify that the dependent variable in the curve in Fig. 1-6-4 does indeed approach a constant value as the independent variable is increased.

All the relevant information regarding the dependence of the recombination coefficient on the gas temperature, the masses of the three species involved, the two Langevin hard-sphere radii and the polarizability of the gas may be readily derived from Table 1-6-4.

E. TEMPERATURE AND INTERACTION

The theory of Thomson provides a useful basis for comparison. It is natural to identify the cross sections appearing in Thomson's formula (1-6-45) with the momentum transfer cross sections. If the interaction is of the Langevin type (1-5-4) the 1–3 momentum transfer cross section is

$$Q_{D,13} = \frac{3\sqrt{2\pi}ep^{\frac{1}{2}}}{8(k\theta)^{\frac{1}{2}}g(A)} \tag{1-6-53}$$

$g(A)$ being a certain tabulated function [13] of the interaction potential and has magnitude 0.5105 in the polarization, or small-ion, limit ($A = 0$) and $A\,g(A)$ tends to 0.75 in the elastic-sphere limit. It may be seen from (1-6-45) that the Thomson approximation to the 1–3 partial recombination coefficient is

$$\alpha_{T,13} = \frac{4\sqrt{2}\pi^2 e^7 p^{\frac{1}{2}} N(Z)}{9(k\theta)^3 (3M_{12})^{\frac{1}{2}} g(A)} \tag{1-6-54}$$

which can, in fact, account for details of the ion-neutral interaction. From (1-6-42) and (1-6-54) the ratio

$$\mathscr{R} = \alpha_{13}/\alpha_{T,13} \tag{1-6-55}$$

$$= \frac{9\sqrt{3}}{8\sqrt{2\pi}} \left(\frac{M_{12}}{M_{13}}\right)^{\frac{1}{2}} g(A)\gamma(a, A) \tag{1-6-56}$$

provides a measure of the effectiveness of the Thomson formula. Figure 1-6-4 shows the variation of $\mathscr{R}$ with the interaction parameter A when $M_1 = M_2 = M_3$ (i.e., $a = \frac{1}{3}$) which is the case for which the Thomson theory was designed. The variation is only slight. This signifies that the Thomson treatment provides a good description of how the recombination coefficient for given species depends on the temperature and on the form of the interaction (at least if this is of the Langevin type). Moreover, the values of $\mathscr{R}$ in Fig. 1-6-4 are quite close to the value, 0.485, eq. (1-4-49), obtained for the special and completely different case of ions recombining in their parent gas (the momentum transfer cross section being taken to be twice the symmetrical resonance charge-transfer cross section).

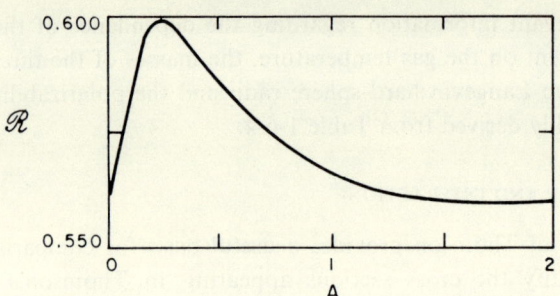

Fig. 1-6-4. Comparison with result given by the Thomson theory. The mass-ratio parameter a is taken to be $\frac{1}{3}$ (corresponding to equal masses); the independent variable is the interaction parameter A; the dependent variable is $\mathscr{R} \equiv \alpha_{13}/\alpha_{T,13}$. (Note: when A is 10, $\mathscr{R}$ is 0.56_2.)

For equal mass species, the Thomson theory overestimates the absolute magnitude of the recombination coefficient by about a factor of 2 which is slight considering that $\alpha_{T,13}$ varies as β^{-3} where

$$\beta = e^2/r_T k\theta. \tag{1-6-57}$$

Thus, agreement with the Bates-Flannery calculations is maintained by arbitrarily putting

$$\beta = 1.5\mathscr{R}^{-\frac{1}{3}} \approx 1.8 \tag{1-6-58}$$

instead of the value 1.5 adopted by Thomson, eq. (1-2-3).

The overall accuracy achieved by the Thomson model is, as Brueckner [16] has emphasized, much better than the accuracies of the various parts of the physical model on which it is based. Figure 1-6-5 provides Bates and Flannery's test of the model. It shows the variation with λ_0 of the ratio

$$P(\lambda_0) \equiv a_{13}(c - \overline{\lambda_0 - \omega})/\alpha_{13} \tag{1-6-59}$$

where $a_{13}(c - \overline{\lambda_0 - \omega})$ is the rate coefficient, determined with the aid of (1-3-9), which describes recombination directly from a free state directly into states with binding energy $\lambda_0 k\theta$ or greater. The model is clearly defective in two respects: (1) the assumed rate coefficient for collisions leading to free-bound transitions of the ion-pair is much too high; (2) the neglected bound-bound transitions are actually very important.

The extent (for equal mass species) to which the errors arising from the above defects cancel is indeed a tribute to the intuition of the originator of

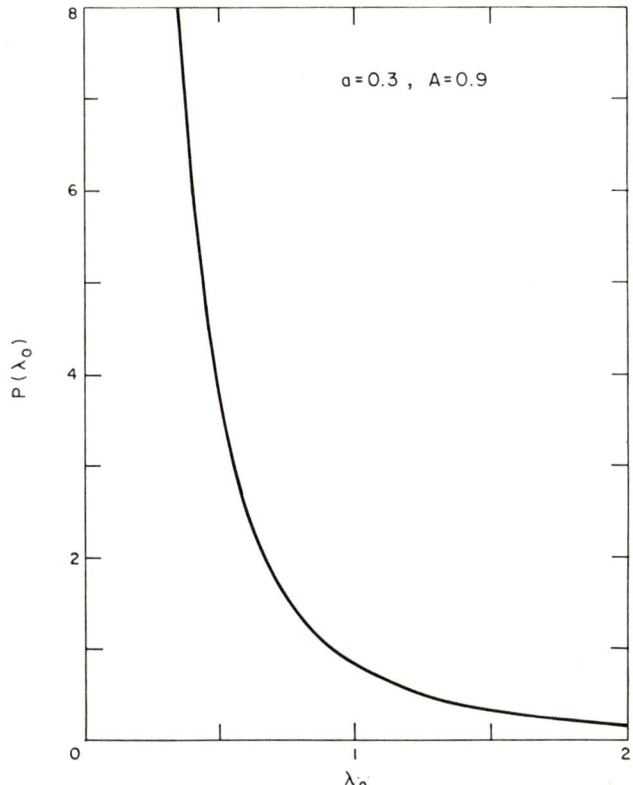

Fig. 1-6-5. Test of details of the Thomson model of recombination $P(\lambda_0)$ as defined in eq. (1-6-59) of text.

the model. See the appendix for a description of the Thomson treatment based on modern terminology.

F. Mass effect

The main features of the dependence of the partial recombination coefficient α_{13} on the masses M_1, M_2 and M_3 may be exhibited by keeping two of the masses fixed and allowing the third to vary. Three cases arise:

Case (i). M_1 and M_2 fixed; M_3 varied.

Let

$$M_1 = eM_2 \qquad (1\text{-}6\text{-}60)$$

then

$$M_3 = \frac{a(1+e)e}{1-ae} M_2 \quad \text{(and } ae \leqq 1\text{)} \tag{1-6-61}$$

with a as defined in (1-5-22).

Case (ii). M_3 and M_1 *fixed;* M_2 *varied.*
Let

$$M_3 = gM_1 \tag{1-6-62}$$

then

$$M_2 = \frac{a(1+g)}{g-a} M_1 \quad \text{(and } a \leqq g\text{)}. \tag{1-6-63}$$

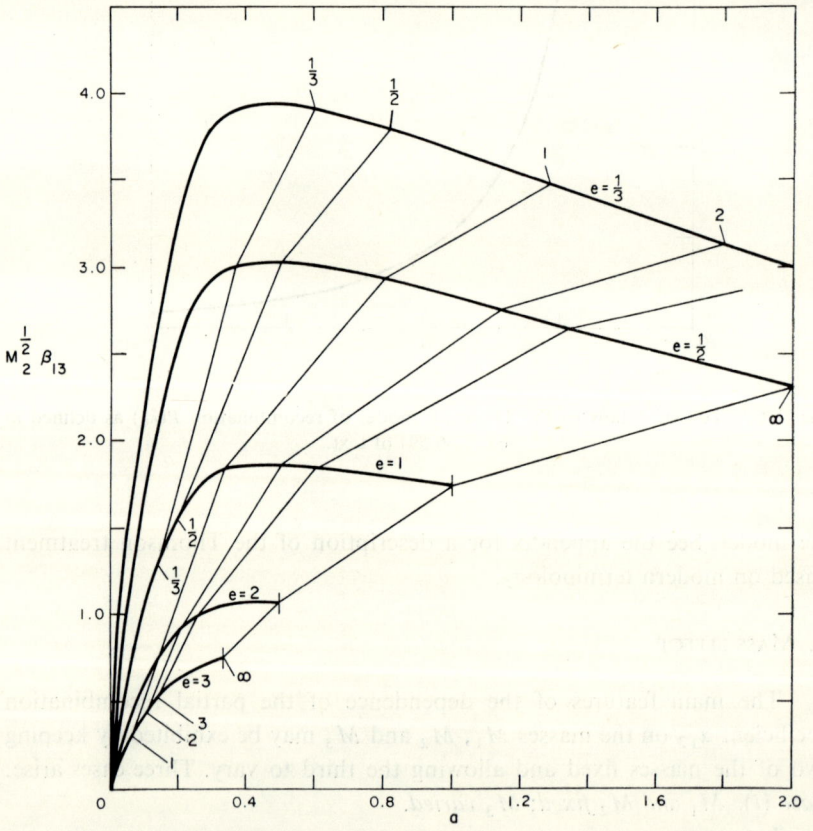

Fig. 1-6-6. Mass effect: M_1 and M_2 fixed; M_3 varied. Dependence of $M_2^{\frac{1}{2}}\beta_{13}$ on a with M_1/M_2 as marked on each curve. Selected values of M_3/M_2 are indicated on the fine straight lines joining corresponding points on the curves.

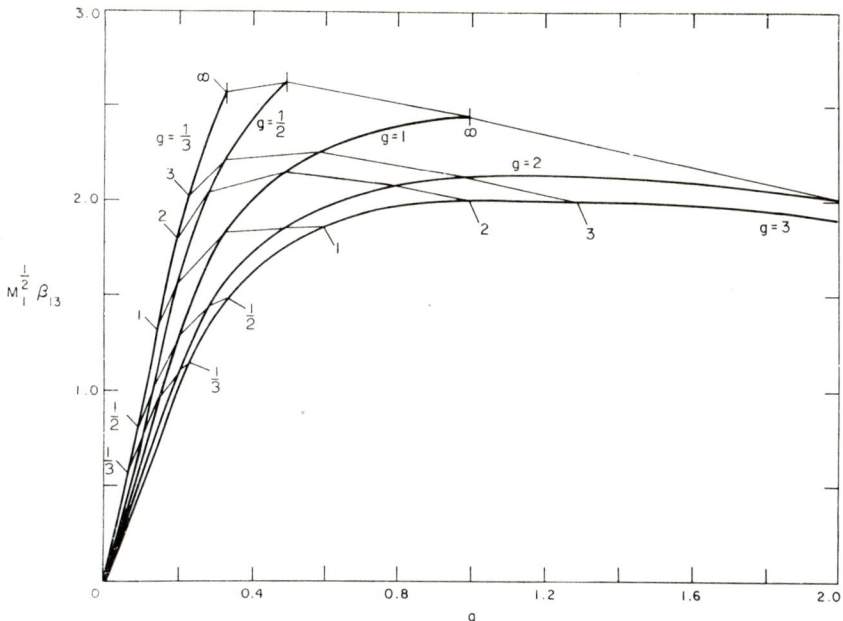

Fig. 1-6-7. Mass effect: M_3 and M_1 fixed; M_2 varied. Dependence of $M_1^{\frac{1}{2}}\beta_{13}$ on a with M_3/M_1 as marked on each curve. Selected values of M_2/M_1 are indicated in the fine straight lines joining corresponding points on the curves.

Case (iii). M_2 and M_3 fixed; M_1 varied.

Let

$$M_2 = fM_3 \tag{1-6-64}$$

then

$$M_1 = \frac{2f}{a(1+f)+\{a^2(1+f)^2+4af\}^{\frac{1}{2}}} M_3. \tag{1-6-65}$$

Figures 1-6-6–1-6-8 show the dependence of the recombination coefficient on the mass ratio parameter a and on the varied masses M_3, M_2 and M_1, respectively. Figure 1-6-6, for example, represents the variation with M_3 for fixed values of M_1/M_2. The fine straight lines join equal values of M_3/M_2 on the curves.

In cases (i) and (ii), $M_2^{\frac{1}{2}}\beta_{13}$ and $M_1^{\frac{1}{2}}\beta_{13}$ are initially increasing functions of M_3 and M_2, respectively; and, having passed through a maximum if e is smaller than or if g is greater than about 2, they tend to non-zero limits

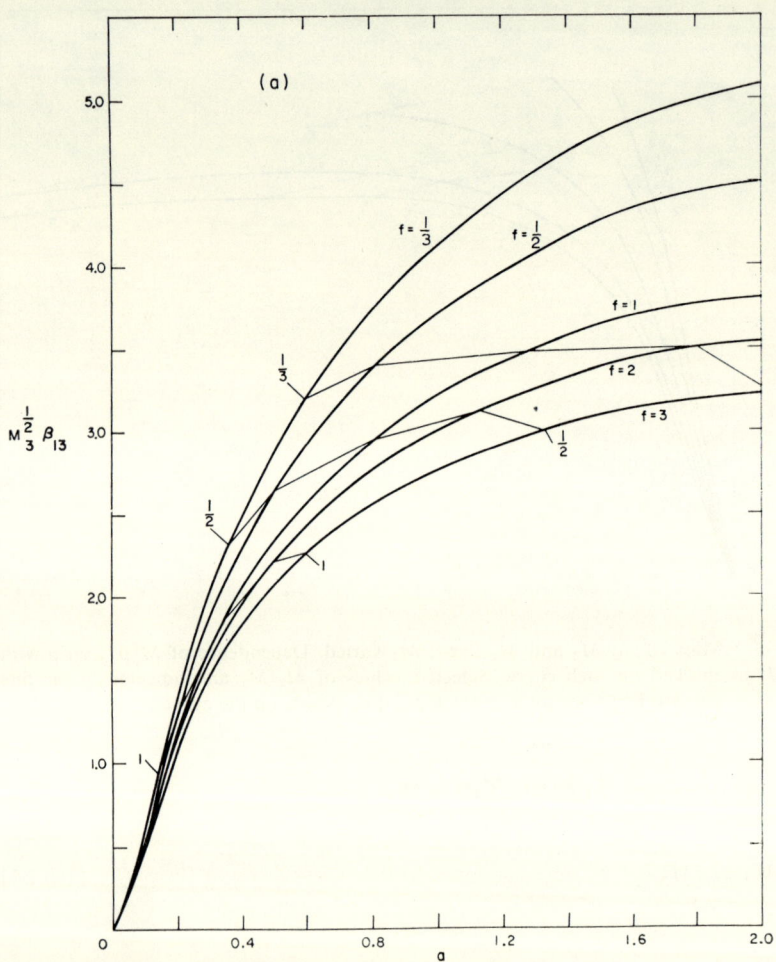

Fig. 1-6-8a. Mass effect: M_2 and M_3 fixed; M_1 varied. Dependence of $M_3^{\frac{2}{3}}\beta_{13}$ on a with M_2/M_3 as marked on each curve. Selected values of M_1/M_3 are indicated on the fine straight lines joining corresponding points on the curves.

as the magnitude of the varied mass tends towards infinity. The maxima (if they exist) are located at values of a which are independent of e and g.

In case (iii), where M_1 is varied, Fig. 1-6-8 demonstrates that $M_3^{\frac{2}{3}}\beta_{13}$ is proportional to $M_1^{\frac{1}{2}}$ in the region where M_1/M_3 is small and for all f falls off as M_1^{-1} in the region where M_1/M_3 is large. The maximum occurs at a value of M_1, which is much less than either M_2 or M_3.

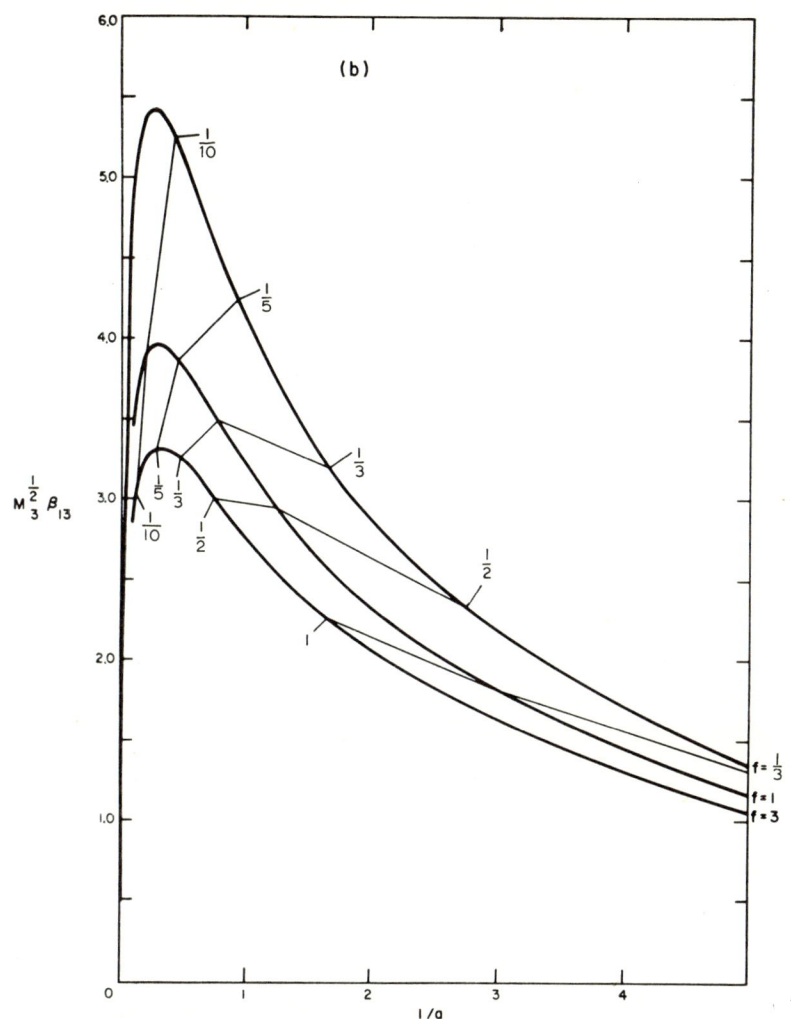

Fig. 1-6-8b. Mass effect: M_2 and M_3 fixed; M_1 varied. Dependence of $M_3^{\frac{1}{2}}\beta_{13}$ on $1/a$ with M_2/M_3 as marked on each curve. Selected values of M_1/M_3 are indicated on the fine straight lines joining corresponding points on the curves.

The Thomson theory, of course, cannot reproduce these effects. Attempts have been made to introduce a simple correction to the Thomson theory to allow for the inequality of the mass. For example, it has been suggested [46] that Q_{13} and Q_{23} of (1-6-45) should be multiplied by the ratio of the smaller to the larger of the two masses involved. Parks [17] has suggested multiplying the Thomson trapping radius r_T in (1-2-3) by $2M_1 M_2 M_3 (M_1+M_2+M_3)/[(M_1+M_2)(M_1+M_3)]^2$. The correction required is clearly much more complicated.

§ 1-7. Theoretical and experimental three-body ionic recombination coefficients

Numerous authors [6, 16, 20] have improperly assumed that the reaction

$$O_2^+ + O_2^- + O_2 \rightarrow [O_4] + O_2 \qquad (1\text{-}7\text{-}1)$$

corresponds to the realistic situation of oxygen ions recombining in oxygen gas. Because of the mass effect, previously discussed, care must be exercised in identification of the possible ions involved. The equilibrium rate constant for

$$O_2^+ + O_2 \rightleftharpoons O_4^+ \qquad (1\text{-}7\text{-}2)$$

has been studied by Yang and Conway [47] whose data show that O_4^+ is the predominant positive ion even at very low neutral densities. In their measurements of mobilities, Snuggs et al. [33] give the ratio of the O_4^+ detected intensity to the O_2^+ intensity to be about $\frac{3}{2}$ at 3 torr and $\frac{3}{1}$ at 6 torr. The results of Voshall, Pack and Phelps [48] for

$$O_2^- + O_2 \rightleftharpoons O_4^- \qquad (1\text{-}7\text{-}3)$$

indicate that at low densities O_2^- is the dominant negative ion. Thus three body recombination at low densities possibly invokes the reaction [49]

$$O_4^+ + O_2^- + O_2 \rightarrow [O_6] + O_2 \qquad (1\text{-}7\text{-}4)$$

rather than (1-7-1). Moreover, at standard number density (Loschmidt's number $N_L = 2.69 \times 10^{19}$ cm^{-3}), the relative concentration of O_4^- to O_2^- is 1:6, which ratio increases with density increase, and so the process

$$O_4^+ + O_4^- + O_2 \rightarrow [O_8] + O_2 \qquad (1\text{-}7\text{-}5)$$

contributes to the recombination, and may well dominate at very high densities. However, in the absence of more explicit experimental evidence, it must be stressed that the above reactions follow from plausible guesses as to the identity of the reactants.

Because the mobility given by the Langevin theory to be

$$K_{Li}(A) = g(A)/(4\pi p M_{i3})^{\frac{1}{2}} N(Z) \qquad (1\text{-}7\text{-}6)$$

with $g(A)$ the function referred to in (1-6-53), varies only slowly with the reduced mass M_{i3} of the ion-neutral system, eq. (1-2-2) predicts in the high-density limit little difference between the recombination coefficients α for (1-7-4) and (1-7-5) so that in this instance the mass effect is not easily recognizable.

Contrary to expectation, symmetrical resonance charge-transfer

$$O_2^- + O_2 \rightarrow O_2 + O_2^- \qquad (1\text{-}7\text{-}7)$$

is *not* the dominant de-activation mechanism responsible for process (1-7-4) since the equilibrium internuclear separations of the molecular ions and neutral differ appreciably enough to make very small vibrational overlap (small Franck-Condon factors) at room temperatures. This results in a small charge-transfer effect which is confirmed by experimental mobility data [33, 50, 51], which show better agreement with the Langevin theoretical treatment of mobility rather than with the theory of mobility [36] for an ion diffusing in its parent gas. Thus (1-7-4) is controlled by the Langevin interaction (hard-sphere repulsion, polarization attraction) and so the theory of Bates and Flannery, rather than that of Bates and Moffett, is applicable.

The relevant formula for the recombination coefficient in the low-density limit is therefore

$$\alpha = 1.3 \times 10^{-8} N(Z) \theta^{-3} e p^{\frac{1}{2}} [M_{12}^{-\frac{1}{2}} \gamma(a, A) + M_{23}^{-\frac{1}{2}} \gamma(b, B)] \quad \text{cm}^3 \text{ sec}^{-1}, \qquad (1\text{-}7\text{-}8)$$

where p is the polarizability of the neutrals Z with number density $N(Z)$, and temperature θ, e is the electronic charge, M_{i3} is the reduced mass of each ion-neutral system, and γ is the tabulated function, Table 1-6-4, of the mass ratio and interaction parameters given by

$$a = M_2 M_3 / M_1 (M_1 + M_2 + M_3), \quad b = (M_1/M_2)^2 a \qquad (1\text{-}7\text{-}9)$$

and

$$A = \frac{(2k\theta)^{\frac{1}{2}}}{p^{\frac{1}{2}} e} S_{13}^2, \quad B = (S_{23}/S_{13})^2 A, \qquad (1\text{-}7\text{-}10)$$

respectively. In the low-density limit with the masses held fixed, the partial recombination coefficients are approximately inversely proportional to the mean free path (or mobility) of each ion in the gas (see (1-2-8) and (1-2-9)). Thus, any departure of the physical ion-neutral interaction from the Langevin interaction can be accounted for by replacing $\gamma(a, A)$ and $\gamma(b, B)$ by

$$\gamma'(a, A) = \frac{K_{L1}(A)}{K_1} \gamma(a, A) \qquad (1\text{-}7\text{-}11)$$

and

$$\gamma'(b, B) = \frac{K_{L2}(B)}{K_2} \gamma(b, B), \qquad (1\text{-}7\text{-}12)$$

respectively, where $K_{1,2}$ are the individual experiment mobilities of the positive and negative ions in the gas, and $K_{L1}(A)$ and $K_{L2}(B)$ are those calculated from the Langevin theory, eq. (1-7-6), with the interaction parameters A and B. Since K_L usually defines an upper limit to the mobility the use of γ' in (1-7-7) will, in general, increase α.

Experimental data on recombination are meager. Indeed, the only low-density results available for meaningful comparison are those of McGowan [30] for the recombination of oxygen ions in oxygen at room temperature and those of Mahan and Person [19] who measured the rate of NO^+ and NO_2^- recombining in various gases at room temperature.

The values of the polarizabilities p and the hard-sphere radii, S_{13} and S_{23}, used in the calculations were obtained from Dalgarno [52] and Hirschfelder, Curtiss and Bird [53]. Table 1-7-1 compares the rates calculated from eq. (1-7-8) with those in columns a and b which were respectively deduced by Landon and Keck [8] and by Mahan and Person [19] from the experimental measurements. The agreement is satisfactory, but it should be noted,

TABLE 1-7-1

NO^+ and NO_2^- recombination coefficients $[N_L/N(Z)] \alpha$ $(10^6 \text{ cm}^3 \text{sec}^{-1})$*

Neutral body Z	Theory	Experiment	
		a	b
H_2	0.76	0.76	1.56
D_2	1.07	0.91	1.64
He	0.69	0.76	1.10
Ne	1.69	2.2	2.79
N_2	3.65	2.7	5.64
Ar	3.69	3.0	3.90
Kr	4.65	3.2	4.70
Xe	5.87	4.6	7.47

* $N(Z)$ is the neutral number density and N_L is Loschmidt's number density ($2.69 \times 10^{19} \text{ cm}^{-3}$).

however, that the experimental situation is complicated by the tendency for heavier ion complexes to form. Also, there is great uncertainty entailed, as demonstrated by the columns a and b of the table, in separating a three-body rate coefficient from the experimental data.

Repeating the calculations by use of Goldsmidt's values of hard-sphere radii [14] resulted in no change greater than a few percent.

TABLE 1-7-2

Recombination coefficient describing $O_4^+ + O_2^- + O_2 \rightarrow [O_6] + O_2$

$N(Z)/N_L$*	Present theory	$\alpha(10^6 \text{ cm}^3\text{sec}^{-1})$ Experiment (McGowan)
0.105	0.43	0.42 ± 0.05
0.147	0.61	0.63 ± 0.07

* $N(Z)$ is the neutral number density and N_L is Loschmidt's number density (2.69×10^{19} cm^{-3}).

Table 1-7-2 displays the excellent agreement between the low-density measurements of McGowan [30] and those calculated from (1-7-8), (1-7-11) and (1-7-12) for the process (1-7-4). The experimental mobilities for O_4^+ and O_2^- used in the calculation were each equal to 2.16 cm^2 volt^{-1} sec^{-1} [33, 51]. In conclusion the Bates-Flannery theory appears to permit the rapid and reliable evaluation of the low-density limit to the three-body ionic rate coefficients.

§ 1-8. Simple treatments of three-body ionic recombination

The concept, first advanced by Pitaevskii [7] in connection with electronic recombination, that recombination be envisaged as diffusion in energy space merited application to ions when Feibelman's study, which followed the lives of individual ion-pairs by a Monte Carlo technique [20], revealed that the fate of an ion-pair could only be predicted until it had survived a large number of collisions with the neutrals, and not just after one collision as assumed by Thomson [2]. Therefore, the Fokker-Planck equation [54] with certain simplifying assumptions was invoked [8] to describe the manner in which the number density of ion-pairs evolves with time and energy. The recombination coefficient α can be then determined from the steady-state solution [7, 8]. Bates and Jundi [23], showed, however, that this procedure meets with limited success.

Flannery [55] correlated the diffusion approach with the treatment discussed in the previous sections by regarding recombination as a Markov process [56]. The Markovian description resulted in equation (1-3-25) for the quasi-equilibrium distribution and also yielded the Fokker-Planck equation when it was assumed that the energy transferred to the neutrals was small in a time sufficiently long for a large number of collisions to occur. Further approximation to the Fokker-Planck equation gave the diffusion-expression of Landon and Keck [8].

While the diffusive approach replaces (1-3-18) by the diffusion equation from which both the steady-state distribution $\rho(\lambda)$ and α can be obtained, Bates and Jundi used an "effective-gradient" approximation for $\rho(\lambda)$ in (1-3-13) to yield a simplified expression for α, which, when compared with the diffusion theory predicts results in closer accord with the full treatment of Bates and Flannery. Therefore, in this section we will attempt to link all these approximate schemes together by finding a common origin.

A. Recombination as a Markov process

Recombination between ions proceeds as a result of a large number of collisions between ion-pairs and third bodies. Let this basic mechanism be described by a probability $P(E, \Delta) \Delta t$ that, in time Δt, an ion-pair with initial internal energy E acquires an energy increment Δ. For these collisions, P does not depend explicitly on the time t so that the distribution of ion-pairs then evolves according to

$$n(E, t) = \Delta t \int n(E - \Delta, t - \Delta t) P(E - \Delta, \Delta) \, d\Delta \qquad (1\text{-}8\text{-}1)$$

which essentially states that the distribution only depends on its immediate previous history. Such an equation defines a *Markov process* [56]. In other words, if a random function assumes the value x (is in state x) at the present time, then its future development is governed by the same rules as the development of the same random function under the condition that it was in state x initially. Therefore, in the definition of the "time-homogeneous" Markov process we consider its evolution for all possible initial states. Such a stochastic process can describe Brownian motion, for example.

1. The Fokker-Planck equation

The interval Δt is assumed to be sufficiently long for one ion-pair to suffer many collisions and yet to be short enough so that the acquired energy Δ is small. Thus, since the collisions can be regarded as being binary, the collision side of the Boltzmann equation [57] can be identified with

$$\left.\frac{\delta n(E, t)}{\delta t}\right]_{\text{collisions}} = \frac{n(E, t) - n(E, t - \Delta t)}{\Delta t}. \qquad (1\text{-}8\text{-}2)$$

Taylor-expansion of (1-8-1) to first order in Δt and to second order in Δ yields

$$n(E, t) = \Delta t \int d\Delta \left[\left\{ n(E, t) P(E, \Delta) - \Delta t\, P(E, \Delta) \frac{\partial n(E, t)}{\partial t} \right\} \right.$$
$$\left. - \left(\Delta \frac{\partial}{\partial E} + \tfrac{1}{2} \Delta^2 \frac{\partial^2}{\partial E^2} \right) \{n(E, t) P(E, \Delta)\} + \ldots \right] \qquad (1\text{-}8\text{-}3)$$

and Δt is long enough that

$$\Delta t \int P(E, \Delta) d\Delta = 1 \qquad (1\text{-}8\text{-}4)$$

that is, the ion-pair with energy E initially must go somewhere in energy space as a result of collisions in this time. Thus

$$\frac{\partial n(E, t)}{\partial t} = -\frac{\partial}{\partial E} \{\tilde{\Delta}_1 n(E, t)\} + \frac{1}{2} \frac{\partial^2}{\partial E^2} \{\tilde{\Delta}_2 n(E, t)\} \qquad (1\text{-}8\text{-}5)$$

where

$$\tilde{\Delta}_n = \int \Delta^n P(E, \Delta) d\Delta \qquad (1\text{-}8\text{-}6)$$

are the moments of energy change with respect to the probability per second that energy changes. Equation (1-8-5) is the Fokker-Planck equation which is valid in the present context when Δt is long enough and Δ is small.

2. Integral equation for the steady-state distribution

When a steady-state is attained, (1-8-1) reduces to

$$n(E) \delta E = \Delta t \int n(E - \Delta) P(E - \Delta, \Delta) \delta E\, d\Delta \qquad (1\text{-}8\text{-}7)$$

where $n(E)\delta E$ is the number of ion-pairs between E and $E+\delta E$, $\Delta t \times P(E-\Delta, \Delta)\delta E$ is the probability that one ion-pair with energy $E-\Delta$ acquires sufficient energy so as to be left in energy interval δE around E after time Δt, and where $n(E-\Delta)d\Delta$ is the number of ion-pairs in interval around $E-\Delta(\equiv E_i)$. From the definition of the rate $K(E_i, E_f)$

$$\Delta t\, P(E-\Delta, \Delta)\,\delta E = \Delta t\, N(Z)\, K(E_i, E)\,\delta E \tag{1-8-8}$$

and hence

$$n(E) = \Delta t\, N(Z) \int_{-E_S}^{\infty} n(E_i)\, K(E_i, E)\, dE_i. \tag{1-8-9}$$

The time Δt is sufficiently long that thermodynamic equilibrium is obtained and hence when we also include in $n(E)$ those ion-pairs which have escaped via the charge-transfer channels we have

$$n_T(E) = \Delta t\, N(Z) \int_{-D}^{\infty} n_T(E_i)\, K(E_i, E)\, dE_i \tag{1-8-10}$$

where $n_T(E)$ is the Saha-Boltzmann number density (1-3-16) and D is the greatest binding energy possessed by an ion-pair. With the aid of detailed-balance (1-3-11), eq. (1-8-10) reduces to

$$\Delta t\, N(Z) \int_{-D}^{\infty} K(E, E_i)\, dE_i = 1 \tag{1-8-11}$$

which, of course, is equivalent to (1-8-4). Inserting Δt above in (1-8-9) and invoking detailed balance again yields

$$\rho(E) \int_{-D}^{\infty} K(E, E_i)\, dE_i = \int_{-E_S}^{\infty} \rho(E_i)\, K(E, E_i)\, dE_i \tag{1-8-12}$$

where the distribution function

$$\rho(E) = n(E)/n_T(E) \tag{1-8-13}$$

and E_S is the binding energy at which the recombination becomes stabilized. Eq. (1-8-12) is identical to the eq. (1-3-25) adopted by Bates, Flannery and Moffett. Moreover, by using (1-8-2) in (1-8-1) together with (1-8-8), it is easily shown in the limit of vanishing Δt that

$$\frac{\partial n(E, t)}{\partial t} = N(Z) \left[\int_{-E_S}^{\infty} n(E_i, t)\, K(E_i, E)\, dE_i - n(E, t) \int_{-D}^{\infty} K(E, E_i)\, dE_i \right]$$

which is the correct equation (1-3-18) that describes the rate of growth of the population of ion pairs-with energy E as the difference of the flow of ion-pairs leaving E from that arriving at this level.

3. Relationship with the diffusion theory

The equation for the ionic distribution used by Pitaevskii [7] and Landon and Keck [8] is

$$n_T(E)\frac{\partial \rho(E, t)}{\partial t} = \frac{1}{2}\frac{\partial}{\partial E}\left[n_T(E)\tilde{A}_2 \frac{\partial \rho}{\partial E}\right] \quad (1\text{-}8\text{-}14)$$

where $\tilde{A}_n$ is defined by (1-8-6) and is therefore given by

$$\tilde{A}_n(E) = N(Z)\int_{-D}^{\infty}(E_f - E)^n K(E, E_f)\,dE_f. \quad (1\text{-}8\text{-}15)$$

Expanding the Fokker-Planck equation (1-8-5) gives

$$n_T(E)\frac{\partial \rho(E, t)}{\partial t} =$$

$$= -\left(\frac{\partial \rho}{\partial E} + \rho\frac{\partial}{\partial E}\right)\left(n_T\tilde{A}_1 - \frac{1}{2}\frac{\partial}{\partial E}(n_T\tilde{A}_2)\right) + \frac{1}{2}\frac{\partial}{\partial E}\left(n_T\tilde{A}_2\frac{\partial \rho}{\partial E}\right) \quad (1\text{-}8\text{-}16)$$

which becomes equivalent to (1-8-14) when the following approximation between the first and second energy-moments

$$n_T(E)\tilde{A}_1(E) \approx \frac{1}{2}\frac{\partial}{\partial E}\{n_T(E)\tilde{A}_2(E)\} \quad (1\text{-}8\text{-}17)$$

is invoked. Thus, not only is the diffusion approach restricted by the fact that Δ must be small but also the above relationship between the energy-moments must be satisfied. The introduction of the symmetrical rate,

$$R(E, E_f) = n_T(E) K(E, E_f) = n_T(E_f) K(E_f, E) \equiv R(E_f, E) \quad (1\text{-}8\text{-}18)$$

demonstrates that we can write

$$R(E, E_f) = \mathcal{R}(\bar{E}, |\Delta|) \quad (1\text{-}8\text{-}19)$$

where $\mathcal{R}$ is an even function of

$$\bar{E} = \tfrac{1}{2}(E + E_f) \quad (1\text{-}8\text{-}20)$$

the mean of the initial and final energies, and a function of

$$\Delta = E_f - E, \quad (1\text{-}8\text{-}21)$$

the energy transferred which is assumed to be small. Inserting (1-8-19) into (1-8-15) with the aid of (1-8-18) and expanding $\bar{E}$ about E, we have

$$n_T(E)\tilde{A}_n(E) = N(Z)\int_{-\infty}^{\infty}\left\{\mathscr{R}(E,|\Delta|)+\tfrac{1}{2}\Delta\left(\frac{\partial\mathscr{R}}{\partial E}\right)_E + \ldots\right\}\Delta^n\,d\Delta \quad (1\text{-}8\text{-}22)$$

for which D has been taken to be effectively infinite. Therefore, as shown by Keck and Carrier [8], the following equations

$$n_T(E)\tilde{A}_1(E) = N(Z)\int_0^{\infty}\left(\frac{\partial\mathscr{R}}{\partial E}\right)_E \Delta^2\,d\Delta + O(\Delta^4) + \ldots \quad (1\text{-}8\text{-}23)$$

and the relation

$$n_T(E)\tilde{A}_2(E) = 2N(Z)\int_0^{\infty}\mathscr{R}(E,|\Delta|)\Delta^2\,d\Delta + O(\Delta^4) + \ldots \quad (1\text{-}8\text{-}24)$$

show that (1-8-17) is satisfied in the limit of vanishing Δ. Hence, in this limit, the Fokker-Planck equation reduces to the diffusion equation.

B. RECOMBINATION COEFFICIENT DEDUCED FROM THE DIFFUSION, EFFECTIVE-GRADIENT AND MODIFIED EFFECTIVE-GRADIENT METHODS

The discussion of the major difficulties encountered in theoretical treatments of the recombination process demands some recapitulation. As before, let us designate the number densities of free positive and negative ions and of third bodies by $N(X^+)$, $N(Y^-)$ and $N(Z)$, respectively. Denote the number density of ion-pairs with internal energy in interval dE about E by $n(E)dE$ when recombination is proceeding and by $n_T(E)dE$ when thermodynamic equilibrium exists. At temperature θ and for negative E, the Saha-Boltzmann equation is

$$\frac{n_T(E)}{N(X^+)N(Y^-)} = \frac{\pi^{\frac{3}{2}}e^6\exp(-E/k\theta)}{2(k\theta)^{\frac{3}{2}}|E|^{\frac{1}{2}}} \equiv f(\theta, E) \quad (1\text{-}8\text{-}25)$$

where e is the electronic charge and k is Boltzmann's constant. The recombination rate $\alpha N(X^+)N(Y^-)$ equals the *net* rate of downflow of ion-pairs past some arbitrary negative energy E_0 and is therefore

$$\alpha N(X^+)N(Y^-) =$$
$$= N(Z)\int_{E_f=-D}^{E_0}dE_f\int_{E_i=E_0}^{\infty}\{n(E_i)K(E_i,E_f)-n(E_f)K(E_f,E_i)\}dE_i \quad (1\text{-}8\text{-}26)$$

where D is the greatest binding energy an ion-pair may have, even transitorily. The application of detailed balance, followed by the introduction of $\rho(E)$ in (1-8-13) yields

$$\alpha = N(Z) \int_{E_f=-D}^{E_0} dE_f \int_{E_i=E_0}^{\infty} f(\theta, E_i)\{\rho(E_i) - \rho(E_f)\} K(E_i, E_f) dE_i \quad (1\text{-}8\text{-}27)$$

which is an alternative way of rewriting eq. (1-3-13).

The two main obstacles which impeded progress in previous theoretical treatments of recombination may now be noted as follows: (i) the correct formulation of an expression for the collisional rate coefficient $K(E_i, E_f)$ and (ii) the solution of (1-8-12) for $\rho(E)$, followed by the double quadrature (1-8-27) for α.

While the calculation of $K(E_i, E_f)$ was simplified by invoking the binary-encounter concept, it was hindered by the difficulty of carrying out correct averages over those regions of multi-dimensional space accessible for a given change in energy. Landon and Keck [8] and Mahan [9], assumed hard-sphere interactions (isotropic scattering) and give approximate expressions for the energy-change moments. The essentially exact rates are those derived by Bates and Flannery (§ 1-5B), for collisions between general mass species obeying Coulombic and Langevin interactions, and by Bates and Moffett (§ 1-4A) for collisions of ions with their parent atoms.

Attempts to avoid the cumbersome procedure inherent in (ii) were made by assuming that the collisional energy changes Δ were sufficiently small ($\ll k\theta$) so as to warrant a diffusion approach, a device that indeed caused great simplification. It must be emphasized, however, that collisions of ions with their parent atoms cannot be treated as diffusive since large energy-transfers occur.

1. *The diffusion method*

The diffusion method assumes that the ionic distribution evolves according to (1-8-14), the steady-state solution of which is given (for negative E) by

$$\rho(E) = 1 - \int_{E}^{0} \frac{dE_0}{n_T(E_0)\tilde{\Delta}_2(E_0)} \bigg/ \int_{-E_s}^{0} \frac{dE_0}{n_T(E_0)\tilde{\Delta}_2(E_0)},$$

$$0 \geqq E \geqq -E_s \quad (1\text{-}8\text{-}28)$$

which satisfies the following boundary conditions (1-3-5)

$$\rho(0) = 1, \quad \rho(-E_s) = 0, \quad \rho(E) \leqq 1 \quad (1\text{-}8\text{-}29)$$

where E_S is the binding energy at which the recombination becomes stabilized by charge-transfer. Thus, the diffusion method predicts $\tfrac{1}{2}n_T(E)\tilde{A}_2(E)\partial\rho/\partial E$, the number of ion-pairs crossing energy E per unit time (the current), to be a constant. Since the ion-pairs have already recombined on reaching $-E_S$, and are then irretrievably lost from the quasi-equilibrium a further boundary condition on the distribution of recombining ion-pairs is therefore

$$\left.\frac{\partial\rho}{\partial E}\right)_{-E_S} = 0. \tag{1-8-30}$$

Integrating (1-8-14) between $-E_S$ and 0 we have

$$\frac{\partial}{\partial t}\int_{-E_S}^{0} n(E)\,dE = \frac{\partial}{\partial t}[X^+ - Y^-] = -\frac{1}{2}\left[\int_{-E_S}^{0}\frac{dE_0}{n_T(E_0)\tilde{A}_2(E_0)}\right]^{-1} \tag{1-8-31}$$

where $[X^+ - Y^-]$ denotes the total number density of ion-pairs in all stages of binding. With the aid of (1-8-25), the expression for the recombination coefficient is

$$\alpha = N(Z)\left[\int_{-E_S}^{0}\frac{dE_0}{\mathcal{D}(E_0)}\right]^{-1} \tag{1-8-32}$$

with

$$\mathcal{D}(E_0) = \tfrac{1}{2}f(\theta, E_0)\int_{-D}^{\infty}(E_f - E_0)^2 K(E_0, E_f)\,dE_f. \tag{1-8-33}$$

We must observe however, that the diffusion approach, while yielding a relatively simple formula for the recombination coefficient α, is valid only when the fractional energy change is small. This criterion is apparently satisfied when a or b in (1-6-29) and (1-6-39) is small i.e.,

$$\begin{aligned}M_2 \ll M_3 \approx M_1\,(a \ll 1)\\ M_1 \ll M_3 \approx M_2\,(b \ll 1)\end{aligned} \tag{1-8-34}$$

and

$$M_3 \ll M_2 \approx M_1\,(a, b \ll 1)$$

where M_1 and M_2 are the ionic masses and M_3 is the mass of the third body. It is important to recognize that the condition $M_3 \gg M_2$ does not ensure a small fractional energy change if $M_1 \approx M_2$. Consequently, the diffusional formulation would have a certain degree of validity only when either the third body or one of the ions is much lighter than the other two particles.

2. The effective-gradient method

Bates and Jundi [23] proposed an alternative procedure which also has the advantage that $\rho(E)$ need not explicitly be determined. Instead of solving either the Fokker-Planck equation or its simpler derivative (the diffusion equation), they reduced the exact expression (1-8-27) by adopting the approximation [58], basic to the *effective-gradient method*,

$$\rho(E_i) - \rho(E_f) \approx (E_i - E_f) \left[\frac{d\rho(E)}{dE} \right]_{E=E_0} \quad (1\text{-}8\text{-}35)$$

to yield

$$\alpha = N(Z) \mathscr{B}(E_0) \left(\frac{d\rho}{dE} \right)_{E_0} \quad (1\text{-}8\text{-}36)$$

with

$$\mathscr{B}(E_0) = \int_{E_f = -D}^{E_0} dE_f \int_{E_i = E_0}^{\infty} f(\theta, E_i)(E_i - E_f) K(E_i, E_f) dE_i. \quad (1\text{-}8\text{-}37)$$

Invoking the boundary conditions (1-8-29) they obtained

$$\alpha = N(Z) \left[\int_{-E_s}^{0} \frac{dE_0}{\mathscr{B}(E_0)} \right]^{-1} \quad (1\text{-}8\text{-}38)$$

for the recombination coefficient which is similar in form to (1-8-32). In the limit of small

$$\Delta = E_f - E_i \quad (1\text{-}8\text{-}39)$$

it may be readily seen that $\mathscr{B}(E_0)$ tends to $\mathscr{D}(E_0)$ when the principle of detailed balance is used to rewrite $\mathscr{B}(E_0)$ as

$$\mathscr{B}(E_0) = \frac{1}{2} \int_{E_f = -D}^{E_0} dE_f \int_{E_i = E_0}^{\infty} \Delta \{ f(\theta, E_i) K(E_i, E_f)$$
$$+ f(\theta, E_f) K(E_f, E_i) \} dE_i. \quad (1\text{-}8\text{-}40)$$

The substitution

$$K(E_i, E_f) dE_f = \Gamma(E_i, \Delta) d\Delta \quad (1\text{-}8\text{-}41)$$

and rearrangement of (1-8-40) gives

$$\mathscr{B}(E_0) = \frac{1}{2} \int_{-D}^{0} \left\{ -\Delta \int_{E_0}^{E_0 - \Delta} f(\theta, E_i) \Gamma(E_i, \Delta) dE_i \right\} d\Delta$$
$$+ \frac{1}{2} \int_{0}^{\infty} \left\{ \Delta \int_{E_0 - \Delta}^{E_0} f(\theta, E_f) \Gamma(E_f, \Delta) dE_f \right\} d\Delta \quad (1\text{-}8\text{-}42)$$

which for Δ small reduces to

$$\mathscr{B}(E_0) \approx \tfrac{1}{2} f(\theta, E_0) \int_{-D}^{\infty} \Delta^2 \, \Gamma(E_0, \Delta) \, \mathrm{d}\Delta \equiv \mathscr{D}(A_0). \tag{1-8-43}$$

Thus, the effective gradient method should give better results than the diffusion method, with which it will agree, of course, for cases involving very small energy-transfers. However, the effective gradient method is more akin to the exact Fokker-Planck equation (1-8-5) than to the diffusion equation (1-8-14) in that the approximation (1-8-17) invoked to reduce (1-8-5) to (1-8-14) is exactly equivalent to the assumption $\mathscr{B} \approx \mathscr{D}$ that introduces accord between (1-8-32) and (1-8-38). This is demonstrated by rewriting condition (1-8-17) as

$$\tfrac{1}{2} n_\mathrm{T}(E_0) \tilde{A}_2(E_0) = \int_{\infty}^{E_0} n_\mathrm{T}(E_\mathrm{i}) \tilde{A}_1(E_\mathrm{i}) \, \mathrm{d}E_\mathrm{i}. \tag{1-8-44}$$

The left-hand side is simply $N(\mathrm{X}^+) \, N(\mathrm{Y}^-) \, N(\mathrm{Z}) \, \mathscr{D}(E_0)$ while the right-hand side is, by (1-8-15)

$$R = N(\mathrm{Z}) \int_{\infty}^{E_0} n_\mathrm{T}(E_\mathrm{i}) \, \mathrm{d}E_\mathrm{i} \int_{-D}^{\infty} (E_\mathrm{f} - E_\mathrm{i}) \, K(E_\mathrm{i}, E_\mathrm{f}) \, \mathrm{d}E_\mathrm{f} \tag{1-8-45}$$

$$= N(\mathrm{Z}) \int_{E_0}^{\infty} n_\mathrm{T}(E_\mathrm{i}) \, \mathrm{d}E_\mathrm{f} \left[\int_{-D}^{E_0} (E_\mathrm{i} - E_\mathrm{f}) \, K(E_\mathrm{i}, E_\mathrm{f}) \, \mathrm{d}E_\mathrm{f} \right.$$

$$\left. + \int_{E_0}^{\infty} (E_\mathrm{i} - E_\mathrm{f}) \, K(E_\mathrm{i}, E_\mathrm{f}) \, \mathrm{d}E_\mathrm{f} \right] \tag{1-8-46}$$

$$= N(\mathrm{X}^+) \, N(\mathrm{Y}^-) \, N(\mathrm{Z}) \, \mathscr{B}(E_0) + N(\mathrm{Z}) \int_{E_0}^{\infty} n_\mathrm{T}(E_\mathrm{i}) \, \mathrm{d}E_\mathrm{i}$$

$$\times \int_{E_0}^{\infty} (E_\mathrm{i} - E_\mathrm{f}) \, K(E_\mathrm{i}, E_\mathrm{f}) \, \mathrm{d}E_\mathrm{f}. \tag{1-8-47}$$

Application of detailed balance demonstrates that the double integral explicit in (1-8-47) is identically zero. Thus, condition (1-8-17) invoked to derive the diffusion equation from the Fokker-Planck equation is valid only when the functions $\mathscr{B}$ and $\mathscr{D}$ are equivalent. We have shown above that $\mathscr{B}$ reduces to $\mathscr{D}$ only in the limit of vanishing Δ. Therefore, the effective-gradient method is more akin to the Fokker-Planck equation than to the diffusion equation, the effectiveness of which can be gauged from the closeness of $\mathscr{B}$ and $\mathscr{D}$, and must become equivalent to the diffusion method only as Δ tends to zero. Moreover, the above argument demonstrates that the effective-gradient method can be alternatively derived from the following equation which is analogous to the diffusion equation (1-8-14),

$$n_T(E)\frac{\partial \rho(E, t)}{\partial t} = \frac{\partial}{\partial E}\left[\frac{\partial \rho}{\partial E}\int_\infty^E n_T(E')\tilde{A}_1(E')\,dE'\right]. \tag{1-8-48}$$

Another approximation was derived by Bates and Jundi also from (1-8-37) by proceeding as before without invoking the principle of detailed balance, with the result

$$\mathscr{B}(E_0) \approx \mathscr{A}(E_0) = f(\theta, E_0)\int_{-D}^0 \Delta^2 \Gamma(E_0, \Delta)\,d\Delta. \tag{1-8-49}$$

Substitution of $\mathscr{A}(E_0)$ for $\mathscr{B}(E_0)$ in (1-8-38) provides a further approximation which will be referred to as the *modified effective-gradient method*.

3. Accuracy of diffusion and effective-gradient methods

As demonstrated in § 1-6C, the recombination coefficient is, to a close approximation, the sum of two partial recombination coefficients: one describing recombination due only to X^+–Z collisions and the other describing recombination due only to Y^-–Z collisions. It is advantageous to confine our attention to these partial recombination coefficients.

By using the expressions for $K(E_i, E_f)$ given by Flannery [45] for selected values of the mass-ratio parameter (1-6-29)

$$a = M_2 M_3/M_1(M_1 + M_2 + M_3) \tag{1-8-50}$$

and the interaction parameter (1-6-30)

$$A = (2k\theta)^{\frac{1}{2}} S^2/p^{\frac{1}{2}} e \tag{1-8-51}$$

Bates and Jundi [23] calculated α_d, α_g and α_m, the partial recombination coefficients given by the diffusion method (1-8-32), the effective-gradient method (1-8-38) and the modified effective-gradient method (1-8-49). Tables 1-8-1–1-8-3 display the dimensionless ratios r_d, r_g and r_m, respectively, defined by

$$r_d \alpha_d = r_g \alpha_g = r_m \alpha_m = \alpha \tag{1-8-52}$$

where α is the accurately computed recombination coefficient of Bates and Flannery (§ 1-6C).

The precision achieved by the diffusion method is rather poor in that r_d varies from about 0.7 for small a to about 0.2 for large a, whereas ideally it should of course be unity. This behaviour is, however, not surprising in view of the fact that the diffusion method is expected to give good results only when small energy transfers occur, i.e., for a small. Thus, the condition for the validity of the diffusion method is so restrictive that any application to cases of ionic recombination met in practice would be meaningless.

TABLE 1-8-1
Accuracy of diffusion method

A	a	$r_d = \alpha/\alpha_d$		
		0.2	0.4	0.9
0.3		0.69	0.39	0.29
0.9		0.71	0.36	0.26
2.0		0.63	0.26	0.19

TABLE 1-8-2
Accuracy of effective-gradient method

A	a	$r_g = \alpha/\alpha_g$		
		0.2	0.4	0.9
0.3		0.85	0.72	0.66
0.9		0.86	0.71	0.64
2.0		0.82	0.62	0.56

TABLE 1-8-3
Accuracy of modified effective-gradient method

A	a	$r_m = \alpha/\alpha_m$		
		0.2	0.4	0.9
0.3		1.15	1.22	1.25
0.9		1.16	1.25	1.30
2.0		1.20	1.33	1.36

Since r_g lies between 0.85 and 0.56 the effective-gradient method is not as much in error, as expected, either with regard to the absolute value of a particular coefficient or with regard to the relative values of different coefficients.

The difference between r_d and r_g serves as a convenient gauge for assessing the merits of the diffusion approach when comparing with not only the effective gradient method but also with the exact Fokker-Planck equation. The closer agreement of α with α_g than with α_d with increasing mass-ratio parameter a simply reflects the fact that α_g is more accurate, being good to second-order in the energy change while α_d is only correct to first-order.

The modified effective-gradient method is as simple to apply as the diffusion method. It appears to be the most accurate of the three methods considered: thus r_m is in the range 1.15 to 1.36, so that 1.25 α_m differs from α by less than 10% for all cases covered by the computations.

The ratios r_d, r_g and r_m vary more rapidly with a than with A. This follows directly from the fact that each approximation is dependent on the value of $\tilde{\Delta}_n$ which in turn, depends more strongly on a than on A. The details of the interaction, determined by A, is contained in the rate coefficient $K(E_i, E_f)$ to which no approximation is made.

§ 1-9. Three-body recombination at moderate and high gas-densities

We conclude this discussion of ionic recombination by considering the implications involved when the density of the gas (third bodies) is raised. Here the physical situation becomes much more complicated than that encountered in the low-density limit. In Thomson's model (§ 1-2A), it is necessary to recognize that the probability that *either* ion experiences a collision when within the trapping sphere can be expressed as the sum of the probabilities w_i for each ion, only if these probabilities are very small, as in the low-density limit when the mean free paths λ_i are much larger than the trapping radius r_T. It is also necessary to recognize that the collisional probability does not continue to increase linearly with the neutral number density as in eq. (1-2-5) but instead must approach unity asymptotically. This difficulty is overcome by noting that (1-2-4) counts the probability $w_1 w_2$ for simultaneous collisions of both ions with a neutral twice so that the well-known Thomson formula for the recombination coefficient as a function of the density is

$$\alpha_T = \pi r_T^2 \tilde{w}_T \left(\frac{3k\theta}{M_{12}}\right)^{\frac{1}{2}} \tag{1-9-1}$$

with

$$\tilde{w}_T = w(x_{T1}) + w(x_{T2}) - w(x_{T1})w(x_{T2}) \tag{1-9-2}$$

in which (cf. Appendix)

$$w(x_{Ti}) = 1 - \{1 - (1 + 2x_{Ti}) \exp(-2x_{Ti})\}/2x_{Ti}^2 \tag{1-9-3}$$

where

$$x_{Ti} = r_T/\lambda_i. \tag{1-9-4}$$

In the low-density region $\lambda_i \gg r_T$ and $w(x_{Ti})$ tends to $4r_T/3\lambda_i$ in agreement with (1-2-5) and for high densities $\lambda_i \ll r_T$ and $w(x_{Ti})$ approaches unity.

It would in practice be a very formidable task to modify the quasi-equilibrium statistical theory of Bates and Flannery correspondingly. The assumption concerning the equality of the populations of accessible elements of phase space cannot be made at moderate and high densities, since it requires the mean time between collisions to be long compared with the orbiting times of the ion-pairs. In consequence, the number of independent variables in the integral equation is three (instead of one as in (1-8-12)) and the final quadrature for the recombination coefficient is six-dimensional (instead of two-dimensional as in (1-8-26)). The computational effort is therefore prohibitive when compared with that involved in the low-density case. Even if computations were feasible, the description of the phenomenon would be incomplete. The rate at which ions of opposite sign approach each other is limited by the speed of random diffusion and by the speed of attractive drift such that at sufficiently high densities the recombination coefficient is given by the Langevin-Harper formula, eq. (1-2-2),

$$\alpha_{LH} = 4\pi e(K_1 + K_2) \tag{1-9-5}$$

K_i being the mobilities (in esu units of cm^2/statvolt-sec; 1 statvolt = 300 V) of the ions in the neutral gas. A unified quasi-equilibrium statistical theory incorporating these effects would be too cumbersome to be useful.

Natanson [6] for the case involving equal-mass species, succeeded in modifying Thomson's model so as to include them. Therefore, Bates and Flannery [12] adjusted the formula of Natanson for the recombination coefficient so as to bring it into accord with the quasi-equilibrium statistical theory in the low-density region. This procedure which reproduces the correct dependence on the mass ratios and interaction strengths will now be discussed.

A. NATANSON'S THEORY

A clear account of the important work of Natanson [6] is to be found in the book of McDaniel [13]. Natanson treated only the special case in which the three species of particles have equal masses and the two ionic species have the same mean free path. By putting a certain parameter β which appears

1-9 RECOMBINATION AT MODERATE AND HIGH GAS-DENSITIES

in eq. (12-4-41) of McDaniel's book, equal to unity (as is suggested), we can write the formula for the recombination coefficient

$$\alpha_N = \tfrac{1\cdot 7}{5}\pi r_N^2 \tilde{w}_N \left(\frac{3k\theta}{M_{12}}\right)^{\frac{1}{2}} C(x_N)\left[1+\frac{17 r_N^2 \tilde{w}_N (3k\theta/M_{12})^{\frac{1}{2}} k\theta\{C(x_N)-1\}}{20(\mathscr{D}_1+\mathscr{D}_2)e^2}\right]^{-1} \tag{1-9-6}$$

where

$$r_N = \tfrac{1}{2}\lambda\left[\left(1+\frac{5e^2}{3k\theta\lambda}\right)^{\frac{1}{2}}-1\right] \rightarrow \begin{cases} \dfrac{5e^2}{12k\theta} & \text{at low densities} \\[6pt] \left(\dfrac{5e^2\lambda}{12k\theta}\right)^{\frac{1}{2}} & \text{at high densities} \end{cases} \tag{1-9-7}$$

λ being the ionic mean free path, is the critical distance replacing r_T of eq. (1-2-3), where

$$\tilde{w}_N = 2w(x_N) - w^2(x_N) \tag{1-9-8}$$

with

$$x_N = r_N/\lambda \tag{1-9-9}$$

is the probability replacing $\tilde{w}_T$ of (1-9-2), where

$$C(x_N) = \exp\left(\tfrac{12}{5}x_N\right) \tag{1-9-10}$$

and where $\mathscr{D}_i$ are the diffusion coefficients of the ions in the gas.

The product $w^2(x_N)$ is simply the probability that both ions, when within the critical sphere of radius r_N, collide simultaneously with the neutral.

Using the standard relation

$$e\mathscr{D}_i = k\theta K_i \tag{1-9-11}$$

the expression for α may be rewritten as

$$\alpha_N^{-1} = \alpha_{TN}^{-1} + \alpha_{LHN}^{-1} \tag{1-9-12}$$

where

$$\alpha_{TN} = \tfrac{1\cdot 7}{5}\pi r_N^2 \tilde{w}_N \left(\frac{3k\theta}{M_{12}}\right)^{\frac{1}{2}} C(x_N) \tag{1-9-13}$$

describes the contribution from individual ion-neutral collisions dominant at low densities and

$$\alpha_{LHN} = \frac{C(x_N)}{C(x_N)-1}\alpha_{LH} \tag{1-9-14}$$

denotes that resulting from the diffusional and mobility effects which limit the recombination at high gas densities, α_{LH} being, as before, the Langevin-Harper recombination coefficient.

Eq. (1-9-13) is similar in form to eq. (1-9-1), the extra factors $\frac{17}{5}$ and $C(x_N)$ arising, respectively, from the curvatures of the trajectories of the ions and from the effect of this Coulomb attraction on their spatial density distributions. Since the *reciprocals* of the rate coefficients of processes which occur in series are additive [58], it is therefore natural to identify α_{LHN} as a rate coefficient associated with the approach of the ions. At very high densities, C tends to infinity and α tends to α_{LH}. At low densities C becomes unity and

$$\alpha_N \to \tfrac{17}{5}\pi r_N^3 (3k\theta/M_{12})^{\frac{1}{2}} 8/3\lambda \tag{1-9-15}$$

so that the ratio

$$\alpha_N/\alpha_T = \tfrac{17}{5} r_N^3/r_T^3 = \tfrac{17}{5}(\tfrac{5}{8})^3 = 0.83. \tag{1-9-16}$$

The parameter C can thus be regarded as being the density-dependent mixing parameter which effectively determines the importance of the effects arising from three-body collisions relative to those arising from random diffusion and attractive drift.

B. Modification of Natanson's Theory

If the mean free paths, instead of having the same value λ as above, have different values, λ_1 and λ_2, then (1-9-7) defines two critical distances which we designate r_{N1} and r_{N2}, respectively. Throughout Natanson's formulae, we arbitrarily replace $C(x_N)$ by the geometric mean

$$\bar{C}(x_{N1}, x_{N2}) = [C(x_{N1})C(x_{N2})]^{\frac{1}{2}} \tag{1-9-17}$$

where

$$x_{Ni} = r_i/\lambda_i, \quad i = 1, 2. \tag{1-9-18}$$

Noting that formula (1-2-4) for $\tilde{w}_N$ is a special case of formula (1-9-2) and recalling the significance of each term of $\tilde{w}_N$, we also replace $r_N^2 \tilde{w}_N$ in (1-9-6) by

$$r_{N1}^2 w(x_{N1}) + r_{N2}^2 w(x_{N2}) - r_{N<}^2 w(y_{N1}) w(y_{N2}) \tag{1-9-19}$$

where $r_{N<}$ is the lesser of r_{N1} and r_{N2} and where

$$y_{Ni} = r_{N<}/\lambda_i. \tag{1-9-20}$$

The low-density limit is then the sum of two partial recombination coefficients

$$a_N(i, 3) = \tfrac{1.7}{5}\pi r_{Ni}^2 w(x_{Ni})(3k\theta/M_{12})^{\frac{1}{2}}, \quad i = 1, 2 \tag{1-9-21}$$

one representing the contribution from 1–3 collisions and the other representing the contribution from 2–3 collisions. This formula must be modified so as to give the limit accurately, in accordance with the predictions of Bates and Flannery [11]. From the several possible methods of modification, we will choose the following procedure.

Let $\alpha_{BF}(i, 3)$ be the partial recombination coefficients calculated by Bates and Flannery for interactions of the Langevin type, and let K_{Li} be the corresponding Langevin mobilities (cf. McDaniel [13]). In the low-density limit the partial-recombination coefficients are approximately inversely proportional to the ionic mean free paths and hence to the mobilities (the masses being held fixed). We can therefore correct for the effect of the difference between the actual and the Langevin interactions by adopting as improved partial recombination coefficients (as in § 1-7)

$$\alpha'_{BF}(i, 3) = (K_{Li}/K_i)\alpha_{BF}(i, 3) \tag{1-9-22}$$

where K_i are the measured mobilities.

Not only does this procedure cause refinement in details of the physical ion-neutral interaction, but also it permits the inclusion of the effects of rotational and vibrational excitation and de-excitation which arise from the collision of either ion with the third body Z, if a molecule. The rate coefficient $K(E_i, E_f)dE_f$ (calculated in § 1-5), considers only the energy-change in the ion-pairs, brought about by *elastic* collisions with the third body. Experimental mobilities [59] include the effects of collisions which leave the neutral in elastic and in all accessible inelastic channels.

To make the low-density limit to (1-9-13) consistent with α'_{BF} we replace r_{Ni}, x_{Ni} and y_{Ni} in (1-9-17) and (1-9-19) by $\rho_i r_{Ni}$, $\rho_i x_{Ni}$, and $\rho_i y_{Ni}$, respectively, with

$$\rho_i^3 = \alpha'_{BF}(i, 3)/a_N(i, 3). \tag{1-9-23}$$

Denoting the modified rate coefficients by the symbol used for the original coefficients with a prime added, we have

$$\alpha'_{TN} = \tfrac{1.7}{5}\pi[\rho_1^2 r_{N1} w(\rho_1 x_{N1}) + \rho_2^2 r_{N2} w(\rho_2 x_{N2}) - \rho_<^2 r_{N<}^2 w(\rho_< y_{N1})$$
$$\times w(\rho_< y_{N2})](3k\theta/M_{12})^{\frac{1}{2}}\bar{C}(\rho_1 x_{N1}, \rho_2 x_{N2}) \tag{1-9-24}$$

and

$$\alpha'_{\text{LH}} = \frac{\bar{C}(\rho_1 x_{\text{N1}}, \rho_2 x_{\text{N2}})}{\bar{C}(\rho_1 x_{\text{N1}}, \rho_2 x_{\text{N2}}) - 1} \alpha_{\text{LH}}. \tag{1-9-25}$$

It is best to adopt the measured values of the mobilities appearing implicitly in these formulae rather than the values calculated on the assumption that the interactions are of the Langevin type; and to adopt also the corresponding values of the mean free paths.

Finally, allowance must be made for the recombination in binary collisions

$$X^+ + Y^- \to X + Y. \tag{1-9-26}$$

Since this process occurs in parallel with the Thomson process we have, if β is its rate coefficient in the low-density region, that

$$(\alpha'_N)^{-1} = (\alpha'_{\text{TN}} + \beta)^{-1} + (\alpha'_{\text{LHN}})^{-1}. \tag{1-9-27}$$

TABLE 1-9-1
Basic information used at 25 °C

(i) Polarizability of O_2: $p = 1.6 \times 10^{-24}$ cm^3 [53].

(ii) Hard-sphere core radii:

$S_{13}(O_4^+ - O_2) = S_{23}(O_4^- - O_2) = 5.08 \times 10^{-8}$ cm*

$S_{23}(O_2^- - O_2) = 3.58 \times 10^{-8}$ cm [53].

(iii) Langevin interaction parameter:

$(2k\theta)^{\frac{1}{2}} S_{13}^2 / p^{\frac{1}{2}} e = \begin{cases} 1.219 \ (O_4^\pm - O_2) \\ 0.581 \ (O_2^- - O_2). \end{cases}$

(iv) Measured zero-field reduced mobilities in oxygen [33] (in cm^2 V^{-1} sec^{-1})

$K(O_4^+) = 2.16$

$K(O_2^-) = 2.16$

$K(O_4^-) = 2.14$.

(v) Deduced values of dimensionless correction factor ρ_i defined by eq. (1-9-23):

For $O_4^+ + O_2^- + O_2$: $\rho_1 = 0.657$, $\rho_2 = 0.978$;

For $O_4^+ + O_4^- + O_2$: $\rho_1 = \rho_2 = 0.826$.

(vi) Binary recombination coefficient β in eq. (1-9-27) $= 2 \times 10^{-7}$ cm^{3}sec^{-1} [30].

---

* O_4 is assumed to be a linear molecule composed of two O_2 molecules and the equilibrium internuclear distance $r_e(O_4)$ is taken as 3×10^{-8} cm [47].

C. Comparison with Experiment

There are few sets of laboratory measurements on ionic recombination that are satisfactory for testing the theory. The most suitable set is that obtained by McGowan [30] in his recent study of ionic recombination in pure oxygen. As previously noted, the three-body recombination processes occurring are probably [49]:

$$O_4^+ + O_2^- + O_2 \to [O_6] + O_2 \qquad (1\text{-}9\text{-}28)$$

at the lower densities and

$$O_4^+ + O_4^- + O_2 \to [O_8] + O_2 \qquad (1\text{-}9\text{-}29)$$

at the higher densities. By using formula (1-9-27) with the basic information contained in Table 1-9-1, we present α'_N for each process as a function of neutral number density. Figure 1-9-1 compares their results with those of McGowan. The agreement is quite good. Figure 1-9-2 continues the theoreti-

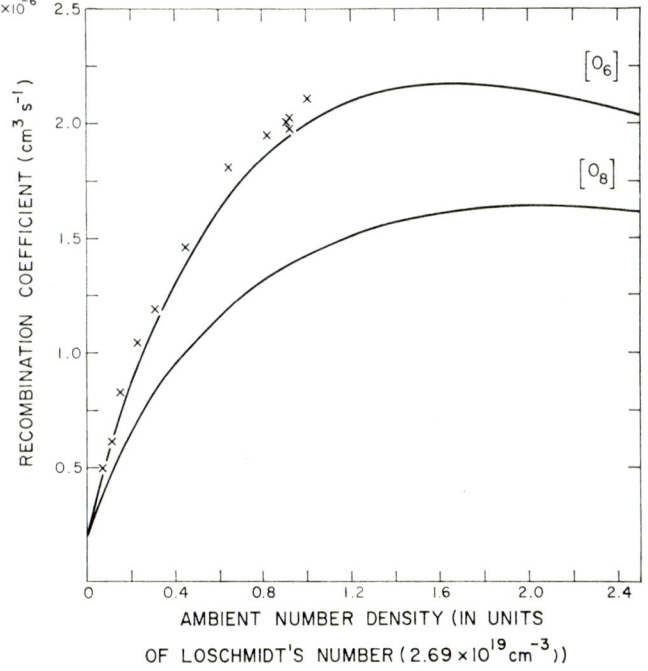

Fig. 1-9-1. Recombination in oxygen at 25 °C. Full curves are calculated α'_N curves, that for process (1-9-28) being marked $[O_6]$ and that for process (1-9-29) being marked $[O_8]$; crosses are measurements by McGowan [30].

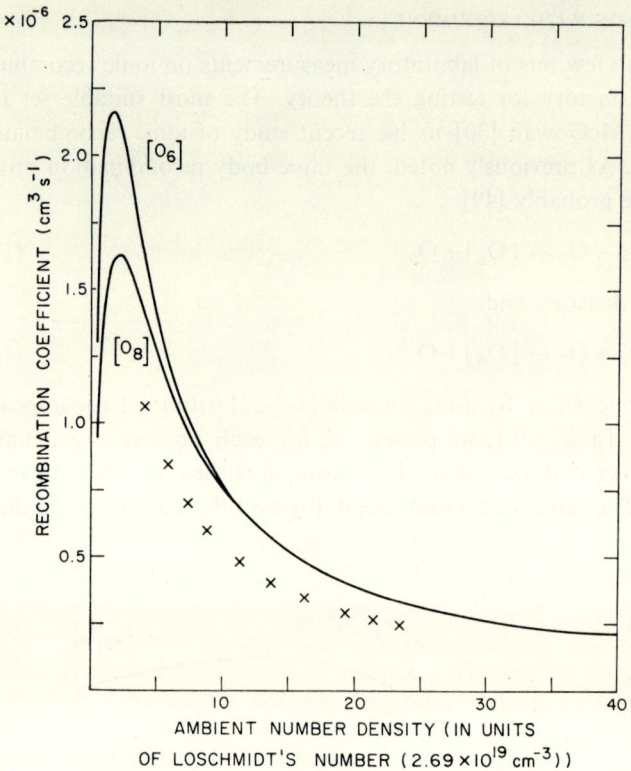

Fig. 1-9-2. Full lines are as indicated in Fig. 1-9-1; crosses denote measurements of Mächler [60] in air.

cal curves into the high-density region to show their form and to illustrate the effect of change of species of ion. Proper comparison data are not available but we include, for interest, the results of the early measurements of Mächler [60] on the recombination of ions (of unknown species) in air. The departure of these theoretical curves from those calculated previously by Bates and Flannery [12] is due to the use of more accurate and up-to-date experimental mobilities [33]. The high density results tend to values in accord with (1-9-5) and, because the measured mobilities for O_4^+, O_2^- and O_4^- are almost equal, the rates for (1-9-28) and (1-9-29) do not manifest any noticeable mass effect and therefore tend to the same high-density limit.

Thus both figures relate the present status between theory and experiment. The situation appears rather encouraging.

Appendix. The Thomson theory

In this appendix, we present, in modern notation, the description of the three-body ionic recombination process as envisaged by Thomson [2]. Each recombination event

$$X^+ + Y^- + Z \to [XY] + Z \tag{1-A-1}$$

is assumed to proceed via the following sequence of channels:

$$X^+ + Y^- \xrightarrow{k_1} (X^+Y^-)^*_{r_T} \tag{1-A-2}$$

$$(X^+Y^-)^*_{r_T} \xrightarrow{k_2} X^+ + Y^- \tag{1-A-3}$$

and

$$(X^+Y^-)^*_{r_T} + Z \xrightarrow{k_3} (X^+Y^-) + Z \to [XY] + Z \tag{1-A-4}$$

where $(X^+Y^-)^*_{r_T}$ denotes ion-pairs formed with a rate k_1 (cm³ sec⁻¹) in states for which the ionic separation is r_T or less. Such ion-pairs can either revert back into their initial incoming channels with a rate k_2 (sec⁻¹) or else suffer de-activation by collisions with the third bodies Z at a rate k_3 (cm³ sec⁻¹) such that the internal energy of each ion-pair is reduced sufficiently for closed orbiting of one ion about the other to occur. The latter possibility is followed by eventual electron-transfer which completes the recombination.

As before, let $N(X^+)$, $N(Y^-)$ and $N(Z)$ be the number densities of the species involved and let N^* denote the number densities of ion-pairs that have ionic separations $\leq r_T$ before collisions with the neutrals occur. The growth of these ion-pairs is described by the equation

$$dN^*/dt = k_1 N(X^+) N(Y^-) - k_2 N^* - k_3 N^* N(Z) \tag{1-A-5}$$

the solution of which evolves with time t according to

$$N^*(t) = \left[\frac{k_1 N(X^+) N(Y^-)}{k_2 + k_3 N(Z)}\right] [1 - \exp(-\{k_2 + k_3 N(Z)\}t)] \tag{1-A-6}$$

in which we assume $N^*(0) = 0$.

When a steady-state is attained, $dN^*/dt = 0$ and the number-density reaches the limit,

$$N^*(t) = \frac{k_1 N(X^+) N(Y^-)}{k_2 + k_3 N(Z)}. \tag{1-A-7}$$

The rate of depletion of the free ions is

$$\frac{dN(X^+)}{dt} = \frac{dN(Y^-)}{dt} = k_2 N^* - k_1 N(X^+) N(Y^-) \qquad (1\text{-A-}8)$$

which, with the aid of (1-A-7) reduces to

$$\frac{dN(X^-)}{dt} = \frac{dN(Y^-)}{dt} = -\frac{k_1 k_3 N(Z)}{k_2 + k_3 N(Z)} N(X^+) N(Y^-) \qquad (1\text{-A-}9)$$

with the result that the coefficient α for recombination is

$$\alpha = -\frac{1}{N(X^+) N(Y^-)} \frac{dN(X^+)}{dt} = \frac{k_1 k_3 N(Z)}{k_2 + k_3 N(Z)}. \qquad (1\text{-A-}10)$$

The positive and negative ions, with velocities v_1 and v_2, respectively, are assumed to be in thermodynamic equilibrium with the surrounding gas. The velocity v_{rel} of the positive ion relative to the negative ion is therefore given by [13]

$$v_{\text{rel}} = (v_1^2 + v_2^2)^{\frac{1}{2}} \qquad (1\text{-A-}11)$$

and hence

$$k_1 = \pi r_T^2 (v_1^2 + v_2^2)^{\frac{1}{2}} \qquad (1\text{-A-}12)$$

is the rate (cm^3 sec^{-1}) of formation of ion-pairs with ionic separations r_T or less. The probability that these ion-pairs suffer collisions with the neutrals is the probability that channel (1-A-4) is followed and is given by

$$w = \frac{k_3 N(Z)}{k_2 + k_3 N(Z)} \qquad (1\text{-A-}13)$$

while

$$1 - w = \frac{k_2}{k_2 + k_3 N(Z)} \qquad (1\text{-A-}14)$$

is the probability that the ions continue their hyperbolic orbits as in (1-A-3), so that from (1-A-10) we deduce the recombination coefficient to be

$$\alpha = \pi r_T^2 w (v_1^2 + v_2^2)^{\frac{1}{2}} \qquad (1\text{-A-}15)$$

in accordance with eq. (1-2-6).

We can calculate w from simple geometrical arguments. Assume that the positive ion crosses the imaginary sphere of radius r_T (centered at the negative ion) at angle of incidence ψ, the angle between the direction of incidence and the normal to the sphere. Then the probability that the ion does not collide with a neutral body while traversing a path length $2r_T \cos \psi$ to arrive at the opposite side is given by the survival equation:

$$P_B = \exp(-2r_T \cos \psi/\lambda^+) \qquad (1\text{-}A\text{-}16)$$

where λ^+ is the mean free path of the positive ion in the gas. The chance dP_A that the ion is incident on the annular element $2\pi r_T^2 \sin \psi \, d\psi$ about ψ is the projected area of this annular element onto the normal of the incident path divided by the target area πr_T^2 of the sphere. Thus

$$dP_A = 2 \sin \psi \cos \psi \, d\psi \qquad (1\text{-}A\text{-}17)$$

such that $\int_{\psi=0}^{\frac{1}{2}\pi} dP_A$ is unity. Thomson [2] inadvertently omitted the factor of $\cos \psi$ in (1-A-17) but Loeb [14] later included it. Therefore, the net probability that the positive ion will cross the surface of the sphere at all angles and then pass through the sphere (in a straight line) without suffering a collision with the neutral is

$$\mathscr{P}^+ = \int P_B \, dP_A = \int_0^{\frac{1}{2}\pi} \sin 2\psi \exp(-2r_T \cos \psi/\lambda^+) \, d\psi$$
$$= \tfrac{1}{2}(\lambda^+/r_T)^2 [1 - \exp(-2r_T/\lambda^+)\{1 + 2r_T/\lambda^+\}]. \qquad (1\text{-}A\text{-}18)$$

Hence, the probability that a positive ion-neutral collision occurs within a distance r_T from the negative ion is

$$w^+ = 1 - \mathscr{P}^+ \qquad (1\text{-}A\text{-}19)$$

and, correspondingly, the probability that a negative-ion neutral collision occurs within a distance r_T from the positive ion is

$$w^- = 1 - \mathscr{P}^- \qquad (1\text{-}A\text{-}20)$$

where $\mathscr{P}^-$ is deduced from (1-A-18) by replacing λ^+ by λ^-, the mean free path of the negative ion. Therefore, the probability w, eq. (1-A-13), that an ion-pair suffers a collision with the neutral is

$$w = w^+ + w^- - w^+ w^- \rightarrow \begin{cases} \tfrac{2}{3} r_T \left(\dfrac{1}{\lambda^+} + \dfrac{1}{\lambda^-} \right) & \text{at low densities} \\ 1 & \text{at high densities} \end{cases} \qquad (1\text{-}A\text{-}21)$$

where $w^+ w^-$, the chance that both ions collide simultaneously with the neutral, has been counted *twice* in the sum $w^+ + w^-$. Substitution of (1-A-21) into (1-A-15) yields the Thomson result.

From eqs. (1-A-2)–(1-A-4) the inadequacies of the Thomson model are immediately obvious and are as follows:

(i) Ion-pairs having ionic separations greater than r_T are ignored from consideration.

(ii) There is no mechanism describing disruption of the ion-pairs after they collide with the third bodies, e.g., the rate, corresponding to the reverse of the forward process with rate k_3 is assumed to be unimportant.

(iii) Only one collision of the ion pairs $(X^+Y^-)^*_{r_T}$ with the neutral is deemed as a necessary and sufficient condition for recombination. The effects of subsequent collisions that could cause further deactivation or disruption are ignored and transitions between bound states of the ion-pair are not acknowledged.

(iv) As the density $N(Z)$ is increased, α tends to k_1, a constant; an incorrect result which arises from the lack of recognition that the speed of approach between the ions is limited at intermediate densities by the speed of random diffusion and by the speed of attractive drift at even higher densities.

Acknowledgments

I am extremely grateful to Professor D. R. Bates for initially introducing me to the subject of ionic recombination in 1964, and also for a countless number of stimulating discussions thereafter. I am also grateful to him for making valuable comments in reading the initial draft of this review. I also acknowledge the benefit of interesting discussions with Professor A. Dalgarno and Dr. C. Bottcher in the course of writing this article.

References

1. P. LANGEVIN, Ann. Chem. Phys. **28** (1903) 289, 433.
2. J. J. THOMSON, Phil. Mag. **47** (1924) 337.
3. W. R. HARPER, Proc. Camb. Phil. Soc. **28** (1932) 219.
4. M. E. GARDNER, Phys. Rev. **53** (1938) 75.
5. J. SAYERS, Proc. Roy. Soc. A (London) **169** (1938) 83.
6. G. L. NATANSON, Soviet Phys.-Tech. Phys. **4** (1959) 1263.
7. L. P. PITAEVSKII, Soviet Phys.-JETP **15** (1962) 919.
8. S. A. LANDON and J. C. KECK, J. Chem. Phys. **48** (1968) 374; J. C. KECK and G. CARRIER, J. Chem. Phys. **43** (1965) 2284.
9. B. H. MAHAN, J. Chem. Phys. **48** (1968) 2629.
10. D. R. BATES and R. J. MOFFETT, Proc. Roy. Soc. A (London) **291** (1966) 1.

11. D. R. BATES and M. R. FLANNERY, Proc. Roy. Soc. A (London) **302** (1968) 367.
12. D. R. BATES and M. R. FLANNERY, J. Phys. B (Atom. Molec. Phys.) [2] **2** (1969) 184.
13. E. W. MCDANIEL, Collision Phenomena in Ionized Gases (Wiley, New York, 1964) Ch. 12 and Ch. 9.
14. L. B. LOEB, Basic Processes of Gaseous Electronics (University of California Press, Berkeley, Los Angeles, 2nd ed., 1955) Ch. 1 and Ch. 6.
15. L. B. LOEB and L. C. MARSHALL, J. Franklin Inst. **208** (1929) 371.
16. K. A. BRUECKNER, J. Chem. Phys. **42** (1964) 439.
17. E. K. PARKS, J. Chem. Phys. **48** (1968) 1483.
18. T. FUENO, H. EYRING and T. REE, Canad. J. Chem. **38** (1960) 1693.
19. B. H. MAHAN and J. C. PERSON, J. Chem. Phys. **40** (1964) 392, 2851.
20. P. J. FEIBELMAN, J. Chem. Phys. **42** (1965) 2462.
21. D. R. BATES, A. E. KINGSTON and R. W. P. MCWHIRTER, Proc. Roy. Soc. A (London) **267** (1962) 297.
22. D. R. BATES, P. B. HAYS and D. SPREVAK, J. Phys. B (Atom. Mol. Phys.) **4** (1971) 962.
23. D. R. BATES and Z. JUNDI, J. Phys. B (Proc. Phys. Soc.) [2] **1** (1968) 1145.
24. J. J. THOMSON and G. P. THOMSON, Conduction of Electricity through Gases, vol. 1 (Cambridge University Press, London, 1928) Ch. 2.
25. L. B. LOEB, Fundamental Processes of Electrical Discharges in Gases (Wiley, New York, 1939) Ch. 2.
26. J. SAYERS, in: Atomic and Molecular Processes, ed. D. R. Bates (Academic Press, New York, 1965) Ch. 8.
27. G. RÜMELIN, Phys. Z. **9** (1908) 659;
G. RÜMELIN, Ann. Physik **43** (1914) 821.
28. S. G. PLIMPTON, Phil. Mag. **25** (1913) 65.
29. L. C. MARSHALL, Phys. Rev. **34** (1929) 618.
30. S. MCGOWAN, Canad. J. Phys. **45** (1967) 439;
S. MCGOWAN, Defence Res. Board of Canada, DRCL Report Nos. 410 (1963), 515 (1966).
31. H. G. EBERT, J. BOOZ and R. KOEPP, Z. Physik **181** (1964) 187.
32. B. H. MAHAN and J. C. PERSON, J. Chem. Phys. **40** (1964) 392;
T. S. CARLTON and B. H. MAHAN, J. Chem. Phys. **40** (1964) 3683;
G. A. FISK, B. H. MAHAN and E. K. PARKS, J. Chem. Phys. **47** (1967) 2649.
33. E. W. MCDANIEL, Case Studies in Atomic Collision Physics 1, eds. E. W. McDaniel and M. R. C. McDowell (North-Holland, Amsterdam, London, 1969) Ch. 1;
R. M. SNUGGS, D. J. VOLZ, J. H. SCHUMMERS, D. W. MARTIN and E. W. MCDANIEL, Phys. Rev. A **3** (1971) 477.
34. L. D. LANDAU and E. M. LIFSHITZ, Statistical Physics (Pergamon Press, London, 1958) Chs. 4, 10.
35. D. R. BATES and T. J. M. BOYD, Proc. Phys. Soc. A **69** (1956) 910.
36. A. DALGARNO, Phil. Trans. Roy. Soc. A (London) **250** (1958) 426.
37. M. R. FLANNERY, Ann. Phys. (N.Y.) **61** (1970) 465.
38. H. R. HASSÉ, Phil. Mag. **1** (1926) 139.
39. E. A. MASON, J. Chem. Phys. **26** (1957) 667.
40. M. R. FLANNERY, Proc. Phys. Soc. **92** (1967) 551.
41. H. GOLDSTEIN, Classical Mechanics (Addison-Wesley, Reading, Mass., 1950) Ch. 3.
42. Z. KOPAL, Numerical Analysis (Chapman and Hall, London, 1955) p. 381.
43. F. J. SMITH, Physica **30** (1964) 497;
H. O'HARA and F. J. SMITH, Computer Journal **11** (1968) 213; **12** (1969) 179.
44. G. GIOUMOUSIS and D. P. STEVENSON, J. Chem. Phys. **29** (1958) 294.
45. M. R. FLANNERY, J. Phys. B (Proc. Phys. Soc.) [2] **1** (1968) 384.

46. H. S. W. Massey and E. H. S. Burhop, Electronic and Ionic Impact Phenomena (Clarendon Press, Oxford, 1952) p. 404.
47. Y. Yang and D. Conway, J. Chem. Phys. **40** (1964) 1729.
48. R. E. Voshall, J. L. Pack and A. V. Phelps, J. Chem. Phys. **43** (1965) 1990.
49. M. R. Flannery, J. Chem. Phys. **50** (1969) 546.
50. L. G. McKnight, Bull. Amer. Phys. Soc. **12** (1967) 228.
51. J. Dutton and P. Howells, Bull. Amer. Phys. Soc. **12** (1967) 228.
52. A. Dalgarno, in: Proc. Conf. on Physical Chemistry in Aerodynamics and Space Flight, Philadelphia, 1959 (Pergamon Press, New York, 1961) p. 217.
53. J. O. Hirschfelder, C. H. Curtiss and R. B. Bird, Molecular Theory of Gases and Liquids (Wiley, New York, 1954) pp. 950, 1110.
54. S. Chandrasekhar, Rev. Mod. Phys. **15** (1943) 1.
55. M. R. Flannery, Ann. Phys. (N.Y.) **67** (1971) 376.
56. P. Mandl, Analytical Treatment of One-Dimensional Markov Processes (Academia, Publishing House of the Czechoslovak Academy of Sciences, Prague, 1968) Ch. 1.
57. E. H. Holt and R. E. Haskell, Foundations of Plasma Dynamics (Macmillan, New York, 1965) Ch. 5.
58. D. R. Bates and A. E. Kingston, Proc. Phys. Soc. **83** (1964) 43.
59. O. H. Crawford, A. Dalgarno and P. B. Hays, Mol. Phys. **13** (1967) 181.
60. W. Mächler, Z. Physik **104** (1936) 1.

CHAPTER 2

PRECISION MEASUREMENTS OF ELECTRON TRANSPORT COEFFICIENTS

BY

M. T. ELFORD

Ion Diffusion Unit,
Research School of Physical Sciences,
Australian National University, Canberra

Contents

	Page
2-1. Introduction	94
2-2. Theory of electron swarms	95
A. The electron energy distribution function	95
1. When only elastic scattering occurs	97
2. When both elastic and inelastic scattering occur	97
B. Transport coefficients	98
1. Electron drift velocity	98
2. The diffusion coefficient	99
3. The characteristic energy	100
4. The magnetic deflection coefficient	102
C. The derivation of cross sections from transport coefficients	103
1. When only elastic scattering occurs	103
2. When both elastic and inelastic scattering occur	106
2-3. The electron drift velocity	107
A. Review of previous methods of measurement	107
1. Oscillographic methods	107
2. Methods using counters	108
3. Electrical shutter methods	109
B. The Bradbury-Nielsen method	111
1. Electrical shutters	111
2. Uniform electric field and geometrical accuracy	112
3. Electron source	114
4. Gas temperature measurement and control	114
5. Gas purity and vacuum requirements	116
6. Pressure measurement	119
7. The determination of the current maxima	119
8. Sources of systematic error	121
9. Corrections for the effect of diffusion	124
10. The linear variation of the drift velocity with gas number density	125
C. Experimental data	126
1. Presentation of data	126
2. Analysis of error	129
3. Results and discussion	129
2-4. The measurement of D_T/μ by the Townsend-Huxley method	137
A. Introduction	137
1. Historical survey	137
B. The solution of the diffusion equation and the current ratio formula	139
1. The current ratio formula	139
2. Effect of the spatial dependence of the electron energy distribution function	141
C. Experimental method	141
1. Diffusion chamber geometry	141
2. Production of uniform electric field	142
3. The collector	145
4. The current ratio	146
5. Temperature measurement and control	147
6. Electron source	147
7. Gaseous impurities	147
D. Sources of error	148

1. Nonuniformity of the electric field	148
2. Finite size source hole	150
3. Incorrect axial alignment	150
4. Presence of negative ions	150
E. Experimental results	151
2-5. Conclusion	155
Acknowledgements	156
References	156

§ 2-1. Introduction

Electrons occur in natural phenomena principally under swarm conditions and the measurement of the coefficients which characterise the macroscopic behaviour of an electron swarm are of obvious significance. A second and increasingly more significant aim of swarm studies is the obtaining of information on electron scattering processes. Despite some earlier successes such as the independent discovery by Townsend of the minimum in the elastic scattering cross section in gases such as argon, i.e. the Ramsauer-Townsend effect, progress towards a detailed knowledge of the microscopic collision processes was hampered by two factors, inadequate theories by which these processes are related to the macroscopic coefficients and inaccurate experimental measurements of these coefficients.

By the early 1950's the first of these difficulties was largely overcome but because of their complexity the theoretical expressions could not be evaluated without making often unreal assumptions (e.g. that the momentum transfer cross section was independent of the energy of the electrons in a swarm or that the distribution of energies of electrons in the swarm had a particular form). Under these circumstances there was little incentive to obtain experimental data of high accuracy. With the increasing use and availability of high speed computers the previous numerical difficulties were overcome and in 1962 Frost and Phelps published the first cross sections for elastic and inelastic electron collision processes (for hydrogen and nitrogen) derived from a numerical solution of the Boltzmann transport equation. The success of this technique in obtaining cross sections over a wide energy range and particularly at very low energies where beam experiments are very difficult to perform, is now well established. Both Phelps [1] in a review of rotational and vibrational excitation and more recently Crompton [2] have emphasised the importance of using reliable and precise data in such analyses. The necessity for high precision has been made particularly evident by Crompton, Gibson and Robertson [3] who show, when discussing the vibrational excitation of hydrogen, how errors in D_T/μ data of only a few percent can lead to very large errors in the derived cross sections for both rotational and vibrational excitation.

This chapter describes the experimental techniques used to obtain precise values of the two coefficients of major interest i.e. the drift velocity W and the characteristic energy eD_T/μ. Other coefficients of significance such as the ionization and attachment coefficients will not be considered here. This chapter is in no sense a review. The experimental techniques

§ 2-2. Theory of electron swarms

The term electron swarm experiments is applied to those experiments in which a large number of electrons move in a gas at a number density such that any electron makes a very large number of collisions before being collected by a surface. The theoretical problem is to relate the observed macroscopic behaviour of the whole electron swarm to the collision processes which occur between an electron and a gas molecule.

A. The Electron Energy Distribution Function

The behaviour of an assembly of electrons (i.e. an electron swarm) in a gas is governed by the cross sections for the various collision processes which occur between the electrons and gas molecules, the gas temperature and number density and the presence of external fields, electric or magnetic. The distribution of the velocities and positions of the electrons at a given instant may be described by a distribution function $F(r, c, t)$ where $F(r, c, t) dr dc$ is the number of electrons within the volume element dr at position r which have velocities lying within the element dc situated at c in velocity space.

By definition

$$\int F(r, c, t) dc = n(r, t) \qquad (2\text{-}2\text{-}1)$$

where $n(r, t)$ is the number density of the electrons in the swarm. The function $F(r, c, t)$ is found by solution of the Boltzmann equation

$$\frac{\partial F}{\partial t} + c \cdot \nabla_r F + a \cdot \nabla_c F = \left(\frac{\partial F}{\partial t}\right)_{\text{coll.}} \qquad (2\text{-}2\text{-}2)$$

where a is the acceleration produced by an external force and $(\partial F/\partial t)_{\text{coll.}}$ is the rate of change of the number of electrons in unit volume of phase space due to collisions of electrons with gas molecules. It will be assumed that electron-electron interactions may be ignored, a condition easily met in the experimental measurements with electron swarms described in § 2-3 and § 2-4. We assume that the only force present is that due to a steady uniform electric field E directed along the Oz axis and that the distribution of electron velocities is constant in time and independent of spatial co-ordinates. This assumption is considered in §2-2B-2.

We write

$$F(\mathbf{r}, \mathbf{c}, t) = f(\mathbf{c}) n(\mathbf{r}, t) \tag{2-2-3}$$

where the function $f(\mathbf{c})$ satisfies the normalising relation

$$\int f(\mathbf{c}) \, d\mathbf{c} = 1. \tag{2-2-4}$$

On substitution of (2-2-3) for $F(\mathbf{r}, \mathbf{c}, t)$ into (2-2-2) we have

$$f(\mathbf{c}) \frac{\partial n(\mathbf{r}, t)}{\partial t} + \mathbf{c} \cdot f(\mathbf{c}) \frac{\partial n(\mathbf{r}, t)}{\partial \mathbf{r}} + \frac{eE}{m} \frac{\partial f(\mathbf{c})}{\partial v_z} \cdot n(\mathbf{r}, t) = \left[\frac{\partial (f(\mathbf{c}) n(\mathbf{r}, t))}{\partial t} \right]_{\text{coll.}} \tag{2-2-5}$$

If it is assumed that the electron number density at any point is constant and that there are no spatial number density gradients, i.e. $\partial n / \partial \mathbf{r} = \partial n / \partial t = 0$ we have the Boltzmann equation for a steady stream of electrons,

$$\frac{eE}{m} \frac{\partial f(\mathbf{c})}{\partial v_z} = \left(\frac{\partial f(\mathbf{c})}{\partial t} \right)_{\text{coll.}} \tag{2-2-6}$$

where v_z is the component of the velocity $\mathbf{c}$ in the Oz direction.

The usual procedure used to solve equation (2-2-6) is to first expand $f(\mathbf{c})$ in terms of spherical harmonics i.e.

$$f(\mathbf{c}) = \sum_l f_l P_l(\cos \theta) = f_0(c) + f_1(c)(\cos \theta) + \ldots \tag{2-2-7}$$

where $P_l (\cos \theta)$ is the Legendre function of order l and θ is the angle between the direction of $\mathbf{c}$ and the polar axis of a spherical coordinate system. The functions f_l are functions of the electron speed c. Only the first two terms of this series are retained (the Lorentz approximation). The neglect of higher order terms has been justified by Holstein [4]. Two equations in $f_0(c)$ and $f_1(c)$ are derived from the Boltzmann equation (2-2-6) by using the Lorentz approximation and performing appropriate integrations over velocity space for electrons of fixed speed c. These equations are then solved to obtain both $f_0(c)$ and $f_1(c)$. It is $f_0(c)$, the isotropic term of the expansion in spherical harmonics, which appears in the formulae for the transport coefficients in § 2-2B. It will be referred to either as the distribution of electron speeds $f_0(c)$ with the normalising relation,

$$\int_0^\infty f_0(c) 4\pi c^2 \, dc = 1 \tag{2-2-8}$$

or as the electron energy distribution $f_0(\varepsilon)$, with the normalising relation

$$\int_0^\infty \varepsilon^{\frac{1}{2}} f_0(\varepsilon) \, d\varepsilon = 1. \tag{2-2-9}$$

1. *When only elastic scattering occurs*

The solution for $f_0(\varepsilon)$ when only elastic collisions occur between electrons and gas molecules and when the motion of the gas molecules is taken into account was first obtained by Davydov [5] and is

$$f_0(\varepsilon) = A \exp\left[-\int_0^\varepsilon \left(\frac{ME^2 e^2}{6mN^2 q_m^2(\varepsilon)\varepsilon} + kT\right)^{-1} d\varepsilon\right] \tag{2-2-10}$$

where M is the molecular mass,
        m is the electron mass,
        N is the gas number density,
        k is the Boltzmann constant,
        T is the absolute gas temperature,
        $q_m(\varepsilon)$ is the energy dependent momentum transfer cross section defined by

$$q_m(\varepsilon) = 2\pi \int_0^\pi I(\Theta)(1 - \cos \Theta) \sin \Theta \, d\Theta \tag{2-2-11}$$

where Θ is the angle of scattering and $I(\Theta)$ is the differential elastic scattering cross section per unit solid angle. The scattering is assumed to be independent of the azimuthal angle. A is a constant determined from the normalising relation (2-2-9).

2. *When both elastic and inelastic scattering occur*

The particular form of the Boltzmann equation (2-2-6) to be solved when both elastic and inelastic collisions occur between electrons and gas molecules has been derived by Holstein [4]. With the inclusion of the term to account for the effect of the molecular energy distribution [6] the equation is [7],

$$\frac{E^2}{3N}\frac{d}{d\varepsilon}\left(\frac{\varepsilon}{q_m(\varepsilon)}\frac{df_0}{d\varepsilon}\right) + \frac{2m}{M}NkT\frac{d}{d\varepsilon}\left(\varepsilon^2 q_m(\varepsilon)\frac{df_0}{d\varepsilon}\right)$$

$$+ \frac{2m}{M}N\frac{d}{d\varepsilon}(\varepsilon^2 q_m(\varepsilon)f_0(\varepsilon))$$

$$+ \sum_{j,k}N_j[(\varepsilon+\varepsilon_{jk})f_0(\varepsilon+\varepsilon_{jk})q_{jk}(\varepsilon+\varepsilon_{jk}) - \varepsilon f_0(\varepsilon)q_{jk}(\varepsilon)]$$

$$+ \sum_{k,j}N_k[(\varepsilon-\varepsilon_{jk})f_0(\varepsilon-\varepsilon_{jk})q_{kj}(\varepsilon-\varepsilon_{jk}) - \varepsilon f_0(\varepsilon)q_{kj}(\varepsilon)] = 0, \qquad (2\text{-}2\text{-}12)$$

where $q_{jk}(\varepsilon)$ is the cross section for the excitation from the jth to the kth state, $q_{kj}(\varepsilon)$ is the cross section for the superelastic transition from the kth to the jth state, ε_{jk} is the difference in internal energy between the jth and kth molecular states and N_j is the number density of molecules in the jth state. Various numerical methods have been employed to solve this equation [7, 8], using an assumed set of cross sections for specific cases.

B. Transport Coefficients

The observed macroscopic properties of an electron swarm moving in a steady uniform electric field are related to the microscopic collision processes, described in terms of collision cross sections, by various integral expressions involving the distribution of electron energies. The two most important of these properties are the electron drift velocity and the ratio of the diffusion coefficient to the electron mobility. A third coefficient is the magnetic deflection coefficient which is measured when a magnetic field is applied transverse to the electric field.

1. Electron drift velocity

Consider a group of electrons travelling in a uniform electric field. The electron drift velocity W is defined as the average electron velocity in the direction of the electric field. Expressed in terms of the velocity distribution function applicable to the situation when $\partial n/\partial r = 0$ the formula for the drift velocity is

$$W = \int f(c)c\,dc. \qquad (2\text{-}2\text{-}13)$$

When the substitution of $f(c)$ by the expansion in spherical harmonics (2-2-7) is made it is found that all terms except that in $f_1(c)$ disappear due to the orthogonal properties of the Legendre function [9]. Thus the expression for W, based on (2-2-13), in terms of $f_1(c)$ is rigorous. In order to express the drift velocity in terms of f_0 it is necessary to obtain a relation between f_1 and f_0. Such a relation is obtained by multiplying equation (2-2-6) by $\cos\theta \sin\theta \, d\theta \, d\phi$ and integrating over all values of θ and ϕ. The angles θ and ϕ are the polar and azimuthal angles of the velocity c. As a result of the assumption that $\partial n/\partial r = 0$, the polar axis of the co-ordinate system lies in the same direction as that of the electric field E.

The drift velocity derived from relation (2-2-13) and in terms of f_0 is [4, 10],

$$W = \frac{E}{N}\left(\frac{2}{m}\right)^{\frac{1}{2}} \frac{e}{3} \int_0^\infty \frac{\varepsilon}{q_m(\varepsilon)} \frac{df_0(\varepsilon)}{d\varepsilon} d\varepsilon. \qquad (2\text{-}2\text{-}14)$$

The Boltzmann equation approach used by Holstein considers the swarm as a whole. An alternative approach is to examine the behaviour of a certain class or group, the so-called mean free path method. This method has been used by Huxley [11] to obtain the same expression (2-2-14), the approximation used in the Huxley method being that the change in electron speed along a free path is very much less than the electron speed, i.e. $\Delta c \ll c$. This approximation is equivalent to the Lorentz approximation used in solving the Boltzmann equation [12]. In a more recent analysis using a free path approach, Cavalleri and Sesta [12] have derived rigorous expressions for the drift velocity without making the assumption $\Delta c \ll c$, but the drift velocity can only be calculated from their formulae for special cases.

2. The diffusion coefficient

The diffusion coefficient is defined in terms of the flux of electrons. Consider a group of electrons travelling in the direction of an electric field E assumed to be in the Oz direction. The net transport of electrons in the Oz direction due to diffusion, across any plane within the group normal to Oz, is defined as $-D_L \partial n/\partial z$, where $\partial n/\partial z$ is the number density gradient at the plane considered. In a direction transverse to E, say Oy, the net transport of electrons across a plane normal to Oy is defined as $-D_T \partial n/\partial y$. It should be noted that the diffusion coefficient is anisotropic. The tensor characteristic of the diffusion coefficient was first discussed by Wannier [13] in connection with the motion of ions in gases and arises from the retention of the spatial gradient term in the Boltzmann equation (2-2-2). Parker and Lowke [14] and Lowke and Parker [15] have shown that when this term is included the distribution of electron energies $f_0(\varepsilon)$ becomes a function of the spatial coordinates, D_L and D_T being in general, unequal. In argon for example, the difference can be as much as a factor of 8.

The transverse diffusion coefficient D_T is unaffected by the inclusion of the spatial gradient term and is given by [9]

$$D_T = \frac{(2/m)^{\frac{1}{2}}}{3N} \int_0^\infty \frac{\varepsilon f_0(\varepsilon)}{q_m(\varepsilon)} d\varepsilon. \qquad (2\text{-}2\text{-}15)$$

A formula for the longitudinal diffusion coefficient D_L has been derived by Parker and Lowke [14].

3. *The characteristic energy*

In the absence of an applied electric field, E, a swarm of electrons in a gas will be in thermal equilibrium with the gas molecules exchanging energy and momentum. The macroscopic motion of the electron swarm will be governed by the electron number density gradients and may be described in terms of the gradient and the diffusion coefficient $D = D_L = D_T$. The mean energy $\langle \varepsilon \rangle$ of the swarm will be that of the gas molecules, i.e. $\frac{3}{2}kT$, and the distribution of electron energies $f_0(\varepsilon)$ will be that given by the Maxwellian distribution function, i.e.

$$f_0(\varepsilon) = A \exp[-\varepsilon/kT] \qquad (2\text{-}2\text{-}16)$$

where A is the normalising constant obtained from (2-2-9). The diffusion coefficient D and electron zero-field mobility μ, defined as

$$\mu = \lim_{E \to 0} (W/E) \qquad (2\text{-}2\text{-}17)$$

are related to the gas temperature T by the Nernst-Townsend (or Einstein) relation [16]

$$D/\mu = kT/e. \qquad (2\text{-}2\text{-}18)$$

When an electric field is applied, the electrons receive energy from the field, an energy balance being established when the rate at which energy is received from the field balances the rate at which energy is transferred to the gas by collisions. The mean energy of the electron swarm in the presence of an electric field will exceed that of the gas molecules, the ratio of the mean electron energy to the mean thermal agitational energy of the gas molecules being defined as the Townsend energy factor k_T, i.e.

$$k_T = \langle \varepsilon \rangle / (\tfrac{3}{2}kT). \qquad (2\text{-}2\text{-}19)$$

The distribution of electron energies is now in general no longer Maxwellian and must be obtained by solution of the Boltzmann equation. The quantity k_T can therefore not be obtained directly by experiment. To take into account the change in mean energy produced by the application of the electric field the relation (2-2-18) can be modified by the introduction of a factor k_1, known as the Townsend energy ratio (not to be confused with the Townsend energy factor k_T) defined in terms of the transverse diffusion coefficient D_T, μ and T as

$$k_1 = (D_T/\mu)e/kT. \tag{2-2-20}$$

Since (D_T/μ) can be measured (§ 2-4) k_1 can be determined. D_T/μ can be written from (2-2-14) and (2-2-15) as

$$eD_T/\mu = -\int_0^\infty \frac{\varepsilon f_0(\varepsilon)\,d\varepsilon}{q_m(\varepsilon)} \bigg/ \int_0^\infty \frac{\varepsilon}{q_m(\varepsilon)}\frac{df_0}{d\varepsilon}\,d\varepsilon, \tag{2-2-21}$$

i.e.

$$eD_T/\mu = \frac{2\langle\varepsilon\rangle}{3F} = k_1 kT \tag{2-2-22}$$

where

$$\langle\varepsilon\rangle = \int_0^\infty \varepsilon^{\frac{3}{2}} f_0(\varepsilon)\,d\varepsilon \tag{2-2-23}$$

and

$$F = -\frac{2}{3}\int_0^\infty \varepsilon^{\frac{3}{2}} f_0(\varepsilon)\,d\varepsilon \int_0^\infty \frac{\varepsilon}{q_m(\varepsilon)}\frac{df_0}{d\varepsilon}\,d\varepsilon \bigg/ \int_0^\infty \frac{\varepsilon}{q_m(\varepsilon)} f_0(\varepsilon)\,d\varepsilon; \tag{2-2-24}$$

F is a dimensionless quantity.

Thus the Townsend energy factor k_T is related to k_1 by the relation

$$k_T = k_1 F. \tag{2-2-25}$$

The value of the parameter F is usually close to 1, but $F = 1$ only in special circumstances. When the electron energy distribution is Maxwellian, $F = 1$, for any energy dependence of q_m. F is also equal to 1 for any energy distribution, provided v_m, the momentum transfer collision frequency, is constant.

Since eD_T/μ provides an estimate of the mean energy of the swarm (e.g. $\langle\varepsilon\rangle = \frac{3}{2}(eD_T/\mu)$ for a Maxwellian distribution) it is known as the characteristic energy ε_K.

4. *The magnetic deflection coefficient*

It has been assumed in the previous sections that the only field present is an electric field in the Oz direction. We now consider the application of a magnetic field B in the Oy direction in addition to the electric field in the Oz direction. Magnetic deflection experiments were first carried out by Townsend and Tizard [17] with the aim of measuring drift velocities. It is now known that the quantity measured in this type of experiment is not the drift velocity (except in special circumstances) but a quantity W_M which is related to the drift velocity by a factor known as the magnetic deflection coefficient χ. W_M is termed the "magnetic drift velocity". Since χ can be at least as large as 1.5 very significant errors can be incurred if W_M is taken to be the drift velocity. The quantity W_M is an independent transport coefficient related to the experimental parameters and cross sections as shown in the formulae below.

In the presence of an electric field and a transverse magnetic field the drift velocity will no longer be directed along the direction of the electric field and can be written

$$W = W_z + W_x \tag{2-2-26}$$

where W_z and W_x are the components of W in the Oz and Ox directions respectively.

The formulae for W_x and W_z in terms of electron speeds c are [9, 18],

$$W_x = -\frac{4\pi}{3}\frac{E}{N}\frac{B}{N}\left(\frac{e}{m}\right)^2 \int_0^\infty \frac{c}{q_m^2\{1+(\omega/v_m)^2\}} \frac{df_0}{dc} dc \tag{2-2-27}$$

$$W_z = -\frac{4\pi}{3}\frac{E}{N}\frac{e}{m} \int_0^\infty \frac{c^2}{q_m\{1+(\omega/v_m)^2\}} \frac{df_0}{dc} dc \tag{2-2-28}$$

where $\omega = eB/m$ and $v_m = Ncq_m(c)$.

In usual laboratory experiments the condition $\omega^2 \ll v_m^2$ holds and the formulae (2-2-27) and (2-2-28) become

$$W_x = -\frac{4\pi}{3}\frac{E}{N}\frac{B}{N}\left(\frac{e}{m}\right)^2 \int_0^\infty \frac{c^2}{q_m^2} \frac{df_0}{dc} dc \tag{2-2-29}$$

$$W_z = -\frac{4\pi}{3}\frac{E}{N}\frac{e}{m} \int_0^\infty \frac{c^2}{q_m} \frac{df_0}{dc} dc = W, \tag{2-2-30}$$

i.e. in the presence of weak magnetic fields the drift velocity in the direction of the applied electric field remains unchanged.

The distribution function $f_0(c)$ in the above expressions has been considered to be unaltered by the presence of the magnetic field. Allis [9] has shown that in the presence of a magnetic field transverse to an electric field E, the electric field in the distribution function should be replaced by an effective field E_{eff}. However, if $\omega^2 \ll v_m^2$ it can be shown that $E \approx E_{\text{eff}}$ and $f_0(c)$ remains unchanged.

The magnetic deflection coefficient χ is defined as

$$\chi = \frac{E}{BW} \frac{W_x}{W_z} \approx \frac{E}{B} \frac{W_x}{W^2} \qquad (2\text{-}2\text{-}31)$$

and the "magnetic drift velocity" W_M as

$$W_M = \frac{E}{B} \frac{W_x}{W_z} = \chi W. \qquad (2\text{-}2\text{-}32)$$

There have been only three recent measurements of W_M, those of Jory [19] for nitrogen, Crompton, Elford and Jory [20] for helium and Creaser [21] for hydrogen and deuterium, all measurements being made at 293 °K. Earlier measurements of W_M by Townsend and his collaborators are summarised by Healey and Reed [22].

Both Jory and Creaser discuss the sources of error in magnetic deflection measurements and show that the accuracy obtained even after close attention to experimental detail is not, in general, as high as that obtained in recent measurements of either D_T/μ or W. For this reason, and the fact that no W_M data are available at low gas temperatures, W_M measurements have not been used as prime data in the derivation of collision cross sections from transport coefficients.

C. The Derivation of Cross Sections from Transport Coefficients

1. *When only elastic scattering occurs*

The electron transport coefficients which are of concern in this chapter, i.e. W and D_T/μ, are functions only of the experimental parameters E/N and T for a given gas. When the electrons make only elastic collisions the transport coefficients are determined, at a given value of E/N and T, only by the momentum transfer cross section. Thus by choosing any one of the transport coefficients, usually the drift velocity, and using an iterative technique it is possible to derive the momentum transfer cross section as a function of electron energy. There are three factors to be considered in this procedure.

(a) The degree of uniqueness of the derived cross section. The distribution of electron energies in helium (77 °K) at $E/N = 8 \times 10^{-3}$ Td and 2.0 Td is shown in Fig. 2-2-1 together with the momentum transfer cross section derived by Crompton, Elford and Robertson [23]. The unit of E/N used is the townsend, defined as 1 townsend (Td) = 10^{-17} V cm^2 [24]. More than 80 % of the electrons in the swarms at these two values of E/N have energies within the energy range shown shaded. It can be seen that at each value of E/N the drift velocity is determined by the variation of q_m with ε over a limited energy range. By varying E/N, different parts of the cross section become significant in determining the drift velocity and if a fit is made to the experimental drift velocity values at a large number of values of E/N, a high degree of uniqueness in the cross section can be achieved provided the experimental scatter in the data is small.

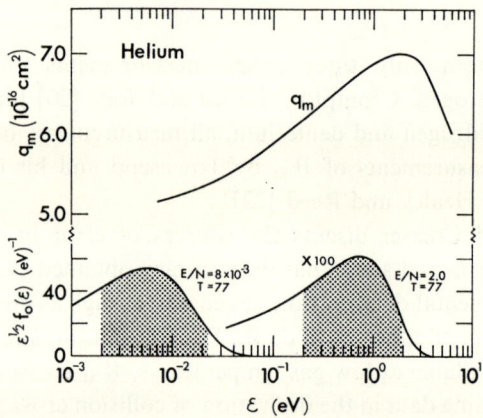

Fig. 2-2-1. The distribution in energy of electrons in helium at 77 °K and at $E/N = 8 \times 10^{-3}$ and 2.0 Td. Note that $\varepsilon^{\frac{1}{2}} f_0(\varepsilon) d\varepsilon$ is the fraction of electrons in the swarm with energies between ε and $\varepsilon + d\varepsilon$. Approximately 80 % of the electrons in the swarm have energies in the region shown shaded. The momentum transfer cross section derived by Crompton, Elford and Robertson [23] is shown in the upper section of the figure.

(b) The error incurred in the derived cross section due to errors in the experimental data or the analysis. Although the energy dependence of the cross section can be determined with a degree of uniqueness governed by the random error of the drift velocity data, the error in the magnitude of the cross section will be determined by the systematic error of the experimental data. The sensitivity of the drift velocity in helium at 77 °K to overall changes in the momentum transfer cross section has been found [23] by increasing the cross section at all energies by 2 % and observing the effect on the computed drift velocities. Over the E/N range used experimentally the calculated drift velocities changed by at least 1%; at the lowest value of E/N the change was 1.6%. Thus in helium at 77 °K, if drift velocity data are available with an error of less than 1%, it is possible to determine the momentum transfer cross section with an error of less than 2% over an energy range which can be determined (section c). Robertson [25] has shown that when the procedure described above is used for neon, the calculated drift velocities at 77 °K are somewhat less sensitive than in helium but a 2% change in the cross section over the full energy range still produced a change of at least 0.8% in the drift velocity over the E/N range used. In the other inert gases the sensitivity is variable due to the presence of the Ramsauer minimum. In helium (when $D_T/\mu \gg kT/e$) the values of D_T/μ are approximately twice as sensitive to changes in the cross section as are the W values, the sensitivity being even greater in other gases in certain energy ranges. In argon Frost and Phelps [26] have shown that the D_T/μ values are much more sensitive than the drift velocity values to changes in q_m over the energy range 0.25 to 0.7 eV. However, since it is usually possible to determine the drift velocity with sufficient precision to offset the difference in sensitivity, W data are used as the prime data for analysis.

The formulae for $f_0(\varepsilon)$ and W used in the analysis have been subject to investigation to determine the magnitude of the error introduced by the assumption that all terms higher than the first two in the expansion of the velocity distribution function (2-2-7) may be neglected. Calculations based on the Cavalleri and Sesta formulae [23] have suggested that the effect of this assumption on the value of W, as computed from (2-2-14), is not significant over the range of E/N values used. Calculations made by Lowke [27], who included the next higher term in the expansion for $f(c)$ and assumed elastic scattering, support this conclusion.

An additional cross check on both the computing procedure and the formulae is provided by using the momentum transfer cross section derived from drift velocity data to predict the measured values of D_T/μ and W_M. The predicted and experimental values were found to agree to within the estimated experimental error [20].

(c) The energy range of the derived cross section. The energy range of the derived cross section is somewhat larger than the range of mean electron energies appropriate to the range of values of E/N for which data are available, due to the electrons having a distributed range of energies about the mean value. The lowest possible mean energy for swarms in a gas at a given temperature is the mean thermal energy of agitation of the gas molecules, i.e. 38 meV at 293 °K and 10 meV at 77 °K. Thus in order to obtain the derived cross section over the widest possible range of energies it is desirable to perform measurements at low gas temperatures and over a wide range of values of E/N. The very low energy region of the derived cross section is of interest as it enables the scattering length to be determined by simple extrapolation to zero energy, the error incurred in the extrapolation being small if the cross section extends to a sufficiently low energy (e.g. 0.008 eV in helium, [23]). Such an extrapolation would appear to be a more accurate method of obtaining the scattering length than the use of modified effective range theory [28] to extrapolate present available total elastic scattering cross sections to zero energy, e.g. [29].

2. *When both elastic and inelastic scattering occur*

When, in addition to elastic scattering, there is only one inelastic process present it is still possible to obtain unique cross sections if data for two transport coefficients are available.

When the energy distribution is governed by more than one inelastic cross section, it is not possible to obtain a unique set of cross sections with data for two transport coefficients only. If two inelastic processes with cross sections $q_1(\varepsilon)$ and $q_2(\varepsilon)$ and thresholds ε_1 and ε_2 occur, then the situation may be summarised as [30],

(1) above a transition energy somewhat below ε_2, neither cross section can be uniquely determined from swarm data alone;

(2) $q_1(\varepsilon)$ can be determined uniquely from threshold to the transition energy. Above this energy, uniqueness in its determination is lost, unless $q_2(\varepsilon)$ is accurately known from another source;

(3) $q_2(\varepsilon)$ can never be determined unless $q_1(\varepsilon)$ is known. On the other hand, provided the separation between the threshold energies ε_1 and ε_2 is sufficiently large $q_2(\varepsilon)$ can be determined within comparatively narrow limits even though $q_1(\varepsilon)$ is not accurately known. This is due to the electron losing as much energy in a single collision of type 2 with energy loss ε_2 as in many collisions with energy loss ε_1.

The simplest case where both elastic and inelastic collisions occur, is para-hydrogen at 77 °K. At this temperature 99.5 % of the molecules are in the $J = 0$ rotational state and 0.5 % are in the $J = 2$ state. Thus for electron swarms at a value of E/N sufficiently low for there to be an insignificant number of electrons that can excite a vibrational transition, there are only two relevant cross sections, one for elastic scattering and the $J = 0 \to 2$ transition. These cross sections can be determined uniquely and have enabled the cross section for the $J = 0 \to 2$ transition to be determined from threshold, 0.044 eV, to 0.3 eV with an error estimated to be less than 5 % [30]. At higher values of E/N the cross section for vibrational excitation (threshold for $V = 0 \to 1$ is 0.516 eV) becomes the dominant inelastic cross section and can be determined over a limited energy range if the $J = 0 \to 2$ rotational transition can be assumed known from a theoretical investigation.

The cross sections derived from the analysis of the data for para-hydrogen can be used to analyse normal hydrogen where an additional cross section, for the $J = 1 \to 3$ transition, has to be included [8]. The next more complex case is that of deuterium [8] and since here there are more unknown cross sections than there are sets of transport coefficient data, it is necessary to use information from theoretical calculations. For other molecular gases the analysis becomes more complex and information from a variety of sources is used to enable a reasonable set of cross sections to be derived, e.g. [31, 32].

The error limits of the derived cross section depend on both the errors of the experimental data used and the fundamental limitations of the method. To limit the first source of errors the data for both the drift velocity and the characteristic energy should be subject to a small absolute error and to small experimental scatter. The techniques used to obtain such data are discussed in § 2-3 and § 2-4.

§ 2-3. The electron drift velocity

A. Review of previous methods of measurement

The wide variety of methods devised to measure the electron drift velocity have been reviewed by Loeb [33] and McDaniel [34] and will only be considered briefly here. The methods used fall into several categories.

1. Oscillographic methods

The method devised by Hornbeck [35] is typical of those of this group. The electrode system consists of two circular and parallel plates. A pulse of photo-electrons is produced at the cathode by a pulse of UV light and drifts to the anode, the drift velocity being obtained from an analysis of the current transient observed with an oscilloscope. The Hornbeck method in a modified form has been used by Bowe [36].

Related methods are those described by Herreng [37], Colli and Facchini [38], Klema and Allen [39], Nagy, Nagy and Dési [40] and Kirschner and Toffollo [41]. The fundamental limitation to the accuracy obtainable in all these methods is the accuracy with which time intervals can be measured from the oscillograph traces. The accuracy of these methods appears to be in general poorer than that obtained by electrical shutter methods.

2. *Methods using counters*

In this technique a Geiger-Mueller counter is used to detect the arrival of electrons at a plane. This method has been developed by Bortner, Hurst and Stone [42], from the original technique of Stevenson [43] and is shown in a more recent form [44] in Fig. 2-3-1. Electrons produced by a pulse of UV light incident on the cathode drift in a uniform electric field to the anode where they are detected by a Geiger-Mueller counter mounted behind an aperture. The time at which the UV light illuminates the cathode is indicated by a signal pulse from a photodiode. This pulse is delayed in

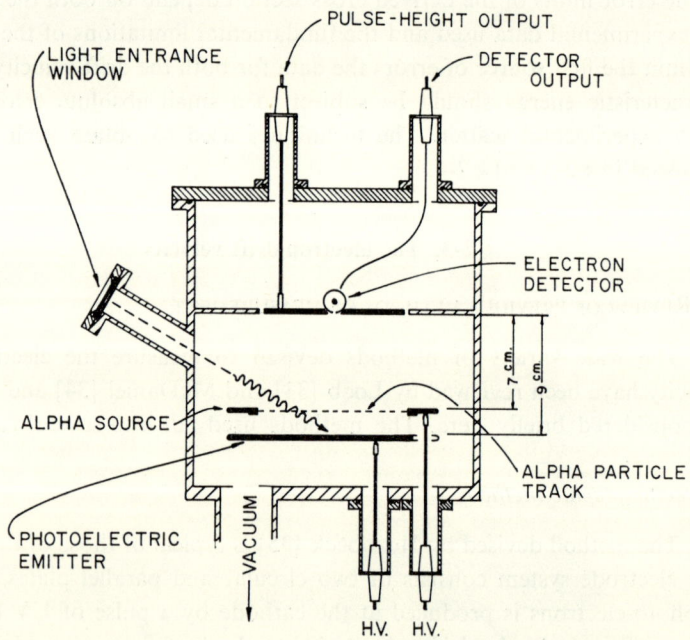

Fig. 2-3-1. The apparatus used by Christophorou, Compton, Hurst and Reinhardt [44] to measure electron drift velocities. The α-particle source was used in a separate experiment to measure the electron attachment coefficient.

time and displayed on a double beam oscilloscope together with the signal pulse from the Geiger-Mueller counter. The transit time is taken as the delay time required to align the two pulses on the oscilloscope screen. Data for a wide range of gases and gas mixtures have been taken with this technique [45, 46].

A modification of this method, to enable both W and D_L/μ to be determined, was devised by Hurst, O'Kelly, Wagner and Stockdale [47]. If the diameter of the hole in the anode and the initial electron intensity are such that there is a negligible probability of more than one electron entering the counter from any one electron pulse it is possible to measure the flight times of individual electrons. The drift velocity and D_L/μ were derived from the distribution of flight times after suitable corrections were made for factors such as counter dead time. An analysis of this and other sources of error in this technique has been made by Hurst and Parks [48]. A further modification by Wagner, Davis and Hurst [49] enabled the technique to be extended to include gases unsuitable for counter operation. The Geiger-Mueller counter was replaced by a detection system incorporating a differentially pumped electron multiplier.

3. *Electrical shutter methods*

The method devised by Bradbury and Nielsen [50] (see also [51]) is shown in Fig. 2-3-2. Electrons produced by photo-emission drift in a uniform electric field to a collector. Two electrical shutters, S_1 and S_2, situated a

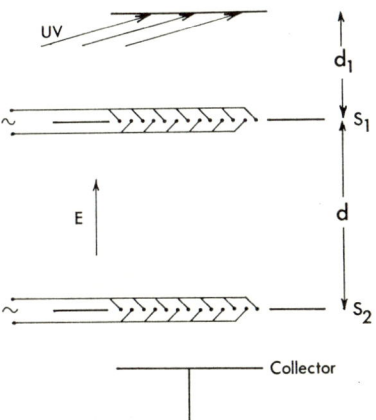

Fig. 2-3-2. Schematic diagram of the Bradbury-Nielsen method for measuring electron drift velocities.

known distance apart are interposed in the path of the electrons. Each shutter consists of a coplanar grid of fine wires, alternate wires being connected together. Alternating potentials 180° out of phase are applied to each set of wires, the signals being of constant and equal amplitude. These signals are superimposed on the d.c. voltage required to maintain the uniformity of the electric field. When the sinusoidal potential between the wires is close to zero electrons are transmitted, but at higher potentials the shutters become opaque to electrons. If the signals applied to both shutters are exactly in phase, a group of electrons which traverses the distance between the shutters in an integral multiple of half periods of the alternating potential will be transmitted to the collector. The current received by the collector plotted against the frequency of the a.c. potential exhibits a series of equally spaced maxima and minima. A typical series of maxima will be shown later in Fig. 2-3-8. If f^n is the frequency at which the nth current maximum occurs and d is the drift distance then we may write

$$W' = 2f^n d/n. \qquad (2\text{-}3\text{-}1)$$

The parameter W' is approximately equal to the drift velocity and will be termed the "measured drift velocity". The corrections to be applied to W' are considered in § 2-3B-9.

A modified form of the Bradbury-Nielsen method has been used by Phelps, Pack and Frost [52], the shutters being opened by pulses of short duration rather than by a sinusoidal potential. The same pulse train was applied to the second shutter as to the first shutter but delayed by a known time interval. The collector current when plotted as a function of delay time exhibited a maximum at a delay time which was taken to be equal to the mean transit time of the electrons. To enable end effects to be eliminated, Pack and Phelps [53] measured the transit time over two different distances d_1 and (d_1+d) (Fig. 2-3-2). Pulsed UV was used and the transit times were measured using the method employed by Phelps, Pack and Frost [52]. The difference of these transit times is the time taken by the electron pulse to travel the distance between the shutters. A second mode of operation, designed to reduce the effect of distortion of the electric field at the shutters was also used. The shutters were held open and closed only for short time intervals. The transit time was now determined from the time delay at which a minimum occurred in the transmitted current. Both modes of operation gave the same drift velocity values to within the experimental scatter.

The original Bradbury-Nielsen method has been extensively examined in papers by Lowke [54, 55], Elford [56], Crompton, Elford and Jory [20],

Crompton, Elford and McIntosh [57], and Crompton, Elford and Robertson [23] and shown to be capable of obtaining data of high precision. The experimental technique developed by these authors and the sources of error in the method are examined in § 2-3B below.

B. THE BRADBURY-NIELSON METHOD

1. *Electrical shutters*

The electrical shutters are required to meet a stringent set of conditions. The wires must be accurately coplanar in order for the length of the drift chamber to be well defined; the distance between wires should be small to enable the transmitted current to be reduced to zero with relatively small potentials between the wires thus enabling the distortion of the main drift field to be kept low; the diameter of the shutter wires should be small relative to the spacing of the wires to reduce the attenuation of the electron current and the insulation between the two sets of wires of a shutter should be high. In addition the shutter must be capable of being outgassed and of being repeatedly cycled from room temperature to low temperatures without damage or loss of mechanical tolerances. Finally the materials used must be compatible with ultra high vacuum practice. A shutter used by Crompton, Elford and McIntosh [57], which meets these conditions, consists of nichrome wires, 0.08 mm in diameter, spaced 0.4 mm apart and supported by a slotted steatite annular ring (outer diameter 5.5 cm, inner diameter 3.0 cm). The surface of the slot was machined flat and parallel to the lower surface of the ring to a tolerance of 0.005 mm. The wires were laid across the surface of the slot and sealed under tension to the ring by Pyroceram frit No. 95 (Corning Glass Co.). After alternate wires were connected together by spot welding to nickel wire leads, the wires were vacuum coated with gold to reduce contact potential differences.

The sinusoidal potential is supplied to the shutters by a resistor-capacitor network designed to ensure that there is a negligible phase shift between the signals applied to the two shutters.

The efficiency of operation of a shutter is determined by measuring the transmitted electron current as a function of the d.c. bias placed between the sets of shutter wires. During this measurement the other shutter is held "open" by maintaining both sets of wires at the d.c. potential appropriate to the plane of the shutter in the electrode system. Typical "cut off curves" are shown in Fig. 2-3-3a for electrons in helium at $E/N = 0.01$ Td and 0.02 Td and in Fig. 2-3-3b for neon at $E/N = 0.02$ Td. The curves for helium are

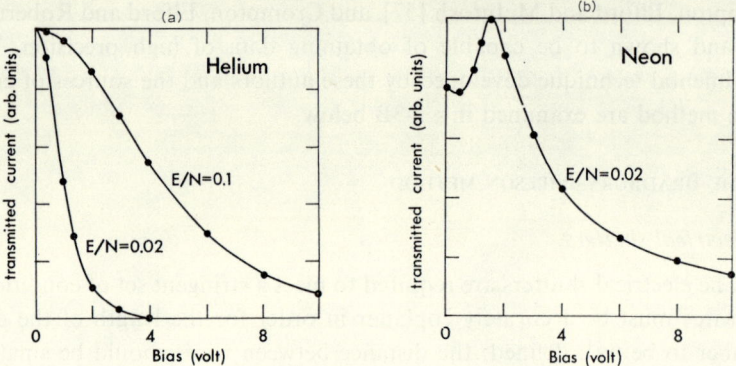

Fig. 2-3-3. The electron current transmitted by a shutter as a function of the d.c. bias between alternate wires, (a) for helium at $E/N = 0.02$ and 0.1 Td, (b) for neon at $E/N = 0.02$ Td.

similar to those for most gases and show that the shutters become increasingly inefficient as E/N increases. This effect is one of the factors which limits the upper value of E/N at which accurate measurements can be made using this method. In neon maximum transmission does not occur at zero bias, this behaviour being apparently due to the rapid decrease in $q_m(\varepsilon)$ with ε which occurs in this gas. A similar effect has been observed in argon [25, 53]. By varying the amplitude of the alternating potential applied to the shutters it has been shown [25] that the value of f^0 is unaffected by the unusual shutter characteristics.

2. *Uniform electric field and geometrical accuracy*

The electrical field in the drift chamber is held uniform by guard electrodes at potentials appropriate to their position in the electrode system, the design of the ring system being governed by the degree of field uniformity required and the method of mounting the shutters. As might be expected the effect of non-uniformities in the electric field of the drift chamber is small. Since the shutter wires form fixed equipotential planes, any increase in the longitudinal component of the electric field in one region of the drift chamber leads to a decrease in an adjacent region. Thus changes in the drift velocity as a result of distortion of the electric field are largely self-cancelling.

Figure 2-3-4 shows the guard electrode system used by Crompton, Elford and Robertson [23]. This system is a modified version of the "thick" electrode system devised by Crompton, Elford and Gascoigne [58] (see § 2-4C-2). The electrodes are held in position by accurately ground glass

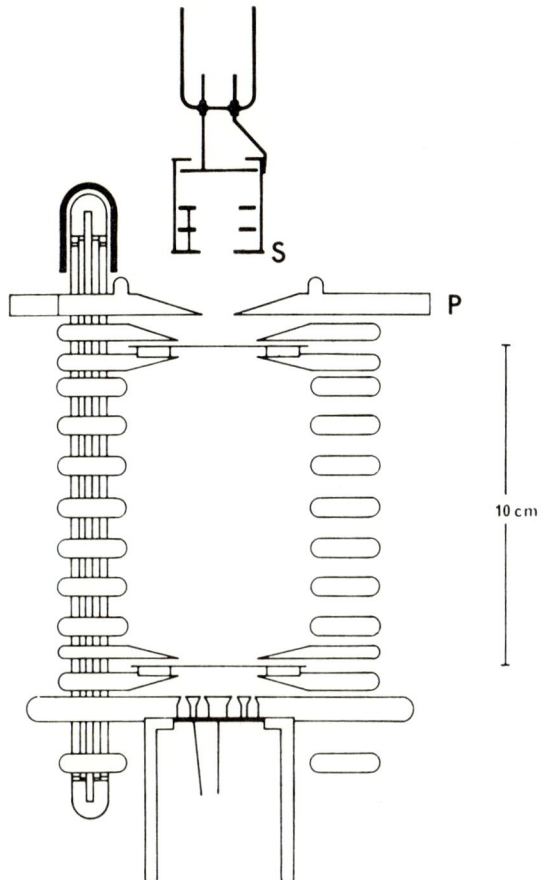

Fig. 2-3-4. Electrode system used by Crompton, Elford and Robertson [23] to measure electron drift velocities.

spacers and sheathed stainless steel tie rods. The electrode potentials were supplied by a highly stabilised power supply and a precision voltage divider, the potential of all electrodes being within 0.1 % of their calculated values. To ensure that the electrons at the top shutter had a distribution of energies which closely approximated that in the drift chamber, the electric field for a distance of 1 cm above the upper shutter was maintained at the same value as that in the drift chamber. The electron current was controlled by varying the field between the source S and the upper electrode P. The electric field strength E was calculated by determining the voltage between the two shutters using a differential voltmeter and dividing by the drift chamber length d.

The chamber length was assumed to be the distance between the mid planes of the shutters, taking into account the finite thickness of the shutter wires. It was determined by direct external measurement taking readings at various points around the circumference, and also by a wire probe method. This length is considered to be known with an error of less than half of a wire diameter, i.e. 0.04 mm. The technique used to measure the length was subject to an error of less than ± 0.001 mm; the uncertainty of 0.04 mm quoted above being due to slight displacements of the wires as a result of the sealing technique. A correction of 0.2 % was made to d when the electrode system was operated at 77 °K to account for thermal contraction.

3. *Electron source*

Although a photo-cathode and UV light have been used as the source of electrons in the majority of drift velocity measurements reported in the literature, at gas pressures in excess of about 10 torr a highly satisfactory source has been found to be one in which electrons are produced by volume ionization using α-particle emission from a silver coated foil of Americium 241 [59]. The foil is contained in a stainless steel cylinder S (Fig. 2-3-4) which has a series of baffles to prevent α-particles entering the drift chamber. This type of source provides a more stable electron current than is obtainable from either a thermionic or photo-electric source and has two particular advantages over platinum filaments used in some earlier measurements, e.g. [56, 57]; no heat is dissipated, a fact of considerable importance in reducing thermal gradients in measurements at low temperatures and reactions which may occur at the hot filament surface are avoided (e.g. the conversion of pure para-hydrogen to the equilibrium para, ortho-hydrogen mixture).

4. *Gas temperature measurement and control*

The gas temperature affects measurements of the drift velocity in two ways. The drift velocity is a function of E/N and the calculation of the gas number density N from the gas pressure requires a knowledge of the gas temperature. The drift velocity is also a function of the gas temperature through the electron energy distribution function. At sufficiently large mean electron energies the influence of the thermal motion of the gas molecules on the electron energy distribution function becomes negligible and the drift velocity becomes a function only of E/N, independent of gas temperature. The temperature must be held highly uniform over the length of the drift chamber to avoid variations in E/N and hence W, and also to reduce uncertainty in the calculated value of N. To maintain temperature stability the experimental tube was immersed in an appropriate liquid contained in a large stainless steel Dewar. The gas temperature was measured by two copper-constantan thermocouples (TC) attached to the electrodes as shown (Fig. 2-3-4).

At room temperature, a water bath was used to maintain temperature stability to within 0.1 °K per hour. When liquid nitrogen baths were used, the temperature as measured by the thermocouples was checked against the bath temperature, the temperature of the liquid nitrogen being found from the barometric pressure and vapour pressure tables. A small correction (< 0.1 °K) was made for the level of oxygen impurity in the liquid nitrogen. The temperature of the bath agreed with that recorded by the thermocouples

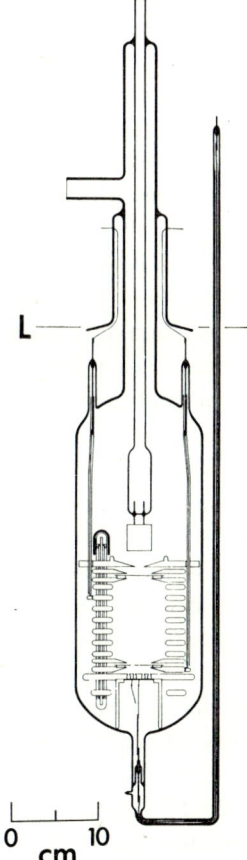

Fig. 2-3-5. Experimental tube used to measure electron drift velocities [23]. For low temperature measurements the tube was immersed in a refrigerant to the level L.

to within the uncertainty of the measurement, i.e. 0.1 °K. No thermal gradients were detectable at 77 °K and the gas temperature was assumed to be that of the bath. One possible cause of thermal gradients is the radiation from surfaces at room temperature to the upper electrode of the system. This effect was made negligible by immersing the tube to the level L shown in Fig. 2-3-5, thus greatly reducing the solid angle subtended to the source by the surface receiving radiation. Even at gas pressures as low as 5 torr at 77° K no thermal gradient was detectable. With these precautions the gas temperature was determined with an error estimated to be less than 0.1 % at 293 °K and 0.2 % at 77 °K.

5. *Gas purity and vacuum requirements*

(1) Inert gases. In the inert gases the effect of a molecular impurity is due almost entirely to inelastic collisions in which the energy lost by the electron can be very much greater than that lost in an elastic collision. The effect of the inelastic collision is to alter the electron energy distribution function and reduce the mean energy of the electrons, usually with a consequent rise in W. The effect of elastic collisions with impurity molecules is very much smaller and for low levels of impurity can be neglected. The effect of small levels of molecular impurities can be large; for example 5 ppm of N_2 has been shown to cause the drift velocity in Ne to increase by up to 1.5 %, while in argon the increase was found to be even larger, up to 2.5 % [25] (see also § 2-3B-8b). The gases used must therefore be of high purity and the outgassing rate of the experimental tube must be sufficiently low that the purity does not change significantly over the time in which measurements are made.

Adequately low outgassing rates can, in general, only be obtained by the use of ultra high vacuum techniques. A block diagram of a typical vacuum system [23] is shown in Fig. 2-3-6. Although the drift tube (Fig. 2-3-5) was designed to be heated to moderate temperatures it was not in general baked before experimental runs as the danger of distorting the electrode structure was considered to be greater than the danger of contamination of the gas from the higher outgassing rate. Before measurements, however, the UHV system was baked at 200 °C for several days. The ultimate pressure of the system was below 10^{-7} torr and in a period of 24 hours the pressure rise was less than 2×10^{-5} torr. Since experiments were generally carried out at pressures greater than 100 torr the contamination from outgassing was less than 0.2 ppm per day. At 77 °K and at higher pressures this rate was even smaller.

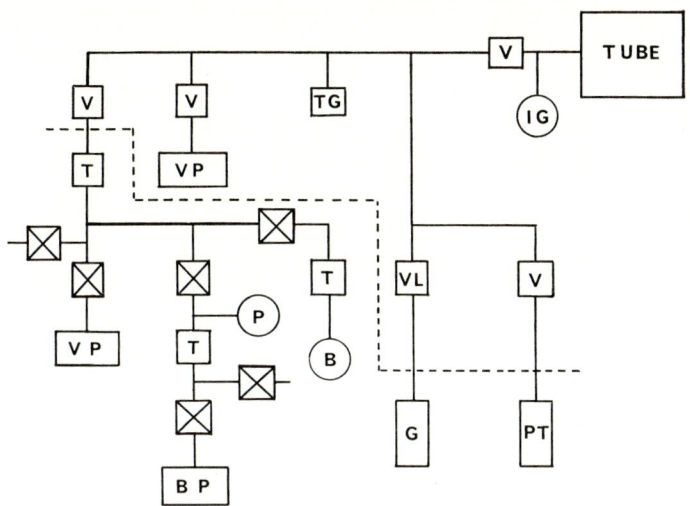

Fig. 2-3-6. Schematic diagram of the vacuum system used in measurements of electron transport coefficients [23]. The section above the dashed line is bakeable.

⊠	Saunders value	T	Liquid nitrogen trap
V	UHV value	IG	Ionization gauge
VL	Variable leak	P	Pirani gauge
VP	8 litre sec^{-1} Vacion pump	TG	Pressure gauge (Texas Instruments Ltd.)
B	Ballast volume	G	Gas cylinder
BP	Rotary backing pump	PT	Silver-palladium osmosis tube

(2) Diatomic gases. In the diatomic gases, as in the inert gases, the effect of impurities is due almost entirely to inelastic collisions but the effect on the transport coefficients of these collisions with impurity molecules will be, in general, much smaller since the fraction of the total energy-transfer to the gas molecules in such collisions is very small. However if the impurity molecules have large inelastic cross sections the sensitivity of the transport coefficients to such impurities can be high. Lowke [55] has shown that 20 ppm of water vapour in nitrogen causes a 0.7 % increase in W at $E/N = 0.48$ Td. Fortunately, maintaining a low water vapour level is a relatively simple experimental procedure and the high sensitivity to water vapour does not present a problem. In general it is possible to tolerate much higher outgassing rates in measurements with diatomic or polyatomic gases.

An additional source of error in measurements in both monatomic and diatomic gases can be the presence of small traces of a gas in which electron attachment occurs. Fig. 2-3-7 shows the effect of the presence of 20 ppm of water vapour in nitrogen at 293 °K, $E/N = 0.48$ Td and at a pressure of

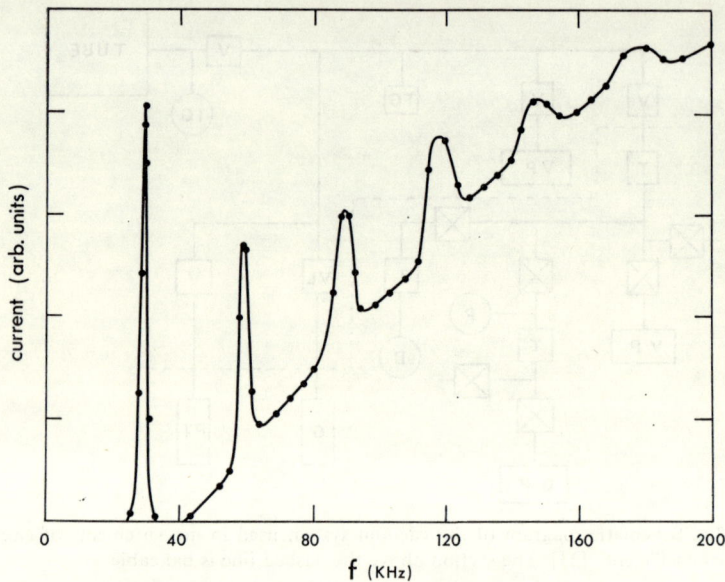

Fig. 2-3-7. A current-frequency curve observed in the presence of a background negative ion current. The curve shown was obtained by Lowke [55] using nitrogen containing 20 ppm of water vapour. The experimental conditions were $T = 293$ °K, $p_{293} = 250$ torr and $E/N = 0.48$ Td.

250 torr [55]. The usual electron current peak system can be seen to be superimposed on a background current of negative ions. The presence of this background current (the cause of which is not understood in this case) results in the electron current maxima being distorted, making accurate measurements of f^n difficult. The rise in background current with increasing frequency is caused by the field between the shutter wires varying sufficiently rapidly that only a fraction of the ions is collected, this fraction decreasing as the frequency increases.

Since the impurities introduced from outgassing of the experimental tube can be reduced to levels which have an insignificant effect on the measurements of transport coefficients, the only significant sources of impurities are the gas inlet system and gas supply. Care is necessary to ensure that the gas inlet system does not add contaminants through the use of pressure regulators employing elastomers in their construction.

Hydrogen and deuterium of adequate purity were obtained by diffusion through a silver-palladium osmosis tube. Other high purity gases were obtained from commercial sources (e.g. Matheson Gas Co. Ltd.) and were supplied in high pressure cylinders. These were connected directly to a UHV leak valve thus avoiding the use of a pressure regulator.

6. *Pressure measurement*

The gas pressures used were from about 5 to 700 torr and were measured with a quartz spiral manometer (Texas Instruments Ltd.). This manometer was calibrated and periodically checked against a CEC type 6201 deadweight tester which has a stated accuracy of 0.015 % of the pressure measured. At pressures less than 15 torr, the lower limit of the piston-cylinder combination of the deadweight tester when used with its reference volume evacuated, linear interpolation was used to calibrate the manometer. The quartz capsule used to measure high gas pressures had a maximum pressure limit of 500 torr. For pressures higher than this, the reference side of the gauge, normally held highly evacuated, was filled with dry nitrogen to a suitable and known pressure. The reference side of the manometer incorporated a large constant-temperature ballast volume to enable the reference pressure to be held stable. The absolute error at all pressures used is considered to be less than 0.1 %.

The gas number density N was usually determined assuming the gas to obey the perfect gas law but in certain gases at high pressures and low temperatures it is necessary to take into account deviations from this law. In terms of the second virial coefficient $B(T)$ the number density is then given by

$$\frac{1}{N} = \frac{1}{p_T} kT \left[1 + B(T) \frac{N}{N_0}\right] \tag{2-3-2}$$

where N_0 is Avogadro's number and p_T is the pressure measured at a gas temperature T. The error incurred by ignoring the second virial coefficient is $< 0.1 \%$ for helium and neon at 77 °K over the range of pressures used (< 700 torr), but in nitrogen at 77 °K and 500 torr the error incurred is almost 2.5 %.

7. *The determination of the current maxima*

Figure 2-3-8 shows a typical plot of the transmitted electron current as a function of the frequency of the signal applied to the shutters. In order to avoid effects caused by space charge repulsion the electron currents used were between 10^{-11} and 10^{-13} A at the collector.

The frequency at which the collector current had a maximum value was determined by measuring the frequencies on either side of the peak at some arbitrary fraction (usually greater than 0.7) of the maximum current.

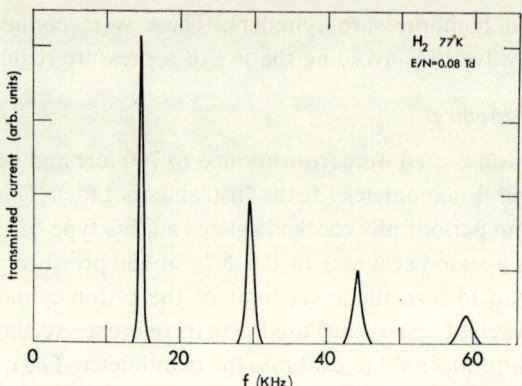

Fig. 2-3-8. A typical electron current-frequency plot obtained using the Bradbury-Nielsen method [71] (for hydrogen at 77 °K and $E/N = 0.08$ Td).

The average of the two frequencies was taken as the frequency at which the current was a maximum, i.e. f^n. Due to the output of the driving amplifier varying slightly with frequency and the transmitted current changing with the amplitude of the a.c. voltage it is necessary to monitor the amplitude to ensure that there is no significant difference in amplitude at the two frequencies. For a given peak a systematic variation in f^n with increasing electron current, indicates that the current peak is distorted. Three estimates of the frequency f^n were made for both first and second maxima and the values averaged to obtain f^1 and f^2.

The accuracy with which the values of f^1 and f^2 can be determined depends on the resolving power of the peak, the electron current stability and the electrometer sensitivity. The resolving power (R.P.) of the peak is defined as the frequency at which maximum current occurs, divided by the frequency spread at half the maximum current. In practice it has been found that a resolving power of 5 or more is adequate to enable f^n to be measured with an error less than 0.1 %.

The resolving power is determined by two factors, the broadening of the group of electrons due to diffusion and the transmission through the shutters. By assuming that the number density of the electrons transmitted by the first shutter has the form of a delta function (i.e. the shutter acts as an "ideal" shutter) Lowke (see [56]) showed that the theoretical resolving power R.P.$_T$ of the group after travelling a distance d is given by

$$\text{R.P.}_T = N^{\frac{1}{2}} \left[\frac{E/N}{D_L/\mu} \frac{d}{2 \ln 2} \right]^{\frac{1}{2}}. \tag{2-3-3}$$

Thus for a given value of E/N (and hence D_L/μ), the value of R.P.$_T$ can be made large by using a sufficiently large gas number density. However at large values of E/N, there is an upper limit imposed on N due to the onset of discharge and this limitation leads to broad current maxima. There are certain gases, e.g. argon, where D_L/μ is large at low values of E/N. In this case even though large values of N can be used, the error in the determination of f'' due to the broad peaks becomes a major part of the error in the drift velocity measurements over this range of E/N. The advantage of using long drift chambers is obvious from expression (2-3-3).

The above analysis is based on ideal shutters. In practice the electron group transmitted by a shutter has a finite distribution in the direction of the electric field, the width of this distribution being a function of the a.c. potential applied to the shutters, the value of E/N and the geometry of the shutters. In general the efficiency of the shutters which is indicated by the d.c. bias or "cut-off curves" (Fig. 2-3-3) decreases as E/N increases. The broadening of the transmitted group of electrons due to this effect can be offset by using larger a.c. potentials between the shutter wires.

High stability of the measured electron current was obtained by using a radioactive source (§ 2-3B-3) and ensuring that all parameters which affect the electron current were held highly stable. In particular, considerable care was taken in shielding and construction to ensure that spurious currents in the electrometer circuit due to microphony, piezo-electric effects etc. were kept very low. Since the measurements of the two frequencies for a chosen current take less than 30 seconds only short term electrometer stability is required. The currents were measured using a 10^{11} ohm resistor and a Vibron Model 62A electrometer.

In general the estimates of f^1 obtained from the first and second peaks agreed to within 0.1 % and for a substantial body of data the agreement was to within 0.05 %.

8. *Sources of systematic error*

Electron drift velocities made using the techniques described in the previous sections are subject not only to systematic errors which arise from the measurement of the experimental parameters e.g. pressure, chamber length (see § 2-3C-2), but from a number of other sources which are now considered.

(a) Non-uniform electric fields and end effects. Distortion of the electric field in the drift chamber caused by incorrect geometry or incorrect potentials of the guard electrodes gives rise to a systematic error which is independent of the electric field established and which cannot therefore be detected by any direct experimental test. One particular drift tube, that used by Crompton, Elford and Jory [20] to measure drift velocities of electrons in helium (293 °K) was designed to have the smallest possible distortion of the electric field but could not be cooled due to the shutter construction. A "thick" guard electrode system (§ 2-4C-2) was used and large aperture shutters were mounted in such a way as to introduce minimum distortion. The experimental tube shown in Fig. 2-3-4 on the other hand was designed to be operated over a wide temperature range and distortion had been unavoidably introduced into the electric field by the shape of the electrodes adjacent to the shutters. However, data taken for helium at 293 °K with this tube has been found to agree with that of Crompton, Elford and Jory to within 0.2 % suggesting that the effect of field distortion is not significant in either tube [25].

An additional source of field distortion is the alternating potential applied between adjacent shutter wires. The distortion from this source however may be ignored, as varying the amplitude of the potential over a wide range has been found to have no observable effect on the measured drift velocity (i.e. < 0.1 %).

Despite the gold coating of the shutter wires, contact potential differences set a lower limit to the potential which can be applied across the drift chamber for reliable measurements. A contact potential difference of ΔV across the drift chamber results in a fractional change in the drift velocity $\Delta W/W$ which is proportional to $\Delta V/V$ at a given value of E/N. To avoid large errors from this source, the potential V across the drift chamber was always greater than 20 volt.

In a previous section (§ 2-3B-2) it was stated that the drift chamber length is the distance between the mid planes of the shutters. Since the behaviour of the electron swarm in the close vicinity of the shutter wires is complex it is not obvious that the effective chamber length is the geometrical length as defined above. A series of measurements in hydrogen using drift chambers of 5 and 10 cm (nominal dimensions) and under conditions where the diffusion correction (§ 2-3B-9) was negligible, agreed to within 0.1 % when the geometrical length was used to calculate the measured drift velocity [60]. It may therefore be assumed that at least over this length range, the geometrical and effective lengths are identical. The fact that the drift velocity is independent of length also confirms that the effect of field distortion produced by the alternating potential applied to the shutters is negligible.

(b) Gas impurities. The most likely impurities to be present in the gas samples in the experimental tubes and which can cause significant changes in measurements in the inert gases are hydrogen and nitrogen. A systematic investigation of the effect of the impurities N_2 and H_2 in the inert gases helium, neon and argon has been carried out by Robertson [25]. In helium it was found that impurity levels greater than 200 ppm of either nitrogen or hydrogen are required for the drift velocity to change by more than 0.1 % at most values of E/N in the experimental range. The analysis supplied with the gas showed impurity levels of 0.5 ppm of N_2 and smaller traces of H_2 and O_2 so that it may be assumed that the data obtained in helium by Crompton, Elford and Robertson [23] is unaffected by impurities.

Robertson [25] has also computed the effects of impurities by assuming a model gas with the atomic weight and momentum transfer cross section of the major constituent gas and the inelastic and super-elastic cross sections of the impurity, multiplied by the fractional population of the impurity molecules. The computing procedure was that of Gibson [8]. A more accurate calculation of the effect of a diatomic gas (hydrogen) on transport coefficients in an inert gas (argon) has been made by Engelhardt and Phelps [61] but the simpler approximate procedure used by Robertson is sufficiently accurate for the study of the effect of small impurity levels. Not only did the measured outgassing rate indicate a negligible rate of contamination but the measurements were stable with time over periods in excess of 24 hours.

TABLE 2-3-1

Measured effect of nitrogen impurities on drift velocities in neon at 77 °K [25]

E/N (Td)	Change in W (%) for 20 ppm N_2
0.016	<0.1
0.024	<0.1
0.04	+0.1
0.08	+0.2
0.12	+0.3
0.16	+0.6
0.24	+2.1
0.32	+4.4

In neon the sensitivity of the drift velocity to impurities is much greater than in helium. Table 2-3-1 shows the effect of 20 ppm of N_2 in neon. It was concluded by Robertson [25] that at higher values of E/N vibrational excitation of nitrogen (probably other than the first vibrational transition) is responsible for the high sensitivity of the measured drift velocities in neon to small traces of nitrogen. Two approaches to the problem of impurities in the inert gases are possible. One is to ensure that the impurity level of the gases used is sufficiently low for the measurements to be unaffected. This presents no difficulty for helium but in the case of N_2 in Ar an impurity level of less than 0.5 ppm is necessary if the error in the drift velocity is to be less than 0.2 % at the higher values of E/N. An alternative approach is to measure the impurity levels and apply a correction based on the known sensitivity.

9. *Correction for the effect of diffusion*

Lowke [54] and more recently Burch [62] have examined the Bradbury-Nielsen method and considered the effects of diffusion both in the drifting electron pulse and at the shutters. In these analyses the diffusion coefficient was assumed to be isotropic. It was found that the measured drift velocity W' is related to the true drift velocity W by the relation

$$W' = W\left[1 + C'\frac{D_L/\mu}{E/N}\right] \qquad (2\text{-}3\text{-}4)$$

or

$$\frac{\Delta W}{W} = C^1\left(\frac{D_L}{D_T}\right)\left(\frac{D_T/\mu}{E/N}\right)\frac{1}{Nd} = C\left(\frac{D_T/\mu}{E/N}\right)\frac{1}{Nd} \qquad (2\text{-}3\text{-}5)$$

where C^1 is a constant and $C = C^1(D_L/D_T)$. The diffusion coefficient D in the original formula of Lowke [54] has been replaced by D_L.

An experimental investigation by Lowke [54] and subsequent sets of data in a number of gases have shown that the deviation of the measured drift velocity from the asymptotic value obtained at high gas number densities obeys the relation above, i.e. $\Delta W/W \propto 1/N$ at a given value of E/N. Since the theoretical value of C was uncertain, the procedure used was to obtain a value of C such that the corrected drift velocities at a given value of E/N were independent of N over a wide range. At high values of E/N the range of gas pressures was limited by the onset of discharge, and it became impossible to obtain a value of C. The value then used in making the correction was that obtained from lower values of E/N. The uncertainty in C and hence in the correction introduced by this procedure was taken into account in estimating the error limits. It should be emphasised that by using high gas number densities and long drift chambers the correction can be made very small. As an example, in the drift velocity data for helium at 77 °K of Crompton, Elford and Robertson [23] which were used to obtain the momentum transfer cross section shown in Fig. 2-3-1 the largest correction made was 0.25 % and in general the correction was less than 0.1 %.

10. *The linear variation of the drift velocity with gas number density*

In 1963 Lowke [55] observed a linear dependence of W on N in N_2 at 77 °K and low values of E/N, the values of W varying by up to 3 % when the pressure was varied between 50 and 500 torr. Similar dependences have been subsequently observed in a number of gases, e.g. [63, 57, 64, 65]. Frommhold [66] has suggested that the linear dependence is due to the formation of temporary negative ions which autoionize after a mean life time which is long compared with the electron mean free time. He has shown that such a hypothesis gives rise to a dependence on the gas number density of the form

$$W^1 \approx W_0 \left(1 - \alpha N\right) \qquad (2\text{-}3\text{-}6)$$

where W^1 is the measured electron drift velocity, W_0 is the "zero density" drift velocity, and α is a coefficient depending on the energy distribution of the electron swarm, the gas and the temperature. The negative ions are assumed to be formed by electrons being trapped in a resonant state slightly below the excitation energy of the inelastic level with which the resonance is associated. An investigation by Crompton and Robertson [64, 65] of density dependent drift velocities in n–H_2, p–H_2, D_2 and He at 77 °K has indicated that the results are in good qualitative agreement with the Frommhold hypothesis and assuming that the resonance is associated with rotational transitions it was found possible to determine the energies at which these resonances should occur.

Typical plots of the deviation of the drift velocity measured at pressure p, i.e. W_p, from the zero pressure value W_0 are shown as a function of gas number density in Fig. 2-3-9 for helium and deuterium [64]. At low values of E/N, low values of E have to be used and the departure from linearity is considered to be due to instrumental effects, in particular those due to surface effects at the wires of the top shutter [60]. These surface effects are only significant at low values of E and may be regarded as negligible in most measurements. In helium it can be seen that there is no observable density dependence in agreement with Frommhold's hypothesis. In the diatomic gases the procedure has been to extrapolate the data to zero number density to obtain the best estimate value of the drift velocity. This extrapolation is

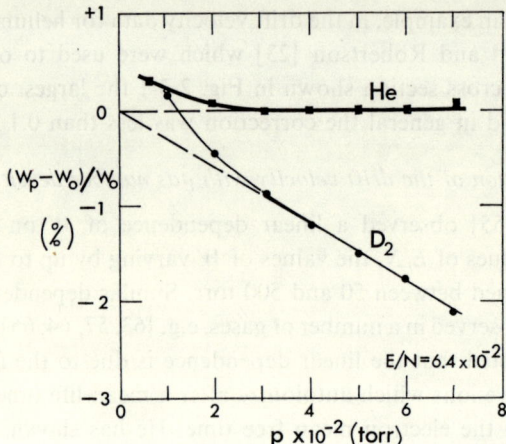

Fig. 2-3-9. The percentage variation of the measured electron drift velocity W_p with pressure in helium and deuterium at $E/N = 6.4 \times 10^{-2}$ Td. W_0 is the drift velocity obtained by extrapolation to zero pressure [65].

considered justified on the grounds that the linearity of W' with N has been well established and that there is strong evidence for the Frommhold hypothesis. The total variation observed in W is small, up to 1.5 % for D_2 over pressures up to 500 torr at 77 °K and about 0.5 % for H_2 under the same conditions. Thus the corrections to obtain the best estimate value are small.

It should be noted that at very high gas number densities Grünberg [67] has observed a pressure dependence of the drift velocity in helium, contrary to the Frommhold hypothesis. It has been suggested that this dependence is probably due to a mechanism different from that postulated by Frommhold since no dependence is found in helium below 8 000 torr. An alternative mechanism applicable to very high gas number densities has been suggested by Legler [68].

C. EXPERIMENTAL DATA

1. *Presentation of data*

The method of presentation of data and the error analysis discussed in this and the following section is that used in published papers of the Electron and Ion Diffusion Unit at the Australian National University.

There are four known effects which can give rise to a dependence of the measured drift velocity W' on gas number density; contact potential differences, diffusion effects, surface effects at the top shutter (peculiar to the

Bradbury-Nielsen method), and the linear variation with pressure discussed in § 2-3B-10. It is therefore essential to tabulate drift velocity data not only as a function of E/N and T but also as a function of p_T, in order to be able to recognise the presence of the effects described above and in some cases to enable corrections to be made. To obtain such a table fixed values of E/p_T (usually E/p_{293} and E/p_{77}) were used and adjustments (usually less than 0.1 %) made to correct the data to the values of E/p_T chosen. These corrections were required to compensate for small variations of the actual gas temperature or pressure from the nominal values. The E/p_T values are related to E/N by

$$E/N = (E/p_T) \cdot 1.0354 \times 10^{-2} T \text{ (Td)} \qquad (2\text{-}3\text{-}7)$$

where E is in V/cm, p_T is in torr, and perfect gas behaviour has been assumed. Check sets of data were taken using different gas samples to ensure that no variation occurred between samples. Portions of the tables of W' obtained in this way are shown in Tables 2-3-2 to 2-3-4.

In the data for helium at 77 °K shown in Table 2-3-2, the first three pressure dependent effects listed above are very small while the fourth pressure effect is absent. A comparison of the data taken over a wide pressure range indicates the order of the internal consistency. By contrast diffusion effects can be seen to be significant in the data for neon at 293 °K (Table 2-3-3). When a correction is made using the relation

$$W' = W(1 + 1.2(D_T/\mu)/V),$$

TABLE 2-3-2

The drift velocities of electrons in helium at 76.8 °K [23]

E/N (Td)	p_{77} (torr)							Best estimate
	700	600	500	400	300	200	100	
0.09543	1.576	1.576	1.575	1.575	1.575	1.575	1.578	
	1.576	1.575	1.575	1.574	1.574	1.574	1.576	1.575
0.1193	1.763	1.763	1.762	1.762	1.762	1.763	1.765	
	1.763	1.762	1.762	1.762	1.761	1.762	1.763	1.762
0.1431	1.930	1.930	1.929	1.929	1.928	1.929	1.932	
	1.930	1.929	1.928	1.928	1.927	1.928	1.929	1.929

The upper entries are the measured drift velocities W' ($\times 10^{-5}$ cm/sec). The lower entries are the measured drift velocities corrected for diffusion effects using the relation $W' = W[1 + 1.5(D_T/\mu)/V]$.

TABLE 2-3-3
The drift velocity of electrons in neon at 293 °K [25]

E/N (Td)	p(torr)				Best estimate
	700	600	500	300	
0.3036	3.463	3.465	3.469		
	3.449	3.449	3.450		3.449
0.4553	4.150	4.163	4.163	4.183	
	4.135	4.145	4.141	4.138	4.140
0.6071		4.721	4.723	4.732	
		4.701	4.700	4.692	4.699

The upper entries are the measured drift velocities W' ($\times 10^{-5}$ cm/sec). The lower entries are the measured drift velocities corrected for diffusion effects using the relation $W' = W[1 + 1.2(D_T/\mu)/V]$.

TABLE 2-3-4
The drift velocity of electrons in para-hydrogen at 77 °K [59]

E/N (Td)	p_{77}(torr)						Best estimate
	350	300	250	200	100	50	
0.06376	1.356	1.357	1.359	1.360	1.362		1.355
0.07970	1.606	1.608	1.611	1.612	1.615	1.614	1.607
0.1435	2.49(8)	2.50(2)	2.50(6)	2.50(8)	2.51(5)	2.52(1)	2.500

The entries are the measured drift velocity W' ($\times 10^{-5}$ cm/sec).

the corrected values are seen to be almost independent of pressure showing that the relation between W' and W above is closely obeyed. Both diffusion effects and contact potential differences between the shutters produce the same form of variation of W' with pressure. However in the results of Table 2-3-3 the smallest potential difference across the drift chamber was 250 volt and thus the effect of contact potential differences may be ignored and the total pressure dependence observed attributed to diffusion effects.

Data for a diatomic gas, para-hydrogen, at 77 °K are presented in Table 2-3-4 to show the linear variation with pressure.

The best estimate values at a given value of E/N were obtained by weighting the data in favour of the values obtained at the highest gas pressures, where contact potential differences, surface effects at the top shutter and diffusion effects are least significant. Where the linear variation with pressure was established as the dominant effect, the best estimate value was found by linear extrapolation to zero pressure.

2. Analysis of error

The error limits were assessed by adding the systematic and random errors. A typical error analysis, in this case for electrons in helium at 77 °K [23] is shown in Table 2-3-5. Since the total systematic error was obtained by adding the contributions arithmetically, the final assigned error may be regarded as the estimated maximum possible error. The estimated random error is approximately one eighth the estimated systematic error in these measurements, a fact of considerable significance in deriving cross sections (§ 2-2C).

TABLE 2-3-5

Sources of systematic error in measurements of electron drift velocities in helium at 77 °K [23]

Source of error	Estimated maximum effect on W (%)
voltage between shutters, V	0.05
temperature	0.2
pressure	0.1
drift distance	0.3
diffusion correction	0.1
Total uncertainty due to systematic errors	0.75
random error	0.1
Total error	0.85
Stated error	1.0 %

3. Results and discussion

The gases considered will be limited to those for which cross sections have been derived by analysis of transport coefficients. It is beyond the scope of this chapter to review all of the very large body of drift velocity data available in the literature for these gases and the papers quoted in Table 2-3-6 are therefore those which report drift velocity data used in the most recent analyses to obtain cross sections using the methods described in § 2-2C. Analyses that have been performed to derive cross sections by these methods are also listed in Table 2-3-6. The analyses listed have been carried out by two laboratories, the Westinghouse Research Laboratories, Pittsburgh and the Ion Diffusion Unit at the Australian National University, Canberra.

It has been shown that the computing procedures used by these two laboratories give the same results when the same input data is used [20, 30]. Thus differences between derived cross sections for a particular gas reflect differences between the data used. These data may differ in either random scatter, systematic error, or both. Consequently it is not desirable to compare derived cross sections from various analyses for a given gas without previous reference to the quality of the transport coefficient data on which these analyses are based.

TABLE 2-3-6

Derivations of cross sections from the analysis of electron transport coefficients

A. Inert gases.

Gas	Cross section derivations	Electron drift velocity measurements		
		T (°K)	Reference	Stated absolute error
He	Frost and Phelps [26] Crompton, Elford and Jory [20] Crompton, Elford and Robertson [23]	77 293	*Crompton, Elford and Robertson [23] *Crompton, Elford and Jory [20]	±1% ±0.5% for $0.018 < E/N < 1.8$, ±1% for $E/N > 1.8$
Ne	Robertson [25]	76.8 293	*Robertson [25] as above	±1% for $E/N < 0.2$, +1%, −2.5% for $E/N > 0.2$ as above
Ar	Frost and Phelps [26] Engelhardt and Phelps (and hydrogen-argon mixtures) [61] Robertson [25]	90 293	*Robertson [25] as above	±1% for $0.01 < E/N < 0.1$, variable for $E/N < 0.01$ but max. error $< 3.0\%$, variable for $E/N > 0.1$ but max. error $< 2\%$ as above
Kr	Frost and Phelps [26]	195 300 368	Pack, Voshall and Phelps [95] as above as above	
Ze	Frost and Phelps [26]	195 300 368	Pack, Voshall and Phelps [95] as above as above	

TABLE 2-3-6 (continued)

B. Molecular gases

Gas	Cross section derivations	Electron drift velocity measurements			Measurements of D_T/μ		
		T (°K)	Reference	Stated absolute error	T (°K)	Reference	Stated absolute error
p-H$_2$	Crompton, Gibson and McIntosh [30]	77	Crompton and McIntosh [59]	±2%	77	Crompton and McIntosh [59]	±2% for $0.4 \leq E/N \leq 6$ ±3% for $E/N < 0.4$
n-H$_2$	Frost and Phelps [7] Engelhardt and Phelps [31] Gibson [8] Crompton, Gibson and Robertson [3]	77	*Robertson [71]	±1%	77	Crompton, Elford and McIntosh [57]	±2%
		293	as above	as above	293	as above	±1%
D$_2$	Engelhardt and Phelps [31] Gibson [8]	77	*Robertson [71]	±1%	77	Crompton, Elford and McIntosh [57]	±2%
		293	*McIntosh [96]	±1% for $0.018 \leq E/N \leq 3.64$ ±2% for $3.64 < E/N \leq 15.2$	293	McIntosh [96]	±1%
N$_2$	Frost and Phelps [7] Engelhardt, Phelps and Risk [97]	77	*Lowke [55]	±2%			
		293	as above	±1%			

TABLE 2-3-6 (continued)

Gas	Cross section derivations	Electron drift velocity measurements			Measurements of D_T/μ		
		T (°K)	Reference	Stated absolute error	T (°K)	Reference	Stated absolute error
CO	Hake and Phelps [32]	77	Pack, Voshall and Phelps [95]		77	Warren and Parker [89]	
		195	as above		288	Skinker and White [98]	
		300	as above				
CO_2	Hake and Phelps [32]	293	Riemann [103]		288	Skinker [100]	
			Errett [101]		195	Rudd (see [22])	
			Bortner, Hurst and Stone [42]			Warren and Parker [95]	
			Frommhold [102]			Cochran and Forester [94]	
		195	Pack, Voshall and Phelps [95]		293	Rees [99]	±1%
		300	as above				
			Levine and Uman [104]				
		293	Elford [56]	±0.5% for $0.3 < E/N < 9.1$; ±1% for $12 < E/N < 21$			
O_2	Hake and Phelps [32]	293	Nielsen and Bradbury [106]		288	Brose [112]	
			Riemann [103]		288	Healey and Kirkpatrick (see [22])	
		293	Doehring [107]		288	Huxley, Crompton and Bagot [109]	±2%
		300	Pack and Phelps [105]		293	Rees [110]	±2%
		293	Frommhold [109]		293	Schlumbohm [111]	

THE ELECTRON DRIFT VELOCITY

Those references marked by an asterisk are to papers which include tables of drift velocity data obtained using the Bradbury-Nielsen method and the techniques described in § 2-3B. To indicate the general behaviour of the drift velocity as a function of E/N and T and the degree of agreement with the data of the papers quoted in Table 2-3-6, four gases helium, neon, hydrogen and nitrogen have been chosen as examples.

In comparing published sets of data a number of difficulties arise. Much of the data is presented in graphical form only (and often on very small

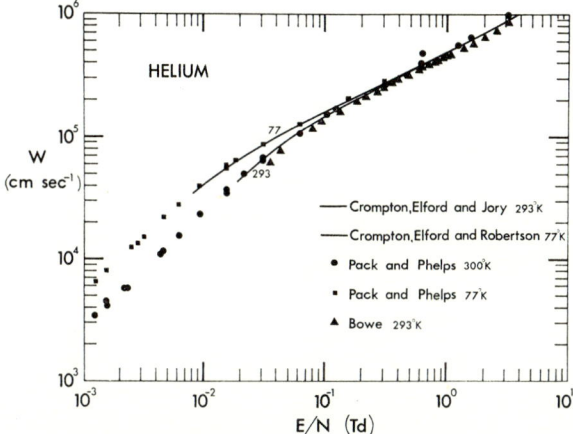

Fig. 2-3-10. The drift velocity of electrons in helium.

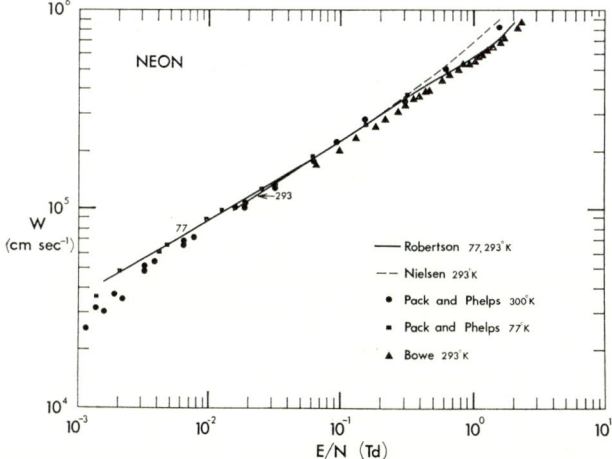

Fig. 2-3-11. The drift velocity of electrons in neon.

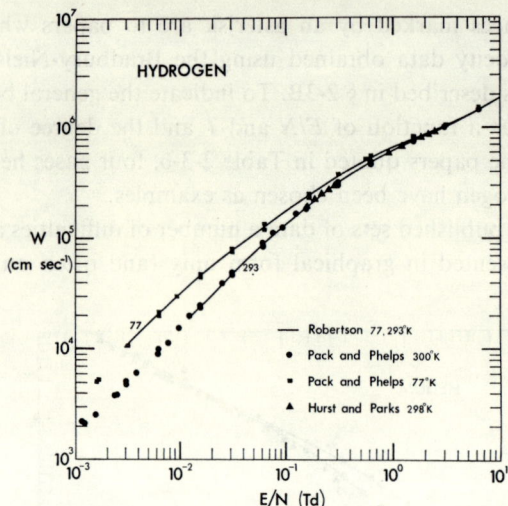

Fig. 2-3-12. The drift velocity of electrons in hydrogen.

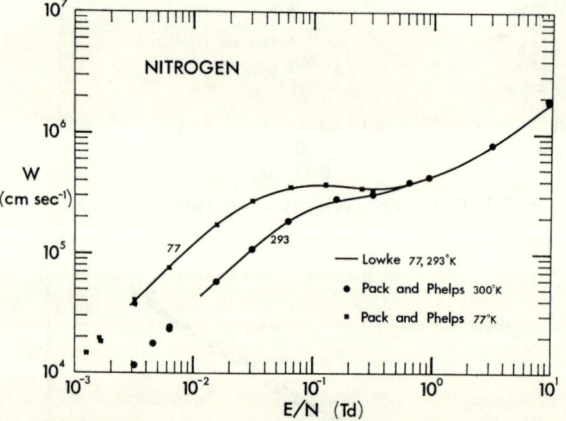

Fig. 2-3-13. The drift velocity of electrons in nitrogen.

graphs), frequently there is no statement of the gas temperature and there is in general no assessment of error, either random or systematic. For room temperature data quoted below, where no gas temperature has been stated, a value of 293 °K has been assumed; this assumption will be indicated where applicable. The data of Crompton, Elford and Jory [20], Crompton, Elford and Robertson [23], Robertson [25] and Lowke [55] are shown in Figs. 2-3-10 to 2-3-13 as continuous lines. No data points of these authors are included as they all lie well within the thickness of the lines.

(a) Helium. The data shown in Fig. 2-3-10 for the drift velocity of electrons in helium are those of Pack and Phelps [53] (tabulated in Pack, Voshall and Phelps [69]) (for 77 and 300 °K), Crompton, Elford and Jory [20] (for 293 °K), Crompton, Elford and Robertson [23] (for 77 °K) and Bowe [36] (for 293 °K ass.). The results of Pack and Phelps are in general agreement with those of both Crompton, Elford and Jory and Crompton, Elford and Robertson for $E/N < 0.5$ Td but at higher values of E/N, they are up to 10 % higher. The data of Bowe lie systematically 5 to 10 % below those of Crompton, Elford and Jory. However, Bowe has subsequently reported data within 2 % of the data of Crompton et al. [20, 70]. Other sets of data for helium at room temperature and not shown in Fig. 2-3-10 have been reported by Wagner, Davis and Hurst [49] (for 293 °K ass.), and Nielsen [51] (for 293 °K).

The data of Wagner et al. lie systematically above the results of Crompton, Elford and Jory by approximately 2–3 % but if the gas temperature were higher than the value of 293 °K which has been assumed, this difference would be reduced. The data of Nielsen lie close to those of Crompton, Elford and Jory. Because of doubtful experimental technique, data published before 1930 has not been considered.

(b) Neon. Figure 2-3-11 shows the data of Robertson [25] (for 77 °K and 293 °K), Pack and Phelps [53] (for 77 °K and 300 °K), Nielsen [51] (for 293 °K) and Bowe [36] (for 293 °K ass.). The data of Bowe lie a systematic 5–10 % below those of Robertson. In view of a similar result in helium (and Bowe's subsequent findings; see section (a) above) it is concluded that the data of Bowe are probably subject to a systematic error of this magnitude [55, 26]. The data of Nielsen lie up to 20 % above those of Robertson.

The data of Pack and Phelps are in general agreement with those of Robertson, but extend to lower values of E/N. However, in this very low region of E/N the experimental scatter of the Pack and Phelps data becomes large.

(c) Hydrogen. The data of Robertson [71] (for 77 and 293 °K), Pack and Phelps [53] (for 77 and 300 °K) and Hurst and Parks [48] (for 298 °K) are shown in Fig. 2-3-12. Other measurements made at room temperature and not shown are those of Lowke [55] (for 77 and 293 °K), Bradbury and Nielsen [50] (for 293 °K) and Wagner, Davis and Hurst [49] (for 293 °K ass.).

The data from all these sources are in good agreement if we take into account an assessment of the experimental scatter for each set of data. The data of Lowke are in particularly close agreement with those of Robertson, the difference being less than 1 % over the common E/N range at both 77 and 293 °K. Measurements of Hurst and Parks, and Wagner, Davis and Hurst, using the method described in § 2-3A-2, i.e. the determination of the drift velocity from the distribution of flight times, agree with each other and those of Robertson to within approximately 2 %.

(d) Nitrogen. In Fig. 2-3-13 the data of Lowke [55] (for 77 and 293 °K) are shown together with the data of Pack and Phelps [53] (for 77 and 300 °K) and Nielsen [50] (for 293 °K). Other room temperature results not shown in Fig. 2-3-13 have been reported by Wagner, Davis and Hurst [49] (for 293 °K ass.), Bowe [36] (for 293 °K ass.), Commetti and Huber [72] (for 303 °K), Nagy, Nagy and Dési [40] (for 293 °K ass.), Colli and Facchini [28] (for 293 °K ass.), Bortner, Hurst and Stone [42] (for 293 °K ass.) and Klema and Allen [39] (for 293 °K ass.).

The results of Wagner et al. are systematically higher than those of Lowke by about 1 % which is within the combined experimental error. The data of Bowe lie a systematic 6 % below those of Lowke in confirmation of the conclusion based on the data for helium and neon that the method used by Bowe was subject to a systematic error of approximately 6 %. Bortner, Hurst and Stone also report data up to 6 % lower than that of Lowke. The data of Commetti and Huber and Nagy, Nagy and Dési agree to within their experimental error with the data of Lowke while those of Klema and Allen and Colli and Facchini are generally higher.

A comparison of the data for these four gases shows that, in general, the data are in agreement with those of the authors listed in Table 2-3-6 to within an experimental error which can reasonably be assigned to the respective measurements.

A detailed explanation of the variation of the drift velocity with both E/N and T requires a discussion of the relevant cross sections for elastic and inelastic collisions and will not be given here. However, two features of the drift velocity curves are worth noting. As E/N is decreased the drift velocity is seen to become proportional to E/N. In general this corresponds to the electron swarms approaching thermal equilibrium with the gas molecules. At sufficiently large values of E/N, the electron energy distribution becomes independent of the thermal motion of the gas molecules and the drift velocity becomes independent of gas temperature. This fact provides a useful check on experimental data taken at different temperatures.

§ 2-4. The measurement of D_T/μ by the Townsend-Huxley method

A. INTRODUCTION

Although a large body of drift velocity data exists, there have been far fewer measurements of the lateral diffusion coefficient D_T.

Three methods have been used to obtain this coefficient directly; microwave methods, the drift-dwell-drift technique of Nelson and Davis [73] and the electron density sampling method of Cavalleri [74]. The microwave methods (see ref. 2 of [74]) in fact measure the ambipolar diffusion coefficient since electron diffusion takes place in the presence of an electric field produced by a high concentration of positive ions. The drift-dwell-drift technique is restricted to thermal energies, i.e. $E/N = 0$ and is complex in that two distinct diffusion processes produce the measured distribution of electron flight times from which the thermal diffusion coefficient is derived. The method of Cavalleri is a particularly significant technique since it enables the diffusion coefficient to be obtained not only at thermal energies but also as a function of E/N. An absolute error of less than $\pm 0.5\%$ is claimed for measurements in helium at 300 °K over an E/N range if 0 to 9 Td and less than 1.5% in the range 9 to 27 Td. Data using this method and with this precision are at present available only for helium.

The most frequently used approach to the study of the diffusion of an electron swarm in an electric field is not the direct measurement of D_T but the measurement of the ratio D_T/μ by the Townsend-Huxley method. The historical development of this method and the techniques necessary to obtain D_T/μ data of high precision are now considered.

1. *Historical survey*

Measurements of the ratio D_T/μ by Townsend and his collaborators in the early years of this century played a significant role in the understanding of the nature of electrical conduction in gases. The apparatus devised by Townsend [75] is shown in Fig. 2-4-1. The electrode system had cylindrical symmetry and the electrons were produced by ionization of the gas by X-rays. After entering the drift chamber through a large hole G, the electrons drifted and diffused in a uniform electric field maintained by guard rings. The ratio of the current collected by the central disc to the total current falling on the anode was measured. This current ratio was related to W/D_T by an expression obtained by solution of the diffusion equation. In 1913 Townsend and Tizard [17] changed the geometry of the anode and cathode,

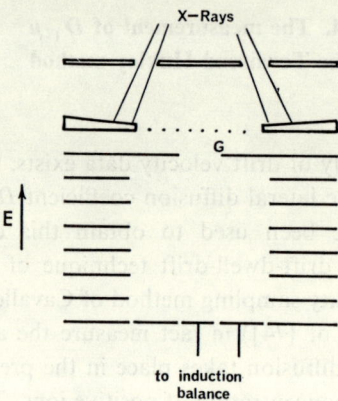

Fig. 2-4-1. Electrode system used by Townsend [75] to measure the ratio D_T/μ.

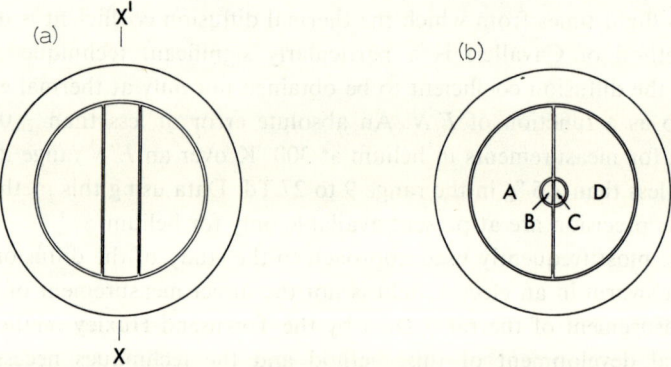

Fig. 2-4-2. Geometry of the collector electrode, (a) of Townsend and Tizard [17], (b) of Huxley [76].

the large circular hole being replaced by a slit, the anode now having the geometry shown in Fig. 2-4-2a. The cathode slit was aligned in the XX' direction. The purpose of this new geometry was to enable both W/D_T and W (but actually W_M) to be measured in the same apparatus but unfortunately the solution of the diffusion equation now became a slowly converging Fourier series that necessitated a large amount of calculation. The solution of the diffusion equation for the Townsend chamber (Fig. 2-4-1) is also slowly convergent.

In 1940 Huxley [76] suggested a change in geometry of the diffusion chamber which led to a series solution of the diffusion equation in terms of Bessel functions which rapidly converged. Electrons enter the diffusion chamber through a small axial hole, approximating a point source of electrons, the anode having the geometry shown in Fig. 2-4-2b. The ratio W/D_T is found from the ratio of the current falling on the inner disc (B+C) to the total current. The magnetic drift velocity W_M is obtained from a measurement of the ratio of the current falling on electrode (A+B) to that falling on (C+D) when a magnetic field is applied transverse to the electric field and in the same direction as the central cut in the collector electrode. Details of the experimental technique used in the measurement of W_M are given by Jory [19] and Creaser [21]. In recent measurements of D_T/μ the slit has been omitted because of the higher geometrical accuracy obtainable in the construction of the simpler collector. The method of measurement of D_T/μ using the general geometry of Townsend but the small source hole and analysis of Huxley has become known as the Townsend-Huxley method and will now be examined. It will be assumed that there is no volume production or loss of electrons in the swarm. The extension of the analysis to the case where either one or more of the processes of attachment, ionization and secondary emission at the electrodes occur has been considered by Huxley [77], Hurst and Huxley [78], Hurst and Liley [79] and Lawson and Lucas [80].

B. THE SOLUTION OF THE DIFFUSION EQUATION AND THE CURRENT RATIO FORMULA

1. *The current ratio formula*

The quantity measured, the current ratio R, is a function of E/N and N. The relation between R and W/D_T is obtained by solution of the time independent diffusion equation

$$\nabla^2 n = 2\lambda \, \partial n/\partial z \quad \text{where} \quad \lambda = W/2D_T \tag{2-4-1}$$

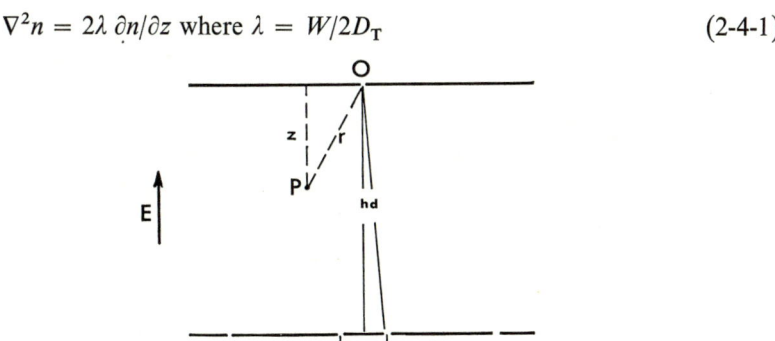

Fig. 2-4-3. Geometry of the Townsend-Huxley diffusion chamber.

on the assumption that the electron energy distribution is spatially independent (see § 2-2B-2). The coordinate system used is shown in Fig. 2-4-3 and n is the electron number density at point $P(r, z)$. The chamber length is h and the diameter of the central collector is $2b$.

Huxley [76] showed that two simple solutions of (2-4-1) are

$$n = A_0 \exp[\lambda z] \exp(-\lambda r)/r \qquad (2\text{-}4\text{-}2)$$

and

$$n = A_1 \frac{z}{r} \exp[\lambda z] \frac{d}{dr}\left\{\frac{\exp(-\lambda r)}{r}\right\}. \qquad (2\text{-}4\text{-}3)$$

The coefficients A_0 and A_1 are determined by the strength of the source.

The question of the appropriate boundary conditions has been discussed at length by Hurst and Liley [79] who suggest that the physical boundary conditions may be complex in that they account for a number of phenomena at or near the boundaries. As a consequence they suggest that the correct ratio formula can only be determined experimentally. Two particular ratio formulae [76] which have been examined using experimental data are

$$R = 1 - \frac{h}{d}\exp[-\lambda(d-h)] \qquad (2\text{-}4\text{-}4)$$

obtained by using (2-4-2) and assuming $n = 0$ at the anode, and

$$R = 1 - \frac{h}{d}\exp[-\lambda(d-h)]\left[\frac{h}{d} - \frac{1}{h\lambda}\left(1 - \frac{h^2}{d^2}\right)\right] \qquad (2\text{-}4\text{-}5)$$

obtained by using (2-4-3) and assuming $n = 0$ at both anode and cathode (except at the source hole). The parameter d is the slant height, given by $d = (h^2 + b^2)^{\frac{1}{2}}$. Huxley and Crompton [81] showed that if formula (2-4-5) were used, the data of Crompton and Sutton [82] for electrons in hydrogen gave the unexpected result that the values of D_T/μ were a function not only of E/N but of N also. On the other hand if formula (2-4-4) were used the values of D_T/μ were found to be a function only of E/N as expected. A more extensive experimental investigation, also in hydrogen, of the validity of formula (2-4-4) has been made by Crompton and Jory [83] using a Townsend-Huxley diffusion chamber of variable geometry. When both d and h were varied widely it was found that formula (2-4-4) gave consistent values of D_T/μ whereas the formula (2-4-5) gave values of D_T/μ which varied with h and p at small values of h. When b/h is small and h is large, formula (2-4-5) approaches formula (2-4-4). For example with $h = 10$ cm, $b = 0.5$ cm the values of k_1 derived by the use of either formula differ by less than 0.2 % for $R > 0.5$. In summary, both formulae give consistent and indistinguishable values of k_1 with changing experimental parameters when h is large and b/h is small but at small values of h and large values of b/h only formula (2-4-4) gives consistent results.

2. *Effect of the spatial dependence of the electron energy distribution function*

Parker [84] has pointed out that all analyses of the behaviour of an electron swarm in a Townsend-Huxley diffusion chamber have been based on the assumption that the spatial gradient term in the Boltzmann equation (2-2-2) can be ignored. When this term is retained Parker found that the electron energy distribution function became a function of spatial co-ordinates and that this caused an error in the value of D_T/μ as derived using formula (2-4-4). The fractional error is given by the approximate formula

$$\frac{\Delta(D_T/\mu)}{(D_T/\mu)} \approx \frac{1}{2}\left(\frac{D_T/\mu}{E/N}\right)\frac{1}{Nh} - \frac{1}{4}\left(\frac{b}{h}\right)^2. \qquad (2\text{-}4\text{-}6)$$

For diffusion chambers with $h = 10$ cm, $b = 0.5$ cm and over the range of values of E/N and N normally used the error in D_T/μ is negligible (< 0.1 %). However, if short chamber lengths and widely divergent streams are used the correction suggested by Parker could be significant. Other analyses of the Townsend-Huxley method based on the Boltzmann equation, including the spatial gradient term, have been carried out by Francey [85] and Desloge and Mitchell [86].

C. Experimental method

Measurements of D_T/μ are essential in the analysis of swarm data to obtain cross sections for inelastic scattering processes (see § 2-2C-2). Since the analysis of swarm data provides the only reliable cross sections in the very low energy regime (less than 0.5 eV approximately), particular emphasis has been placed in recent measurements, e.g. [57], on obtaining data of high precision at low gas temperatures and over a wide range of values of E/N. The techniques used by Crompton and his colleagues to obtain such data will now be discussed.

1. *Diffusion chamber geometry*

In order to obtain accurate D_T/μ data at low values of E/N and low gas temperatures it is necessary to choose the geometrical parameters of the diffusion chamber with care. A number of conditions restrict the values which these parameters may have. The current ratios should lie between 0.2 and 0.9 for accurate measurement, b/h should be small and h large to ensure that the current ratio formulae should approach the asymptotic form, and the potential difference across the chamber should be adequately high to reduce uncertainty caused by the presence of contact potential differences. A lower limit to the size of b is set by assumptions concerning the annular gap. The effective diameter of the central disk $2b$ is normally taken as the actual diameter plus the width of the gap between disk and annulus on the assumption that the electrons arriving at the gap divide equally between disk and annulus. To reduce the uncertainty caused by this assumption the gap width is made small compared to the diameter of the central disk. In practice a value of $b = 0.5$ cm, annular gap width $= 0.005$ cm and $h = 10$ cm has been found to enable D_T/μ to be measured over a wide range of values of E/N at low gas temperatures, and fulfil the conditions stated above. However, even with this geometry it is found that for thermal electrons in hydrogen at 77 °K the current ratio lies between 0.2 and 0.9 only if the potential difference across the diffusion chamber is less than 25 volt. Under these conditions contact potential differences become significant. Thus the state of the surfaces of anode and cathode set the limit to the accuracy obtainable at very low values of E/N.

The inner diameter of the diffusion chamber is chosen to ensure that a negligible fraction of the electron swarm reaches the walls. For example, in the measurements of Crompton, Elford and McIntosh in hydrogen, less than 0.004 % of the electron swarm is found outside a radial distance of 4 cm at the anode. The 5 cm radius system used ensured that the presence of the radial boundary had a negligible effect on the electron swarm. Thus in theoretical analyses of the motion of electron swarms in diffusion chambers of the geometry described above, radial boundary conditions may be ignored.

2. *Production of a uniform electric field*

The degree of uniformity required in the electric field for accurate D_T/μ measurements is very high since small radial electric fields can cause significant changes in the current distribution over the collector. As there is a significant electron number density only in a restricted volume of the diffusion chamber, it is necessary to achieve a highly uniform field only over this volume. An electrode geometry which produces such a field is the "thick" electrode system of Crompton, Elford and Gascoigne (Fig. 2-4-4).

2-4 THE MEASUREMENT OF D_T/μ

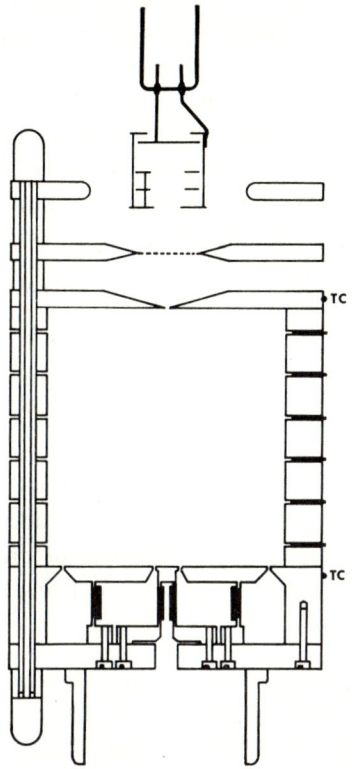

Fig. 2-4-4. Electrode system used to measure D_T/μ [57].

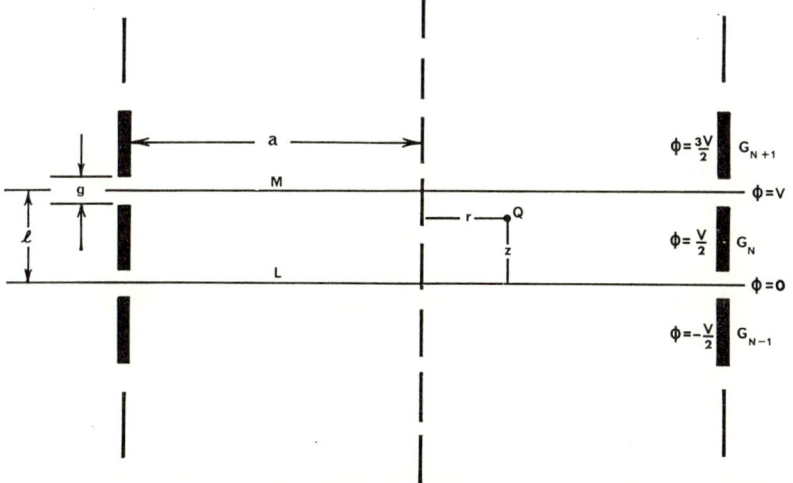

Fig. 2-4-5. Coordinate system used in the derivation of the potential field of a "thick" guard electrode system.

The potential field generated by a system of identical guard electrodes spaced at regular intervals can be examined by considering the field within any module of the system. The planes L and M of the system shown in Fig. 2-4-5 define such a module. If the potentials of the electrodes G_{N-1}, G_N and G_{N+1} are $-\frac{1}{2}X$, $\frac{1}{2}X$ and $\frac{3}{2}X$ respectively, the potentials of the planes L and M will be 0 and X respectively and the solution to the Laplace equation with these boundary conditions is [87]

$$\phi = \frac{Xz}{l} + \sum_N \frac{2X}{N\pi} \frac{I_0(N\pi X/l)}{I_0(N\pi a/l)} \sin(N\pi z/l) \frac{\sin(N\pi g/2l)}{N\pi g/2l}$$

where ϕ is the potential of the point Q at distances z from L and r from the axis of the system,

- l is the length of the module,
- a is the inner radius of the electrode system,
- g is the thickness of the gap between electrodes,
- I_0 is the modified Bessel function of zero order, and
- N is a summation index, the summation being carried out for all even values of N.

With $l = \frac{10}{6}$ cm and $a = 5$ cm, i.e. the values used in the system shown in Fig. 2-4-4, it is found that the maximum deviation of the potential from the value corresponding to a uniform field is less than 0.05 % within an axial cylindrical volume 6 cm in diameter. For values of the current ratio $R > 0.2$, 99.9 % of the electron stream arrives at the anode within a radius of 3 cm. The condition of high uniformity throughout the cylindrical volume stated above and used in the design of the electrode system is in fact more stringent that is required. The effect of field distortion is to deflect electrons which in a uniform field would have been collected on the central disk, to the outer annulus, or vice versa. The deflection of electrons by field distortion over either the inner disk or outer annulus cannot affect the current ratio. The field is therefore required to be highly uniform only over a cylindrical volume much smaller in diameter than the 6 cm quoted above.

The thick electrode system has a high degree of mechanical rigidity and acts as an electrostatic shield for the chamber. The thick electrode system produces a high degree of uniformity over a restricted volume. By contrast the more conventional ring system while theoretically providing uniformity over a larger volume in an apparatus of the same dimensions, is unlikely to produce a high degree of uniformity anywhere because of geometrical distortion of the electrodes.

The electric field above the source hole is maintained at the same value as in the drift chamber to ensure that when the electrons enter the chamber, they have approximately the distribution of electron energies appropriate to the value of E/N used.

3. *The collector*

The collector must have high geometrical precision, maintain its dimensions over a large temperature range and have a high resistance between sections as leakage currents are required to be less than 10^{-15} A. The deviation of the collector surface from a plane is critical. In one early series of measurements, reported by Crompton, Elford and Gascoigne, a lip approximately 0.0004 cm high at the inner edge of the outer annulus was found to cause a systematic error of 1.5% through the production of a radial field in the vicinity of the gap.

A collector which has successfully met the conditions described and which has been cycled many times between 293°K and 77°K is shown in Fig. 2-4-6. The inner disk and outer annulus are supported on thin walled copper cylinders vacuum brazed to alumina tubes. The cylinders are supported by a heavy copper base plate and differential contraction is taken up in distortion of the thin copper cylinder walls (thickness 0.04 cm).

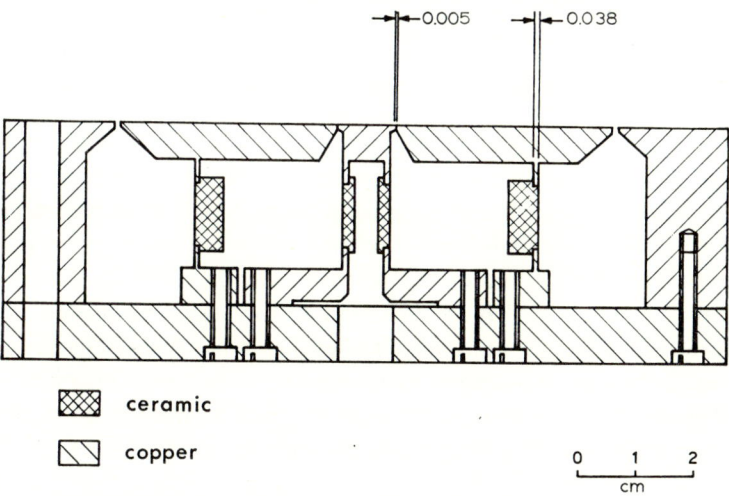

Fig. 2-4-6. The construction of a collector designed to be varied over a wide temperature range without significant loss of geometrical precision [57].

4. *The current ratio*

The currents used in the experiments were small ranging from 10^{-12} to 10^{-11} A to avoid introducing effects caused by Coulomb repulsion. One method of measuring the ratio of two currents each of this order of magnitude has been described by Crompton, Elford and Gascoigne. A simplified circuit of the double induction balance which they used is shown in Fig. 2-4-7. Two displacement currents i_1 and i_2 are generated by applying potentials V_1 and V_2, both of which vary at a constant rate, to the capacitors C_1 and C_2 which have a capacitance of approximately 30 pF and which are matched to within 0.01 % [82]. The ratio of the displacement currents is determined by the setting of the potentiometer P_2, which has a resolution of 1 in 1000.

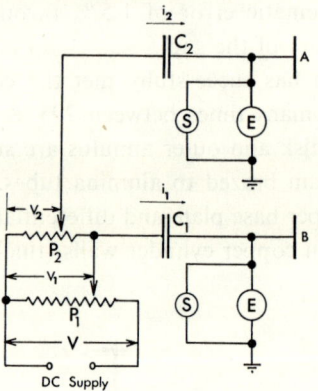

Fig. 2-4-7. Simplified circuit of the double induction balance used to measure current ratios.

To measure the ratio of the currents received by the collectors A and B, first voltage V and then P_2 is adjusted until A and B are maintained at earth potential to within 0.2 mV. The ratio of the collected currents is then obtained from the setting of the potentiometer P_2. The deviations from earth potential of A and B are detected by two valve electrometers E. The particular advantage of using valve rather than other types of electrometers for this application is discussed by Crompton, Elford and Gascoigne.

The required precision is obtained by making the balance technique an integrating one. The integration commences when $V_1 = V_2 = 0$ and terminates when $V_1 = V$, the collectors being unearthed and earthed at the correct instants by the use of a cam, and electromagnetic earthing switches S. The integration period is 55 seconds.

The leakage rate of the collector and electrometer system was less than 10^{-16} A, while the zero drift was less than ± 0.2 mV over periods of several hours. Current ratios can be measured with an error of less than 0.1 % when $R = 0.5$ and each current is 5×10^{-13} A.

5. *Temperature measurement and control*

Measurements of D_T/μ, like those of W, are affected by the gas temperature both through the determination of N and through the electron energy distribution. At very low values of E/N, where the mean electron energy is close to that of the mean thermal agitational energy of the gas molecules, the value of D_T/μ approaches kT/e, and hence an error of 1 °K in gas temperature measurement at 293° K can lead to an error of approx. 0.3 %, the error incurred at 77 °K being greater than 1 %. At low values of E/N, D_T/μ varies slowly with E/N (e.g. Fig. 2-4-9) so that errors in the determination of N are less serious than at high values of E/N.

The gas temperature was measured by copper-constantan thermocouples (TC) attached to the electrode system as shown in Fig. 2-4-4. The temperature was held stable by use of the techniques described in § 2-3B-4.

6. *Electron source*

In earlier measurements of D_T/μ in hydrogen, nitrogen and helium made by Crompton and Elford [88] and Crompton, Elford and Jory [20] the electrons were obtained by thermionic emission from a heated platinum filament. The heat dissipated by the filament was removed by a cooling jacket, through which water was circulated in a closed system by a pump. The disadvantages of this type of electron source compared with a radioactive source have already been discussed in § 2-3B-3. It was found however that when the radioactive source was used some electrons were produced in the chamber by weak γ radiation from the Americium 241. If not taken into account, these electrons which had not entered the chamber through the source hole, would give rise to a false distribution of electron current at the collector. A compensation procedure for this effect is described by Crompton and McIntosh [59]. At the highest gas pressures used, the current to the annular ring, for which compensation was made, was only a few percent of the current to be measured. The background current to the central disk was almost negligible.

7. *Gaseous impurities*

The effect of gas impurities on electron transport coefficients has been discussed in § 2-3B-5. In the case of the inert gases, D_T/μ data are not used as prime data in deriving cross sections and hence the problem of obtaining data where very low levels of diatomic impurities are significant can be avoided. In helium, the inert gas for which D_T/μ values are least sensitive to diatomic impurities, it has been shown [58] that 100 ppm of H_2 or N_2 has a negligible effect on the measurements. For gases, other than the inert gases, adequate purity can be obtained by using the techniques described in §2-3B-5.

D. SOURCES OF ERROR

1. *Nonuniformity of the electric field*

Radial electric fields arising from field nonuniformity cause an incorrect distribution of electrons over the collector and can be introduced in three ways; by errors in the geometry of the guard electrode structure (or errors in the guard electrode potentials), by deviations of the collector surface from a plane and by contact potential differences particularly over the collector surface in the vicinity of the annular gap.

Although the required degree of field uniformity can be theoretically achieved by choosing suitable geometrical parameters for the "thick" guard electrode system (§ 2-4C-2), it is necessary to construct the system to a high degree of precision if significant distortion is to be avoided. For example, in the guard ring system shown in Fig. 2-4-4 the full guard rings were 1.616 ± 0.003 cm in height and the glass spacers between rings were 0.051 ± 0.002 cm in thickness. These tolerances and the special care taken to ensure a plane collector surface made the effect of incorrect geometry on the results taken with this system, e.g. [20, 57] negligible.

The more difficult problem is that of contact potential differences between sections of the electrode system or between different points on the collector. To reduce these effects both anode and cathode were vacuum coated with gold and the guard rings were gold plated. In an alternative procedure used by Warren and Parker [89, 90] to reduce contact potential differences all interior surfaces of the diffusion chamber were coated with colloidal graphite. However, the scatter in the experimental data of these workers prevents a comparison being made between the effectiveness of gold and colloidal graphite as methods of reducing contact potential differences in this application. Tests using an apparatus, described by Crompton, Elford and Gascoigne, to measure contact potential differences over a surface up to 10 cm in diameter showed that unless considerable care was exercised, contact potential differences of between 30 and 100 mV existed between various points on a surface vacuum coated with gold. The quality of various collector surfaces from the point of view of contact potential differences was found both by direct measurement to determine the variation of the contact potential difference and by using the collector in measurements of D_T/μ at varying values of E but a constant value of E/N. After a number of attempts a collector with a maximum variation of 8 mV in contact potential difference between any two small areas 6 mm in diameter (the diameter of the probe used in the direct measurement of the contact potential difference) was obtained, the variation over the central region being not greater than 3 mV. This collector was found to give the smallest variation of D_T/μ with varying E (approximately 1 %) and the results were more stable with time and previous history of the tube than other collector surfaces investigated. The remaining discrepancy was attributed to the existence of a contact potential difference between the plated guard electrodes and the vacuum coated collector surface. Tests using the apparatus for measuring contact potential differences showed that differences of the order of 100 mV could exist between such surfaces and accordingly a compensating potential was placed between the set of guard electrodes and the anode and cathode.

In measurements of D_T/μ at room temperature the exact value of this compensating potential was determined by measuring D_T/μ for potassium ions at sufficiently low values of E/N that $D_T/\mu = kT/e$ to within 0.2 %. Thus in hydrogen results were taken at values of $E/N < 1.5$ Td and the compensating potential set such that the measured value of D_T/μ was kT/e. Crompton, Elford and Gascoigne have given a detailed justification for setting the compensating potential in this way and also a method for detecting and identifying errors caused by geometrical inaccuracy or contact potential differences.

At 77 °K, the compensating procedure using potassium ions has not been used as it was found [57] that a positive ion current on the collector caused the value of D_T/μ for ions to increase over a period of several hours until the value was in error by more than 10 %, the initial value being only 1 % too high. Measurements of D_T/μ with electrons, following the ion measurements, showed similar behaviour, the value also increasing with time, indicating that a contact potential difference is not responsible. When only electron currents were used the results were stable and the compensating potential was set by choosing the value for which D_T/μ became constant for a given value of E/N, when E was varied as widely as possible.

Although this unknown surface effect could be avoided a second surface effect was found to occur in measurements at 77 °K. After measurements of D_T/μ at high values of E/N, subsequent measurements at low values of E/N were found to be increased and Crompton, Elford and McIntosh have suggested that this effect is due to the formation of an insulating layer on the collector surface. The published results of these experiments are however not subject to error from this cause. It should be emphasised that the compensating procedure discussed above to take into account contact potential differences eliminates an error of less than 1 % at the lowest field strengths used. At higher field strengths the effect of contact potential differences becomes negligible.

2. *Finite size source hole*

In the theory of a Townsend-Huxley diffusion chamber it has been assumed that the source hole can be regarded as a point source of electrons. In practice, a hole less than 0.1 cm in diameter transmits insufficient current for accurate current ratio measurements. The effect of the finite size of the source hole has been investigated by Crompton and Jory [83] and for $h = 10$ cm and $b/h = 0.05$ the effect of the 0.1 cm diameter hole is estimated to cause ratios measured at high gas pressures (and hence large current ratios) to be just significantly larger than those measured at smaller pressures. However, it is considered that the calculations overestimate the effect since the electron number density was assumed constant across the source hole. The absence of a systematic pressure dependence in the experimental data suggests that in fact the error due to the finite size of the hole has been overestimated.

3. *Incorrect axial alignment*

Misalignment of the source hole over the central disk leads to an error in the current ratio which increases with the value of the ratio. Calculations made by Crompton and Jory for the case where $b = 0.5$ cm and $h = 10$ cm show that a misalignment of 0.02 cm causes an error of approximately 0.1 % at $R = 0.9$. Such a misalignment is much greater than that incurred when the alignment is made by the procedure described by Crompton, Elford and Gascoigne. By employing a Taylor-Hobson Micro-Alignment Telescope it is estimated that the maximum error in alignment is less than 0.005 cm.

4. *Presence of negative ions*

In 1963 Crompton and Elford [88] reported that the measured values of k_1 were dependent on the electron currents used, even though the magnitude of these currents was of the order of 10^{-11} to 10^{-12} A. Crompton,

Elford and McIntosh [57] investigated this dependence further both at 293 °K and 77 °K. These authors found that the measured values of D_T/μ were linearly dependent on the total current and the values of D_T/μ shown in their tables were found by linear extrapolation to zero current from measurements at 1×10^{-12} A and 2×10^{-12} A. The difference between the extrapolated value of D_T/μ and the value recorded at 1×10^{-12} A was nearly always less than 1 % for room temperature experiments and decreased rapidly with increasing E/N. Crompton, Elford and McIntosh have produced evidence that the effect is due to space charge repulsion enhanced by the presence of negative ions, presumably O_2^-, formed by 3 body attachment [91]. The dependence of the current ratio on the total electron current was found to be in excellent qualitative agreement with the theoretical analysis of Liley [92] and measurements with hydrogen samples containing known impurity levels of oxygen are in semi-quantitative agreement. The current dependence observed in the D_T/μ measurements in hydrogen and deuterium [57] was found to be caused by an oxygen impurity level of less than 3 parts in 10^{10}. It should be emphasised that although in this instance the current dependence observed did not cause a significant error in the measured values of D_T/μ it is obviously essential to use gases with an extremely low oxygen impurity level.

E. Experimental results

Contact potential differences, the effect of the finite size of the source hole and the error suggested by Parker (§ 2-4B-2) all give rise to values of D_T/μ which are a function of the gas pressure as well as E/N at a particular gas temperature. It is therefore necessary to take data over the widest possible range of pressures at each value of E/N and compare such results. Part of a set of D_T/μ data, in this case for hydrogen at 293 °K [57] is shown

TABLE 2-4-1

Experimental values of D_T/μ for electrons in hydrogen at 293 °K [57]
(entries are in volt)

E/N (Td)	p_{293} (torr)						Best estimate
	500	400	300	200	150	100	
0.04553	0.0275(1)	0.0275(2)	0.0275(1)	0.0274(2)			0.0274(9)
0.06071	0.0286(1)	0.0285(8)	0.0285(9)	0.0285(2)	0.0284(7)		0.0285(5)
0.07589		0.0297(3)	0.0297(1)	0.0296(6)	0.0296(2)		0.0296(8)
0.09106			0.0308	0.0308	0.0308	0.0307	0.0308

in Table 2-4-1. It can be seen that there is no significant systematic variation of D_T/μ with pressure. An error analysis carried out in a similar manner to that shown in Table 2-3-5 for drift velocity data, resulted in an absolute error of $\pm 1\%$ being placed on these data. In this case the analysis was based on the sources of error discussed by Crompton and Jory [83] and Crompton, Elford and Gascoigne [58].

References to measurements of D_T/μ used in the most recent analyses to obtain cross sections are given in Table 2-3-6. Measurements in helium and hydrogen have been chosen as examples of the variation of D_T/μ with E/N and T and to indicate the degree of agreement between measurements made using the techniques described in this section and those of other workers. Data for helium have been included even though the momentum transfer cross section for helium is derived from drift velocity data and not measurements of D_T/μ. A comparison of the experimental values of D_T/μ and those predicted using a momentum transfer cross section derived from drift velocity data, provides an important check on both the analysis and experimental data. With the exception of the results of Townsend and Bailey all D_T/μ data quoted in sections (a) and (b) below were obtained using the Townsend-Huxley method.

(a) Helium. The data of Crompton, Elford and Jory [20] (for 293 °K) and Warren and Parker [89] (for 77 and 300 °K) are shown in Fig. 2-4-8. The curve of best fit through the D_T/μ data of Crompton et al. agrees to within 1% with that calculated using the momentum transfer cross section of Crompton, Elford and Robertson [23] derived from their drift velocity data at 77 °K. The two curves are therefore indistinguishable in this figure. The data of Warren and Parker agree with those of Crompton, Elford and Jory at 293 °K and with the predicted values of D_T/μ at 77 °K to within the experimental scatter of their measurements of a few percent. Robertson [60] has also shown that the cross section of Crompton, Elford and Robertson for helium predicts the $D_T N$ values of Cavalleri [74] to within 1% for $E/N < 1$, a satisfactory result in view of the very different type of measurement involved in the Cavalleri experiment.

(b) Hydrogen. It has been shown that it is possible to obtain cross sections for elastic scattering and for inelastic scattering particularly near threshold with a high degree of uniqueness in the case of hydrogen. There has been therefore strong interest in obtaining accurate D_T/μ data for this gas.

The results of Crompton, Elford and McIntosh [57] (for 77 and

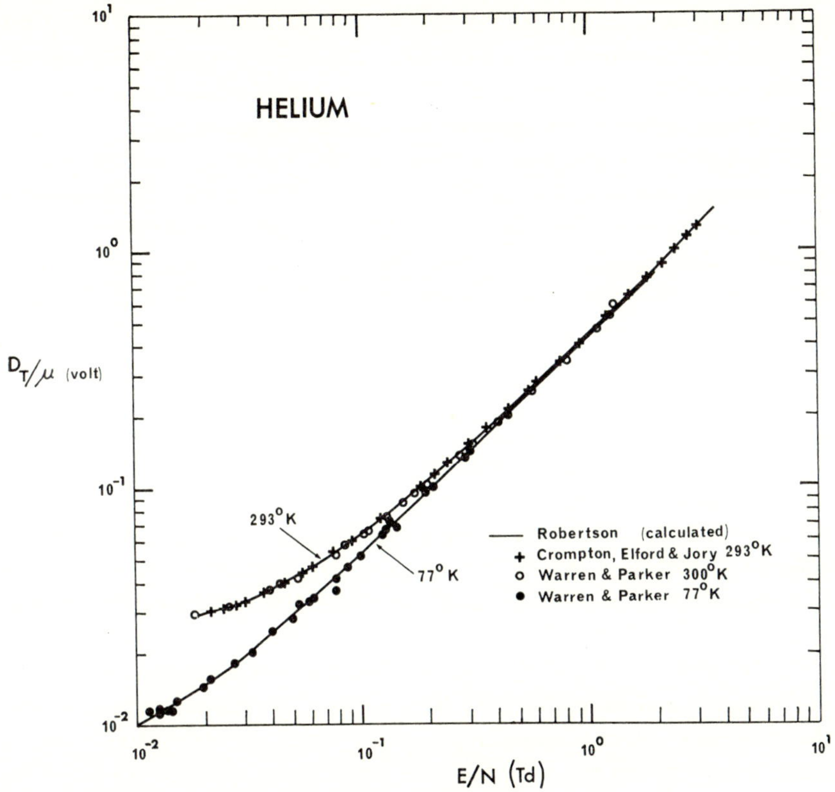

2-4 THE MEASUREMENT OF D_T/μ

Fig. 2-4-8. Measurements of D_T/μ in helium.

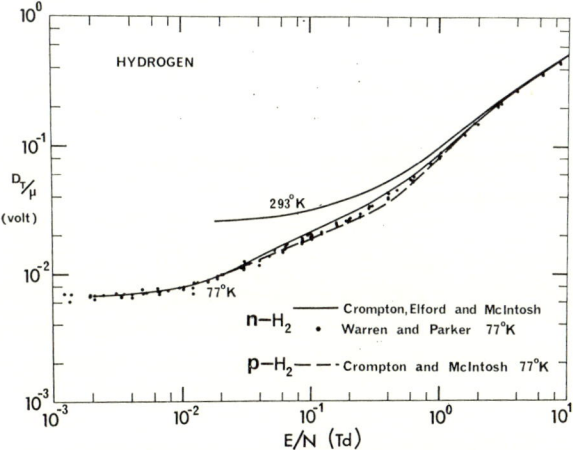

Fig. 2-4-9. Measurements of D_T/μ in hydrogen.

293 °K) and Warren and Parker [89] (for 77 °K) are shown in Fig. 2-4-9. The data of Crompton et al. are shown as a continuous line, all experimental points lying within the thickness of the line. Other measurements of D_T/μ in hydrogen at room temperature but not shown are those of Townsend and Bailey [93], Crompton and Sutton [82], Crompton and Jory [83], Cochran and Forrester [94] and Crompton and Elford [88]. The agreement of the data of Crompton, Elford and McIntosh with that of Crompton and Elford is within 0.5 % over the common range of E/N values (i.e. $0.018 \leqslant E/N \leqslant 0.3$). While the data of Crompton and Sutton (for $E/N > $ 3Td) and Townsend and Bailey (assuming a temperature of 288 °K) are in fair agreement with those of Crompton, Elford and McIntosh the data of Crompton and Sutton at lower values of E/N are significantly different. A discussion of this difference and also the large discrepancy with the results of Cochran and Forrester is given by Crompton and Jory.

The data of Warren and Parker show a large experimental scatter (often in excess of 5 %) some points disagreeing with the data of Crompton, Elford and McIntosh (for 77 °K) by up to 20 %. Probable sources of error in the measurements of Warren and Parker have been discussed by Crompton et al. [57].

Also shown in Fig. 2-4-9 are the data for para hydrogen (for 77 °K) of Crompton and McIntosh [59]. Because para and normal hydrogen have the same momentum transfer cross section, electron swarms when close to thermal equilibrium will have the same value of D_T/μ in both gases. If the value of E/N is sufficiently large that the dominant inelastic energy transfer process is vibrational excitation then again the values of D_T/μ will be the same in both normal and para hydrogen. The difference between the values of D_T/μ for these gases at intermediate values of E/N which can be seen in Fig. 2-4-9, stems from the difference in the statistical weights of the rotational levels of the two gases. At 77 °K, 99.5 % of the para-hydrogen molecules are in the $J = 0$ rotational state and the only rotational excitation process that contributes significantly to the energy loss by the swarm is the $J = 0 \rightarrow 2$ transition. However, in normal hydrogen at 77 °K, 75 % of the molecules occupy the $J = 1$ level and only 25 % the $J = 0$ level. Since the cross sections for the $J = 0 \rightarrow 2$ and $1 \rightarrow 3$ transitions are different (the thresholds are 0.045 and 0.075 eV respectively) the rotational energy losses in normal and para hydrogen will differ, giving rise to differences in transport coefficients.

Although the effect of various inelastic processes can be seen in hydrogen, because of the separation in energy of the thresholds for the low energy inelastic processes, in the heavier diatomic gases where there are a number of inelastic processes with closely spaced thresholds no simple discussion of the kind given for hydrogen is possible.

Two features of the D_T/μ curves are the same for all gases. At sufficiently low values of E/N, D_T/μ approaches a constant value kT/e, i.e. the swarm approaches thermal equilibrium with the gas molecules while at sufficiently high values of E/N, the mean electron energy becomes independent of gas temperature and as in the case of the drift velocity, the values of D_T/μ become independent of temperature.

§ 2-5. Conclusion

The particular contribution of electron swarm studies to the understanding of very low energy electron collision phenomena and their advantages over beam techniques in this energy regime, has been discussed at length by Crompton [2] and will not be repeated here. It is sufficient to state that below about 1 eV, cross sections derived from electron transport coefficients are believed to be the most accurate available, while below about 0.1 eV they are the only cross sections available.

Three factors determine the accuracy of cross sections derived from electron transport coefficients; the degree of uniqueness, the precision of the experimental data and the reliability of the method of analysis. This chapter has been principally concerned with the second of these factors. The accuracy now obtainable in measurements of W and D_T/μ has been shown to be due not only to improved measurement and control of experimental parameters but also to an understanding of the effects which limit the accuracy which may be obtained under a given set of experimental conditions. A typical example is the effect of diffusion on the drifting electron group in experiments to measure the drift velocity. From a knowledge of the functional dependence of these effects on the experimental parameters it has been possible to redesign the experiments and thus reduce the errors caused by these effects to the desired degree.

As the precision of the transport coefficient measurements has improved, attention has been focused increasingly on the third of the factors mentioned above, the reliability of the method of analysis. The form of the Boltzmann equation used and the assumptions involved in the method of solution have been and continue to be the subject of considerable theoretical interest. As a result of these studies there is emerging a better understanding of the behaviour of electrons under swarm conditions and a higher degree of confidence in cross sections derived from electron transport coefficients.

Acknowledgements

The author wishes to record his appreciation of many helpful discussions with Dr. R. W. Crompton. The comments of Dr. J. J. Lowke and members of the Ion Diffusion Unit, especially Prof. Sir Leonard Huxley, Dr. D. K. Gibson and Dr. A. G. Robertson are also gratefully acknowledged. He also wishes to thank Dr. Crompton and Dr. Robertson for permission to use their data in advance of publication.

References

1. A. V. Phelps, Phys. Rev. **40** (1968) 399.
2. R. W. Crompton, Advan. Electron. Electron Phys. **27** (1969) 1.
3. R. W. Crompton, D. K. Gibson and A. G. Robertson, Phys. Rev. **A2** (1970) 1386.
4. T. Holstein, Phys. Rev. **70** (1946) 367.
5. B. Davydov, Phys. Zeits. Sowjetunion **8** (1935) 59.
6. H. Margenau, Phys. Rev. **69** (1946) 508.
7. L. S. Frost and A. V. Phelps, Phys. Rev. **127** (1962) 1621.
8. D. K. Gibson, Australian J. Phys. **23** (1970) 683.
9. W. P. Allis, in: Handbuch der Physik, Vol. **21**, ed. S. Flugge (Springer-Verlag, Berlin, 1956) p. 383.
10. P. M. Davidson, Proc. Phys. Soc. B **67** (1954) 159.
11. L. G. H. Huxley, Australian J. Phys. **13** (1960) 578.
12. G. Cavalleri and G. Sesta, Phys. Rev. **170** (1968) 286.
13. G. H. Wannier, Bell System Tech. J. **32** (1953) 211.
14. J. H. Parker Jr. and J. J. Lowke, Phys. Rev. **181** (1969) 290.
15. J. J. Lowke and J. H. Parker Jr., Phys. Rev. **181** (1969) 302.
16. E. W. McDaniel, Collision Phenomena in Ionized Gases (Wiley, New York, 1964) Ch. 10.
17. J. S. Townsend and H. T. Tizard, Proc. Roy. Soc. A **88** (1913) 336.
18. L. G. H. Huxley, Australian J. Phys. **13** (1960) 718.
19. R. L. Jory, Australian J. Phys. **18** (1965) 237.
20. R. W. Crompton, M. T. Elford and R. L. Jory, Australian J. Phys. **20** (1967) 369.
21. R. P. Creaser, Australian J. Phys. **20** (1967) 547.
22. R. H. Healey and J. W. Reed, The Behaviour of Slow Electrons in Gases (Amalgamated Wireless (Australasia) Limited, Sydney, 1941).
23. R. W. Crompton, M. T. Elford and A. G. Robertson, Australian J. Phys. **23** (1970) 667.
24. L. G. H. Huxley, R. W. Crompton and M. T. Elford, Bull. Inst. Physics and Physical Society **17** (1966) 251.
25. A. G. Robertson, to be published.
26. L. S. Frost and A. V. Phelps, Phys. Rev. **136** (1964) A 1538.
27. J. J. Lowke, private communication.
28. T. F. O'Malley, Phys. Rev. **130** (1963) 1020.

29. D. E. GOLDEN, Phys. Rev. **151** (1966) 48.
30. R. W. CROMPTON, D. K. GIBSON and A. I. MCINTOSH, Australian J. Phys. **22** 1969 (715).
31. A. G. ENGELHARDT and A. V. PHELPS, Phys. Rev. **131** (1963) 2115.
32. R. D. HAKE JR. and A. V. PHELPS, Phys. Rev. **158** (1967) 70.
33. L. B. LOEB, Basic Processes of Gaseous Electronics (University of California Press, Berkeley, 1955) Ch. 3.
34. E. W. MCDANIEL, Collision Phenomena in Ionized Gases (Wiley, New York, 1964) Ch. 11.
35. J. A. HORNBECK, Phys. Rev. **83** (1951) 375.
36. J. C. BOWE, Phys. Rev. **117** (1960) 1411.
37. P. HERRENG, Compt. Rend. **217** (1943) 75.
38. L. COLLI and U. FACCHINI, Rev. Sci. Instr. **23** (1952) 39.
39. E. D. KLEMA and J. S. ALLEN, Phys. Rev. **77** (1950) 661.
40. T. NAGY, L. NAGY and S. DÉSI, Nucl. Instrum. **8** (1960) 327.
41. E. J. M. KIRSCHNER and D. S. TOFFOLLO, J. Appl. Phys. **23** (1952) 594.
42. T. E. BORTNER, G. S. HURST and W. G. STONE, Rev. Sci. Instr. **28** (1957) 103.
43. A. STEVENSON, Rev. Sci. Instr. **23** (1952) 93.
44. L. G. CHRISTOPHOROU, R. N. COMPTON, G. S. HURST and P. W. REINHARDT, J. Chem. Phys. **43** (1965) 4273.
45. L. G. CHRISTOPHOROU, G. S. HURST and A. HADJIANTONIOU, J. Chem. Phys. **44** (1966) 3506.
46. L. G. CHRISTOPHOROU and A. A. CHRISTODOULIDES, J. Phys. B. (Atom. Molec. Phys.) **2** (1969) 71.
47. G. S. HURST, L. B. O'KELLY, E. B. WAGNER and J. A. STOCKDALE, J. Chem. Phys. **39** (1963) 1341.
48. G. S. HURST and J. E. PARKS, J. Chem. Phys. **45** (1966) 282.
49. E. B. WAGNER, F. J. DAVIS and G. S. HURST, J. Chem. Phys. **47** (1967) 3138.
50. N. E. BRADBURY and R. A. NIELSEN, Phys. Rev. **49** (1936) 388.
51. R. A. NIELSEN, Phys. Rev. **50** (1936) 950.
52. A. V. PHELPS, J. L. PACK and L. S. FROST, Phys. Rev. **117** (1960) 470.
53. J. L. PACK and A. V. PHELPS, Phys. Rev. **121** (1961) 798.
54. J. J. LOWKE, Australian J. Phys. **15** (1962) 39.
55. J. J. LOWKE, Australian J. Phys. **16** (1963) 115.
56. M. T. ELFORD, Australian J. Phys. **19** (1966) 629.
57. R. W. CROMPTON, M. T. ELFORD and A. J. MCINTOSH, Australian J. Phys. **21** (1968) 43.
58. R. W. CROMPTON, M. T. ELFORD and J. GASCOIGNE, Australian J. Phys. **18** (1965) 409.
59. R. W. CROMPTON and A. I. MCINTOSH, Australian J. Phys. **21** (1968) 637.
60. A. G. ROBERTSON, Ph.D. thesis, Australian National University, unpublished.
61. A. G. ENGELHARDT and A. V. PHELPS, Phys. Rev. **133** (1964) A375.
62. D. S. BURCH, private communication.
63. R. GRÜNBERG, Z. Physik **204** (1967) 12.
64. R. W. CROMPTON and A. G. ROBERTSON, Bull. Am. Phys. Soc. **14** (1969) 259.
65. R. W. CROMPTON and A. G. ROBERTSON, Australian J. Phys., in press.
66. L. FROMMHOLD, Phys. Rev. **172** (1968) 118.
67. R. GRÜNBERG, Z. Naturforsch. **24A** (1969) 1838.
68. W. LEGLER, Phys. Letters **31A** (1970) 129.
69. J. L. PACK, R. E. VOSHALL and A. V. PHELPS, Research Rep. 62-928-113-R1, Westinghouse Research Laboratories (1962).
70. J. C. BOWE, Bull. Am. Phys. Soc. **12** (1967) 232.

71. A. G. Robertson, Australian J. Phys. **24** (1971) 445.
72. A. Commetti and P. Huber, Helv. Phys. Acta **33** (1960) 911.
73. D. R. Nelson and F. J. Davis, J. Chem. Phys. **51** (1969) 2322.
74. G. Cavalleri, Phys. Rev. **179** (1969) 186.
75. J. S. Townsend, Proc. Roy. Soc. A **81** (1908) 464.
76. L. G. H. Huxley, Phil. Mag. **30** (1940) 396.
77. L. G. H. Huxley, Australian J. Phys. **12** (1959) 172.
78. C. A. Hurst and L. G. H. Huxley, Australian J. Phys. **13** (1960) 21.
79. C. A. Hurst and B. S. Liley, Australian J. Phys. **18** (1965) 521.
80. P. A. Lawson and J. Lucas, Proc. Phys. Soc. **85** (1965) 177.
81. L. G. H. Huxley and R. W. Crompton, Proc. Phys. Soc. B **68** (1955) 381.
82. R. W. Crompton and D. S. Sutton, Proc. Roy. Soc. A **215** (1952) 467.
83. R. W. Crompton and R. L. Jory, Australian J. Phys. **15** (1962) 451.
84. J. H. Parker Jr., Phys. Rev. **132** (1963) 2096.
85. J. L. A. Francey, J. Phys. B. (Atom. Molec. Phys.) **2** (1969) 669 and 680.
86. E. A. Desloge and R. D. Mitchell, Australian J. Phys. **23** (1970) 497.
87. C. A. Hurst, private communication (cited in [58]).
88. R. W. Crompton and M. T. Elford, Proc. 6th Intern. Conf. Ioniz. Phen. Gases (Paris 1963) p. 337.
89. R. W. Warren and J. H. Parker Jr., Phys. Rev. **128** (1962) 2661.
90. J. H. Parker Jr. and R. W. Warren, Rev. Sci. Instr. **33** (1962) 948.
91. L. M. Chanin, A. V. Phelps and M. A. Biondi, Phys. Rev. **128** (1962) 219.
92. B. S. Liley, Australian J. Phys. **20** (1967) 527.
93. J. S. Townsend and V. A. Bailey, Phil. Mag. **42** (1921) 873.
94. L. W. Cochran and D. W. Forrester, Phys. Rev. **126** (1962) 1785.
95. J. L. Pack, R. E. Voshall and A. V. Phelps, Phys. Rev. **127** (1962) 2084.
96. A. I. McIntosh, Australian J. Phys. **19** (1966) 805.
97. A. G. Engelhardt, A. V. Phelps and C. G. Risk, Phys. Rev. **135** (1964) A1566.
98. M. F. Skinker and J. V. White, Phil. Mag. **46** (1932) 630.
99. J. A. Rees, Australian J. Phys. **17** (1964) 462.
100. M. F. Skinker, Phil. Mag. **44** (1922) 994.
101. D. Errett, Ph.D. thesis, Purdue University, 1951, unpublished.
102. L. Frommhold, Z. Physik **160** (1960) 554.
103. W. Riemann, Z. Physik **122** (1944) 216.
104. N. E. Levine and M. A. Uman, J. Appl. Phys. **35** (1964) 2618.
105. J. L. Pack and A. V. Phelps, J. Chem. Phys. **44** (1966) 1870 and **45** (1966) 4316.
106. R. A. Nielsen and N. E. Bradbury, Phys. Rev. **51** (1937) 69.
107. A. Doehring, Z. Naturforsch. **7** (1952) 253.
108. L. Frommhold, Fortschr. Physik **12** (1964) 597.
109. L. G. H. Huxley, R. W. Crompton and C. H. Bagot, Australian J. Phys. **12** (1959) 303.
110. J. A. Rees, Australian J. Phys. **18** (1965) 41.
111. H. Schlumbohm, Z. Physik **184** (1965) 492.
112. H. L. Brose, Phil. Mag. **50** (1925) 536.

CHAPTER 3

DIFFERENTIAL CROSS SECTIONS IN ELECTRON IMPACT IONIZATION

BY

H. EHRHARDT, K. H. HESSELBACHER, K. JUNG and K. WILLMANN

Department of Physics,
University Trier-Kaiserslautern,
Kaiserslautern, Germany

Contents

	Page
3-1. Introduction	161
3-2. Kinematics and notations	162
3-3. Different types of cross sections	163
3-4. Choice of target gas and collision variables	166
3-5. The apparatus	167
A. Description of the electron impact spectrometer	167
B. Detection electronics and measurement procedures	172
C. Intensities	176
D. Background problems	178
E. Stray fields	179
F. Modes of operation	180
G. Normalization problems	182
3-6. Experimental Results	182
A. Double differential cross sections	183
B. The triple differential cross sections	188
3-7. Comparison with theory	196
3-8. Measurements close to threshold	202
Acknowledgement	206
References	206

§ 3-1. Introduction

During the last fifty years a large number of publications have appeared on the experimental determination of cross sections for electron impact ionization of atoms and molecules. Most of the data obtained were of phenomenological character and in connection with radiation chemistry, mass spectrometry, plasma physics or phenomena in ionized gases. Often no attempt is made to determine the final electronic state of the ions produced and in some experiments it was not possible to discriminate between the different charge states.

Only a few experiments have been made in which atoms have been studied in ionizing transitions from a well known initial state to a final state which is determined in the experiment. On the other hand, the demand for such detailed data on ionizing collisions of low energy electrons with atoms has increased very rapidly during the last decade, since such data are of considerable interest for the physics of laboratory, terrestrial and extraterrestrial plasmas, which are not in thermodynamical equilibrium. It is a severe problem to satisfy this demand. Experimentally it is already difficult to determine the absolute total cross sections for ionizing transitions from the ground state of an atom to a given final state of the ion; it may be done by observation of the optical transitions of the ions after the collision, if inner shell or double excitation or cascading prior to the observed optical transition can be excluded. If cross sections are to be measured for the electron impact ionization of atoms or ions in an excited state to a specified state of the ionized target particle, then the experimental technique of today does not yet allow such measurements except for a very few cases (ionization of atoms in metastable states). (Cf. the article by Rundel and Stebbings, this volume.) Also the demand for accurate and detailed cross section data for low energy electron impact ionization can not be satisfied by theory. Unfortunately, the theory is not yet in the position to offer accurate approximations in the low energy region. The reasons for this are twofold: i) even the simplest electron impact ionization process, namely that of atomic hydrogen, is already a three particle problem and ii) the experimental data for a thorough comparison with theoretical results do not exist or are not detailed enough (see § 3-3). But such experimental data are needed for comparison with theory, since it is nearly impossible to quote error limits after having made a sequence of complicated approximations in the calculations.

These arguments were the starting point for the research work which is described in this article. Our main concern is to present and discuss double and triple differential cross sections for the electron impact ionization of helium and to a lesser extent, of argon. Such cross sections are functions of several parameters, namely of the collision energy, of the scattering angles of the two outgoing electrons, and of the energy of one of the two outgoing electrons. These functions represent, we think, the detailed information which is needed for comparison with theory, since the comparison can be made before integration to the total cross section. Each integration step has an averaging effect and therefore should be avoided. Before integration, it should be possible to purse the consequences of certain approximations more easily than after integration. This argument holds especially, if the approximation is made to simplify the mathematics and not for physical reasons. In this case the physical meaning of the approximation may be cleared up by comparison with differential cross sections rather than integrated data. On the other hand, quite often a collision model is used, with the assumption that the true collision problem has similar properties. This procedure has the advantage of physical transparency but requires as much detailed information about the actual collision process as possible. In this case, the structures in the double and triple differential cross sections may give valuable hints for models suitable for theoretical approximations. From the total cross section it seems impossible to get such information, since generally it is a smooth curve without significant structures.

§ 3-2. Kinematics and notations

If an electron of impact energy E_0 (see Fig. 3-2-1) collides with an atom in the state Ψ_i and if E_0 exceeds the potential IP for single ionization,

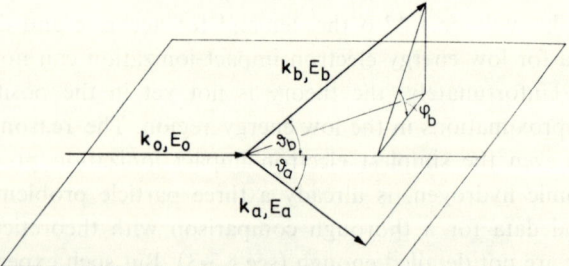

Fig. 3-2-1. Schematic diagram of the kinematics of an ionizing electron collision with an atom. The incoming electron has kinetic energy E_0 and momentum k_0, the scattered and the ejected electrons possess energy E_a and E_b, resp., and momentum k_a and k_b, resp.. Since the atomic electron before the collision has momentum and since during the collision the ion can take up momentum the three vectors k_0, k_a and k_b need not be in one plane.

then it is possible that two electrons with the energies E_a and E_b leave the collision region with the scattering angles ϑ_a and ϑ_b with respect to the direction of the electron beam. The ion is left behind in the state Ψ_f. Since the atomic electron before the collision has momentum, and since in addition to that, the ion may take up some momentum during the collision, the trajectories of the incoming electron and of the two outgoing electrons need not be in a plane. The angle between the planes (k_0, k_a) and (k_0, k_b) is called φ_b, where k_0, k_a, k_b are the momenta of the electrons with the energy E_0, E_a, E_b. Energy conservation requires:

$$E_0 = E_a + E_b + IP - E_{ex,i} + E_{ex,f}. \qquad (3\text{-}2\text{-}1)$$

If the atom before the collision and the ion after the collision are in their ground states, the excitation energies $E_{ex,i}$ (atom) and $E_{ex,f}$ (ion) are zero and therefore

$$E_0 = E_a + E_b + IP. \qquad (3\text{-}2\text{-}2)$$

This is the case for most of the collisions which are discussed in this study. For fixed E_0, equation (3-2-2) requires that any change in energy E_a is compensated by that in E_b, or in other words, the excess energy $E_0 - IP$ is distributed between the two outgoing electrons.

§ 3-3. Different types of cross sections

As in § 3-2 we assume that during the electron collision with the atom a transition occurs from a defined state of the target particle into a defined state of the ion. The most detailed description of such an ionizing collision by electron impact is the triple differential cross section

$$\frac{d^3\sigma}{dE\,d\Omega_a\,d\Omega_b} = f_3(E_0, E_a, \vartheta_a, \vartheta_b, \varphi_b), \qquad (3\text{-}3\text{-}1)$$

which is a function of the kinematic parameters E_0, E_a, ϑ_a, ϑ_b and φ_b. In eq. (3-3-1), E_a may be replaced by E_b by use of eq. (3-2-2), i.e. E_a and E_b are not both free parameters of the collision. The measurement of both energies E_a and E_b together with the measurement of E_0 determines the value of IP, i.e. the type of collision. For example, if IP is the first ionization potential, then we know that the ionization of an outer shell electron is studied.

The triple differential cross section has to be measured in a coincidence experiment, in which the energies and angles of both outgoing electrons are to be measured for one single collision process. A full graphical representation is impossible, since it would require a 6-dimensional space. Two different cuts in different subspaces have been measured, one which is described in this review, for helium in the energy range E_0 from 30 eV to 260 eV, and another by Amaldi et al. [1] for carbon and oxygen in the keV range. The present authors in their experiments varied ϑ_b, whereas Amaldi et al. varied E_0. In both types of experiments all the other parameters are kept constant. Therefore, the first mentioned experiments give the angular correlation between the two outgoing electrons, which contains sensitive information on the type of collision, i.e. whether little or much momentum is transferred to the residual ion during the collision between the incoming and the atomic electron (see ref. [2] and § 3-6B). The experiments of Amaldi et al. yield the momentum distribution of the atomic electron before the collision in a good approximation [3].

The double differential cross section

$$\frac{d^2\sigma}{dE\,d\Omega} = f_2(E_0, E, \vartheta) = \int d\Omega_b f_3(E_0, E, \vartheta, \vartheta_b, \varphi_b) \qquad (3\text{-}3\text{-}2)$$

is lower by two orders with respect to the number of kinematic parameters. In the equivalent experiments the energy E and angle ϑ of one of the outgoing electrons is measured disregarding the scattering angle of the second electron. This is equivalent to the integration of $d^3\sigma$ over all values of the angles ϑ_b and φ_b. In the measurements of the double differential cross sections presented in this chapter, one of the three parameters E_0, E and ϑ is varied, whereas the two remaining parameters have been kept constant during one experiment. Consequently angular distributions (E_0 and E fixed) or energy loss spectra of the first kind (E_0 and ϑ are constant) and of the second kind (E and ϑ are constant) have been obtained.

Double differential cross sections, namely angular distributions and energy loss spectra of the first kind have also been measured by other authors. Most of these experiments were carried out in the years from 1930 to 1936. Although the main results obtained in these years are in good agreement with present day data, the low energy measurements show large errors, sometimes even of an order of magnitude. Most of the errors seem to be due to incomplete magnetic field shielding and to poor transmission properties of the electron optical devices used (see § 3-5A). In the last few years angular distributions of electrons from ionizing collisions of atoms, with different energy losses of high energy electrons, have been measured by Lassettre et al. and Simpson et al. (see § 3-6A).

The single differential cross section $d\sigma/d\Omega$ seems to be without obvious physical interpretation and therefore has not been measured.

The single differential cross section

$$\frac{d\sigma}{dE} = f_1(E_0, E) = \int d\Omega \, f_2(E_0, E, \vartheta) \tag{3-3-3}$$

is a function of the collision energy E_0 and the energy E of the detected electrons. This cross section represents the energy distribution (integrated over all angles) of the scattered and ejected electrons after the ionization process. $d\sigma/dE$ is of great practical importance, e.g. energy loss of photoelectrons in the earth ionosphere, degradation spectrum in radiation chemistry etc.. In principle, it has a shape as given in Fig. 3-3-1. The energy distribution of the ionization electrons is of course symmetrical with respect to $\frac{1}{2}(E_a+E_b)$, since to each fast electron belongs a slow electron of complementary energy. From calculations and measurements of double differential cross sections it is known that $d\sigma/dE$ has maxima close to the energies E_0-IP and zero. If E_0 is large enough, then the two groups S and E are fairly well separated and may be called the scattered (S) and the ejected (E) electrons, although the two particles (outgoing from *one* ionization process) are in principle undistinguishable. This terminology is supported by the different angular behaviour of the fast (S) and slow (E) electrons (see § 3-6A).

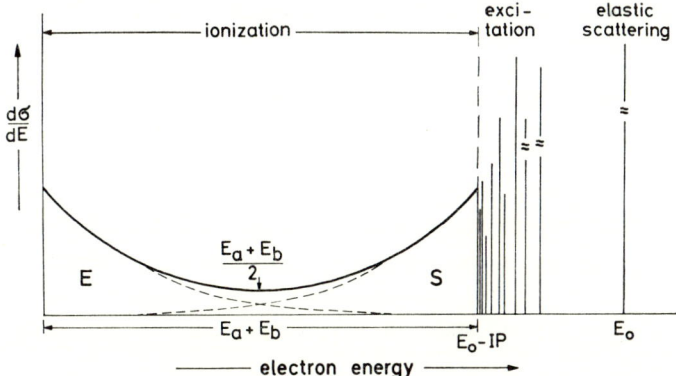

Fig. 3-3-1. Schematic diagram of the energy loss spectrum $d\sigma/dE$ (integrated over all scattering angles) of electrons after collisions with atoms, the ionization potential of which is *IP*. For large values of E_0 ($E_0 \gtrsim 2 \times IP$) two groups of electrons are found (E and S), which are fairly well separated and which have intensity distributions symmetric with respect to the middle energy $\frac{1}{2}(E_a+E_b)$. The fast ionization electrons (S) are called the "scattered", whereas the slow (E) are called the "ejected" electrons. The two groups show very different angular distributions (see § 3-6A).

The integral over $d\sigma/dE$ gives twice the value of the total cross section for the impact energy E_0:

$$\sigma = f_0(E_0) = \tfrac{1}{2} \int_0^{E_0 - IP} dE \, f_1(E_0, E). \tag{3-3-4}$$

In view of the many different experimental results which can be obtained for one single value of E_0 (differential and total cross sections), a comparison with theory looks very promising.

§ 3-4. Choice of target gas and collision variables

The measurements presented in this article are intended to give a fairly complete knowledge about the kinematics of electron impact ionization and as much detailed experimental data as possible for a collision system which is as simple as possible in order to be of value for comparison with theory. Therefore we have choosen helium as target gas. Of course, atomic hydrogen would have been more suitable, but the neutral beam intensities which can be obtained for H-atoms are too low for reasonable coincidence count rates. Argon has been used only in a few cases for comparison with the results on helium.

Most measurements deal with the single ionization of helium atoms by electron impact for collision energies E_0 from threshold up to about 260 eV. The atoms and the ions produced are in their ground states. Therefore the investigated collisions are

$$e^- + He(1s^2) \rightarrow He^+(1s) + 2e^-. \tag{3-4-1}$$

For energies E_0 from ionization threshold up to 50 eV no other processes are possible, since these energies are too low for the simultaneous ionization of the atom and the excitation of the ion. This is not true for $E_0 = 58$ eV and higher. Although in the coincidence experiments the energies of both outgoing electrons are measured and therefore it is insured that the ions finally are in the ground state, the following sequence of processes may occur:

$$e^-(E_0) + He(1s^2) \rightarrow He^{**}(nl, n'l') + e^-(E = E_0 - E^{**}) \tag{3-4-2}$$

$$He^{**}(nl, n'l') \rightarrow He^+(1s) + e^-(E = E^{**} - E_{He^+}(1s)). \tag{3-4-3}$$

The energy positions of the doubly excited states in helium and their f-values are well known. If the energy transferred to the helium atom during the collision is in the range from approximately 58 to 65 eV or between 70 and 72 eV, Rydberg series with $n = 2$ or $n = 3$ can be excited, and the ionization continuum is very much affected by the interference of the electrons of the direct (non resonant) ionization process (eq. (3-4-1)) with the autoionization electrons (eqs. (3-4-2) and (3-4-3)). But outside these two energy ranges the f-values are so small (considerably less than 10^{-3}) compared to the direct ionization continuum at the relevant energies, that the direct process is not influenced by interference. This has been assured in the present experiment by the choice of the energies of the outgoing electrons. They are not at values given by the eqs. (3-4-2) and (3-4-3). For some measurements a collision energy of about 260 eV was chosen in order to have simpler theoretical conditions than for low impact energies. At 260 eV, exchange effects should play an unimportant role. The collisions are also less complicated than for low impact energies, since, in most cases, one electron is still fast after the collision and it does not very much affect the remaining electron–ion system. These arguments encouraged us to compare the experimental results for $E_0 = 260$ eV with a few well known approximations. The findings are described in the section § 3-7.

§ 3-5. The apparatus

A. Description of the electron impact spectrometer

A schematic diagram of the electron impact spectrometer used by the authors [4–7] is shown in Fig. 3-5-1. The arrangement is operated in a

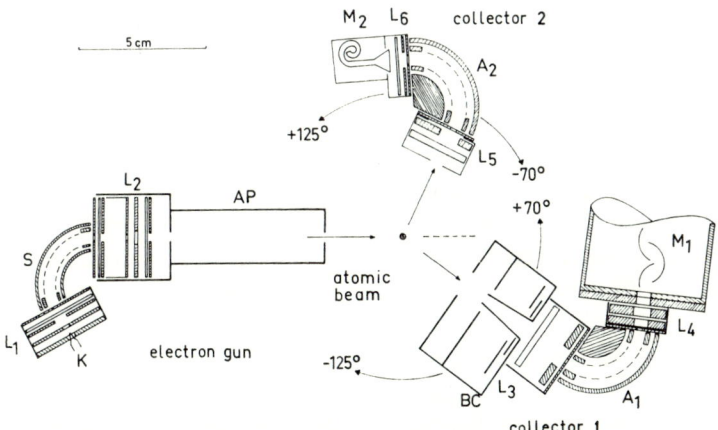

Fig. 3-5-1. The electron impact spectrometer, described in § 3-5.

stainless steel vacuum chamber of 50 cm diameter. The electrons are emitted from the tip of the hairpin cathode K, made from thoriated tungsten wire, and their energy has approximately a Maxwellian distribution with an energy half width of about 300 to 400 meV. Since the electrons are leaving only from the tip of the cathode, the lens system L_1 is essentially focusing a point source which is easier to deal with experimentally than an extended source, and this constitutes the advantage of a hairpin cathode. Thoriated tungsten wire is operated at about 1300° K and therefore has a relatively long life time and a small energy distribution for the emitted electrons. Furthermore only small amounts of matter are evaporated from the cathode and, therefore the amount deposited on the surfaces of the electron optical systems is minimal.

A 127° electrostatic electron selector S [8] is used to obtain an energy resolution for the electron beam small compared to the excess energy E_0–IP. Taking into account the fringing field correction as calculated by Herzog [9] the effective deflection angle is only 119°. The energy resolution is of great importance, if the primary energy E_0 is only somewhat greater (ca. 1 eV or less) than the ionization potential IP of the investigated target atoms. For the present results the energy resolution is not too important, since for most measurements the lowest energy of the incident electrons is several eV above threshold. The measured energy width is of the order of a few 100 meV.

The lens system L_1 has a twofold function, it focuses the electron coming from K onto the entrance slit of the selector S and it also accelerates them to its transmission energy (ca. 8 eV). This lens system as well as all the other lens systems denoted by $L_i (i = 2, \ldots, 6)$ contain two pairs of split plates to deflect the position of the focused image in two directions perpendicular to each other.

The lens system L_2 following the exit slit of the selector S focuses the electron beam onto the atomic beam in the scattering centre and also accelerates the electrons to the desired primary energy, which can be varied from a few eV to about 400 eV. The lower limit is determined by space charge and by electrostatic and magnetic stray fields. For an energy resolution of ca. 300 meV an electron beam current of some 10^{-7} A has been obtained.

The box AP is at ground potential and its purpose is twofold: it stops the electrons that have left the gun diffusely and it also shields the well collimated beam in a field free path of about 10 cm between the end of L_2 and the scattering centre from possible stray fields. The whole gun itself is housed in a tight box to prevent electrons from escaping from the cathode region and causing undesirable background.

All components of the electron optics (lenses, electrostatic, cylinders) are machined from non-magnetic stainless steel and are gold plated in order to reduce contact potentials and insure clean surfaces. Other authors [8, 10, 11] use different materials to reduce the same surface effects. We believe that the essential point to stress here is to preserve clean surfaces. This can be achieved by baking out the spectrometer periodically. In our case the gun is continuously heated to ca. 200 °C and the collector systems are baked out about every six weeks.

The target gas flows into the scattering centre through a gold plated tube of ca. 5 mm in length and of a diameter of 0.4 mm. The pressure of ca. 10^{-6} torr in the vacuum chamber rises to about 10^{-4} torr when the gas is flowing through the gas tube. The electron beam crosses the gas beam about 2 mm above the gas inlet tube. In this area (scattering center) the gas pressure is at least one order of magnitude higher than in other parts of the vacuum chamber. In this pressure range (10^{-3} torr) the scattering rate increases linearly with the pressure, i.e., the shape of the gas beam is independent of the pressure. This observation is important if double differential cross sections are measured but of not too much importance for the measurement of triple differential cross sections (angular correlations between the two ionization electrons, see below).

The electrons leaving the reaction center are analyzed with respect to their energy and scattering angle in collector 1 or 2 and are detected in coincidence. Both collector systems can be rotated from 70° on one side of the primary beam to 125° on the other side (see Fig. 3-5-1) independently of each other. In order to make the mechanical arrangement simple the azimuth φ_b (see Fig. 3-2-1) is chosen to be zero degrees, i.e. measurements of out of plane scattering are not possible. The angular resolution of collector 1 is about $\pm 1.5°$ whereas that of analyzer 2 amounts to ca. $\pm 2°$. For both systems the scattering angle is adjustable to better than $\pm 0.5°$.

The lens systems L_3 and L_5 focus the electrons outgoing from the scattering centre on the entrance slits of the analyzers A_1 and A_2 and also accelerate or retard the electrons to the transmission energy (ca. 5 eV) of the cylindrical condensers. For the measurements presented in this chapter the energy resolution of the two analyzers was about 0.5 eV for A_1 and ca. 1.5 eV for A_2. A_1 and A_2 are constructed in the same way as the selector S of the electron gun. The electrostatic cylindrical condensers are formed from grids of high transmission (ca. 80%). Behind the grids are positively charged plates to absorbe the stray electrons that have the wrong energy.

The lens systems L_4 and L_6 following the exit slits of A_1 and A_2 resp. focus the electrons on the multiplier M_1 or M_2. M_1 is a 17 stage electron multiplier and M_2 a channeltron. The channeltron has a relatively small gain for large counting rates [12]; however its terminals have only a small magnetic field (a few 10 mG) and its small size makes it possible to cover a greater angular range.

BC is an electron beam eliminator. The electrons scattered by the target gas get to collector 1 through a 26 mm long cone which is surrounded by a grounded box with a rectangular slit parallel to the scattering plane (not shown in the diagram). If collector 1 recorded scattered electrons at small angles (3°–15°) the primary electron beam goes through the slit of the box BC and is accelerated towards a positively charged electrode at the end of the box. Because of the retarding potential *only* the elastically scattered electrons that bounce off from this electrode have a small chance of leaving the box BC. This system is very effective in reducing the electron background by several orders of magnitude.

The electron optical systems are adjusted with the electrons scattered from the gas beam. Because of the small counting rate (see below), maximum transmission of the systems is desired. This is obtained by adjusting for maximum counting rate at a fixed detector energy.

In Fig. 3-5-2 a schematic diagram of another collector system is shown. This collector was developed in order to have a measurable and constant transmission of electrons in a relatively large energy range. This is of im-

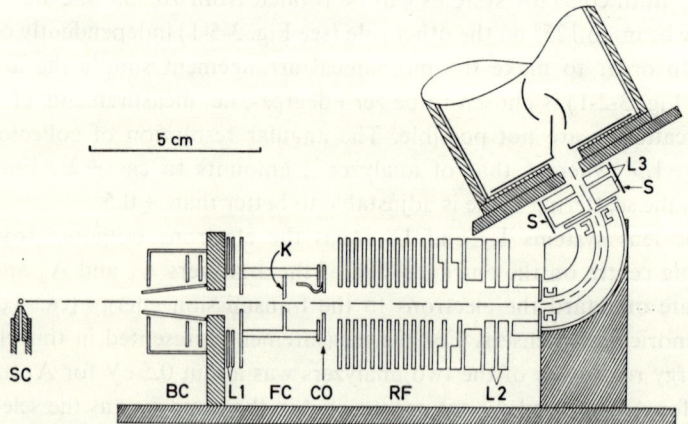

Fig. 3-5-2. Electron collector system used for the measurements of energy loss spectra with constant transmission with respect to the electron energy. For details see § 3-5A.

portance for measurements of energy distributions at a fixed scattering angle. The problem of constant transmission was solved by using a slightly varying retarding field connected to a 127° electrostatic cylindrical condenser. Constant transmission over a large energy range could be achieved by allowing only electrons traveling nearly parallel to the optical axis of the retarding field to enter. For this reason the system detects only electrons scattered into a small solid angle ($\lesssim 1/5000$), and consequently the signal was weak. This system was therefore suitable for recording energy loss spectra but it was not useful for coincidence measurements.

The electrons pass through a field free region (cone of the beam collector, Faraday cage FC), and a collimating region (lens system L_1) into the retarding field, where the electron energy is reduced to 5 eV. As in the case of Fig. 3-5-1 here also all lens systems L_i ($i = 1, 2, 3$) contain two pairs of split plates to deflect the position of the focused image. Behind the retarding field the Einzel-lens L_2 produces an image on the entrance slit of the 127° electrostatic analyzer. Following the exit slit of this analyzer there is the lens system L_3 that produces an image on the first dynode of the electron multiplier. As in Fig. 3-5-1, also in Fig. 3-5-2 the sizes of the different parts of the system can be determined with the scale shown in the figure.

Typical transmission values were $(30\pm5)\%$ over a large energy range at high energies (for example $60 \leq E \leq 250$ eV) and also over a somewhat smaller energy range at lower energies (for example $40 \leq E \leq 120$ eV). The transmission is measured by positioning the collector in a zero degree direction, so that the primary electron beam fully enters the system. About 80% of the electron beam passes through the electrode K in the centre of the Faraday cage FC. The intensity I_0 of the transmitted electron beam is measured by applying a negative voltage on the exit electrode CO of FC, so that I_0 is trapped in the Faraday cage. After opening electrically the cage FC, I_0 enters the retarding field, passes through the 127° electrostatic analyzer onto the first dynode of the electron multiplier. The current I_D which reaches the first dynode of the multiplier is measured with an electrometer. Most of the current I_0 is lost in the entrance and exit slits of the electrostatic analyzer. If the energy resolution of the 127° condenser is not smaller than the energy half width of the primary electron beam, the transmission T of the system is given by $T = I_D/I_0$.

In general the transmission T depends somewhat on the energy resolution of the 127° analyzer. In principle it is of little importance, if the value of the transmission T obtained is somewhat lower or higher. Of great importance, however, for the measurements of the energy distribution of ionization electrons is constancy of transmission with variable detection energy. This is achieved by varying the potentials of the retarding field and of the lens system L_2 as a function of the detector energy automatically in such a way that all the electrons within the energy band that can be transmitted by the 127° cylindrical condenser are exactly focused at the entrance slit.

The box BC is also an electron beam eliminator when the analyzer is positioned at small angles (3°–15°), and it operates in the same way as discussed above in connection with the description of Fig. 3-5-1. Angular measurements below 3° are impossible, since the number of background electrons becomes too high. For measurements above 15° the primary electron beam misses the collector. The negatively charged screening box S prevents any stray electrons from reaching the first dynode of the multiplier and from being detected.

B. DETECTION ELECTRONICS AND MEASUREMENT PROCEDURES

A schematic diagram of the whole electronic arrangement is shown in Fig. 3-5-3. Pulses produced on the anode of each multiplier are coupled by a capacitor to a cathode follower for matching these to a 50 ohm cable

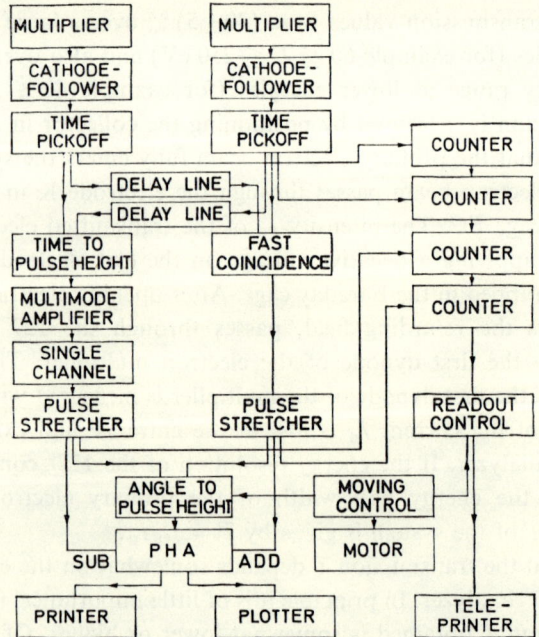

Fig. 3-5-3. Block diagram of the electronic arrangement.

that carries the signal outside the vacuum chamber. For counting rates of ca. 10^4 pulses per second after the cathode followers the pulse amplitude amounts to about 20 mV, if the applied voltages for the 17 stage multiplier is 2–2.5 kV and for the channeltron 2.5–3 kV. The matched pulses of each detector system go into separate preamplifiers with discriminators (time pickoff) and then are fed into counters. With a good adjustment the rate of dark pulses is about 1 pulse per 10 minutes or less.

For recording energy loss spectra and also double differential angular distributions only one collector system is used. For this purpose a ratemeter is connected (not shown in the figure) parallel to one of the two upper counters of Fig. 3-5-3. The ratemeter gives a d.c.-output signal whose amplitude is proportional to the counting rate. This signal is fed into an X-Y-recorder to obtain the measured distribution directly. To measure the triple differential cross section of the electron impact ionization the whole arrangement shown in Fig. 3-5-3 is necessary.

To test the adjustment of the entire apparatus a time spectrum is recorded before every measurement. For this purpose the pulses of the two separate electronic circuits are fed into the start- or stop-input of the time-to-pulse height converter (TPH). The pulse amplitude on the output of the TPH is proportional to the time shift between the start- and stop-input pulse. The output signal of the TPH is shaped by a multimode amplifier and stored in a 400 channel pulse height analyzer (PHA) connected parallel to the single channel analyzer (not drawn in). In this way the channel numbers give a time scale.

An example of a time spectrum is plotted in Fig. 3-5-4. The target gas is argon and the primary energy $E_0 = 110$ eV. The coincidence rate for

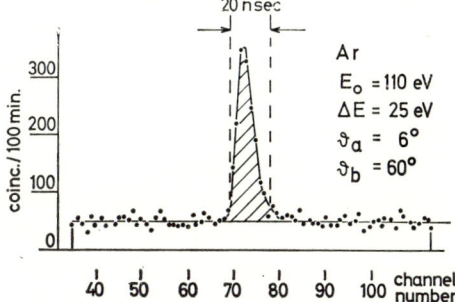

Fig. 3-5-4. Example of a time spectrum for ionizing collisions of 110 eV electrons on argon. The shaded area represents the number of the ionization processes. The electrons have the energies $E_a = 85$ eV and $E_b = 9$ eV and are detected at the indicated angles. This number is called the rate of the true coincidences.

10 minutes is recorded as a function of the channel number. The energy loss of the scattered electrons is $\Delta E = 25$ eV, and they are therefore left with an energy $E_a = 85$ eV after collision and are then collected at an angle $\vartheta_a = 6°$. The ejected electrons are collected at $\vartheta_b = 60°$ and their energy is $E_b = 9.2$ eV (see eq. (3-2-1), $IP = 15.8$ eV, $E_{ex,i} = E_{ex,f} = 0$ eV). The two ionization electrons are detected on different sides with respect to the primary electron beam.

The constant background on the recording represents the accidental coincidences, of which the rate is given by

$$N_a = N_1 \times N_2 \times 2\tau \qquad (3\text{-}5\text{-}1)$$

where N_1 and N_2 signify the count rates of the separate electronic circuits connected to collector 1 and 2 resp. and τ is the time resolution of the coincidence unit used. The maximum count rate, occurring between channel numbers 70 and 80, represents the true coincidences. With the help of a delay line (see Fig. 3-5-3) of about 150 nsec on the stop-input of the time-to-pulse height converter this maximum was shifted to the centre of the time region stored by the TPA. The time resolution is about 2.5 nsec per channel. The coincidence maximum has a time width of ca. 20 nsec. With a good adjustment of the apparatus this time can be improved to about 12 nsec for the chosen collision parameters (see below). Half of this time spread is probably due to the velocity uncertainty of the detected electrons due to the energy resolution of the analyzers A_1 and A_2 (see above) as well as by the fact that the scattering centre is not a point source for the electrons contributing to the true coincidence count rate but instead an extended source. Other contributions to the time spread of the coincidence peak of the time spectrum result from the walk (different amplitudes of the multiplier pulses) and the jitter (noise of the signal and the discriminator circuit) inherent in this electronic equipment.

Both ionization electrons collected by the analyzers 1 and 2 are registered on the input of the fast coincidence unit (see Fig. 3-5-3) with a definite time shift caused by the mostly very different velocities of the two ionization electrons leaving the reaction centre and the different long leads of the separate electronic circuits. This time shift is compensated for by a delay line, so that the fast coincidence unit indicates the true coincidence rate plus the accidental coincidences that occur in this time interval.

Accidental coincidences can be registered by a single channel analyzer with an adjusted window width equal to the resolution time of the fast coincidence unit (10–15 nsec). The threshold of the single channel analyzer is tuned so that its window does not indicate any true coincidence. The difference between the count rates measured with the fast coincidence unit and the single channel analyzer gives the real coincidence count rate.

The electronic modules, which have a good time resolution (in particular the time pickoff), are very sensitive devices. The time pickoff is triggered by pulse rises with the effect that the pulses may be of high impedance if they are steep enough. For this reason the electronic equipment is very sensitive to "stray" pulses especially if they rise fast. As an undesirable consequence of this sensitivity, fast oscillations of power supplies caused by momentarily overloading the network are counted individually (more than 100 coincidences per msec). For this reason it was found necessary to prestabilize all important power supplies. Much care was taken to insure good screening of all leads. It was found convenient to use pulse stretchers (see Fig. 3-5-3) for shaping the output pulses of the fast coincidence unit and the single channel analyzer. The pulse stretchers give only one output signal within a 10 msec interval. All pulses which appear within a shorter time interval are not registered. In this way only one in some 100 false coincidences is counted. The large dead time is not disadvantageous, since the actual coincidence count rate ranges from ca. 0.5 to 50 events per minute (see below).

In order to improve the statistics of a small number of counts, long measuring times are required and the instrumentation has to be fully automated. To record the angular correlation between the two ionization electrons leaving the scattering centre, the angular position ϑ_b (see Fig. 3-2-1) of one collector is varied continuously by a motor, while the angle ϑ_a of the other analyzer is kept constant. Furthermore all the energies are also maintained, the energy E_0 of the primary electron beam and the energies E_a and E_b of the detected particles.

The standard pulses made by the pulse stretchers are transformed by the angle-to-pulse height converter (APH) to pulses whose amplitudes are functions of the angle ϑ_b. With the help of the moving control (see Fig. 3-5-3) a d.c.-voltage which is proportional to the value of the angular position ϑ_b of the collector system is produced. This d.c.-voltage is fed into the APH as reference voltage for the amplitude of the output pulses. The 400 channel PHA adds pulses coming from the fast coincidence unit and subtracts those coming from the single channel analyzer. In this way the true coincidence counting rate is stored on the PHA as a function of the angle ϑ_b. In contrast to the time spectrum in this case, the channel numbers of the PHA represent an angular scale instead of a time scale. The results stored in the memory of the PHA are recorded on a printer and/or a plotter.

The moving control also monitors a motor that sweeps the collector back and forth between fixed values of ϑ_b, which can be preset. At the turning points the moving control operates the readout control and the count rates stored in the four counters are typed out by a teleprinter. The two upper counters (Fig. 3-5-3) store the count rates of the two separate electronic circuits connected to collectors 1 and 2. One of the two lower counters stores the accidental coincidences while the other one stores the combined count rates of the accidental coincidences plus the true coincidences. The four printed count rates should remain approximately constant throughout the time of measurement and any discrepancies would reveal a malfunctioning of the apparatus and the measurement will have to be discontinued.

In general several days of continuous operation are needed to obtain the necessary statistics and produce one useful experimental curve. To eliminate long term fluctuations (variations in the intensity of the primary electron beam, pressure variations, etc.) it is advantageous to sweep through the chosen angular region of ϑ_b many times during the measurements for one curve. In the results presented below ca. 20 minutes were required for one complete sweep through an angular range of 100°, which would correspond to some 100 to 200 complete sweeps for the total time to obtain one curve.

C. INTENSITIES

The rate of the accidental coincidences is given by eq. (3-5-1) and can in a good coincidence experiment be of the same order of magnitude as the count rate of the true coincidences. However they should not exceed it very much. The number of true coincidences N_c is given by

$$N_c = P_1 \times P_2 \times N_T \tag{3-5-2}$$

where N_T is the total number of "coincidence" processes in the scattering centre, P_1 and P_2 are the detection probabilities of the particle collectors 1 and 2. For an estimation of N_c, the probabilities P_i may be written as products of several functions, namely

$$P_i = W_i(\Omega_i) \times \Delta\Omega_i \times W_i(E_i) \times \Delta E_i \times W_{iT} \times W_{iA}, \quad i = 1, 2. \tag{3-5-3}$$

Here, $W(\Omega_i)$ is the probability of a particle being scattered into the direction Ω_i, and $\Delta\Omega_i$ is the acceptance angle of the detector i. $W(E_i) \times \Delta E_i$ is the probability that the electron has an energy in the range between E_i and $E_i + \Delta E_i$, W_{iT} is the transmission of the electron optical system of the detector i (the energy dependence of this function shall be neglected here). W_{iA}

stands for the detection efficiency of the multiplier M_i and the ancillary electronic circuit.

All these numbers may be very small and should be carefully estimated before planning a coincidence experiment. Taking into account the possible angular distributions occurring in low energy electron scattering by atomic systems one will find that the factor $W_1(\Omega_1) \times W_2(\Omega_2) \times \Delta\Omega_1 \times \Delta\Omega_2$ may be of the order of 10^{-6} or smaller (in the apparatus described in this review, $\Delta\Omega_i$ is of the order of 10^{-3}). $W(E_i)$ is given by the energy dependence [5, 13] of the double differential cross section for a fixed scattering angle, i.e. $W(E_i)$ denotes the probability of the splitting of the excess energy $E_0 - IP$ by the two ionization electrons. ΔE_i is the energy resolution of the detector i.

If, for $E_0 < IP + E_{\text{ex}, i}$ two energies are known, the third can be determined from eq. (3-2-1). For this reason, if an energy selected electron beam is used in a coincidence experiment it is sufficient to have only one collector as an energy analyzing system. Collector 2 can only be an angular measuring device for one of the two ionization electrons [4]. A considerable disadvantage of such an arrangement is the fact that the number N_2 of eq. (3-5-1) will be very large since all the elastically and inelastically scattered electrons as well as the background electrons with different energies will be detected. It is better, therefore, to use a second energy analyzing system as in the apparatus described here.

In the first instance the best adjustment seems to be $\Delta E_1 = \Delta E_2$. If the energy resolutions ΔE_i are of the order of magnitude of 1 eV and the energy of the primary electron beam amounts to about 100 eV (in this energy region, often the total cross section for the electron impact ionization has a maximum [14, 15]) the factor $W_1(E_1) \times W_2(E_2) \times \Delta E_1 \times \Delta E_2$ will have a magnitude ranging from 10^{-2} to 10^{-1}. Electron optical transmission probabilities range from 10^{-2} to 1 and W_{iA} from 10^{-1} to 1. Thus, we find that the product $P_1 \times P_2$ of eq. (3-5-2) is of the order of 10^{-9} or smaller.

The total number of coincidence events N_T in the scattering centre cannot be increased beyond certain limits. N_T is proportional to the intensity of the primary electron beam; however, the number of accidental coincidences N_a is proportional to the square of this intensity and therefore increases much faster than N_T. The limit arises at the moment that it becomes impossible to decide whether two particles arise from one or two different collision processes. The limit is given by the time resolution which is determined by several factors as explained above and amounts to ca. 10^{-8} sec.

For the following estimation we will assume that N_a does not become greater than N_c. The average number R of ionizing impacts for one electron over the length S through a gas with the cross section Q at the pressure P is given by

$$R = 3 \times 10^{16} \times Q[\text{cm}^2] \times P[\text{torr}] \times S[\text{cm}]. \tag{3-5-4}$$

From R we obtain N_T by multiplying R by the number of electrons in the primary electron beam. Setting $Q \approx 10^{-17} \text{ cm}^2$, $P \approx 10^{-3}$ torr and $S \approx 3 \times 10^{-1}$ cm, and using an intensity for the electron beam of ca. 10^{-7} A, we find $N_T \approx 10^8$ events per second. From eq. (3-5-2) we obtain for N_c a value of about 10^{-1} true coincidences per second, i.e. ca. 10 events per minute. Using eq. (3-5-1) we calculate the same number for the rate of accidental coincidences N_a with $\tau \approx 10^{-8}$ sec. Taking into account the background electrons detected by the collectors, the ratio between true coincidences and accidental coincidences is not 1 : 1 but a little smaller, say 1 : 1.3. For the results reported here the true coincidence rate ranged from about 0.5 to 50 coincidences per minute, and the number of accidental coincidences was about 2 to 5 times greater. If the number of true coincidences drops below 1 or 0.1 events per minute, then the adjustment of the apparatus and the measurements become very tedious or impossible.

D. BACKGROUND PROBLEMS

In every electron impact spectrometer the electron background is a serious problem, since slow background electrons are easily produced, but difficult to handle without affecting the true scattered electrons. Too large a background poses the following two problems. First, it is essential to determine exactly the background and to subtract this from the count rate. Second, all results are affected by the statistical error of the background, which can be greater than the error of the true signal or even greater than the true signal itself.

In order to reduce the background it is necessary to know its sources. We have to distinguish two types of background: the one that has the same energy as the primary electrons and the other with energies of a few eV (< 5 eV). The first one is of importance in elastic scattering. However, for scattering angles greater than ca. $10°$ it is usually much smaller than the true scattering rate (this angle will be a function of the electron impact cross section of the target gas). If one is looking for inelastic scattering, using energy analyzing devices, the background is important only for very small angles ($\vartheta \lesssim 3°$). Of course, this angle is also a function of the magnitude of the investigated cross section.

The low energy background is more troublesome. It is mainly caused by electrons of the primary electron beam inelastically scattered by walls. For this reason walls should not be in the vicinity of the electron beam. The electron impact spectrometer described here possesses an electron beam "eliminator", which is connected to the analyzer. The primary beam enters into this eliminator, if small angle scattering is to be measured. For large angle scattering ($> 15°$), the electron beam misses the walls of the analyzer. In this case a small electron collector moves automatically into the 0° position to eliminate the primary beam. In addition, the low energy background can be reduced by a positively charged mesh close to the walls of the vacuum chamber. The scattering centre of course has to be properly shielded from this charged surface.

Another source of low energy background electrons ($E < 5$ eV) is the gun itself, especially the area around the cathode. This background can be prevented from entering the scattering chamber by housing the entire gun in a tight box. The only opening is for the exit of the primary beam. Still another source of slow background electrons from the gun may be contained in the primary electron beam itself. This contribution may be in the order of 0.1 to 1 % of the total beam intensity. It is produced by inelastic scattering of fast beam electrons from the solid surfaces of the lens systems. In the interaction region the slow electrons can be scattered elastically from the target gas, and if the cross section for the investigated inelastic process is very small, as it is in ionization processes, the contribution of the mentioned elastic processes to the count rate can easily be of the same order of magnitude as that of the main process. In the present apparatus the number of slow electrons in the beam was very efficiently reduced by introducing a negatively charged aperture in the AP box (see Fig. 3-5-1). The aperture is placed in the middle of this box and the potential on it was about two thirds of that of the primary beam energy. It has a sufficiently large opening, ca. 6 mm in diameter, to avoid further inelastic collisions.

It is also important to enclose the complete detector system into a tight box to avoid stray electrons from reaching the multiplier. Any small gaps, even as small as $\frac{1}{10}$ mm, should be closed, since slow electrons of some 100 meV can make very many wall collisions and zig-zag their way into the electric field of the multiplier and produce a signal.

E. STRAY FIELDS

It is of considerable importance for low energy electron scattering to reduce stray fields as much as possible. For example an electrostatic stray field in the scattering centre will falsify the primary electron impact energy by an unknown amount and, more seriously, angular and energy distributions of slow ejected electrons from ionization collision may be totally changed. To eliminate these errors in the electron impact spectrometer used in our laboratory, the scattering region has been carefully screened from all electrical potentials, by surrounding all optical systems and all the leads with grounded metal boxes. The scattering centre is enclosed in a cage of mesh wire (transmission 98 %) with a gap of ca. 1.5 cm in the scattering plane to guarantee free trajectories for the scattered and ejected electrons to the collectors.

The earth's magnetic field is compensated by Helmholtz coils and μ-metal. μ-metal is also used to reduce the magnetic stray field due to the cathode current and magnetic metal parts of the electron multiplier. The first aperture in front of the cathode K (see Fig. 3-5-1) is made out of μ-metal and the multiplier is placed in a μ-metal tube. With these precautions the remaining magnetic field in the scattering centre has been measured to be smaller than 3 mG but increases to some 10 mG at a distance of ca. 5 cm from the centre. Since the electron energies are always greater than 1 eV at this distance from the scattering centre, the residual magnetic field cannot seriously affect the experiment. The calculated radius of curvature of the detected electrons is greater than 1 m.

F. Modes of operation

There are three main possible modes of operation for the investigation of ionization processes with an electron impact spectrometer of the type discussed here: first, the energy loss spectra, second, angular dependence measurements and third, coincidence measurements. Energy loss spectra and angular dependences are obtained by using only one of the analyzing systems.

The most important factor for reliable energy loss spectra is the knowledge of the transmission as a function of the detector energy. A collector system with measurable transmission is shown in Fig. 3-5-2 and has been described in § 3-5A. Energy loss spectra and angular dependences are double differential cross sections, $d^2\sigma/d\Omega\,dE = f_2(E_0, E_1\vartheta)$. An energy loss spectrum of the first kind is obtained for fixed values of E_0 and ϑ by varying the energy E. For an angular dependence measurement the count rate is plotted for fixed energies E_0 and E as a function of the varying scattering angle ϑ. For a good angular dependence measurement it is essential to know the angular correction function of the apparatus. Quite often a simple sin ϑ-correction is used. For a scattering chamber such a correction is sufficient; however, if a gas beam is used, the correction function will be between sin ϑ and a constant (i.e. no correction). Since in our experiment the gas beam produces a rather high background pressure, and since signal electrons have been scattered by particles, which belong to the background, we used an experimentally obtained correction function of the angular intensity distribution. It has been obtained by the following procedure: First, the angular distribution of the scattered electrons for a given E_0 and E is recorded for the case of the gas beam on. Holding the same energy parameters constant a second angular dependence is measured for a diffuse gas flow into the vacuum chamber instead of a gas flow through the beam tube in the scattering centre. In both runs the background pressure is the same. By subtracting the second count rate from the first for each scattering angle the correct angular distribution follows. All angular distribution data presented in this article have been obtained by this procedure. The method will give reasonable results only if the full gas beam is in the field of view of the collector system. If this is not the case the angular correction function will change for different collector energies E, since the electron optical properties vary with the detection energy. The errors of the measured angular distributions are not larger than about 20 %.

In the third mode of operation of the electron impact spectrometer for ionization processes both ionization electrons are detected in coincidence after having been analyzed with respect to their energy and scattering angle. In this mode the triple differential cross section $d^3\sigma/d\Omega_a d\Omega_b dE = f_3(E_0, E_a, \vartheta_a, \vartheta_b, \varphi_b)$ is measured. $d^3\sigma$ can be explored as a function of the angle ϑ_b of the ejected electrons by keeping E_0, E_a (and E_b), ϑ_a and φ_b constant. In such a measurement the angular correlation between the two ionization electrons is determined.

Since the electrons contributing to the count rate of the true coincidences have to come from a region of the scattering centre in the field of view of both collector systems, the error of the angular distribution is much smaller than in the case of a double differential angular dependence experiment discussed above. By rotating one collector system while the other is held constant, the scattering region in the field of view of both systems does not change appreciably (the error amounts at the most to about 10 %).

Other modes for the investigation of the triple differential cross section $d^3\sigma$ are possible, for example, by varying E_0, E_a and E_b keeping ϑ_a and ϑ_b constant or by varying E_a and E_b (see eq. (3-2-2)) keeping E_0, ϑ_a and ϑ_b constant. No measurements using these operational mode have been made in our laboratory, although with the present experimental equipment it can be done.

G. NORMALIZATION PROBLEMS

Since the transmission characteristics of the detector systems given in Fig. 3-5-1 are not accurately known it is impossible at the moment to normalize to a common intensity scale the angular dependences (double and triple differential cross sections) obtained at different detection energies for the same primary energy E_0. Another unsolved problem is the normalization of the results obtained for the same detector energy but different primary electron energies E_0. The knowledge of the intensity of the primary beam is not sufficient since it is also essential to know exactly the focusing characteristics of the primary beam in the scattering centre as a function of its energy and to know the pressure distribution of the gas beam in this region.

The coincidence results obtained for the same energy parameters E_0 and $E_a (E_b)$ but for different scattering angles can be normalized to each other without great difficulties. The angular distributions for positive and negative values of ϑ_b (ϑ_a being constant, see below) are normalized to each other by changing the angle ϑ_a on alternating sides with respect to the primary electron beam while keeping the angle ϑ_b constant. The error in the normalizing procedure amounts to about $\pm 10 \%$ at the maximum of the curves. Curves for different values of ϑ_a are normalized to each other in an identical way with about the same error.

The reported energy values were measured with 127° electrostatic analyzers, but calibrated with a retarding field method. For the calibration an energy loss spectrum was plotted on a x-y-recorder for a given primary energy. From this plot the zero of the energy scale was established and the values of the primary energy as well as the detector energies were read with a digital voltmeter. An error of about ± 0.5 eV was estimated.

§ 3-6. Experimental results

During the last decade, reliable experimental results have been obtained of total cross sections for the single ionization by electron impact of neutral atoms by several investigators [16–20]. In some cases errors as low as 2 % have been quoted (see for example Rapp and Englander-Golden [21]). The range of collision energy covered by such experiments reaches from threshold up to several keV. These data can be regarded as good experimental material for comparison with theory, especially, if the intial state of the atom and the final state of the ion is known. This is the case for the ionization of hydrogen atoms and the ionization of helium atoms in the energy range from threshold up to about 60 eV. For helium, which is of most interest for our investigations, a large number of calculations for the single ionization into the ground state of the ion have been made [22, 23] but all approximations predict too high values of the total cross section in the range of collision energies from threshold to about 500 eV. The latest calculations from Economides and McDowell [24] however show a much better agreement in this region.

A. DOUBLE DIFFERENTIAL CROSS SECTIONS

For meaningful measurements of double differential cross sections $d^2\sigma/dEd\Omega$ it is necessary to know the initial state of the atom before the collision and to analyze the ionization electrons after the collision with respect to their energies and scattering angles. Both possible dependences of the double differential cross section for ionizing collisions of electrons with rare gas atoms have been investigated by Hughes and McMillen (1932) [25, 26], Tate and Palmer (1932) [27], Mohr and Nicoll (1934) [28], Goodrich (1937) [29] and by others. The angular range of detection extended from about 0° to 170°, the range of collision energies from about 50 eV to several hundreds eV, and the range of energies of the detected electrons from close to zero up to $E_0 - IP$.

It is known, now, that some of these measurements have large errors; especially the reported energy and angular distributions of very slow electrons are questionable, since they are very sensitive to stray magnetic and electric fields and to transmission properties of electron optical devices. Nevertheless, many valuable results have been obtained with rather simple experimental equipment. The most important results are the following.

1. The higher the energy of the scattered electrons, the more is the intensity peaked in forward direction.

2. The angular distribution of very low energy electrons is nearly isotropic.

3. The angular distributions of electrons of medium energy often show an intensity maximum close to 60°.

4. Energy distributions show two intensity maxima, one for fast electrons (energy close to $E_0 - IP$) and one for electrons of nearly zero energy. In between (energy around $\frac{1}{2}(E_0 - IP)$) the energy distribution has a minimum of intensity.

Goodrich (1937) [29] has compared his experimental results with calculations by Wetzel. Wetzel applied a simplified form of the Born-approximation in which not only the incident and scattered electron but also the ejected electron were described by plane waves. The lack of orthogonality between final and initial state made it necessary to include the interaction between the incident electron and the ion into the perturbation potential. The theory reproduces the general characteristics of the measurements. Moreover, the high energy group of the energy distribution $d\sigma/dE$ is explained as being due to the colliding electrons, which have suffered during the collision an energy loss that is somewhat larger than the ionization potential IP of the target atom. The intensity maximum in the low energy region of the energy distribution is identified with the atomic electrons, which have been ejected during the collision. The maxima in the angular distributions which appear for medium energy electrons around 60° have been explained by Massey et al. [22] by energy and momentum conservation during the ionizing collision.

During the last few years several investigations have been made, namely by Lassettre et al. [30] and by Simpson et al. [31], on angular distributions of high energy electrons (several hundreds of eV) in a relatively small angular range close to the forward direction. The main purpose of these measurements was the determination of the generalized oscillator strength in the ionization continuum and the comparison with high energy approximations. The investigation of single ionization through multiple excitation of the target atom and of multiple ionization has made good progress lately, but the discussion of these phenomena would be outside the scope of this article.

Double differential cross sections for the electron impact ionization of helium and argon have been measured by the present authors, since with present experimental techniques it can be hoped to obtain more precise data. Special care has been taken to avoid magnetic and electric stray fields as well as electric fields due to space charge. If high currents are used, in the collision centre space charge potentials of several volts may very easily arise which drastically change the energy and angular distributions of slow ionization electrons. In our experiment, counting techniques are applied and therefore all currents are low enough to reduce space charge potentials down to a few mV. For the present experiments a range of collision energy from 25 eV to 260 eV and an angular range from 3° to 125° was chosen. The errors are approximately 10 %, if no error bars are shown in the figures.

Figure 3-6-1 shows examples of energy spectra of the ionization electrons from helium and argon for the collision energy of 200 eV. The curves

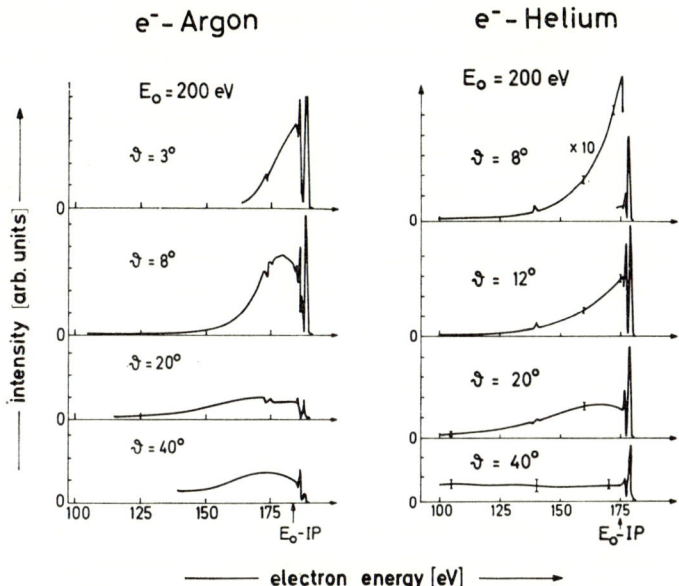

Fig. 3-6-1. Examples of energy dependences of scattered electrons after collisions with argon and helium for a primary energy $E_0 = 200$ eV. The onset of the ionization continuum (at the energy $E_0 - IP$) is marked by arrows. These spectra are measured at angles which are indicated in the figure by ϑ. The direct ionization continuum is affected by double excitation and negative ion resonances.

are reproductions from x-y-recordings. The peaks for energies above 175.5 eV for helium and above 184.2 eV for argon result from excitation processes. It can be seen how the ionization continuum connects to the excitation region. The little humps in the ionization continuum (below the ionization onset) are due to short lived negative ion states (resonances) or to double excitation and subsequent autoionization [30, 31]. For argon the same general behaviour is found as in the case of helium. Also in the case of argon, the curves are very much affected by double excitation and resonances. In addition, considerable contributions from double ionization [14, 41, 32] and from ionization of electrons from deeper shells [33] may be present. The intensity from the excitation processes drops more rapidly with scattering angle than the intensity in the ionization continuum.

The symmetry of the energy spectrum $d\sigma/dE$ with respect to the energy value $\frac{1}{2}(E_a + E_b)$ (see Fig. 3-3-1) is no longer observed for the double differential cross sections in Fig. 3-6-1. Instead, the group of the fast electrons (scattered electrons) shows a very rapid decrease of intensity with increasing scattering angle. For the two rare gases measured, the energy dependences for $\vartheta = 40°$ get very flat and the intensity is very low. Because of the small intensity, no measurements above $\vartheta = 40°$ were made. Energy loss spectra for helium and incident electrons with 100 eV impact energy have been compared [5] with calculations using the approximation, given by the authors earlier [4]. In this approximation the two outgoing electrons are described by plane waves and for the helium atom before the collision Hartree-Fock solutions are used. The theory overestimates the intensity of the scattered electrons for large scattering angles and underestimates the scattered intensity for angles lower than about 15°. But the gross features are reproduced by the theory, namely the rapid decrease of intensity with increasing angle and decreasing energy of the scattered electrons and the nearly constant intensity of the ejected electrons with respect to the ejection angle. Calculations by Peach [49] using the normal Born-approximation with orthogonal wave functions in the initial and final state give better agreement with experiment for fast electrons and for small scattering angles. For large angles ($\vartheta > 60°$) the intensities are much too low, especially for smaller energies of the ejected electrons.

An interesting behaviour of the scattered electrons is shown in Fig. 3-6-2. Plotted are energy spectra of the scattered electrons on the same energy scale, with $X = E_0 - IP$ but for different collision energies E_0 and for the scattering angle $\vartheta = 12°$. Each curve is normalized to 100 at its maximum. The figure shows that a large energy transfer from the colliding electron to the ejected electron gets more probable if the scattering angle and/or the

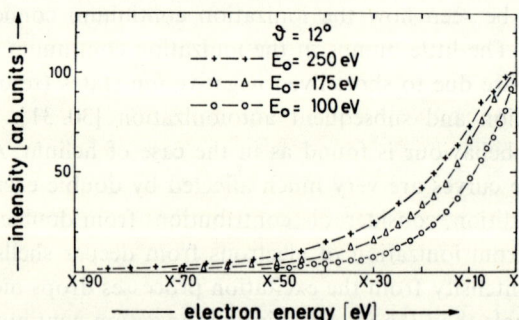

Fig. 3-6-2. Energy loss spectra (of the first kind) for helium taken at the scattering angle $\vartheta = 12°$ for different collision energies E_0. All curves are on the same energy scale. X is the excess energy ($X = E_0 - IP$, $IP = 24.45$ eV is the ionization potential of helium). The curves are normalized to 100 at their maxima. The diagram shows that the half widths of the curves increase with increasing collision energy E_0.

collision energy increase. This is to be expected from a type of collision, for which the binary interaction of the incoming electron with one of the atomic electrons dominates. The important role of binary collisions for ionization of atoms by electron impact has been shown by Ehrhardt et al. [6, 7] and by Amaldi et al. [1] by means of coincidence experiments and by the quite successful application of the binary encounter theory to the results of these experiments by Vriens [2].

Figure 3-6-3 shows a set of angular dependences for ejected electrons of different energies E. The collision energy E_0 is 256.5 eV. Measurements on helium for different collision energies (200, 125, 100, 80, 50, 40, 30 eV) have quite similar results. The measurements for $E_0 = 40$ eV are shown in Fig. 3-6-4. In all the measurements of angular dependences the following observations are made:

1. Ejected electrons of energy $E \leq 3$ eV may have a slight and broad minimum in the angular range from approximately 30° to 90°.

2. Ejected electrons in the energy range from about 3 to 6 eV are nearly flat in the whole angular range measured.

3. Ejected electrons with energy $E \geq 10$ eV have a maximum at about 60°. For measurements with $E_0 \leq 80$ eV this maximum vanishes more and more if E_0 decreases.

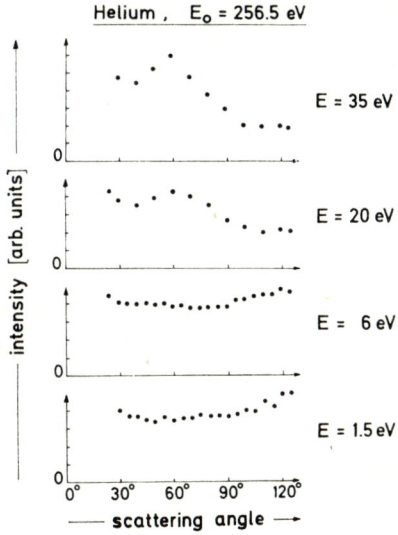

Fig. 3-6-3. Angular dependences for the slow electrons after ionizing collisions of 256.5 eV electrons with helium. taken for several energies E, indicated in the figures.

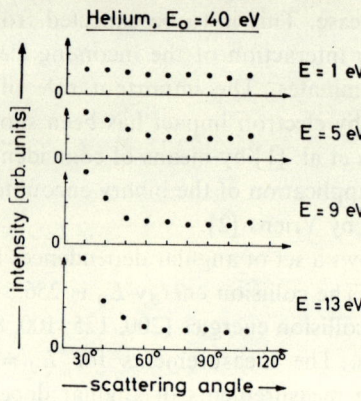

Fig. 3-6-4. Angular dependences of the electrons after ionizing collisions of 40 eV electrons with helium. The energies of the detected electrons are indicated in the diagrams by E.

The maximum is a consequence of the binary character of the collision (see § 3-6B). Since most of the scattered electrons are scattered into the forward direction the maximum shows that the angle between the scattered and the ejected electron for most collision processes is between 50° and 90°. This is roughly in agreement with the expectations from a two-body collision and the predictions from the binary encounter theory. The theory is best for relatively high energies of the ejected electrons. This statement seems to be confirmed by the observation of the intensity maximum only for electrons with $E \geq 10$ eV. The constant intensity distribution with respect to the angle of the ejected electrons with $E \leq 10$ eV is compatible with the results of the coincidence experiments of this group. These results are:
i) If the energy of the ejected electron is low (a few eV), then the intensities in the "recoil" and in the "binary encounter" peak are nearly the same;
ii) If the scattering angle of the faster electron is varied by only a small amount then the position of the "binary" maximum increases very much. From these findings we can expect an angular distribution which is symmetrical about 90°: Most of the electrons detected under smaller angles come from collisions of a binary type whereas those detected under larger angles (above 90°) have suffered a collision of the recoil type.

B. THE TRIPLE DIFFERENTIAL CROSS SECTION

The triple differential cross section $d^3\sigma/dE d\Omega_a d\Omega_b$ gives the most detailed informations about the ionization process by electron impact. Since it is a function of several parameters one can examine quite different dependences and therefore make very detailed comparisons with theory. In our experiment we have fired electrons of a fixed energy E_0 onto helium atoms. After the ionizing collision in the reaction centre, two electrons leave the collision region. The scattered electron with the energy E_a is detected by the collector 1 at the angle ϑ_a, the value of which is maintained during the experiment. The detector 2 rotates around the centre and detects only those electrons which processes the energy E_b.

The process of ionization without excitation requires:

$$E_0 = E_a + E_b + IP$$

where IP is the ionization potential of the target gas. The energies E_a and E_b are chosen so that only transitions from the ground state of the atom into the ground state of the residual ion can be observed. The two outgoing electrons are detected in coincidence. The dependence of the coincidence rate with respect to the scattering angle of the ejected slow electron ϑ_b has been investigated.

For primary energies E_0, where E_0 is large with respect to the ionization potential (e.g. $E_0 = 250$ eV), one must choose small angles ϑ_a and energies E_a in a region close to $E_0 - IP$, because otherwise the count rate would be too small. For such a choice of parameters most of the fast electrons after the collision are scattered into a small cone around the forward direction (ϑ up to about 10°). Within this angular range the intensity decreases very rapidly by several orders of magnitude as ϑ increases. In addition, the largest probability exists for collisions in which one electron leaves the collision centre with energy close to $E_0 - IP$ and the other electron is ejected with the complementary energy (according to the energy conservation law) in the range from zero to about 15 eV with its maximum in the region of a few eV. It is very improbable that both outgoing electrons have the same energy $\frac{1}{2}(E_a + E_b) = \frac{1}{2}(E_0 - IP)$ (see Fig. 3-3-1). Such a limitation in the choice of collision parameters is not given for low collision energy, i.e. $E_0 \lesssim 40$ eV.

The experimental results of the number of the coincidence events versus the angle ϑ_b of the slow electrons are given in polar diagrams. All collisional parameters which have been kept constant in the individual experiment, namely E_0, E_a, E_b and ϑ_a are marked in the figures. The arrows from the bottom of the figures to the centre indicate the incoming electrons of energy E_0, the arrows from the centre into the direction ϑ_a indicate the direction in which the fast electrons are detected after the collision. The slow electrons are detected in the direction ϑ_b. The distances of the points from the centre of the diagram represent the probability of finding the slow electron at the angle ϑ_b if the corresponding fast electron has been detected at the angle ϑ_a. Both collectors move in a plane, that means, that the azimuth φ_b (see Fig. 3-2-1) is 0° or 180°. In the diagrams ϑ_b has positive sign if φ_b is zero, and it has negative sign if φ_b is 180°. The angle ϑ_a has always been assigned a positive value. If ϑ_b has a positive value, it means that both electrons are detected on the same side of the primary beam.

Most figures of angular correlation spectra show the points actually obtained and therefore the statistics, whereas the full and dotted curves in the Fig. 3-6-5 represent, for the sake of clearness, the averaged experimental

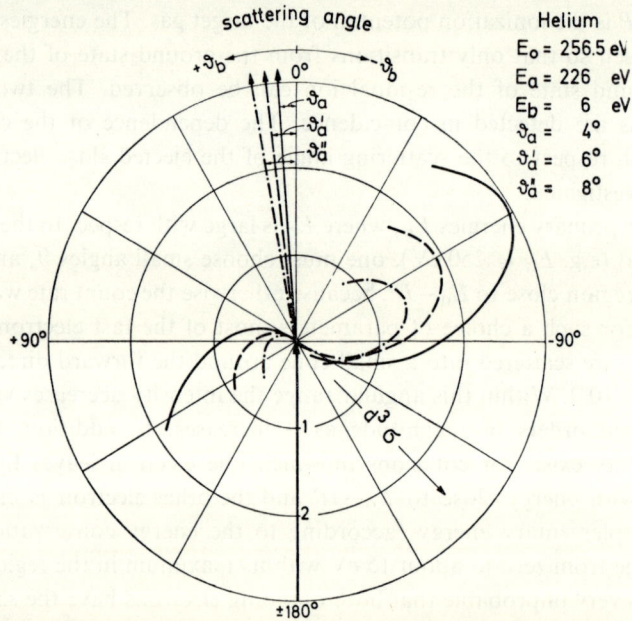

Fig. 3-6-5. Triple differential cross section for electron impact ionization of helium as a function of the scattering angle ϑ_b of the slow electrons. This diagram is intended to show the trend of the position and the height of the binary peak, if the angle ϑ_a is varied.

results. The fluctuation of the actual points in these measurements is similar to that of the others. All intensities are plotted in arbitrary units. If several curves are shown in one figure then they are normalized to each other, i.e. they are on the same intensity scale (excepting Figs. 3-7-1, 3-7-2, 3-7-3, where the measurements are fitted to calculations which will be discussed below, § 3-7).

The most obvious experimental result is that a strong angular correlation exists between the two ionization electrons (see for example Fig. 3-7-1), i.e. the ejected electron leaves the collision complex preferentially in a direction which is dependent on the angle at which the scattered electron is detected. One can distinguish two separate angular regions where the slow electrons are found, namely in forward direction, often around $\vartheta_b = -60°$, and in backward direction, approximately $\vartheta_b = +140°$.

By comparison with theory it was shown [2], that in those collisions where the slow electrons are ejected in forward direction ($\vartheta_b \approx -60°$), only little momentum is transferred to the remaining ions. To a first approximation only two particles, the scattered and the ejected electron, take part in the collision. Therefore the forward peak is called the binary encounter peak. The two outgoing electrons are found on opposite sides of the primary beam. In collisions which give rise to backscattering of the slow electrons a large amount of momentum is transferred to the ions. The ejected electron has been recoiled in the rising area of the Coulomb potential of the ion. Therefore this peak is called recoil peak. For this type of collision both outgoing electrons are mostly detected in the same half plane (in the polar diagrams the left-hand side half plane). If the explanation for the recoil peak is right, the ions produced must be pushed into a direction opposite to the direction of the ejected electrons. Therefore the angular dependence of the intensity distribution of the ions should be anisotropic. Very recently, McConkey et al. [34] have measured the ionic angular distribution for different impact energies below 30 eV for helium and argon. For each value of the collision energy E_0 the authors found a broad maximum in the expected direction. The angular position of the peak decreases with increasing E_0. These results justify the explanation given for the recoil peak.

The binary encounter peak is characterized by its symmetrical shape around the axis of maximum intensity. This angle may be called $\vartheta_{b\,max}^{bin}$. The angular position of the binary encounter peak depends strongly on the choice of the scattering angle ϑ_a of the fast electron and on the choice of the ratio E_a/E_b. The value of $\vartheta_{b\,max}^{bin}$ increases with increasing ϑ_a and decreasing energy of the slow electron E_b. Figure 3-6-5 gives experimental results for the angular correlation distributions. These curves show the trend of $\vartheta_{b\,max}^{bin}$ for different values of ϑ_a (4°, 6°, 8°). All other parameters (E_0, E_a, E_b) are identical for the three curves. The same figure also shows that the intensity of the binary and the recoil peak decreases with increasing angle ϑ_a.

Collisions with only little momentum transfer to the nucleus, but large momentum transfer to one of the atomic electrons, show several properties which are typical for a classical binary encounter between two hard spheres of the same mass. In such a classical collision the angle between the two particles after the encounter is 90° whereas in the electron impact ionization $|\vartheta_a| + |\vartheta_{b\,max}^{bin}|$ is smaller than 90°, mostly around 70° to 80°. The qualitative interpretation for deviation from 90° is that the binding energy of the atomic electron has to be taken into account. Also, the experimental results of this review, namely that $\vartheta_{b\,max}^{bin}$ increases as ϑ_a increases and as E_b decreases, can be explained by a simple momentum diagram. Of course, this statement does not imply that the classical binary encounter model is able to describe the binary encounter peak in all details at low electron impact energies.

In nearly all measurements the binary encounter peak seems to have symmetric shape around its maximum $\vartheta_{b\,max}^{bin}$. A few measurements make an exception. In such cases, the binary encounter peak is disturbed at its large angle side by an intensity increase which seems to belong to the recoil peak, so that in this angular range only a minimum of the intensity is reached instead of an intensity zero. The symmetric intensity distribution of the binary encounter peak enables one to measure its full angular width at half height of the maximum intensity. It is interesting to note that this angular half width is (within experimental accuracy) not dependent on the energy of the colliding electron in the range from $E_0 = 30$ eV up to 260 eV, but there is a definite dependence on the energy of the ejected electron E_b. If E_b is small, e.g. 2.5 eV, the average half width is about 55°, whereas for relatively high E_b, e.g. 15.5 eV, the half width has decreased to about 47°. It has been shown theoretically by Blassgold et al. [3] and by Vriens [2] for very high impact energies that the shape of the binary encounter peak depends strongly on the momentum distribution of the atomic electrons before their ejection into the continuum. The present experiments show that this statement seems to be still valid for low impact energies.

Another group of the slow electrons is preferentially ejected into the backward direction (recoil peak), and from the experimental results it seems to be clear that the angular range of ejection is between $\vartheta_b = +90°$ and about $\vartheta_b = -120°$ (through 180°) for positive ϑ_a. The experiment does not allow continuous scanning through this angular range, but by comparison with theory, it has been shown that in this angular range there is only one peak if helium is chosen as target gas. Results with argon indicate more than one peak in the forward direction. This has been predicted for high impact energies by Vriens [2]. Since the angular range of observation is limited in the experiment to angles smaller than 125° in both half planes, for helium only a partial investigation of the recoil peak is possible. In fact, we could not find the intensity maximum of the recoil peak, and we also do not know whether its intensity distribution is symmetric with respect to $\vartheta_{b\,max}^{recoil}$. In spite of this a few interesting statements can be made concerning the recoil peak. In several measurements (see Figs. 3-6-6, 3-6-8 and 3-8-4), the onset of the recoil peak in the half plane of the negative ϑ_b (right-hand sides of the figures) in the range from $\vartheta_b = -100°$ to $\vartheta_b = -125°$ is clearly visible. If one assumes that the recoil peak also has symmetric shape (as can be seen from some theoretical calculations) then the measurements indicate that the angular half width of the recoil peak is considerably larger than the half width of the binary peak (possibly 90° or more). The recoil peak often has larger intensity than predicted by different theoretical approximations (see

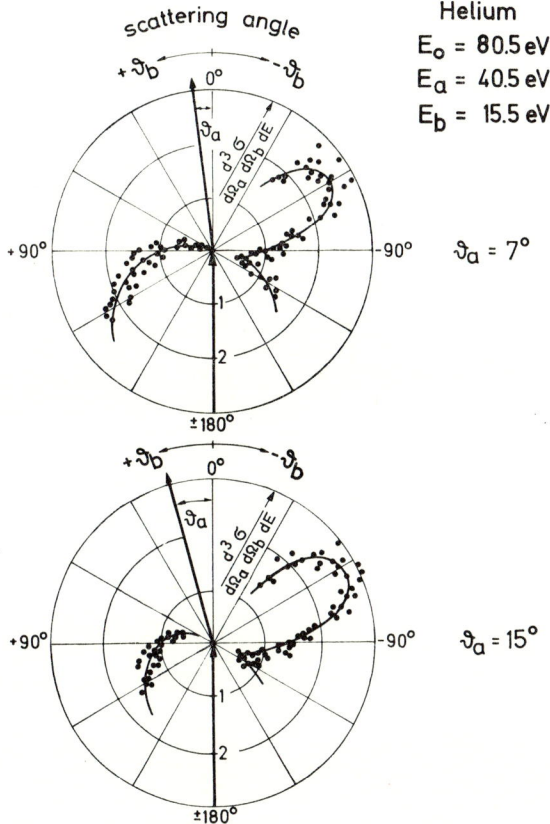

Fig. 3-6-6. The triple differential cross section as a function of the scattering angle ϑ_b of the slow ionization electrons for two different detection angles of the fast electron. All parameters which have been kept constant during the experiment are indicated in the figure. The full line is the average over the actually obtained measurements (arbitrary units).

Figs. 3-7-1 and 3-7-3). With the same assumption (symmetric shape) one concludes that for low impact energies the possible position of the intensity maximum of the recoil peak may be close to $\vartheta_b = +160°$. This seems to be important since it is equivalent to the statement that the symmetry axis of the binary encounter peak does not coincide with the symmetry axis of the recoil peak. In other words, the two maxima are not in opposite directions.

In some measurements we have observed (i.e. for certain sets of kinematic parameters E_0, E_a, ϑ_a), that the total intensity which contributes to the recoil peak is larger than that of the binary encounter peak (see Figs. 3-6-7 and 3-6-8) whereas for other parameters the opposite behaviour is

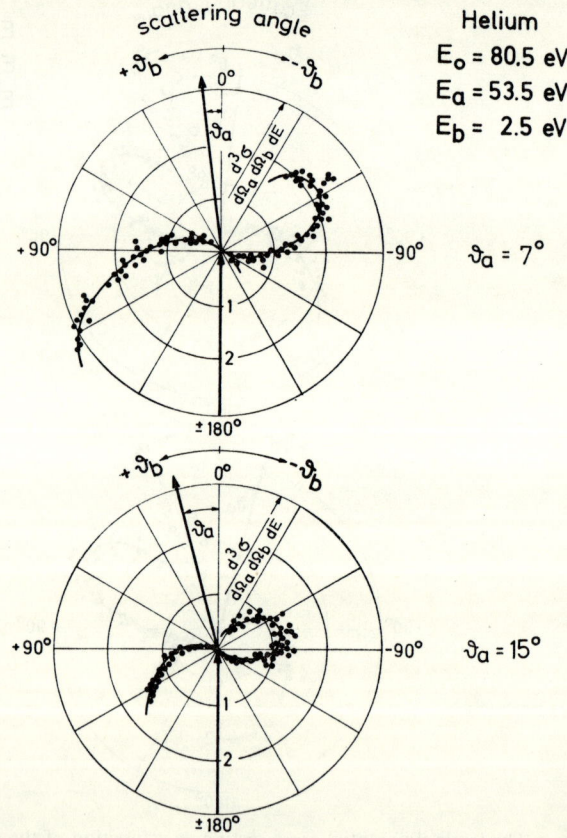

Fig. 3-6-7. Triple differential cross section for helium as a function of the scattering angle ϑ_b of the slow ionization electrons, if the corresponding fast electrons are detected at the angle $\vartheta_a = 7°$ (upper figure) and $\vartheta_a = 15°$ (lower figure). The values are given in arbitrary units. The full curves are averages of the actually obtained points (arbitrary units).

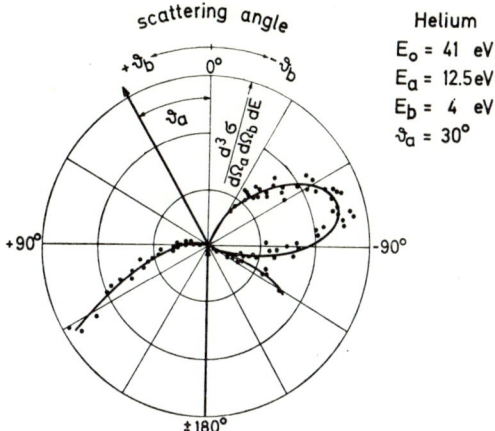

Fig. 3-6-8. Triple differential cross section for collisions of 41 eV electrons with helium. The parameters which have been kept constant during the experiment are indicated in the diagram. The points give the actually obtained values (in arbitrary units); the full line is an averaged curve. On the right-hand side the onset of the recoil peak can be seen.

obtained (see for example the Fig. 3-6-6 for $E_0 = 80.5$ eV and $\vartheta_a = 15°$). For $E_0 = 256.5$ eV, in none of our measurements does the intensity in the recoil peak exceed the intensity in the binary peak.

A qualitative model for the explanation of the origin of the recoil peak has first been given by Vriens [2]. Before the collision the atomic electron is bound in the potential well of the nucleus. During the collision the atomic electron gains energy and makes an upward transition into the continuum and starts to travel into the direction of the binary peak. Some electrons proceed this path and therefore contribute to the binary peak. Others are reflected in the region of varying potential caused by the attracting field of the ion and therefore contribute to the recoil peak, and at the same time they transfer momentum to the remaining ion. This model implies, i) that the axis of symmetry of the recoil and the binary peak are the same, ii) that the angular half width of the recoil peak is larger than the half width of the binary peak and iii) the intensity of the recoil peak is lower than or equal to the intensity of the binary peak, but never higher. Obviously, the first and last implications are not in agreement with our experiments. However if the model is slightly improved by the effect on the potential due to the electron which is scattered into the angle ϑ_a, then this altered model is able to explain qualitatively the experimental results. The presence of the electron "a" increases the potential in the direction of ϑ_a and therefore the axis of symmetry is bent, i.e. both electrons, those which are ejected in the forward direction (binary peak) and those which are ejected in the backward direction (recoil peak) are repelled to larger angles. The influence of the electron "a" on the potential also explains the increasing intensity of the recoil peak for decreasing energy of the electron "b".

Mostly we have studied the cross section $d^3\sigma(E_0, E_a, E_b, \vartheta_a, \vartheta_b, \varphi_b)/dE_i d\Omega_a d\Omega_b$ as a function of the angle ϑ_b of the slow (ejected) electrons. It is also possible to investigate other dependences, for example, the variation of $d^3\sigma$ with the angle ϑ_a of the fast electrons. If the *slow* electron is detected at a *fixed angle* ϑ_b, the corresponding fast electrons mostly are found in forward direction: but the maximum of this coincidence peak seems not to be located directly at the angle $\vartheta_a = 0°$. For some collision parameters we were able to detect the maximum, but it was impossible to measure its half width because it is not possible to scan the collector through the region of small angles below 3°. In this region the primary beam would enter the collector. But one can say that the shape, especially the slope of the peak, is strongly dependent on the scattering parameters. Amaldi et al. [1] have investigated also in a coincidence experiment another dependence of the cross section $d^3\sigma$. This group used electrons in the keV region for the ionization of atoms in very thin foils. The two outcoming electrons are detected in coincidence. The detection energies and the angular positions of the collectors are kept constant during one experiment at $E_a = E_b = 7.3$ keV and $\vartheta_a = \vartheta_b = 45°$. The primary energy E_0 is varied. The maximum of the coincidence distribution gives the binding energy of the electron removed from the atom, whereas the distribution itself is directly correlated to the square of the momentum distribution of the atomic electron before the collision.

§ 3-7. Comparison with theory

In this section some experimental results will be compared with results obtained by simple quantum mechanical calculations. For energies of the incident electrons which are about 10 times larger than the ionization potential, the agreement between theory and measurements of the *total* cross section is relatively good [35]. It will be seen from this section, that this is not so for differential cross sections. Differential cross sections represent more critical test material.

For the scattering of an electron by a helium atom, the Hamiltonian for the problem is

$$H = T_1 + (T_2 + T_3 + V_{23} + V_{24} + V_{34}) + V_{12} + V_{13} + V_{14} \tag{3-7-1}$$

where "1" refers to the incident electron, "2" and "3" refer to the two electrons of the helium atom, and "4" stands for the helium nucleus. The T_i ($i = 1, 2, 3$) are the kinetic energies of the electrons, and $V_{i,k}$ ($i, k = 1, 2, 3$, $i \neq k$) are the interaction potentials between the charged particles. The nucleus is at rest ($T_4 = 0$).

Neglecting the amplitude for electron capture and simultaneous ejection of the two atomic electrons, the expression for the triple differential cross section has the form [35]

$$d^3\sigma = 2 \times (2\pi)^{-5} \frac{k_a \cdot k_b}{k_0} \{\tfrac{1}{4}|f+g|^2 + \tfrac{3}{4}|f-g|^2\}$$

where $k_j = (2mE_j/\hbar^2)^{\frac{1}{2}}$ with $j = 0$, a, b. The factor 2 indicates that helium has two equivalent electrons. Choosing the target nucleus as the origin of the coordinate system, in the Born approximation the direct scattering amplitude f and the exchange amplitude g are given by

$$f = \int d\tau_{123} \exp(-i\mathbf{k}_a \cdot \mathbf{r}_1) \cdot \varphi_f^*(k_b, \mathbf{r}_2) \cdot \varphi_{He^+}^*(\mathbf{r}_3) \cdot W$$
$$\cdot \exp(-i\mathbf{k}_0 \cdot \mathbf{r}_1) \cdot \varphi_{He}(\mathbf{r}_2, \mathbf{r}_3)$$
$$g = \int d\tau_{123} \varphi_f^*(k_b, \mathbf{r}_1) \cdot \exp(-i\mathbf{k}_a \cdot \mathbf{r}_2) \cdot \varphi_{He^+}^*(\mathbf{r}_3) \cdot W$$
$$\cdot \exp(i\mathbf{k}_0 \cdot \mathbf{r}_1) \cdot \varphi_{He}(\mathbf{r}_1, \mathbf{r}_2)$$

(3-7-2)

where W is the perturbation potential, $\varphi_{He}(\mathbf{r}_2, \mathbf{r}_3)$ is the wave function for He in the ground state, and φ_{He^+} is the hydrogenic wave function for the He^+ ion, i.e.

$$\varphi_{He^+}(\mathbf{r}_3) = 2 \times (2/\pi)^{\frac{1}{2}} \cdot e^{-2r_3}.$$

Different Born approximations use different expressions for W and φ_f, where φ_f is the second continuum wave function.

The simplest theory is the plane wave Born approximation (see for example Glassgold and Ialango [3]). In this calculation φ_f is represented by $\varphi_f = \exp(i\mathbf{k}_b \cdot \mathbf{r})$, i.e. both outgoing electrons are described by plane waves. In the potential W, the interaction of the incident electron with both atomic electrons is taken into account, whereas the interaction with the target nucleus is ignored, i.e.

$$W = W_1 = 1/r_{12} + 1/r_{13}. \tag{3-7-4}$$

This assumption is made in analogy to the Born approximation for excitation to bound states below the ionization threshold. Due to the orthogonality of the wave functions for the case of excitation, the interaction between the incident electron and the target nucleus does not contribute to the cross section. But this is not the case for ionizing collisions. In order to calculate the cross section (3-7-2) a Hartree-Fock expression is used as the wave function for the helium atom:

$$\varphi_{He}(r_2, r_3) = \eta(r_2) \cdot \eta(r_3), \qquad (3\text{-}7\text{-}5)$$

where

$$\eta(r) = N_0(e^{-\gamma r} + c \cdot e^{-2\gamma r})$$

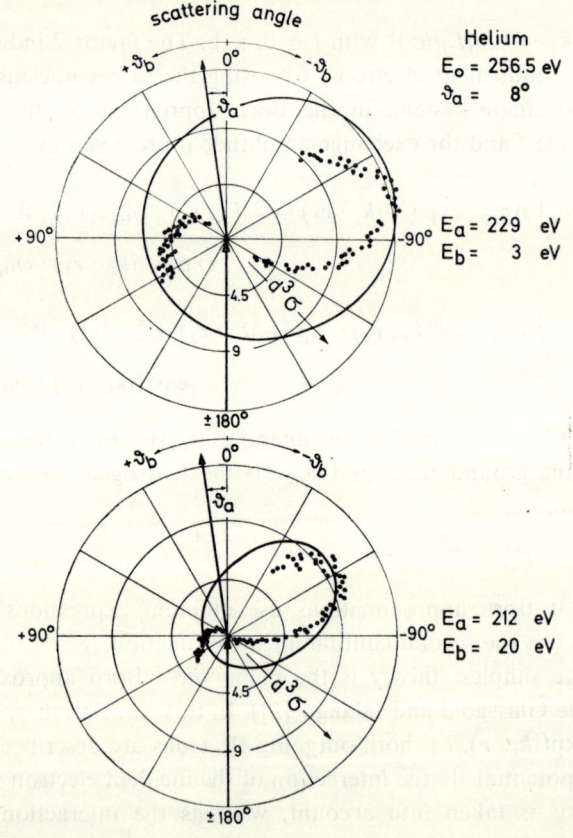

Fig. 3-7-1. Triple differential cross section for electron impact ionization of helium as a function of the detection angle ϑ_b of the slow electrons. The dots are the actually obtained values, whereas the full lines represent the result of calculations in the plane-wave-Born approximation (see § 3-7). The measured curves are fitted to the theory at the binary peak. The values are given in atomic units (a_0^2/Ry).

with $N_0 = 0.837$, $\gamma = 1.456$ and $c = 0.6$ [35]. In Fig. 3-7-1 theoretical curves (full lines) and experimental results (dots) are lotted. They are normalized to each other at the intensity maximum in the forward direction. Atomic units are chosen. For both graphs the primary energy is $E_0 = 256.5$ eV, and the fast electron is detected at $\vartheta_a = 8°$. The two plots differ from each other by the detection energies of the two outgoing electrons. Considerable discrepancies between theory and experiment are found. The theoretical curves show a minimum of $d^3\sigma$ in the backward direction, whereas the experimental data indicate a second maximum (recoil peak) of the coincidence cross section. The two zeros between the binary peak and the backscattering peak in the measurements cannot be reproduced by this theory.

For large values of E_0 and E_b the relative intensity in backward direction decreases (see above), so that the agreement between theory and experiment becomes considerably better. For such energy values, close agreement between the plane wave Born approximation and the binary encounter theory, developed by Vriens, is found, apart from some slight differences [2]. The binary encounter theory will not be applied to the present measurements because it has already been discussed by Vriens.

In the plane–plane approximation [4, 37, 38] the interaction between the incident electron and the target nucleus is taken into account, i.e. the perturbation (3-7-4) is replaced by

$$W = W_2 = -2/r_1 + 1/r_{12} + 1/r_{13}, \qquad (3\text{-}7\text{-}6)$$

whereas the wave functions are the same as used in the plane wave Born approximation. The negative sign of the nucleus interaction causes the scattering amplitude to change its sign with respect to the scattering angle. This implies zeros in $d^3\sigma$ for certain directions.

In Fig. 3-7-2 results of the plane–plane approximation are plotted together with experimental curves for two sets of collision parameters. In both cases the detection angle for the scattered electron is $\vartheta_a = 4°$. The energies of the scattered and the ejected electrons are $E_a = 229$ eV and $E_b = 3$ eV for the upper part of the figure and $E_a = 212$ eV and $E_b = 20$ eV for the lower part. Again atomic units are used.

In the lower picture a very good agreement between theory and experiment is found. The measured maxima of $d^3\sigma$ as well as the minima are reproduced by the calculation. The magnitude of the binary peak and the recoil peak are well reproduced. Such good agreement between experiment and the plane–plane approximation is always found for results for which the energy E_b of the ejected electron is about 20 eV or larger. Since the scattering

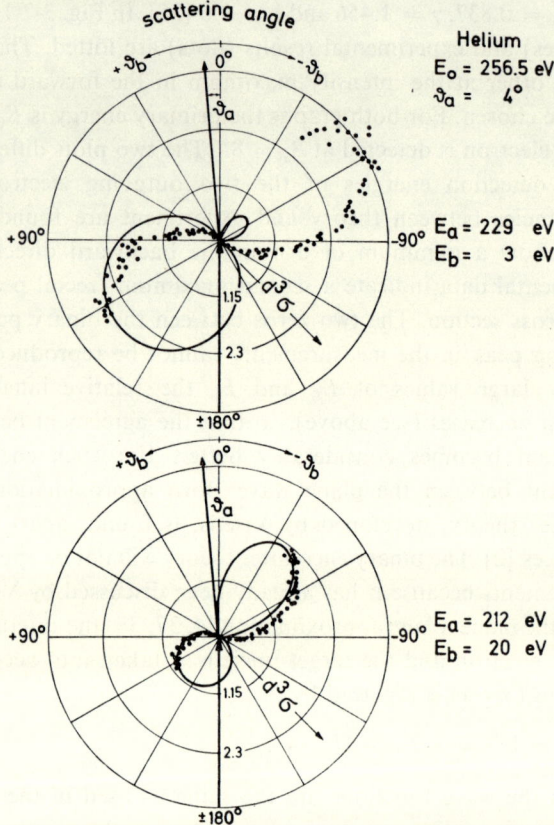

Fig. 3-7-2. Triple differential cross section for electron impact ionization of helium as a function of the detection angle of the slow electron. The full lines are results of calculations using the plane-plane approximation (see § 3-7). The dots give the measured values, fitted on the theoretical curves at $\vartheta_b = +120°$. Intensities are given in atomic units (a_0^2/Ry).

intensity in the backward direction in this approximation is caused by the interaction of the incident electron with the target nucleus, this distribution is called the recoil peak. At the scattering angle of the intensity maximum of the recoil peak, the momentum transfer of the ejected electron to the remaining ion has a maximum. For smaller values of E_b the agreement between the plane–plane approximation and experiment is not good. Such an example is given in the upper part of Fig. 3-7-2 where $E_b = 3$ eV. Theoretical and experimental data are normalized to each other at large positive angles ϑ_b. In this case, the theory overestimates the intensity of the recoil peak. The discrepancy is not surprising, since a plane wave cannot be a good description for an electron moving slowly in a Coulomb field.

This lack of the theory is removed in the (normal) Born approximation. Here the slow electron is represented by a Coulomb function

$$\varphi_f(\mathbf{k_b}, \mathbf{r}) = e^{\frac{1}{2}\pi n} \Gamma(1+in) \cdot \exp(i\mathbf{k_b} \cdot \mathbf{r}) \cdot {}_1F_1(-in, 1; -i(k_b r + \mathbf{k_b} \cdot \mathbf{r})) \quad (3\text{-}7\text{-}7)$$

where $n = z/k_b$, and z is the effective charge acting on the electron. Using the expressions for the direct scattering amplitude f, given by Landau and Lifshitz, the coincidence cross section was calculated in this approximation. In Fig. 3-7-3 results are presented for two sets of collision parameters. Since

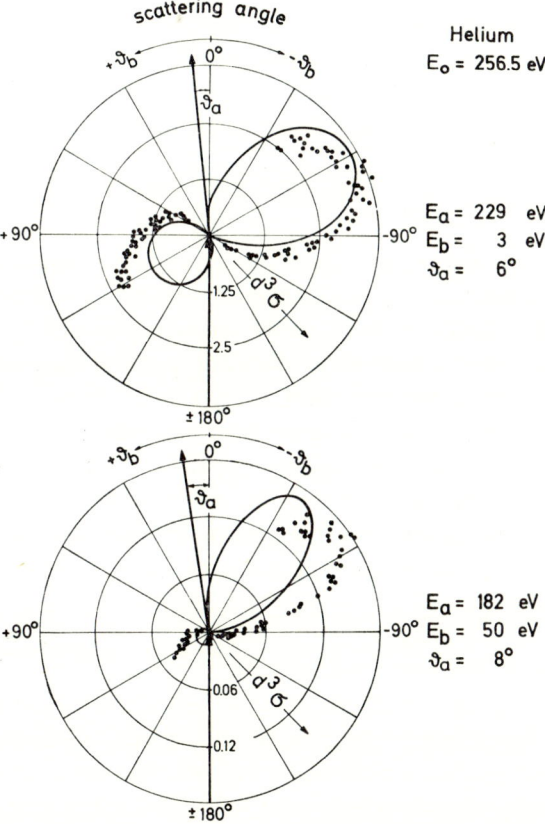

Fig. 3-7-3. Triple differential cross section for electron impact ionization of helium as a function of the detection angle of the slow electrons. The full lines represent results of the Born-approximation using a plane wave and a Coulomb wave. The dots represent the measurements. They are fitted to the theory at the binary peak. Atomic units (a_0^2/Ry) are used.

the scattered electron is faster than the ejected electron, it is assumed that the slow electron is moving in the field of the ion. For this reason $z = 1.7$ was chosen for the calculations. The interaction of the incident electron with the target nucleus was neglected. The agreement between theory and experiment is better for small energies E_b (upper graph) than in the plane–plane approximation but for $E_b = 50$ eV (lower diagram) the theory shows a recoil peak which is much too small. Moreover there is a considerable discrepancy between theory and experiment with regard to the angular position ϑ_b of the binary peak.

Taking into account the interaction of the incident electron with the target nucleus, the shape of the theoretical curve is not very much affected. Neither the angular positions of the maxima and minima of $d^3\sigma$ nor the ratio of the intensities of the two maxima are in better agreement with the experimental results. Similarly, for a few curves, the inclusion of the exchange amplitude has been tested by use of the expression given by Peterkop [38]

$$g(k_a, k_b) = f(k_b, k_a).$$

For the collision parameters of our experimental results the exchange effect on $d^3\sigma$ is about 10 % for the primary energy $E_0 = 256.5$ eV and increases up to 30 % for $E_0 = 30$ eV; in the symmetrical case $(E_a = E_b) f$ and g have the same order of magnitude. Calculating $d^3\sigma$ in the Geltman [39] or in the Vainshtein approximation [40] a very similar shape for the angular dependence has been obtained as in the Born approximation. Better agreement between theory and experiment can possibly be expected by taking into account the correlation between the two outgoing electrons caused by their Coulomb interaction, i.e. in the continuum wave functions a correlation term should be introduced. Such calculations are in progress.

§ 3-8. Measurements close to the ionization threshold

During the last 20 years many experiments have been made on the energy dependence of the total ionization cross section for electron impact close to the ionization threshold [43–48]. The latest measurements indicate a non-linear threshold behaviour. Of course, these measurements are very difficult and the effective deviation from non-linearity is small and therefore the errors relatively large. It seems possible that experimental results of double and triple differential cross sections in connection with theory give valuable contributions for the solution of the ionization threshold problem.

In this paragraph are represented a few data of this kind, which have been obtained by the present authors.

From the measurement of double differential electron impact ionization cross sections it is known that for sufficiently high values of the primary energy E_0 two groups of electrons are found after the collision: i) the fast electrons, which mainly travel into the forward direction and ii) the slow electrons, which have a nearly spherical angular distribution. It is possible (of course within limits) to call the fast electrons "scattered" and the slow electrons "ejected", in other words the particle exchange and the momentum exchange is small at sufficiently high collision energies. This changes significantly if E_0 is lowered below 50 eV. For impact energies around 30 eV and lower the two groups are no longer distinguishable, and all energies in the range from zero to $E_0 - IP$ are present with nearly equal probability.

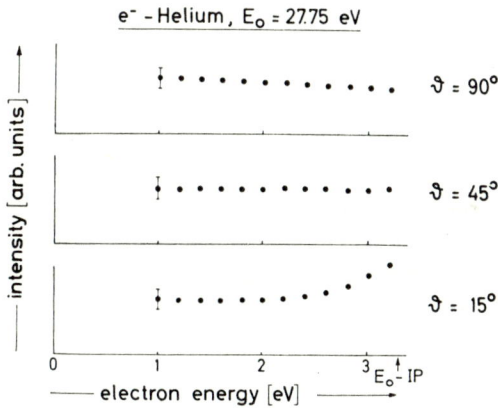

Fig. 3-8-1. Examples of energy loss spectra (of the first kind) from helium for ionizing collisions of electrons with energy E_0 near the ionization threshold for three scattering angles ϑ. The energy of detection ranges between 1 eV and $E_0 - IP$. The intensity distributions taken at large angles ($\vartheta = 45°$ and $90°$) indicate no or only a slight variation with the electron energy.

In Fig. 3-8-1 an example is given of energy loss spectra of the first for $E_0 = 27.75$ eV in the energy range of 1 eV to $E_0 - IP = 3.3$ eV. The intensity distributions measured at large angles ($\vartheta = 45°$ and $90°$) indicate no or only a slight variation with the electron energy. At the scattering angle $\vartheta = 15°$ there are about 1.6 times as many fast electrons as those of medium or low energy. One can assume that the cross section $d\sigma/dE$ (integrated over all angles) varies only very smoothly with respect to the energy of the detected electrons.

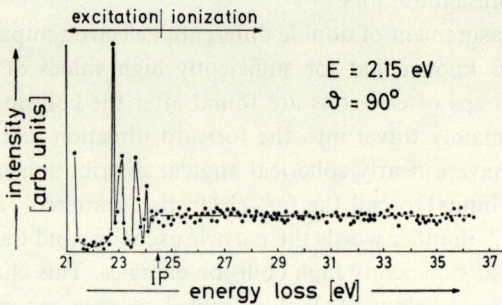

Fig. 3-8-2. The figure shows an energy loss spectrum of the second kind for helium, i.e. the energy E of the detected electrons and the angle ϑ are kept constant whereas the energy of the incoming electrons E_0 (primary energy) is varied. The cross section is plotted versus the energy loss $\Delta E = E_0 - E$ of the electrons. The onset of the ionization is marked by an arrow. In the ionization range near the threshold this type of cross section is constant within the accuracy of the measurement.

Fig. 3-8-2 shows an energy loss spectrum of the second kind, i.e. the detection energy E and the angle are kept constant whereas E_0 varies. The cross section in the ionization range is constant within the errors of the measurement and similar results have been obtained for other angles ϑ and energies E. This is also the same statement as has been found by [41, 42] who used the SF_6-scavenger method and the trapped electron method.

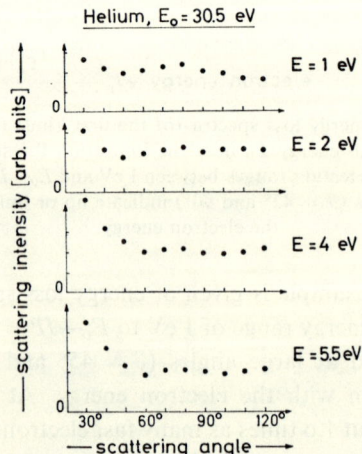

Fig. 3-8-3. Angular distributions of ionization electrons after collisions with helium. The impact energy amounts to 30.5 eV (6 eV above the ionization threshold). The energy of the detected electrons is indicated in the diagrams by E.

Also the angular distributions of electrons of all different energies are nearly spherical (see Fig. 3-8-3). Therefore it seemed interesting to us to measure angular correlations in the low impact energy region. In Fig. 3-8-4 ($E_0 = 41$ eV) the binary and the recoil peak are both still visible, although the binary peak has become very small and most collisions are of the recoil type. Both electrons are chosen equally fast, so that the mutual influence on each other should be largest. For impact energies as low as 30.5 eV, i.e. 6 eV above *IP*, it seems surprising to find still well developed binary peaks as is seen from Fig. 3-8-5. Of course, at such low energies the expressions "binary" and "recoil" may loose their applicability.

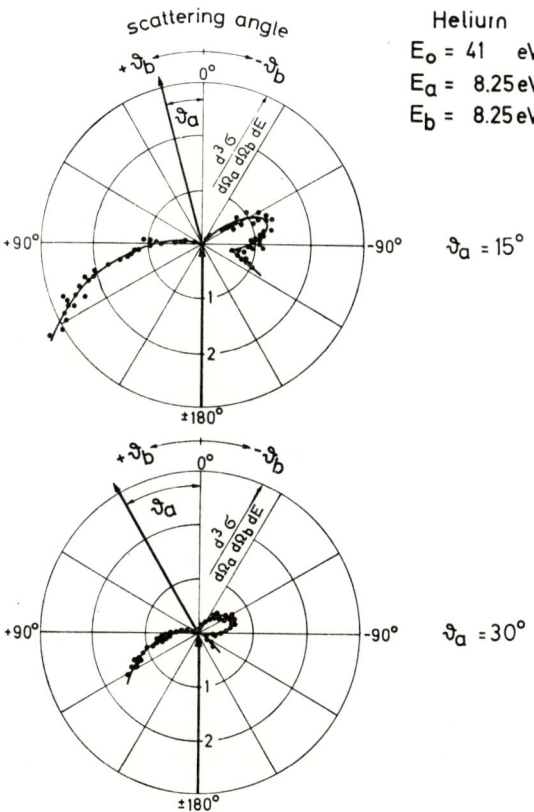

Fig. 3-8-4. Triple differential cross section for helium as a function of the scattering angle ϑ_b of the ionization electrons, if the corresponding electrons are detected at $\vartheta_a = 15°$ (upper diagram) and $\vartheta_a = 30°$ (lower diagram). Both outgoing electrons have the same energy (symmetrical case). The full line is the average over the actually obtained measurements (arbitrary units).

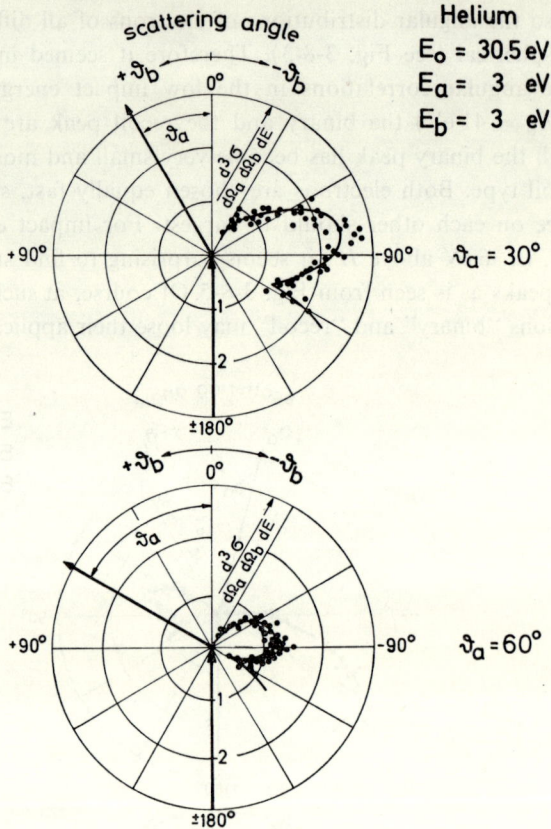

Fig. 3-8-5. Triple differential cross section for helium as a function of the scattering angle ϑ_b of the ionization electrons, if the corresponding electrons are detected at $\vartheta_a = 30°$ (upper diagram) and $\vartheta_a = 60°$ (lower diagram). Both outgoing electrons have the same energy of $E_a = E_b = 3$ eV (symmetrical case). All parameters which have been kept constant during the experiment, are indicated in the figure. The full line is the average over the actually obtained measurements (arbitrary units).

Acknowledgement

The authors acknowledge very much theoretical discussions with M. Schulz and L. Vriens as well as the financial support from the Deutsche Forschungsgemeinschaft.

References

1. U. AMALDI, A. EGIDI, R. MARCONERO and G. PIZELLA, Rev. Sci. Instr. **40** (1969) 1001.
2. L. VRIENS, Physica **45** (1969) 400; **47** (1970) 267;
   L. VRIENS and T. P. M. BONSEN, J. Phys. B **1** (1968) 1123.

3. A. E. Glassgold and G. Ialongo, Phys. Rev. **175** (1968) 151.
4. H. Ehrhardt, M. Schulz, T. Tekaat and K. Willmann, Phys. Rev. Letters **22** (1969) 89.
5. H. Ehrhardt, K. H. Hesselbacher, K. Jung, T. Tekaat, M. Schulz and K. Willmann, Z. Physik, to be published.
6. H. Ehrhardt, K. H. Hesselbacher, K. Jung, T. Tekaat and K. Willmann, to be published.
7. H. Ehrhardt, K. H. Hesselbacher, K. Jung, M. Schulz and K. Willmann, to be published.
8. L. Kervin and P. Marmet, Can. J. Phys. **38** (1960) 787.
9. R. Herzog, Z. Physik **97** (1935) 586.
10. G. J. Schulz, Phys. Rev. **125** (1962) 229.
11. C. E. Brion, J. Chem. Phys. **40** (1964) 2995.
12. A. Egidi, R. Marconero, G. Pizella and F. Sperli, Rev. Sci. Instr. **40** (1969) 88.
13. L. J. Kieffer and G. H. Dunn, Rev. Mod. Phys. **38** (1966) 1.
14. M. R. Rudge, Rev. Mod. Phys. **40** (1968) 1.
15. I. H. Sloan, Proc. Phys. Soc. **85** (1965) 435.
16. W. L. Fite and R. T. Brackmann, Phys. Rev. **112** (1958) 1141.
17. R. K. Asundi and M. V. Kurepa, J. Electron. Control **15** (1963) 41.
18. B. L. Schram, F. J. de Heer, M. J. van der Wiel and J. Kistemaker, Physica **31** (1964) 94.
19. R. K. Asundi, J. D. Craggs and M. V. Kurepa, Proc. Phys. Soc. **82** (1963) 967.
20. B. A. Tozer and J. D. Craggs, J. Electron. Control **8** (1960) 103.
21. D. Rapp and P. Englander-Golden, J. Chem. Phys. **43** (1965) 1464.
22. H. S. W. Massey and E. H. S. Burhop, Electronic and Ionic Impact Phenomena, Vol. 1 (Oxford 1969).
23. G. Peach, Proc. Phys. Soc. **87** (1966) 389
24. D. G. Economides and M. R. C. McDowell, J. Phys. B **2** (1969) 1323.
25. A. L. Hughes and J. H. McMillen, Phys. Rev. **39** (1932) 585.
26. A. L. Hughes and J. H. McMillen, Phys. Rev. **41** (1932) 39.
27. J. T. Tate and R. R. Palmer, Phys. Rev. **40** (1932) 731.
28. C. B. O. Mohr and F. H. Nicoll, Proc. Roy. Soc. A **144** (1934) 596.
29. M. Goodrich, Phys. Rev. **52** (1937) 259.
30. S. M. Silvermann and E. N. Lassettre, J. Chem. Phys. **40** (1964) 1265.
31. J. A. Simpson, G. E. Chamberlain and S. Mielczarek, Phys. Rev. **139** (1965) A1039.
32. Th. M. El-Sherbini, M. J. van der Wiel and F. J. de Heer, Physica **48** (1970) 157.
33. G. Peach, J. Phys. B **3** (1970) 328.
34. J. McConkey, W. R. Newell and A. Grove, Angular Distributions of Ions Resulting from Electron Impact on the Rare Gases, preprint.
35. M. R. H. Rudge, Rev. Mod. Phys. **40** (1968) 564.
36. N. F. Mott and H. S. W. Massey, The Theory of Atomic Collisions (London, Oxford 1965).
37. W. W. Wetzel, Phys. Rev. **44** (1933) 25.
38. R. P. Peterkop, Phys. Rev. **164** (1967) 1220.
39. S. Geltman, Phys. Rev. **120** (1956) 171.
40. L. Vainstein, L. Presnyakov and J. Sobel'man, Sov. Phys. JETP **18** (1964) 1383.
41. J. T. Grissom, W. T. Naff, W. R. Garret, R. N. Compton and J. A. Stockdale, Second Intern. Conf. on Atomic Physics, Oxford, July 1970.
42. C. E. Brion and L. A. Olsen, J. Phys. B **3** (1970) 1020.
43. R. E. Fox, W. H. Hickam, T. Kjeldaas and D. J. Grove, Phys. Rev. **84** (1951) 859; J. Chem. Phys. **20** (1952) 1055.
44. R. E. Fox, W. M. Hickam and T. Kjeldaas, Phys. Rev. **96** (1954) 63.

45. D. C. Frost and C. A. McDowell, J. Am. Chem. Soc. **80** (1958) 6183.
46. C. E. Brion and G. E. Thomas, Phys. Rev. Letters **20** (1968) 241.
47. P. Marchand, C. Paquet and P. Marmet, Phys. Rev. **180** (1969) 123.
48. J. W. McGowan and E. M. Clarke, Phys. Rev. **167** (1968) 43.
49. G. Peach, private communication.

CHAPTER 4

INTERPRETATION OF SPECTRAL INTENSITIES FROM LABORATORY AND ASTROPHYSICAL PLASMAS

BY

A. H. GABRIEL AND CAROLE JORDAN

Astrophysics Research Unit, Culham Laboratory, Abingdon, Berkshire, England

February 1971

Contents

	Page
4-1. Introduction	211
4-2. Theoretical methods	212
A. Statistical equilibrium for excited states	212
B. Effects of metastable levels	214
C. Collisional excitation	215
D. Levels above the first ionization limit	219
E. Line ratio measurement of electron temperature	222
F. Line ratio measurement of electron density	223
4-3. Collision rate experiments	224
A. General methods	224
B. Electron beam excitation	225
C. Laboratory plasma experiments	225
D. Theta-pinches	227
4-4. Lithium-like ions	229
A. General description	229
B. Energy levels	231
C. Transition probabilities	231
D. Collisional excitation rates	232
E. Statistical equilibrium	234
F. Line intensities in laboratory sources	235
G. Solar line intensities	237
H. Dielectronic satellite lines	241
4-5. Beryllium-like ions	242
A. General description	242
B. Energy levels	243
C. Transition probabilities	244
D. Collisional excitation rates	244
E. Equations of statistical equilibrium	249
F. Laboratory plasmas	252
G. Solar line intensities	255
H. Low density objects, quasars, and planetary nebulae	264
I. Recent developments	265
4-6. Helium-like ions	265
A. General description	265
B. Energy levels	267
C. Transition probabilities	268
D. Collisional excitation rates	270
E. Low density plasmas	274
F. High density plasmas	277
G. Intermediate densities	281
H. Line ratios, N_e and T_e	283
I. Dielectronic satellite lines	286
J. Absolute intensities	286
K. Recent developments	287
Acknowledgements	288
References	288

§ 4-1. Introduction

This chapter aims to consider the present state of knowledge concerning the intensities of the principal spectral lines arising from some simple types of ions. Line intensities, although straightforward to measure, can depend on many complex interactions among the particles in a plasma, and it is only recently that their interpretation has become well understood. There are plasmas in which these interactions or collisions occur in such a way that the transition rate due to each is equal to the rate of the precisely inverse transition. In this case, the problem reduces to one of thermodynamic equilibrium. The populations of the states can be found from the equations of statistical mechanics and no further knowledge of individual interaction rates is required. In the more general case, transitions may be balanced by processes which are not precisely their inverse, or may not be balanced at all, so that the steady-state solution is not reached and the populations vary with time.

For a fuller description of the overall problem involving departures from thermodynamic equilibrium and consideration of ionization and excitation models in steady-state and transient plasmas, the reader is referred to an article by McWhirter [1]. In the present chapter we are concerned with one particular type of model, which in practice represents a situation that is found in a wide range of laboratory and astrophysical plasmas. In this model the distribution of ionization stages is either steady-state or transient. However, if transient, the time scale for changes in ionization is sufficiently long that the relative population of excited states within each ion is determined by steady-state considerations. Furthermore, the excited states are in general well removed from thermodynamic equilibrium populations.

The steady-state equations for excited state populations are very simple in the case of a degenerate hydrogenic system. Such a system has a smoothly varying energy level spacing and no metastable levels. When looking at more complex ions, other factors must be taken into account, particularly when considering the lower-lying excited levels which give rise to the more intense spectral lines. We shall consider therefore the problem for two, three and four electron ions. Hydrogenic ions will not be considered, although it is important to realise that they also are subject to some complicating factors. In particular, for the more highly-charged members of the sequence, 2s and 2p levels can no longer be considered as degenerate, and the metastable 2s level must be taken into account.

Since the emphasis here is on the atomic physics, no account will be taken of optical transfer problems, all plasmas being regarded as optically thin. Transfer effects are a property of the bulk dimensions of the plasma and cannot be considered entirely in terms of individual atomic interactions. Furthermore only the lower excited states will be considered. The problem of intensities along Rydberg series is common to all types of ion, and can be found in more general texts.

Many properties of the ions considered vary smoothly along isoelectronic sequences, but with a more rapid change at the lower end towards the neutral atom. Since we are not concerned here with low temperature plasmas, those properties of neutrals and singly charged ions, which are not characteristic of the isoelectronic sequence, will not be considered further.

§ 4-2. Theoretical methods

A. Statistical equilibrium for excited states

Considering now only cases in which the changes in population of each ion stage are slow compared with excited state redistribution, each ion stage can be treated independently and assumed to have a fixed total population. Therefore we consider at first only those processes leading to transfer between the excited states. Other processes involving ionization and recombination can be included later as a perturbation for those cases where they have a direct influence on excited state populations. The rate of change of the population density N_i of a state i can be expressed as a function of all the other excited states j,

$$\frac{dN_i}{dt} = \sum_j N_j N_e C(j \to i) - \sum_j N_i N_e C(i \to j) \\ + \sum_{j>i} N_j A(j \to i) - \sum_{j<i} N_i A(i \to j). \quad (4\text{-}2\text{-}1)$$

Here $C(i \to j)$ is the rate coefficient for collision with electrons (density N_e) producing the transition i to j, and $A(i \to j)$ is the spontaneous radiative decay rate i to j. It is assumed in eq. (4-2-1) that the plasma is optically thin so that photoexcitation is absent and that electrons are the only particles capable of impact-exciting an ion. Excitation of ions by impact with other heavy particles (ions or protons) is not in general an efficient mechanism. For some transitions between close-lying levels, however, it can become important, and then additional terms must be added to the equation. Although we are limiting the treatment to optically thin plasmas, these will on occasion be exposed to exciting photons from an intense external source, so that this would also require additional terms in eq. (4-2-1).

Since we are concerned with the time-independent solution, the left-hand side of eq. (4-2-1) can be equated to zero, and the four terms on the right-hand side regrouped to give

$$N_i = \frac{N_e \sum_j N_j C(j \to i) + \sum_{j>i} N_j A(j \to i)}{N_e \sum_j C(i \to j) + \sum_{j<i} A(i \to j)}. \qquad (4\text{-}2\text{-}2)$$

For the simplest cases at low electron density, this equation is dominated by the first term in the numerator and the second term in the denominator, with the main contribution to the summation over j coming from the ground state g. It then reduces to the so-called coronal excitation equation, first derived by Elwert [2] to account for excitation in the solar corona:

$$N_i = \frac{N_e N_g C(g \to i)}{A(i \to g)}. \qquad (4\text{-}2\text{-}3)$$

On the other hand, it is clear that at sufficiently high density, the collisional terms in eq. (4-2-2) will dominate over the radiative terms, giving

$$N_i \sum_j C(i \to j) = \sum_j N_j C(j \to i). \qquad (4\text{-}2\text{-}4)$$

Since these processes are precisely inverse, this is a sufficient condition for establishing a thermodynamic equilibrium population distribution between states i and j. We shall rarely be concerned with the conditions of eq. (4-2-4) except in some special cases involving two or three nearby excited levels only. High temperature plasmas are generally closer to eq. (4-2-3), but for ions having some long-lived excited states it is necessary to involve some of the other terms of eq. (4-2-2).

The general solution of eq. (4-2-2) involves the solution of n simultaneous equations, one for each state involved, where i and j range from 1 to n. In a practical case, the problem is to reduce the number of states (and hence equations) by considering a closed group of states, such that the populations of those outside have little effect on those inside. For example, since both $C(i \to j)$ and $A(j \to i)$ decrease rapidly as the level j increases in principal quantum number, it is often justifiable to exclude all the higher levels when considering the lower excited states. For many ions the majority of the ion population will be in the ground state. This can also lead to a useful approximation by treating N_g as a constant, and equal to the ion population. In some cases, a group of close-lying levels (e.g. the fine structure levels of an LS term) can be shown to be in thermodynamic equilibrium among themselves, due to very large rates for collisional transfer between them. Their relative population is then statistical and they can be treated for the purposes of eq. (4-2-2) as a single level, using appropriately averaged rate coefficients and transition probabilities. The effective rate coefficient for the general case of transitions between two such mixed states a and b, is obtained by summing over the final states, and averaging over the initial states; i.e.

$$\bar{C}(a \to b) = \frac{\sum_a \sum_b g_a C(a \to b)}{\sum_a g_a} \qquad (4\text{-}2\text{-}5)$$

and

$$\bar{A}(a \to b) = \frac{\sum_a \sum_b g_a A(a \to b)}{\sum_a g_a} \qquad (4\text{-}2\text{-}6)$$

where g_a are the statistical weights of the initial states a. For $\bar{C}$ or $\bar{A}$ to be meaningful the initial states must be statistically distributed, though not necessarily the final states.

After solving the appropriate equations and obtaining the population densities N_i, the intensity I of any line $i \to j$ can be obtained simply from the relationship

$$I(i \to j) = N_i A(i \to j) \qquad (4\text{-}2\text{-}7)$$

with I in photons emitted in all directions, or

$$I'(i \to j) = \frac{h\nu}{4\pi} N_i A(i \to j) \qquad (4\text{-}2\text{-}8)$$

where I' is now in erg cm^{-3} sec^{-1} sterad^{-1}. It should be noted that if eq. (4-2-7) is applied to the special case of a resonance line represented by eq. (4-2-3), it is no longer necessary to know the value of $A(i \to j)$, since

$$I(i \to g) = N_e N_g C(g \to i). \qquad (4\text{-}2\text{-}9)$$

B. Effects of Metastable Levels

Starting with a simple ion, such as can be represented by eq. (4-2-3), it is clear that the immediate effect of introducing a metastable or long life-time level k, is to build up N_k to a much higher population than similar levels with an allowed decay. This then affects other level populations, since in eq. (4-2-2) for level i, numerator terms with $j = k$ become important in addition to those involving the ground. It is then necessary to return to a fuller set of equations to obtain a solution. In more physical terms, the level k having an excess population begins to transfer to other excited states rather than decaying to the ground, and thus affects their populations.

In addition to the above, metastable levels can cause another effect which disturbs the approximation made in separating ionization and excitation processes. While the greater part of each ion population is confined to its ground state, and while ionization and recombination occur only between successive ground states, it is valid to treat each ion as a separate system. The existence of a metastable level with a long lifetime and excess population leads to the importance of ionization or recombination directly to or from metastable levels. This effect does not depend on whether the ionization balance is steady state or transient. When such processes are of moderate magnitude, they can be estimated and allowed for by additional terms in eq. (4-2-2). When they are large, it may be necessary to solve eq. (4-2-2) simultaneously for levels of more than one ion stage, or even in the case of transient ionization, to obtain the full time dependent solution to eq. (4-2-1) with the addition of ionization and recombination terms.

C. Collisional excitation

The above considerations emphasize the importance of the collisional excitation rate coefficients C in determining the spectral intensities. C is related to the collision cross-section Q by

$$C = \langle vQ \rangle \qquad (4\text{-}2\text{-}10)$$

where the product is averaged over the distribution function of v, the relative velocity of the colliding particles. So that for electron collisions

$$C = \int_{E_0}^{\infty} vQ f(E) \, dE \qquad (4\text{-}2\text{-}11)$$

where v and E are the electron velocity and energy, $f(E)$ is the distribution function, and E_0 is the threshold for excitation or zero for de-excitation. For most practical cases, $f(E)$ is a Maxwellian distribution with an electron temperature T_e.

A precise evaluation of Q is difficult, although a number of approximations is available. In order to examine some of the general properties of rate coefficients, it is helpful to consider one of the simpler approximate forms commonly used for optically allowed transitions, given by

$$Q = \frac{8\pi}{\sqrt{3}} \frac{f}{EE_0} \bar{g} \pi a_0^2 \tag{4-2-12}$$

where f is the absorption oscillator strength, a_0 is the Bohr radius and $\bar{g}$ is a slowly varying semi-empirical correction factor tabulated by Van Regemorter [3]. The energies are in Rydberg units. For neutral atoms $\bar{g}$ is zero at threshold but for ions it takes a value of 0.2 from threshold up to twice threshold. This expression is good to a factor of 3. The formula is often used as a means of expressing other more accurate computations, by using $\bar{g}$ as an adjustable parameter in place of Van Regemorter's empirical values. A combination of eqs. (4-2-11) and (4-2-12) gives

$$C = 8\sqrt{\frac{\pi}{3}} \frac{ha_0}{m} \frac{f}{E_0 T_e^{\frac{1}{2}}} P \exp(-E_0/T_e) \tag{4-2-13}$$

with E_0 and T_e in Rydberg units, or

$$C = 1.7 \times 10^{-3} \frac{f}{E_0 T_e^{\frac{1}{2}}} P \exp(-E_0/kT_e) \tag{4-2-14}$$

with E_0 in eV and T_e in °K. Here P is the average value of $\bar{g}$, given by

$$P = \exp(E_0/kT) \int_{E_0/kT_e}^{\infty} \bar{g} \exp(-x) \, dx; \tag{4-2-15}$$

P is also tabulated by Van Regemorter, and for $E/kT_e > 1$ tends to the value of $\bar{g}$ at threshold, i.e. 0.2.

Theoretical cross-sections are often expressed in terms of collision strengths Ω rather than Q. These are related by

$$\Omega = \frac{1}{\pi a_0^2} Q g_1 E \tag{4-2-16}$$

where g_1 is the statistical weight of the initial level and E is in Rydberg. The main advantage is that by taking out the $1/E$ variation, close to threshold Ω is a constant with respect to E. Thus eq. (4-2-12) becomes

$$\Omega = \frac{8\pi}{\sqrt{3}} \frac{f}{E_0} g_1 \bar{g} \tag{4-2-17}$$

which is constant if $\bar{g}$ is constant. If eq. (4-2-11) for C is now written in terms of Ω, then

$$C = 1.36 \frac{1}{g_1} \frac{1}{T_e^{\frac{3}{2}}} \int_{E_0/kT_e}^{\infty} \Omega \exp(-E/kT_e) \, dE. \tag{4-2-18}$$

If now Ω is taken to be constant at some value $\bar{\Omega}$, this becomes

$$C = 8.65 \times 10^{-6} \frac{1}{g_1} \frac{1}{T_e^{\frac{1}{2}}} \bar{\Omega} \exp(-E_0/kT_e), \tag{4-2-19}$$

so that detailed calculations for collisions can be expressed in terms of Q, Ω, $\bar{\Omega}$ or $\bar{g}$ and can be related to the rate coefficient C through eq. (4-2-11), (4-2-17), (4-2-18) or (4-2-19).

Collisional excitation and de-excitation rates are related simply, so that if one is known, the other can be readily derived. The relation can be obtained from the Principle of Detailed Balance, or alternatively derived directly from a quantum mechanical approach. For rate coefficients (j is the upper level),

$$C(j \rightarrow i) = C(i \rightarrow j) \frac{g_i}{g_j} \exp(E(ij)/kT_e). \tag{4-2-20}$$

For cross-sections

$$Q_{E'}(j \rightarrow i) = Q_E(i \rightarrow j) \frac{g_i}{g_j} \frac{E}{E'} \tag{4-2-21}$$

where

$$E' = E - E(ij),$$

and for collision strengths

$$\Omega_{E'}(j \rightarrow i) = \Omega_E(i \rightarrow j). \tag{4-2-22}$$

For the excitation of optically allowed transitions, the cross-sections depend approximately on the oscillator strength f as shown by eq. (4-2-12). At high electron energies the factor $\bar{g}$ takes the asymptotic form

$$(\sqrt{3}/2\pi) \log_e(E/E_0)$$

so that the cross-section has an asymptotic form $E^{-1} \log_e(E/E_0)$. This is true also of more precise calculations, whenever the transition is optically allowed. For optically forbidden transitions, the excitation cross-section cannot of course be approximated by an expression of this type and depends on more detailed calculations. However the values can be as large or even larger than for similar optically allowed transitions. At high energies, the cross-sections decrease asymptotically as E^{-1} for Δl-forbidden transitions, or as E^{-3} for spin-forbidden transitions. This rapid fall-off for spin change excitation arises since such transitions can only occur through electron exchange, a process which becomes very unlikely at high impact energies.

For a review of the theoretical methods available for electron impact excitation cross-sections, the reader is referred to Moiseiwitsch and Smith [4], and to Bely and Van Regemorter [5]. Most methods require detailed calculations to be carried out for the particular ion and transition used. Two methods are available for the optically allowed transitions which allow the user to derive cross-sections using only the oscillator strength and energy levels of the transition. The first of these is the formula due to Van Regemorter which was discussed above and reproduced in eq. (4-2-12). The other is a generalization to positive ions by Burgess [6] of the impact parameter method developed by Seaton [7]. This is not in a simple analytic form and the more complex functions are tabulated, but it can be appplied readily to a number of transitions. The formulation referred to by Burgess as "weak coupling" has been used for a number of the cases which are dealt with later. It gives an improved accuracy, usually better than a factor of 2, when compared with eq. (4-2-12). An exception arises in the case of the helium-like ion resonance lines, where eq. (4-2-12) appears to give significantly better values.

For greater accuracy, and for optically forbidden transitions one has to rely on more specific calculations. The collision problem can be formulated exactly in quantum theory, including in full the effects of exchange. The resulting set of coupled equations involves all levels of the ion, not only the initial and final states. The extent of this set must then be limited before solution, so that the computational effort does not become prohibitive. This approach, known usually as the "close coupling" method has been carried out by a number of workers for a range of transitions. Its validity depends largely on whether the practical restriction in the number of levels included is a realistic physical approximation. Levels close to the initial and final states couple strongly during the collision, and it is important to include these. A further simplification is possible if the exchange part of the equations can be treated in a more approximate form. In many practical cases, close-coupling calculations, limited by computational capacity, will give inferior results to other approximations.

As most of the ions of interest are highly charged, approximations based on a Coulomb potential have the advantage that they become more exact as Z tends to infinity. The simplest of such approximations is the Coulomb–Born. This is the Born approximation with allowance for the influence of the Coulomb field on the incoming electron wave. Because of the dominance of the Coulomb terms, it gives better results for ions than the Born approximation does for neutrals. Exchange can be included in which case it becomes the Coulomb–Born–Oppenheimer approximation. For low Z the exchange treatment becomes unreliable. This method has been widely used by many authors. (See for example Bely [8], Burgess et al. [9] and Blaha [10].) A more exact matrix algebra leads to the Coulomb–Born II or Coulomb–Born–Oppenheimer II, methods which it is claimed account to some extent for coupling with other states. This is often referred to as the "unitarized" approximation, and satisfies certain conservation conditions, which are otherwise violated. However, it appears that this extension is by no means consistent in improving the accuracy.

A better method has been developed by Seaton and collaborators for use with positive ions. As part of a general purpose computer program they are applying a distorted wave approximation which involves an improved treatment of exchange, together with some coupling of additional states. This has been formulated to facilitate its simple application to a wide range of ions, and should prove to be of considerable value over the next few years. An important feature of this approach is the provision of configuration mixing in the description of the target ion. This can be very important for some ions. Indeed, Johnston and Kunze [11] have pointed out that for allowed transitions, good rate coefficients may be obtained by inserting improved oscillator strengths, based on configuration mixing, in the Van Regemorter formula of eq. (4-2-14).

D. Levels above the first ionization limit

In all but the one-electron ion, there exist many series of energy levels above the lowest ionization potential. These arise from configurations in which more than one electron is in an excited state or in which an inner electron is excited. Many of these states can autoionize into adjacent continua, although a few have no continuum of suitable parity and L-value available. For the populating mechanism of all these states four possible processes can be considered, although some may not be permitted depending on the particular configuration involved. These processes are

1) impact excitation of an inner-shell electron,
2) impact (or photon) ionization of an inner-shell electron,

3) simultaneous impact excitation of two electrons,
4) dielectronic capture.

Of these, the first three are normally such small effects as to be negligible so long as the departure from steady-state ionization is not very large. Process 1) can become significant in transient plasmas, e.g. sparks and solar flares. The fourth process is the inverse of autoionization and can have important effects on the intensities of observed spectral lines. There are two main effects; enhancement of intensity of the resonance lines and the introduction of weak "satellite" lines. These will be considered in turn.

Dielectronic capture can be regarded as the first part of the overall process of dielectronic recombination. This was first worked out in detail by Burgess [12, 13] who regards dielectronic recombination as a 3-part process:

1) The colliding electron at below threshold energy excites an ion to a resonance transition, but itself gets trapped in an outer orbital. If at this point autoionization occurs we return to the starting point and no recombination has taken place.

2) Alternatively, the doubly excited ion stabilizes by radiative decay of the inner excited electron. This emits a photon of resonance radiation of the recombining ion, shifted slightly to longer wavelength by the influence of the additional outer electron.

3) The now singly excited ion cascades to its ground state.

The overall result is recombination plus the emission of a photon close to the resonance line of the recombining ion. Burgess summed over all possible states of both the inner and outer excited electron and showed that the main contribution to recombination (and therefore photon emission) arises from states with the outer electron having high principal quantum numbers, typically between 50 and 200. For these high levels, in fact for any $n \gtrsim 4$, the emission will be indistinguishable in wavelength from the equivalent resonance line.

Burgess carried out detailed calculations for a number of specific ions, and then derived a semi-empirical formula [14] which he claimed valid for a wide range of ions. The recombination rate coefficient α due to this process is given by

$$\alpha = \frac{8.2 \times 10^{-4}}{T_e^{\frac{3}{2}}} \frac{Z^{\frac{1}{2}}(Z+1)^2}{(Z^2+13.4)^{\frac{1}{2}}} \sum_j \frac{f(i \to j) E_0^{\frac{1}{2}}}{(1+0.105x+0.015x^2)} \exp(-E_0/kT)$$

(4-2-23)

where $x = 0.0735 E_0/(Z+1)$, E_0 is in eV, T_e in °K. E_0 is the energy and f the oscillator strength for the resonance transition $i \rightarrow j$ in the recombining ion. Z is the charge on recombining ion, so that for recombination from say oxygen VII to oxygen VI, Z is equal to 6. The expression is summed over the strong resonance levels j to obtain a total recombination rate. Since we are concerned here with the contribution to the effective excitation of an individual line, we need consider only one term of the j summation. The value of α_j for each term in the equation is then equal to the effective excitation rate coefficient for the dielectronic contribution to the resonance line. Each such term in eq. (4-2-23) contains implicitly the summation over all states of the outer recombining electron.

The ratio of the dielectronic to the direct contribution to a line is thus given by

$$\frac{\alpha_j}{C(i \rightarrow j)} = 0.48 \frac{Z^{\frac{1}{2}}(Z+1)^2}{(Z^2+13.4)^{\frac{1}{2}}} \frac{E_0^{\frac{3}{2}}}{T_e P} \frac{1}{(1+0.105x+0.015x^2)} \quad (4\text{-}2\text{-}24)$$

derived from eq. (4-2-14) and one j-term from eq. (4-2-23). When the Z-dependence of E_0 and T_e are taken into account this is a slowly increasing function of Z, and can for some ions and transitions reach values of the ratio in excess of 0.1.

Burgess's theory has been developed for the large number of states which contribute to the dielectronic recombination process, involving as they do states with the outer electron in high principle quantum numbers. For the lowest of such states, the approximations become poor and an alternative formulation is advisable. This was first proposed by Gabriel et al. [15] to account for the intensities of satellite lines produced by dielectronic recombination, and well separated on the long wavelength side of resonance lines. In this case we are concerned with only one j-term of the summation in eq. (4-2-23), only the lowest term in the implicit summation over the outer electron states, and furthermore only one LS term of the configuration resulting from this selection. The effective recombination rate resulting in one particular satellite line is then

$$\alpha_s = \frac{2.06 \times 10^{-16}}{T^{\frac{3}{2}}} \frac{g_s}{g_i} A_r \left(\frac{A_a}{A_a + A_r}\right) \exp(-E_s/kT_e) \quad (4\text{-}2\text{-}25)$$

where E_s is the energy of the doubly excited state s of statistical weight g_s, above that of the ground state of the recombining ion of statistical weight g_i. A_a and A_r are the transition probabilities for decay of the state s by autoionization and stabilizing radiation respectively. For many of the satellites, $A_a/(A_a+A_r) \sim 1$. The above equation then reduces to the effective rate that would apply if the state s had a thermodynamic equilibrium population with respect to the recombining ion. This is understandable, since there is then a balance between dielectronic capture and autoionization, two precisely inverse processes. The ratio of satellite intensity to impact excited resonance line can now be derived from eqs. (4-2-14) and (4-2-25) to be

$$\frac{\alpha_s}{C(i \rightarrow j)} = \frac{1.2 \times 10^{-13}}{T_e} \frac{g_s}{g_i} \frac{E(ij)A_r}{Pf(i \rightarrow j)} \left(\frac{A_a}{A_a+A_r}\right) \exp\left((E(ij)-E_s)/kT_e\right). \quad (4\text{-}2\text{-}26)$$

This ratio varies approximately as Z^4 on account of the terms $E(ij)A_r/T_e$, so that the satellites become relatively more prominent for highly ionized ions in the soft X-ray region. In applying the above equation to any particular case allowance must be made for possible contributions to the resonance line from dielectronic recombination according to eq. (4-2-24) and also for contributions to the satellite levels by direct impact excitation.

A third influence of dielectronic recombination on the intensities of spectral lines, arises from the effect of the increased total recombination rate on the ionization equilibrium or time-history. This is the effect that has provided the main incentive for the initial work on the process. Since the ionization distribution is outside the scope of this chapter, it will not be considered further.

E. Line Ratio Measurement of Electron Temperature

In § 4-2A we showed that the intensity of a spectral line can in general be represented by a complex function of electron temperature and density. In spite of this, a wide range of spectral intensities can be approximated by the coronal excitation equation (4-2-3) or (4-2-9). If we consider the ratio of photon intensities of two lines g–i and g–k, both excited from the ground and decaying to the ground, it can be seen that

$$\frac{I(i \rightarrow g)}{I(j \rightarrow g)} = \frac{C(g \rightarrow i)}{C(g \rightarrow j)}. \quad (4\text{-}2\text{-}27)$$

Using the expression (4-2-14) for the C's,

$$\frac{I(i \rightarrow g)}{I(j \rightarrow g)} = \frac{f(g \rightarrow i)}{f(g \rightarrow j)} \frac{E(gj)}{E(gi)} \frac{P(gi)}{P(gj)} \exp(-(E(gi)-E(gj))/kT_e). \quad (4\text{-}2\text{-}28)$$

It is clear that the ratio is independent of electron density, and that the dependence on electron temperature is given almost entirely by the exponential term. This term derives from the Maxwellian electron energy distribution (see eq. (4-2-11)), so that we are in effect examining those portions of the distribution which have energies above $E(gi)$ and $E(gj)$. From eq. (4-2-28) it can be seen that the ratio is a sensitive function of temperature when $|E(gi) - E(gj)| \gg kT_e$ and insensitive when $|E(gi) - E(gj)| \ll kT_e$. For the sensitive measurement of electron temperature we therefore choose two lines of one ion, excited from the ground state, with their upper levels well separated in energy. The choice of a simple ion for which eq. (4-2-3) is valid is helpful since this avoids a dependence on electron density. The use of more precise values for the collisional excitation rates will improve the accuracy without affecting any of the above considerations.

One practical disadvantage arises when the two lines are both transitions to the ground. The requirement $|E(gi) - E(gj)| \gg kT_e$ implies that the lines will be well separated in wavelength. This can raise severe problems in relative intensity calibration of the measuring equipment over a large spectral range.

F. Line Ratio Measurement of Electron Density

A density dependence is introduced into line ratios as soon as the coronal equation (4-2-3) breaks down for one of them. Consider as a simple example the above case of two transitions $g \to i$ and $g \to k$, where k now has a long lifetime (low $A(k \to g)$) and an alternative collisional loss rate $C(k \to m)$, where m represents any or all other levels. In this situation the coronal equation is not valid.

$$N_i = \frac{N_e N_g C(g \to i)}{A(i \to g)} \tag{4-2-29}$$

as before, and

$$N_k = \frac{N_e N_g C(g \to k)}{A(k \to g) + N_e C(k \to m)} \tag{4-2-30}$$

giving the ratio

$$\frac{I_i}{I_k} = \frac{C(g \to i)}{C(g \to k)} \left(1 + \frac{N_e C(k \to m)}{A(k \to g)}\right). \tag{4-2-31}$$

The first term in eq. (4-2-31) is the simple coronal result, while the second term introduces a density dependence. This becomes sensitive to density when

$$A(k \to g) \sim N_e C(k \to m). \tag{4-2-32}$$

For small densities it is insensitive, and for much larger densities the intensity ratio of eq. (4-2-31) will become too large to measure I_k reliably.

§ 4-3. Collision rate experiments

A. General methods

It will be seen from the discussion in § 4-2C that the problem of calculating reliable excitation rates without prohibitive computational effort is in many cases unsolved. This then opens the field to worthwhile experimental measurements. Since the quantities required for the analysis of spectral intensities are the excitation rate coefficients $C(i \to j)$, it is clear that there are two fundamental approaches available. One is to measure $C(i \to j)$ directly as a function of the electron temperature T_e. The second is to measure the equivalent excitation cross-section $Q(i \to j)$ as a function of electron energy E. In the latter case, the measured Q can be compared directly with theoretical values, and can be readily integrated over Maxwellian electron energy distributions to obtain values of C. The converse is not possible, and measured rate coefficients C cannot in general be unfolded to give Q as a function of E.

These two approaches, the measurement of Q or C, are reflected in two different experimental methods. In the first, to measure Q, a monoenergetic beam of electrons is incident on a target of ions. This can be in the form of a gas (for neutral atoms), or an atomic beam (for neutrals or ions) travelling at an angle to, and intersecting, the electron beam. The efficiency of excitation is then determined either by detecting the photons emitted, the excited atoms produced (for metastables) or the energy-loss from the incident electron beam. This method is limited by the difficulty of producing sufficiently intense beams of multiply-ionized atoms. In the alternative method, the target ions are produced in a plasma source, the exciting electrons being those occurring naturally in the same source. Excitation is detected by the photons emitted and the measurement gives directly the excitation rate coefficient, rather than the cross-section.

In both of these experimental methods, photon emission does not depend only on the collisional rate coefficient to the upper excited level. Complications arise due to alternative spontaneous decay routes and cascade contributions from higher levels. In addition, beam methods suffer from the effects of anisotropic emission, and plasma methods from the effects of metastable states. Because of these problems, some workers measure what they refer to as optical excitation functions, which are in effect cross-sections for the production of a particular photon, irrespective of the route. Such quantities are not in general of direct use in the present type of analysis. Therefore experiments should be designed to provide individual excitation rates.

B. ELECTRON BEAM EXCITATION

For reviews of electron beam methods, the reader is referred to Moiseiwitsch and Smith [4], Harrison [16] and Heddle and Keesing [17]. Methods using gas targets are limited to neutrals, and are not therefore relevant to the present article.

One of the classic crossed beam experiments is that carried out by Fite and Brackmann [18] in 1958, for the excitation of hydrogen Lyman α. An electron beam from a gun was crossed at right angles with an atomic beam from a furnace. Photons were detected by means of an ionization chamber. To cope with a large background and small signal, the atomic beam was chopped and phase-sensitive detection used to record only those counts which synchronized with the beam chopping frequency. Many problems arise in interpreting the signals from such an experiment. Some of the background is modulated by the chopper. The geometry of the beam intersection is difficult to establish and can vary with the electron energy.

For ionized targets, a crossed beam experiment has been performed on helium II by Dance et al. [19] at Culham, who measured excitation to the 2s state by detecting the metastables produced. Interference effects between the two beams and the detector system were important and the experimental technique was therefore quite complex.

Extension of these techniques to more highly charged ions has not so far proved feasible. It is difficult to produce sufficiently intense beams of ions, and even if this were possible the space charge effects would be serious. Detection of photons is also a problem. The intensities available might be adequate for large aperture ionization chambers to be used, but would not be sufficient if a spectrometer were required to provide the necessary wavelength resolution to reject other nearby spectral lines. For these reasons, plasma excitation remains the principle experimental technique for measuring excitation rates in ionized atoms.

C. LABORATORY PLASMA EXPERIMENTS

As was shown in § 4-2A, for an ion with simple coronal excitation, the intensity of a resonance line is given from eq. (4-2-9) by $N_e N_g C(g \to i)$.

This is proportional to N_e^2 and a function of T_e. Similarly, from eq. (4-2-28), the ratio of two resonance lines is given by a function of T_e only. Thus, measurements of spectral intensities from a plasma can be interpreted from two viewpoints. If the coefficients C are known as a function of temperature, the intensities can be used to determine T_e and possibly N_e. If on the other hand T_e and N_e are known from some other measurements, then the spectral intensities can be used to determine the excitation rate coefficients C. Non-spectroscopic techniques are available for T_e and N_e in laboratory plasmas, that cannot be used for astrophysical plasmas. We can thus think of the laboratory measurement which measures the C's as calibrating the technique for determining T_e and N_e for astrophysical plasmas. This is in general an oversimplification, and in a realistic case all the competing atomic processes must be evaluated for the two very different density regimes involved. Furthermore, since pulsed laboratory plasmas exhibit a transient ionizing condition, the temperature at which a particular ion exists may range from 1.3 to 3 times its ionization equilibrium temperature. The coefficients measured must then be scaled over this range before being applied to an astrophysical plasma.

The non-spectroscopic technique mentioned above is the measurement of the profile of ruby laser radiation after scattering by the free electrons in the plasma. This technique has been perfected over the last ten years and can now be regarded as a routine diagnostic for laboratory plasmas. It is described in review articles by Kunze [20] and by Evans and Katzenstein [21]. Since the scattered radiation is wavelength shifted by the Doppler effect of the free electron, this is a particularly direct method of determining T_e. We are concerned here with only the simplest form of Thomson scattering, since under these conditions the profile of the scattered radiation gives T_e while the intensity gives N_e. Since the laser beam can be brought to a focus within the plasma, and viewed at a scattering angle of 90°, it is possible by this method to measure T_e and N_e for any small volume element of the plasma.

There is now a wide range of plasma sources available for spectroscopic studies, as reviewed recently by Gabriel [22]. However only two types have so far been used for the simultaneous study of spectral intensities and laser scattering, thus enabling a quantitative measurement of excitation rates. The first of these was ZETA, a toroidal device built at Harwell for research into controlled fusion. This has been used by Boland et al. [23] for the measurement of excitation rates in lithium-like nitrogen v. The plasma under the conditions chosen had an electron temperature of 2.1×10^5 °K and a density $\sim 4 \times 10^{14}$ cm^{-3}. The temperature was measured by laser scattering, but the density was determined from the absolute intensity of the visible continuum. The laser gave a 300 MW pulse of 15 ns duration. The radiation scattered at 90° was dispersed and then split in the focal plane of the spectrometer into four photomultipliers covering different points on the Gaussian profile. Spectral intensities were recorded using a grazing-incidence spectrometer, which was intensity calibrated using the branching ratio technique [24]. The overall accuracy for the collisional rate coefficients was estimated to be $\pm 20\%$ for relative values and $\pm 50\%$ for absolute values. The results obtained will be discussed further in § 4-4F.

D. Theta-pinches

More recent measurements of excitation rates have involved the use of theta-pinch devices. Such machines have been developed in many laboratories over the last decade for the study of controlled fusion. The technique involves the discharge of a high-voltage capacitor bank into a single-turn cylindrical coil. This closely surrounds a quartz tube containing the plasma, and coupling of the electrical discharge to the plasma can be represented as a 1 : 1 transformer. The plasma is heated by a combination of shock-heating, Joule-heating, and adiabatic compression. In more sophisticated devices, further capacitor banks are used to pre-heat the plasma and to provide an initial bias magnetic field. The theta-pinch is useful for plasmas in the density range 10^{15} to 10^{17} cm^{-3} with temperatures up to 4×10^6 °K.

It was shown by Green et al. [25] that, due to cooling by axial thermal conduction, and the steep dependence of conduction on electron temperature, theta-pinches exhibit a maximum value of T_e, effectively stabilized over a wide range of energy input rates, and relatively insensitive to changes in the circuit parameters. A significant drop in T_e can be obtained by loss of efficiency, for example in the pre-heater, but only at a cost of reproducibility. For quantitative measurements, the temperature range is thus limited to 1.5×10^6 to 4×10^6. Variation of density in a particular device also presents difficulties. Changing the initial gas pressure is only partly effective, resulting mainly in a change in the final compressed plasma diameter. For larger variations it is necessary to alter the tube and coil diameters, which alters the magnetic compression field, and consequently the plasma pressure and density. This is clearly not an easy parameter to adjust, and only one device has been provided with such a readily-changeable tube and coil system. It will be clear from § 4-2, that in general one desires to work at the lowest possible electron density. However, when higher-order density effects become significant, it is important to be able to vary the density in order to evaluate them.

The first theta-pinch to be designed specifically for this type of work was built at the University of Maryland by DeSilva and Kunze [26]. It used a 9 kJ, 40 kV capacitor bank together with preheater and bias banks. Since it aimed primarily at low densities the coil diameter was 15 cm, i.e. about twice that conventionally used, and 32 cm long. For typical operating conditions, the filling pressure of 12 mtorr of hydrogen gave $T_e \sim 3 \times 10^6$ °K with $N_e \sim 3 \times 10^{15}$ cm^{-3}. A laser scattering system was used to measure both T_e and N_e. The beam from a 200 MW ruby laser passed axially along the discharge and was focused at the mid-plane of the coil. Radiation scattered at 90° was passed into a multichannel spectrometer, which allowed simultaneous measurements of seven portions of the spectral profile. The laser, optics and detecting system were mounted on a common carriage, permitting easy scanning of the scattering volume along a diameter of the plasma.

The Maryland theta-pinch has been used by Kunze et al. [27] for measurements on helium-like ions and by Kunze and Johnston for measurements on lithium-like ions [28] and beryllium-like ions [11]. The results will be discussed later in the appropriate sections. For some measurements, the bank has been increased to 15 kJ, enabling the density to increase to $\sim 6.5 \times 10^{15}$ cm^{-3}. Also a low density condition was established in work by Kunze and Gabriel [29], by filling with 1 mtorr of methane, which gave $N_e \sim 10^{15}$ cm^{-3}. For all of these experiments, the spectroscopic studies in the ultraviolet region were carried out by viewing the plasma axially from the ends of the tube.

A theta-pinch has now been constructed for spectroscopic studies at the Astrophysics Research Unit, Culham. Although similar in many respects to the Maryland machine, it has a higher energy storage of 48 kJ enabling it to produce plasmas of higher density. The temperature range is from 2.0×10^6 °K to 3.5×10^6 °K. It has been provided with interchangeable coil and tube assemblies of 15 cm and 8 cm diameters. These give density ranges of 3 to 15×10^{15} cm^{-3} and 4 to 15×10^{16} cm^{-3} respectively. Both tubes are provided with side arms at their mid-planes so that the spectroscopic measurements can be made radially at the centre of the plasma. The layout is shown in Fig. 4-3-1. The coil is shown slotted at its mid-plane and also in two other positions. This, and the slots in the collector plates, facilitate removing the end sections, thereby reducing the coil length by a factor of 2. The laser scattering is similar to that used at Maryland. However, the

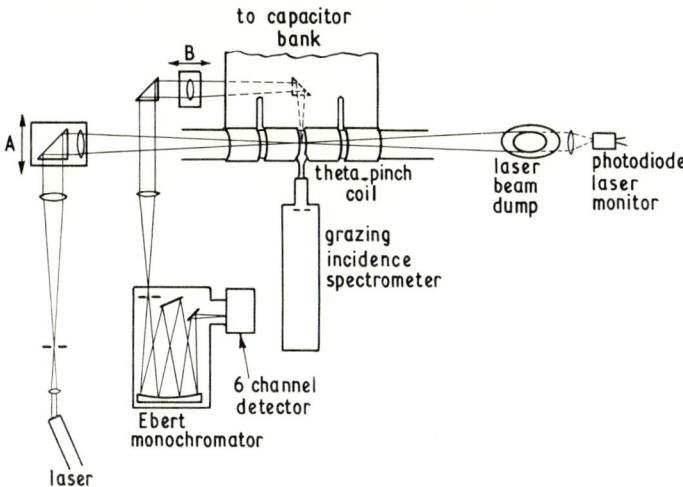

Fig. 4-3-1. Layout of the theta-pinch device at the Astrophysics Research Unit, showing the laser scattering system and ultraviolet spectroscopic apparatus.

method of scanning the viewing volume radially is somewhat different. The two tables marked A and B are mounted on calibrated slides and are each moved in the directions indicated, by the same distance. The side-arm viewing offers two advantages. Since the path length in the plasma is reduced, the problem of self-absorption of lines in the plasma is much less. Furthermore, since the viewing direction is the same line-of-sight as the radial scan of the scattering volume, it is possible to make precise corrections for any changes in T_e or N_e along this line. In practice such variations are small and are often ignored. The A.R.U. theta-pinch has now been used for a number of experiments. Tondello and McWhirter [30] have studied beryllium-like neon VII spectra, Gabriel and Paget [31] have studied dielectronic satellite lines from lithium-like nitrogen and oxygen and Gabriel and Paget have carried out further experiments in collaboration with the Maryland group on helium-like carbon, nitrogen and oxygen [32]. All these are considered later in their appropriate sections.

§ 4-4. Lithium-like ions

A. General description

Apart from the hydrogenic ions the lithium-like ions are the simplest to consider, both from the point of view of theoretical calculations and of

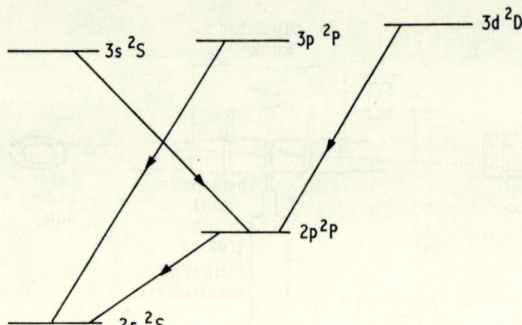

Fig. 4-4-1. Partial term scheme for lithium-like ions, showing transitions observed in solar spectra.

laboratory experiments. As for hydrogenic ions there is only one electron outside the closed shell. However the absence of a metastable level (2s is now the ground state) makes the lithium-like ions even simpler than the hydrogenic ions, and leads to the coronal excitation model (given by eq. (4-2-3)), being valid for levels with $n = 2$ and $n = 3$ for densities up to about 10^{14} cm^{-3} (in nitrogen v). Figure 4-4-1 shows the term diagram for levels in configurations with $n = 2$ and $n = 3$. Transitions observed in solar spectra are marked.

The lithium-like ions are, in principle, particularly suitable for making measurements of electron temperatures. The wide separation in energy of the levels with $n = 2$ and $n = 3$ gives good temperature sensitivity to the relative intensities of lines from these levels (see § 4-2E). In view of the accuracy of the theoretical atomic data in this iso-electronic sequence, further intensity data for the lines from the lithium-like ions in the solar spectrum would be valuable.

The excitation cross-sections for these ions have been calculated in several approximations. Laboratory experiments have confirmed the theoretical excitation rates between the $n = 2$ and $n = 3$ levels within the experimental error of a factor of two, but more accurate measurements would be useful.

The lithium-like ions have important series above their first ionization limit in which one of the 1s electrons is excited. These levels give rise to prominent satellite spectra adjacent to the resonance lines of the helium-like ions. Such levels can be populated by dielectronic capture, and in some high density laboratory sources direct inner shell excitation also contributes.

B. ENERGY LEVELS

The energies of levels with $n = 2, 3$ are known in lithium-like ions of all elements as far along the isoelectronic sequence as chlorine XV. The 2p ^{2}P levels are given in Moore [33] for ions of lithium to oxygen. The neon VIII 2s ^{2}S–2p ^{2}P transitions have been classified by Fawcett et al. [34] and by Bokasten et al. [35]. Recent work by Fawcett [36] has extended the classification of the ^{2}P levels to chlorine XV. The 2s ^{2}S–2p ^{2}P transitions in neon VIII, magnesium X and silicon XII were first identified in the solar spectrum by Detwiler et al. [37] using wavelengths predicted by Edlén [38, 39].

The $n = 3$ levels are complete in Moore [33] for ions as high as aluminium XI, except for neon VIII, for which the energy levels are given by Fawcett et al. [40] and House and Sawyer [41]. The $n = 3$ levels are now known as far along the sequence as chlorine XV, from the recent work by Tondello [42] on silicon XII, by Fawcett et al. [43] on phosphorous XIII and sulphur XIV and by Fawcett [36] on chlorine XV. More accurate wavelengths for some of the transitions between the levels with $n = 2$ and $n = 3$ have been given by Feldman et al. [44].

Transitions from levels above the ionization limit which give rise to satellite spectra have been observed for many years [45]. In particular, transitions from the 1s 2s 2p and 1s 2p^2 configurations have recently been classified by Gabriel and Jordan [46] in ions from carbon IV to aluminium XI. Extrapolations along the iso-electronic sequence, based on the above data, have been used to propose classifications for lines from higher ions observed in the solar spectrum [47, 48, 49].

C. TRANSITION PROBABILITIES

The tabulations of Wiese et al. [50] give the transition probabilities for transitions between levels with $n = 2$ and $n = 3$ for ions as far as neon VIII in the lithium isoelectronic sequence. The values he quotes are those calculated by Weiss [51] using a Hartree-Fock self consistent field method. The results are expected to be accurate to within 10%.

Data for ions between sodium IX and phosphorous XIII are given by Wiese et al. [52]. For the 2s ^{2}S–2p ^{2}P transition the values quoted are from calculations by Cohen and Dalgarno [53]. For other transitions the Coulomb approximation has been used; these values should be accurate to within $\pm 10\%$ because there is only one electron outside the closed shell and the stage of ionization is high.

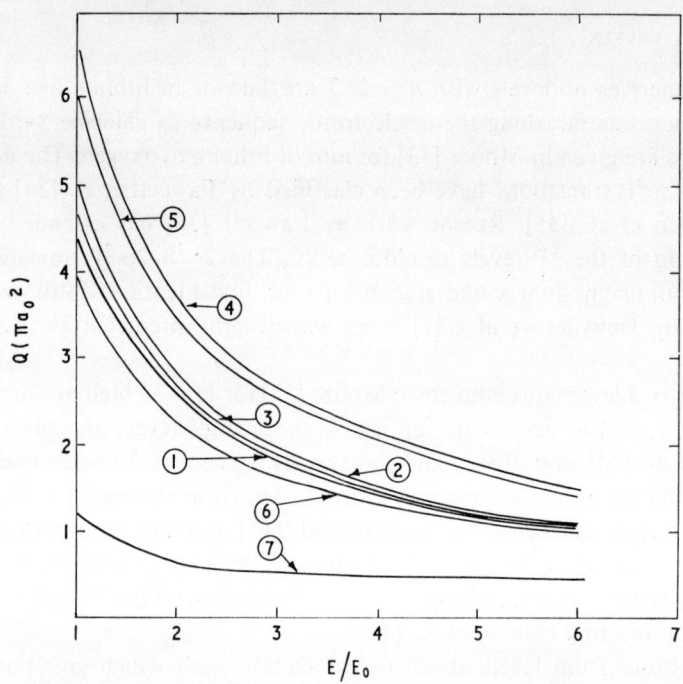

Fig. 4-4-2. The 2s → 2p cross-section as a function of incident energy, calculated in the following approximations; (1) close-coupling [54], (2) Coulomb–Born I [8], (3) Coulomb–Born II [8], (4) Coulomb-Bethe I [8], (5) Coulomb–Bethe II [8], (6) the impact-parameter method [6], (7) the $\bar{g}$ approximation [3].

D. Collisional excitation rates

The collision cross-sections for the 2s–2p transition in lithium-like ions have been calculated by several methods, and the values obtained using the more reliable approximations suggest that these results are accurate to within 10%. Several calculations of the cross-section as a function of the energy of the incident electron, for the 2s–2p transition in nitrogen V are shown in Fig. 4-4-2. The results shown are from the close-coupling calculations by Burke et al. [54]; from Bely's [8] calculations in the Coulomb–Born I and II and Coulomb–Bethe I and II approximations; from the impact parameter method by Burgess [6], and from the $\bar{g}$ approximation given by Van Regemorter [3]. It is apparent from Fig. 4-4-2 that the Coulomb–Born I and II and impact parameter methods give values which agree well with those from the close coupling calculations. However, the $\bar{g}$ approximation underestimates the cross-section by a factor of about four near threshold.

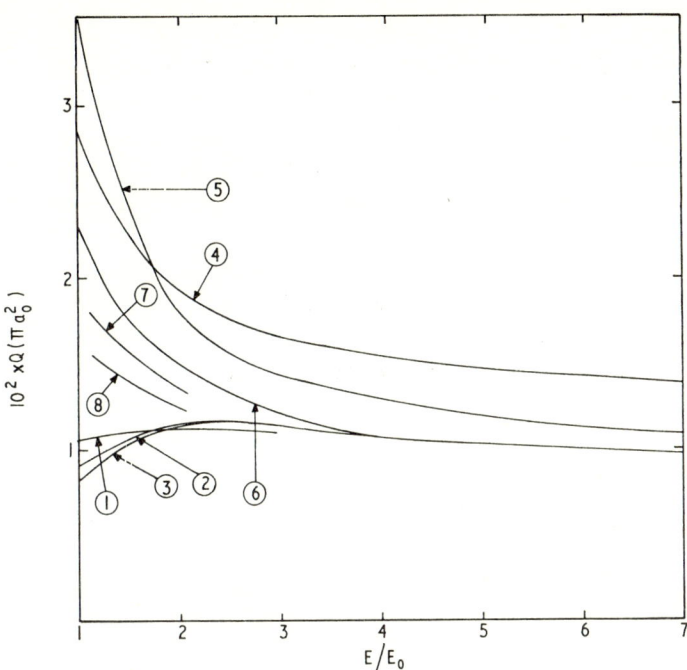

Fig. 4-4-3. The 2s → 3p cross-section as a function of incident energy, calculated in the following approximations; (1) strong-coupling, no exchange [54], (2) Coulomb–Born I [8], (3) Coulomb–Born II [8], (4) the impact-parameter method [6], (5) the $\bar{g}$ approximation [3], (6) the "best estimate" of Burke et al. [54], (7) close-coupling, no exchange [54], (8) strong-coupling, with exchange [54].

Figure 4-4-3 shows the 2s→3p cross-section calculated using various approximations. The results illustrated are from the Coulomb-Born I and II calculations by Bely [8], from the impact-parameter and $\bar{g}$ approximations [6, 3] and from several calculations by Burke et al. [54]. Burke et al. have calculated the cross-section using strong coupling (i.e. only initial and final states included) with and without exchange; using close coupling (all $n = 2$ and $n = 3$ levels included) with and without exchange and have made a "best estimate" of the cross-section by adding the effects of close coupling and exchange to the values obtained using the strong-coupling no exchange approximation. There is obviously greater uncertainty (about a factor of 2 at threshold) in the 2s→3p cross-section than in the 2s→2p cross-section.

Burke et al. [54] have also calculated the 2s → 3s, 3d cross-sections in nitrogen V using the same approximations as above, and the 2p → 3s, 3p, 3d; 3s → 3p, 3d and 3p → 3d cross-sections using the Coulomb–Born approximation, since for this latter group of transitions the different approximations give very similar results. They suggest that their results are accurate to within 20 % near threshold, except for the 2p–3s transition where the uncertainty is larger. Bely and Petrini [55] have recently pointed out that the calculations of Burke et al. for the 2p–3p and 2p–3d transitions contain a small computational error, and the results of Bely and Petrini (see below) should therefore be used for these transitions.

Bely [56] and Bely and Petrini [55] have calculated the 2s → ns, nd and 2p → ns, np, nd cross-sections, where $n \geq 3$, and have also used the Coulomb–Born approximation. They give results for beryllium II, nitrogen V, neon VIII and $z = \infty$ so that the cross-sections for any ion in the iso-electronic sequence can be derived. As several authors have pointed out, it is interesting that the permitted transitions do not always have the largest cross-section; in particular the 2s → 3s and 2s → 3d cross-sections are larger than those for 2s → 3p, and the 2p → 3p cross-sections are larger than those for 2p → 3s and are the same order of magnitude as those for 2p → 3d. It should be noticed that in eq. (4-2-16), i.e.

$$\Omega = Q g_1 E / \pi a_0^2,$$

Bely [8, 56] uses, for the statistical weight of an LS term, $g_1 = (2L+1)$ and not as is more usual (Allen [57]) $g_1 = (2L+1)(2S+1)$.

Recently, Flower [58] has used the general computer programs developed by Eissner and Nussbaumer [59, 60], to calculate the collision strengths for transitions between the $n = 2$ and $n = 3$ levels in nitrogen V and silicon XII in the distorted wave approximation (see also Saraph et al. [61]). In nitrogen V his results for 2s–2p agree almost exactly with those of Burke et al [54]. For 2s–3p his values agree to within a few percent with those of Bely [8]. For other transitions there are differences of up to 30 % between Flower's results and those of Bely [56] and Bely and Petrini [55]. For silicon XII Flower compares his computed values with those obtained from Bely [8, 56] and Bely and Petrini [55] using their suggested interpolation procedure. Two sets of results agree within 30 %, except for the 2s–3s transition for which Flower's value is about 40 % larger than that obtained by Bely and Petrini.

E. STATISTICAL EQUILIBRIUM

For $N_e \lesssim 10^{14}$ cm^{-3} the simple coronal model can be applied to the 2p levels and the levels with $n = 3$; i.e. the population of each excited level is determined by a balance between the excitation rate for collisions from the ground term and the spontaneous radiative decay to all levels.

At higher densities ($N_e \gtrsim 10^{14}$ cm^{-3} for nitrogen v) or in conditions where the 2s–2p transition becomes optically thick, collisions from 2p to 3s, 3p, 3d and collisions between the levels with $n = 3$ must also be included.

At temperatures close to that at which an ion has its maximum steady-state population, cascades from levels with $n = 3$ do not make a significant contribution to the 2p population. However, at higher temperatures (say 10 times that at which the ion has its maximum steady-state population), cascades from levels with $n = 3$ can contribute about 10 % of the total rate to the 2p level.

The large energy difference between the levels with $n = 2$ and $n = 3$ leads to a strong temperature dependence in the relative intensities of lines from the 2p level and $n = 3$ levels (see § 4-2E, eq. (4-2-28)). These line intensity ratios are therefore suitable for making measurements of the electron temperature, particularly in laboratory plasmas which have a uniform electron density distribution. The temperature and density gradients in the solar atmosphere introduce difficulties which reduce the sensitivity of the method (see § 4-4G).

F. Line intensities in laboratory sources

The absolute intensities of transitions between the levels with $n = 2$ and $n = 3$ can be measured in a laboratory plasma under conditions where the electron temperature and density are uniform. If the electron density is low enough for the coronal model to be applicable then the actual value of N_e need not be known in order to interpret relative intensities. Two approaches can be applied to the interpretation of the line intensities. The first has been discussed in § 4-2E and this is to use the theoretical calculations of the excitation cross-sections to determine the electron temperature. The second approach is the converse, which is to assume that the electron temperature and density are known and hence determine the excitation rate coefficients.

The former approach was first used by Heroux [62, 63] to measure the electron temperature in the ZETA plasma device, from the intensity ratio of the 2s–2p and 2s–3p transitions in nitrogen v. The actual temperature results are now superseded because the cross-sections he used ($\bar{g} = 0.20$ in eq. (4-2-12)) have been shown to be incorrect.

Two groups of experimenters have used the converse approach and have made laboratory measurements of excitation rate coefficients. Boland et al.

[23] have measured the excitation rate coefficients for the 2s–2p; 2s–3s, 3p, 3d and 2s–4p transitions in nitrogen v and Kunze and Johnston [28] have measured the coefficients for the 2s-2p; 2s–3s, 3p, 3d and 2s–4s, 4p, 4d transitions in nitrogen v, oxygen vi and neon viii. The experimental methods used by these groups have been described in § 4-3C and § 4-3D.

Boland et al. [23] used the plasma device ZETA, in which $N_e \sim 10^{14}$ cm^{-3} and $T_e \sim 2 \times 10^5$ °K. Under these conditions the coronal excitation model can be used, except for the 3d ^{2}D levels. For 3d ^{2}D the additional process of collisional excitation from 2p ^{2}P must be included, even if the transition 2s ^{2}S–2p ^{2}P is optically thin. In the above experiment the optical depth for the 2s ^{2}S–2p ^{2}P transition was 0.38 at the line centre, which implies a probability of escape for the radiation of 0.75 and an enhancement of the 2p ^{2}P population by the reciprocal of this factor. Boland et al. find that 29 % of the excitations to 3d ^{2}D originate from the 2p levels. A comparison between their deduced rates for the 2s → 2p; 2s → 3s, 3p, 3d and 2s → 4p transitions in nitrogen v, the theoretical calculations of Burke et al. [54] and Taylor and Lewis [64], and the $\bar{g}$ approximation (from Van Regemorter [3]) is shown in Fig. 4-4-4. The results of Boland et al. are indicated by

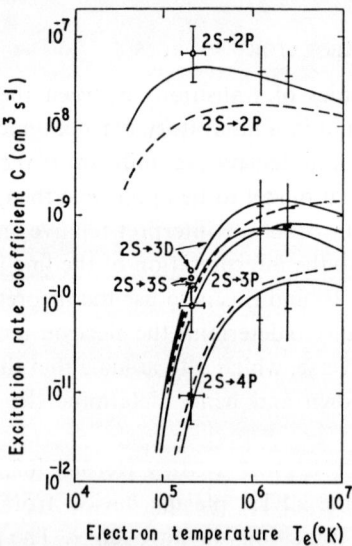

Fig. 4-4-4. The observed and calculated excitation rate coefficients for transitions in nitrogen v. Circles denote the experimental results of Boland et al. [23], crosses denote the experimental results of Kunze and Johnston [28]. The full lines are the theoretical rates from the cross-sections of Burke et al. [54] and Taylor and Lewis [64] (for the 2s–4p only); the dashed lines are the theoretical rates using the $\bar{g}$ approximation [3].

circles. When the experimentally determined excitation rate coefficient for the 2s → 2p transition is normalized to the theoretical value, the measured rates for the other transitions agree with the calculations of Burke et al. and of Taylor and Lewis to within the expected relative error of $\pm 20\,\%$. The uncertainty in the absolute experimental values is given as $\pm 50\,\%$.

Kunze and Johnston [28] used a theta-pinch device in which the temperature and density were considerably higher; $N_e \sim 4 \times 10^{15}\,\mathrm{cm}^{-3}$ and $T_e \sim 10^6\,°\mathrm{K}$. The measured total excitation rates to the excited levels therefore include contributions from several processes. However, since observations were made at a number of densities and temperatures the contributions from the various processes could be studied. In particular, the 2s–2p, 2s–3p and 2p–3d transitions in oxygen VI suffered from self-absorption. The high 2p ^{2}P population and large N_e lead to significant loss rates from the 2p ^{2}P levels to the 3s, 3p and 3d levels. The high values of T_e lead to large 2s → 3s, 3p, 3d excitation rates which give significant contributions to the 2p ^{2}P population through cascades. When considering the 2p ^{2}P population the above terms tend to cancel each other. Cascades from levels with $n \geq 4$ were found to be negligible (less than 7 % for nitrogen V).

For nitrogen V, the lowest ion studied, and at the highest densities used, $N_e \sim 6 \times 10^{15}\,\mathrm{cm}^{-2}$, collisions between levels of the same principal quantum number are comparable or even larger than the radiative rates. When this occurs the excitation rates to the individual levels cannot be deduced because the latter tend towards relative populations proportional to their statistical weights. Thus for nitrogen V only the average rates from 2s plus 2p to 3p plus 3d can be found. The experimental results of Kunze and Johnston [28] are compared with the theoretical calculations of Burke et al. [54] and Taylor and Lewis [64] in Fig. 4-4-4, and are indicated by crosses.

G. Solar Line Intensities

The method of using "known" excitation rates to determine electron temperatures has been applied by Heroux and Cohen [65] to observed solar intensities. They have compared temperatures derived from the observed relative intensities of the 2s–2p, 2s–3p, 2p–3s and 2p–3d transitions in the ions oxygen VI, neon VIII and magnesium X, using the theoretical excitation cross-sections of Burke et al. [54] and Bely [8, 56], with those predicted from ionization equilibrium calculations. The main experimental problems in the method are (i) to establish an accurate intensity calibration over the wide range of wavelengths (50–1250 Å) needed to include the relevant transitions, and (ii) to measure the intensity of the weak 2–3 transitions and resolve them from the many other lines between 50 Å and 150 Å. The observed intensity ratios are claimed to be accurate to within ±40 %.

Because the electron density in the solar atmosphere where the above ions are formed is low ($N_e \lesssim 10^{10}$ cm^{-3}), the simple coronal model can be applied to the excited levels. However, because the electron density and temperature vary through the solar chromosphere and corona, eq. (4-2-28) should not be applied directly to the solar relative intensities. Instead, an integration over the whole atmosphere should be used. The relative intensity of two lines, for example the 2s–2p and 2s–3p transitions, is then given by

$$\frac{I(i \to g)}{I(j \to g)} = \frac{\int_V C(g \to i)(N(\text{ion})/N(E))N_e^2 \, dV}{\int_V C(g \to j)(N(\text{ion})/N(E))N_e^2 \, dV} \tag{4-4-1}$$

where $N(\text{ion})/N(E)$ is the fraction of the element in the relevant stage of ionization, and the integration is over the whole volume V of the atmosphere. At low densities (the limit depends on the stage of ionization) and in the steady state situation, $N(\text{ion})/N(E)$ is independent of the electron density and is given by a balance between the collisional ionization rate and the radiative plus di-electronic recombination rate.

At the densities found in the solar transition region and corona, $N_e \sim 5 \times 10^9$ to 2×10^8 cm^{-3}, the di-electronic rate given by Burgess [14] should be reduced because many of the levels to which the recombination takes place (see § 4-2D) are above the thermal limit, i.e. collisions to higher levels take place more rapidly than decay to lower levels. Thus a density term enters the values of $N(\text{ion})/N(E)$. Jordan [66] has calculated $N(\text{ion})/N(E)$ as a function of temperature for all elements abundant in the sun, for both the low density case and also for the density dependent case using a model of the solar atmosphere [67].

Heroux and Cohen have derived temperatures from their observed line intensity ratios by assuming that the lines from a given ion are formed in the same volume of the atmosphere. They have allowed for the temperature variation over this volume but not for the variation of density or the temperature gradient. They put (from eq. (4-4-1))

$$\frac{I(i \to g)}{I(j \to g)} = \frac{\int (C(g \to i)/C(g \to j))(N(\text{ion})/N(E)) \, dT_e}{\int (N(\text{ion})/N(E)) \, dT_e} \tag{4-4-2}$$

thereby assuming that $N_e^2 \, dV/dT_e$, or $N_e^2 \, dh/dT_e$ is constant over the temperature range in which each line is formed. Hence they calculate T_e using the theoretical values of Burke et al. [54] and Bely [8, 56] for the ratio of excitation rates. The temperatures they derive are given in the last column of Table 4-4-1.

TABLE 4-4-1

Ion	Transition	T_m (°K) for solar N_e	T pred. (°K)	T_{obs} (°K) Heroux and Cohen [65] (taking ratio to 2s–2p)
O VI	2s–2p	2.9×10^5	3.1×10^5	
	2s–3p	3.2×10^5	$[0.32–1.4 \times 10^6]$*	4.0×10^5
	2p–3s	3.2×10^5	$[0.32–1.4 \times 10^6]$*	3.7×10^5
	2p–3d	3.2×10^5	$[0.32–1.4 \times 10^6]$*	3.3×10^5
Ne VIII	2s–2p	5.9×10^5	8.0×10^5	
	2p–3s	6.6×10^5	1.0×10^6	1.1×10^6
Mg X	2s–2p	1.10×10^6	1.18×10^6	
	2s–3p	1.18×10^6	1.26×10^6	1.3×10^6
	2p–3s	1.18×10^6	1.26×10^6	1.4×10^6
	2p–3d	1.18×10^6	1.26×10^6	1.5×10^6

* No clear peak – see text.

It is interesting that the observed temperatures derived from the $\Delta n = 1$ transitions are in good agreement. Heroux and Cohen used collision strengths for the 2s–3p transitions based on the "best estimate" of Burke et al. [54] (see Fig. 4-4-3). The observations therefore tend to support a value for the 2s–3p collision strength which at threshold is a factor of two larger than that derived from the Coulomb–Born approximation.

When considering the temperature at which lines in the solar extreme ultraviolet spectrum are formed it has been usual (Pottasch [68]) to assume that each line is formed predominantly at the temperature, T_m, at which the function

$$g(T) = T_e^{-\frac{1}{2}} \frac{N(\text{ion})}{N(\text{E})} \exp\left(-E(gi)/kT_e\right) \qquad (4\text{-}4\text{-}3)$$

has its maximum value for the relevant line, because the function falls steeply at temperatures below or above T_m. The values of T_m for the observed lines are also given in Table 4-4-1. The integral over volume in eq. (4-4-1) is then restricted to a volume ΔV, corresponding to the region where $\Delta T = 1.4\, T_m$

to 0.70 T_m. The distribution of $\int_{\Delta T} N_e^2 (dh/dT_e) dT_e$ with T_e can be found by using many lines. The value of $\int_{\Delta T} N_e^2 (dh/dT_e) dT_e$ is much greater in the corona than in the transition region, and this contribution should be included in a summation over the whole atmosphere. Thus it would be expected that a line is formed predominantly at the temperature, T_{pred}, at which the function

$$g(T) N_e^2 (dh/dT_e) \Delta T_e \tag{4-4-4}$$

has its maximum value. The model by Jordan [67] has been used for the variation of N_e^2 and dh/dT_e with temperature. This model has a quiet coronal temperature of 1.4×10^6 °K. As can be seen from Table 4-4-1 the "observed" temperatures are higher than the average values of T_m for the lines concerned. The cause of the difference between T_m and T_{pred} is the large increase in $\int_{\Delta T} N_e^2 (dh/dT_e) dT_e$ between the transition region ($T_e \sim 10^5$ °K) and the corona ($T_e \sim 10^6$ °K), combined with the slow fall off in $N(\text{ion})/N(\text{E})$ with temperature, for the lithium-like ions. This contribution from coronal material is not very important for the 2s–2p transition in oxygen VI, but it has a large effect on the $n = 2$ to $n = 3$ transitions. Instead of showing a clear peak at $T_e \approx 3.2 \times 10^5$ °K, the computed emission is roughly constant between 3.2×10^5 °K and 1.4×10^6 °K. Since the "observed" temperature is an average from the 2s–2p and $n = 2$ to $n = 3$ transitions it will depend critically on the amount of material present in the corona as a function of temperature. If the term $N_e^2 \, dh/dT_e$ is included in the equation used by Heroux and Cohen the temperatures derived from their observations should be *reduced*, since smaller values of $C(g \to i)/C(g \to j)$ would be needed to give the observed intensity ratios. The agreement between observed and predicted temperatures would then be improved. The coronal contribution to the neon VIII emission is important for both the 2s–2p and 2p–3s transitions. The temperature at which the magnesium X lines would have maximum emission in a uniform atmosphere is close to the mean quiet coronal temperature so that coronal contribution dominates both the 2s–2p and $n = 2$ to $n = 3$ transitions. Thus the magnesium X line ratios should indicate the quiet coronal temperature. The lines from higher ions such as silicon XII would be sensitive to the amount of material present in active regions.

From the above discussion it can be seen that apart from the lines of magnesium X the solar lithium-like ion lines do not provide a good method of measuring the electron temperature unless it can be assumed that a model of the transition region and corona is known. As discussed in § 4-5G the line ratios in beryllium-like ions are more suitable for measuring solar electron temperatures. However, if the electron temperature distribution in the solar atmosphere could be established from the beryllium-like ions, the observed line ratios from the lithium-like ions could be used to check the values of $N_e^2 \, dh/dT_e$ as a function of temperature in the available models.

H. Dielectronic Satellite Lines

The mechanism for the production of satellites to resonance lines by dielectronic recombination was described for the general case in § 4-2D. Conditions favourable for the production of these depend on the individual term systems and on changes in ionization energy with ion stage. As a consequence, recombination from helium-like to lithium-like ions provides the most prominent satellites, and these appear as weak features to the long wavelength of the helium-like resonance lines. Classification of the states involved, arising from the configurations 1s 2s 2p and 1s 2p² was carried out by Gabriel and Jordan [46], who also proposed dielectronic capture as the populating mechanism. The quantitative theory was presented by Gabriel et al. [15] who pointed out that for these, the lowest levels involved in dielectronic capture, the autoionizing rate A_a is much larger than the radiative decay rate A_r. Equation (4-2-26) for the intensity ratio of satellite to resonance line then reduces to

$$\frac{\alpha_s}{C(i \to j)} = \frac{1.2 \times 10^{-13}}{T_e} \frac{g_s}{g_i} \frac{E(ij)A_r}{Pf(i \to j)} \exp\left((E(ij) - E_s)/kT_e\right) \quad (4\text{-}4\text{-}5)$$

so that the intensities are now independent of A_a. This approximation $A_a \gg A_r$ is valid for the ions carbon IV to oxygen VI as observed in laboratory plasmas, but not for higher ions in the sequence.

A laboratory experiment to measure the satellite intensities has been carried out by Gabriel and Paget [31] using the A.R.U. theta-pinch apparatus described in § 4-3D. For the ions nitrogen V and oxygen VI, for which LS coupling applies, the two intense satellites arise from the transitions $1s^2 \, 2s \,^2S\text{-}1s \, 2p(^1P)2s \,^2P^0$ and $1s^2 \, 2p \,^2P^0\text{-}1s \, 2p^2 \,^2D$, having approximately equal intensities. The predicted relative intensities of the $^2S\text{-}^2P^0$ satellite to the resonance line, for nitrogen and oxygen, are shown in Fig. 4-4-5. These are derived from eq. (4-4-5), with a correction for the dielectronic contribution to the resonance line, according to eq. (4-2-24). The observed ratios are also indicated and can be seen to be in agreement with theory to within the experimental accuracy of ±20%. Since the measurements were carried out at densities varying by a factor of 5, the agreement with theory also supports the absence of any density dependence in the ratio.

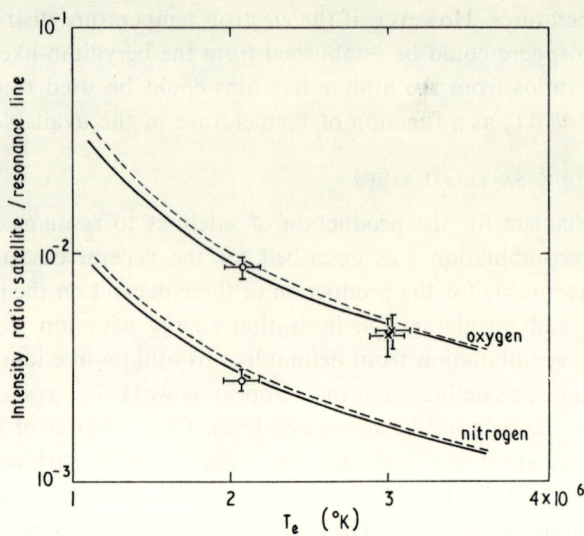

Fig. 4-4-5. Comparison of observed and calculated values for the intensity ratio of the ^{2}S–^{2}P^0 satellite to the ^{1}S–^{1}P^0 helium-like resonance line [31]. The dashed curves are computed from eq. (4-4-5), while the full curves have been corrected for the di-electronic contribution to the resonance line, according to eq. (4-2-24). The points marked by circles were measured at $N_e = 1 \times 10^{16}$ cm^{-3}, and those marked by crosses at $N_e = 5 \times 10^{16}$ cm^{-3}.

For higher ions in the sequence measurements of satellite ratios have been made in solar flares. Walker and Rugge [49] have observed silicon XII and sulphur XIV while Neupert and Swartz have observed iron XXIV. For these ions, the assumptions of LS coupling, non-relativistic mechanics, and $A_a \gg A_r$ will break down, and a more complete theory is required. If however, the above simple theory is used as an approximation, then satellite to resonance line ratios are expected to be of the order of 0.1 for silicon, 0.3 for calcium and 1 for iron. These ratios are broadly consistent with those observed.

§ 4-5. Beryllium-like ions

A. General description

The additional 2s electron causes the beryllium-like ions to have a more complicated term structure than that of the lithium-like ions. Figure 4-5-1 shows the term diagram for the ground and first two excited configurations. The major additional complication is the existence of the metastable 2s 2p ^{3}P$_{0,1,2}$ levels. However, as discussed below and in § 4-2G, these metastable levels do introduce the possibility of observing relative line intensities which are sensitive to the electron density in the emitting plasma.

As in the lithium-like ions, the levels with $n = 2$ and $n = 3$ are widely separated and the relative intensities of lines originating from these levels provide a means of measuring electron temperatures.

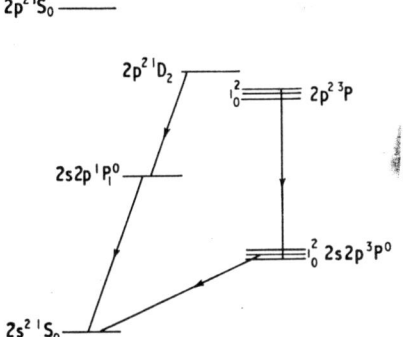

Fig. 4-5-1. Partial term diagram for the beryllium-like ions, showing transitions observed in solar spectra.

Many new atomic data for the beryllium-like ions have become available in the past few years, although work is still in progress on energy levels and collision strengths.

B. Energy Levels

The energy levels of nearly all terms of interest have been known for many years in several ions of astrophysical interest, i.e. carbon III, nitrogen IV, oxygen V and magnesium IX (Moore [33]). Kelly [69] lists further classifications in neon VII and silicon XI, and the situation in these and higher ions has been considerably improved by the recent work of Tondello and Paget [70] on neon VII; Tondello [42] on silicon XI; Fawcett et al. [43] on phosphorous XII and sulphur XIII, and of Fawcett [36] on sodium VII–chlorine XIV. The absolute position of the triplet terms, and in particular the wavelength of the intercombination transition $2s^2$ $^1S_0$–$2s\,2p$ $^3P_1$ has been determined in carbon III, nitrogen IV, oxygen V and silicon XI from the solar observations of Burton et al. [71] and of Burton and Ridgeley [72]. Edlén et al. [73] have confirmed the identifications in carbon III to oxygen V by laboratory experiments. Tondello and Paget [70] have recently observed the neon VII transition in a laboratory plasma. However, the absolute position of the 2s 2p ^{3}P term is still unknown in sodium VIII, magnesium IX, sulphur XIII and higher ions.

C. Transition Probabilities

The tabulations of Wiese et al. [50] provide oscillator strengths for most of the permitted transitions of interest in ions up to oxygen V along the iso-electronic sequence, but the tabulations are less complete for neon VII. Transition probabilities for forbidden transitions within the 2s 2p configuration are also given in ref. [50]. The more recent publication by Wiese et al. [52] gives some permitted oscillator strengths for singlet transitions in magnesium IX and silicon XI, but includes only one triplet transition. However, ref. [52] does include transition probabilities for transitions within the 2s 2p configuration and for the $2s^2\ ^1S_0$–2s 2p $^3P_{1,2}$ transitions.

The $2s^2\ ^1S_0$–2s 2p $^3P_1$ transition probabilities for carbon III to neon VII have been calculated by Garstang and Shamey [74] and by Naqvi [75]. Values for the $2s^2\ ^1S_0$–2s 2p $^3P_2$ transition in carbon III to oxygen V have been calculated by Osterbrock [76] using the method given by Garstang [77]. The values for neon VII and magnesium IX may be interpolated between oxygen V and silicon XI using the scaling of Z^3 given in ref. [77].

Oscillator strengths for the resonance line and transitions between excited levels in carbon III, nitrogen IX and oxygen V have recently been calculated by Friedrich and Trefftz [78] including configuration mixing and further values may soon be available from the program of Eissner and Nussbaumer [60]; so far only results for nitrogen IV have been published (Nussbaumer [79]). See also § 4–5I.

D. Collisional excitation rates

Osterbrock [76] has computed the collision strengths for the $2s^2\ ^1S$–2s 2p ^1P and $2s^2\ ^1S$–2s 2p ^3P transitions in boron II to oxygen V and neon VII, by solving the coupled integro-differential equations by an iterative procedure. He used the formulation of Saraph et al. [61] for the coupled equations. The ion wave functions were calculated from the program of Eissner and Nussbaumer [60], which includes configuration mixing. The main approximation in the solution of the equations is the neglect of interaction between states other than those in the $2s^2$ and 2s2p configuration. For the $2s^2\ ^1S$–2s 2p ^3P transition, results are given for energies above and below the ^1P threshold. In the latter case, the collision strengths have been averaged over the resonances below the 2s 2p ^1P level using the method of Gailitis [80].

The only other precise calculations available for comparison with the results of Osterbrock are those of Eissner [81], who has used the distorted wave approximation. Eissner has calculated the collision strengths including (a) two configurations only, $2s^2$ and $2s\,2p$, (b) three configurations, $2s^2$, $2s\,2p$ and $2p^2$, and (c) for neon VII, six configurations, as above plus $2s\,3s$, $2s\,3p$ and $2s\,3d$. Eissner and Osterbrock have both investigated the effect of including a different number of partial wave-contributions (see below).

A less accurate method of calculating the collision rates for permitted transitions is to use the weak-coupling impact-parameter approximation [5] which makes no allowance for exchange and includes only the final and initial states. Blaha [10] has used his simplified Coulomb–Born approximation to obtain effective gaunt factors for the permitted resonance transitions in beryllium-like ions. Blaha's results are 10–20 % larger than those derived using the impact-parameter method, in accordance with his conclusion that his method tends to overestimate the gaunt factors by that amount.

Figures 4-5-2 to 4-5-6 show comparisons between several sets of calculations of Ω, the collision strength, for the $2s^2\,{}^1S-2s\,2p\,{}^1P$ and $2s\,2p\,{}^3P-2p^2\,{}^3P$ transitions in carbon III, oxygen V and neon VII. They show (i) Eissner's results including two and three configurations, and the effect of including states with $l \leq 5$ or all l states. (ii) Osterbrock's results including two configurations, with $l < 7$ and all l states. (iii) Results using the impact-parameter tables of Burgess. (iv) Results using the $\bar{g}$ approximation given by Van Regemorter [3].

It can be seen that close to threshold the inclusion of states with $l > 5$ hardly changes the values of Ω, since these states become important only at high energies. However, the inclusion of the third configuration $2p^2$ has a marked effect on the $2s^2\,{}^1S-2s\,2p\,{}^1P$ collision strength. The three configuration values are between factors of 1.5 and 1.7 smaller than the two-configuration values. The impact-parameter method, essentially a two-configuration calculation, gives results up to 30 % smaller than the two-configuration results of Eissner. The impact-parameter method *overestimates* the collision strength compared with Eissner's three-configuration results by a factor of 1.6 in carbon III, but only by about 25 % in higher ions. The values of Ω near threshold calculated using the values of $\bar{g}$ given by Van Regemorter are between factors of 1.4 and 2 smaller than those from Eissner's three configuration calculations.

For the $2s\,2p\,{}^3P-2p^2\,{}^3P$ transition the agreement between Eissner's three configuration calculations and the impact-parameter approximation is quite good, with the values of Ω near threshold agreeing within 20 %.

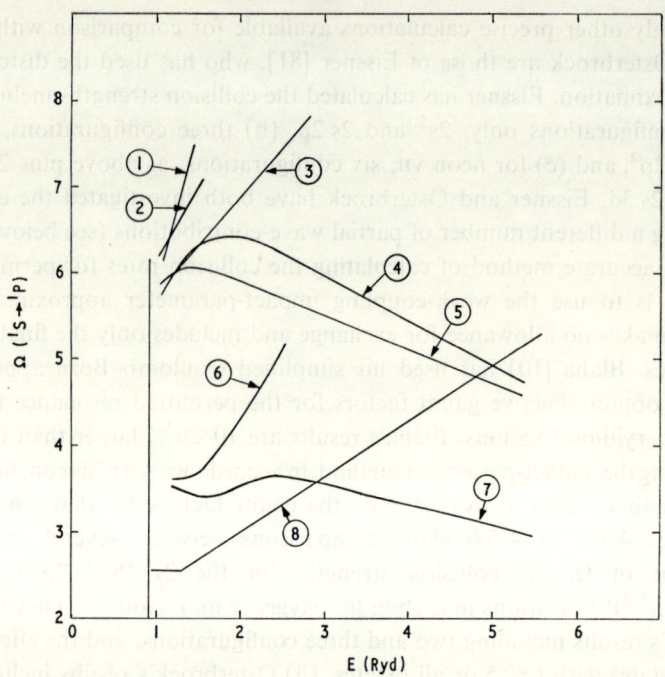

Fig. 4-5-2. The collision strength for the $2s^2$ ^1S–$2s\,2p$ ^1P transition in carbon III, as a function of incident energy, calculated in the following approximations: (1) distorted wave, two configurations, $\Sigma\,l = \infty$ [81], (2) coupled equations, two configurations, $\Sigma\,l = \infty$ [76], (3) the impact-parameter method [6], (4) distorted wave, two configurations $l \leq 5$ [81], (5) coupled equations, two configurations $l \leq 7$ [76], (6) distorted wave, three configurations, $\Sigma l = \infty$ [81], (7) distorted wave, three configurations, $l \leq 6$ [81], (8) the $\bar{g}$ approximation [3].

The results of Eissner and Osterbrock for the $2s^2$ ^1S–$2s\,2p$ ^3P transition are in good agreement, taking Osterbrock's values prior to the inclusion of resonances below the $2s\,2p$ ^1P level. These resonances are important in carbon III under solar conditions and increase the excitation rate by a factor of about two. They must also be taken into account in nitrogen IV and oxygen V but are unimportant in neon VII and higher ions.

Eissner has also calculated collision strengths for transitions between $2s^2$ ^1S, $2s\,2p$ ^3P and the higher configuration, $2s\,3s$, $2s\,3p$ and $2s\,3d$ in neon VII. The comparison between these theoretical calculations and the experimental results of Tondello and McWhirter [30] and Johnston and Kunze [11] is given in § 4-5F.

In the calculations of solar line intensities discussed in § 4-5G, Eissner's cross-sections have been used except for the $2s^2$ ^1S–$2s\,2p$ ^3P transitions and

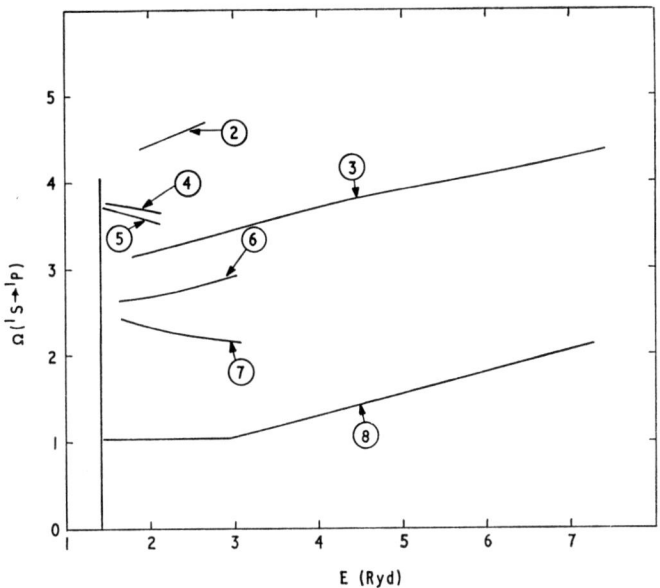

Fig. 4-5-3. As for Fig. 4-5-2 for the $2s^2\ {}^1S\text{–}2s\ 2p\ {}^1P$ transition in oxygen v.

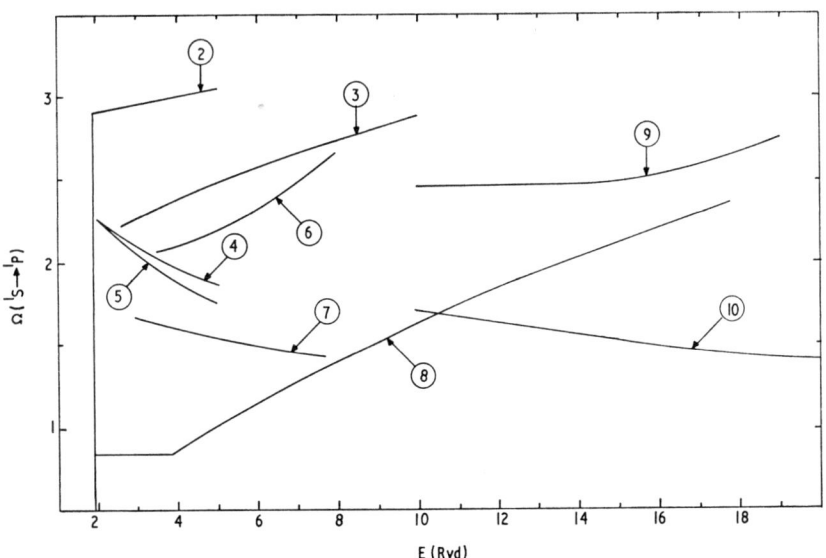

Fig. 4-5-4. As for Fig. 4-5-2 for the $2s^2\ {}^1S\text{–}2s\ 2p\ {}^1P$ transition in neon VII and also (9) distorted wave, six configurations, $\Sigma l = \infty$ [81] and (10) distorted wave, six configurations, $l \leq 6$ [81].

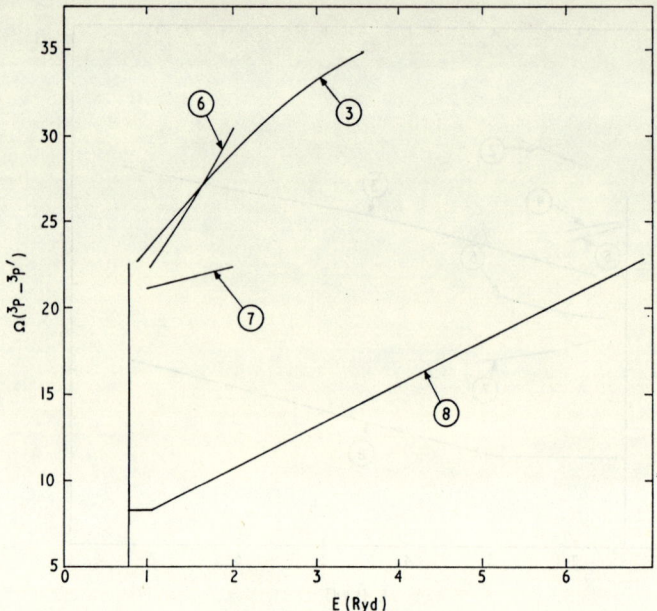

Fig. 4-5-5. The collision strength for the 2s 2p ^{3}P–2p^2 ^{3}P transition in carbon III. (Key to approximations given with Fig. 4-5-2.)

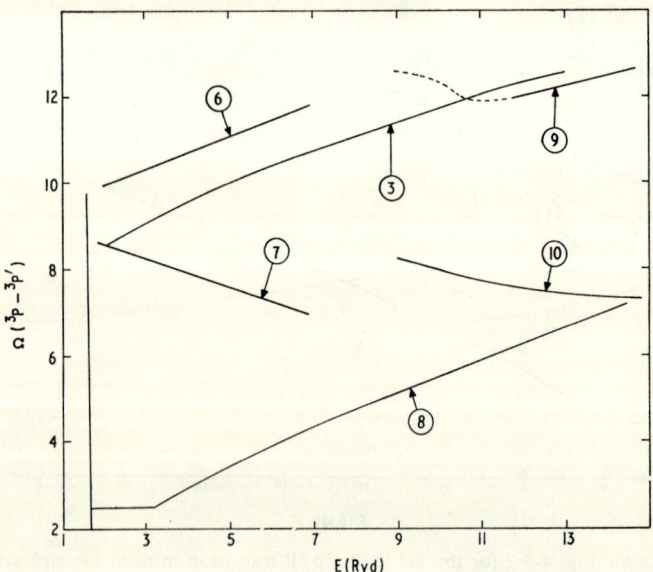

Fig. 4-5-6. As for Fig. 4-5-5 for the 2s 2p ^{3}P–2p^2 ^{3}P transition in neon VII.

those between the fine structure levels of 2s 2p $^3P_{0,1,2}$. For the former, Osterbrock's calculations including resonances have been adopted. When the calculations presented in § 4-5G were made no precise values of the collision strengths for transitions between the levels of the 2s 2p ^3P term were available. However, they have been estimated by using the results of Saraph et al. [61] for the $2p^2$ ^3P term in an ion of the same charge. Recently, Eissner has calculated the collision strengths for the fine structure transitions between the 2s 2p ^3P levels in oxygen V. His values agree within 10 % with those used in the present calculations.

No calculations are as yet available for the proton excitation rates between the ^3P fine structure levels. Bely and Faucher [82] have calculated the proton excitation rates for positive ions in the iso-electronic sequences with 2p and $2p^5$ ground state configurations. Some estimate of the relative importance of proton and electron impact rates may be made by comparing the results of Bely and Faucher with the electron excitation rates for the same transition calculated by Blaha [83]. For low ions, e.g. carbon III, the proton impacts can probably be neglected but for oxygen V and higher ions the proton impact rate could be as large as the electron impact rate, at the optimum temperature for the emission of the resonance and intercombination transitions. However, because the mixing of the 2s 2p ^3P levels by collisional transfer between 2s 2p ^3P and $2p^2$ ^3P is comparable with the mixing by collisional transfer between the 2s 2p ^3P fine structure levels themselves, adding a proton impact rate which is equal to the electron impact rate changes the resulting level populations by only 24 % in oxygen V and 9 % in silicon XI.

E. EQUATIONS OF STATISTICAL EQUILIBRIUM

The equation of statistical equilibrium can be set up for each excited level and can be solved as a function of density and temperature. When considering levels in the 2s 2p and $2p^2$ configurations, the exclusion of levels with $n \geq 3$ is justified by the much smaller excitation rates to these levels.

Although the relative importance of the different populating and depopulating mechanisms changes along the isoelectronic sequence, for the conditions extant in the solar atmosphere the following general conclusions can be drawn. All excitations originate from the ground state of the beryllium-like ion (although excitations from 2s 2p ^3P occur, the 2s 2p ^3P levels are themselves excited from the ground) and recombination and ionization into the excited levels are both negligible. For 2s 2p ^1P, excitation from $2s^2$ ^1S dominates and the contributions from cascades following excitations from $2s^2$ ^1S and 2s 2p ^3P to $2p^2$ ^1D, ^1S, and from excitations from 2s 2p ^{3}P, are together less than 20 % of the total rate. For 2s 2p ^{3}P, excitation from 2s^2 ^{1}S is also the most important process, and cascades following excitation from 2s^2 ^{1}S to 2p^2 ^{1}D, ^{1}S and ^{3}P contribute less than 10 % of the total rate.

The transitions 2s 2p ^{1}P–2p^2 ^{1}D and 2s 2p ^{3}P–2p^2 ^{3}P differ in that 2p^2 ^{1}D is excited from the 2s^2 ^{1}S and 2s 2p ^{3}P levels and not from 2s 2p ^{1}P, whilst 2p^2 ^{3}P is excited predominantly from 2s 2p ^{3}P. For levels from which permitted transitions occur, i.e. 2s 2p ^{1}P, 2p^2 ^{1}D and 2p^2 ^{3}P, the dominant de-population process is spontaneous radiative decay. The depopulation of the 2s 2p ^{3}P as a whole proceeds, in carbon III, by collisional de-excitation to 2s^2 ^{1}S, collisional excitation to 2s 2p ^{1}P, and spontaneous radiative decay from the ^{3}P$_1$ level; in higher members of the iso-electronic sequence collisional de-excitation becomes unimportant, but at very low densities ($\leq 10^8$ cm^{-3}) the radiative decay of the ^{3}P$_2$ level must also be included. The relative populations of the fine structure levels of the ^{3}P term depend on several exciting and de-exciting mechanisms. These are, collisions to and from 2s^2 ^{1}S, collisions between the ^{3}P$_{0,1,2}$ levels, redistribution by collisions to 2p^2 ^{3}P followed by radiative decay back to 2s 2p ^{3}P, collisional de-population by excitations to 2s 2p ^{1}P, 2p^2 ^{1}D and 2p^2 ^{1}S, and radiative decay of the 2s 2p ^{3}P$_1$ and ^{3}P$_2$ levels.

In silicon XI the additional processes of radiative decay, photoexcitation and stimulated emission between the fine structure levels must also be included.

In carbon III and nitrogen IV where the metastable ^{3}P population is comparable with that of the ground state the singlet levels in the 2s 3l configurations will be populated both from the ground state and from the ^{3}P levels. The triplet levels in the 2s 3l configuration will however be populated predominantly from the metastable triplet. In higher ions where the metastable population is lower the situation will reverse; the singlet levels will be populated from the ground state and the triplet levels from both the ground state and metastable triplet.

In transient, ionizing laboratory plasmas the additional processes of ionization into the 2s 2p ^{3}P levels from the boron-like ions and ionization from the 2s 2p ^{3}P levels to the lithium-like ions must also be included in the calculations of these triplet level populations. The higher densities ($N_e \sim 10^{14}$ cm^{-3}) in laboratory plasmas increase the importance of the collisional processes i.e. collisions from 2s 2p ^{3}P to other excited levels, collisions to the ground state, and collisions between ^{3}P fine structure levels, relative to the radiative decay from the ^{3}P$_1$ level.

The relative importance of excitations from the metastable triplet and ground terms to the singlet and triplet levels in the 2s 3l configurations is the same as for the low ions under solar conditions, as discussed above.

The equations of statistical equilibrium for levels in the 2s 2p and $2p^2$ configurations are given below, as specific examples of the general equation (4-2-2). The notation is the same as that in § 4-2A, with the addition of R_{nm} as the branching ratio for a transition from a level with $J = n$ in the $2p^2\ ^3P$ term to a transition with $J = m$ in the 2s 2p ^3P term and where $\sum_m R_{nm} = 1$. If individual J-values are not given the rate is that for the whole multiplet. The $2p^2\ ^3P$ term is designated by $^3P'$ to distinguish it from the 2s 2p ^3P. The equations are

$$N(^1P) = \frac{N_e[N(^1S)C(^1S \to\ ^1P) + N(^3P)C(^3P \to\ ^1P) + N(^1D)A(^1D \to\ ^1P)]}{A(^1P \to\ ^1S)} \quad (4\text{-}5\text{-}1)$$

$$N(^1D) = \frac{N_e[N(^1S)C(^1S \to\ ^1D) + N(^3P)C(^3P \to\ ^1D)]}{A(^1D \to\ ^1P)} \quad (4\text{-}5\text{-}2)$$

$$N(^3P') = \frac{N_e[N(^3P)C(^3P \to\ ^3P') + N(^1S)C(^1S \to\ ^3P')]}{A(^3P' \to\ ^3P)} \quad (4\text{-}5\text{-}3)$$

$$N(^3P) = \frac{N_e N(^1S)C(^1S \to\ ^3P) - N(^3P_1)A(^3P_1 \to\ ^1S) + N(^3P_2)A(^3P_2 \to\ ^1S)}{N_e[C(^3P \to\ ^1S) + C(^3P \to\ ^1P) + C(^3P \to\ ^1D)]} \quad (4\text{-}5\text{-}4)$$

$$N(^3P_1) = N_e[N(^1S)C(^1S \to\ ^3P_1)$$
$$+ \sum_{j=0,2} N(^3P_j)(C(^3P_j -\ ^3P_1) + \sum_{k=1,2} (^3P_j -\ ^3P'_k)R_{k1})]/$$
$$[A(^3P_1 -\ ^1S) + N_e(C(^3P_1 \to\ ^1S) + \sum_{k=0,2} C(^3P_1 -\ ^3P_k)$$
$$+ C(^3P_1 \to\ ^1P) + C(^3P_1 \to\ ^1D) + \sum_{k=1,2} C(^3P_1 -\ ^3P'_k)(1 - R_{k1}))] \quad (4\text{-}5\text{-}5)$$

with similar equations for $^3P_0$ and $^3P_2$.

The relative intensities of any two lines from levels j and k can then be found by applying eq. (4-2-7), i.e.

$$\frac{I(j \to i)}{I(k \to i)} = \frac{\lambda_{ik}}{\lambda_{ij}} \frac{N_j A(j \to i)}{N_k A(k \to i)} \quad (4\text{-}5\text{-}6)$$

where I is measured in erg cm^{-2} sec^{-1}, and λ is the wavelength of the transition. Equation (4-5-6) should only be applied to lines which are formed in the same region of the plasma. Otherwise the intensities should be integrated over the whole volume of the plasma.

F. LABORATORY PLASMAS

Analyses of laboratory spectra of beryllium-like ions can be made using methods similar to those used for lithium-like ions. Either known excitation rates can be used to determine the temperature in a laboratory source, or the observed absolute intensities and known values of the electron density and temperature can be used to determine the excitation rates.

Tondello and McWhirter [30] have used the latter approach to find the excitation rates for eighteen transitions in neon VII, these being between levels in the $2s^2$, $2s\,2p$, $2s\,3l$, $2p\,3l$, $2s\,4l$ and $2s\,5d$ configurations. Their experimental procedure is described in more detail in § 4-3D; the basis of the experiment was to make accurate measurements of the absolute intensities of the lines emitted by a theta-pinch plasma for which the electron density and temperature were measured by laser scattering techniques.

Because of the existence of the metastable triplet level the observed intensity in a given line depends on the relative populations of the ground state and metastable level and also on several excitation rates. The experimentally determined quantity is not strictly the excitation rate for the observed transition but is the total excitation rate to the upper level from all possible routes. The equation for the intensity of a singlet transition, terminating on the ground state, i, can be written as

$$I(i \to j) = \text{const.} \times N_e^2 \frac{N(E)}{N_e} \frac{N_i}{N(E)} \left[C(i \to j) + \frac{N_m}{N_i} C(m \to j) \right] \quad (4\text{-}5\text{-}7)$$

and for a triplet level, terminating on the metastable level, m, as

$$I(m \to j) = \text{const.} \times N_e^2 \frac{N(E)}{N_e} \frac{N_m}{N(E)} \left[C(m \to j) + \frac{N_i}{N_m} C(i \to j) \right]. \quad (4\text{-}5\text{-}8)$$

Tondello and McWhirter calculated the ion population $(N_i + N_m)/N(E)$ from the transient ionization equations. They calculated the experimental excitation rates from the observed intensities, density and temperature assuming that the triplet levels are populated only from the metastable triplet and that the singlet levels are populated only from the ground state. Tondello and McWhirter estimated that the metastable population should be within a factor of two of the Boltzmann value and took $N(^3P)/N(^1S) = 8.3$, the Boltzmann population at 2.1×10^6 °K. This value was used to find

$N_i/N(E)$ and $N_m/N(E)$, factors needed in the derivation of the experimental rates, $C(i \to j)$ or $C(m \to j)$ which are given in Table 4-5-1. The theoretical rates listed are the total rates to the excited level from both the ground state and metastable level, calculated using the collision-strengths of Eissner [81]. The assumed Boltzmann population also enters these theoretical rates. The calculated percentage contribution from either the metastable level or ground state is also given in Table 4-5-1. It can be seen that the assumption that the

TABLE 4-5-1

Experimental and theoretical excitation rates for neon VII transitions

Transition		λ (Å)	Excitation coefficient ($cm^3 sec^{-1}$)		Percentage excitation from metastable, or ground level*
			Experimental	Theoretical	
2s 2p ^{3}P–2p^2 ^{3}P		561	4.8×10^{-9}	7.9×10^{-9}	99.997
–2s 3s ^{3}S		115	3.4×10^{-11}	2.6×10^{-11}	90
–2s 3d ^{3}D		106	1.6×10^{-10}	9.1×10^{-10}	98.4
2s^2 ^{1}S	–2s 2p ^{1}P	465	3.1×10^{-8}	1.6×10^{-8}	92
	–2s 3s ^{1}S	128	2.3×10^{-10}	3.5×10^{-10}	92.7
	–2s 3d ^{1}D	117	4.3×10^{-10}	7.5×10^{-10}	71
	–2s 3p ^{1}P	97	3.1×10^{-10}	3.6×10^{-10}	67

* From the metastable for the triplet transitions, from the ground level for the singlet transitions.

triplets are excited predominantly from the metastable triplet is justified. The assumption that 2s 3d ^{1}D and 2s 3p ^{1}P are populated predominantly from the ground state is less accurate but a lower triplet population would improve the approximation. It can be seen from Table 4-5-1 that the theoretical and experimental results agree within the quoted experimental accuracy of a factor of 2 for absolute values and $\pm 50\%$ for the relative values, with the exception of the 2s 2p ^{3}P^0–2s 3d ^{3}D transition where there is a discrepancy of about a factor of six, and for which there is no apparent explanation. Tondello and McWhirter have estimated that the effects of cascade, stepwise excitation (other than from the metastable level) and opacity are all less than the experimental uncertainty. Also, they have measured the intensity of the intersystem 2s^2 ^{1}S$_0$–2s 2p ^{3}P$_1$ transition, and find that the population of the metastable level is 50% of the Boltzmann value, although the accuracy is only a factor of 2 because of the weakness of the intersystem line. They conclude that theoretical calculations for further transitions and experiments with increased accuracy are desirable.

Since Tondello and McWhirter made their calculations, further theoretical excitation rates for transitions from the 2s 2p ^{3}P level to the 2s 2p ^{1}P, 2p^2 ^{1}S and ^{1}D levels have become available (Eissner [81]). The new rates are larger than previous estimates and lead to a calculated metastable population of $N(^3P)/N(^1S) = 4.2$, a factor of two smaller than the Boltzmann value. Use of this lower population would have the following effects on the rates given in Table 4-5-1. The experimental rates for the triplet transitions would be increased by 10%, while those for the singlet transitions would be reduced by 80%. The theoretical rates remain the same to within 20%.

An unsatisfactory aspect of the rates given in Table 4-5-1 is that the experimental ratio of the rates for the 2s 2p ^{3}P–2p^2 ^{3}P and 2s^2 ^{1}S–2s 2p ^{1}P transitions is a factor of three smaller than the theoretical ratio. The lines observed are strong and are close in wavelength, and it would be expected that the experimental ratio should be good. The lower metastable population would reduce this discrepancy to a factor of 1.7.

Further measurements of excitation rates in beryllium-like ions have been made by Johnston and Kunze [11]. They have used the Maryland thetapinch (see § 4-3D) to determine excitation rates for singlet and triplet transitions between levels with $n = 2$, 3 and 4 in the ions nitrogen IV, oxygen V, neon VII and silicon XI. Electron densities in the range $(1.8–9) \times 10^{15}$ cm^{-3} and electron temperatures in the range $(0.8–3.0) \times 10^6$ °K were used. For oxygen V and neon VII it was possible to make measurements at several densities and temperatures. Johnston and Kunze also observed the ^{1}S–^{3}P intersystem line in neon VII and derive the population of the metastable level from its intensity.

Johnston and Kunze compared their experimental total rates with those calculated by Osterbrock [76] and the $\bar{g}$ approximation of Van Regemorter [3]. They did not include collisional excitation from 2s 2p ^{3}P to 2p^2 ^{1}S and 2p^2 ^{1}D as rates were not available. Using the values calculated by Eissner [81] would lead to a metastable population of $N(^3P)/N(^1S) = 2.6$ rather than the value of 3.2 calculated by Johnston and Kunze. The value observed by Johnston and Kunze was 0.92, but their experimental error could include values of up to 2.5.

Johnston and Kunze used their experimentally determined value of $N(^3P)/N(^1S) = 0.92$ in calculating the theoretical total excitation rates to each level. For the 2s^2 ^{1}S–2s 2p ^{1}P and 2s 2p ^{3}P–2p^2 ^{3}P transitions their experimental rates are about a factor of two larger than those calculated using the $\bar{g}$ approximation and should therefore agree quite well with those calculated by Eissner (see Figs. 4-5-2 to 4-5-6). In neon VII, Johnston and

Kunze find that the ratio of the experimental rates for the 2s 2p ^{3}P–2p^2 ^{3}P and 2s^2 ^{1}S–2s 2p ^{1}P transitions is 30% larger than the theoretical value. Use of Eissner's cross-sections and the higher metastable population, $N(^3P)/N(^1S) = 2.6$, would give a theoretical value 10% larger than the observed value.

G. Solar line intensities

Some or all of the four transitions shown in Fig. 4-5-1 are observed from each of the ions, carbon III, nitrogen IV, oxygen V and silicon XI in the solar spectrum. The available useful intensity data are given in Table 4-5-2 and are taken from the following sources; Hall et al. [84], Burton et al. [85] and Freeman and Jones [86]. The data in refs. [84] and [86] refer to whole sun intensities, those in ref. [85] refer to a region of the quiet disk.

Equations (4-5-1) to (4-5-5) and the similar equations for $N(^3P_0)$ and $N(^3P_2)$ have been solved as a function of density and temperature and the relative intensities of the lines have been calculated using eq. (4-5-6). The computed and observed relative intensities are shown in Figs. 4-5-7 to 4-5-10. In carbon III, nitrogen IV and oxygen V, the 2s^2 ^{1}S–2s 2p ^{1}P, 2s^2 ^{1}S–2s 2p ^{3}P$_1$ and 2s 2p ^{3}P–2p^2 ^{3}P′ transitions are all observed, and the ratios $I(^1S-^3P_1)/I(^3P-^3P')$ and $I(^3P-^3P')/I(^1S-^1P)$ are plotted. In silicon

TABLE 4-5-2

Intensity data for beryllium-like ions observed in the sun

Ion	λ	Intensity (erg cm^{-2} sec)	
		Refs. [85, 86]	Ref. [84]
C III	977	0.023	0.081
	1909	(0.015)	–
	1175	0.012	0.051
N IV	765	–	0.006
	1487	0.0013	–
	923	–	0.0032
O V	630	0.024	0.045
	1218	0.0011	–
	760	–	0.003
	1371	(0.0004)	–
Si XI	303	0.030	–
	581	–	0.0018

Values for ref. [85] in parentheses are deduced from limb intensities and other limb/disk ratios.

XI, only the $2s^2\ {}^1S$–$2s\ 2p\ {}^3P_1$ and $2s^2\ {}^1S$–$2s\ 2p\ {}^1P$ transitions are observed. The ratio $I({}^1S$–${}^3P_1)/I({}^1S$–${}^1P)$ is therefore plotted instead of $I({}^3P$–${}^3P')/I({}^1S$–${}^1P)$. The population ratios $\sum N({}^3P)/N({}^1S)$ and $(N({}^3P_1)/N({}^1S))/N_e$ for carbon III, nitrogen IV and oxygen V have been published previously (Jordan [87]).

The atomic data used, the sources of which have been discussed in § 4-5B, C and D, are given in Table 4-5-3. The collision strengths for nitrogen IV were obtained by interpolation, and those for silicon XI by extrapolation.

From Figs. 4-5-7 to 4-5-10 it can be seen that the intensity ratios in carbon III depend on both the electron density and temperature, whilst in higher ions, the dependence on these parameters decreases. It is therefore difficult to derive the electron density from the line ratio in ions other than carbon III, and it is more significant to compare the observed line ratios directly with those computed.

As discussed in § 4-4G, the temperature at which a line has its maximum emission can usually be taken as the temperature at which the function

$$g(T) = T_e^{-\frac{1}{2}} \frac{N(\text{ion})}{N(\text{E})} \exp\left(-E(gi)/kT_e\right) \qquad (4\text{-}5\text{-}9)$$

has its maximum value. However, in the region of the solar atmosphere where the carbon III lines are formed the variation of the density and temperature gradient with height is rapid and must also be taken into account. Thus the temperature at which the function

$$g(T)N_e^2 dh/dT_e \qquad (4\text{-}5\text{-}10)$$

has its maximum value should be used. This value is $5.6 \times 10^4\ °K$ compared with $7 \times 10^4\ °K$ derived from eq. (4-5-9). In Figs. 4-5-7 to 4-5-9 the optimum temperatures are indicated by T_m.

Table 4-5-4 gives the observed intensity ratios; the calculated intensity ratios at the given values of T_m and N_e, which are from Jordan [66, 67] and where possible, the values of N_e deduced from the observed line ratios.

From Table 4-5-4 it can be seen that for carbon III the observed and predicted values of $I({}^3P$–${}^3P')/I({}^1S$–${}^1P)$ are in fair agreement. The measured relative intensities of these transitions should be accurate to within a factor of two since the lines are not widely separated in wavelength and are of comparable intensity. The electron density deduced from the observed intensity ratio is about a factor of two smaller than that given by the model, and this can be considered as satisfactory agreement. The observed disk

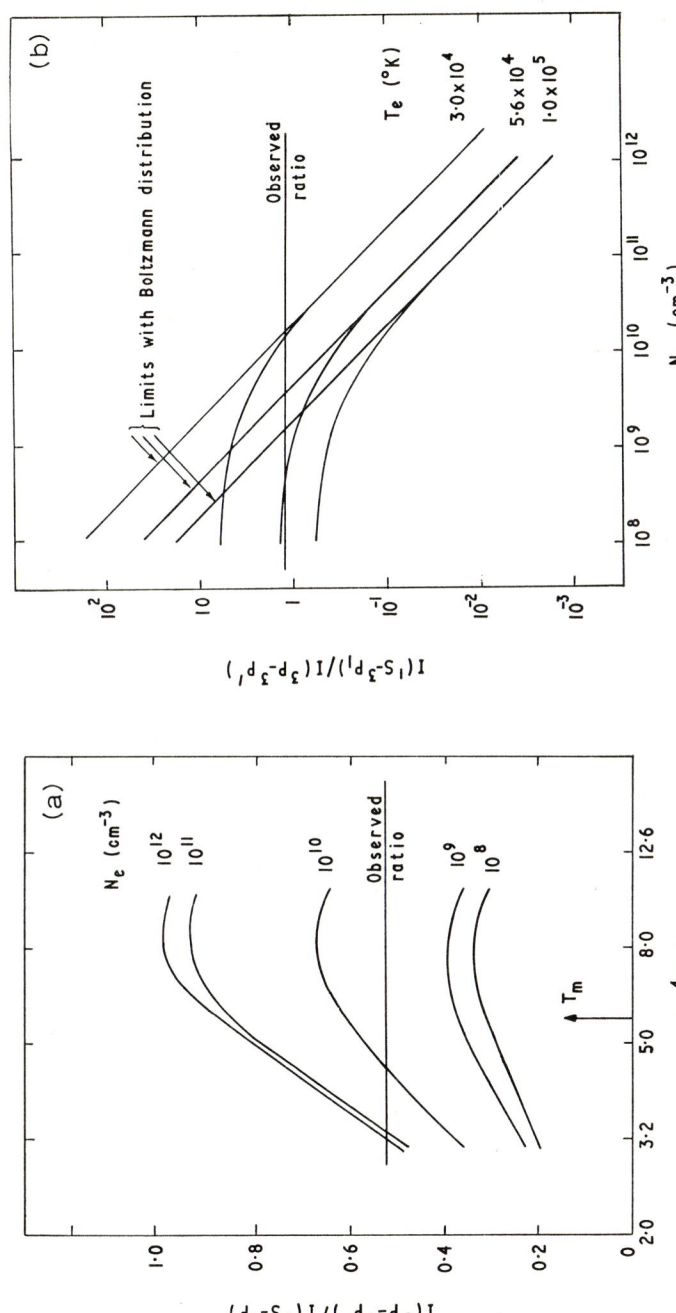

Fig. 4-5-7. The computed and observed values of the ratios (a) $I(^3P-^3P')/I(^1S-^1P)$ and (b) $I(^1S-^3P_1)/I(^3P-^3P')$ in carbon III as a function of density and temperature. The limits refer to a Boltzmann distribution among the fine structure levels of the ^3P term.

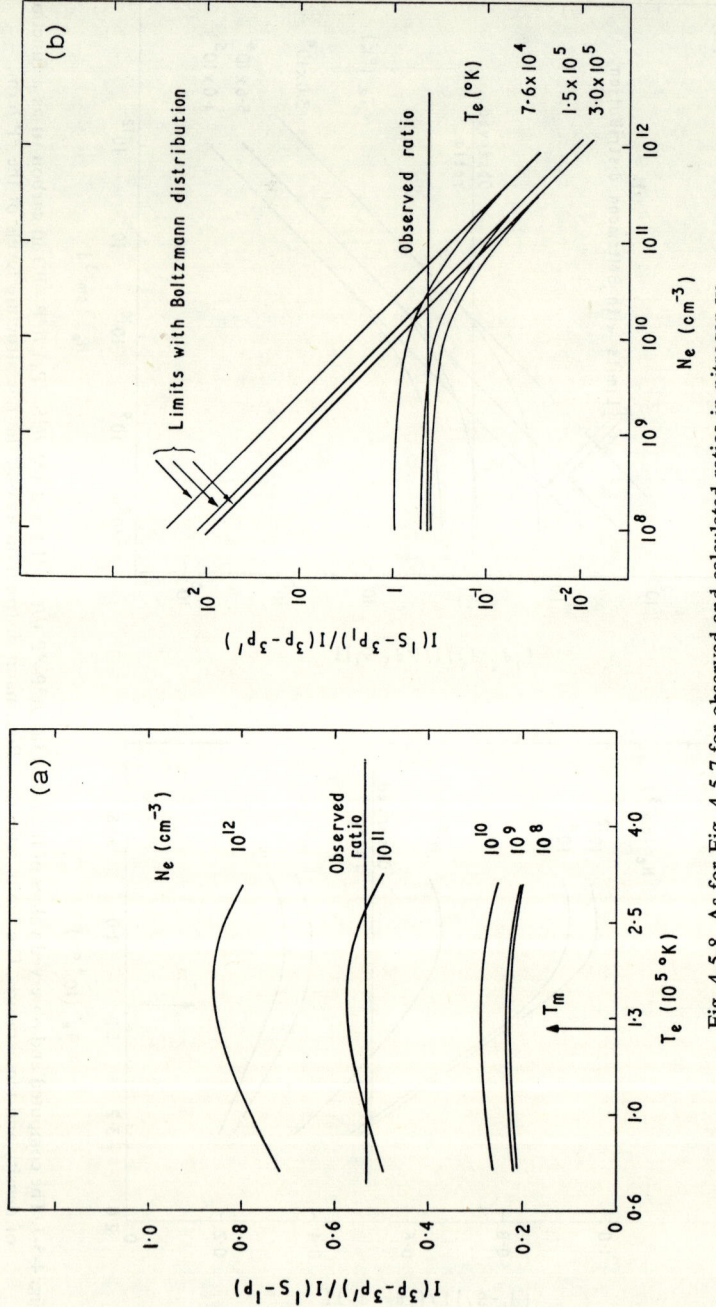

Fig. 4-5-8. As for Fig. 4-5-7 for observed and calculated ratios in nitrogen IV.

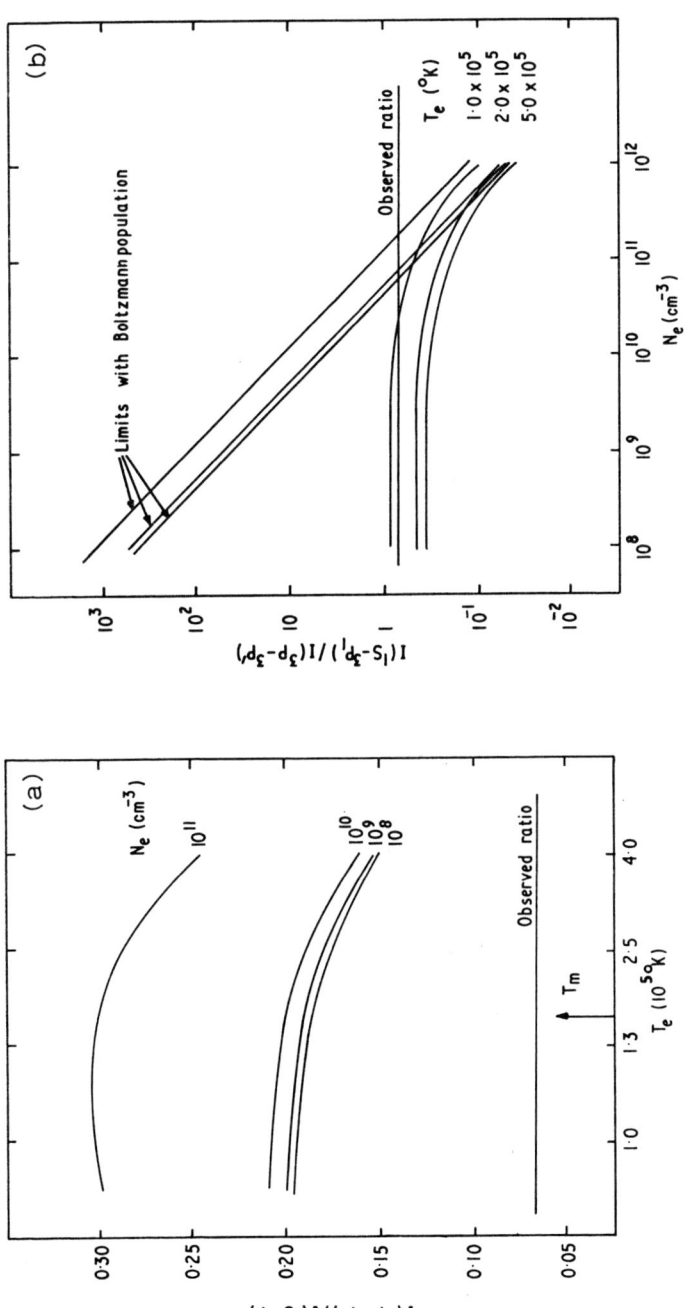

Fig. 4-5-9. As for Fig. 4-5-7 for observed and calculated ratios in oxygen v.

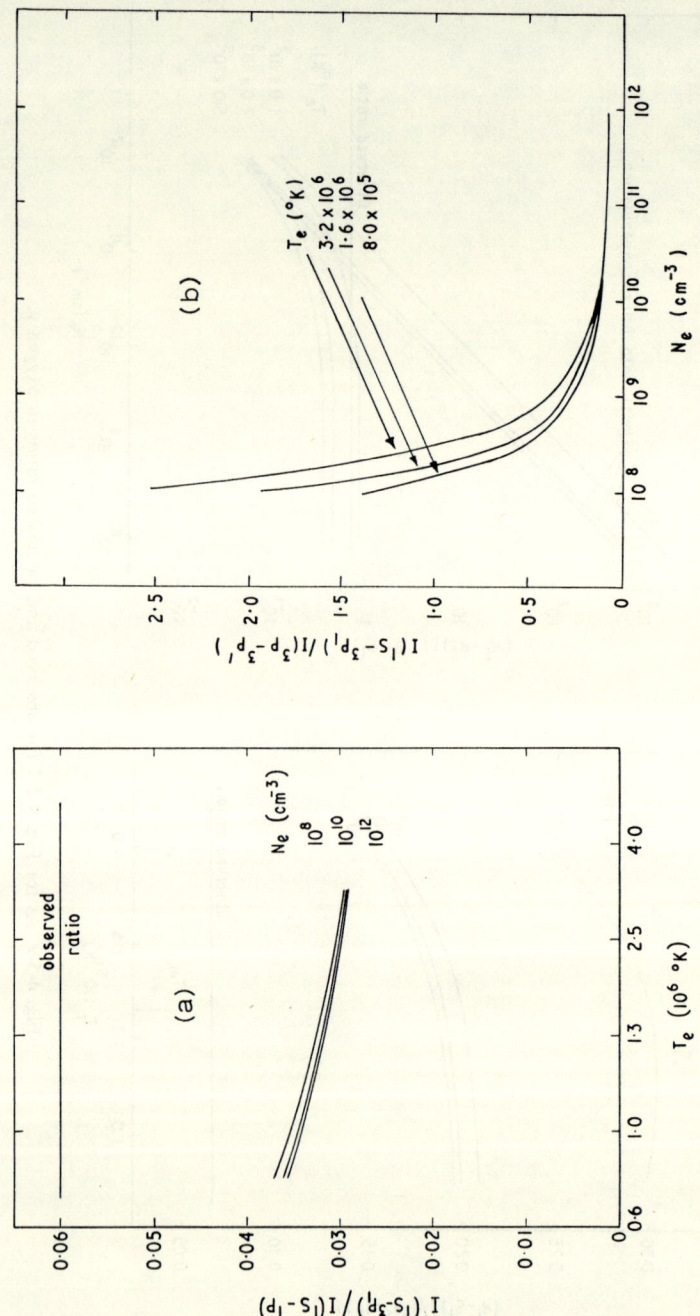

Fig. 4-5-10. The computed and observed values of the ratios (a) $I(^1S-^3P_1)/I(^1S-^1P)$ and (b) $I(^1S-^3P_1)/I(^3P-^3P')$ in silicon XI as a function of density and temperature.

TABLE 4-5-3. Atomic data for beryllium-like ions

C III

Transition	λ (Å)	W (eV)	Ω	A (sec⁻¹)
$2s^2\ ^1S\ -2s\ 2p\ ^1P^0$	977	—	3.75	—
$-2s\ 2p\ ^3P^0_0$	1909	—	1.1*	1.90×10^2
$-2s\ 2p\ ^3P^0_2$	1907	—	—	3.1×10^{-3}
$-2p^2\ ^1D$	—	18.0	0.26	—
$-2p^2\ ^3P$	—	16.2	0.017	—
$2s\ 2p\ ^3P^0-2p^2\ ^3P$	1176	—	23.3	—
$-2s\ 2p\ ^1P^0$	—	6.2	2.16	—
$-2p^2\ ^1D$	—	11.5	1.19	—
$2s\ 2p\ ^1P-2p^2\ ^1D$	—	5.4	—	—
$2s\ 2p\ ^3P_0-2s\ 2p\ ^3P_1$	$\sim 3 \times 10^6$	—	0.38	—
$^3P_0-\ ^3P_2$		—	0.21	2.2×10^{-7}
$^3P_1-\ ^3P_2$		—	0.95	2.5×10^{-6}

N IV

Transition	λ (Å)	W (eV)	Ω	A (sec⁻¹)
$2s^2\ ^1S\ -2s\ 2p\ ^1P^0$	765	—	3.50	—
$-2s\ 2p\ ^3P^0_0$	1487	—	0.55*	9.2×10^2
$-2s\ 2p\ ^3P^0_2$	1483	—	—	6.6×10^{-3}
$-2p^2\ ^1D$	—	23.3	0.13	—
$-2p^2\ ^3P$	—	21.7	0.013	—
$2s\ 2p\ ^3P^0-2p^2\ ^3P$	923	—	17.7	—
$-2s\ 2p\ ^1P^0$	—	7.8	1.06	—
$-2p^2\ ^1D$	—	15.0	0.70	—
$2s\ 2p\ ^1P-2p^2\ ^1D$	—	7.2	—	—
$2s\ 2p\ ^3P_0-2s\ 2p\ ^3P_1$	$\sim 10^6$	—	0.22	—
$^3P_0-\ ^3P_2$		—	0.16	4.6×10^{-6}
$^3P_1-\ ^3P_2$		—	0.64	4.0×10^{-5}

O V

Transition	λ (Å)	W (eV)	Ω	A (sec⁻¹)
$2s^2\ ^1S\ -2s\ 2p\ ^1P^0$	630	—	3.07	—
$-2s\ 2p\ ^3P^0_1$	1218	—	0.35*	3.6×10^3
$-2s\ 2p\ ^3P^0_2$	1214	—	—	1.3×10^{-2}
$-2p^2\ ^1D$	—	28.	0.078	—
$-2p^2\ ^3P$	—	26.4	0.0091	—
$2s\ 2p\ ^3P^0-2p^2\ ^3P$	760	—	16.7	—
$-2s\ 2p\ ^1P^0$	—	9.4	0.71	—
$-2p^2\ ^1D$	—	18.4	0.46	—
$2s\ 2p\ ^1P-2p^2\ ^1D$	1371	—	—	—
$2s\ 2p\ ^3P_0-2s\ 2p\ ^3P_1$	$\sim 5 \times 10^5$	—	0.24	—
$^3P_0-\ ^3P_2$		—	0.12	4.6×10^{-5}
$^3P_1-\ ^3P_2$		—	0.58	3.9×10^{-4}

Si XI

Transition	λ (Å)	W (eV)	Ω	A (sec⁻¹)
$2s^2\ ^1S\ -2s\ 2p\ ^1P^0$	303	—	1.45	—
$-2s\ 2p\ ^3P^0_1$	581	—	0.052*	6.5×10^5
$-2s\ 2p\ ^3P^0_2$	565	—	—	0.21
$-2p^2\ ^1D$	—	61.4	0.027	—
$-2p^2\ ^3P$	—	55.4	0.0013	—
$2s\ 2p\ ^3P^0-2p^2\ ^3P$	365	—	5.63	—
$-2s\ 2p\ ^1P^0$	—	19.7	0.11	—
$-2p^2\ ^1D$	—	40.1	0.031	—
$2s\ 2p\ ^1P-2p^2\ ^1D$	—	20.4	—	—
$2s\ 2p\ ^3P_0-2s\ 2p\ ^3P_1$	$\sim 3 \times 10^4$	—	0.050	0.255
$^3P_0-\ ^3P_2$		—	0.025	—
$^3P_1-\ ^3P_2$		—	0.12	1.95

* The collision strength is for the total $^1S-^3P$ transition.

TABLE 4-5-4

Comparison of observed and calculated values of intensity ratios and electron densities

Ion	I	Observed ratios	Calculated ratios	T_e (°K)	N_e (cm^{-3}) EUV model	N_e (cm^{-3}) from observed ratio
C III	$\dfrac{I(^3P-^3P')}{I(^1S-^1P)}$	0.52	0.60	5.6×10^4	1.0×10^{10}	5.2×10^9
	$\dfrac{I(^1S-^3P_1)}{I(^3P-^3P')}$	(1.25)	0.36	5.6×10^4	1.0×10^{10}	4.0×10^8
N IV	$\dfrac{I(^3P-^3P')}{I(^1S-^1P)}$	0.53	0.25	1.5×10^5	3.7×10^9	8.5×10^{10}
	$\dfrac{I(^1S-^3P_1)}{I(^3P-^3P')}$	0.41	0.43	1.5×10^5	3.7×10^9	no solution at T_m
O V	$\dfrac{I(^3P-^3P')}{I(^1S-^1P)}$	0.067	0.19	2.3×10^5	2.5×10^9	no solution
	$\dfrac{I(^1S-^3P_1)}{I(^3P-^3P')}$	0.69	0.45	2.3×10^5	2.5×10^9	no solution
	$\dfrac{I(^1P-^1D)}{I(^1S-^1P)}$	(0.017)	0.0092	2.3×10^5	2.5×10^9	–
Si XI	$\dfrac{I(^1S-^3P_1)}{I(^1S-^1P)}$	0.060	0.021	1.6×10^6	2.0×10^8	–

Values in parentheses are deduced from limb intensities and other limb/disk ratios.

intensity listed for the $^1S-^3P_1$ transition has been deduced from the limb intensity and the limb to disk ratios of other optically thin transitions formed at similar temperatures (see ref. [85]). The agreement between the observed and calculated ratios for $I(^1S-^3P_1)/I(^3P-^3P')$ is not satisfactory, there being a discrepancy of a factor of 3.5 between the two values. Figure 4-5-7 shows that if $T_e = 5.6 \times 10^4$ °K and $N_e \sim 10^{10}$ cm^{-3}, a satisfactory solution cannot be found by increasing the mixing between the triplet levels, because the intensity ratio is already close to that expected from a Boltzmann distribution among the ^3P fine structure levels. The atomic parameters $A(^3P_1 \to {}^1S)$ and $C(^3P \to {}^3P')$, which appear in eqs. (4-5-3) and (4-5-5), should not be in error by more than 50%. The most likely source of error is the intensity data, since the two lines are at widely different wavelengths. The intensities [85] should be accurate to within a factor of two at any wavelength, so a factor of 4 does represent an upper limit to the expected error in relative intensities. To obtain a fit to all three lines in carbon III by altering the temperature, a value of $T_e < 3 \times 10^4$ °K would be needed, and from the model used this appears to be unlikely. Nor are optical depth effects expected to be important. The computed optical depth in the ^1S–^1P transition at the centre of the disk is $\tau \approx 0.063$ [85]; the optical depth in the ^3P–$^3P'$ transition will be even less.

In nitrogen IV the observed intensity ratio $I(^3P$–$^3P')/I(^1S$–$^1P)$ is about a factor of two larger than that predicted by the calculations and the EUV model. Considering the weakness of the ^3P–$^3P'$ multiplet and the possible blending with hydrogen Lyman θ this disagreement is not surprising. The nitrogen IV lines do not provide a sensitive method of determining electron density. In order to compare the intensity of the ^1S–$^3P_1$ transition with that of the ^3P–$^3P'$ transition two sets of data must be combined. The ratio is therefore liable to wavelength dependent errors in either set of data. The agreement between the observed and calculated intensities is fortuitously good.

The observed ratio $I(^3P$–$^3P')/I(^1S$–$^1P)$ in oxygen V is a factor 3 smaller than that predicted by the calculations and the EUV model. Since the ^3P–$^3P'$ multiplet is weak the origin of the difference is probably again the intensity data. The observed ratio $I(^1S$–$^3P_1)/I(^3P$–$^3P')$ is only a factor 1.5 larger than that predicted by the EUV model which can be regarded as satisfactory agreement. The transition 2s 2p ^1P–2p^2 ^1D is also observed in oxygen V in the limb spectrum [85], and a value for the disk can be deduced from other limb to disk ratios. The observed ratio $I(^1P$–$^1D)/I(^1S$–$^1P)$ is about a factor of two larger than that calculated.

Only one ratio $I(^1S$–$^3P_1)/I(^1S$–$^1P)$ is observed in silicon XI. The observed value is about a factor of 3 larger than that calculated, which is within the possible error in the intensity data since the 2s ^1S–2s 2p $^3P_1$ transition is very weak.

In summary, the intensity data at present available are the limiting factors when comparing observed and computed line ratios. More accurate ratios in carbon III would be useful since relative intensities in this ion can be used to determine electron densities.

The calculations shown in Figs. 4-5-7 to 4-5-10 can also be used to interpret the variations in line ratios observed from the Orbiting Solar Observatory (OSO) IV by Noyes et al. [88]. For the two active regions for which data have been published, the ratio $I(^3P$–$^3P')/I(^1S$–$^1P)$ changes in carbon III by a factor of about 1.34, and in O V by about a factor of 1.1. Interpreted as density changes at constant temperature, these variations in intensity ratios correspond to density increases of 4.6 and 6 respectively. Active region data from OSO VI have not yet been published; it will be of interest to apply the results of the present calculations to further intensity ratio variations.

The method of measuring electron temperatures from the relative intensities of lines from levels of widely separated energies, described in § 4-2E can in principle be applied to the beryllium-like ions. However there are not as yet sufficient intensity data obtained from one flight for the method to be applied. For oxygen V and higher ions in the sequence the metastable population is sufficiently low for the singlet levels to be populated predominantly from the ground state alone. The electron temperature can then be determined from singlet transitions without knowledge of the electron densities. The $n = 3$ triplet levels will be excited from both the ground state and metastable level, and it may be necessary to assume a model of the electron density to compute the metastable population. Because the value of $N(\text{ion})/N(\text{E})$ for beryllium-like ions falls rapidly with increasing temperature the lines should be formed over a restricted temperature range and the variation of $N_e^2 dh/dT_e$ need not be included in the integration over temperature.

H. Low density objects, quasars, and planetary nebulae

Osterbrock [89] has used his calculations of collision strengths for carbon III, nitrogen IV and oxygen V to predict the populations of the 2s 2p $^3P_1$ and $^3P_2$ levels as a function of electron density. He points out that there are three density regions for the above populations. These are

(i) $N_e < N_2$
(ii) $N_2 < N_e < N_1$
(iii) $N_1 < N_e$

where N_2 and N_1 are the values of N_e at which the radiative decay of $^3P_2$ and $^3P_1$ respectively, are equal to the collisional de-excitation rates. For $N_e < N_2$, 99 % of collisions to the ^3P levels result in a photon in one of the two lines, with a comparable number of photons in each line. If $N_2 < N_e < N_1$, the $^3P_2$ line is very much weaker than the $^3P_1$ line. Hence if both lines were observed, and the electron density in the source was $\sim 10^5$–10^6 cm^{-3} it would be possible to measure the electron density from the line ratio. These values of electron density are comparable with those suggested by quasar models (Osterbrock and Parker [90]). However, the width of lines in quasar spectra will make such observations difficult since the separation of the two lines in carbon III is only 2 Å, and the oxygen V lines will be blended with H Ly-α. In the range $N_2 < N_e < N_1$, about 85% of the excitations to the ^3P levels result in emission of a photon from $^3P_1$, the high rate arising because the collisions between the three triplet levels are comparable with the collisional de-excitation rates. The lines from both the $^3P_1$ and $^3P_2$ levels in carbon III and nitrogen IV should be visible in the spectra of planetary nebulae where $N_e \sim 10^4$ cm^{-3}. Osterbrock's calculations are all for $T_e = 10^4$ °K, but may be scaled to apply to other temperatures.

I. RECENT DEVELOPMENTS

Since the calculations described above were completed, Nussbaumer [91] has computed the transition probabilities for the $2s^2\ ^1S_0$–$2s\ 2p\ ^3P_1$ transition in the beryllium-like ions including configuration mixing. He finds that the transition probabilities are about a factor of two smaller than those calculated by Garstang and Shamey [74]. A factor of two reduction in the $2s^2\ ^1S_0$–$2s\ 2p\ ^3P_1$ decay rate would have the following effects on calculated line ratios. In carbon III, the calculated values of $I(^3P-^3P')/I(^1S-^1P)$ will be 20% greater, leading to a value of N_e a factor of two smaller than that given in Table 4-5-4. The calculated value of $I(^1S-^3P_1)/I(^3P-^3P')$ will be 0.57 its previous value. Thus the discrepancies between calculated and observed values are *increased*. In oxygen V and higher ions the population increase is offset by the decrease in A-value and the calculated line ratios remain the same.

The calculations of Osterbrock [89] for low density objects would also be modified, in that the density limit, N_1, at which collisional de-excitation equals radiative decay for the $^3P_1$ level would also be reduced by a factor of two. This would not alter his general conclusion that the lines from $^3P_1$ and $^3P_2$ could be used to measure electron densities in the range $N_e \sim 10^5$–10^6 cm^{-3}.

Because of the high density in laboratory plasmas ($N_e \sim 10^{15}$ cm^{-3}) the change in the A-value does not have a significant effect on the results discussed in § 4-5F.

§ 4-6. Helium-like ions

A. GENERAL DESCRIPTION

The term structure of the two-electron ions is characterized by the high ionization potential of the 1s electrons as well as the relatively high excitation potential to the first excited $n = 2$ terms. The scheme is shown in Fig. 4-6-1 for carbon v, for terms below the first ionization potential. Since the lowest two-electron excited configuration $2s^2$ lies well above the first ionization potential, all the terms shown in Fig. 4-6-1 are due to one-electron excitation of the type 1s nl. Each such configuration gives singlet and triplet terms 1s nl ^1L and 1s nl ^3L, which will be abbreviated here as n ^1L and n ^3L. The ground configuration is the exception giving only a singlet term 1 ^1S. Spin-orbit interaction is small in the two-electron ion, so that for practical purposes the J-splitting is readily observable only in the 2 ^3S–2 ^3P transition. In strict LS coupling, there are two metastables 2 ^1S and 2 ^3S, and the intercombination line 1 ^1S–2 ^3P is forbidden. However, the following

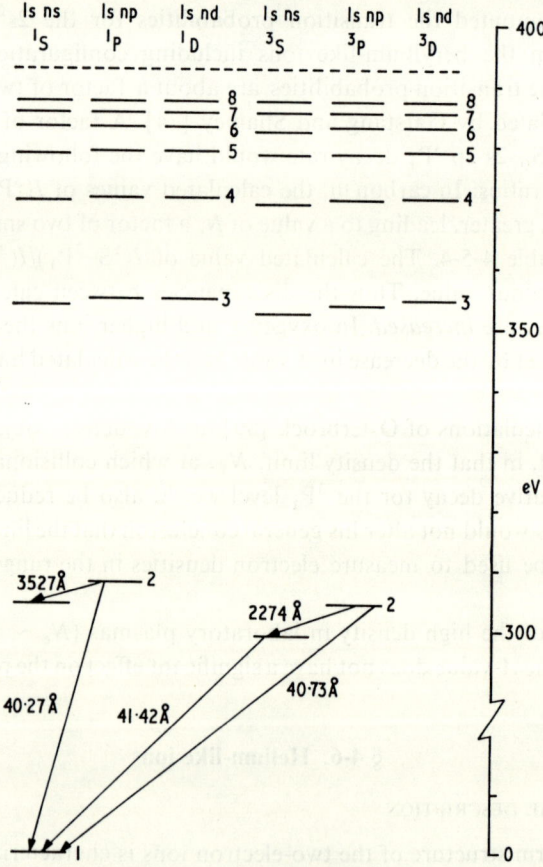

Fig. 4-6-1. Energy levels for carbon v.

forbidden transitions do occur, in increasing order of probability. The 1 ^{1}S–2 ^{3}S transition proceeds through relativistic magnetic dipole radiation with a rate scaling as Z^{10}; the 1 ^{1}S–2 ^{1}S transition occurs by two-photon electric dipole emission with a rate scaling as Z^6; and the 1 ^{1}S–2 ^{3}P$_1$ transition proceeds through spin-orbit interaction with a rate scaling as Z^9. (For large Z the 1 ^{1}S–2 ^{3}P$_2$ magnetic quadrupole transition becomes significant with a rate scaling as Z^8.) Since the allowed electric dipole rates scale only as Z^4 (or as Z for $\Delta n = 0$ transitions), the forbidden transitions become increasingly important at larger Z.

The excited states are all grouped relatively close to the ionization limit, so that it is not possible to observe ratios of intensities from spectral lines having very different excitation potentials. These ions are not therefore a suitable choice for measurements of electron temperature. However, the existence of the three long-lived terms 2 ^{3}S, 2 ^{1}S and 2 ^{3}P give rise to some complex variations of intensities with electron density, thereby making density determinations possible.

In the limit of low density, all the excited states are populated from the ground by electron impact, and decay directly or by cascade to the ground through the emission of radiation. Relative intensities of spectral lines are then independent of density. As the density increases, the long-lived states are depleted additionally by collisions to nearby states. This produces changes in relative intensities in the density region where

$$N_e C \sim A \qquad (4\text{-}6\text{-}1)$$

where C is the coefficient for collisional loss and A is the spontaneous decay rate. The analysis for these ions is fortuitously simplified by the fact that the regions in which N_e is of the order given by eq. (4-6-1) are well separated for each of the three main forbidden transitions. As N_e increases from the low density limit, collisions 2 ^{3}S → 2 ^{3}P first depopulate 2 ^{3}S so that the 1 ^{1}S–2 ^{3}S line intensity falls to zero. Then at intermediate densities 2 ^{1}S → 2 ^{1}P collisions depopulate the 2 ^{1}S term so that intensity is transferred from the two-photon continuum to the 1 ^{1}S–2 ^{1}P line. Finally at higher density, the now collisionally mixed 2 ^{3}S and 2 ^{3}P terms suffer collisional loss to the excited singlet terms as well as by stepwise ionization, resulting in a loss of 1 ^{1}S–2 ^{3}P intensity relative to the 1 ^{1}S–2 ^{1}P line.

B. Energy levels

Most of the important terms below the first ionization limit are listed for several of the ions by Moore [33] and by Kelly [69]. Detailed studies of particular ions are also given by Edlén [92], Fawcett and Irons [93] and for carbon v by Edlén and Löfstrand [94].

A particular problem concerns the position of the $2\,^1S$ term, which remained undetermined until very recently in ions above beryllium III. This is because of the very small branching ratio for decay of $2\,^1P$ excitation through the $2\,^1S$–$2\,^1P$ line, which therefore has a very low intensity. Calculated wavelengths for this line are given by Wiese et al. [51] for several members of the sequence. Prasad and El-Menshawy [95] made intensity measurements on a line at 3540.8 Å, the computed wavelength [50] in carbon v. However measurements in a laser-produced polyethelyne plasma by Boland et al. [96] showed a line at 3526.7 Å which they identified as this transition. Subsequent detailed calculations by Accad et al. [97] confirmed this latter identification. The line has not yet been observed in higher ions of the sequence.

Configurations with both electrons excited lie above the first ionization limit, for all ions of the sequence. Detailed calculations have been carried out by several authors for helium (see e.g. Burke [98]) and these have been supported by some experiments. Perrott and Stewart [99, 100] have calculated energy levels for $2p^2\,^1D$ and $2s^2\,^1S$ and Chan and Stewart [101] for $2s\,2p\,^{1,\,3}P$ for ions up to boron IV. These levels all autoionize and are therefore very difficult to produce in emission spectra for low Z. However, $p^2\,^3P$ and $p \cdot p'\,^{1,\,3}P$ levels are non-autoionizing in LS coupling and are expected to produce emission lines more readily. Some of these have been calculated by Drake and Dalgarno [102] up to neon IX. The $1s\,2p\,^3P^0$–$2p^2\,^3P$ transition in helium was seen in emission by Kruger [103] in 1930. The observation and classification of lines from both autoionizing and non-autoionizing levels have been carried out in beryllium III [104] and carbon v [105]. Similar transitions are also identified in the solar spectrum for magnesium XI and silicon XIII, by Walker and Rugge [49].

C. TRANSITION PROBABILITIES

f-values for optically allowed transitions are tabulated by Wiese et al. [50]. A number of calculations have been carried out in the last few years for other transitions. The transition rate for the intercombination line $1\,^1S$–$2\,^3P_1$ arises through the progressive break-down of LS coupling along the iso-electronic sequence. Rates calculated by Elton [106] and Garstang [77] have now been superseded by the more complete calculations of Drake and Dalgarno [107]. A direct measurement of this rate has now been carried out in oxygen VII using beam-foil spectroscopic techniques [108], and gives results in agreement with theory. At high Z, the 1 ^{1}S–2 ^{3}P$_2$ transition becomes important through magnetic quadrupole radiation [77]. These rates have now been calculated by Drake [109], for all ions up to argon XVII. Beam foil measurements of this transition rate have been made [110] in argon XVII, and are in agreement with theory.

The 2 ^{1}S term decays by two-photon emission. This produces a continuum, with its threshold at the 1 ^{1}S–2 ^{1}S energy interval, rising to a maximum on a wavelength scale at $\sim$ 1.24 times the threshold and falling away slowly to longer wavelengths. The rates for ions up to neon IX have been calculated by Drake et al. [111]. An experimental observation of this continuum has been reported by Elton et al. [112] in neon IX, using a theta-pinch plasma, but the conditions of the experiment did not allow the measurement of the transition rate. At the time of writing there has been no positive observation of such a continuum from an astrophysical source.

The 2 ^{3}S term was, until recently, also expected to decay by two-photon emission, and there have been several calculations of the transition probability. The early calculation by Mathis [113] has been shown to contain a serious error, corrected by the later work of Bely and Faucher [114] and Drake et al. [111]. In the case of the triplet, the rate is much smaller than the singlet, and the continuum peaks sharply, close the threshold, so that it could be mistaken for a broad line. A study of recently

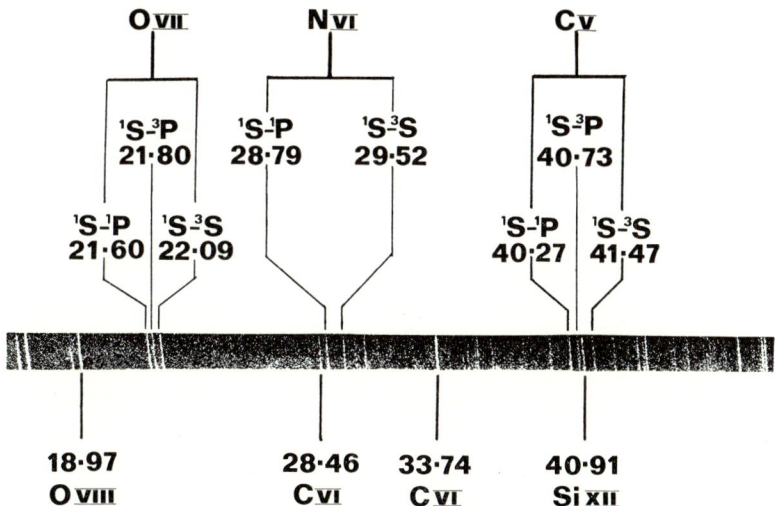

Fig. 4-6-2. Solar soft X-ray spectrum recorded by Freeman and Jones [86], showing 1 ^{1}S–2 ^{3}S forbidden lines in carbon V, nitrogen VI and oxygen VII.

available solar soft X-ray spectra [115, 116, 117] however showed no such continua. However, an additional strong line to the longwave side of the resonance $1\,^1S$–$2\,^1P$ and intercombination $1\,^1S$–$2\,^3P_{1,2}$ lines appeared for each of the helium-like ions. Figure 4-6-2 is a recent solar spectrum by Freeman and Jones [86], showing these transitions in carbon, nitrogen and oxygen. Gabriel and Jordan [46] concluded that there was in fact a single-photon magnetic-dipole transition from the $2\,^3S$ level which dominates over the two-photon rate, and they were able to deduce a minimum rate of 10 sec^{-1} for oxygen VII. Griem [118] then showed that using Dirac mechanics, a large relativistic magnetic dipole probability could be calculated, that had been previously overlooked in helium-like ions. His scaling as Z^8, giving a rate for oxygen of 32 sec^{-1}, was too small however as a result of an algebraic error in the derivation. Following suggestions by other workers [119, 120, 121], Gabriel and Jordan [122] were able to locate the discrepancy and showed that the scaling should be as Z^{10} [123]. The constant they adopted was derived semi-empirically from solar observations and was a factor of two smaller than that originally used by Griem, giving a rate for oxygen of 830 sec^{-1}. The theory used was based upon hydrogenic approximations, but was expected to be accurate to within a factor of 2. More recent calculations by Drake [124], taking account of electron–electron interaction, support this estimate, giving rates $\lesssim 25\%$ larger than the semi-empirical values of Gabriel and Jordan. Drake's value for oxygen VII is 1044 sec^{-1}. Furthermore this rate has recently been measured directly for argon XVII in a beam-foil experiment [125]. It agrees to within 20% with Drake's calculation.

D. COLLISIONAL EXCITATION RATES

The impact excitation processes which are important in determining the intensities of lines from $n = 2$ terms are a) excitation from the ground to each of the four $n = 2$ terms, b) excitation transfer between the $n = 2$ terms, and c) ionization from the $n = 2$ triplet levels. Collisions between the J levels of the $2\,^3P$ term are never important; levels become statistically mixed through the intermediate $2\,^3S$ level, before the density is high enough for direct mixing.

At the time of writing, the most complete theoretical cross-sections available use Coulomb–Born–Oppenheimer methods and suffer from the usual uncertainties concerning the validity of the exchange contributions. Bely [126] carried out calculations for oxygen VII. More complete data can be derived from the C–B–O calculations of Burgess et al. [9] for hydrogenic ions of infinite nuclear charge, which can be expected to be valid approxima-

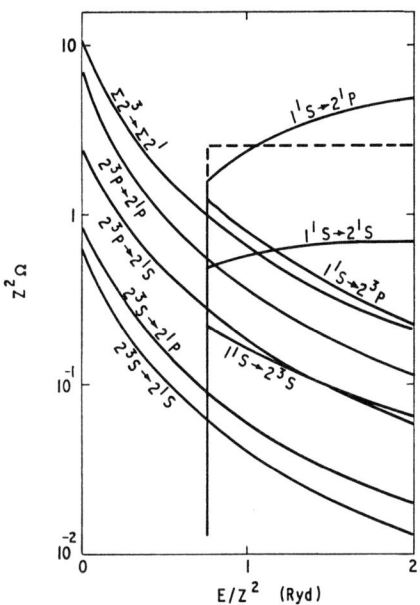

Fig. 4-6-3. Collision strengths for helium-like ions of infinite charge, derived from Burgess et al. [9]. The dashed curve is the value for $1\,^1S \to 2\,^1P$ from the effective gaunt factor approximation [3].

tely for helium-like ions above carbon v, except for the above exchange uncertainties. Some of the transitions required here are derived in their paper; the remainder can be calculated [127] from their tabulated R-matrices. The collision strengths for each of the above excitation processes are shown in Fig. 4-6-3 as a function of the impact energy. These are plotted in "reduced" form of $Z^2\Omega$ as a function of E/Z^2 where $(Z-1)$ is the charge on the ion. The total strength for $n = 2$ triplet to singlet transitions is based on the assumption that the initial levels ($2\,^3S_1$, $2\,^3P_{0,1,2}$) are populated statistically. Also shown as a dashed line is the effective Gaunt factor approximation of Van Regemorter [3] for the resonance transition $1\,^1S \to 2\,^1P$. In this ion where coupling of states is not large it can be expected to give a good value, and in fact agrees well with the Coulomb–Born result. Burgess's weak-coupling impact parameter approximation [6] is found to give a result which is a factor of 2 larger.

Fig. 4-6-4 shows the excitation rate coefficients for excitations from the ground, plotted as a function of electron temperature, derived by integrating the above collision strengths over the Maxwellian electron energy distribution. Also shown are four ratios of various excitation rates, which are

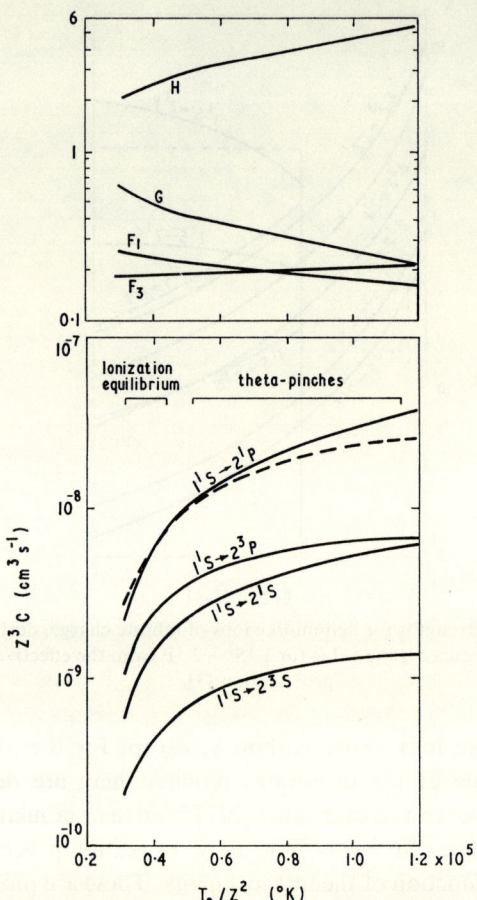

Fig. 4-6-4. Excitation rate coefficient of helium-like ions, derived from the data in Fig. 4-6-3. The upper curves show ratios of some of these rates, defined as follows:

$$F_1 = C(1\ ^1S \to 2\ ^1S)/C(1\ ^1S \to 2\ ^1P),\ F_3 = C(1\ ^1S \to 2\ ^3S)/C(1\ ^1S \to 2\ ^3P),$$
$$G = [C(1\ ^1S \to 2\ ^3S) + C(1\ ^1S \to 2\ ^3P)]/C(1\ ^1S \to 2\ ^1P),$$
$$H = [C(1\ ^1S \to 2\ ^1S) + C(1\ ^1S \to 2\ ^1P)]/[C(1\ ^1S \to 2\ ^3S) + C(1\ ^1S \to 2\ ^3P)].$$

relevant to the later intensity analysis. The reduced temperature T/Z^2 at which each helium-like ion would be the dominant ion stage is indicated both for conditions of ionization equilibrium [66], as expected in low density astrophysical plasmas, and for a typical transient ionizing plasma, such as a theta-pinch.

Bely [126] has looked at the contribution to be expected from cascade, i.e. excitation from the ground to an $n = 2$ level via an intermediate higher level. He claims that excitation to the $n = 2$ triplet levels can be increased by up to 40 % by this effect, whereas the singlets are increased by only a few per cent. This is not included in the calculated rates reproduced here.

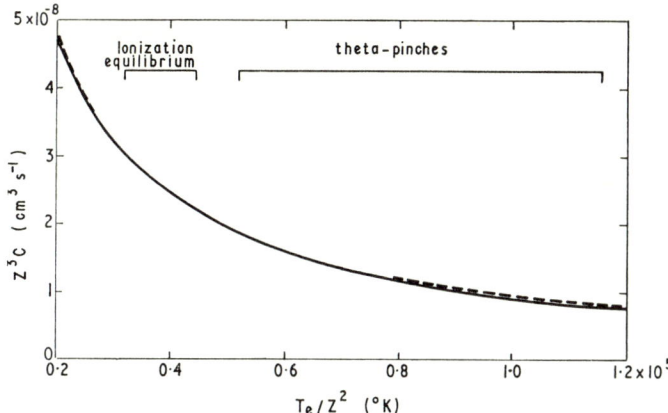

Fig. 4-6-5. Excitation rate coefficient for $n = 2$ triplet to singlet transfer in helium-like ions, derived from the data in Fig. 4-6-3. The dashed curve represents the analytic approximation to this rate, given by eq. (4-6-2).

For the total $n = 2$ triplet to singlet collision rate, integration leads to the result shown in Fig. 4-6-5. A good approximation to this is given by the analytic expression

$$Z^3 C(2^3 \to 2^1) = 9.7 \times 10^{-4} (T/Z^2)^{-1} \qquad (4\text{-}6\text{-}2)$$

which is shown as a dashed curve in the figure. Here 2^3 implies the combined $n = 2$ triplet levels, etc. For the optically allowed transitions $2\,^1S \to 2\,^1P$ and $2\,^3S \to 2\,^3P$ good results are to be expected from either the Van Regemorter formula or the Burgess impact parameter method. These two methods are found to give rates in good agreement, and the former has been adopted in the intensity analysis which follows later. (But see § 4-6K.) It should be noted that these rates are larger than those given by eq. (4-6-2) for the optically forbidden transitions by a factor $\gtrsim 150$.

For ionization from the $n = 2$ triplet terms, there are Coulomb–Born calculations by Moores [128] for carbon v. Moores' results, which are essentially the same for $2\,^3S$ and $2\,^3P$, are found to agree within 10 % both with Wilson and White (as used by McWhirter [1]) and with a new semi-empirical formula by Kunze [124]. Here, we modify the Wilson and White expression to obtain a best fit to Moores' data. This gives

$$S(2^3) = 0.61 \times 10^{-5} E^{-\frac{3}{2}} \left(\frac{kT}{E}\right)^{\frac{1}{2}} \left(2.54 + \frac{kT}{E}\right)^{-1} \exp\left(-\frac{E}{kT}\right), \quad (4\text{-}6\text{-}3)$$

with E and kT in eV, and where E is the ionization energy of the $n = 2$ triplet levels. This expression does not require scaling for other values of Z and can be applied throughout the sequence. It reproduces Moores' result to within 5% over the range $0.2 < kT/E < 6$.

E. Low density plasmas

In the quiet solar corona, helium-like ions can be treated in the low-density limit, but in hot, dense active regions other processes must be included. Collisions $2\,^3\text{S} \to 2\,^3\text{P}$ then begin to compete with spontaneous radiation

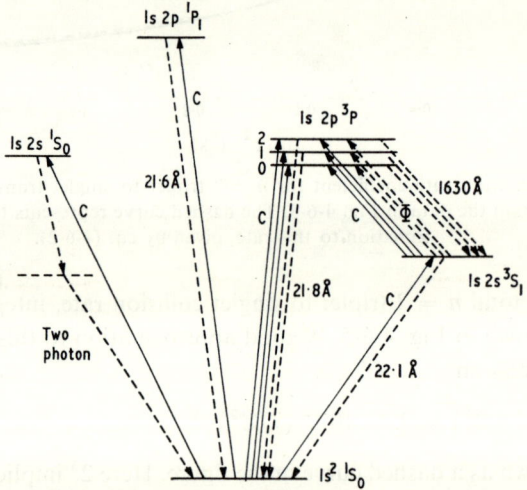

Fig. 4-6-6. Schematic representation of the lower energy levels of the helium-like ion showing those processes taken into account in the analysis for low-density plasmas [130] (see § 4-6E). The wavelengths indicated relate to the oxygen VII ion.

for decay of the $2\,^3\text{S}$ level, and intensity is transferred from the forbidden to the intercombination line. The theory of this process has been worked out by Gabriel and Jordan [130]. The equations of statistical equilibrium were solved for the ground state and the six $n = 2$ levels. The levels and the processes taken into account are shown in Fig. 4-6-6. The intensity ratio R of the forbidden to intercombination line intensities is found to be given by

$$R = \frac{A(2\,^3\text{S} \to 1\,^1\text{S})}{[N_e C(2\,^3\text{S} \to 2\,^3\text{P}) + \Phi](1+F_3) + A(2\,^3\text{S} \to 1\,^1\text{S})} \left(\frac{1+F_3}{B_3} - 1\right)$$

$$(4\text{-}6\text{-}4)$$

where

$$F_3 = C(1\,^1S \to 2\,^3S)/C(1\,^1S \to 2\,^3P)$$

and the effective branching ratio

$$B_3 = \frac{1}{3}\frac{A(2\,^3P_1 \to 1\,^1S)}{A(2\,^3P_1 \to 1\,^1S)+A(2\,^3P \to 2\,^3S)}$$
$$+ \frac{5}{9}\frac{A(2\,^3P_2 \to 1\,^1S)}{A(2\,^3P_2 \to 1\,^1S)+A(2\,^3P \to 2\,^3S)}. \quad (4\text{-}6\text{-}5)$$

N_e is the electron density and Φ (in sec^{-1}) is the photo-excitation rate from $2\,^3S$ to $2\,^3P$, which in the sun is important only for carbon v. The rates $C(1\,^1S \to 2\,^3P)$ and $C(2\,^1S \to 2\,^3P)$ are assumed to be distributed among the three $2\,^3P$ levels in the ratio of their statistical weights. The rates $C(1\,^1S \to 2)$ include contributions from cascade. The temperature T_e used to calculated $C(2\,^3S \to 2\,^3P)$ is taken as that value at which the resonance line has its maximum emission. F_3 has been taken as 0.35 which fits the observations better than the 0.2 indicated by Fig. 4-6-4 but of course now includes cascade. The resulting density is insensitive to the choice of values for T_e or F_3. It is only necessary to assume that F_3 does not vary significantly with T_e. Using eq. (4-6-4), the observed ratio R can now be interpreted in terms of N_e, whenever R differs significantly from its low density-limit R_0, given by

$$R_0 = (1+F_3)/B_3 - 1. \quad (4\text{-}6\text{-}6)$$

It is the variation of R between observations, rather than its absolute value, that is used to confirm that $R > R_0$, thus guarding against possible errors due to errors say in F_3.

The interpretation of observed ratios given in ref. [130] are based upon the early erroneous values of $A(2\,^3S \to 1\,^1S)$. This was later repeated [131] with the new semi-empirical A-values. The intensity ratio is shown as a function of N_e in Fig. 4-6-7, this time plotted in the form $1/R$ in order to give a linear plot. The effect of putting $\Phi = 260$ sec^{-1}, its computed value in carbon v, is shown as the dashed line in the figure. $1/R_0$ for carbon v is then increased from 0.09 to 0.75. Densities derived for some solar active regions [131] range up to $\sim 10^{13}$ cm^{-3} using the observations reported. It is most important in interpreting such observations to ensure that the measured ratios do vary significantly between observations, taking into account the accuracy of measurement. Otherwise all the densities may be effectively those which give the low-density limit ratio R_0.

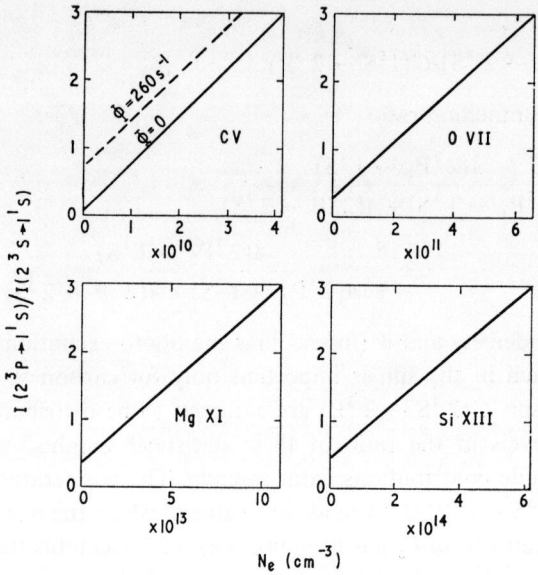

Fig. 4-6-7. Computed values of the intensity ratio $(1\,^1S\text{–}2\,^3P)/(1\,^1S\text{–}2\,^3S)$ as a function of density for several helium-like ions [131]. The dashed curve for carbon v shows the effect of including photo-excitation $2\,^3S \to 2\,^3P$ in the sun.

Table 4-6-1 lists the atomic coefficients for a number of ions together with the limiting ratio $1/R_0$ and the lower limit for density measurement N_e^*. This is arbitrarily defined as the density for which $R = 1.1 \times R_0$ (with $\Phi = 0$). It is given by

$$N_e^* = 0.111 \frac{A(2\,^3S \to 1\,^1S)}{(1+F_3)C(2\,^3S \to 2\,^3P)}. \tag{4-6-7}$$

It can be readily seen from Fig. 4-6-6, that if we define an observed intensity ratio G by

$$G = \frac{I(2\,^3S \to 1\,^1S) + I(2\,^3P \to 1\,^1S)}{I(2\,^1P \to 1\,^1S)} \tag{4-6-8}$$

then throughout this density regime, G is given by

$$G = \frac{C(1\,^1S \to 2\,^3S) + C(1\,^1S \to 2\,^3P)}{C(1\,^1S \to 2\,^1P)}. \tag{4-6-9}$$

It is seen [130, 131] that over a wide range of observations, $G \sim 1.1$, this supporting the assumption that ratios of ground state excitation rates are insensitive to T_e or Z. Notice again that the observed value of G differs from the calculated value in Fig. 4-5-4, but that the observed rate coefficients include cascade.

TABLE 4-6-1

Atomic coefficients, and limiting values of ratios and densities, used in the analysis of low-density plasmas

Ion	T_e (°K)	$A(2\,^3S \to 1\,^1S)$ (sec^{-1})	$C(2\,^3S \to 2\,^3P)$ (10^{-9} cm^3 sec^{-1})	B_3	$\dfrac{1}{R_0}$	N_e^* (cm^{-3})
C v	1.0 (6)	4.86 (1)	28.0	0.11	0.090	1.4 (8)
N vi	1.25 (6)	2.53 (2)	17.0	0.22	0.20	1.2 (9)
O vii	1.95 (6)	1.04 (3)	11.6	0.29	0.28	7.3 (9)
Ne ix	3.5 (6)	1.09 (4)	5.82	0.34	0.33	1.5 (11)
Mg xi	5.8 (6)	7.24 (4)	3.28	0.38	0.39	1.8 (12)
Al xii	7.4 (6)	1.66 (5)	2.57	0.41	0.43	5.3 (12)
Si xiii	8.9 (6)	3.56 (5)	2.16	0.45	0.50	1.3 (13)
S xv	1.55 (7)	1.41 (6)	1.29	0.56	0.71	8.9 (13)
Ar xvii	1.8 (7)	4.71 (6)	0.95	0.68	1.0	4.9 (14)
Ca xix	2.2 (7)	1.38 (7)	0.69	0.77	1.3	1.6 (15)
Fe xxv	4.0 (7)	2.00 (8)	0.32	0.87	1.8	5.1 (16)

Numbers in parentheses represent powers of ten for multiplying factors.

In view of the much higher densities predicted by this theory with the corrected transition probabilities, it is important to re-examine critically the solar ratios measured, to ensure that after taking full allowance for observational errors, the ratio R does vary significantly, and departs significantly from its low density limit. If this critical check is carried out for the ions considered in ref. [131], neon ix and magnesium xi are the ions that continue to show real departures from the low-density limit. More recently, new data has been recorded for silicon xiii [49], and in flares for calcium xix [132] and iron xxv [48]. All of these give ratios which are consistent with the values of R_0 listed in Table 4-6-1. This is not surprising, since the density N_e^* required for significant change in R increases rapidly with nuclear charge, reaching 5×10^{16} cm^{-3} for iron xxv.

F. High density plasmas

The majority of measurements on laboratory plasmas have been carried out in the higher density regime in which both two-photon emission from $2\,^1S$ and the forbidden line from $2\,^3S$ are completely suppressed. These two levels then decay by excitation to $2\,^1\mathrm{P}$ and $2\,^3\mathrm{P}$ respectively followed by radiation to the ground. At this point, which for the purposes of this section is a new low-density limit, the resonance line intensity is given by the sum of excitation to $2\,^1\mathrm{S}$ and $2\,^1\mathrm{P}$ and the intercombination line by the sum of excitation to $2\,^3\mathrm{S}$ and $2\,^3\mathrm{P}$. Furthermore since collisional exchange between $2\,^3\mathrm{S}$ and $2\,^3\mathrm{P}$ has become comparable with radiative processes, these two terms are approaching statistical relative populations. As the density is increased the excited level populations increase relative to the ground. The mixed $2\,^3\mathrm{S}$ and $2\,^3\mathrm{P}$ levels have an exceptionally high population due to their slow decay through the intercombination line, and this leads to the onset of other collisional loss mechanisms from these levels. Kunze et al. [27] developed a theory for this process which took account of two such mechanisms; direct ionization $S(2^3)$ from the level, and collisional transfer $C(2^3 \rightarrow 2^1)$ to the $n = 2$ singlet levels. They derived for the ratio of resonance to intercombination line

$$\frac{I(2\,^1\mathrm{P} \rightarrow 1\,^1\mathrm{S})}{I(2\,^3\mathrm{P} \rightarrow 1\,^1\mathrm{S})} = R_1 + \frac{N_e}{A(2^3 \rightarrow 1)}[(R_1+1)C(2^3 \rightarrow 2^1) + R_1 S(2^3)] \quad (4\text{-}6\text{-}10)$$

where

$$R_1 = \frac{C(1\,^1\mathrm{S} \rightarrow 2\,^1\mathrm{S}) + C(1\,^1\mathrm{S} \rightarrow 2\,^1\mathrm{P})}{C(1\,^3\mathrm{S} \rightarrow 2\,^3\mathrm{S}) + C(1\,^1\mathrm{S} \rightarrow 2\,^3\mathrm{P})} \quad (4\text{-}6\text{-}11)$$

is the new low-density limit ratio defined in this section, and $A(2^3 \rightarrow 1)$ is the intercombination line transition probability averaged over the initial 2^3 levels, so that

$$A(2^3 \rightarrow 1) = \tfrac{1}{4} A(2\,^3\mathrm{P}_1 \rightarrow 1\,^1\mathrm{S}). \quad (4\text{-}6\text{-}12)$$

The expression (4-6-10) is generally valid if $C(2^3 \rightarrow 2^1)$ includes all 2^3 loss processes that lead to emission of a resonance line photon, and $S(2^3)$ includes all those that do not. Equation (4-6-10) can be rewritten

$$\frac{I(2\,^1\mathrm{P} \rightarrow 1\,^1\mathrm{S})}{I(2\,^3\mathrm{P} \rightarrow 1\,^1\mathrm{S})} = R_1 + PN_e \quad (4\text{-}6\text{-}13)$$

where P is a constant with respect to N_e and only a slow function of T_e. P is thus the slope of the straight line plot obtained from $I(2\,^1\mathrm{P} \rightarrow 1\,^1\mathrm{S})/I(2\,^3\mathrm{P} \rightarrow 1\,^1\mathrm{S})$ versus N_e.

Kunze et al. [27] measured this ratio as a function of N_e for carbon v in a theta-pinch plasma. Their results are shown in Fig. 4-6-8. The low

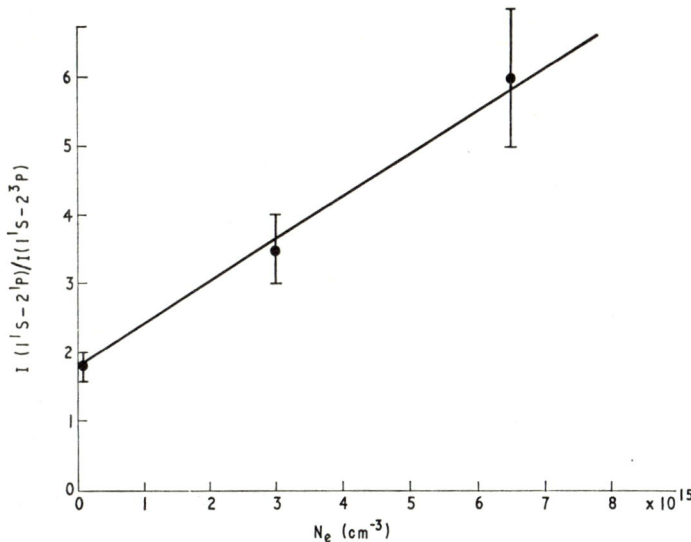

Fig. 4-6-8. Line ratios for carbon v measured as a function of electron density by Kunze et al. [27].

density point was in fact derived from an oxygen VII measurement but its use has been justified by subsequent low density, carbon V measurement. This gives a value of 1.8 for R_1 and 6.4×10^{-16} cm^3 for P. For oxygen they obtained only a value for R_1 of 1.8. All the measurements were carried out close to $T_e = 3 \times 10^6$ °K. R_1, as defined by eq. (4-6-11), is in effect the rate coefficient ratio labelled as H in Fig. 4-6-4, when the effects of cascade are included. Bely [126] has calculated this ratio for oxygen VII and shown that cascades make a large contribution to the triplets, which increases with temperature. This has the effect of lowering H and reducing its temperature dependence. The above experimental results show that R_1 does not vary significantly between carbon V and oxygen VII, while the reduced temperature T_e/Z^2 varies by a factor of 2. This suggests that Bely has underestimated the triplet cascade contribution, and that this contribution has the result of effectively removing the dependence of R_1 on T_e.

Return now to a consideration of the slope P, from Fig. 4-6-8. Since $A(2^3 \to 1)$ increases rapidly with the charge Z, P will decrease with Z. For higher ions it is therefore, necessary to work at higher densities in order to measure P. The higher densities are difficult to achieve, so that measurements are good for carbon V, becoming less reliable for nitrogen VI and oxygen VII. A single measurement of the intensity ratio ($= 2.5$) was made for oxygen VII, by Elton and Koppendorfer [133] at $N_e = 6.2 \times 10^{16}$ cm^{-3}. Recently a series of collaborative experiments have been carried out by Kunze at the University of Maryland and Gabriel and Paget at the A.R.U. Culham [32]. Both groups used theta-pinches. The Maryland machine covers the range $N_e = 1$ to 15×10^{15} cm^{-3}, while the A.R.U. machine can operate up to over 10^{17} cm^{-3}, thereby including a very wide range between the two. Experiments have been carried out on carbon, nitrogen and oxygen, and where possible at different temperatures. $P \sim 6$ to 7×10^{-16} cm^3 has been confirmed for carbon V up to the higher density, for $T \sim 3 \times 10^6$ °K. Values for nitrogen VI are $\sim 8 \times 10^{-17}$ cm^3 and for oxygen VII $\sim 1.2 \times 10^{-17}$ cm^3. These results are tentative as the experiments are still in progress at the time of writing. Measurements carried out by Prasad and El-Menshawy [95] for carbon V at 4×10^{17} cm^{-3} must unfortunately be disregarded. This is because they made measurements on what they claimed to be the 2 ^{1}S–2 ^{1}P line at 3541 Å, and this is now believed to be an incorrect identification (see § 4-6B).

Both Kunze et al. [27], and Elton and Koppendorfer [133] assumed that P was dominated by the first term in the square bracket in eq. (4-6-10). The second term, ionization from the 2^3 levels, they corrected for from theory, in order to derive a value for $C(2^3 \to 2^1)$. This value turned out to be ~ 20 times larger than that given by the Coulomb–Born calculations. Bely [126] claimed that this interpretation was incorrect. He argued that $C(2^3 \to 2^1)$ was negligible and that P was determined entirely by $S(2^3)$. It now appears that he was partly correct and that $S(2^3)$ was underestimated in the earlier analyses. However, recent improved calculations of $S(2^3)$ as detailed in § 4-6D can account for only $\frac{2}{3}$ of P, and the remaining $\frac{1}{3}$ still requires $C(2^3 \to 2^1)$ to be ~ 7 times larger than theory. A further complication arises from the variation with T_e. The experiments indicate that P increases slowly with T_e. This trend is supported by $S(2^3)$ (see eq. (4-6-3)) on account of the exponential term, but not by $C(2^3 \to 2^1)$ (see eq. (4-6-2)) which has the reverse trend. Furthermore, a $\frac{1}{3}$ contribution from $C(2^3 \to 2^1)$ would be sufficient to reverse the T_e variation in P. A possible explanation lies in the suggestion that an additional process accounts for $\frac{1}{3}$ of P while $S(2^3)$ accounts for $\frac{2}{3}$, $C(2^3 \to 2^1)$ being negligible. The additional process suggested is the collisional transfer from the 2^3 levels to the *singlet* levels with $n \geq 3$, followed by cascade to the ground. This would require to be $\sim \frac{1}{5}$ of the excitation to the *triplet* $n \geq 3$ levels and would have the correct T_e dependence on account of the re-introduction of the exponential term. The higher proportion of spin transfer collisions for these as compared to the 2 → 2 collisions can be understood, since more of the excitation will occur close to threshold. A rate coefficient of the form

$$C(2^3 \to \geq 3^1) \sim \frac{3.5 \times 10^{-5}}{Z^2 T_e^{\frac{1}{2}}} \exp\left(-\frac{E}{kT}\right) \qquad (4\text{-}6\text{-}14)$$

is suggested, in which E is the order of the energy interval $n = 2$ to 3 in the ion concerned. Equation (4-6-10) should then be rewritten

$$\frac{I(2^1P \to 1^1S)}{I(2^3P \to 1^1S)} = R_1 + PN_e$$

where

$$P = \frac{R_1}{A(2^3 \to 1)}[S(2^3) + \alpha C(2^3 \to \geq 3^1)] \qquad (4\text{-}6\text{-}15)$$

and

$$\alpha = 1 + \beta/R_1. \qquad (4\text{-}6\text{-}16)$$

β is the proportion of $C(2^3 \to > 3^1)$ collisions resulting in the emission of a $1\,^1S$–$2\,^1P$ photon. In the absence of any more detailed knowledge of these collisions, we assume $\beta = \frac{2}{3}$, the value we would get if only $n = 3$ singlet levels are involved and if the rate is shared statistically between them. If this is in error, it results in a somewhat smaller error in the constant derived for eq. (4-6-14).

If in the above experiments, a measurement is made in addition of the absolute intensity of the $2\,^3S$–$2\,^3P$ multiplet in the long wavelength ultraviolet region, it is possible to obtain the absolute value of $C(1\,^1S \to 2\,^1P)$. This was carried out in references [27] and [133] to experimental accuracies of 40% and a factor of 2, respectively. The results were found to agree to within the experimental errors with rates calculated using eq. (4-2-14).

G. Intermediate Densities

Before proceeding to derive the overall variation of line ratios with density, it is necessary to look at the region intermediate between the low density and high density regions dealt with in § 4-6 E and F. The 2^3 levels are behaving here as a single system, decaying through the intercombination line. The $2\,^1S$ level is, however, decaying partly by two-photon emission and partly, depending on the density, by excitation to $2\,^1P$. This region has not been dealt with quantitatively elsewhere either by theory or experiment.

The theory is obtained quite simply by adapting the model already derived for the triplets in eqs. (4-6-4) and (4-6-5). Thus

$$\frac{I(2\,^1S \to 1\,^1S)}{I(2\,^1P \to 1\,^1S)} = \frac{A(2\,^1S \to 1\,^1S)}{N_e C(2\,^1S \to 2\,^1P)(1+F_1)+A(2\,^1S \to 1\,^1S)} \left(\frac{1+F_1}{B_1}-1\right)$$

(4-6-17)

where

$$F_1 = \frac{C(1\,^1S \to 2\,^1S)}{C(1\,^1S \to 2\,^3P)}$$

and

$$B_1 = \frac{A(2\,^1P \to 1\,^1S)}{A(2\,^1P \to 1\,^1S)+A(2\,^1P \to 2\,^1S)}$$

(4-6-18)

≈ 1 for all ions.

In these equations, $2\,^1S \to 1\,^1S$ represents the two-photon process. F_1 can be derived by combining the observed solar ratios $F_3 = 0.35$ and $G = 1.1$, with the laboratory result $R_1 = 1.8$. These lead to $F_1 = 1.0$ a result rather insensitive to changes in F_3. This value for F_1 might be considered too high, from a theoretical viewpoint. An error might arise through combining solar and laboratory ratios, measured at different temperatures and possibly having different cascade contributions. However, in the absence of any better data we must for now assume the value $F_1 = 1.0$.

The intensity ratio given in eq. (4-6-17) is of the two-photon continuum to the resonance line and is not easily measured. However the equation does give the proportion of the total singlet excitation which decays via the resonance line. Thus, by definition (eq. (4-6-8)), the observed ratio G is given by

$$G = \frac{I(2\,^3S \to 1\,^1S)+I(2\,^3P \to 1\,^1S)}{I(2\,^1P \to 1\,^1S)}$$

$$= \frac{C(1\,^1S \to 2\,^3S)+C(1\,^1S \to 2\,^3P)}{C(1\,^1S \to 2\,^1P)} \times \frac{1}{1+F_1} \times \left(1+\frac{I(2\,^1S \to 1\,^1S)}{I(2\,^1P \to 1\,^1S)}\right).$$

(4-6-19)

In conjunction with eq. (4-6-17) this gives the observed ratio G throughout the intermediate density region. The atomic coefficients used in this region are listed in Table 4-6-2.

TABLE 4-6-2
Atomic coefficients used for intermediate density region

Ion	T_e (°K)	$A(2\ ^1S \to 1\ ^1S)$ (sec^{-1})	$C(2\ ^1S \to 2\ ^1P)$ (10^{-8} cm^3 sec^{-1})
C v	1.0 (6)	3.31 (5)	3.6
	3.1 (6)		2.8
N vi	1.25 (6)	9.43 (5)	2.53
	3.1 (6)		1.78
O vii	1.95 (6)	2.31 (6)	1.48
	3.1 (6)		1.30
Ne ix	3.5 (6)	1.00 (7)	0.75

Numbers in parentheses represent powers of ten for multiplying factors.

H. LINE RATIOS, N_e AND T_e

It is now possible to calculate the variation of the three $2 \to 1$ lines and the two-photon continuum as a function of N_e throughout all three regions considered. It must be realised that a number of uncertainties of detail still exist in the theory so that these variations should be considered approximate in some respects. The relative intensities of these four emissions are plotted in Fig. 4-6-9 for carbon v, nitrogen vi, oxygen vii and neon ix. For carbon v the effect of photoexcitation $2\ ^3S \to 2\ ^3P$ for the sun is also shown. For carbon v to oxygen vii the appropriate portions of the curves are plotted at two different temperatures, corresponding to the temperatures expected for these ions, in ionization equilibrium as well as in a typical ionizing laboratory plasma.

An experimental study of the intermediate region will be a problem, since it is difficult to measure the relative intensity of a line to a continuum of comparable total intensity. One experimental observation of the two-photon continuum has been reported. This was by Elton et al. [112] in neon ix in a theta-pinch at a density of 10^{16} cm^{-3}. Further work is required in this area of study. An alternative method of studying this region would be to observe the change in the intensity ratio of resonance to intercombination line as the density is varied. No quantitative data is available, although it is instructive to recall that early photographic spectra of the toroidal pinch Zeta taken in the grazing-incidence region [134] showed comparable intensities for these two lines at densities $\sim 10^{14}$ cm^{-3}, for oxygen vii. This is broadly consistent with the analysis in Fig. 4-6-9 and lends support to a value for $F_1 \sim 1.0$.

As stated earlier, the helium-like ion is not ideally suited for temperature

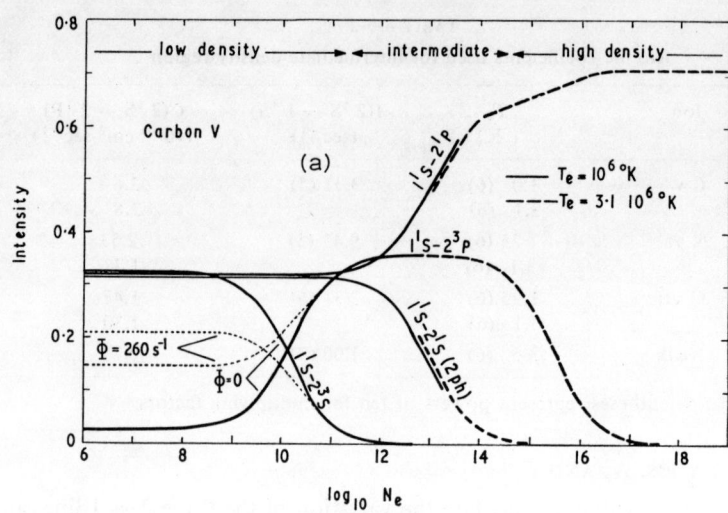

(a)

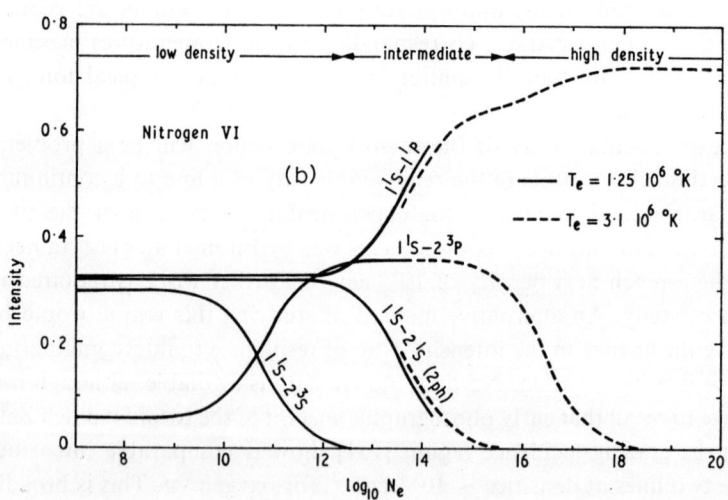

(b)

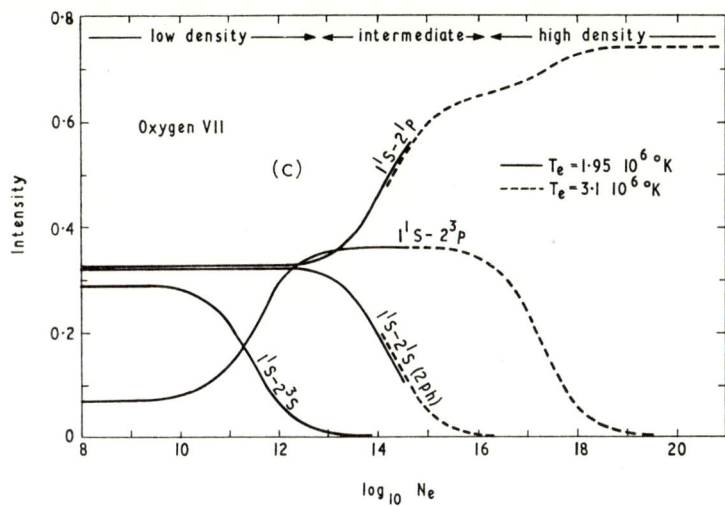

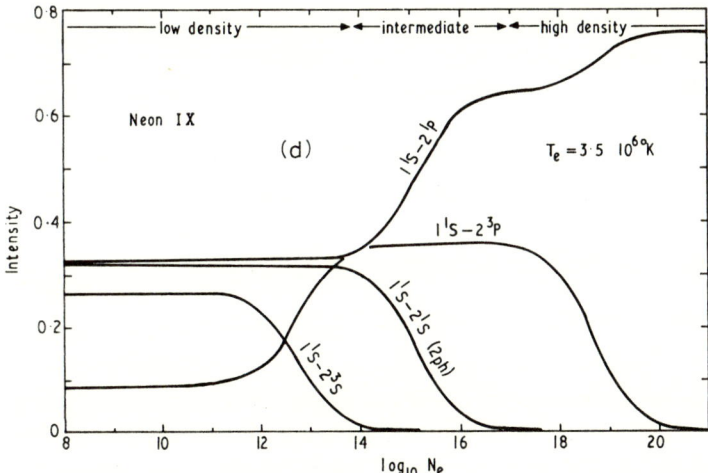

Fig. 4-6-9. Predicted line intensities for helium-like ions over a wide range of densities. They are plotted on an intensity scale as a proportion of the total $1 \to 2$ excitation rate, including cascade. The three density regions dealt with in the text are indicated. See also § 4-6K.

measurement from line intensity ratios. However, two further methods have been proposed in the literature and should be mentioned briefly. The first is a proposal by Williams and Kaufman [135] to use the absolute intensity of the $2\,^3S$–$2\,^3P$ carbon v multiplet. In the light of the more recent studies of this ion, we can now see that their assumption of a Boltzmann population for the $2\,^3S$ level was incorrect. Furthermore, the cross-section they adopted for $1\,^1S \rightarrow 2\,^3P$ excitation was an overestimate by a factor ~ 300. A corrected version of this theory was derived by Kunze et al. [136], but the method cannot be recommended for ease of application. A better method has been proposed in this same paper [136] for use with transient ionizing plasmas, i.e. most pulsed laboratory plasmas. It makes use of the time history of the $2\,^3S$–$2\,^3P$ intensity in order to derive an ionization rate for the ion, and thence the electron temperature. For this particular type of analysis the helium-like ion offers some unique advantages, on account of the large increase in ionization potential between the lithium-like and helium-like ion. This causes the helium-like ion to persist over a wide temperature range (or long duration in the case of transient plasmas), and to achieve at its peak over 80% of the population density of the element concerned.

I. Dielectronic satellite lines

Emission spectra from two-electron excited states have been observed both in the laboratory and from the sun, as detailed in § 4-6B. As with the three-electron system, these also can be produced by dielectronic recombination, which is likely to be the dominant mechanism except for very high density plasmas. The general considerations of § 4-2D and § 4-4H will then apply also to these transitions. Walker and Rugge [49] have applied such a theory to line ratios of satellite to hydrogenlike resonance lines measured from magnesium XI and XII in the sun. They obtained agreement between observation and theory to within a factor of 3, which was probably limited by the values available for the radiative transition probability A_r.

J. Absolute intensities

The interpretation of helium-like ion spectra outlined above has emphasized relative intensities of $2 \rightarrow 1$ transitions and their relation to relative collision rates $1 \rightarrow 2$. The only direct measurements of absolute excitation rates are those indicated at the end of § 4-6F, in particular that of Kunze et al. [27] who measured in effect the sum of $C(1\,^1S \rightarrow 2\,^1P)$ and $C(1\,^1S \rightarrow 2\,^1S)$ for carbon v. This gave a value for fP (as in eq. (4-2-13)) of $0.23 \pm 40\%$ compared with a theoretical predication (see § 4-6D) for $C(1\,^1S \rightarrow 2\,^1P)$

alone of 0.13. The measurement of satellite to resonance line ratios by Gabriel and Paget [31], described in § 4-4H can also be interpreted as a measurement of the sum of excitation rates to the $2\,^1S$ and $2\,^1P$ levels. In this case a value of fP is found of $0.13\pm20\%$. These two experimental results are consistent within the errors quoted. If they are to agree with the theoretical estimate for $2\,^1P$ alone, then the contribution from excitation to $2\,^1S$ must be small. This suggests that the ratio $C(1\,^1S \to 2\,^1S)/C(1\,^1S \to 2^1P)$, refered to above as F_1, is less than 0.2 rather than the value of 1.0 adopted in § 4-6G.

K. Recent developments

The curves shown in Fig. 4-6-9 suggest a valuable method of measuring electron densities over ten order of magnitude, merely from the relative intensities of three close lines in one ion. However, uncertainties remain and it will be most important to improve the collision rates used, over the next few years.

The rates used for collisions $2\,^3S \to 2\,^3P$ and $2\,^1S \to 2\,^1P$ were obtained from eq. (4-2-13) and it appears now that these may be too low. Recent calculations by Blaha [10] using a Coulomb–Born approximation suggest increases of a factor of 2 in the ions quoted here, rising to 3 or more for ions of higher charge, when proton impacts become important. This will have the effect of sliding the low and intermediate portions of the curves in Fig. 4-6-9 to the left by a factor of 2 in density.

The change in F_1 to 0.2, suggested in the above section, will, if true, have the effect of reducing the intensity change in the $1\,^1S$–$2\,^1P$ line in the intermediate density section of these curves.

Engelhardt et al. [137] have pointed out that the assumption of statistical mixing of the $^3P_{0,1,2}$ levels at the lowest end of the high density range may not be fully justified. If this is allowed for in the interpretation of the results of Kunze et al. [24] it can be shown (Gabriel et al. [29]) that a lower value is obtained for the ratio

$$R_1 = \frac{C(1\,^1S \to 2\,^1S)+C(1\,^1S \to 2\,^1P)}{C(1\,^1S \to 2\,^3S)+C(1\,^1S \to 2\,^3P)}.$$

This will have only a small effect on the ratios shown in Fig. 4-6-9. Engelhart et al. [137] also develop a model which could be used for measuring electron density in the region between the above "intermediate" and "high" density regions. This is based on a measurement of the relative line intensities within the 2^3S–$2\,^3P$ multiplet.

Acknowledgements

The authors are grateful to many members of the Astrophysics Research Unit, whose helpful comments have been of much assistance in the writing of this chapter. We would also like to thank those who have allowed us to make use of their work in advance of publication, in particular Dr. W. Eissner of University College London.

References

1. R. W. P. McWhirter, in: Plasma Diagnostic Techniques, eds. R. H. Huddlestone and S. L. Leonard (Academic Press, New York, 1965) p. 201.
2. G. Elwert, Z. Naturforsch. **7a** (1952) 432.
3. H. van Regemorter, Astrophys. J. **136** (1962) 906.
4. B. L. Moiseiwitsch and S. J. Smith, Rev. Mod. Phys. **40** (1968) 238.
5. O. Bely and H. Van Regemorter, in: Annual Review of Astronomy and Astrophysics, Vol. 8 (1970) 329.
6. A. Burgess, Proc. Symp. on Atomic Collision Processes in Plasmas, U.K. Atomic Energy Authority Rep. AERE-R 4818 (1964) p. 63.
7. M. J. Seaton, Proc. Phys. Soc. (London) **79** (1962) 1105.
8. O. Bely, Ann. Astrophys. **29** (1966) 131.
9. A. Burgess, D. G. Hummer and J. A. Tully, Phil. Trans. Roy. Soc. London **266** (1970) 225.
10. M. Blaha, Bull. Amer. Astr. Soc. **3** (1971) 246.
11. W. D. Johnston and J. H. Kunze, Phys. Rev. A **4** (1971) 962.
12. A. Burgess, Astrophys. J. **139** (1964) 776.
13. A. Burgess, Ann. Astrophys. **28** (1965) 774.
14. A. Burgess, Astrophys. J. **141** (1965) 1588.
15. A. H. Gabriel, C. Jordan and T. M. Paget, in: Proc. 6th Intern. Conf. on Physics of Electronic and Atomic Collisions (M.I.T. Press, Massachusetts, 1969) p. 558.
16. M. F. A. Harrison, in: Methods of Experimental Physics, ed. L. Marton (Academic Press, New York, 1968) Vol. 7A, p. 95.
17. D. W. O. Heddle and R. G. W. Keesing, in: Advances in Atomic and Molecular Physics, eds. D. R. Bates and I. Estermann (Academic Press, New York, 1968) Vol. 4.
18. W. L. Fite and R. T. Brackmann, Phys. Rev. **112** (1958) 1151.
19. D. F. Dance, M. F. A. Harrison and A. C. H. Smith, Proc. Roy. Soc. London **A290** (1966) 74.
20. H. J. Kunze, in: Plasma Diagnostics, ed. Lochte-Holtgreven (North-Holland, Amsterdam, 1968) p. 550.
21. D. E. Evans and J. Katzenstein, Rep. Progr. Phys. **32** (1969) 207.
22. A. H. Gabriel, Nucl. Instr. and Methods **90** (1970) 157.
23. B. C. Boland, F. C. Jahoda, T. J. L. Jones and R. W. P. McWhirter, J. Phys. B. **3** (1970) 1134.
24. W. G. Griffin and R. W. P. McWhirter, Proc. Conf. on Optical Instruments and Techniques, ed. K. J. Habell (Chapman and Hall, London, 1962) p. 14.
25. T. S. Green, D. L. Fisher, A. H. Gabriel, F. J. Morgan and A. A. Newton, Phys. Fluids **10** (1967) 1663.
26. A. W. DeSilva and H. J. Kunze, J. Appl. Phys. **39** (1968) 2458.
27. H. J. Kunze, A. H. Gabriel and H. R. Griem, Phys. Rev. **165** (1968) 267.
28. H. J. Kunze and W. D. Johnston, Phys. Rev. A **3** (1971) 1384.

REFERENCES

29. H. J. Kunze and A. H. Gabriel, Phys. Fluids **11** (1968) 1216.
30. G. Tondello and R. W. P. McWhirter, J. Phys. B **4** (1971) 715.
31. A. H. Gabriel and T. M. Paget, J. Phys. B. **5** (1972) in press.
32. A. H. Gabriel, T. M. Paget and H. J. Kunze, to be published.
33. C. E. Moore, N.B.S. Circular No. 467 (U.S. Govt. Printing Office, Washington D.C., 1949).
34. B. C. Fawcett, B. B. Jones and R. Wilson, Proc. Phys. Soc. **78** (1961) 1223.
35. K. Bockasten, R. Hallin and T. P. Hughes, Proc. Phys. Soc. **81** (1963) 522.
36. B. C. Fawcett, J. Phys. B **3** (1970) 1152.
37. C. R. Detwiler, J. D. Purcell and R. Tousey, Mém. Soc. Roy. Liège 5me Ser. **4** (1961) 254.
38. B. Edlén, Rept. Progr. Phys. **26** (1963) 181.
39. B. Edlén, Handbuch der Physik **27** (1963) 80.
40. B. C. Fawcett, A. H. Gabriel, W. G. Griffin, B. B. Jones and N. J. Peacock, Proc. Phys. Soc. **84** (1964) 257.
41. L. L. House and G. A. Sawyer, Astrophys. J. **139** (1964) 775.
42. G. Tondello, J. Phys. B **2** (1969) 727.
43. B. C. Fawcett, R. A. Hardcastle and G. Tondello, J. Phys. B **3** (1970) 564.
44. U. Feldman, L. Cohen and W. Behring, J. Opt. Soc. Am. **60** (1970) 891.
45. B. Edlén and F. Tyrén, Nature **143** (1939) 940.
46. A. H. Gabriel and C. Jordan, Nature **221** (1969) 947.
47. J. F. Meekins, G. A. Doschek, H. Friedman, T. A. Chubb and R. W. Kreplin, Solar Phys. **13** (1970) 198.
48. W. M. Neupert and M. Swartz, Astrophys. J. **160** (1970) L189.
49. A. B. C. Walker and H. R. Rugge, Aerospace Report No. TR-0059 (9260-02)-2 (1970) and Astrophys. J. **164** (1971) 181.
50. W. L. Wiese, M. W. Smith and B. M. Glennon, NSROS-NBS 4 (U.S. Govt. Printing Office, Washington D.C., 1966).
51. A. W. Weiss, Astrophys. J. **138** (1963) 1262.
52. W. L. Wiese, M. W. Smith and B. M. Miles, NSROS-NBS 22 (U.S. Govt. Printing Office, Washington D.C., 1969).
53. M. Cohen and A. Dalgarno, Proc. Roy. Soc. London **A280** (1964) 258.
54. P. G. Burke, J. H. Tait and B. A. Lewis, Proc. Phys. Soc. **87** (1966) 209.
55. O. Bely and D. Petrini, Astron. and Astrophys. **6** (1970) 318.
56. O. Bely, Ann. Astrophys. **29** (1966) 683.
57. C. W. Allen, Astrophysical Quantities (Univ. of London, the Athlone Press, 2nd ed. 1963).
58. D. R. Flower, in: Highlights of Astronomy 1970, ed. C. de Jager (D. Reidel Publ. Co., Dordrecht, Holland) p. 512, and J. Phys. B **4** (1971) 697.
59. W. Eissner, to be published.
60. W. Eissner and H. Nussbaumer, J. Phys. B **2** (1969) 1028.
61. H. E. Saraph, M. J. Seaton and J. Shemming, Phil. Trans. Roy. Soc. London **A264** (1969) 77.
62. L. Heroux, Nature **198** (1963) 1291.
63. L. Heroux, Proc. Phys. Soc. **83** (1964) 121.
64. H. J. Taylor and B. A. Lewis, AERE-R-5061 Harwell (1965).
65. L. Heroux and M. Cohen, Phil. Trans. Roy. Soc. London A **270** (1971) 99.
66. C. Jordan, Mon. Not. Roy. Astr. Soc. **142** (1969) 501.
67. C. Jordan, Ph. D. Thesis, University of London (1965).
68. S. R. Pottasch, Astrophys. J. **137** (1963) 945.
69. R. L. Kelly, N.R.L. Report 6648 (U.S. Govt. Printing Office, Washington D.C., 1968).

70. G. Tondello and T. M. Paget, J. Phys. B **3** (1970) 1757.
71. W. M. Burton, A. Ridgeley and R. Wilson, Mon. Not. Roy. Astr. Soc. **135** (1967) 207.
72. W. M. Burton and A. Ridgeley, Solar Phys. **14** (1970) 3.
73. B. Edlén, H. P. Palenius, K. Bockasten, R. Hallin and J. Bromander, Solar Phys. **9** (1969) 432.
74. R. H. Garstang and L. J. Shamey, Astrophys. J. **148** (1967) 665.
75. A. M. Naqvi, Proc. Seventh Intern. Conf. on Phenomena in Ionized Gases, Belgrade, 1965 (Gradevinska Knjiga Pub. House, Belgrade, 1966) Vol. 2, p. 558.
76. D. E. Osterbrock, J. Phys. B **3** (1970) 149.
77. R. H. Garstang, Astrophys. J. **148** (1967) 579.
78. H. Friedrich and E. Trefftz (Max Planck Inst. Report, MPI-PHE (Astro 29), München, 1969).
79. H. Nussbaumer, Mon. Not. Roy. Astr. Soc. **145** (1969) 141.
80. M. Gailitis, Soc. Phys. - JETP **17** (1963) 1328.
81. W. Eissner, to be published.
82. O. Bely and P. Faucher, Astron. and Astrophys. **6** (1970) 88.
83. M. Blaha, Astron. and Astrophys. **1** (1969) 42.
84. C. A. Hall, K. R. Damon and H. E. Hinteregger, Space Research **3** (1963) 745.
85. W. M. Burton, C. Jordan, A. Ridgeley and R. Wilson, Phil. Trans. Roy. Soc. Soc. London A **270** (1971) 81.
86. F. F. Freeman and B. B. Jones, Solar Phys. **15** (1970) 288.
87. C. Jordan, in: Highlights of Astronomy 1970, ed. C. de Jager (D. Reidel Publ. Co., Dordrecht, Holland) p. 519.
88. R. W. Noyes, G. L. Withbroe and R. P. Kirshner, Solar Phys. **11** (1970) 388.
89. D. E. Osterbrock, Astrophys. J. **160** (1970) 25.
90. D. E. Osterbrock and R. A. R. Parker, Astrophys. J. **143** (1966) 268.
91. H. Nussbaumer, Astron. and Astrophys. **16** (1972) 77.
92. B. Edlén, Arkiv Fysik **4** (1952) 441.
93. B. C. Fawcett and F. E. Irons, Proc. Phys. Soc. London **89** (1966) 1063.
94. B. Edlén and B. Löfstrand, J. Phys. B **3** (1970) 1380.
95. A. N. Prasad and M. F. El-Menshawy, J. Phys. B **1** (1968) 471.
96. B. C. Boland, F.E. Irons and R. W. P. McWhirter, J. Phys. B **1** (1968) 1180.
97. Y. Accad, C. L. Pekeris and B. Schiff, Phys. Rev. **183** (1969) 78.
98. P. G. Burke, in: Advances in Atomic and Molecular Physics, Vol. 4 (Academic Press, New York, 1968) p. 173.
99. R. H. Perrott and A. L. Stewart, J. Phys. B **1** (1968) 381.
100. R. H. Perrott and A. L. Stewart, J. Phys. B **1** (1968) 1226.
101. Y. M. C. Chan and A. L. Stewart, Proc. Phys. Soc. London **90** (1967) 619.
102. G. W. F. Drake and A. Dalgarno, Phys. Rev. A. **1** (1970) 1325.
103. P. G. Kruger, Phys. Rev. **36** (1930) 855.
104. S. Goldsmith, J. Phys. B. **2** (1969) 1075.
105. U. Feldman and L. Cohen, Astrophys. J. **158** (1969) L169.
106. R. C. Elton, Astrophys. J. **148** (1967) 573.
107. G. W. F. Drake and A. Dalgarno, Astrophys. J. **157** (1967) 459.
108. I. A. Sellin, M. Brown, W. W. Smith and B. Donnally, Phys. Rev. A **2** (1970) 1189.
109. G. W. F. Drake, Astrophys. J. **158** (1969) 1199.
110. R. Marrus and R. W. Schmieder, Phys. Rev. Letters **25** (1970) 1689.
111. G. W. F. Drake, G. A. Victor and A. Dalgarno, Phys. Rev. **180** (1969) 25.
112. R. C. Elton, L. J. Palumbo and H. R. Griem, Phys. Rev. Letters **20** (1968) 783.
113. J. Mathis, Astrophys. J. **125** (1957) 318.
114. O. Bely and P. Faucher, Astron. and Astrophys. **1** (1969) 37.

115. G. Fritz, R. W. Kreplin, J. F. Meekins, A. E. Unzicker and H. Friedman, Astrophys. J. **148** (1967) L133.
116. H. R. Rugge and A. B. C. Walker, in: Space Research 8 (North-Holland, Amsterdam, 1968) p. 439.
117. B. B. Jones, F. F. Freeman and R. Wilson, Nature **219** (1968) 252.
118. H. R. Griem, Astrophys. J. **156** (1969) L103.
119. R. H. Garstang, private communication.
120. S. Woosley, private communication.
121. I. P. Grant, private communication.
122. A. H. Gabriel and C. Jordan, Phys. Letters **32A** (1970) 166.
123. H. R. Griem, Astrophys. J. **161** (1970) L155.
124. G. W. F. Drake, Phys. Rev. A **3** (1971) 908.
125. R. W. Schmieder and R. Marrus, Phys. Rev. Letters **25** (1970) 1254.
126. O. Bely, Phys. Letters **26A** (1968) 408.
127. J. A. Tully, private communication.
128. D. L. Moores, private communication.
129. H. J. Kunze, Phys. Rev. A **3** (1971) 937.
130. A. H. Gabriel and C. Jordan, Mon. Not. Roy. Astron. Soc. **145** (1969) 241.
131. F. F. Freeman, A. H. Gabriel, B. B. Jones and C. Jordan, Phil. Trans. Roy. Soc. London A **270** (1971) 127.
132. G. A. Doschek and J. F. Meekins, Solar Phys. **13** (1970) 220.
133. R. C. Elton and W. W. Köpendorfer, Phys. Rev. **160** (1967) 194.
134. B. C. Fawcett, A. H. Gabriel, W. G. Griffin, B. B. Jones and R. Wilson, Nature **200** (1963) 1303.
135. R. V. Williams and S. Kaufman, Proc. Phys. Soc. London **75** (1960) 329.
136. H. J. Kunze, A. H. Gabriel and H. R. Griem, Phys. Fluids **11** (1968) 662.
137. W. Engelhardt, W. Köppendörfer and J. Sommer, to be published.

CHAPTER 5

ATOMIC PROCESSES IN ASTROPHYSICAL PLASMAS

BY

VALERIE P. MYERSCOUGH

Department of Mathematics, Queen Mary College, London, England

and

G. PEACH

Department of Physics, University College, London, England

Contents

	Page
5-1. Introduction	295
5-2. Stellar atmospheres	296
A. General theory of radiative transfer	296
B. The construction of model atmospheres	300
C. Line radiation	302
D. Departures from local thermodynamic equilibrium	305
5-3. The continuous spectrum	312
A. General formulae	312
B. Absorption by neutral atoms and positive ions	316
C. Photodetachment of negative ions	322
D. Coherent scattering	328
5-4. The line spectrum	329
A. Transition probabilities	329
B. The line profiles – general formulae	333
C. Pressure broadening of spectral lines	336
5-5. Collisional excitation and ionization	354
A. The quantum mechanical theory – general formulae and approximations	354
B. Classical and semi-classical approximations	361
5-6. Interpretation of observations	366
A. Curve of growth analysis	366
B. The solar atmosphere	368
C. The early-type stars and planetary nebulae	373
D. The late-type stars molecular absorption	379
E. Stars of non-solar composition	381
1. Weak helium line stars	381
2. Hydrogen deficient stars	383
3. Carbon stars	384
5-7. Problems requiring further study	387
Acknowledgments	389
References	389

§ 5-1. Introduction

The radiation emitted by a star yields important information concerning its temperature, surface gravity and chemical composition, and one of the major problems in astrophysics is the interpretation of such stellar spectra. The observed radiation is compared with models of known characteristics, and the degree of agreement between the observed and model spectra enables the structure and composition of a particular star to be deduced.

Most stars exhibit a continuous spectrum together with a large number of dark absorption lines, and such stars are grouped into classes according to the appearance of their visible spectra. These classes O, B, A, F, G, K and M form a sequence which is directly related to surface temperature (see e.g. Aller [1]). The hottest (O and B) stars are intrinsically blue in colour, and both emission and absorption line objects are observed; prominent features of O stars are lines of HI, HeI, HeII, OIII, NIII and CIII. As the stellar temperature decreases, the degree of ionization of the elements is lessened, and so in class B the lines of HeII disappear and those of FeIII and MgII begin to occur. The A and F stars have strong lines of HI and singly ionised metals, while the major features of class G, of which the sun is a member, are the resonance lines of CaII, and lines of such neutral metals as FeI, TiI, MgI and SiI. In class K the metallic lines strengthen, and the cool red stars of class M have spectra dominated by molecular bands. The temperature of a black body radiating the same amount of energy as the star varies from about 2×10^5 °K for the central stars of some planetary nebulae to about 2000 °K for class M objects; the sun has a temperature of approximately 6000 °K on this scale. There are some stars, however, which do not fall into any of these categories, for example, the hot Wolf-Rayet stars which exhibit extremely strong, broad emission lines of carbon or nitrogen, and the cool R and N stars which show bands of carbon compounds.

A major problem in the analysis of a particular stellar spectrum is to determine whether, irrespective of the temperature of the star, the relative abundances of the elements present in the atmosphere are in solar (or cosmic) proportions (see e.g. Allen [2]). Such analyses are critically dependent not only on the physical assumptions of a model atmosphere, but also on the atomic data used in its construction. The theoretical approach is outlined in § 5-2, while in § 5-3, § 5-4 and § 5-5 the relevant atomic data is considered, together with an indication of its accuracy. In § 5-6 we discuss the application of these results to the interpretation of observations; the ultimate test of the validity of a model is the comparison of the predicted spectrum with that observed for a particular star or class of stars. Finally in § 5-7 we consider some of the atomic parameters which are needed more accurately, and indicate a few of the outstanding problems still to be resolved.

§ 5-2. Stellar atmospheres

A. General Theory of Radiative Transfer

We shall consider here only the outer layers of a star which determine the emergent spectrum; at sufficiently great depths in the star the temperature is high enough for nuclear processes to occur and rather different physical problems arise. The outer layers, or atmosphere, whose geometric depth is small compared with the radius of the star, may be considered to be stratified in plane parallel layers. The matter in these layers absorbs, scatters and re-emits radiation and, in order to examine the characteristics of this radiation, we must define its intensity and flux, and also the coefficients of emission and absorption (see e.g., Chandrasekhar [3]).

We consider a cylindrical element of volume dV of cross sectional area $d\sigma$ and length dr, at geometric depth z in a layer of density ρ. The orientation is such that the axis of the cylinder is inclined at an angle θ to the outward normal to the stellar surface and ϕ is the corresponding azimuthal angle (see

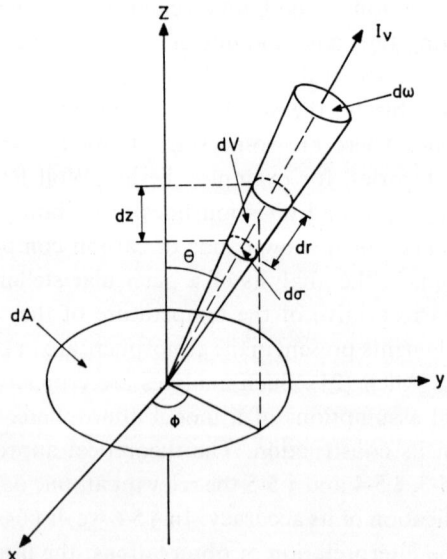

Fig. 5-2-1. Schematic diagram for the derivation of the transfer equation.

Fig. 5-2-1). The specific intensity $I_\nu(\theta, \phi)$ at frequency ν, in the direction (θ, ϕ), is defined as the amount of energy per second flowing through unit cross sectional area perpendicular to the direction (θ, ϕ), per unit solid angle and in unit frequency interval. Thus $I_\nu(\theta, \phi)$ has units erg sec^{-1} ster^{-1} Hz^{-1} cm^{-2}, and the amount of energy flowing through area $d\sigma$ in frequency interval $d\nu$ in the cone of solid angle $d\omega$ about the direction (θ, ϕ) is

$$dE_\nu = I_\nu(\theta, \phi) d\sigma \, d\nu \, d\omega \quad (\text{erg sec}^{-1}). \tag{5-2-1}$$

The energy per second crossing the area dA perpendicular to z due to the intensity in the direction (θ, ϕ) is given by

$$dE_\nu = I_\nu(\theta, \phi) \cos\theta \, dA \, d\nu \, d\omega, \tag{5-2-2}$$

and the total amount of radiation per second per unit frequency interval crossing unit area perpendicular to z is the flux at that depth,

$$\begin{aligned} F_\nu &= \int I_\nu(\theta, \phi) \cos\theta \, d\omega \\ &= \int_0^\pi \int_0^{2\pi} I_\nu(\theta, \phi) \cos\theta \sin\theta \, d\theta \, d\phi \quad (\text{erg cm}^{-2} \text{ sec}^{-1} \text{ Hz}^{-1}). \end{aligned} \tag{5-2-3}$$

The mean intensity, J_ν, is defined by

$$\begin{aligned} J_\nu &= \frac{1}{4\pi} \int I_\nu(\theta, \phi) \, d\omega \\ &= \frac{1}{4\pi} \int_0^\pi \int_0^{2\pi} I_\nu(\theta, \phi) \sin\theta \, d\theta \, d\phi. \end{aligned} \tag{5-2-4}$$

For the case of a plane parallel stratified medium, $I_\nu(\theta, \phi)$ is independent of ϕ, and the integrations over ϕ in (5-2-3) and (5-2-4) are trivial; the following results are all for this particular case.

Provided that a steady state exists, radiation flowing through the volume element dV is scattered, or absorbed and re-emitted by the material. We consider for simplicity the change in intensity I_ν arising from pure absorption and emission processes only. The loss in intensity as radiation flows through dV in the direction (θ, ϕ) due to absorption is proportional to ρ, the density of the material, and to the intensity of the incident radiation. Thus

$$dI_\nu = -\kappa_\nu \rho I_\nu \, dr \tag{5-2-5}$$

where $\kappa_\nu (\text{cm}^2 \text{ g}^{-1})$ is the mass absorption coefficient and we include in κ_ν

contributions from stimulated emission (or negative absorptions). The mass emission coefficient j_ν (erg sec^{-1} ster^{-1} Hz^{-1} g^{-1}) is the energy emitted per second, per unit solid angle, per unit frequency interval and for unit mass of material. The increase in the intensity in the direction (θ, ϕ) is then

$$dI_\nu = \rho j_\nu dr. \tag{5-2-6}$$

The coefficient j_ν depends only on spontaneous emission processes and hence is independent of direction. Equations (5-2-5) and (5-2-6) combine to give the net change in the intensity

$$dI_\nu = (j_\nu - \kappa_\nu I_\nu)\rho\, dr \tag{5-2-7}$$

which is the equation of radiative transfer. In the plane parallel model, the intensity I_ν is a function of z and θ only, where

$$dz = \mu\, dr \tag{5-2-8}$$

and $\mu = \cos \theta$. We introduce a dimensionless parameter, the monochromatic optical depth τ_ν, defined by

$$d\tau_\nu = -\kappa_\nu \rho\, dz \tag{5-2-9}$$

so that τ_ν increases inwards from zero at the surface of the star. The transfer equation (5-2-7) then becomes

$$\mu \frac{dI_\nu}{d\tau_\nu} = I_\nu - S_\nu \tag{5-2-10}$$

where the intensity is now written as $I_\nu \equiv I_\nu(\tau_\nu, \mu)$ and the quantity $S_\nu \equiv S_\nu(\tau_\nu)$ is the source function, defined according to

$$S_\nu = j_\nu / \kappa_\nu. \tag{5-2-11}$$

If the material is considered to radiate as a black body at temperature T then $I_\nu \equiv B_\nu(T)$ where $B_\nu(T)$ is the Planck function

$$B_\nu = \frac{2h\nu^3}{c^2} (e^{h\nu/KT} - 1)^{-1} \quad (\text{erg sec}^{-1}\ \text{ster}^{-1}\ \text{Hz}^{-1}\ \text{cm}^{-2}). \tag{5-2-12}$$

In this simple case, since $dB_\nu/dr = 0$, it follows from equation (5-2-7) that

$$j_\nu = \kappa_\nu B_\nu(T). \tag{5-2-13}$$

In the general case, however, the source function S_ν defined by (5-2-11) depends on the emission and absorption processes occurring in the material, and hence on the relative populations of the individual levels of the atoms and ions at depth τ_ν. Here, since at the densities occurring in astrophysical plasmas the velocity distribution of free electrons is Maxwellian, an electron termperature T can always be assigned. If, further, this temperature can be assumed to characterise the local matter (so that the level populations are given by the Boltzmann and Saha laws), then $S_\nu \equiv B_\nu(T)$. The validity of this assumption of local thermodynamic equilibrium (LTE) is discussed in § 5-2D.

Equation (5-2-10) may be solved formally by making the substitution

$$I_\nu(\tau_\nu, \mu) = C(\tau_\nu, \mu) \exp(\tau_\nu/\mu) \tag{5-2-14}$$

to obtain

$$C(\tau_\nu, \mu) - C(0, \mu) = -\int_0^{\tau_\nu} \exp(-t_\nu/\mu) S_\nu(t_\nu) dt_\nu/\mu. \tag{5-2-15}$$

The boundary condition at the surface is

$$I_\nu(0, \mu) = 0 \quad \text{for} \quad \mu < 0 \tag{5-2-16}$$

since there can be no inward flow of radiation from outside the star. Further, from consideration of the conservation of energy, $I_\nu(\tau_\nu, \mu)$ must remain finite for $\mu > 0$ and increasing τ_ν, so that

$$C(\tau_\nu, \mu) \to 0 \quad \text{as} \quad \tau_\nu \to \infty. \tag{5-2-17}$$

Combining equations (5-2-14) and (5-2-15) with conditions (5-2-16) and (5-2-17) we have

$$I_\nu(\tau_\nu, \mu) = \exp(\tau_\nu/\mu) \int_{\tau_\nu}^\infty \exp(-t_\nu/\mu) S_\nu(t_\nu) dt_\nu/\mu, \quad \mu > 0$$

$$I_\nu(\tau_\nu, \mu) = -\exp(\tau_\nu/\mu) \int_0^{\tau_\nu} \exp(-t_\nu/\mu) S_\nu(t_\nu) dt_\nu/\mu, \quad \mu < 0. \tag{5-2-18}$$

We use definitions (5-2-3) and (5-2-4) together with equations (5-2-18), to obtain the mean intensity

$$J_\nu(\tau_\nu) = \frac{1}{2} \int_0^\infty S_\nu(t_\nu) E_1(|\tau_\nu - t_\nu|) dt_\nu, \tag{5-2-19}$$

and the flux

$$F_\nu(\tau_\nu) = 2\pi \int_{\tau_\nu}^\infty S_\nu(t_\nu) E_2(t_\nu - \tau_\nu) dt_\nu - 2\pi \int_0^{\tau_\nu} S_\nu(t_\nu) E_2(\tau_\nu - t_\nu) dt_\nu \tag{5-2-20}$$

where the exponential integral $E_n(x)$ is defined by

$$E_n(x) = \int_1^\infty \frac{e^{-xy}}{y^n} \, dy. \tag{5-2-21}$$

B. THE CONSTRUCTION OF MODEL ATMOSPHERES

Some additional information to the theory of § 5-2A is required before we can relate the surface temperature, surface gravity g and chemical composition to the observed spectrum of the star. We assume that the atmosphere is in

(a) radiative equilibrium,
(b) hydrostatic equilibrium, and
(c) a steady state.

Condition (a) implies that there are no energy sources present and that the transport of energy through the atmosphere only takes place by radiation. Thus the net flux $F(\tau)$ integrated over frequency must be independent of optical depth τ. We can define an effective temperature T_e of the star by

$$F(\tau) = F = \sigma_R T_e^4 \tag{5-2-22}$$

where σ_R is the Stefan–Boltzmann constant; thus T_e is the temperature of a black body with integrated flux equal to F. Condition (b) implies that the effects of turbulence and convection are neglected, so that

$$dP/dz = -g\rho \tag{5-2-23}$$

where P is the pressure at depth z. In general both the gas pressure P_g and the radiation pressure P_r contribute to P, although P_r is usually negligible in comparison to P_g for stars cooler than class B. We may solve eq. (5-2-23) in conjunction with (5-2-19) and (5-2-20) using initially the simple T–τ relation

$$T^4 = \tfrac{3}{4} T_e^4 (\tau + q(\tau)) \tag{5-2-24}$$

derived from radiative transfer theory by Milne and Eddington assuming κ_ν to be independent of frequency (see e.g. [3]). Here τ is a mean optical depth defined by

$$d\tau = -\kappa\rho \, dz \tag{5-2-25}$$

where κ is a mean continuous opacity. In the Milne–Eddington approximation, $q(\tau) = \tfrac{2}{3}$, while expressions for the frequency dependent case have been given by Mark [4] and Henyey [5]. κ is usually obtained from the Rosseland mean [3]

$$\frac{1}{\kappa}\int_0^\infty \frac{\partial B_\nu}{\partial T}\,d\nu = \int_0^\infty \frac{1}{\kappa_\nu}\frac{\partial B_\nu}{\partial T}\,d\nu. \qquad (5\text{-}2\text{-}26)$$

On combining (5-2-23) and (5-2-25), we then have

$$\frac{dP_g}{d\tau} + \frac{dP_r}{d\tau} = \frac{g}{\kappa}. \qquad (5\text{-}2\text{-}27)$$

Thus κ_ν and hence κ are functions of the electron pressure P_e, temperature T and the relative abundances of the atmospheric constituents; the detailed derivation of κ_ν is illustrated in § 5-3 and § 5-4. The monochromatic optical depth τ_ν is given by

$$\tau_\nu = \int_0^\tau \frac{\kappa_\nu(t)}{\kappa(t)}\,dt \qquad (5\text{-}2\text{-}28)$$

and the temperature–optical depth law is modified until condition (a) is satisfied, that is, at each optical depth τ,

$$\int_0^\infty F_\nu\,d\nu = F \equiv \sigma_R T_e^4 \qquad (5\text{-}2\text{-}29)$$

(see e.g. [6]). The parameters g and T_e are chosen initially and then may be subsequently altered to produce a solution for the emergent flux which is in improved agreement with observations.

The third assumption (c) involves solving a full set of equations for the relative populations of the energy levels of each atomic constituent, taking into account all the important photo and collisional excitation, ionization, de-excitation and recombination processes. If, however, the collisional rates are much more important than the radiative rates in determining the level populations, the additional assumption of LTE is a good approximation. This leads to an enormous simplification of the problem, since the relative populations of the levels are related to the local kinetic temperature, T, by the Boltzmann and Saha laws, see § 5-2D, and the source function S_ν is given by (5-2-13) to be $B_\nu(T)$. The LTE approximation appears to be satisfactory for sufficiently large optical depths, where the electron number density, N_e, is of the order of 10^{15} cm^{-3}. In the less dense, extreme outer layers, however, this assumption always breaks down.

At some depth in the atmosphere, dependent on the type of star considered, the transport of energy by convection becomes important. However, a comparison of theoretical and observed energy distributions indicates that, for most cases, the layers which determine the character of the emergent spectrum have temperature gradients determined by radiative, rather than convective, transfer so that condition (a) is reasonable. The effects of turbulent motion are found in the extended atmospheres of stars with low surface gravities. In such cases, since blobs of gas are moving randomly with a variety of velocities, assumption (b), which presupposes a static atmosphere, does not hold and neither does the assumption of plane parallel stratification.

In view of the above comments, and due to the wealth of observational material, it is clear that more accurate models can be constructed for the sun than for any other star. Additional information is obtained by studying the intensity at different positions on the solar disc. From equation (5-2-18), the emergent intensity is

$$I_\nu(0, \mu) = \int_0^\infty S_\nu(t_\nu) \exp(-t_\nu/\mu) \, dt_\nu/\mu. \tag{5-2-30}$$

Thus in the centre of the disc ($\mu = 1$), the contributions to I_ν come from layers with optical depth $0 < \tau_\nu < 1$, whereas as we approach the limb ($\mu \to 0$), the intensity is only determined by the extreme outer layers. Pagel [7], and Lambert and Pagel [8] have shown that, in the sun, the LTE approximation is sufficiently accurate to describe the continuum emission, the weak lines in the visible, and the wings of strong absorption lines which are formed at large enough optical depths. In a typical photospheric (atmospheric) region with $T = 6000\ °K$ and electron density $N_e \sim 10^{14}\ cm^{-3}$, the proportion of neutral hydrogen in the $n = 3$ level is given closely by the Boltzmann relation and absorption by hydrogen in this state contributes to the visible continuum spectrum. The main source of opacity, however, arises from the processes

$$H^- + h\nu \to H + e$$

and (5-2-31)

$$H + e + h\nu \to H + e.$$

This was first suggested by Wildt [9], since although the ratio of the number density of H^- ions to that of neutral hydrogen is only about 10^{-7}, this is greater than the corresponding ratios for the populations of the $n = 2$ to $n = 1$ and $n = 3$ to $n = 1$ levels in hydrogen at optical depths $\tau \lesssim 1$ in the solar atmosphere. Since the Balmer lines are very strong in the sun, H^- is also important in determining the quality of the emergent radiation, and, in fact, continuous absorption by H^- and free–free absorption by electrons in the field of neutral hydrogen are the major sources of opacity in normal stars for which $4000\,°K \leq T_e \leq 7000\,°K$. The dominant process governing the amount of H^- is the associative detachment reaction of H^- with H, $[H^- + H \to (H_2^-) \to H_2 + e]$, see [7], and calculations of the opacity due to processes (5-2-31) are discussed in § 5-3. Continuous absorption by neutral hydrogen and the metallic constituents is only important in the ultraviolet and detailed LTE continuum models are now available (e.g., Carbon and Gingerich [10]).

For the higher temperature A and B stars, hydrogen is the dominant source of opacity, while in the hotter stars absorption by HeI and HeII is also important. For stars hotter than class A, the scattering of radiation by free electrons must also be considered. A series of models which include all the major sources of continuum opacity is available [11]; these have been fully iterated by the standard techniques [6] to obtain strict radiative equilibrium. More detailed results of particular cases are given in § 5-6.

C. Line Radiation

The continuous spectrum of a star is sufficient to determine the general structure of the atmosphere, but information on abundances comes from an analysis of individual absorption lines. The total absorption coefficient κ_ν at line frequency ν may be written as

$$\kappa_\nu = \kappa_c + l_\nu \qquad (5\text{-}2\text{-}32)$$

where κ_c is the continuous absorption coefficient, and l_ν represents only absorption in the line; the detailed form of l_ν is discussed in § 5-4. κ_c may be assumed to be independent of frequency across the width of the line and thus the transfer equation (5-2-10) for the line becomes (see e.g. [1] and [12]),

$$\mu \frac{dI_\nu}{d\tau_c} = (1+\eta_\nu)I_\nu - S_c - \eta_\nu S_\nu \qquad (5\text{-}2\text{-}33)$$

where

$$\eta_\nu = l_\nu/\kappa_c \qquad (5\text{-}2\text{-}34)$$

and τ_c is the optical depth in the continuum. The source functions in the continuum and the line are S_c and S_ν respectively. Equation (5-2-33) can be solved to give a result analogous to (5-2-30)

$$I_\nu(0, \mu) = \int_0^\infty (S_c + \eta_\nu S_\nu) \exp\{-(t_\nu + \tau_c)/\mu\} d\tau_c/\mu \qquad (5\text{-}2\text{-}35)$$

where

$$dt_\nu = \eta_\nu d\tau_c. \qquad (5\text{-}2\text{-}36)$$

The residual intensity is defined as the ratio of the intensity in the line to that in the continuum and is thus given by

$$r_\nu(0, \mu) = I_\nu(0, \mu)/I_c(0, \mu) \qquad (5\text{-}2\text{-}37)$$

where $I_c(0, \mu)$ is the emergent continuum intensity. For stars other than the sun, we can observe only the integrated intensity, so we then use a definition of r_ν analogous to (5-2-37) in terms of the emergent flux. The equivalent width of the line on a wavelength scale is defined by

$$W_\lambda = \int_{-\infty}^{+\infty} (1 - r_\nu) d(\Delta\lambda) \qquad (5\text{-}2\text{-}38)$$

where $\Delta\lambda$ is the wavelength separation from the line centre, or on a frequency scale by

$$W_\nu = \int_{-\infty}^{+\infty} (1 - r_\nu) d(\Delta\nu). \qquad (5\text{-}2\text{-}39)$$

A simplification of (5-2-35) is possible if S_c can be replaced by a black body function. If both the continuum and the line can be treated using LTE, (5-2-35) reduces to

$$I_\nu(0, \mu) = \int_0^\infty B_\nu(t_\nu) \exp(-t_\nu/\mu) dt_\nu/\mu \qquad (5\text{-}2\text{-}40)$$

where t_ν is the total optical depth in the line. Equation (5-2-40) can only be used when the populations of the atomic energy levels are determined by collisional processes. The opposite extreme is that in which level populations of the atoms absorbing the line radiation are completely dominated by radiative processes, usually called monochromatic radiative transfer (MRT). The scattering of the radiation is assumed to be coherent, that is, scattering does not cause any change in frequency, and the line source function, S_ν, then becomes equal to the mean intensity J_ν. For many lines formed in the solar photosphere, neither the LTE nor MRT approximations are justified [13].

Both processes can be allowed for (see [1] and [12]) by writing

$$S_v = \varepsilon B_v + (1-\varepsilon) J_v \qquad (5\text{-}2\text{-}41)$$

in equation (5-2-33), where the probability of pure absorption is ε and of coherent scattering is $1-\varepsilon$. However, non-coherent scattering of radiation is important in the formation of the strong resonance lines and has been considered by Hummer in a series of papers (see [14] and [15]).

D. Departures from Local Thermodynamic Equilibrium

In the outer layers of a stellar atmosphere, where the electron density is usually less than 10^{14} cm^{-3}, the radiation field is often sufficiently strong that radiative processes are important in the population and depopulation of the atomic energy levels of interest. Thus the assumption of LTE is invalid, and Thomas and others (see e.g. [16]) have developed non local theories (NLTE) to analyse the problem. These may be used to study the magnitude of departures from LTE, which are important for the interpretation of strong absorption lines in the sun and also for the analysis of both the strong lines and the continuum emission in the hotter stars.

If the populations of the atomic energy levels are in a steady state, then the rate of change with time of the number of atoms in each level is zero and the various processes populating and depopulating a given level must be considered individually. Processes involving the emission or absorption of radiation can be described in terms of the Einstein coefficients A_{ji}, B_{ji} and B_{ij} where i and j are the lower and upper levels of the transition respectively. The coefficients B_{ij} and B_{ji} are defined with respect to the mean intensity of the radiation so that

$$A_{ji} = \frac{2h v_{ij}^3}{c^2} B_{ji} \quad (\text{sec}^{-1}) \qquad (5\text{-}2\text{-}42)$$

and

$$\omega_i B_{ij} = \omega_j B_{ji} \qquad (5\text{-}2\text{-}43)$$

where v_{ij} is the frequency of the radiation emitted in the transition $j \to i$ and ω_i and ω_j are the statistical weights of levels i and j. Strictly, the number of atoms per cm^3 making the transition $i \to j$ per second depends on the actual intensity, which is a function of angle (cf. § 5-2A). However, this introduces more complexity into an already difficult problem and as yet no one has attempted to find a solution for this case.

B_{ij} may be expressed in terms of the oscillator strength f_{ij} (see § 5-4A) giving

$$B_{ij} = \frac{\pi e^2}{mc} \frac{4\pi}{h\nu_{ij}} f_{ij} \quad (\sec^2 g^{-1}). \tag{5-2-44}$$

We define the function $\phi_{ij}(\nu')$ to be the line shape for the transition $i \to j$ in which radiation of frequency ν' is absorbed. $\phi_{ij}(\nu')$ is normalized according to

$$\int_{-\infty}^{+\infty} \phi_{ij}(\nu') \, d(\Delta\nu) = 1 \tag{5-2-45}$$

where $\Delta\nu = \nu' - \nu_{ij}$ is the frequency separation from the line centre. The absorption cross section $a_{ij}(\nu')$ is given by

$$a_{ij}(\nu') = \frac{\pi e^2}{mc} f_{ij} \phi_{ij}(\nu') \quad (\text{cm}^2) \tag{5-2-46}$$

or, alternatively, on using (5-2-44) we obtain

$$a_{ij}(\nu') = \frac{h\nu_{ij}}{4\pi} B_{ij} \phi_{ij}(\nu'). \tag{5-2-47}$$

In equations (5-2-44) and (5-2-46), e and m are the electronic charge and mass respectively.

The coefficient C_{ij} is defined to be the number of transitions $i \to j$ per second due to collisions and we shall consider here only those transitions caused by electron impact. The rate coefficients are related to each other through the result obtained from detailed balancing, i.e.,

$$\omega_i C_{ij} = \omega_j C_{ji} \exp(-h\nu_{ij}/KT) \tag{5-2-48}$$

where T is the electron temperature. C_{ij} may be expressed in terms of the cross-section for excitation, $Q_{ij}(v)$ (cm^2), (see § 5-5), by

$$C_{ij} = N_e \int_{v_0}^{\infty} v f(v) Q_{ij}(v) \, dv \quad (\sec^{-1}). \tag{5-2-49}$$

Here $f(v)$ is a distribution function of initial velocity v, which can be taken to be Maxwellian, since electron–electron collisions produce a much more rapid redistribution of electron energy than do radiative or inelastic collision processes. The lower limit of the integral in (5-2-49) is v_0, where $\tfrac{1}{2}mv_0^2$ is the threshold energy for the excitation, and N_e is the number density of the free electrons.

The definitions of A_{ji}, B_{ij} and C_{ij} may be extended to include the case in which j is a continuum level, denoted here by c. As $j \to c$, the frequency $v_{ij} \to v$, where v is a continuous frequency variable and

$$\int a_{ij}(v')\,dv' \to v_0\, a_v(i), \qquad (5\text{-}2\text{-}50)$$

where $a_v(i)$ (cm^2) is the usual photoionization cross-section for level i (see § 5-3A). The frequency v_0 is given by

$$h v_0 = \frac{z^2 e^2}{2 a_0} \qquad (5\text{-}2\text{-}51)$$

where a_0 is the Bohr radius and $(z-1)$ is the net charge on the absorbing atom (see § 5-3A). Thus hv_0 is the ionization potential of the ground state of a hydrogenic atom of nuclear charge z. The relation (5-2-50) between the bound-bound and bound-free cross-sections can be expressed in terms of the Einstein B coefficients as

$$B_{ij} \to B_{ic}$$

where B_{ic} is a continuous function of v and

$$a_v(i) = \frac{hv}{4\pi v_0} B_{ic}. \qquad (5\text{-}2\text{-}52)$$

The coefficient C_{ic} is given by

$$C_{ic} = N_e \int_{v_0}^{\infty} v f(v) Q_{ic}(v)\, dv \quad (\text{sec}^{-1}) \qquad (5\text{-}2\text{-}53)$$

where $Q_{ic}(v)$ is the cross-section for collisional ionization (see § 5-5) and $\tfrac{1}{2}mv_0^2 = I_i$ is now the ionisation potential of level i.

The populations of the levels i and j, N_i and N_j respectively, may be expressed in terms of the number density of ions of net charge z, N_+, and the electron density N_e. We have

$$\frac{N_j}{N_i} = \frac{b_j \omega_j}{b_i \omega_i} \exp(-h v_{ij}/KT) \qquad (5\text{-}2\text{-}54)$$

and

$$N_i = b_i N_e N_+ \left(\frac{h^2}{2\pi m KT}\right)^{\tfrac{3}{2}} \frac{\omega_i}{2 U_+} \exp(I_i/KT) \qquad (5\text{-}2\text{-}55)$$

where U_+ is the partition function for the residual ion. The factors b_i and b_j are dimensionless and represent departures from thermodynamic equilibrium. If $b_i = b_j = 1$ in (5-2-54) and (5-2-55), the Boltzmann and Saha formulae are obtained. Equations (5-2-43) and (5-2-54) give

$$\frac{N_j}{b_j} B_{ji} = \frac{N_i}{b_i} B_{ij} \exp\left(-h\nu_{ij}/KT\right) \tag{5-2-56}$$

and since we assume that the continuum states are statistically populated, $b_j \to 1$ as the transition is made from a bound state j to a continuum state c, while

$$N_j B_{ji} \to \frac{N_i}{b_i} B_{ic} \exp\left(-h\nu/KT\right). \tag{5-2-57}$$

Also, using relations (5-2-42) and (5-2-57), we have

$$N_j A_{ji} \to \frac{2h\nu^3}{c^2} \frac{N_i}{b_i} B_{ic} \exp\left(-h\nu/KT\right) \tag{5-2-58}$$

as $j \to c$. Relations (5-2-57) and (5-2-58) give the number per second of stimulated and spontaneous recombinations to level i. Finally, as $j \to c$

$$\sum_j N_j C_{ji} \to \frac{N_i}{b_i} C_{ic} \tag{5-2-59}$$

and this gives the number of 3-body recombination processes which are the inverse to the processes of collisional ionization.

In order to derive the source function S_ν (see eq. (5-2-11)) in the general case, we require the detailed behaviour of the mass absorption coefficient κ_ν and the mass emission coefficient j_ν. We consider firstly the opacity $\kappa_\nu(i)$ arising from line and continuous absorption by atoms in level i, then

$$\rho \kappa_\nu(i) = \sum_{j=i+1}^{\infty} \frac{h\nu_{ij}}{4\pi} \{N_i B_{ij} \phi_{ij}(\nu) - N_j B_{ji} \phi_{ji}(\nu)\}$$
$$+ \frac{h\nu}{4\pi\nu_0} N_i B_{ic} \left\{1 - \frac{1}{b_i} \exp\left(-h\nu/KT\right)\right\} \tag{5-2-60}$$

where the states of the atom are labelled by $j = 1, 2, \ldots$ in order of increasing energy, and the stimulated emission processes have been treated as negative absorptions. The second term in (5-2-60) represents the continuous absorption due to bound–free transitions (cf. eq. (5-3-7)) and only contributes to $\kappa_\nu(i)$ if $h\nu \geq I_i$, while the total opacity is given by

$$\kappa_\nu = \sum_i \kappa_\nu(i). \tag{5-2-61}$$

The summation in (5-2-61) may be extended to include the continuum states $i = c$ and hence the free–free opacity $\kappa_\nu(c)$; if $i = c$, there is no contribution to $\kappa_\nu(i)$ from the first term in (5-2-60). The line profile for absorption is the same as for stimulated emission, therefore

$$\phi_{ij}(\nu) = \phi_{ji}(\nu) \tag{5-2-62}$$

for all values of ν.

In a similar way, we define the emission coefficient $j_\nu(i)$ to be

$$\rho j_\nu(i) = \sum_{j=1}^{i-1} \frac{h\nu_{ij}}{4\pi} N_i A_{ij} \psi_{ij}(\nu) \tag{5-2-63}$$

where $\psi_{ij}(\nu)$ is the profile for emission in a transition $i \to j$. This profile is not in general the same as $\phi_{ij}(\nu)$ since $j_\nu(i)$ contains a contribution from the scattering of photons of frequencies within the lines $i \to j$. $\psi_{ij}(\nu)$ is normalized according to the rule (5-2-45) and the total emission coefficient j_ν is given by

$$j_\nu = \sum_i j_\nu(i) \tag{5-2-64}$$

where the summation includes continuum states. The contribution to (5-2-64) from continuum states $i = c$ can be evaluated using (5-2-58) in equation (5-2-63), that is, as $i \to c$ so $\nu_{ij} \to \nu$ and

$$\frac{h\nu_{ij}}{4\pi} N_i A_{ij} \int \psi_{ij}(\nu') d\nu' \to \frac{h\nu}{4\pi} \frac{2h\nu^3}{c^2} \frac{N_j}{b_j} B_{jc} \exp(-h\nu/KT). \tag{5-2-65}$$

The only terms in the sum over j in (5-2-63) which then contribute are those for which $h\nu \geq I_j$. If j is a continuum level also, its energy E_j is determined by the relation $h\nu = E_i - E_j$ where E_i is the energy of level i.

Equations (5-2-61) and (5-2-64) give the source function S_ν involved in the equation of radiative transfer (5-2-10). However, extra information is still required in order to solve the problem; this is obtained from the equations of steady state for each level i in the atom which determine the number densities N_i (or departure coefficients b_i) of atoms in levels i ($i = 1, 2, \ldots, \infty$). The steady state equation for the ith level is

$$N_i \sum_{j=1}^{\infty}{}' B_{ij} \int_0^{\infty} \phi_{ij}(v) J_v \, dv + \frac{N_i}{hv_0} \int_{I_i}^{\infty} B_{ic}(v) J_v \, d(hv)$$

$$+ N_i \sum_{j=1}^{\infty}{}' C_{ij} + N_i C_{ic} + N_i \sum_{j=1}^{i-1} A_{ij} = \sum_{j=1}^{\infty}{}' N_j B_{ji} \int_0^{\infty} \phi_{ij}(v) J_v \, dv$$

$$+ \frac{1}{hv_0} \frac{N_i}{b_i} \int_{I_i}^{\infty} B_{ic}(v) J_v \exp(-hv/KT) \, d(hv) + \sum_{j=1}^{\infty}{}' N_j C_{ji} + \frac{N_i}{b_i} C_{ic}$$

$$+ \sum_{j=i+1}^{\infty} N_j A_{ji} + \frac{2}{c^2 v_0} \frac{N_i}{b_i} \int_{I_i}^{\infty} v^3 B_{ic}(v) \exp(-hv/KT) \, d(hv). \quad (5\text{-}2\text{-}66)$$

Here, J_v is the intensity I_v averaged over all directions (see eq. (5-2-4)) and the prime (') on a summation means that the state $j = i$ is to be omitted. Equation (5-2-66) is the full steady state equation for a level i in a single atomic species and for a single charge state of that species. The problem is of course even more complicated if other atoms and ions are taken into account, but we may consider certain approximations to the general equations (5-2-10) and (5-2-66) that have already been solved.

In a low density astrophysical plasma such as a gaseous nebula ($N_e \sim 10^4$ cm^{-3}, $T \sim 10^4$ °K) there are departures from thermodynamic equilibrium even for levels of very high excitation. The intensity of the radiation is extremely low (see [17]) and hence all the terms in (5-2-66) which involve absorption and stimulated emission may be neglected. This results in a considerable simplification of the problem, since the solution of (5-2-66) for the level populations is uncoupled from the solution of the radiative transfer equation. A nebula is predominantly composed of atomic hydrogen and the best solution of (5-2-66) for hydrogen has recently been obtained by Brocklehurst [18] (see § 5-6C) who solves the infinite set of equations by a matrix condensation technique [19] using the best available collision cross-sections (see § 5-5). In the solar corona ($N_e \sim 10^7$ cm^{-3}, $T \sim 2 \times 10^6$ °K, see the review by Seaton [20]), where the interpretation of the emission spectrum of multiply ionized ions is important, a contribution from dielectronic recombination must be added to the ordinary radiative recombination terms [21]. In this process, an ion A^{+s+1} and an electron make a radiationless transition to a doubly excited state of A^{+s} and this state then decays to a lower level with the emission of a photon.

The problem of departures from LTE in stellar atmospheres is more difficult. The solutions of (5-2-10) and (5-2-66) are now coupled together by the radiation field, since the terms in (5-2-66) involving J_v cannot be neglected. Deviations from LTE are apparent in the solar photosphere and chromosphere (see § 5-6B) and in most A and B stars (see § 5-6C). Models of early type stars have been obtained in which the major constituents, hydrogen and helium have been represented by N discrete levels plus the continuum (see e.g. [22]). Of these N levels $n(\ll N)$ have been allowed to depart from LTE and typical values of n and N are $5 < n < 10$ and $16 < N < 32$. Departures from LTE in the line radiation are due to non-equilibrium populations in both the upper and lower levels of the line, whereas departures in the continuum radiation depend only on the populations of the bound levels of the system from which a photon $h\nu$ can produce photoionization. In practice expressions (5-2-60) and (5-2-63) for $\kappa_\nu(i)$ and $j_\nu(i)$ are more simple than they appear, since in general, for a given frequency ν, at most one or two lines will contribute effectively to the sums over i and j.

Finally, we consider the form of the source functions S_ν for the simple model in which only two levels, labelled 1 and 2 say, are taken into account (see e.g. [14]). We then have

$$\rho \kappa_\nu = \frac{h\nu_{12}}{4\pi} \{N_1 B_{12} - N_2 B_{21}\} \phi_{12}(\nu) \tag{5-2-67}$$

and

$$\rho j_\nu = \frac{h\nu_{12}}{4\pi} N_2 A_{21} \psi_{21}(\nu_,). \tag{5-2-68}$$

The equation (5-2-66) becomes for $i = 1$

$$N_1 B_{12} \int_0^{+\infty} \phi_{12}(\nu) J_\nu d\nu + N_1 C_{12} = N_2 B_{21} \int_0^{+\infty} \phi_{12}(\nu) J_\nu d\nu$$
$$+ N_2 C_{21} + N_2 A_{21}. \tag{5-2-69}$$

Thus using equations (5-2-11), (5-2-67), (5-2-68) and (5-2-69), we have for the line source function

$$S_\nu = \left\{ \varepsilon B_{\nu_{12}} + (1-\varepsilon) \int_0^{+\infty} \phi_{12}(\nu) J_\nu d\nu \right\} \frac{\psi_{21}(\nu)}{\phi_{12}(\nu)}, \tag{5-2-70}$$

where $B_{\nu_{12}}$ is the Planck function (5-2-12) and, using relation (5-2-48)

$$\varepsilon = \frac{C_{21}}{C_{21} + A_{21}[1 - \exp(-h\nu_{12}/KT)]^{-1}}. \tag{5-2-71}$$

If κ_ν and j_ν have the same frequency dependence, an approximation that is valid for Doppler and collision broadening (see § 5-4B), but not for natural broadening, we can write

$$\psi_{21}(v) = \phi_{12}(v) \tag{5-2-72}$$

for all values of v. Then ε is the probability per scattering that the photon is absorbed and that subsequently collisions redistribute the energy. The second term in S_v represents the incoherent scattering of the photon within the line, that is the photon is absorbed and then re-emitted in a different direction with a slightly different frequency due to the finite breadth of the line. If the state 1 is infinitely narrow the scattering is always coherent, i.e., there is no change of frequency. Complete coherence has been assumed in all the calculations for multi-level atoms outlined above and discussed in § 5-6. This special case (cf. (5-2-41)) can be obtained from (5-2-70) by replacing $\phi_{12}(v)$ under the integral sign by $\delta(v-v_{12})$. Hummer [23] has found that the assumption $\phi_{12}(v) = \psi_{21}(v)$ gives good values of S_v near the line centre, but is less accurate in the line wings. We note that in addition to this approximation we must have $C_{21} \gg A_{21}$ and hence $\varepsilon \sim 1$ in order to recover the LTE result that $S_v = B_v$ for the line radiation. Much work remains to be done on the solutions of the coupled radiative transfer–steady state problem.

§ 5-3. The continuous spectrum

A. General formulae

The gas pressure P_g(dyn. cm^{-2}) at each depth in the atmosphere due to atoms and atomic ions, is given by

$$P_g = NKT; \quad N = \sum_A N(A) \tag{5-3-1}$$

where $N(A)$ is the number density of atomic species A. If $N(H)$ is the density of hydrogen atoms, we may write

$$N = N(H) \sum_A \frac{N(A)}{N(H)}; \tag{5-3-2}$$

the abundance of species A relative to hydrogen is an input parameter for any model atmosphere calculation. The opacity κ_v (cm^2 g^{-1}), first introduced in equation (5-2-5) is given by

$$\rho \kappa_v = \sum_A N(A) m(A) \kappa_v(A) \quad (\text{cm}^{-1}) \tag{5-3-3}$$

where

$$\rho = \sum_A N(A) m(A) \quad (\text{g cm}^{-3}). \tag{5-3-4}$$

The atom A has a mass $m(A)$ (gram) and an associated opacity $\kappa_\nu(A)$, which is determined by bound-free and free-free absorption processes. We write

$$\rho \kappa_\nu(A) = m(A) \sum_s N(A^{+s}) \kappa_\nu(A^{+s}) \tag{5-3-5}$$

where A^{+s} denotes the atom A when it is s times ionized. If $s = 0$, the atom is neutral and if $s = -1$ it is a negative atomic ion. The opacity $\kappa_\nu(A^{+s})$ is given by

$$m(A, \kappa_\nu(A^{+s})) = k_\nu^B + k_\nu^F \quad (\text{cm}^2) \tag{5-3-6}$$

where k_ν^B and k_ν^F are absorption coefficients per atom A^{+s} for the bound-free and free-free absorption of radiation of frequency ν by the atom A^{+s}. We have

$$N(A^{+s}) k_\nu^B = \sum_i N_i a_\nu(i) \left\{ 1 - \frac{1}{b_i} \exp(-h\nu/KT) \right\} \tag{5-3-7}$$

where the summation is over all bound states i of the atom A^{+s} from which a photon of energy $h\nu$ can detach an electron. The number density of atoms A^{+s} in state i is N_i and the factor $\{1-(1/b_i) \exp(-h\nu/KT)\}$ allows for the inverse process of induced recombination. The departure coefficient b_i is defined in § 5-2D. The cross-section $a_\nu(i)$ (cm^2) is that for the photoionization process

$$A^{+s}(i) + h\nu \rightarrow A^{+s+1} + e \tag{5-3-8}$$

where the atom A^{+s+1} is in its ground configuration. Radiation can also be absorbed by an electron moving in the field of the atom A^{+s+1} causing a transition to a continuum state of A^{+s} of higher energy. This process can be written as

$$A^{+s+1} + e + h\nu \rightarrow A^{+s+1} + e. \tag{5-3-9}$$

We define a quantity $a_\nu(E) \, dN_e$ (cm^2) which is a cross-section for absorption of radiation of frequency ν by a continuum state of A^{+s} for which the number density of electrons with energies in the range $(E, E+dE)$ is dN_e. We assume that the initial distribution of energy E of the free electrons is Maxwellian, and so the probability that the electrons of number density

N_e have an energy in the range $(E, E+dE)$ is

$$\frac{dN_e}{N_e} = 2\pi(\pi KT)^{-\frac{3}{2}} E^{\frac{1}{2}} \exp(-E/KT) dE. \tag{5-3-10}$$

Thus the total cross-section per unit electron density is

$$\int a_\nu(E) \frac{dN_e}{N_e} = 2\pi(\pi KT)^{-\frac{3}{2}} \int_0^\infty a_\nu(E) E^{\frac{1}{2}} \exp(-E/KT) dE \tag{5-3-11}$$

and the absorption coefficient k_ν^F is given by

$$N(A^{+s}) k_\nu^F = N(A^{+s+1}) \int a_\nu(E, dN_e \{1 - \exp(-h\nu/KT)\} \tag{5-3-12}$$

allowing for bremsstrahlung, the inverse of process (5-3-9). The number density $N(A^{+s})$ in equations (5-3-7) and (5-3-12) is given by

$$N(A^{+s}) = \sum_i N_i \tag{5-3-13}$$

and $N(A^{+s+1})$ is the number density of atoms A^{+s+1} in the ground configuration only. The absorption coefficients for transitions in which the atom A^{+s+1} is left in an excited configuration can be neglected in comparison with competing processes. If local thermodynamic equilibrium holds, the atomic energy levels are statistically populated, the coefficients $b_i = 1$ for all values of i and the number densities $N(A^{+s})$, N_i and $N(A^{+s+1})$ are related through the Boltzmann and Saha equations (cf. (5-2-54) and (5-2-55)).

The photoionization cross-section, $a_\nu(i)$, is given by

$$a_\nu(i) = \frac{8\pi^3 e^2 \nu}{3c\omega_i} \sum \left| \int \Psi_j^* R \Psi_i d\tau \right|^2; \quad R = \sum_p r_p \tag{5-3-14}$$

where Ψ_i is the total wave function for the initial atomic system A^{+s} and Ψ_j is the total wave function for the final state $(A^{+s+1} + e)$. The position vector of the pth electron relative to the nucleus is r_p and $d\tau$ is an element of volume in the multidimensional space spanned by the vectors r_p. The statistical weight of the state i is ω_i and the summation is over all the degenerate sublevels of the initial and final states of the system. For many cases of interest we can use wave functions Ψ_i and Ψ_j which are anti-symmetrized combinations of one electron orbitals and in which the wave function for the ejected electron is assumed to be orthogonal to all the orbitals of the atom A^{+s}. If state i is such that the active electron has quantum number nl, eq. (5-3-14) reduces to

$$a_v(nl) = \tfrac{4}{3}\pi\alpha a_0^2 D(I_{nl}+k'^2) \sum_{l'=l\pm 1} C_{l'} \left| \int_0^\infty P_{nl}(r) r G_{k'l'}(r)\,dr \right|^2 \qquad (5\text{-}3\text{-}15)$$

where r is now in atomic units, that is $|r| = ra_0$, and $P_{nl}(r)$ is the radial function for the nl electron. This electron is ejected with energy k'^2 rydberg and has a final state continuum wave function $G_{k'l'}(r)$, with asymptotic amplitude $k'^{-\frac{1}{2}}$. The energy required to remove the nl electron is I_{nl} rydberg, and therefore

$$\frac{z^2 v}{v_0} = I_{nl} + k'^2, \qquad (5\text{-}3\text{-}16)$$

where v_0 is given by (5-2-51). α is the fine structure constant and D is a factor of the order of unity which depends only on the overlap of the one electron states of the atom A^{+s+1} with the corresponding (non-active) states of the atom A^{+s}. The coefficients $C_{l'}$ are coupling coefficients, see for example [24]. For an electron outside a closed shell

$$C_{l'} = \frac{l_>}{2l+1}; \quad l_> = \max(l, l'). \qquad (5\text{-}3\text{-}17)$$

The cross-section for free–free absorption can also be derived from (5-3-14) in which Ψ_i now denotes the initial continuum state of A^{+s}. If Ψ_i and Ψ_j are expressed in terms of one electron orbitals, we obtain

$$a_v(E, l) = \frac{2^4 \pi^2 \alpha a_0^5}{3} \frac{(k'^2 - k^2)}{k} \sum_{l'=l\pm 1} C_{l'} \left| \int_0^\infty G_{kl}(r) r G_{k'l'}(r)\,dr \right|^2 \qquad (5\text{-}3\text{-}18)$$

where

$$E = (k^2/z^2) h v_0. \qquad (5\text{-}3\text{-}19)$$

Exact expressions for $a_v(i)$ and $a_v(E)$ exist if the absorbing system is hydrogenic (cf. [25]). If the nuclear charge is z, the photoionization cross-section from a level i of principal quantum number n is

$$a_v(n) = \frac{2^6 \alpha \pi a_0^2}{3\sqrt{3} z^2} \frac{g(n, k')}{\sigma^3 n^5} \qquad (5\text{-}3\text{-}20)$$

where

$$g(n, k') = \frac{\pi\sqrt{3} \exp\left[(-4z/k')\tan^{-1}(nk'/z)\right]}{(1-\exp(-2\pi z/k'))\sigma^{\frac{1}{2}}} |\Delta(n, k')| \tag{5-3-21}$$

and

$$\sigma = v/v_0.$$

The free–free absorption cross-section can be obtained from

$$a_\nu(E) = \frac{2^7\pi^3 a_0^5 \alpha}{3\sqrt{3}z^4} \frac{g(k, k')}{\sigma^3 k} \tag{5-3-22}$$

where

$$g(k, k') = \frac{\pi\sqrt{3}(k'+k)}{\sigma z} \exp(-2\pi z/k')[1-\exp(-2\pi z/k)]^{-1}$$
$$\times [1-\exp(-2\pi z/k')]^{-1}|\Delta(k, k')|. \tag{5-3-23}$$

The quantity Δ which appears in (5-3-21) and (5-3-23) is given by

$$\Delta = [{}_2F_1(1-c, -d, 1; x)]^2 - [{}_2F_1(1-d, -c, 1; x)]^2 \tag{5-3-24}$$

where for $\Delta \equiv \Delta(n, k')$

$$c = iz/k'; \quad d = n; \quad x = -4nizk'/(iz-nk')^2 \tag{5-3-25}$$

and for $\Delta \equiv \Delta(k, k')$

$$c = iz/k'; \quad d = iz/k; \quad x = -4kk'/(k'-k)^2. \tag{5-3-26}$$

Similar expressions for transitions in which the orbital angular momentum is specified are given by Burgess [26].

B. Absorption by Neutral Atoms and Positive Ions

In order to evaluate the bound–free absorption coefficient k_ν^B, we need the photoionization cross-sections $a_\nu(i)$ for all the contributing levels. For some of the atomic systems of astrophysical interest, cross sections have been measured for the removal of an electron from the ground state, but there is very little experimental data available on either the photoionization of atoms in excited states or the absorption by positive atomic ions. Biberman and Norman [27] have reviewed the continuous spectra from atomic plasmas and a recent bibliography of photoabsorption cross-section data [28] lists the experimental and theoretical work available up to the end of 1969.

Most of the observational material available on the emergent flux from stars covers the visible region of the spectrum. At these wavelengths continuous absorption by neutral atoms or positive ions occurs by photoionization from excited states and consequently it is necessary to develop theoretical methods in order to evaluate k_v^B. Early calculations assumed that all the excited levels could be treated as though they were hydrogenic. If we write the residual charge $s+1 = z$, the absorption by a complex atom or positive ion, A^{+s} ($s = 0, 1, 2, \ldots$) can be treated in the same way as that produced by a hydrogenic atom with nuclear charge z. The formula (5-3-15) can then be used with $P_{nl}(r)$ and $G_{k'l'}(r)$ taken to be hydrogenic, and a value for the ionization potential I_{nl} deduced from spectroscopic data. For highly excited levels, where this data is not available the position of the levels must be estimated. This approximation when used to evaluate k_v^B can give results which are in error by as much as an order of magnitude.

More recently methods have been developed [24, 29, 30, 31] which allow to some extent for the complexity of the atom. The single channel quantum defect method given by Seaton [32] is used to obtain approximate wave functions for the system. The effective quantum number n^* and quantum defect μ for a series of levels nl (for which only n is allowed to vary), are defined by

$$I_{nl} = z^2/n^{*2} \; ; \quad n^* = n - \mu. \tag{5-3-27}$$

The quantum defect method relates μ to the phase shift for the elastic scattering of electrons with orbital angular momentum quantum number l. Thus μ, which is usually a slowly varying function of energy, can be obtained from spectroscopic data at discrete points corresponding to the energy levels, and then extrapolated to positive energies to give a continuous function corresponding to the phase shift. The photoionization cross-section can then be expressed in the form

$$a_v(nl) = \frac{8\pi a_0^2 n^{*3}}{3z^2 \zeta(n^*, l)} (1+\varepsilon' n^{*2})^{-3} \times$$
$$\times \sum_{l'=l\pm 1} C_{l'}[G(n^*l; \varepsilon'l') \cos\{\pi(n^* + \mu'(\varepsilon') + \chi(n^*l; \varepsilon'l'))\}]^2 \tag{5-3-28}$$

where

$$\varepsilon' = k'^2/z^2 \tag{5-3-29}$$

and $\mu'(\varepsilon')$ is the extrapolated quantum defect for the series of levels nl'. The quantity $\zeta(n^*, l)$ can be taken to be unity for all but the lowest members of the series and the functions $G(n^*l; \varepsilon'l')$ and $\chi(n^*l; \varepsilon'l')$ are independent of the charge s of the atom. Tabulations of G and χ as functions of n^* and ε' have been made for $(l, l') = (0, 1), (1, 0), (1, 2), (2, 1), (2, 3)$ and $(3, 2)$, see [29].

The absorption coefficients k_ν^B may now be evaluated in the following way. Since the levels nl become hydrogenic for $l > l_0$ where $l_0 = 2$ or 3, we use (5-3-28) to provide a correction factor to the exact hydrogenic result, for the levels with $0 \leq l \leq l_0$. The coefficient k_ν^B may be written in the form (cf. (5-3-7))

$$N(A^{+s})k_\nu^B = \{\sum_n N_n a_\nu^H(n) + \sum_{l=0}^{l_0} \sum_n N_{nl}(a_\nu(nl) - a_\nu^H(nl))\}$$
$$\times \left(1 - \frac{1}{b_i} \exp(-h\nu/KT)\right) \quad (5\text{-}3\text{-}30)$$

where $a_\nu^H(n)$ is given by (5-3-20) and $a_\nu(nl)$ and $a_\nu^H(nl)$ are both given by (5-3-28). The cross-section $a_\nu^H(nl)$ is determined by setting $n^* = n$ and $\mu' = 0$ in (5-3-28) and the tabulated values of G and χ are such that the exact value of the cross section for the photoionization of a hydrogenic state nl is then obtained. The populations of the hydrogenic and non-hydrogenic states are given by N_n and N_{nl} respectively and the summations are over all those states for which

$$\nu \geq \nu_0/n^{*2}. \quad (5\text{-}3\text{-}31)$$

The individual cross-sections $a_\nu(nl)$ have, in general, finite positive values at threshold, and therefore the coefficient k_ν^B contains a series of discontinuities each caused by the onset of absorption by an atomic level.

A similar approach may be made to the calculation of k_ν^F, the free–free absorption coefficient, which is a smoothly varying function of ν. The contribution of k_ν^F to the opacity is particularly important at longer wavelengths where the only terms contributing to k_ν^B are those corresponding to very highly excited levels. The coefficient k_ν^F may be written as (see [33] and eq. (5-3-12)),

$$N(A^{+s})k_\nu^F = N(A^{+s+1})\left\{\int a_\nu^H(E)\,dN_e + \sum_{l=0}^{l_0} \int (a_\nu(E, l) - a_\nu^H(E, l))\,dN_e\right\}$$
$$\times (1 - \exp(h\nu/KT)) \quad (5\text{-}3\text{-}32)$$

where $a_\nu^H(E)$ is given by (5-3-22). The cross-section $a_\nu(E, l)\,dN_e$ may be obtained from

$$a_v(E, l) = \frac{16\pi^2 \alpha a_0^5}{3z^5 \sigma^3 \varepsilon^{\frac{1}{2}}} \sum_{l'=l\pm 1} C_{l'} \left[G(\varepsilon l; \varepsilon' l') \cos \{\pi(\mu'(\varepsilon') - \mu(\varepsilon) + \chi(\varepsilon l; \varepsilon' l'))\} \right]^2$$

(5-3-33)

where

$$\varepsilon = k^2/z^2 \tag{5-3-34}$$

and $G(\varepsilon l; \varepsilon' l')$ and $\chi(\varepsilon l; \varepsilon' l')$ are tabulated in [34]. The functions G and χ are smoothly varying as $-1/n^{*2} \to +\varepsilon$ and $a_v^H(E, l)$ is evaluated from (5-3-33) by setting $\mu = \mu' = 0$.

Results for the total absorption coefficient, $k_v^B + k_v^F$ are tabulated in [31] for 21 atoms and ions of astrophysical interest, using the methods outlined above. The work of [31] does not include the factor allowing for stimulated emission and it assumes that the relative populations of the energy levels are given by the Boltzmann and Saha laws (cf. (5-2-54) and (5-2-55)). Where there are either experimental or superior theoretical data available for $a_v(nl)$, these cross-sections have been used in preference to those obtained using the quantum defect method. In general, the quantum defect method is more accurate for excited states than for the ground state of an atomic system, especially if the ground state contains a number of equivalent electrons in its outer shell. Another limitation is that the quantum defects μ and μ' can only, under the most favourable conditions, be reliably extrapolated into the continuum for ε and ε' in the range $0 \leq \varepsilon, \varepsilon' \lesssim 0.5$, and so (5-3-28) cannot be expected to give a reasonable estimate of $a_v(nl)$ far away from threshold. Travis and Matsushima [35] have also evaluated k_v^B for some of the elements treated in [31] in which they use the earlier bound–free general formula [24] and assume that the factor $g(n, k')$ in $a_v^H(n)$ (see (5-3-21)) is unity. They conclude that when the quantum defects as determined from experiment do not vary smoothly with energy, variations in the way they are extrapolated can lead to variations in individual cross-sections of about a factor of 2. In Figure 5-3-1 a comparison is made of the results of [30, 31] for the total absorption by nitrogen with the experimental work of Boldt [36] in the visible region of the spectrum; it can be seen that the overall agreement is quite good. The difference between the two sets of theoretical results arises because in [30], l_0 is taken as 2 in eq. (5-3-30), while in [31] $l_0 = 3$. In principle the results of [31] should be more accurate, although in this case those of [30] are closer to the experimental values.

Other theoretical methods of evaluating $a_v(nl)$ are possible in certain cases. If the outer shell of the atom in its ground configuration contains p electrons, the dominant contribution to $a_v(np)$ usually comes from the

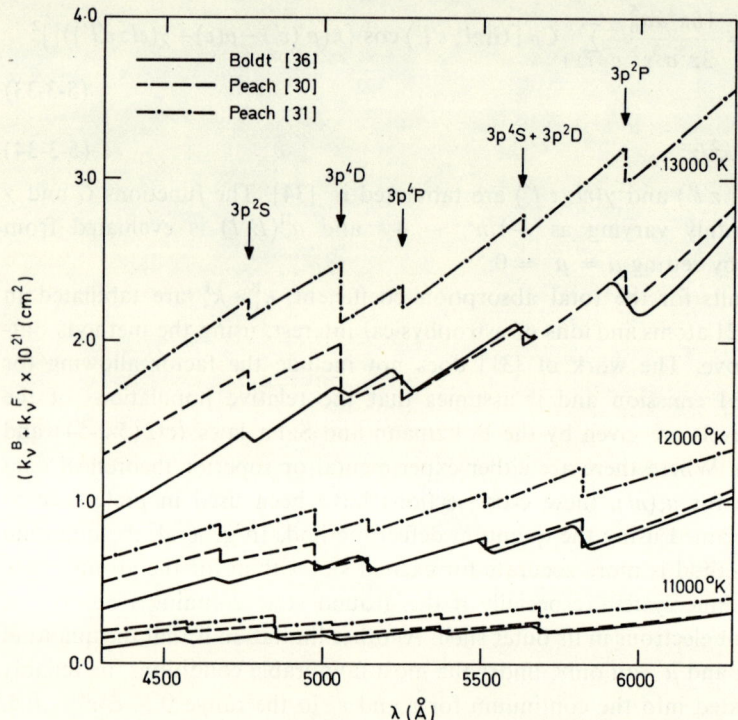

Fig. 5-3-1. Total continuous absorption coefficient for nitrogen; no allowance has been made for stimulated emission.

transition $np \to k'd$ and the phase shift for the elastic scattering of electrons with $l' = 2$ is usually small. Clementi [37] has calculated Hartree–Fock wave functions, $P_{nl}(r)$, for the ground configurations of many complex atoms and isoelectronic atomic ions. These functions are presented in a convenient analytic form and if we assume that the phase shifts are zero, $G_{k'l'}(r)$ in equation (5-3-15) becomes a regular Coulomb function, hence the integral can be evaluated analytically. Tabulations of values of $a_\nu(nl)$ for atoms and positive ions in the isoelectronic sequences boron to neon and aluminium to argon using this simple approximation are given by Peach [38].

Calculations have been carried out for all the neutral atoms in the sequence helium to xenon by McGuire [39]* using a modified Herman–

---

* The results for silicon shown in Fig. 5-3-3 are taken from the report rather than from the published paper which gives incorrect results for this case, and the threshold is taken to be the experimental rather than the theoretical one which was used by the author.

Skillman potential to describe each atomic system. Also Moores [40] has presented photoionization cross-sections for the ground states of C^+, N^+, N^{+2}, O^+ and O^{+2} obtained using bound and free state wave functions from the work of Eissner and Nussbaumer [41]. These functions are the solutions for the atomic system when it is described by a Thomas–Fermi potential.

The best calculations of photoionization cross-sections so far are those in which close coupling wave functions are used to describe the continuum state. The work of Norcross [42] for $He(1s^2)$ ^1S is in excellent agreement with experiment [43], while for $He(1s)(2s)$ $^{1,\,3}S$ his results agree quite well with those derived using the quantum defect method. Henry and Williams [44] have calculated cross-sections for N, N^+, O, O^+ and O^{+2} and Conneely et al. [45] have used close coupling functions for the photoionization of Al, Si and S. The results for Si compare well with the experimental data of Rich [46] and the experimental work on N and O [47, 48], is in generally good agreement with the data in [44]. A computer program which evaluates photoionization cross-sections for atoms with configuration $(np)^q$, $q = 1, 2, \ldots, 6$; $n = 2, 3$ using the close coupling method has recently been

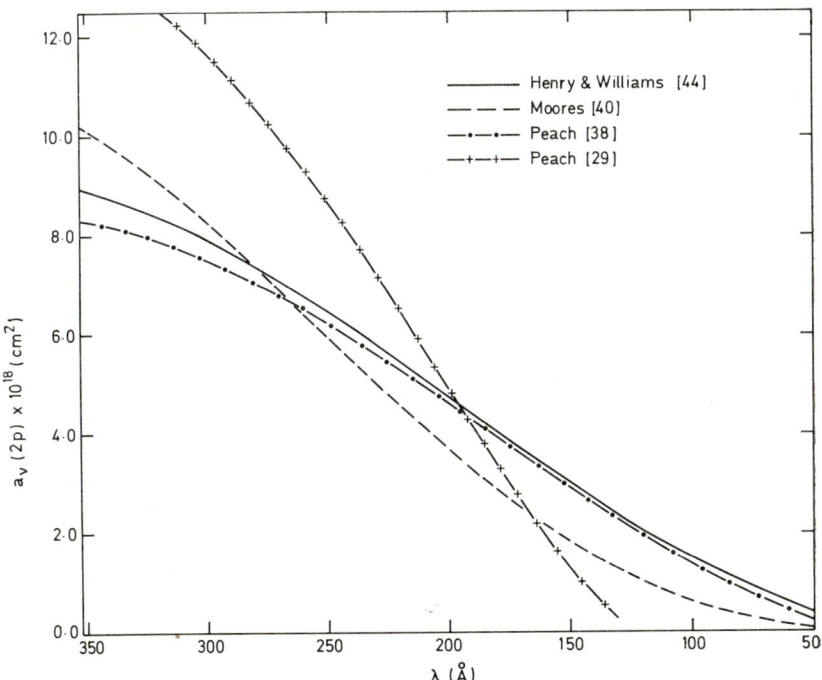

Fig. 5-3-2. Photoionization of the ground state of ionized oxygen.

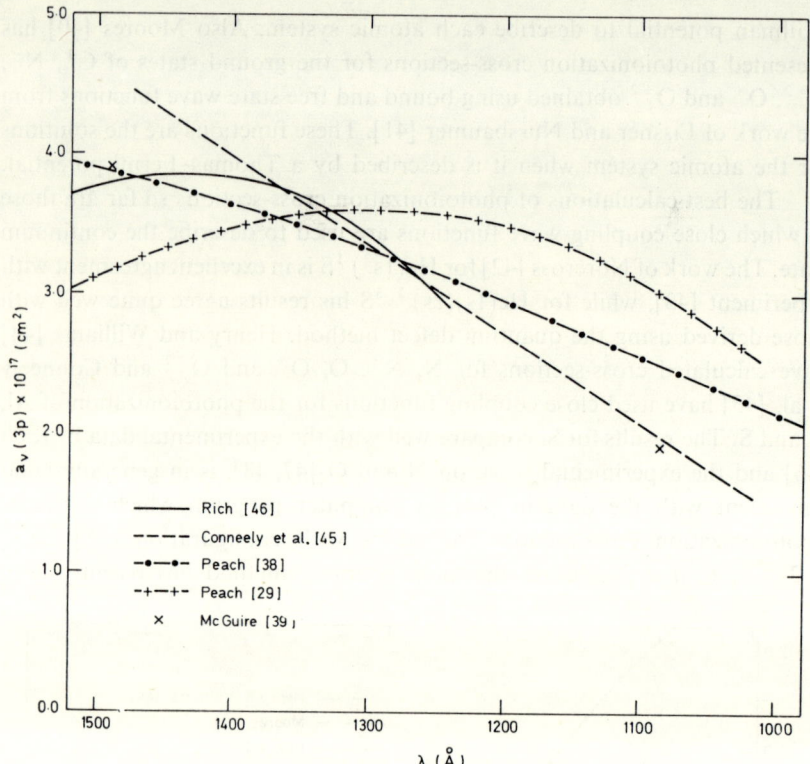

Fig. 5-3-3. Photoionization of the ground state of silicon.

published by Conneely et al. [49]. Figures 5-3-2 and 5-3-3 show comparisons of the results of different approximations for the photoionization of the ground states of O^+ and Si; in estimating $\zeta(n^*, l)$ for the photoionization cross-sections given by the general formula [29] it has been assumed that $\mu = n - n^*$ takes the values 1 and 2 for O^+ and Si respectively at $n^* = 1$.

Much work remains to be done to improve our knowledge of photo-absorption by the elements that are most abundant in stellar atmospheres (see Table 5-6-1 and § 5-6).

C. PHOTODETACHMENT OF NEGATIVE IONS

Absorption by both atomic and molecular negative ions is important in determining the emergent spectrum of stars in classes G, K or M. In this section only detachment from negative atomic ions is discussed; the available data on molecular negative ions is presented in Chapter 8 of this volume. Photodetachment has been reviewed by Branscomb [50] and more recently by Biberman and Norman [27], and the bibliography [28] lists the results available.

The bound–free absorption coefficient, k_ν^B for a negative atomic ion has a somewhat different behaviour as a function of frequency from that for a neutral atom or positive ion. One reason for this is that in this case $z = 0$, and hence the active electron is moving in a field which is determined at large distances by the centrifugal contribution to the potential rather than by the Coulomb potential. Secondly, a negative ion only has one or two true bound states instead of the infinite number for a neutral system. The binding energies of these states vary from 0.0056 rydberg for $H^-(1s)^2$ ^1S to about 0.27 rydberg for $Cl^-(2p)^6$ ^1S.

We may derive the form of the detachment cross section $a_\nu(nl)$ at threshold quite generally, without specifying a particular atomic system. We consider the simple approximation in which the free electron in the final state is described by a plane wave, i.e.,

$$G_{k'l'}(r) = (\tfrac{1}{2}\pi r)^{\frac{1}{2}} J_{l'+\frac{1}{2}}(k'r) \tag{5-3-35}$$

where $J_{l'+\frac{1}{2}}(k'r)$ is the Bessel function given by

$$J_{l'+\frac{1}{2}}(k'r) = (\tfrac{1}{2}k'r)^{l'+\frac{1}{2}} \sum_{m=0}^{\infty} \frac{(-1)^m (\tfrac{1}{2}k'r)^{2m}}{m!\,\Gamma(l'+m+\tfrac{3}{2})}. \tag{5-3-36}$$

For small values of k', the main contribution to the integral in the expression (5-3-15) for $a_\nu(nl)$ comes from large values of r, for which the choice of wave function $G_{k'l'}(r)$ in (5-3-35) should be reasonable. This choice of course neglects the phase shift which occurs in the true solution, caused by the presence of a short range field due to the neutral atom. It is clear that if we use (5-3-35) and (5-3-36) in the evaluation of $a_\nu(nl)$ in (5-3-15), we obtain near threshold that the cross-section varies as

$$a_\nu(nl) \propto (I_{nl}+k'^2) k'^{2l'+1}(a_0+a_1 k'^2+ \ldots) \tag{5-3-37}$$

where $l' = l-1$ for $l' \neq 0$ and $l' = 1$ for $l = 0$ and $a_0, a_1, \ldots$, are constants. This behaviour is independent of the choice of the bound state wave function, $P_{nl}(r)$, and even if the form of $G_{k'l'}(r)$ is modified to take account of polarization and short range effects in the atomic potential, the leading term in the expansion (5-3-37) is unaltered. When a polarization potential, $V(r) = \alpha_p/r^4$ is included in the differential equation to be satisfied by $G_{k'l'}(r)$, it has been shown by O'Malley et al. [51] that $G_{k'l'}(r)$ may be expressed in terms of Mathieu functions. The use of this wave function in expression (5-3-15) produces a threshold law which contains additional terms depending on odd as well as even powers of k', and for the case $l' = 0$, it also contains terms dependent upon log k'.

The free–free absorption coefficient, k_v^F, for negative ions is a smoothly varying function and at sufficiently long wavelengths will become the dominant contribution to the total absorption coefficient, $k_v^B + k_v^F$. In the most simple approximation to k_v^F, we can evaluate $a_v(E, l)$ from equation (5-3-18) by using expression (5-3-35) for $G_{kl}(r)$ and $G_{k'l'}(r)$ and then use (5-3-12) to evaluate k_v^F by summing over all the angular momentum states l. Only the ratio $N(A)/N(A^-)$ will depend on the properties of the particular atom (A) concerned. In a better approximation, $G_{kl}(r)$ and $G_{k'l'}(r)$ incorporate the effects of scattering and then for $l, l' \lesssim 2$, $a_v(E, l)$ will depend directly on the characteristics of the atomic field.

The first negative ion which was found to be of astrophysical importance was the H^- ion (see § 5-2B and § 5-6B). Many calculations of the bound–free absorption coefficient have been carried out, see [28], since those of Chandrasekhar [52, 53] showed that H^- was the major source of continuous absorption at visible wavelengths in the sun. Recent calculations of Doughty et al. [54] employ Hartree–Fock eigenfunction expansions for the continuum wave functions in which the 1s, 2s, 2p, 3s, 3p and 3d states of atomic hydrogen have been included, and the seventy-parameter bound state function of Schwarz for H^-. Bell and Kingston [55] have used the same bound state function together with a continuum function obtained using the method of polarized orbitals. The binding energy of H^- is very small and a consequence of this is that many parameters must be included in the bound wave function in order to obtain a good value of this energy and also of the photodetachment cross-section. The most recent calculations of the free–free absorption coefficient for H^- are those of Doughty and Fraser [56] and Stilley and Callaway [57]. The approximations used in [56] are similar to those of [54] and in [57], the polarized orbital method is used (see Chapter 7). All these authors evaluate k_v^B and k_v^F using both the dipole length form of the matrix element, see (5-3-14) and the dipole velocity version which gives results identical to the dipole length expression if and only if exact initial and final total wave functions are used. A comparison of the two sets of results provides some check on the consistency of the calculation. Figure 5-3-4 shows the results for the total absorption coefficient derived from the dipole length results of [54–57], for a temperature $T = 5600\,°K$ and assuming $N(H)/N(H^-)$ is given by the Saha equation. The experimental results for k_v^B which are also included, are taken from the work of Smith and Burch [58], [59] and Bohm and Rehder [60].

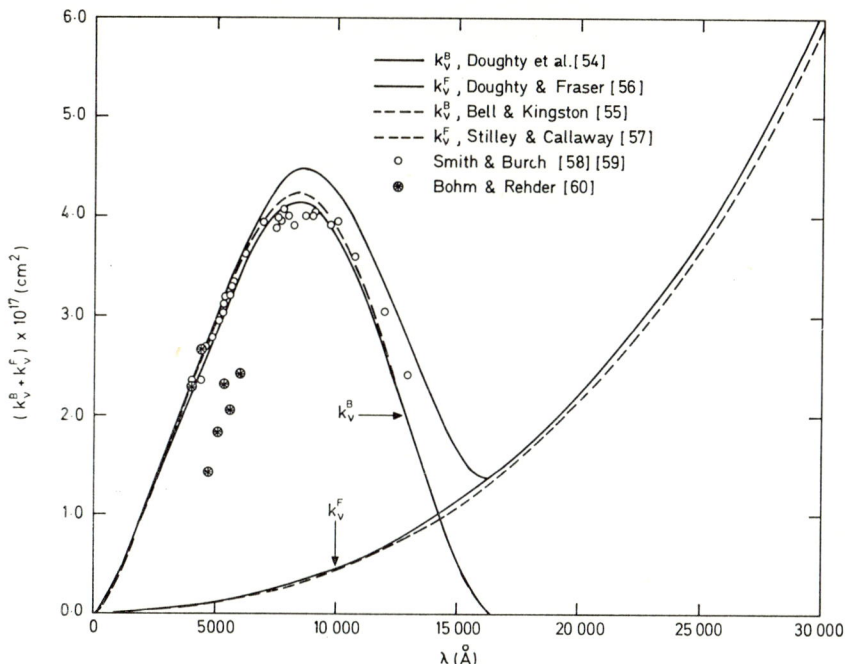

Fig. 5-3-4. Total absorption coefficient of H^- for $T = 5600$ °K; no allowance has been made for stimulated emission. The relative measurements [58, 59], have been normalized at 5280 Å to the results of [55].

Branscomb and Pagel [61] have considered the possible importance of negative ions other than H^- in the stellar atmospheres of late type stars and more recently Vardya [62] has improved and extended their work. Of the atomic negative ions considered, they conclude that C^-, S^- and Cl^- may contribute significantly to the absorption of radiation in late type stars, but that O^- and Si^- are much less likely to be important.

Experimental work has been carried out on the photodetachment of some of the lighter negative ions which are of interest. Seman and Branscomb [63] have studied C^- and the most recent work on O^- is that of Branscomb et al. [64] in which they observe absorption corresponding to the oxygen atom being left in both the $(2p)^4\ ^3P$ and $(2p)^4\ ^1D$ states. Detachment from F^- has also been observed [65, 66], and data have been obtained in three different experiments [67–69] for the transitions in Cl^- in which the neutral chlorine atom is left in the $^2P_{\frac{1}{2}}$ or $^2P_{\frac{3}{2}}$ state. In this case the fine structure splitting of the ^2P level is sufficiently large that it should be taken into account explicitly in a calculation of the cross-section. Cooper and Martin

[70] have calculated photodetachment cross-sections for C^-, O^-, F^- and Cl^- using an approximation in which both bound and free wave functions are calculated in the same central potential. The form of this potential has been chosen to allow approximately for exchange and the long range effects of polarization. A somewhat similar choice of potential is adopted by Robinson and Geltman [71] in their calculations for C^-, O^-, F^-, Si^-, S^- and Cl^- and for the cases where a direct comparison can be made the results of [70, 71] are not very different. Myerscough and McDowell have calculated both k_v^B and k_v^F for C^- [72, 73], using an analytic Hartree–Fock bound state function similar to that given in [37] and numerical solutions of the Hartree–Fock equations for the electron moving in the field of neutral carbon. More recently, Henry has calculated cross-sections for the removal of an electron from C^-, N^- and O^- [74–76], using the polarized orbital method of Temkin [77] for the continuum state. Schneider et al. [78] and Garrett and Jackson [79] have also evaluated absorption cross-sections for O^- and the programme described in [49] has been used to obtain photodetachment cross-sections for Si^-, S^- and Cl^- by Conneely et al. [45]. Moskvin [80] has also studied C^-, F^- and Cl^-, but his evaluation of the matrix element for the p → s transition is incorrect. In all the calculations outlined above, it has been found that if both the dipole length and dipole velocity forms of the matrix element have been evaluated, the agreement between them is not very good and that if there is experimental data available for comparison, in some cases the dipole length results, and in other cases the dipole velocity results, are in the best agreement. A simple estimate of the cross-sections, using the dipole length matrix element, can be obtained for negative ions of configuration $(np)^q$, $q = 1, 2, \ldots, 6$; $n = 2, 3$ by adopting plane waves to describe the free electron states k's and k'd and using the Hartree–Fock functions of [37] for the bound states [81]. The integrals in (5-3-15) can then be carried out analytically and the cross-section is evaluated using the experimentally determined electron affinities. The results obtained appear to be no less good than those obtained by other methods, although the method certainly represents a very simple approximation to the exact problem. Figures 5-3-5 and 5-3-6 show the various results of experiment and theory for the photodetachment of C^- and Cl^-.

A process not considered by either Branscomb and Pagel [61] or Vardya [62] is that of absorption of radiation by a free–free transition in He^-. Helium is the most abundant element in stellar atmospheres apart from hydrogen and in certain stars the helium abundance considerably exceeds that of hydrogen. Initially, estimates of k_v^F for He^- were made by Somer-

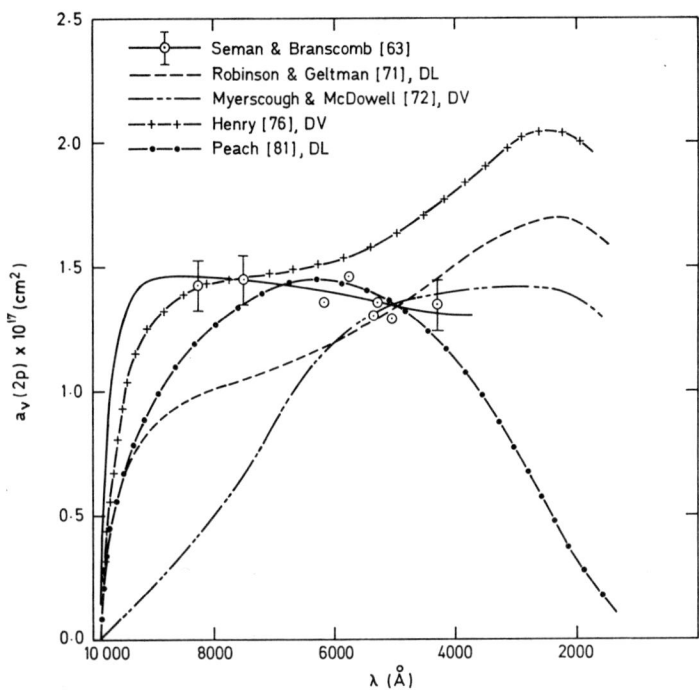

Fig. 5-3-5. The photodetachment of C^-; DL = dipole length, DV = dipole velocity.

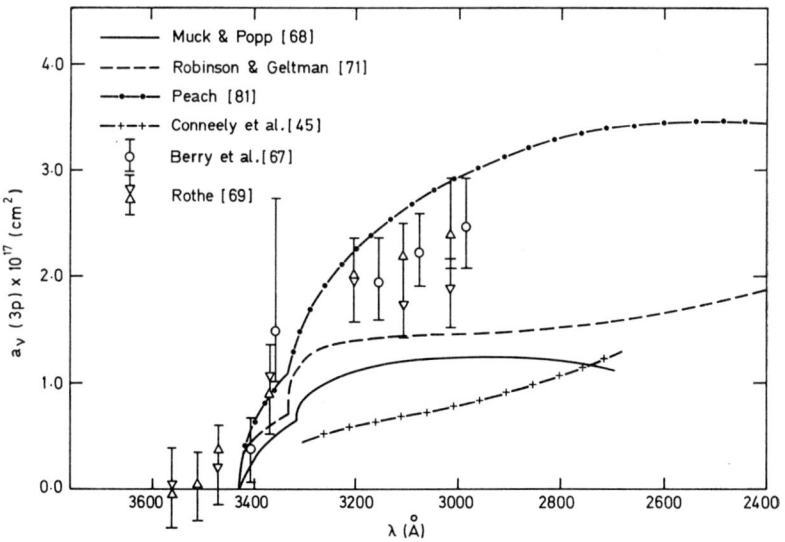

Fig. 5-3-6. The photodetachment of Cl^-.

ville [82] and the most recent calculations are those of John [83] in which he finds differences of between 2% and 24% with the earlier calculations of McDowell et al. [84].

The role of negative atomic and molecular ions in determining the emergent spectrum of cool stars is discussed in §5-6.

D. COHERENT SCATTERING

The coherent scattering of radiation can be a very important factor in certain types of stellar atmospheres. In eq. (5-2-41) the form of the source function is given when both pure absorption and coherent scattering both contribute to the opacity. If σ is the cross-section for coherent scattering, the probability for the occurrence of a pure absorption process is given by

$$\varepsilon = \frac{\rho \kappa_v}{\rho \kappa_v + N_c \sigma} \tag{5-3-38}$$

where κ_v and ρ are defined in § 5-3A and N_c is the number density of particles which are coherently scattering the radiation. The two forms of such scattering are Thomson and Rayleigh scattering. In the early type stars O and B, the hydrogen is almost completely ionized and the free electrons produced scatter photons coherently; this process is described by the Thomson cross-section (cf. [1], p. 169);

$$\sigma = \frac{8\pi e^4}{3m^2 c^4} \quad (\text{cm}^2). \tag{5-3-39}$$

In the atmospheres of the very hot stars for which $T_e \sim 10^5 \, °K$, the contribution to the total opacity from electron scattering dominates that from pure absorption processes [85].

Coherent scattering is not very important for stars in classes A, F or G, but in the cool stars the scattering of radiation by atoms and molecules becomes significant at the shorter wavelengths. The Rayleigh scattering cross-section for such processes is given by

$$\sigma = \frac{2^7 \pi^5 v^4 \alpha_p^2}{3c^4} \quad (\text{cm}^2) \tag{5-3-40}$$

where α_p is the polarizability of the atom or molecule. The polarizability of an atom in a state i can be written as

$$\alpha_p = \frac{2}{3} \left| \sum_{i' \neq i} \frac{(E_{i'} - E_i)|M|^2}{(E^2 - (E_{i'} - E_i)^2)} \right| \quad (\text{cm}^3) \tag{5-3-41}$$

where $E = hv$ is the energy of the incident photon and the summation is over all states i' with energy $E_{i'}$, except the one for which $E_{i'} = E_i$, the energy of the state i. If $E \ll |E_{i'}-E_i|$ for all the states i' which contribute appreciably to the sum in (5-3-41) the E^2 term in the denominator of (5-3-41) can be neglected and then the usual frequency independent expression for α_p is obtained. The matrix element M for allowed transitions $i \to i'$ is given by

$$M = \int \Psi_{i'}^* R \Psi_i \, d\tau; \quad R = \sum_j r_j \qquad (5\text{-}3\text{-}42)$$

(cf. (5-3-14)) where Ψ_i and $\Psi_{i'}$ are the total wave functions for states i and i' and R is the sum over the position vectors of all the electrons in the atom. The summation in (5-3-41) includes terms in which i' is a continuum level and must be averaged over the degenerate levels of state i. It is clear that the calculation of α_p from (5-3-41) involves estimates for numerous oscillator strengths $f_{ii'}$ (see § 5-4A) and the photoionization cross-section, $a_v(i)$ which is given by equation (5-3-14). Tarafdar and Vardya [86] have calculated the Rayleigh scattering cross-sections for He, C, N and O which may make a significant contribution to the ultraviolet opacity in carbon stars, helium stars and other stars of peculiar composition (see § 5-6E). Their calculations are based on the use of the quantum defect method and hence their final results can only be expected to be approximately correct.

It should be noted that whereas the Thomson scattering is independent of frequency, Rayleigh scattering has a strong dependence on wavelength, ($\sigma \propto \lambda^{-4}$). Both types of scattering are non-isotropic and hence the scattered radiation is polarized.

§ 5-4. The line spectrum

A. Transition probabilities

In this section a brief summary is given of the formulae used in calculating transition probabilities, but no attempt is made to give a comprehensive review of the data available. Atomic transition probabilities for all the elements hydrogen to calcium have been published in two volumes by Wiese et al. [87, 88]. The authors give an assessment of the accuracy of each value listed and the tables contain data on allowed and forbidden transitions for both neutral atoms and positive ions. General reviews on allowed and forbidden transitions have been published by Nicholls and Stewart [89] and Garstang [90] respectively.

The oscillator strength f_{ij}, for the absorption of radiation of frequency v_{ij} in an allowed transition $i \to j$, is given by

$$f_{ij} = \frac{8\pi^2 m v_{ij}}{3h\omega_i} \sum \left| \int \Psi_j^* R \Psi_i \, d\tau \right|^2; \quad R = \sum_p r_p, \tag{5-4-1}$$

where the summation is over all the degenerate states of i and j and Ψ_i and Ψ_j are the total initial and final wave functions respectively, for the atomic system. If Ψ_i and Ψ_j are expressed in terms of anti-symmetrized combinations of one electron functions and states i and j have the quantum numbers nl and $n'l'$ respectively, (5-4-1) reduces to

$$f_{nl,n'l'} = \frac{Dz^2 v_{ij}}{3v_0} C_{l'} \left| \int_0^\infty P_{nl}(r) \, r \, P_{n'l'}(r) \, dr \right|^2 \tag{5-4-2}$$

where ra_0 is the length of the vector r and v_0 is defined by (5-2-51). The quantities D and $C_{l'}$ are defined in § 5-3A and as before $(z-1) = s$ is the net charge on the absorbing atom A^{+s}. The orbital angular momenta obey the rule $l' = l \pm 1$, and since r behaves like $1/z$ along an isoelectronic sequence, eq. (5-4-2) is clearly not explicitly dependent upon z.

Exact expressions for the oscillator strength exist for the case of a transition $n \to n'$ in a hydrogenic system of nuclear charge z, similar to those given in (5-3-20) and (5-3-22). It may be shown (cf. [25]) that

$$f_{nn'} = \frac{(2nn')^6}{3\sqrt{3}\pi\omega_n} (n'^2 - n^2)^{-3} g(n, n') \tag{5-4-3}$$

where

$$g(n, n') = \pi\sqrt{3} \frac{nn'}{(n'-n)} \left(\frac{n'-n}{n'+n} \right)^{2n+2n'} |\Delta(n, n')| \tag{5-4-4}$$

and $\Delta(n, n')$ is given by (5-3-24) where

$$c = n'; \quad d = n; \quad x = \frac{-4nn'}{(n'-n)^2}. \tag{5-4-5}$$

Similar expressions, in which the orbital angular momentum is specified, can be derived; Green et al. [91] list oscillator strengths for transitions $nl \to n'l'$ with $2 \leq n \leq 20$ and $0 \leq l \leq n-1$. The oscillator strength $f_{nn'}$ is related to $f_{nl,n'l'}$ by

$$f_{nn'} = \sum_{l=0}^{n-1} \sum_{l'=l\pm 1} \frac{(2l+1)}{n^2} f_{nl,n'l'} \tag{5-4-6}$$

and extensive tabulations of $f_{nn'}$ have recently been published by Goldwire [92] and Menzel [93] in which $2 \leq n' \leq 900$ and $1 \leq n'-n \leq 50$.

A general sum rule is obeyed by the oscillator strengths f_{ij}. If the atom or ion has N electrons then

$$\sum_j f_{ij} = N \tag{5-4-7}$$

where the summation is over all bound and continuum states to which an atom in level i may make a transition. (A reasonable approximation may sometimes be obtained by considering only these transitions involving valence electrons; in this case N is the number of valence electrons.) Equation (5-4-7) contains emission as well as absorption oscillator strengths as defined by (5-4-1); these are related through the additional general relation

$$\omega_i f_{ij} = -\omega_j f_{ji} \tag{5-4-8}$$

where the emission oscillator strength is negative. The summation over the continuum states in (5-4-7) can be expressed as an integral over the energy of the free electron. In the hydrogenic case $N = 1$ and as the level $n'l'$ changes from a bound state to the continuum, $-z^2/n'^2 \to k'^2$ and

$$\tfrac{1}{2} n'^3 \sum_{l'=l\pm 1} f_{nl, n'l'} \to \frac{z^2}{4\pi^2 \alpha a_0^2} a_\nu(nl) \tag{5-4-9}$$

where $a_\nu(nl)$ is given by (5-3-15).

For a complex atom or ion expression (5-4-2) may be evaluated using the Coulomb approximation of Bates and Damgaard [94], which corresponds to the quantum defect method used for bound-free and free-free transitions and described in § 5-3B. This approximation holds for the same type of transitions for which the general formulae (5-3-28) and (5-3-33) are valid, that is for resonance transitions in systems with only one valence electron and for transitions between excited states, and has been widely used in the compilation of tables [87, 88]. However, the program used by Wiese et al. [87] to compute oscillator strengths using the Coulomb approximation was supplied by Griem [95] and it has been noted by Froese and Underhill [96] that the results of Griem do not always agree with those of other workers using the same approximation. Tables from which the integral in (5-4-2) can be evaluated for given values of n^*, l, n'^* and l' are given in [94] where n^* and n'^* are the effective quantum numbers for the levels nl and $n'l'$ respectively. The conclusion of Bates and Damgaard that a suitable independent variable for their tabulation was $(n^* - n'^*)$ is in accord with the form adopted for $a_v(nl)$ in (5-3-28) in the bound–free case, in which the main variation in the matrix element is determined by the variable $(n^* + \mu')$.

New techniques, both theoretical and experimental, have recently been developed in order to improve the scope and accuracy of the existing data on oscillator strengths. The beam-foil technique developed by Bashkin and others [97] provides a new method for obtaining radiative lifetimes directly; the ions are excited by passing them through a thin foil of beryllium or carbon and then allowing them to decay. Oscillator strengths for many lines, particularly those from highly ionized species are greatly needed in astrophysics. Recently, Ervens and Berg [98] have measured oscillator strengths for visible lines of NIII and Hofmann [99] has studied lines of SiI, SiII and SiIII in the wavelength region 1100–2600 Å. Warner, in a series of papers, see for instance [100], has used a scaled Thomas–Fermi potential in which the radial functions are calculated and use is made of observationally determined energy levels to determine the scaling parameter. Some allowance for configuration mixing is made and his results for SiI agree reasonably with those of [99] in the cases where direct comparison can be made. Nussbaumer [101] has also used the Thomas–Fermi method and has developed a program in which allowance may be made for the interaction of many different configurations. His results for NIV should be considerably more accurate than those tabulated in [87]. Ab initio calculations which use Hartree–Fock wave functions and the method of superposition of configurations have been carried out for many transitions (see e.g. [96, 102] in which the boron and carbon isoelectronic sequences are studied). For highly ionized ions, the neglect of relativistic effects and spin–orbit coupling can lead to large discrepancies between experimentally and theoretically determined energy levels [103]. Although oscillator strengths for allowed transitions are not radically affected, the strengths of forbidden lines can be greatly enhanced because the strong coupling between the levels may cause the introduction of an allowed component into the transition. Westhaus and Sinanoğlu [104] have developed methods in which effects of electron correlation in both ground and excited states are included and have applied their theory to the calculation of oscillator strengths for transitions of the type

$$1s^2 2s^2 2p^q \to 1s^2 2s 2p^{q+1}; \quad 1 \leq q \leq 5, \qquad (5\text{-}4\text{-}10)$$

in ions of carbon, nitrogen, oxygen, fluorine and neon. They obtain encouraging agreement with experiment. It has been pointed out by Sinanoğlu [105], that in a transition involving equivalent electrons such as (5-4-10), allowance should be made for the fact that the 1s, 2s and 2p orbitals of the initial state are somewhat different from those of the final state, and that agreement between the dipole length (5-4-1) and dipole velocity forms for the matrix element determining the oscillator strength, is no guarantee that the calculation is accurate. Stewart [106] has shown that for optical dipole transitions between states which are degenerate in the limit of infinite nuclear charge, the dipole length matrix element is likely to give better results than the dipole velocity formula. The results of [104] support this conclusion.

Forbidden transitions, for which $l = l'$ or $l = l' \pm 2$, are of importance in the solar corona and in planetary nebulae (see § 5-6). The developments in the theory of transition probabilities outlined above will also undoubtedly be widely used in the next few years to add to the existing data on forbidden transitions. A recent review of the available data is given in [107].

B. THE LINE PROFILES - GENERAL FORMULAE

In this section we discuss the various processes which determine the basic shape of a line profile, $\phi_{ij}(v)$, which was first introduced in § 5-2D. We shall not attempt to interpret the observed shapes of spectral lines from a particular star, since this, in general, involves a knowledge not only of the line profile, but also the solution of the radiative transfer equation (5-2-10) together with the steady state equations (5-2-66).

There are four causes of spectral line broadening which should be taken into account. Natural broadening of the line occurs because the energy levels of an atom have a finite width, or alternatively, there is a finite lifetime for which the atom can occupy an energy state before spontaneous decay takes place. Secondly, any motion of the emitting (or absorbing) atom in the line of sight results in a shift of the observed frequencies from their true values through the Doppler effect. Thirdly, the pressure broadening of a line is caused by the effects of the field due to surrounding particles (or perturbers) on the atoms which are emitting (or absorbing) the line radiation. These perturbers may be free electrons, atoms, molecules and atomic or molecular ions. Lastly, the instrument that is used to observe the spectral line has itself a finite resolving power and hence there is an instrumental profile of finite width which adds to the intrinsic width of the line. The instrumental profile can be determined and its effects extracted from the observations so that the actual profile produced by the laboratory or astrophysical plasma is obtained.

The shape of a line representing the transition $i \to j$ and broadened purely by spontaneous decay processes in the atom is given by

$$\phi_{ij}(v) = \frac{\gamma_{ij}}{(\omega-\omega_{ij})^2 + (\tfrac{1}{2}\gamma_{ij})^2}\,;\quad \omega = 2\pi v, \tag{5-4-11}$$

where we have introduced the angular frequency ω. The circular frequency of the unbroadened line is ω_{ij} and is defined by

$$\hbar\omega_{ij} = |E_i - E_j| \tag{5-4-12}$$

where E_i and E_j are the energies of levels i and j respectively. The quantities ω and ω_{ij} in (5-4-11) should not be confused with the statistical weights ω_i and ω_j defined in § 5-2D. The line shape (5-4-11) is called a Lorentz profile and throughout this and the following section, the quantity $\phi(v)$ denotes a profile normalized according to relation (5-2-45). The full width of the line at half the maximum intensity is γ_{ij} which may be evaluated from

$$\gamma_{ij} = \sum_{i'=1}^{i-1} A_{ii'} + \sum_{j'=1}^{j-1} A_{jj'}, \tag{5-4-13}$$

where $A_{ii'}$ and $A_{jj'}$ are Einstein coefficients as defined in § 5-2D and the energy levels are labelled 1, 2, ..., in order of increasing energy. If the radiation density is large, the lifetimes of levels i and j will be decreased by the effects of stimulated emission and absorption of radiation upon their populations and then (5-4-13) must be replaced by

$$\gamma_{ij} = \sum_{i'=1}^{i-1} A_{ii'} + \sum_{j'=1}^{j-1} A_{jj'} + \sum_{i' \neq i} B_{ii'} I_v + \sum_{j' \neq j} B_{jj'} I_v. \tag{5-4-14}$$

Here the summations in the third and fourth terms of (5-4-14) are over all levels including the continuum except those for which $i' = i$ and $j' = j$, and the intensity of radiation at frequency v in the line is I_v.

If the motion of the emitting or absorbing atom in the line of sight is completely random, so that the velocity distribution of the atom is Maxwellian, then the profile for a line which is broadened purely by the Doppler effect is

$$\phi_{ij}(v) = 2\pi \int_{-\infty}^{+\infty} \delta\left(\omega - \omega_{ij} - \frac{v\omega_{ij}}{c}\right) f_1(v)\,dv \tag{5-4-15}$$

where v is the thermal velocity of the emitter in the line of sight and

$$f_1(v) = \left(\frac{M}{2\pi KT}\right)^{\tfrac{1}{2}} \exp(-Mv^2/2KT);\quad \int_{-\infty}^{+\infty} f_1(v)\,dv = 1. \tag{5-4-16}$$

On combining (5-4-15) and (5-4-16) we have the Doppler profile

$$\phi_{ij}(v) = \frac{2\sqrt{\pi}}{\xi\omega_{ij}} \exp\left\{-\left(\frac{\omega-\omega_{ij}}{\xi\omega_{ij}}\right)^2\right\}; \quad \xi = \left(\frac{2KT}{Mc^2}\right)^{\frac{1}{2}}, \quad (5\text{-}4\text{-}17)$$

where T is the kinetic temperature, M is the mass of the emitter and ξ is dimensionless. The half width of the line is

$$\gamma_D = 2\xi\omega_{ij}\sqrt{\ln 2} \quad (5\text{-}4\text{-}18)$$

where, as in eq. (5-4-13), γ_D is in units of the angular frequency. If the motion of the emitting atoms in the line of sight is influenced by turbulence, that is by large scale motions of the plasma, the effect can be allowed for approximately, by introducing a "turbulent" velocity v_t. The distribution function $f_1(v)$ in (5-4-16) becomes

$$f_1(v) = \frac{1}{\zeta c\pi^{\frac{1}{2}}} \exp(-v^2/c^2\zeta^2) \quad (5\text{-}4\text{-}19)$$

where ζ replaces ξ and is defined by

$$\zeta^2 = \xi^2 + v_t^2/c^2. \quad (5\text{-}4\text{-}20)$$

The theory of turbulence in stellar atmospheres is not well understood and the procedure outlined in (5-4-19) and (5-4-20) for obtaining the line profile is only very approximate. Thus results obtained for v_t by matching the line shapes given by (5-4-17) and (5-4-20) with those derived from observation, must be regarded as extremely inaccurate, see [1]. Additional Doppler broadening of lines is also observed in the spectra of rotating stars, the line shape being the result of a superposition of line profiles from different parts of the star. In rapidly rotating stars, lines which are normally strong may be broadened to the extent that they are indistinguishable from the background continuum, and thus are inappropriate for analysis.

The finite resolving power of the instrument used for the observation is not dependent upon the intrinsic shape of the spectral line observed. Thus if $\phi_A(v)$ is the apparatus profile and $\phi_{ij}(v)$ the actual profile of the line, the shape of the observed profile is given by

$$\phi_{obs}(v) = \int \phi_{ij}(v-v')\phi_A(v')dv'. \quad (5\text{-}4\text{-}21)$$

The "folding" integral in (5-4-21) can be used for any two profiles formed by independent processes.

The Lorentz profile (5-4-11) not only represents a naturally broadened line, but with an appropriate value of the half width, can give the shape of a line broadened by certain types of perturbers (see § 5-4C). In many cases, the Doppler effect and the processes that give rise to a line broadened with a Lorentz profile, can be treated as independent. Thus we may use the expression (5-4-21) to fold the two profiles together and obtain for the line shape

$$\phi_{ij}(v) = \frac{\gamma_{ij}}{\xi \omega_{ij} \pi^{\frac{3}{2}}} \int_0^\infty [(\omega - \omega' - \omega_{ij})^2 + (\tfrac{1}{2}\gamma_{ij})^2]^{-1} \exp\left\{-\left(\frac{\omega' - \omega_{ij}}{\xi \omega_{ij}}\right)^2\right\} d\omega' \tag{5-4-22}$$

which is often referred to as the Voigt profile. Extensive tables of this profile have been compiled by Hummer [108].

The line shape observed depends on the temperature and density of the plasma and on the unperturbed frequency of the line (see (5-4-17)) as well as on the finite lifetimes of the atomic energy levels. In many cases, however, the effects of natural broadening can be neglected in comparison with the pressure effects described in the next section.

C. PRESSURE BROADENING OF SPECTRAL LINES

The theories of the pressure broadening of spectral lines fall into two main categories. Firstly, the quasi-static theories in which the effects of all the perturbers acting simultaneously on the emitting or absorbing atom are taken into account but no motion of the perturbers is explicitly included. Secondly, the situation in which the perturbers are moving sufficiently fast, so that the broadening can be considered as arising from a series of binary collisions between the atom and a perturber, is described by an impact theory of line broadening.

A purely classical impact theory was developed by Lindholm [109, 110] and independently by Foley [111]; the treatment of the theory presented here was first given by Traving [112]. The radiation emitted in the line is represented by a classical oscillator of angular frequency ω and the effect of each collision between an atom and a perturber is to cause a rapid change of phase in the wave train. The amplitude of the oscillation is given by

$$f(t) = \exp\left[i \int_0^t \omega(t) dt\right] \equiv \exp\left[i\omega_{ij} t + i\eta(t)\right] \tag{5-4-23}$$

where ω_{ij} is the unperturbed frequency of the oscillation and $\eta(t)$ is the sum of all phase changes in the time interval 0 to t. The line profile may be written as

$$\phi_{ij}(\nu) = 2\,\mathrm{Re} \int_0^\infty C(s) e^{-i\omega s}\,ds \qquad (5\text{-}4\text{-}24)$$

where "Re" denotes the real part, and $C(s)$ is the correlation function, which is related to the amplitude $f(t)$ by

$$C(s) = \lim_{T\to\infty} \frac{1}{T} \int_{-\frac{1}{2}T}^{+\frac{1}{2}T} f^*(t) f(t+s)\,ds. \qquad (5\text{-}4\text{-}25)$$

Alternatively, on using (5-4-23)

$$C(s) = \exp(i\omega_{ij}s)\,\langle \exp(i\eta(t+s) - i\eta(t))\rangle \equiv \exp(i\omega_{ij}s)\,\tilde{C}(s) \quad (5\text{-}4\text{-}26)$$

where $\tilde{C}(s)$ denotes the average $\langle \exp(i\eta(t+s) - i\eta(t))\rangle$ of the quantity $\exp(i\eta(t+s) - i\eta(t))$ over time t. If we consider the change in $\tilde{C}(s)$ in time Δs and make the impact approximation, which assumes that the additional phase change η' in the time interval Δs is independent of the momentary phase, then η' and $\eta(t)$ are independent functions and the time average may be split into two factors. We further assume that a time difference Δs which satisfies this condition, is also sufficiently small, that the finite difference Δs may be replaced by an infinitesimal difference ds. The function $\tilde{C}(s)$ then satisfies the differential equation

$$\tilde{C}(s+ds) - \tilde{C}(s) = d\tilde{C}(s) = \langle \exp(i\eta(t+s) - i\eta(t))\rangle \langle e^{i\eta'} - 1\rangle. \qquad (5\text{-}4\text{-}27)$$

The ergodic hypothesis, which states that the average over the time history of the radiation of one oscillator can be replaced by the average over the whole assembly of radiating oscillators at any given instance of time, may be used together with (5-4-26) to replace (5-4-27) by

$$d\tilde{C}(s) = \overline{(e^{i\eta'} - 1)}\,\tilde{C}(s) \qquad (5\text{-}4\text{-}28)$$

where the bar over the factor in (5-4-28) denotes an ensemble average. An additional approximation must now be made in order to integrate (5-4-28), the result of which is to reduce the correlation function to a quantity depending only on a series of binary collisions between the atom and the perturber. If ρ is the impact parameter for a collision we may define a critical impact parameter ρ_0, or alternatively a phase shift $\eta_0 \equiv \eta(\rho_0)$, such that a weak collision is one for which $\rho > \rho_0$ or $|\eta(\rho)| < |\eta_0|$. Here $\eta(\rho)$ denotes the phase shift caused by one collision at an impact parameter ρ and η_0 is chosen to be less than unity. If only weak collisions occur in the time interval ds, such that the sum of all the phase shifts in time ds is also less than unity, then

$$\overline{(\exp(i\eta')-1)} = \sum_k \overline{(\exp(i\eta_k)-1)}; \quad |\eta'| < |\eta_0|, \quad |\eta_k| < |\eta_0| \qquad (5\text{-}4\text{-}29)$$

where η_k is the phase shift caused by the kth perturber. It is assumed that strong collisions, that is those for which $\rho \leq \rho_0$ or $|\eta(\rho)| \geq \eta_0$, are infrequent and that at most one occurs in the time ds. If a strong collision does occur, the effects of all the weak collisions in the same time interval are neglected and

$$\overline{(\exp(i\eta')-1)} = \overline{(\exp(i\eta_s)-1)} \qquad (5\text{-}4\text{-}30)$$

where $\eta_s(\rho)$ is the phase shift caused by the strong collision. We assume finally that the perturbers move past the emitting atom in parallel straight line paths at an average velocity $\bar{v}$. Thus, if N is the number density of the perturbers, the number that collide with the atom in time ds at impact parameters in the range $(\rho, \rho + d\rho)$ is

$$2\pi N \bar{v} \rho \, d\rho \, ds. \qquad (5\text{-}4\text{-}31)$$

On using (5-4-29) for all $\rho > \rho_0$, (5-4-30) for $\rho \leq \rho_0$ and expression (5-4-31) for the number of perturbers, eq. (5-4-27) becomes

$$d\tilde{C}(s) = 2\pi N \bar{v} \int_0^\infty (e^{i\eta(\rho)} - 1) \rho \, d\rho \, ds, \qquad (5\text{-}4\text{-}32)$$

where $\eta(\rho)$ is the phase shift caused by one collision only. Thus finally, we can express the line shape as

$$\phi_{ij}(v) = \frac{2w}{(\omega - \omega_{ij} - d)^2 + w^2} \qquad (5\text{-}4\text{-}33)$$

where

$$w + id = 2\pi N \bar{v} \int_0^\infty (1 - e^{-i\eta}) \rho \, d\rho. \qquad (5\text{-}4\text{-}34)$$

The profile (5-4-33) is a Lorentz profile of half width $2w$ and has a shift in the position of the central frequency from the unperturbed frequency ω_{ij}, of amount d.

The Lindholm–Foley theory does not allow for any interchange of energy between the atoms emitting the line radiation and the perturbers. If the width and shift of the line are to be evaluated from (5-4-33), $\eta(\rho)$

must be related to the properties of the atom and its interaction with the perturber. We may write

$$\eta(\rho) = \int_{-\infty}^{+\infty} (\omega(t) - \omega_{ij}) \, dt \qquad (5\text{-}4\text{-}35)$$

and on using perturbation theory to determine the interaction energy when the atom is in level i and in level j, we have

$$\omega(t) - \omega_{ij} = C_p/r^p \qquad (5\text{-}4\text{-}36)$$

where r is the atom–perturber distance. The parameter p depends only on the nature of the atom–perturber interaction, while C_p depends on this interaction and on the internal properties of the atom, but not on r. It is assumed in (5-4-36) that r is always much greater than the mean radius of the atom and for a perturber moving in a straight line,

$$r = (\rho^2 + v^2 t^2)^{\frac{1}{2}} \qquad (5\text{-}4\text{-}37)$$

where the perturber is at its distance of closest approach at time $t = 0$. We may use (5-4-35), (5-4-36) and (5-4-37) to evaluate $\eta(\rho)$ and then substitution in (5-4-34) gives

$$w = \pi N \bar{v} \left[\frac{\beta_p C_p}{\bar{v}} \right]^{2/(p-1)} \Gamma\left(\frac{p-3}{p-1}\right) \cos\left(\frac{\pi}{p-1}\right) \qquad (5\text{-}4\text{-}38)$$

where

$$\beta_p = \frac{\sqrt{\pi} \Gamma(\frac{1}{2}(p-1))}{\Gamma(\frac{1}{2}p)}. \qquad (5\text{-}4\text{-}39)$$

A similar expression to (5-4-38) exists for the shift. A finite result for the width is obtained from (5-4-38) for $p \geq 3$, since, as $p \to 3$,

$$\Gamma\left(\frac{p-3}{p-1}\right) \cos\left(\frac{\pi}{p-1}\right) \to \tfrac{1}{2}\pi. \qquad (5\text{-}4\text{-}40)$$

The cases $p = 3$, 4 and 6 correspond to resonance, quadratic Stark, and Van der Waals broadening respectively. For the linear Stark effect for which $p = 2$, (5-4-38) becomes infinite, but a finite result is obtained if all collisions for which $\eta(\rho) < \eta_{\min}$ are neglected, where $\eta_{\min}$ is the phase shift for the most distant effective collision. This difficulty occurs because the screening effects produced by the perturbers have not been allowed for in (5-4-36); $\eta_{\min}$ is often chosen to correspond to an impact parameter equal to the Debye radius.

The basic assumptions of the impact and binary collision approximations are also present in the semi-classical impact theory of Griem and Kolb (cf. [95]) and in the work of Baranger [113–115] who has developed an impact theory for the general case of a line composed of overlapping components, using both semi-classical and fully quantum mechanical arguments. In the semi-classical theory, the perturbers are treated classically, but the emitting (or absorbing) atom is described quantum mechanically. This corresponds to a quantum mechanical evaluation of C_p in the Lindholm-Foley theory. The ensemble average is carried out by making the binary collision assumptions already described, but a further average is made over all possible orientations of the perturber–atom collision in space. Thus if the impact parameter vector ρ is in the direction defined by the atom and the point of closest approach of the perturber, the average is over all directions of ρ and the velocity v of the perturber, subject to the condition $\rho \cdot v = 0$. It is also assumed that no change in the velocity v is caused by a collision, an assumption which is implicit in equation (5-4-37) in the Lindholm–Foley theory. The width and shift of an isolated (or one-component) line with upper state i and lower state j is (cf. [116])

$$w + id = 2\pi N \int v f(v) \, dv \int \{1 - S_{ii}(\rho, v) S_{jj}^*(\rho, v)\} \rho \, d\rho, \qquad (5\text{-}4\text{-}41)$$

a relation which is analogous to (5-4-34). In equation (5-4-41), $S_{ii}(\rho, v)$ and $S_{jj}(\rho, v)$ are elements of the scattering matrix describing an elastic perturber–atom collision, and for which the total energy of the perturber–atom system is $E_i + \tfrac{1}{2}mv^2$ and $E_j + \tfrac{1}{2}mv^2$ respectively. Here, m is the mass of the perturber and in (5-4-41), $f(v)$ is a Maxwellian distribution given by

$$f(v) \, dv = f_1(v_x) f_1(v_y) f_1(v_z) \, dv_x \, dv_y \, dv_z \qquad (5\text{-}4\text{-}42)$$

where v_x, v_y and v_z are the cartesian components of the velocity. The function $f_1(v)$ in (5-4-42) is given by (5-4-16) in which the mass M has been replaced by m. A comparison of equations (5-4-34) and (5-4-41) show that the quantity $e^{-i\eta}$ has been replaced by $S_{ii}(\rho, v) S_{jj}^*(\rho, v)$, and explicit expressions for $S_{ii}(\rho, v)$ and $S_{jj}(\rho, v)$ are given in § 5-5.

If the perturber motion is treated quantum mechanically, equation (5-4-41) becomes

$$w + id = 2\pi N \int v f(v) \, dv \, \frac{\hbar^2}{2m^2 v^2} \sum_l (2l+1)\{1 - S_{ii}(l, v) S_{jj}^*(l, v)\} \qquad (5\text{-}4\text{-}43)$$

in which the integral over ρ has been replaced by a summation over l, the angular momentum quantum number for the lth partial wave of the perturber. An alternative way of writing this result for the half width is (cf. [115]),

$$2w = N \int v f(v) \, dv \left\{ \sum_{i' \neq i} Q_{ii'}(v) + \sum_{j' \neq j} Q_{jj'}(v) \right.$$
$$\left. + \int |f_i(v, \theta) - f_j(v, \theta)|^2 \sin \theta \, d\theta \, d\phi \right\} \quad (5\text{-}4\text{-}44)$$

where the summations are over all the inelastic cross-sections $Q_{ii'}$ and $Q_{jj'}$ for collisional transitions $i \to i'$ and $j \to j'$ respectively, and $f_i(v, \theta)$ and $f_j(v, \theta)$ are the elastic scattering amplitudes for scattering by the atom in level i and level j. The collision cross-sections and amplitudes are discussed in § 5-5. The first two terms in (5-4-44) are the probabilities per unit time that collisions cause the decay of levels i and j respectively, and are analogous to the two terms in (5-4-13) which give the radiative transition probabilities. Equation (5-4-44) is valid only when, on average, a large number of collisions take place which broaden the energy levels (and hence shorten their lifetimes), before a radiative transition $i \to j$ takes place.

The calculation of a profile for a line composed of several overlapping components is more difficult and involves the inversion of a matrix of the order of the product of the statistical weights of the upper and lower groups of levels (cf. [95]). A deficiency of these theories is that they do not allow for the occurrence of simultaneous strong collisions and for $|\omega - \omega_{ij}|$ sufficiently large the theory always breaks down, since to cause a large frequency shift, a strong collision of long duration is required. This violates the basic assumptions made in deriving the results (5-4-41) and (5-4-43). Nevertheless, the impact theories have been very successful in predicting the broadening of lines by electron collisions and they have also been used to study certain types of atom–atom broadening.

The conditions necessary for the impact approximation to be valid are summarized in [116]. If τ is the time of duration of a typical collision, then

$$w\tau \ll 1 \quad (5\text{-}4\text{-}45)$$

since the time between collisions is of the order of w^{-1}. The condition for a collision to be weak can be written in terms of the average interaction energy $\bar{V}$, which depends on an average impact parameter $\bar{\rho}$ and the dimensions of the radiating atom. Thus

$$\bar{V}\tau/\hbar \ll 1 \quad (5\text{-}4\text{-}46)$$

where $\bar{V}$ can be expressed in terms of $\bar{\rho}$ which may be taken approximately equal to the nearest neighbour distance. The requirement that strong collisions are well separated in time may be written as

$$w_s \tau_s \ll 1 \tag{5-4-47}$$

where τ_s is the duration time for a strong collision and w_s^{-1} is the average time between strong collisions. If Q_s is the cross-section for collisions in which $\rho \leq \rho_0$, then w_s is given by

$$2w_s = N \int v f(v) Q_s(v) \, dv. \tag{5-4-48}$$

The validity of (5-4-45) can only be ascertained after the calculation for the profile has already been performed, unless some prior estimate of w can be obtained from elsewhere.

In the simplest form of a quasi-static treatment of the line broadening problem, the radiating atom is surrounded by a cluster of identical perturbers, each of which affects the atom, but interactions between the perturbers themselves are neglected. The motion of the perturbers is not included explicitly, but an intensity distribution is obtained by averaging over all possible positions of the perturbers. Within this basic approximation two possibilities exist. Either a field distribution is obtained by summing the fields at the emitting atom due to all the perturbers; a frequency distribution may then be derived from the ordinary theory of the Stark effect. Alternatively, the frequency shift $(\omega - \omega_{ij})$ in the emitted radiation, caused by each perturber, may be used to obtain the distribution of frequencies in the line directly.

We suppose that the lth perturber has position vector r_l relative to the atom and define the probability that the total field F at the emitting atom lies in the range $(F, F+dF)$ to be $W(F)$; the atom and perturbers are enclosed in a volume V containing a total number of perturbers Q. Then, if $P(r_1, r_2, \ldots, r_Q)$ is the probability for finding the Q perturbers at positions specified by the ranges (r_1, r_1+dr_1), $(r_1, r_2+dr_2), \ldots, (r_Q, r_Q+dr_Q)$, $W(F)$ is given by

$$W(F) = \int \delta(F - \sum_l F_l) P(r_1, r_2, \ldots, r_Q) \, dr_1 \, dr_2 \cdots dr_Q \tag{5-4-49}$$

which we may write as

$$W(F) \equiv \frac{1}{(2\pi)^3} \int \exp(i\mathbf{k} \cdot \mathbf{F}) C(\mathbf{k}) \, d\mathbf{k}, \tag{5-4-50}$$

where F_l is the field due to the lth perturber at the emitter and (5-4-50) is obtained from (5-4-49) by representing the delta function by an integral. If the assumption is now made that the perturbers do not interact with each other, the probability $P(r_1, r_2, \ldots, r_Q)$ may be written as

$$P(r_1, r_2, \ldots, r_Q) = V^{-Q} \tag{5-4-51}$$

if the emitter is a neutral particle, and hence $C(k)$ reduces to

$$C(k) = \left[\frac{1}{V}\int_V \exp(-i\mathbf{k} \cdot \mathbf{F}_1) d\mathbf{r}_1\right]^Q$$

$$= \left[1 - \frac{N}{Q}\int(1-\exp(-i\mathbf{k} \cdot \mathbf{F}_1)) d\mathbf{r}_1\right]^Q \tag{5-4-52}$$

where, as before, N is the number density of the perturbers. Since in the physical situations of interest the volume V contains many perturbers and is very large compared with atomic dimensions, we take the limit of Q, $V \to \infty$ while $N = Q/V$ remains finite, in (5-4-52). Thus, in the limit

$$C(k) = \exp\left\{-N\int(1-\exp(-i\mathbf{k} \cdot \mathbf{F}_1)) d\mathbf{r}_1\right\}, \tag{5-4-53}$$

and this integral can be evaluated for different types of fields F_1. If the perturbers are singly charged positive ions

$$F_1 = -(e/r_1^3)r_1 \tag{5-4-54}$$

and

$$C(k) = \exp\left\{-\tfrac{4}{15}N(2\pi ek)^{\frac{3}{2}}\right\}. \tag{5-4-55}$$

The field distribution is then obtained from (5-4-50), a result originally derived by Holtsmark [117]. For the case of a charged radiating atom, an additional factor occurs in the integrand of (5-4-53) in order to account for the long range Coulomb repulsion between the ion and the ionic perturber. If $\beta = F/F_0$ is a dimensionless variable, where F_0 is given by

$$F_0 = 2\pi e(\tfrac{4}{15}N)^{\frac{2}{3}}, \tag{5-4-56}$$

then the distribution (5-4-49) becomes

$$W(\beta) = \frac{2\beta}{\pi}\int_0^\infty \sin(\beta\eta)\exp(-\eta^{\frac{3}{2}})\eta\,d\eta; \quad \int_0^\infty W(\beta)\,d\beta = 1. \tag{5-4-57}$$

For perturbers which have dipole or quadrupole fields, eq. (5-4-49) must be generalized to include an average over all possible directions of the Q identical dipoles with moments $\mu_1, \mu_2, \ldots, \mu_Q$. The dipole field

$$F_1 = \frac{\mu_1}{r_1^3} - \frac{3r_1(r_1 \cdot \mu_1)}{r_1^5} \tag{5-4-58}$$

gives, after a lengthy calculation,

$$C(k) = \exp\left\{-\tfrac{1}{3}\pi^2 \left(1 + \frac{1}{4\sqrt{3}} \ln(7+4\sqrt{3})\right) N\mu_1 k\right\} \tag{5-4-59}$$

and hence the field distribution

$$W(\beta) = \frac{4}{\pi} \frac{\beta^2}{(1+\beta^2)^2}; \quad \int_0^\infty W(\beta)\,d\beta = 1. \tag{5-4-60}$$

In equation (5-4-60), $\beta = F/F_0$ as before, while

$$F_0 = \tfrac{1}{3}\pi^2 \left(1 + \frac{1}{4\sqrt{3}} \ln(7+4\sqrt{3})\right) \mu_1 N \tag{5-4-61}$$

and the result (5-4-60) was first obtained by Holtsmark [117]. More details of the derivations of (5-4-57) and (5-4-60) are given by Breene [118].

The distribution of intensity in the line emitted by the atom may now be calculated from

$$\phi(v) = 2\pi \int_0^\infty \delta(\omega - \omega_{ij} - d_t \beta^t) W(\beta)\,d\beta; \quad \int \phi(v)\,dv = 1 \tag{5-4-62}$$

where $t = 1$ for an atom whose levels are split according to the linear Stark effect and $t = 2$ for the quadratic Stark effect. The constant d_t depends on t and the displacements of the upper and lower levels of the line by the field are contained in the term $d_t \beta^t$.

The quasi-static approximation from which the distribution of frequency shifts results directly, is obtained by writing

$$\phi(v) = 2\pi \int \delta\left(\omega - \omega_{ij} - \sum_l \Delta\omega_l\right) P(r_1, r_2, \ldots, r_Q)\,dr_1\,dr_2 \cdots dr_Q \tag{5-4-63}$$

$$\equiv \int_{-\infty}^{+\infty} \exp\{ik(\omega - \omega_{ij})\} D(k)\,dk \tag{5-4-64}$$

where $\Delta\omega_l$ is the shift in angular frequency in the line caused by the lth

perturber. Equations (5-4-63) and (5-4-64) are analogous to equations (5-4-49) and (5-4-50) except that the distribution of a scalar rather than a vector quantity is considered; the function $D(k)$ is determined in exactly the same way as $C(\mathbf{k})$. Equations (5-4-63) and (5-4-64) lead to the result

$$D(k) = \exp\left(-N\int (1-\exp(-ik\Delta\omega_1))\,d\mathbf{r}_1\right). \tag{5-4-65}$$

If the interaction between atom and perturber is of the Van der Waals' type, we may write

$$\Delta\omega_1 = -|C_6|/r_1^6 \tag{5-4-66}$$

(cf. (5-4-36)), so that

$$D(k) = \exp\left(-\tfrac{2}{3}\pi(2\pi k|C_6|)^{\frac{1}{2}}(1-i)N\right); \quad k \geqq 0 \tag{5-4-67}$$

and

$$D(-k) = D^*(k). \tag{5-4-68}$$

The profile $\phi(v)$ can then be evaluated from (5-4-64) and this method was first introduced by Margenau [119]. Similar distributions can be evaluated for other types of interaction, but it is important to realise that by expressing $\phi(v)$ according to eq. (5-4-63), we have assumed that the frequency shifts are additive. This assumption was also made in establishing the impact theory (cf. (5-4-23)), but here it represents a much more drastic approximation since these shifts occur simultaneously. For example if we consider the effect of linear Stark broadening produced in the emitting atom by perturbers which are singly charged positive ions, then

$$\Delta\omega_1 = C_2/r_1^2 \tag{5-4-69}$$

and the integral in (5-4-65) diverges. In this case the assumption that the frequency shifts are additive corresponds to taking the fields $\mathbf{F}_l$, $l = 1, 2, \ldots, Q$, all in the same direction, which is clearly unrealistic. It should be noted further that all these quasi-static results are invalid if the atom and perturber are too close together, since only the leading term in the atom–perturber interaction has been retained in the calculation of the distribution.

The criteria for the validity of the quasi-static theories are essentially the reverse of those given for the impact approximation. Although there have been attempts to derive a unified theory of line broadening [120, 121], which reproduces the results of the impact approximation in the high energy limit and those of the quasi-static theory in the low energy limit, there is some doubt as to whether this theory really incorporates all the features of the complete profile in either the impact or quasi-static limits [122]. Recently, some progress has been made in extending the quasi-static theory to include to first order the fluctuations in time of the field F [123]. This involves obtaining a distribution function $W(\mathbf{F}, \mathbf{f})$ where $\mathbf{f} = d\mathbf{F}/dt$ is included explicitly and the results are useful both in bridging the gap between the quasi-static and impact regimes and in estimating the validity of the quasi-static theory. In Fig. 5-4-1 we compare the results of the Lindholm–Foley impact theory, with those obtained from the quasistatic and modified quasi-static theories, for two components of $H\beta$ broadened by proton perturbers. The smaller the value of $|C_2|$, the greater is the deviation of the modified-quasi-static approximation from its quasi-static limit.

The results (5-4-55) can easily be generalized to the case where the perturbers are multiply charged and where perturbers of different charge states are present in the plasma together [124]. For the case of singly charged perturbers, calculations have been presented in which some account has been taken of the perturber–perturber interactions, so that the probability $P(r_1, r_2, \ldots, r_Q)$ is no longer given by (5-4-51). Baranger and Mozer [125, 126] have treated the cases of electron perturbers with a uniform neutralizing background and positive ion perturbers shielded by electrons, in which they include pair correlations between the perturbers. Hooper [127, 128] later improved this work by including all correlations to a high degree of accuracy.

Throughout this discussion of quasi-static theory, classical statistics have been used. A review article which studies classical methods for obtaining line shapes in some detail has recently been published [129].

Much theoretical and experimental work has been carried out to obtain line profiles of Balmer lines broadened by protons and electrons. Recent experimental studies of the broadening of $H\beta$, $H\gamma$ and $H\delta$ have been made by Hill and Gerardo [130] for electron (and proton) densities in the range $(1.3–7.1) \times 10^{16}$ cm^{-3} and a plasma temperature of $\sim 2.3 \times 10^4$ °K, while for higher densities ($\sim 2 \times 10^{17}$ cm^{-3}), measurements have been published by Morris and Krey [131] and Shumaker and Popenoe [132] for $H\beta$. Experiments on $H\gamma$ by Bengtson et al. [133] for electron densities in the range $(2–20) \times 10^{16}$ cm^{-3} and a temperature in the range $(1.2–2.3) \times 10^4$ °K give half widths for $H\gamma$ that agree to within about 10 % with the calculations of Kepple and Griem [134]; while the experimental $H\beta$ profiles agree with the results of [134] to within about 4 %, except in the region of the line centre. The calculations in [134] are made using the semi-classical impact approxima-

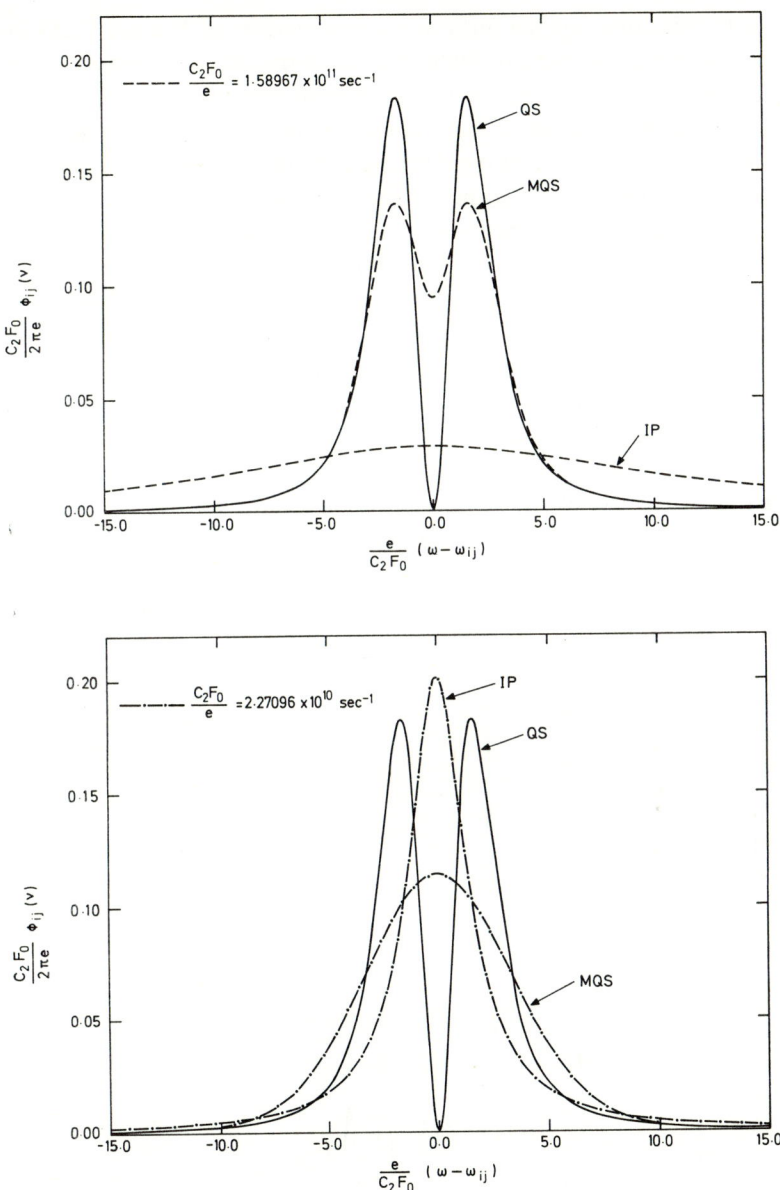

Fig. 5-4-1. Profiles for two components of Hβ at $T = 10^4$ °K and proton density $N = 1.259 \times 10^{14}$ cm^{-3}. QS = Holtsmark theory, MQS = modified quasi-static theory, IP = Lindholm–Foley theory.

tion for the electrons and the quasi-static distributions of Hooper [128] for the protons. Since the levels of hydrogen are degenerate with respect to the orbital angular momentum, the ion field splits the two levels into their components through the linear Stark effect, and the observed hydrogen lines are composed of a number of overlapping components. Hence calculations become more lengthy as the principal quantum numbers of the upper and lower levels increase. The program devised for the computations in [134] has been used to obtain profiles for H_6–H_{12} (the suffix indicates the principal quantum number for the upper level) by Bengtson et al. [135] to obtain results for comparison with their experiments at a temperature of 1850 °K and an electron density of 1.2×10^{13} cm^{-3}. The ratio of experimental to theoretical half width varies between about 0.84 and 0.90. A program has recently been developed, Peach [136], which calculates hydrogen line profiles using the same basic approximations as those used in [134], but with minor differences in the use of the semi-classical collision data for the electron-hydrogen collisions, see § 5-5. Reasonable agreement is obtained with the results of Kepple and Griem for $H\alpha$, $H\beta$, $H\gamma$ and $H\delta$ and with the experiments [135] for H_6–H_{12}. The ratio of experimental to theoretical half-

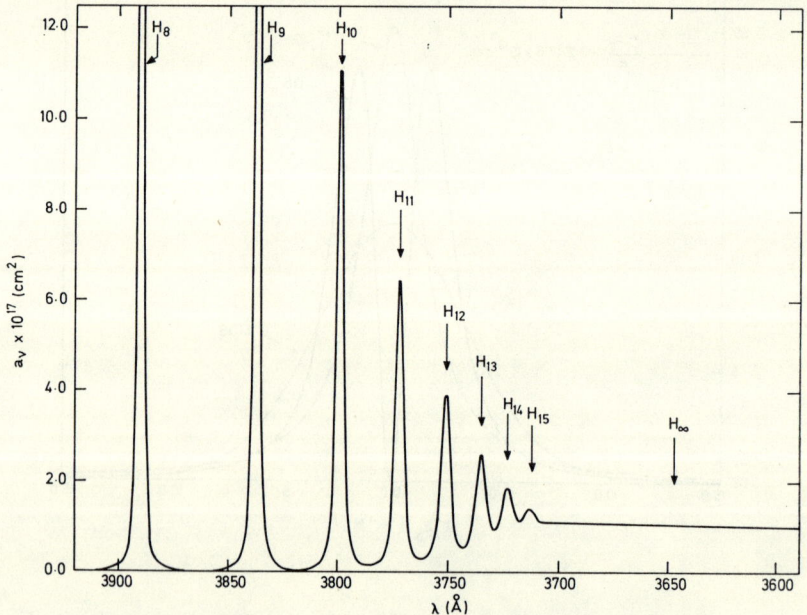

Fig. 5-4-2. The absorption cross-section for the Balmer series of hydrogen at $T = 10^4$ °K and electron density $N_e = 1.259 \times 10^{14}$ cm^{-3}.

widths for the high Balmer lines varies from 0.87 to 0.91. Fig. 5-4-2 shows the absorption cross-section for the Balmer series in hydrogen for an electron density of 1.259×10^{14} cm^{-3} and a temperature of 10^4 °K, which are thought to be typical of conditions in a main sequence A0 star, Pagel [136]. These results have been obtained using the program described in [136] and the formulae (5-4-3) and (5-3-20) for the oscillator strengths and the continuous absorption cross-section respectively. It is seen that there is reasonable agreement with the observations of the spectra of A0 stars, which show that the last resolvable Balmer line is $H_{n'}$ where $n' = 15$. The validity of the quasi-static approximation used to estimate the broadening by protons of these Balmer lines is indicated by Fig. 5-4-1.

The agreement between experiment and theory is not so satisfactory for the Lyman lines. Experiments have been carried out by Boldt and Cooper [137] for Lyα and by Elton and Griem [138] for Lyα and Lyβ. In Fig. 5-4-3 we compare the variation with frequency in the wing of Lyα which is obtained from experiment and theory [95, 121, 136, 139]. The results of [136] represent an improvement on those of [95, 139] for $\Delta\lambda < 3$ Å, but for larger values of $\Delta\lambda$ they become increasingly bad, as no attempt is made in [136] to incorporate, even empirically, the effects of the transition from the impact broadening approximation to the quasi-static approximation for the electrons.

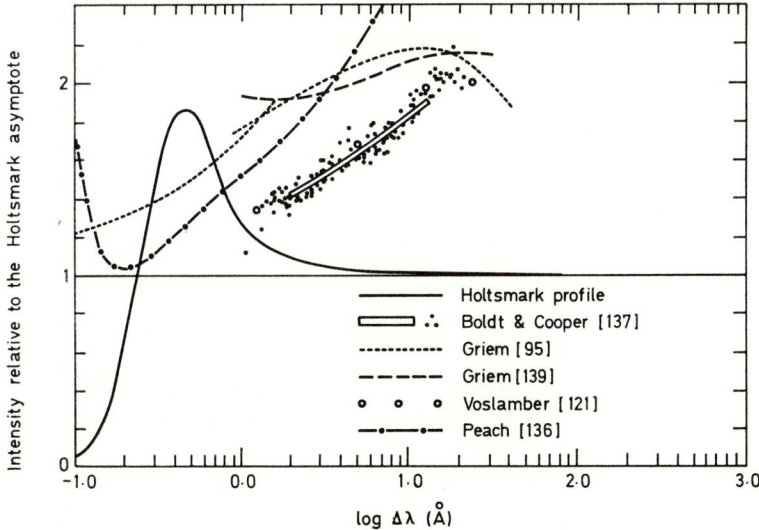

Fig. 5-4-3. The ratio of the Lyα profile to its asymptotic Holtsmark value at $T = 1.22 \times 10^4$ °K and electron density $N_e = 8.4 \times 10^{16}$ cm^{-3}.

This effect is included in [121] and for $\Delta\lambda > 20$ Å, the intensity ratio in Fig. 5-4-3 has its ultimate value, 2. Recent calculations by Vidal, Cooper and Smith [120], using their unified theory, give an intensity ratio which is also a little higher than that given by experiment. They are unable to reproduce the results of [121] and this discrepancy has still to be resolved. No calculations have yet been made for Ly α using the accurate quantum-mechanical collision data for the electron-hydrogen problem, which is already available (see § 5-5).

The Stark broadening of the radio frequency transitions in hydrogen of the type $n \to n+1$, has been calculated by Griem [140]. He estimates the half widths from the behaviour of the profiles in the line wings, and finds that only electron collisions which produce a change in principal quantum number effectively contribute to the broadening (cf. (5-4-44)). These calculations have recently been repeated [141], using more refined data for the inelastic cross sections (see § 5-5) and in which complete line profiles are computed. Griem's conclusions are shown to be essentially correct, and results obtained for the half widths are 20–30 % larger than those given in [140]. This a significant result for the interpretation of emission lines in HII regions (see § 5-6C).

Edmonds et al. [142] have calculated profiles for the Lyman, Balmer, Paschen and Brackett series in hydrogen, for temperatures and densities of astrophysical interest, using the theory of [125, 126]. They claim that the use of the quasi-static theory for both electron and proton perturbers gives better agreement with laboratory experiments for the lower densities common in stellar atmospheres, than does the use of the impact theory for the electrons and the quasi-static theory for the protons. The validity of this claim is uncertain in view of the results of [135] and their agreement with theory.

Some recent measurements on lines of HeII in the visible and near ultraviolet [143] considerably extend the data available on the broadening of ionized helium lines. They provide a good test of the theory, since at the temperature $T \sim 2.3 \times 10^5$ °K, the helium plasma is almost completely ionized and so the positive ion perturbers are structureless particles of charge $Z = 2$. The program [136] may also be used to calculate profiles for lines of hydrogenic positive ions, which are broadened by electrons and by positive ions of multiple charge. The half width indicated for the 4686 Å line is considerably smaller than that measured and it seems likely that at this temperature and electron density ($\sim 10^{17}$ cm^{-3}), the quasi-static approximation for the broadening by the ions is not completely valid.

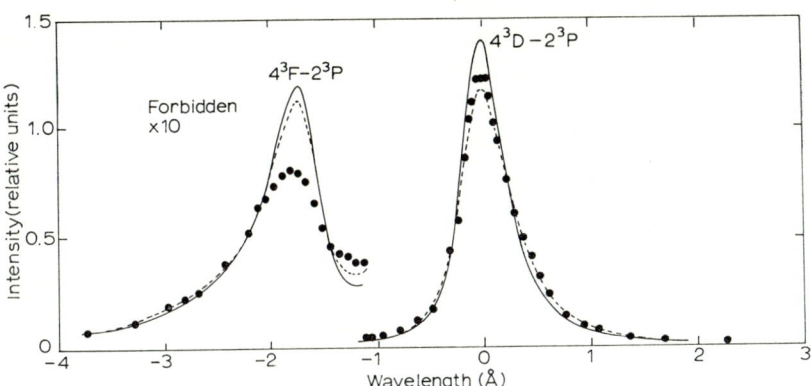

Fig. 5-4-4. The HeI line $(2p)\,^3P$–$(4d)\,^3D$ and its forbidden component $(2p)\,^3P$–$(4f)\,^3F$ at $T = 10^4\,°K$ and electron density $N_e = 10^{15}\,cm^{-3}$; ——— Barnard et al. [146], - - - Griem [143], ● Burgess and Cairns [144]. [D. D. Burgess and C. J. Cairns, J. Phys. B 3 (1970) L67. Reprinted by permission of the Institute of Physics and the Physical Society; © 1970 by the Institute of Physics and the Physical Society.]

The HeI line $(2p)\,^3P$–$(4d)\,^3D$ at 4471 Å and its nearby forbidden component $(2p)\,^3P$–$(4f)\,^3F$ at 4470 Å are of considerable astrophysical interest (see § 5-6C). Very recently, Burgess and Cairns [144] have measured the profiles of these lines for an electron density $\sim 10^{15}\,cm^{-3}$ and a temperature of $10^4\,°K$ and find, that although the calculations of Griem [145] and Barnard et al. [146] agree quite well with their results for the allowed component and in the far wings of the forbidden line, there is a discrepancy in the intensity near the centre of the forbidden component, which cannot be accounted for by experimental error (see Fig. 5-4-4). This discrepancy may well be due to the use of a quasi-static distribution for the ion field at a temperature and density where significant departures from the quasi-static limit can be expected. Until this difficulty is resolved, the use of the theoretical profiles in calculations to detect NLTE effects in helium in stellar atmospheres, and hence to gain information on surface density and helium abundance, can only be extremely uncertain. Further calculations for the allowed and forbidden lines $2p$–nl ($n \geq 5$, $l = 1, 2, \ldots, n-1$) using similar methods to those of [145], have been carried out by Gieske and Griem [147].

Many semi-classical calculations of half widths using the impact approximation have been made for isolated lines in neutral and ionized complex atoms. Tables of widths for lines broadened by electron impact are available [95] and they contain data for neutral atoms and singly ionized ions in the sequence helium to caesium. Experiments have been carried out on the 1335 Å CII line and the 1085 Å NII line for an electron density $\sim 3.3 \times 10^{17}$ cm^{-3} [148] and Roberts and Eckerle [149] have studied several CII lines in the region 2992–3920 Å and also the resonance lines of CaII. Meyer and Beck [150] have recently measured the shift and width of two lines of neutral silicon, 3905 Å ((4s) ^{1}P–(3p)^2 ^{1}S) and 2881 Å ((4s) ^{1}P–(3p)^2 ^{1}D), produced by both perturbing electrons and neutral hydrogen atoms. The agreement between the experimental and theoretically determined half widths is much better for electron broadening than for the neutral atom broadening; the reason for this will be discussed later. Sahal–Bréchot has calculated half widths for lines of HeI, OI, CsI, CII, NII, MgII, SiII, SiIII, SII, SIII, ArII and CaII [151, 152] broadened by electrons and singly charged ions. Roberts [153] has also used the semi-classical impact approximation, in which he has included the effects of the resonant scattering of electrons by the radiating ions and electron ion quadrupole excitation, for the widths of several lines of NII and ArII. Where it is possible to make a direct comparison with the earlier work of Sahal–Bréchot, the widths are larger by factors of ~ 1.1–1.7, but both results are within the error bars on the experimental points (see [151, 153] for references to the experimental work available). Until quite recently, no fully quantum mechanical calculations using eq. (5-4-44) had been carried out, because of the lack of the necessary electron–atom collision data. However, half widths for the resonance lines of CaII and MgII broadened by electron impact have now been calculated [154–156] using quantum mechanical data for the scattering matrix (see § 5-5). The agreement with the results of semi-classical methods is quite good, but the agreement with experiment [149, 152, 157] is much less satisfactory. More work, both experimental and theoretical, is needed to resolve this discrepancy. The calculated shifts quoted in the work above are unlikely to be very reliable, for they depend largely on the effects of very distant collisions and hence on the polarization of both the levels of the line. It is difficult to incorporate the correct polarization of the upper level of a transition into the electron–atom collision data, but further approaches are currently being considered (see § 5-5).

The problems that arise in the study of atom–atom broadening are somewhat different. Many measurements have been made of the broadening and shift of spectral lines broadened by a "foreign" gas, and from this information attempts have been made to deduce the form of the interatomic potential. A Lennard–Jones potential is assumed, giving a frequency shift

$$\omega - \omega_{ij} = C_{12}/r^{12} - C_6/r^6, \tag{5-4-70}$$

and the width and shift of the line are calculated using the Lindholm–Foley theory with (5-4-70) in place of (5-4-36). The second term in (5-4-70) represents the Van der Waals interaction but the first term is somewhat arbitrary. Recent calculations [158] show that experimental data can be fitted more successfully by introducing an additional term, $-C_8/r^8$ into (5-4-70). It is concluded that calculations of line widths using only the leading, Van der Waals term, in the interaction potential almost always gives incorrect results. The widths are sensitive to the precise form of the interaction at small interatomic distances, since the radiating atom and the perturber must approach quite closely in order to cause an appreciable shift of frequency. This was first mentioned by Anderson [159] as long ago as 1949 in his paper on the impact approximation and its uses. Thus the numerous calculations of line widths, see for instance [160–162] which are based on the Van der Waals interaction only, and which have been used in the interpretation of solar and other infra-red lines, should be treated with considerable caution. In the sun, the broadening of the sodium resonance lines is largely caused by neutral hydrogen and helium atoms. A very recent calculation [163] in which an attempt is made to derive a more realistic form for the short-range potential, has been carried out for the broadening of the sodium lines by helium. Although the results are considerably larger than those obtained using the Van der Waals interaction alone, they are still over a factor of 2 smaller than the widths measured in the latest experiment [164] by Burgess and Grindley. The work of [163] confirms that the short range part of the potential is the dominant factor in the determination of the width.

Finally, it should be realized that the use of formulae for line profiles, which are based on either the impact theory or the quasi-static theory, eventually break down if information on the frequency variation extremely far away from the line centre is required. An example of this recently received quite a lot of attention. It was suggested by Cuny [165] that the resonance broadening of Lyα might appreciably affect the solar opacity at wavelengths over 1000 Å from the line centre. However, for such large frequency shifts to occur, the system of atom+perturber, in this case H+H, is such that the two atoms are very close and behave like a quasi-molecule. No other perturber is important, since the probability for two perturbers to be so close to the atom at the same time is negligible. These quasi-molecular effects have been studied by Sands et al. [166] who pointed out the error in the assumptions and conclusions of [165] and by Mies [167].

§ 5-5. Collisional excitation and ionization

A. The quantum mechanical theory - general formulae and approximations

In this section a summary is given of the formulae used in calculating excitation and ionization cross-sections, together with a brief outline of the existing approximations and results. The main emphasis will be on excitation and ionization by electron impact, but proton impact is also considered. Proton impact can be effective in producing transitions in the highly ionized ions present in the solar corona, and collision data is also required if the broadening of a spectral line by protons is treated using the impact approximation (see eq. (5-4-44), § 5-4C).

No attempt is made to give a complete review of the data available. There are many articles which discuss the theoretical work on the excitation of atoms by electron impact, including those by Seaton [168], Peterkop and Veldre [169], Burke [170] and Moiseiwitsch and Smith [171]. Experimental methods and results have also been presented in [171], and by Fite [172] and Heddle and Keesing [173]. Electron collisions with positive ions have been reviewed by Seaton [174] and two articles by Kieffer and Dunn [175] and Rudge [176] deal with collisional ionization from the experimental and theoretical viewpoints respectively. The most recent review of excitation and ionization by electron impact is that of Bely and Van Regemorter [177] and the Oak Ridge Atomic and Molecular Information Center issue periodically an up-to-date bibliography of atomic and molecular processes.

We consider for simplicity, the case of an atom (or positive ion) with one valence electron outside a closed shell. If the valence electron and the scattered electron have coordinates r_1 and r_2 respectively, the total wave function for the system of atom + electron may be expressed as*

$$\Psi(r_1, r_2) = \sum_i \psi(i|r_1) F(i|r_2) \tag{5-5-1}$$

where $\psi(i|r_1)$ is the eigenfunction of the valence electron in the unperturbed atom with the associated quantum numbers denoted by i. The summation in (5-5-1) is over a complete set of states and thus includes an integration involving the continuum. If spin effects are included, (5-5-1) must be replaced by an explicitly symmetrized function

$$\Psi^\pm(r_1, r_2) = \frac{1}{\sqrt{2}} \sum_i [\psi(i|r_1) F^\pm(i|r_2) \pm \psi(i|r_2) F^\pm(i|r_1)] \tag{5-5-2}$$

* See, however, Chapter 6 for an alternative approach.

where the plus sign denotes a singlet state of the electron–atom system and the minus sign a triplet state. The wave functions $\psi^{\pm}(r_1, r_2)$ must satisfy the Schrödinger equation

$$H(r_1, r_2)\Psi^{\pm}(r_1, r_2) = E\Psi^{\pm}(r_1, r_2) \tag{5-5-3}$$

where $E\,e^2/a_0$ is the energy of the whole system, and is therefore conserved throughout the collision, and $H(r_1, r_2)$ is the Hamiltonian, which is given by

$$H(r_1, r_2) = -\tfrac{1}{2}\nabla_1^2 - \tfrac{1}{2}\nabla_2^2 - \frac{Z}{r_1} - \frac{Z}{r_2} + \frac{1}{r_{12}} + U(r_1) + U(r_2). \tag{5-5-4}$$

In (5-5-4), r_{12} is the interelectronic distance, Z is the net charge on the atomic core and $U(r)$ is a short range potential which arises if the core contains electrons, but is zero if the unperturbed atom is hydrogenic. The functions $F^{\pm}(i|r_2)$ describing the scattered electron must be determined in each case by a numerical solution of the coupled equations

$$\left[\nabla^2 + \frac{2(Z-1)}{r} + k_i^2\right] F^{\pm}(i|r) = 2\sum_j [V_{ij}(r) \pm W_{ij}(r)] F^{\pm}(j|r) \tag{5-5-5}$$

which are derived from (5-5-2) and (5-5-3), together with the equation satisfied by $\psi(i|r)$, that is,

$$[-\tfrac{1}{2}\nabla^2 - Z/r + U(r)]\psi(i|r) = E_i\psi(i|r). \tag{5-5-6}$$

In (5-5-6) $E_i\,e^2/a_0$ is the energy of the state i of the unperturbed atom while in (5-5-5) $\tfrac{1}{2}k_i^2\,e^2/a_0$ is the energy of the scattered electron, given by

$$E_i + \tfrac{1}{2}k_i^2 = E. \tag{5-5-7}$$

The operators $V_{ij}(r)$ and $W_{ij}(r)$ in (5-5-5) are defined by

$$V_{ij}(r_2) = \int \psi^*(i|r_1)\left[-\frac{1}{r_2} + \frac{1}{r_{12}} + U(r_1)\right]\psi(j|r_1)\,dr_1 \tag{5-5-8}$$

and

$$W_{ij}(r_2)F^{\pm}(j|r_2) = \int \psi^*(i|r_1)[H(r_1, r_2) - E]F(j|r_1)\,dr_1\,\psi(j|r_2). \tag{5-5-9}$$

Unless the atom (or positive ion) is hydrogenic no exact solution of (5-5-6) exists, and so approximate solutions, usually Hartree–Fock wave functions, are used, although recently Norcross [42] has used wave functions from the work of Eissner and Nussbaumer [41].

For $Z = 1$, the asymptotic form of the total wave function is

$$\Psi^{\pm}(r_1, r_2) \underset{r_2 \to \infty}{\sim} \frac{1}{\sqrt{2}} \left\{ \psi(i|r_1) \exp(i k_i \cdot r_2) \right.$$
$$\left. + \sum_j \psi(j|r_1) f^{\pm}(\hat{k}_j, \hat{k}_i) \frac{\exp(i k_j \cdot r_2)}{r_2} \right\} \quad (5\text{-}5\text{-}10)$$

where $f^{\pm}(\hat{k}_j, \hat{k}_i)$ is the scattering amplitude for an atomic transition $i \to j$. If $Z \neq 1$, a similar expansion exists in terms of Coulomb waves. The cross-section for this transition is then given by an expression which depends only on the boundary conditions

$$Q_{ij}(v) = \frac{1}{4\pi \omega_i} \frac{k_j}{k_i} \sum \int \tfrac{1}{4} \{|f^+(\hat{k}_j, \hat{k}_i)|^2 + 3|f^-(\hat{k}_j, \hat{k}_i)|^2\} \, d\hat{k}_i \, d\hat{k}_j \quad (5\text{-}5\text{-}11)$$

where we have integrated over all angles of scattering and averaged over all the directions of the incident electron. The summation is over all degenerate sublevels of the states i and j and ω_i is the statistical weight of level i. Equation (5-5-11) is the correct expression for a random orientation of the spins of the target atom and the incident electron, whose velocity v is related to k_i by

$$k_i = m a_0 v / \hbar$$

where m is the electron mass.

The solution of (5-5-5) involves an infinite number of coupled integro-differential equations and so in practice in the close coupling method only the first few terms in (5-5-2) are considered. Most work has been carried out on the electron–hydrogen problem, in which states nl up to and including $n = 3$ have been retained in (5-5-2), see [178], but unfortunately this expansion does not converge very rapidly. A method has been developed in which "pseudo-states" are added to the terms retained in (5-5-2) to replace all the neglected terms which involve the higher excited states and the continuum [179–181]. Seaton [182] is investigating a more economical method of solving large systems of coupled equations, using a matrix inversion technique.

An expansion (5-5-1) for the total wave function may also be made for the atom+proton problem. Although the wave function is not complicated by the inclusion of the symmetry properties in (5-5-2), the expansion is very slowly convergent. However, Cheshire et al. [183] have shown that good results for excitation of the 2s, 2p states in hydrogen may be obtained by retaining the 1s, 2s and 2p states in (5-5-1) and then adding pseudo-states which simulate the molecular properties of the system at small separations.

If a partial wave analysis is carried out for the scattered electron, the scattering amplitude may be expressed in terms of normalized spherical harmonics $Y_{lm}(\hat{k})$ as

$$f^{\pm}(\hat{k}_j, \hat{k}_i) = \frac{2\pi i}{(k_i k_j)^{\frac{1}{2}}} \sum_{lml'm'} i^{l-l'} Y_{lm}^*(\hat{k}_i) Y_{l'm'}(\hat{k}_j) T^{\pm}(jl'm', ilm) \qquad (5\text{-}5\text{-}12)$$

where the scattered electron has quantum numbers lm and $l'm'$ before and after the collision. The quantities $T^{\pm}(jl'm', ilm)$ are functions of the total energy Ee^2/a_0 and depend on the quantum numbers of the atom and the scattered electron before and after collision. The matrix of the quantities $T^{\pm}$ for all possible states ilm and $jl'm'$ is the transition matrix T and the scattering matrix S is defined by

$$S = 1 - T \qquad (5\text{-}5\text{-}13)$$

where I denotes the unit matrix. T can be partitioned, so that

$$T = \begin{pmatrix} T^+ & 0 \\ 0 & T^- \end{pmatrix} \qquad (5\text{-}5\text{-}14)$$

where 0 is the null matrix and S is unitary and symmetric, that is

$$S^{\dagger} S = 1; \quad S = S'. \qquad (5\text{-}5\text{-}15)$$

The submatrices $S^{\pm}$ also satisfy separately the equations (5-5-15).

The elements of the scattering matrix which appear in eq. (5-4-43) for the width and shift of a spectral line due to an elastic perturber–atom collision are given by

$$S_{ii}(l, v) = \frac{1}{\omega_i} \sum_{mm'} S(ilm', ilm) \qquad (5\text{-}5\text{-}16)$$

together with a similar expression for $S_{jj}(l, v)$. It should be noted that in the derivation of (5-4-43), which is the strict quantum mechanical equivalent of (5-4-41), any possible change in the angular momentum of the scattered electron caused by the collision has been neglected. In a more exact derivation for an allowed line, the series which is summed over l in (5-4-43) should be replaced by

$$\tfrac{1}{4} \sum_{l'L_i^T L_j^T} (-1)^{l+l'}(2L_i^T+1)(2L_j^T+1) \begin{Bmatrix} L_i & L_i^T & l \\ L_j^T & L_j & 1 \end{Bmatrix} \begin{Bmatrix} L_i & L_i^T & l' \\ L_j^T & L_j & 1 \end{Bmatrix} (X^+ + 3X^-) \qquad (5\text{-}5\text{-}17)$$

where $\begin{Bmatrix} a & b & c \\ d & e & f \end{Bmatrix}$ indicates a 6-j symbol, and

$$X^\pm = \int_0^\infty \frac{f(v)}{v} dv \{\delta_{ll'} - S_{ii}^\pm(L_i l' L_i^T, L_i l L_i^T) S_{jj}^{\pm*}(L_j l' L_j^T, L_j l L_j^T)\}, \quad (5\text{-}5\text{-}18)$$

$f(v) dv$ being the Maxwellian velocity distribution (5-4-42). In (5-5-17) and (5-5-18) the scattered electron has angular momenta l and l' before and after the collision, and the total angular momenta L_i^T and L_j^T are conserved during a collision with the atom in its angular momentum states L_i and L_j respectively. The double suffices on the scattering matrix elements in (5-5-18) have been retained in order to emphasize the fact that S_{ii} and S_{jj} are elements of the scattering matrix evaluated at different total energies. A completely general formula for the L–S coupling case is given in [154]. If, alternatively, we express the line half width w in terms of excitation cross-sections through eq. (5-4-44), then the cross-section for a transition $i \to j$ may be written in terms of the transition matrices (5-5-14) as

$$Q_{ij}(v) = \frac{\pi a_0^2}{k_i^2} \frac{1}{4(2L_i+1)} \sum_{ll'L} (2L+1)[|T^+(L_j l'L, L_i lL)|^2 + 3|T^-(L_j l'L, L_i lL)|^2]. \quad (5\text{-}5\text{-}19)$$

Returning to equation (5-5-10) the scattering amplitude is given by

$$f^\pm(\hat{\mathbf{k}}_j, \hat{\mathbf{k}}_i) = -\frac{1}{2\pi} \int \psi^*(j|\mathbf{r}_1) \exp(-i\mathbf{k}_j \cdot \mathbf{r}_2) \left\{ -\frac{1}{r_2} + \frac{1}{r_{12}} \right\}$$
$$\times \Psi^\pm(\mathbf{r}_1, \mathbf{r}_2) d\mathbf{r}_1 d\mathbf{r}_2, \quad (5\text{-}5\text{-}20)$$

and if $\Psi^\pm(\mathbf{r}_1, \mathbf{r}_2)$ is replaced by $\psi(i|\mathbf{r}_1) \exp(i\mathbf{k}_i \cdot \mathbf{r}_2)$ we obtain the Born approximation which is valid at high energies. For scattering by positive ions the plane waves are replaced by Coulomb waves with charge $z = Z-1$. In the high energy limit the Born approximation to the cross-section for an allowed transition tends to the Bethe formula, that is,

$$Q_{ij}(v) = \frac{4\pi^2 a_0^2}{k_i^2 \sqrt{3}} \frac{f_{ij}}{(E_j - E_i)} g(k_i, k_j) \quad (5\text{-}5\text{-}21)$$

where f_{ij} is the oscillator strength defined in (5-5-1). Equation (5-5-21) may be obtained directly from (5-5-11) and (5-5-20) in which the Born approximation to $\Psi^\pm(\mathbf{r}_1, \mathbf{r}_2)$ has been made. If we introduce the relative momentum vector $\mathbf{K} = \mathbf{k}_i - \mathbf{k}_j$, in the limit of high energies only small values of K effectively contribute to the cross section. Thus the factor $\exp(i\mathbf{K} \cdot \mathbf{r})$ in (5-5-20) may be replaced by $(1+i\mathbf{K} \cdot \mathbf{r})$ for small values of K and the integrals over $\hat{\mathbf{k}}_i$ and $\hat{\mathbf{k}}_j$ in (5-5-11) transformed into an integral over $d\mathbf{K}$. Hence the function $g(k_i, k_j)$ in (5-5-21), which is a Gaunt factor analogous to that defined in (5-3-23) for hydrogen, is given by

$$g(k_i, k_j) = \frac{\sqrt{3}}{\pi} \ln \left(\frac{K_0}{|k_i - k_j|} \right), \qquad (5\text{-}5\text{-}22)$$

where the integral over K has been cut off at an upper value, K_0. The introduction of K_0 is necessary since the expansion of the integrand in powers of K is not valid over the whole range of values of K up to $(k_i + k_j)$. There is no simple way of determining K_0, which is a constant for any given transition. This can only be done by making a comparison between the Bethe result and the full Born approximation at high values of the incident energy. Finally, equations (5-5-21) and (5-5-22) lead to the result

$$Q_{ij}(v) = \frac{A}{k_i^2} \ln(k_i^2) + \frac{B}{k_i^2} \qquad (5\text{-}5\text{-}23)$$

where A and B are constants for a given transition and A is independent of the value of K_0. Van Regemorter [177] has derived values of an effective Gaunt factor for optically allowed transitions from a comparison of theoretical and experimental collision data. This approximation, which has been widely used by astrophysicists, should be treated with caution as it can only be expected to give results correct to about a factor of two, and can be in error by as much as an order of magnitude.

The scattering amplitude for ionization is given by (5-5-20), but with the wave function $\psi(j|r_1)$ for the state j replaced by a continuum function. If $Q_{ic}(E_c, v)$ is the differential cross-section for a transition to a continuum level c, with energy in the range $(E_c, E_c + dE_c)$ atomic units, then the total cross-section is

$$Q_{ic}(v) = \int_0^{\frac{1}{2}E} Q_{ic}(E_c, v) \, dE_c. \qquad (5\text{-}5\text{-}24)$$

In the high energy limit

$$Q_{ic}(v) = \frac{1}{k_i^2 \alpha \sqrt{3}} \int_0^{\frac{1}{2}E} \frac{a_v(i) g(k_i, k_c)}{(I + E_c)} \, dE_c$$

where $\qquad (5\text{-}5\text{-}25)$

$$\frac{a_0 h v}{e^2} = I + E_c = \tfrac{1}{2}(k_i^2 - k_c^2)$$

and $a_v(i)$ is the photoionization cross-section defined by (5-3-14). The ionization potential of the atom is $e^2 I/a_0$ and in the high energy limit $Q_{ic}(v)$ also has the form (5-5-23). For the collisional excitation of a forbidden transition, $Q_{ic}(v)$ behaves as

$$Q_{ic}(v) = A'/k_i^2 \qquad (5\text{-}5\text{-}26)$$

in the limit of high energies, where A' is a constant. The excitation and ionization cross-sections satisfy

$$n_j^3 \sum_{l_j=0}^{n_j-1} Q_{ij}(v) \to Q_{ic}(E_c, v) \qquad (5\text{-}5\text{-}27)$$

as the bound states $n_j l_j$ go over to continuum states $E_c l_j$. The relation (5-5-27) is analogous to the expression (5-4-9) for radiative processes and, in the limit of high incident energies, they are essentially equivalent.

Some recent work [184] on the ionization of Li$^+$ and Mg$^+$, shows that the limiting form of the cross-section, as given by (5-5-23) is not valid until $k_i^2 \gtrsim 40I$, a much higher value than previously thought. Many calculations have been carried out for excitation and ionization using the Born approximation, see Tables 1 and 2 in reference [177], and calculations have now been made for ionization by electron and proton impact of atoms and positive ions with outer electrons $(2p)^q$ and $(3p)^q$ ($q = 1, 2, \ldots, 6$), [185, 186, 40]. The work of [40] indicates that the Born cross-section for the ejection of a 2p electron tends to the form (5-5-23) at about $k_i^2 \approx 10I$. At higher energies the contribution to the cross-section from the ejection of a 2s or 3s electron becomes important, and since it has recently been discovered by the author that the results given in [185] should all be divided by a factor of approximately two [187], the agreement with experiment for neon and argon is now very good. The scaled cross-sections, which should not vary greatly along an iso-electronic sequence,

$$\bar{Q}_{ic}(v) = \frac{(2I)^2}{q} Q_{ic}(v) \qquad (5\text{-}5\text{-}28)$$

agree well with those obtained in [40], when this error is taken into account.

Recently, cross-sections and collision rates, C_{ij} and C_{ic} have been evaluated for electrons incident on hydrogen (cf. (5-2-49) and (5-2-53)) using semi-empirical methods [188] and making use of the continuity rule (5-5-27) as a check for consistency between different sets of collision data. These calculations are for use in NLTE problems which involve the solution

B. Classical and Semi-Classical Approximations

The article by Burgess and Percival [189] gives a comprehensive review of the classical theory of atomic scattering. In this section we concentrate on the methods which can give, relatively simply, estimates of a wide variety of collisional excitation cross-sections.

If the motion of the colliding particle is treated classically, so that its orbit can be specified in terms of an impact parameter ρ, the cross section for the $i \rightarrow j$ transition is given by

$$Q_{ij}(v) = 2\pi \int_0^\infty P_{ij}(\rho)\rho\,d\rho \tag{5-5-29}$$

where $P_{ij}(\rho)$ is the probability that the transition $i \rightarrow j$ occurs. Although the possibility that the incident particle gains energy during the collision is taken into account, we neglect the effects on the orbit of the scattered particle of a corresponding loss of energy. Thus if the target system is a neutral atom, the orbit is assumed to be a straight line, but if it is a positive ion, the orbit is hyperbolic; it has been shown in the context of line broadening calculations [116] that the assumption of a straight line path in the case of a charged target can considerably underestimate the excitation cross-section. If we use first-order time-dependent perturbation theory, we obtain

$$P_{ij}(\rho) = \frac{1}{\hbar^2\omega_i}\sum \left| \int_{-\infty}^{+\infty} V_{ij}(t) \exp(i\omega_{ij}t)\,dt \right|^2 \tag{5-5-30}$$

where $\hbar\omega_{ij}$ is the transition energy, the summation is over all the degenerate sublevels of states i and j, and ω_i is the statistical weight of level i. The quantity $V_{ij}(t)$ is given by

$$V_{ij}(t) = e^2 \int \psi^*(j|r_1) \left[-\frac{1}{r_2(t)} + \frac{1}{|r_2(t)-r_1|} \right] \psi(i|r_1)\,dr_1 \tag{5-5-31}$$

where the notation is that of the previous section. The corresponding expression for the element of the scattering matrix $S_{ii}(\rho, v)$ which occurs in (5-4-41) is

$$S_{ii}(\rho, v) = \frac{1}{\omega_i}\sum(1 - T_{ii}) \tag{5-5-32}$$

where the summation is over all the degenerate states of i and j, and T_{ii} is given by

$$T_{ii} = \frac{i}{\hbar}\int_{-\infty}^{+\infty} V_{ii}(t)\,dt + \frac{1}{\hbar^2}\sum_j \int_{-\infty}^{+\infty} V_{ij}(t)\exp(i\omega_{ij}t)\,dt$$

$$\times \int_{-\infty}^{t} V_{ji}(t')\exp(-i\omega_{ij}t')\,dt'. \qquad (5\text{-}5\text{-}33)$$

The summation in (5-5-33) is over a complete set of states j of the unperturbed atom and higher order terms in $V_{ij}(t)$ have been neglected. From eqs. (5-5-30), (5-5-32) and (5-5-33), we obtain

$$\sum_j P_{ij}(\rho) = 2\,\text{Re}\,(1 - S_{ii}(\rho, v)). \qquad (5\text{-}5\text{-}34)$$

If the transition $i \to j$ is allowed ($\Delta l = 1$), (5-5-31) can be approximated by

$$V_{ij}(t) = e^2 \int \psi^*(j|r_1) \frac{r_1 \cdot r_2(t)}{r_2^3(t)} \psi(i|r_1)\,dr_1 \qquad (5\text{-}5\text{-}35)$$

where we have assumed $r_2(t) > r_1$ at all points of the orbit. For the case of a neutral atom, Seaton [190] has evaluated (5-5-30) using expression (5-5-35) for $V_{ij}(t)$ and he obtains

$$P_{ij}(\rho) = \frac{2e^4}{m\hbar^2 \rho^2 v^2 \omega_{ij}} f_{ij}\beta^2 [K_0^2(\beta) + K_1^2(\beta)] \qquad (5\text{-}5\text{-}36)$$

where m is the electron mass. If proton impact is considered, $P_{ij}(\rho)$ must be multiplied by m/M where M is the proton mass. The quantity β is dimensionless and is given by

$$\beta = \omega_{ij}\rho/v \qquad (5\text{-}5\text{-}37)$$

and the $K_n(\beta)$ are modified Bessel (or Bassett) functions. For small values of ρ the use of (5-5-35) becomes invalid and a cut-off factor must be introduced in evaluating the cross-section (5-5-29). We follow the method of [190] and evaluate $Q_{ij}(v)$ in two different ways; either

$$Q_{ij}(v) = 2\pi \int_{\rho_0}^{\infty} P_{ij}(\rho)\rho\,d\rho \qquad (5\text{-}5\text{-}38)$$

or

$$Q_{ij}(v) = \tfrac{1}{2}\pi\rho_1^2 + 2\pi \int_{\rho_1}^{\infty} P_{ij}(\rho)\rho\,d\rho. \qquad (5\text{-}5\text{-}39)$$

The best estimate of the cross-section for a given incident velocity v is then taken to be whichever is the smaller of (5-5-38) and (5-5-39). The cut-off impact parameter ρ_0 is the smaller of the mean radii of states i and j, and ρ_1 is defined by

$$P_{ij}(\rho_1) = \tfrac{1}{2}. \tag{5-5-40}$$

In the limit of high incident energies, ρ_0 corresponds directly to the cut off in relative momentum K_0 introduced in equation (5-5-22) (see Seaton [190]).

An explicit expression for $Q_{ij}(v)$ in (5-5-38) is given by

$$Q_{ij}(v) = \frac{4\pi a_0^2}{k_i^2(E_j - E_i)} f_{ij}[\beta_0 K_0(\beta_0) K_1(\beta_0)]; \quad E_j - E_i = \frac{a_0 \hbar}{e^2} \omega_{ij} \tag{5-5-41}$$

and this should be compared with the Bethe formula (5-5-21). In (5-5-41), β_0 corresponds to the cut-off parameter ρ_0.

A similar treatment is possible for the excitation of positive ions by electron impact and has been carried out by Burgess [191] based on an analysis for nuclear problems by Alder et al. [192]. In this case, the probability for an allowed transition $i \to j$ in a positive ion, on which there is a net charge z, is

$$P_{ij}(\rho) = \frac{2e^4}{m\hbar \rho^2 v^2 \omega_{ij}} f_{ij} e^{\pi\xi} \beta^2 \left[[K'_{i\xi}(\beta)]^2 + \frac{\varepsilon^2 - 1}{\varepsilon^2} [K_{i\xi}(\beta)]^2 \right] \tag{5-5-42}$$

where here

$$\xi = \frac{ze^2}{mv^3} \omega_{ij} \tag{5-5-43}$$

and ε denotes the eccentricity of the hyperbolic orbit. The prime on $K'_{i\xi}(\beta)$ in (5-5-42) denotes the derivative with respect to β, and β itself is now given by

$$\beta = \xi\varepsilon. \tag{5-5-44}$$

The modified Bessel functions $K_{i\xi}(\beta)$, which depend on the imaginary parameter $i\xi$ are dealt with fully in [193]. The cut-off procedure is similar to that already described, except in this case the distance of closest approach R_c is not equal to ρ, but is given by

$$R_c = \frac{ze^2}{mv^2}(\varepsilon - 1); \quad \rho = R_c\left(R_c + \frac{2ze^2}{mv^2}\right) \tag{5-5-45}$$

and so β_0 is defined in terms of an eccentricity ε_0 which is obtained from

(5-5-45) with R_c equal to the smaller of the mean radii of states i and j. Thus the cross-section derived from (5-5-38) is

$$Q_{ij}(v) = \frac{4\pi a_0^2}{k_i^2(E_j - E_i)} f_{ij} e^{\pi\xi} [\beta_0 K'_{i\xi}(\beta_0) K_{i\xi}(\beta_0)]. \tag{5-5-46}$$

If, as is the case in hydrogenic ions, allowed transitions can take place in which there is no energy change, the relation (5-5-41) reduces to

$$Q_{ij}(v) = \frac{4\pi a_0^2}{k_i^2(E_j - E_i)} f_{ij} \ln\left(\frac{\rho_{\max}}{\rho_0}\right) \tag{5-5-47}$$

where now the integral for $Q_{ij}(v)$ also diverges for large ρ. The upper cut-off parameter, $\rho_{\max}$, may be chosen to be the Debye shielding distance, if the collision occurs in a reasonably dense plasma, or it can be chosen to allow for the decay of the states i and j by spontaneous emission if the density is low; this point is discussed in [194]. A similar relation holds for positive ions, and in the limit $\omega_{ij} \to 0$, (5-5-46) reduces to

$$Q_{ij}(v) = \frac{4\pi a_0^2}{k_i^2(E_j - E_i)} f_{ij} \ln\left(\frac{\varepsilon_{\max}}{\varepsilon_0}\right) \tag{5-5-48}$$

where $\varepsilon_{\max}$ can be chosen in a similar way to $\rho_{\max}$ for the neutral case, that is, the various choices of cut-off are tried for the distance of closest approach, R_c.

The methods outlined above are most reliable when the oscillator strength for the transition is large. Stauffer and McDowell [195] have used the impact parameter method for the electron excitation of electric quadrupole transitions in neutral atoms and find that it gives reasonable results for transitions in which $\Delta l = 2$, but not for those with $\Delta l = 0$. Moiseiwitsch [196] has shown that, for any transition $i \to j$ produced by heavy-particle impact, the wave and impact parameter treatments of the problem are equivalent at high incident energies.

If collision cross-sections are needed in which the levels i and j are both highly excited, use may be made of the correspondence principle of Bohr. Percival and Richards [197] have used this to evaluate cross sections for transitions $n \to n'$ in hydrogenic ions where

$$(n' - n)/n \ll 1. \tag{5-5-49}$$

If

$$k_i^2 \gg Z^2/n^2 \tag{5-5-50}$$

where $Z = (z+1)$ is the nuclear charge of the hydrogenic ion and M is the mass of the incident proton or electron ($M = m$ for electron impact), the incident particle may be supposed to move along a straight line path in all cases. They obtain the result (5-5-41) in which β_0 is determined from

$$\rho_0 = n^2/Z \tag{5-5-51}$$

and the oscillator strengths f_{ij} for these transitions are given by the asymptotic formulae and tables of [93]. This agrees with the results of Saraph [198] for high values of n and is valid for $k_i^2/Z^2 \gtrsim 1$. These results have now been extended [199] to include the region where

$$1/n < k_i^2/Z^2 < 1 \tag{5-5-52}$$

and they obtain for this range of energy

$$Q_{ij}(v) = \frac{2A}{k_i^2} \ln(k_i^2) + \frac{B'}{k_i^2} \tag{5-5-53}$$

where A and B' are constants for a given transition. Eq. (5-5-53) must be compared with the result for $k_i^2/Z^2 > 1$, which can be written in the form

$$Q_{ij}(v) = \frac{A}{k_i^2} \ln(k_i^2) + \frac{B}{k_i^2} \tag{5-5-54}$$

where the same constant A appears in both (5-5-53) and (5-5-54). The correspondence principle has also been used recently to evaluate cross-sections for collisions between hydrogen atoms and heavy particles [199].

Finally, for calculations of the broadening of lines emitted by hydrogenic ions, it can be useful (see [136]) to use wave functions for the hydrogenic system which depend on parabolic coordinates, and which maintain the separability of the Schrödinger equation for an atom in an electric field. Hughes [200] has given the general formula for the transformation from this Stark representation to the usual representation in spherical polar coordinates and quantum numbers nlm. In the Stark representation the angular momentum l is replaced by another quantum number k such that

$$n = k + k' + |m| + 1 \tag{5-5-55}$$

where k and k' are integers greater than or equal to zero. Then if $\phi(nkm)$ represents a wave function in this representation, Hughes* shows that

$$\phi(nkm) = \sum_{l=|m|}^{n-1} (-1)^K (2l+1)^{\frac{1}{2}} \begin{pmatrix} N & N & l \\ M_1 & M_2 & -m \end{pmatrix} \psi(nlm) \tag{5-5-56}$$

* The phase factor in (5-5-56) is the corrected one given by Pfennig, see [120].

where $\psi(nlm)$ is in the usual spherical polar wave function for hydrogen and $\begin{pmatrix} a & b & c \\ d & e & f \end{pmatrix}$ is a 3-j symbol. The quantities K, N, M_1 and M_2 in (5-5-56) are defined by

$$K = \tfrac{1}{2}(2k'+|m|+m)+1; \quad N = \tfrac{1}{2}(n-1) \tag{5-5-57}$$

and

$$M_1 = \tfrac{1}{2}(m+k'-k); \quad M_2 = \tfrac{1}{2}(m+k-k'). \tag{5-5-58}$$

§ 5-6. Interpretation of observations

A. Curve of Growth Analysis

In order to establish the effective temperature and gravity of a star and the relative abundances of its atmospheric constituents, it is necessary to construct reasonably accurate physical models which reproduce the observed spectrum of the star. The simplest method for obtaining the relative abundances of the elements in a stellar atmosphere is a curve of growth analysis in which one has a precise relationship between equivalent width and the number of atoms which produce the line. A simple model of the atmosphere and a mechanism of line formation is assumed in order to calculate the ratio r_ν of line to continuum emission; the suitability of any particular approach will vary with the type of line studied. We consider here the Schuster–Schwarzschild model in which the rate of change of intensity in the line of sight is due only to absorption in the line, all emission processes in the region of formation of the line being neglected. In this case we can write

$$dI_\nu = -l_\nu \rho I_\nu \, dr \tag{5-6-1}$$

which is a special case of equation (5-2-5). If the optical depth in the line of sight is t_ν [cf. (5-2-36)], (5-6-1) may be integrated to give

$$I_\nu = I_c e^{-t_\nu} \tag{5-6-2}$$

where $I_\nu \equiv I_\nu(0, \mu)$, and $I_c \equiv I_c(0, \mu)$ is the emergent intensity in the continuum. The equivalent width of the line is then given by (5-2-38) or (5-2-39) where

$$r_\nu = e^{-t_\nu}. \tag{5-6-3}$$

Consider now the line absorption caused by a transition $i \to j$, where $N_i(\text{cm}^{-3})$ is the number density of atoms in level i and f_{ij} is the oscillator strength (cf. § 5-4). Then the optical depth may be written

$$t_\nu = \int l_\nu \rho \, dr = \frac{\pi e^2}{mc} \mathcal{N}_i f_{ij} \phi(\nu) \tag{5-6-4}$$

where

$$\mathcal{N}_i = \int N_i \, dr \quad (\text{cm}^{-2}) \tag{5-6-5}$$

is the number of atoms in level i in a column of unit cross-sectional area in the line of sight. The depth of such a column is determined by the thickness of the layers which are effective in producing the line. The profile of the line is given by $\phi(\nu)$, where

$$\int_{-\infty}^{+\infty} \phi(\nu) \, d(\Delta \nu) = 1 \tag{5-6-6}$$

and is discussed in detail in § 5-4. For weak lines, $t_\nu \ll 1$ and (5-2-39) gives

$$W_\nu \approx \int t_\nu \, d\nu = \frac{\pi e^2}{mc} \mathcal{N}_i f_{ij} \tag{5-6-7}$$

that is, the equivalent width varies linearly with $\mathcal{N}_i f_{ij}$. For small values of $\mathcal{N}_i$, $\phi(\nu)$ is determined completely by Doppler broadening [see (5-4-17)], but as $\mathcal{N}_i$ increases, collisional and radiative effects become important, so that $\phi(\nu)$ becomes a Voigt profile (5-4-22) and finally tends to a Lorentz profile (5-4-11). For the very strong lines we can write

$$W_\nu = \int_{-\infty}^{+\infty} [1 - \exp\{-\alpha/((2\pi \Delta \nu)^2 + (\tfrac{1}{2}\gamma)^2)\}] \, d(\Delta \nu) \tag{5-6-8}$$

which may be approximated by

$$W_\nu \approx \frac{\sqrt{\alpha}}{2\pi} \int_0^\infty y^{-\frac{3}{2}}(1 - e^{-y}) \, dy \tag{5-6-9}$$

where a Lorentz profile has been used for $\phi(\nu)$ in (5-6-4) and

$$\alpha = \frac{\pi e^2}{mc} \mathcal{N}_i f_{ij} \gamma. \tag{5-6-10}$$

The equivalent width is determined here by the wings of the line and hence the factor $(\tfrac{1}{2}\gamma)^2$ in eq. (5-6-8) is neglected in (5-6-9). Equations (5-6-9) and (5-6-10) show that for strong lines $W_\nu \propto (\mathcal{N}_i f_{ij} \gamma)^{\frac{1}{2}}$. Thus if the observed equivalent widths of lines, arising from the same level i of a particular species, are tabulated against the corresponding relative f-values, an empirical curve is derived. A comparison with the theoretical curve of growth obtained from simple models then determines $\mathcal{N}_i$.

An alternative approach is a purely differential one (see e.g. Aller [1]), resulting from comparison of the curve of growth for the star with that for the sun or another standard star. A recent review of abundance determinations from stellar spectra has been presented by Cayrel and Cayrel de Strobel [201].

B. THE SOLAR ATMOSPHERE

The first solar curve of growth was constructed by Minnaert and Mulders [202], and more detailed approaches have been given by Menzel [203] and Unsöld [204]. The most elaborate calculations for theoretical curves of growth are given by Wrubel [205], and one particularly important application of these results was the determination of the solar metal abundances by Goldberg, Muller and Aller [206], using the model atmosphere of Aller and Pierce [207]. This work gave reasonable abundances for most of the chemical elements observed, with the exception of iron, for which the (essentially NLTE) coronal values obtained later [208, 209], were significantly higher. In order to determine how much of this discrepancy was due to the simplifying assumptions of the curve of growth analysis, a detailed LTE investigation of the solar photosphere was subsequently carried out by Lambert and Warner [210–212], using a model atmosphere whose temperature distribution was derived semi-empirically from a study of the solar continuous spectrum between 5000 Å and 13μ. The oscillator strengths for all the transitions involved were normalized wherever possible to experimental results (e.g. Richter [213, 214] for carbon and nitrogen; Solarski and Wiese [215] for oxygen), or to the results of the Coulomb approximation obtained from the work of Bates and Damgaard [94]. The iron group oscillator strengths were taken for the neutrals from extensive compilations [216, 217] of both experimental and theoretical results, and for the once ionized species from the results of Warner [218]. The effects of pressure broadening of the spectral lines were taken into account [see § 5-4C], and a uniform microturbulent velocity of 1.8 km sec^{-1} was assumed. Lambert and Warner obtained results which showed good internal consistency for the metal abundances derived from the equivalent widths and wings of lines of different multiplets, and which did not differ substantially from those obtained in [206]. In particular, the abundance of iron derived from the analysis of both the FeI and FeII lines was about a factor of ten smaller than the coronal value, and it seemed unlikely [7] that this could be purely a result of NLTE effects in the photosphere.

This discrepancy appears now to be resolved due to recent new determinations of the oscillator strengths for visual multiplets of FeI and FeII by the Kiel group [219–221], who point out that earlier experimental work by King and King [222] contains a systematic error dependent on the excitation temperature of the species under consideration. In particular, the f-values derived for FeI are significantly lower than the results tabulated by Corliss and Bozmann, and are in good agreement with preliminary values obtained by Bridges and Wiese [223]. A re-analysis of the solar iron spectrum [220, 221], using the empirical model of Holweger [224] gives a consistent picture. On the scale $\log N(H) = 12.00$, values of $\log N(Fe) = 7.60$ and 7.63 were derived from the FeI and FeII lines respectively, in excellent agreement with the coronal result of about 7.6. Preliminary values of $\log N(Mg) = 7.5$ and $\log N(Si) = 7.6$ have been reported by Unsöld [225]; these do not differ substantially from earlier photospheric results. Thus the assumption of LTE appears to be adequate for the determination of abundances in solar type stars from lines in the visual region of the spectrum. Table 5-6-1 lists the currently accepted abundances of the major constituents of the solar photosphere; the helium value is not derived directly, since helium shows no detectable lines in the visual, but is the result obtained from early-type stars (see § 5-6C).

TABLE 5-6-1

Relative abundances of the major constituents of the solar photosphere [206, 210, 211, 212, 220, 221, 225]

Element	$\log[N(El)]$	Element	$\log[N(El)]$
H	12.00	Mg	7.50
He	11.00	Al	6.40
C	8.55	Si	7.55
N	7.93	S	7.21
O	8.77	Ca	6.15
Na	6.18	Fe	7.60

The behaviour of the solar photosphere at wavelengths less than about 3000 Å is less easily understood. Rocket and satellite data are available (e.g. [226, 227]), but many of the absorption lines are strong resonance transitions, and are formed high in the photosphere where the assumption of LTE is extremely unreliable. Semi-empirical models [224, 228] indicate a temperature rise in the high photosphere, from a minimum of about 4200 °K at optical depth of about 10^{-4}. New models, including the region of this temperature minimum, have been reported (see e.g. the Harvard Smithsonian Reference Atmosphere presented by Gingerich et al. [229] who incorporate bound–free absorption by MgI, SiI, AlI and, in particular, FeI, in the ultraviolet and consider the effect of CO formation). Also, a detailed study of the behaviour of the Lyman continuum across the limb has been deduced from NLTE models by Kalkofen [230]. The present theoretical results, however, fail to predict the observed ultraviolet continuum; this may not only be because of NLTE effects, but also because the observations refer to a "quasi-continuum", heavily blanketed by metallic lines, while the predictions refer to a real continuum. Figures 5-6-1 and 5-6-2 illustrate the behaviour of the continuous opacity in the visual in solar-type models [231, 35] which is dominated by H^-, while Fig. 5-6-3 gives a comparison of the ultraviolet

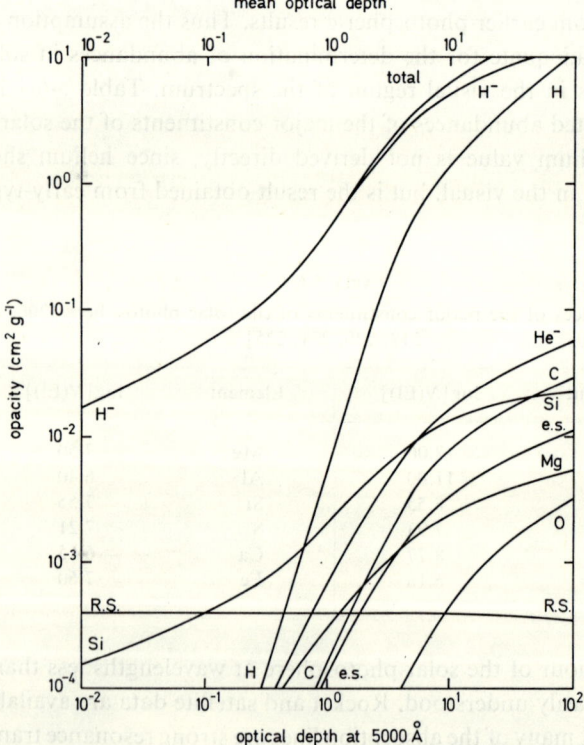

Fig. 5-6-1. Schematic behaviour of the opacity as a function of optical depth in a model atmosphere with $T_e = 6000$ °K, $\log g = 4$ and solar [206] abundances, [231].

intensity from a theoretical solar model [232] with the observations of Bonnet [226] and Parkinson and Reeves [227].

Above the temperature minimum in the solar atmosphere there is a dilute but extended medium; the chromosphere, in which the temperature rises from about 4000 °K to 10^6 °K in less than 5000 km, and the corona, which exhibits a far ultraviolet and X-ray spectrum of emission lines from highly ionized species. Almost all our current knowledge of the low chromosphere comes from interpretation of the CaII H and K lines (3968 Å, 3933 Å)

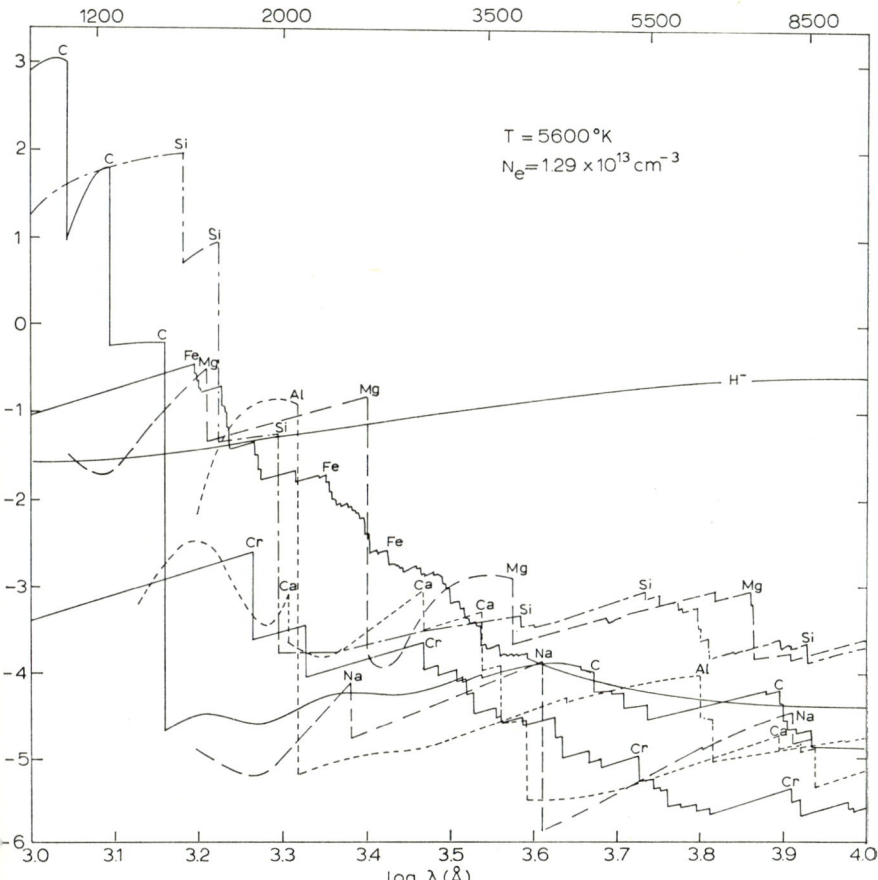

Fig. 5-6-2. Logarithms of the mass absorption coefficients for the individual metals in solar [206] abundances, [35]. The values of the temperature and electron density correspond approximately to optical depth 0.5 in a solar model. [L. D. Travis and S. Matsushima, Astrophys. J. **154** (1968) 689. Reprinted by permission of University of Chicago Press; © 1968 by the University of Chicago.]

and Hα (6563 Å). In this region of the atmosphere the radiative equilibrium condition is not valid; although a small portion of the temperature rise may be explained in purely radiative terms [233], the major part is due to mechanical heating, and so chromospheric models must incorporate such energy source terms. The most recent detailed investigation of the H and K lines has been presented by Linsky [234], who has developed a method for determining the line source functions S_ν arising when the lines are coupled together in a four level representation of the calcium ion, and when the ionization equilibrium is governed by the solar ultraviolet radiation field. The levels specifically included were the ground term $4\,^2S_{\frac{1}{2}}$, the upper levels $4\,^2P_{\frac{1}{2},\frac{3}{2}}$ of the H and K lines, and the $3\,^2D$ level, since the infrared multiplet $4\,^2P$–$3\,^2D$, which depopulates the $4\,^2P$ states, can radically alter the H and K line source

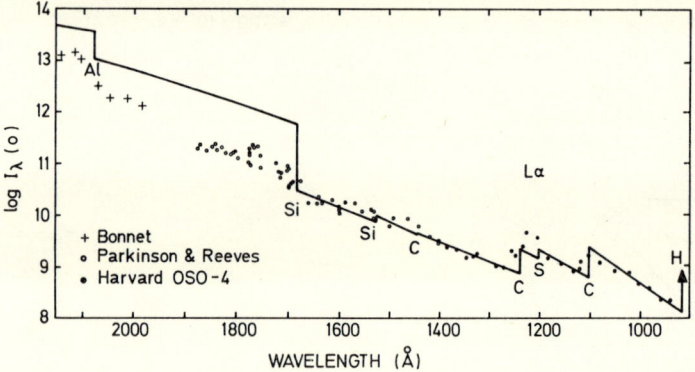

Fig. 5-6-3. Comparison of the observed solar ultraviolet intensity [226, 227] with results from a preliminary version [232] of the Harvard Smithsonian Reference Atmosphere [229].

functions [235]. The statistical equilibrium equations (5-2-66) then give the populations of these levels in terms of the collisional and radiative rates between the levels and the continuum, and were solved simultaneously with the transfer equation. Models for the low chromosphere were then constructed assuming hydrostatic equilibrium, and LTE representation of the metal ionization, and a NLTE formulation for the ionization equilibria of hydrogen and calcium. A detailed consideration of the relative importance of the various collision processes in determining the ionization equilibria is given. The results of the model chromospheres developed for $N(\text{He})/N(\text{H}) = 0.1$ and temperature distributions modified to yield the observed K line profiles predict ratios for the central intensities of the H and K lines in good agreement with recent observational data [234, 236].

The coronal emission spectrum and its interpretation is discussed in Chapter 4. Much of the data on collision cross-sections for the excitation of highly ionized species has been produced by Seaton and his collaborators (see e.g. [20, 237, 238]). Detailed coronal models may only be evaluated if reliable results for the collision rates occurring in the statistical equilibrium equations are available.

C. THE EARLY-TYPE STARS AND PLANETARY NEBULAE

Photoionization of neutral hydrogen and coherent scattering of radiation by electrons are the major sources of opacity in A and B stars, while in the O stars absorption by neutral and ionized helium becomes important. The effective temperature and surface gravity of most normal A and B stars can be determined from model atmosphere calculations, by comparison of the predicted slope of the Paschen continuum, the depth of the Balmer discontinuity, and the wings of the hydrogen lines with observations. The continuum in the visual is virtually unaltered by NLTE effects, except in stars

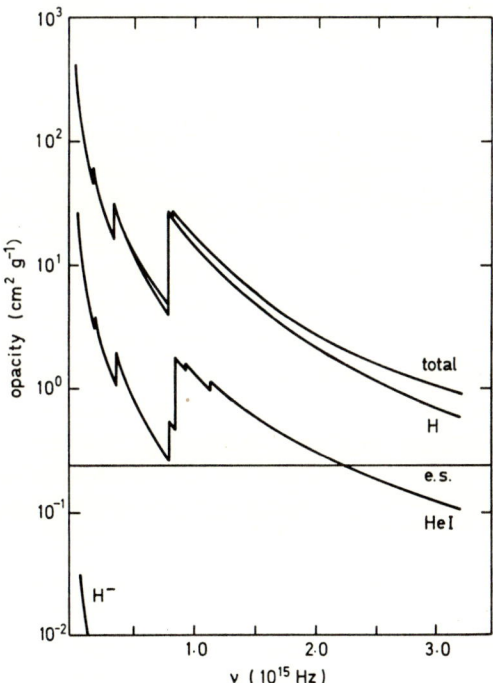

Fig. 5-6-4. Schematic behaviour of the opacity at optical depth unity for a model with $T_e = 15000\ °K$, $\log g = 4$ and solar abundances [243].

of very low surface gravity. Detailed LTE models are available, for example, [239–242], and the fluxes derived from a series of models with effective temperature in the range 10^4 °K to 5×10^4 °K have been tabulated [11]. Figure 5-6-4 illustrates the behaviour of the continuous opacity at optical depth unity for a model with $T_e = 15\,000$ °K and $\log g = 4$, [243].

It is important for the correct prediction of the ultraviolet flux in B stars to include the blanketing effect of absorption lines, see for example [244–246], in order that meaningful comparison with rocket observations [247] may be obtained. This problem is discussed in the comprehensive review by Wilson and Boksenberg [248], and it has been estimated that errors of up to 2000 °K may occur in the effective temperatures of hot stars obtained from the Balmer jumps of unblanketed models [249]. There is now reasonable agreement between the predicted ultraviolet fluxes from line blanketed LTE models and observations of B-stars, and detailed models are available [250].

Departures from LTE in the hydrogen lines of B stars can be quite substantial. The suggestion by Kalkofen [251, 252] that the bound–bound transitions in hydrogen can be assumed to be in detailed balance has been used by Peterson and Strom [253, 254] in a study of the Balmer lines. Below a certain depth, the chief non-equilibrium processes controlling the populations of the hydrogen bound states were assumed to be bound–free radiative transitions, and it was estimated that this would give good results for the wings of the Balmer lines. A grid of model atmospheres for $10\,000$ °K $\leq T_e \leq 15\,000$ °K and $2.5 \leq \log g \leq 4.0$ was computed, with the first five levels of hydrogen in NLTE and the remainder, plus the continuum, in LTE. Modified Bethe cross-sections given by Mihalas [255] were used and the line profiles were obtained from the results of [256, 142], (see § 5-4C). The profiles derived for Hα are quite different from those calculated assuming LTE, and the ratios of the equivalent widths of Hα to Hγ are consistent with observations of B stars. Subsequent calculations by Auer and Mihalas [257] on the effects of Lyα and Hα on the temperature structure of NLTE models indicate a temperature rise near the surface, and these authors have recently presented models based on a sixteen level hydrogen atom in which the first five levels are allowed to depart from LTE in a self-consistent manner [22]. Here the Lyman lines only were assumed to be in detailed balance, all transitions between the levels $n = 2, 3, 4$ and 5 being specifically included. With such detailed NLTE calculations the entire line profile can be predicted, in particular, the line core. The wings of Hα appear weakened, in agreement with earlier results [253, 254], but the core is considerably strengthened; this is consistent with observations of B stars. Hγ is little affected by departures from LTE, but the Paschen lines are strengthened; Fig. 5-6-5 shows the predicted profiles of Hα and Pα for the model with $T_e = 15\,000$ °K and $\log g = 3$. The flux in the continuum is virtually unaltered by departures from LTE, except in the low gravity models and

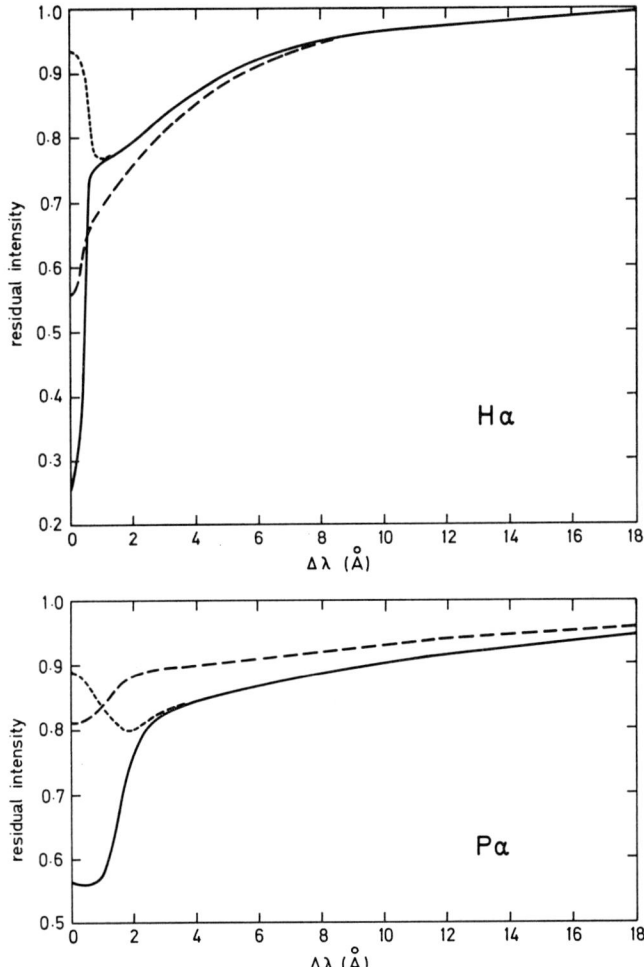

Fig. 5-6-5. Profiles of Hα and Pα for a model atmosphere consisting entirely of hydrogen with $T_e = 15\,000$ °K, $\log g = 3$, [22]. - - - LTE calculation, . . . NLTE for the continuum only (detailed balancing in the lines), ——— NLTE model including the lines. [L. H. Auer and D. Mihalas, Astrophys. J. **160** (1970) 233. Reprinted by permission of the University of Chicago Press; © 1970 by the University of Chicago.]

in the unobservable region beyond the Lyman limit. Mihalas and Auer [258] have also discussed the effects of rotational broadening on the line profiles and show that the central depth of Hα is a sensitive indicator of stellar rotation velocity.

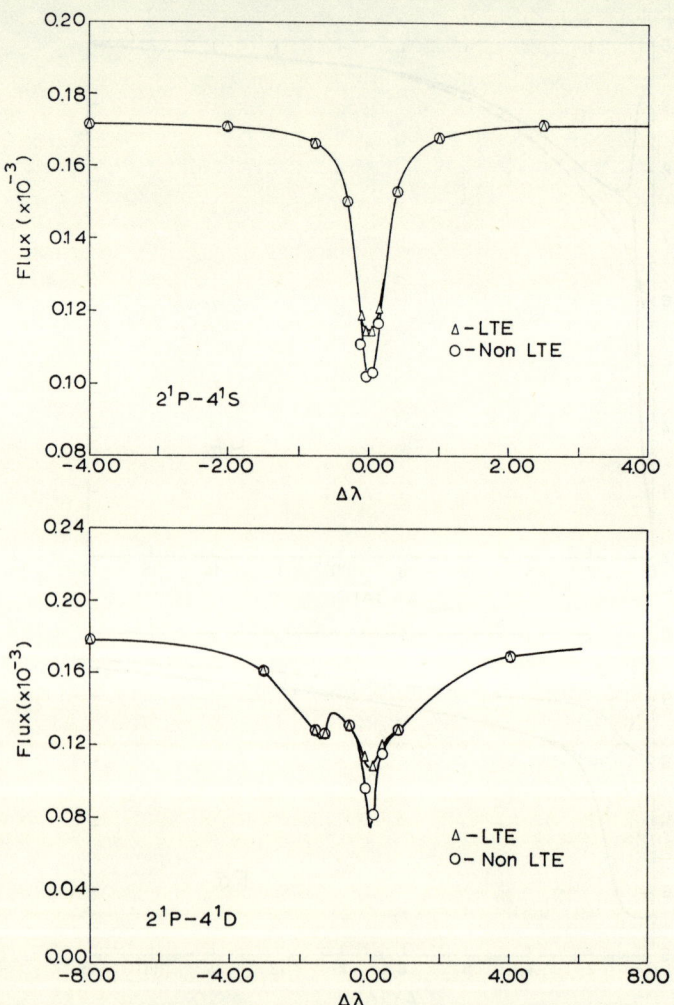

Fig. 5-6-6. Theoretical and profiles for He I lines in a B-star model atmosphere with $T_e = 20000$ °K, $\log g = 4$, [261]. The $2\,^1P$–$4\,^1S$ profile is typical of those having no forbidden component, while the $2\,^1P$–$4\,^1D$ is typical of those having a forbidden component. [A. Poland, Astrophys. J. **160** (1970) 609. Reprinted by permission of the University of Chicago Press; © 1970 by the University of Chicago.]

Helium lines in B stars are important for abundance determinations, since in the cooler stars very few helium lines are strong enough to be detected. Norris [259] has carried out an LTE analysis of twelve He I lines in normal sharp line B stars using profiles taken from [145] and [147]. For all the lines except 4009 Å(2^1P–$7\,^1$D), 5047 Å($2\,^1$P–$4\,^1$S), 5875 Å($2\,^3$P–$3\,^3$D), and 6678Å($2\,^1$P–$3\,^1$D) the equivalent widths determined yield a unique helium abundance given by $N(\text{He})/N(\text{H}) = 0.09 \pm 0.015$. This supports the conclusions of Johnson and Poland [260] and Poland [261], who consider departures from LTE in the transitions $2\,^{1,3}$S–$4\,^{1,3}$P, $2\,^{1,3}$P–$4\,^{1,3}$S, $2\,^{1,3}$P–$5\,^{1,3}$S, and $2\,^{1,3}$P–$4\,^{1,3}$D, and conclude that although the cores of the lines are generally strengthened the equivalent widths are essentially unchanged, so that they provide reasonable data for LTE abundance determinations, see Fig. 5-6-6. The "helium anomaly" (the observed strengthening of the triplets with respect to the singlets along the spectral sequence from B3 to O6) is consequently explained by both Norris and Poland as a natural result of the LTE curve of growth, resulting from comparison of strong lines (triplet series) with weak ones (singlet series).

The analysis of very strong helium lines is more difficult. The abundance determined by Shipman and Strom [262] from an essentially LTE analysis of the line 4471Å($2\,^3$P–$4\,^3$D) using the broadening results of Griem [145] and Barnard et al. [146] agree well with those derived from observations of the weak triplet 4713 Å($2\,^3$P–$4\,^3$S) for stars later than B3. However, as discussed by Hearn [263], departures from LTE in the lines 4471Å and 5875Å($2\,^3$P–$3\,^3$D) can be quite substantial and consequently the abundance of helium can only be obtained correctly by detailed comparison of many theoretical and observed profiles. Further investigation by Snijders and Underhill [264] indicates that the LTE profiles of the triplet 4471Å predicted by Shipman and Strom show a marked discrepancy with observations of B stars. The cores of the observed lines are shallower than those of the predicted profiles and the theory significantly overestimates the intensity of the forbidden component at 4470Å($2\,^3$P–$4\,^3$F). A similar discrepancy exists between the observed profile of the forbidden component in a laboratory plasma [144] and the best available theory (see Fig. 5-4-4 and § 5-4C). Until the line broadening problems are resolved, it is difficult to estimate NLTE effects in the strong helium lines in B stars, and hence the use of such lines for abundance determinations is rather uncertain.

The hot O stars are less well understood than the A and B stars; the predicted continuum flux in the ultraviolet is larger than is observed. NLTE is very important in the lines since, for example, the observed width of Hβ

remains approximately constant for the spectral types B0 to O4 [265], while LTE calculations would predict a sharp decrease. Mihalas and Auer [266] and Auer [267] have constructed models with 25 000 °K $\leq T_e \leq$ 50 000 °K which allow for departures from LTE in H_I, He_I and He_{II} and predict substantial changes in the far ultraviolet flux. The Balmer discontinuity is significantly increased for the models with $T_e > 35\,000$ °K implying a higher temperature scale for the O stars. The intensities of Hα and Hβ appear much stronger than LTE results would suggest and it is also likely that the assumed gravities for O stars should be reduced by about $\Delta(\log g) = 0.5$, which improves agreement with values derived from stellar evolution.

Some of the hottest stars are the central stars of planetary nebulae, and it is important to have reasonably accurate determinations of the flux in the H_I, He_I and He_{II} continua of such stars in order to interpret the spectra produced in the surrounding nebula. Models have been computed by Gebbie and Seaton [268], Böhm and Deinzer [269], Gebbie [85], Böhm [270], and Hummer and Mihalas [271]. The most detailed models [271] are in hydrostatic, radiative and local thermodynamic equilibrium, with 3×10^4 °K $\leq T_e \leq 2 \times 10^5$ °K and $3.4 \leq \log g \leq 7.5$. The ration $N(\text{He})/N(\text{H})$ was taken as 0.16, somewhat higher than the currently accepted value, and the metal content was slightly varied. The usual sources of opacity, namely, H, H_2^+, He, He^+ and electron scattering, were included, together with contributions from one or more states of ions of C, N, O and Ne. However, Böhm [270] has pointed out that deviations from hydrostatic equilibrium in very hot stars are quite likely, and line blanketing and NLTE effects can be considerable in the ultraviolet continuum.

The electron temperature of most gaseous nebulae is about 10^4 °K while the electron density is about 10^4 cm^{-3}, and consequently the interpretation of the spectra produced by such nebulae depends strongly on collision processes. The ultraviolet radiation, emitted by the hot central star in the case of a planetary nebula, or in dilute form from many stars in the case of an interstellar H_{II} region, causes photoionization of the major nebula constituents and an emission line spectrum is exhibited. The lines of hydrogen and helium are produced by electron–ion recombination followed by cascade, and are particularly prominent at radio frequencies in H_{II} regions, while in the optical regions there occur forbidden lines of [O$_{II}$] caused through collisional excitation of metastable states. From observations of the relative intensities of the emission lines, the electron density and temperature of the nebula can be determined; the detailed formulation of the problem has been given by Aller [272] and Seaton [273], and the atomic collision processes of importance are discussed by Seaton [17]. The electron density is normally obtained from the ratio of the intensity of the 3729 Å $^2D_{\frac{5}{2}}-^4S_{\frac{3}{2}}$ component to the 3726 Å $^2D_{\frac{3}{2}}-^4S_{\frac{3}{2}}$ component of the OII ($2p^3$) doublet (Seaton and Osterbrock [274]). Saraph and Seaton [275] have recently reported detailed results for the electron densities of eleven planetaries derived from this ratio in [OII], and also from similar ratios in [SII], [ClIII], [ArIV] and [KV] with configuration ($3p^3$). The most accurate atomic data available [276, 277] was used for the collision strengths. At radio frequencies, since the discovery of recombination lines of hydrogen [278] in HII regions, much work has been concentrated on methods of obtaining theoretical intensities for such lines. These are observed as $n+1 \to n$ transitions for very large values of n, where the populations of the levels differ only marginally from their LTE values. However Goldberg [279] has shown that very slight departures of the populations of the upper and lower levels from their equilibrium values can lead to inversion, and hence amplification of the lines by stimulated emission. Since the radiation field is relatively weak in an HII region, the statistical equilibrium equations (5-2-66) may be uncoupled from the radiative transfer problem, and detailed calculations for the departure coefficients b_n have been presented [18] for $n \leq 300$. Collisions between all levels, n and n', were allowed for, using the rates obtained from the correspondence principle [199]. The results are given for a wide range of electron temperatures and densities and the interpretation of recent observations and the Stark broadening of the lines is discussed by Duprée and Goldberg [280].

The related problem of the helium abundance in planetary nebulae has been investigated by Seaton [273] and Harman and Seaton [281]. A survey by Kaler [282], based on theoretical results and on observations obtained using photographic photometry, leads to a mean $N(\text{He})/N(\text{H})$ for planetaries of 0.14, with a range of 0.09 to 0.20. However, the most recent determination of $N(\text{He})/N(\text{H})$ is 0.10 [283]. This was obtained using the same theoretical data, but with measurements derived from photoelectric photometry; the difference in the two mean values is due solely to errors in the earlier photographic photometry. The present mean value for planetaries agrees well with the data from optical observations of extragalactic nebulae [284], while radio frequency results from the relative intensities of H 109α and He 109α in HII regions [285], give an average of 0.084 ± 0.003.

D. THE LATE-TYPE STARS; MOLECULAR ABSORPTION

The stars in classes K and M exhibit absorption spectra strongly dominated by molecular bands; a comprehensive review of the infrared spectra of these objects has been presented by Spinrad and Wing [286]. The abundant molecules H_2, H_2O and CO can be observed only in the infrared, while molecules such as CN have strong transitions in this region, and thus the interpretation of infrared spectra provides information on the abundances of the light elements. A recent review of the composition of cool stars has been given by Pagel [287], with particular reference to theories of nucleosynthesis and evolution.

Progress in constructing model atmospheres for stars with effective temperatures less than about 4000 °K has been hindered, mainly by the difficulties incurred in the treatment of molecular line blanketing and convection. The most important molecular source of opacity is H_2O, whose lines completely cover the spectrum for $\lambda > 0.8\,\mu$. Auman [288] has calculated the opacity due to this molecule by obtaining the strengths and positions of the individual spectral lines and summing their contributions, and similar results are available for CO [289] and CN [290]. Initial attempts to calculate model atmospheres for cool stars [291–295], assumed purely radiative equilibrium, and allowed only approximately for the opacity due to water vapour. Subsequently results have been presented by Auman [296] which include the effects of convection and the detailed behaviour of the H_2O opacity, together with contributions from H, H^-, H_2^-, He^-, H_2^+, the major metals, and scattering processes. Absorption by H_2O is dominant throughout most of the high gravity (dwarf) models, but H^- becomes significant in the deeper layers of the low gravity (giant) stars. Consequently the emergent flux behaves differently in the two cases; only in the giant atmospheres is there a sharp maximum at $1.6\,\mu$ due to the minimum in the H^- opacity. These models have been applied to a discussion of the molecular abundances in K and M stars [297] with particular reference to the apparently low value for H_2O found in M stars [298, 299]. However, it is likely that departures from LTE will be important in giant and supergiant atmospheres, and the contribution to the opacity due to spectral lines of other molecules such as CO, CN, OH, TiO, H_2S, SiO may be significant [296, 297].

Little theoretical work has been carried out on late-type stars of non-solar composition. The old evolved stars certainly have a tendency to be metal poor, but it has been suggested by Spinrad and Taylor [300] that the K giants in the old galactic clusters M67 and NGC 188 may be super metal rich, with abundances of Ca, Mg and Na of three or four times the solar value. (This effect may alternatively be due to a high helium content.) In the cool carbon stars it is likely that the bound–free continua of negative ions such as Cl^-, C^-, C_2^-, CN^- will be important [62], in addition to the usual atomic and molecular opacities. Unfortunately little is known about the behaviour of photo-detachment cross-sections for negative molecular ions other than H_2^-, O_2^- and OH^- (see Chapter 8), but C^- (see § 5-3C) is a major source of opacity in R Coronae Borealis stars (see § 5-6E). A further problem in the study of cool stars is the formation of solid particles [301, 302]; a recent consideration of stellar atmospheres with $T_e \lesssim 3000\,°K$ by Krishna Swamy [303] indicates that extinction due to graphite grains is comparable with absorption by atomic and molecular sources.

Molecular band spectra are also important in the very low density HI regions of interstellar space, where the mean kinetic temperature is about 100 °K. Absorption lines of NaI (D-lines), KI, CaI, CaII (H and K), TiII, FeI and molecular lines due to CN, CH and CH^+ have long been observed in the optical spectrum of HI regions, and another characteristic feature is the 21 cm (1428 MHz) fine structure transition in hydrogen. Information concerning the intensity of the cosmic microwave background can be determined from the CN, CH and CH^+ lines [304, 305], yielding an equivalent black-body temperature of 2.7 °K. Recently many more molecular lines have been identified in interstellar space at infrared and radio frequencies. The four Λ doublet transitions of OH at 1720, 1667, 1665, 1612 MHz have been observed in both emission and absorption (see e.g. [306]) and appear in many cases to show anomalous intensity ratios and strong linear and circular polarisation (see e.g. [307]), with possible secular variations. NH_3 and H_2O have also been observed (see e.g. [308]), while interstellar formaldehyde, H_2CO, has been detected in absorption [309] at 4830 MHz. Recent surveys (see e.g. [310]) indicate that H_2CO in absorption is well correlated in both direction and velocity with OH and HI in absorption. In cooler dark dust clouds, however, OH appears in normal thermal emission [311], while H_2CO is found in absorption [312]. A detailed discussion of the observations of many unusual molecules found in interstellar space, together with the problems of their theoretical interpretation, is given by Somerville [313].

E. Stars of non-solar composition

1. *Weak helium line stars*

The majority of early-type stars have a helium abundance given approximately by $N(He)/N(H) = 0.1$, but there are groups of stars in which the helium lines appear relatively weak compared with normal stars of similar temperature. Some of these are the Ap stars, which possess extremely strong magnetic fields and have slow rotation rates; in particular those which show an excess of silicon or manganese [314]. There is also a group of weak helium line stars [315] which may be an extension of the Ap stars to slightly higher temperatures. The abundance anomalies of these stars appear to depend strongly on effective temperature [316]; oxygen is most underabundant in the cooler stars with 7 700 °K $< T_e <$ 11 700 °K, while silicon and manganese are most overabundant in the ranges 11 200 °K $< T_e <$ 18 000 °K and 11 000 °K $< T_e <$ 14 400 °K respectively. Another unusual feature of the Ap stars is that some have variable spectra; in particular large changes are often seen in the strengths of lines due to the rare earths, chromium, strontium and silicon.

The number density of the metallic elements relative to hydrogen is very important in determinations of the observed spectrum of these stars in the visual and ultraviolet. Silicon is a strong source of opacity throughout this region of the spectrum in silicon-rich Ap stars [317, 249] and neglect of this can result in a serious overestimate for the effective temperature deduced from the Balmer discontinuity [318]. Detailed studies of a few Ap stars are available, in particular the manganese star 53 Tau, for which model atmosphere analyses have been presented by Auer et al. [319] and Strom [320]. The results of the first authors, which were based on an LTE analysis of the relatively weak lines using the model atmosphere technique of Mihalas [242], allowed for broadening by electrons and ions, together with natural damping, and assumed the turbulent velocity to be zero, a choice indicated by previous curves of growth [321]. Most of the metals lighter than titanium appeared to have roughly normal abundances, but severe overabundances existed for the heavier elements, with the exception of iron and chromium. Helium was found to be five times underabundant, and manganese two hundred times overabundant. However, the results of Strom, which were derived by adopting different values of the turbulent velocity in the visual and ultraviolet, in order to obtain the same abundance from both the strong and weak lines of an element, differ substantially from those of [319], by factors of up to four. It is difficult to estimate at present how much these discrepancies are due to the assumptions made in an LTE analysis, how much to inaccuracies in the opacity and how much to the parameters included in the broadening theory.

It has been suggested by Michaud [322] that diffusion processes are responsible for most of the abundance anomalies in Ap stars. If the atmospheres are sufficiently stable for such processes to be important, gravitational settling of the predominantly neutral elements He, N, O would lead to underabundances of these elements in stars with the observed ranges of effective temperature, while radiation pressure would lead to overabundances of certain heavy elements in the stars where they are observed. This effect appears to be consistent with the results of Norris [259], who investigated twelve of the hotter weak helium line stars having effective temperatures between about 14 000 °K and 20 000 °K. Apparent deficiencies of helium of factors of two to fifteen were found in these stars, some of which were silicon rich and others phosphorus rich, while the ranges of effective temperature obtained for the anomalies were in accordance with the hypothesis of Michaud. It has been shown further by Peterson [323] that variations in the abundance of metals such as silicon, across the surface of Ap stars, lead to spectral variations associated with rotational period; this may provide the mechanism for their variability. The related problem of diffusion processes and abundance anomalies in metallic (Am) stars has recently been discussed by Watson [324].

2. Hydrogen deficient stars

Some of the most interesting problems in the study of astrophysical plasmas occur in the analysis of the spectra of white dwarfs [325], for these are stars at a late stage of evolution, and are dense, compact objects. Many appear to be deficient in hydrogen; for example the analysis by Weidemann [326] of Van Maanen indicated an effective temperature of about 5870 °K, a surface gravity given by $\log g = 8$, and an estimate for $N(\text{H})/N(\text{He})$ of 0.0178. Consequently other sources of opacity rather than H or H^- are frequently important in the determination of the flux from such stars.

A comprehensive study of the atmospheres of hotter white dwarfs has recently been presented by Strittmatter and Wickramasinghe [327]. Models were constructed with effective temperatures between 10 000 °K and 25 000 °K and surface gravities in the range $6 \leq \log g \leq 9$; the approximation of LTE is valid due to the high densities involved. Various values of the helium abundance were considered, and the effects of Lyman and Balmer line blanketing were included. Line profiles were calculated for He I 4471Å ($2\,^3P-4\,^3D$), 4713Å ($2\,^3P-4\,^3S$), 5875Å ($2^3P-3\,^3D$), Mg II 4481Å ($3^2D-4\,^2F$), Si III 4129Å ($3^2D-4\,^2F$), C II 4267Å ($3\,^2D-4\,^2F$), and the H and K lines of Ca II. The line broadening mechanisms specified were Doppler broadening and radiation damping, together with collisional broadening by He and H and quadratic Stark broadening; these latter two effects are very significant in dwarf atmospheres. On consideration of the roles of accretion, convection, diffusion and magnetic fields in such models, Strittmatter and

Wickramasinghe suggest that a consistent picture of white dwarf spectra may be obtained if most of these stars have helium rich envelopes which accrete a surface layer of hydrogen. Under normal conditions this produces the type of spectrum shown by the wide range of white dwarf DA stars, namely, one in which only lines of hydrogen are observed. Within the temperature range $15\,000\,°K \leq T_e \leq 18\,000\,°K$ the upper layer of helium is convectively unstable, and the accretion products are mixed down, producing a typical DB spectrum (that is one consisting primarily of He I lines), with the observed temperature of such a white dwarf. The hydrogen abundance in the surface layers of these stars must be reduced by at least a factor of 10^5 relative to helium compared with its value in normal stars. Consequently He$^-$ appears to be the major source of opacity in DB stars and the metal abundances indicated are significantly lower than the solar values.

A fairly comprehensive list of stars extremely deficient in hydrogen has been given by Hill [328]. Besides the white dwarfs discussed in this section, this also includes some of the carbon-rich helium stars mentioned below.

3. *Carbon stars*

Probably the most interesting class of carbon stars are the R Coronae Borealis variables, of which the prototype R CrB is a supergiant of approximately solar effective temperature, which suffers seemingly random large fluctuations in visual intensity on the time scale of a few years. During these periods the absorption spectrum appears heavily veiled, and many metallic emission lines are observed [329]. The major features of the normal spectrum at maximum intensity are the strength of the numerous neutral carbon lines, the relative weakness of the C_2 and CN bands, and an apparent deficiency of hydrogen. A detailed equivalent width analysis by Searle [330] led to a carbon to iron ratio about twenty five times the solar value, and a carbon to hydrogen ratio of about ten, while among the metals there appeared to be no abundance anomalies exceeding a factor of two. These results have been essentially confirmed [331, 332], and the southern R CrB variable, RY Sgr, exhibits similar behaviour [333].

A grid of model atmospheres in LTE has been constructed for R CrB by Myerscough [334], using the abundances indicated by Searle, in order to determine the effective temperature and gravity by comparison with the observed visual continuum [335] and electron pressure [332]. Figure 5-6-7 shows the behaviour of the opacity at 5000 Å for a typical model, and illustrates the importance of absorption by He$^-$, C$^-$ and Si in the atmo-

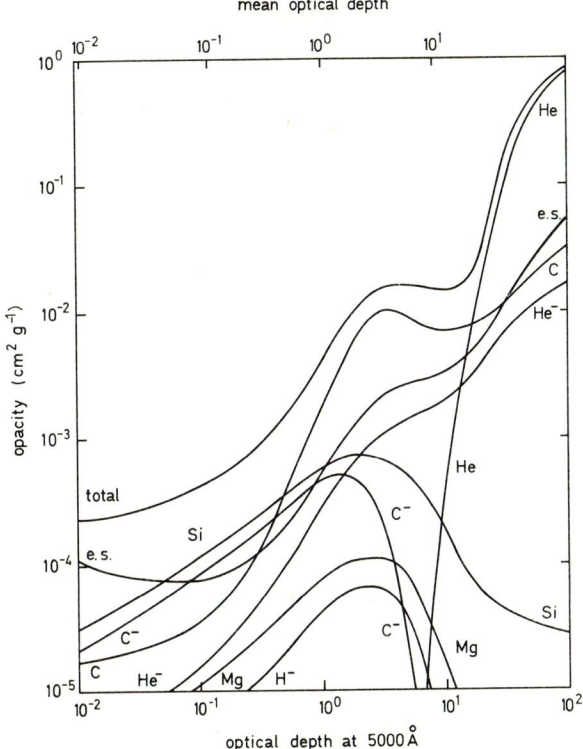

Fig. 5-6-7. Schematic behaviour of the opacity for a model with $T_e = 6000\,°K$, $g = 1.0$ [231] and the observed abundances for R CrB [330].

spheres of such hydrogen deficient carbon stars. The best fit to the continuum data for R CrB was obtained for $T_e \sim 5800\,°K$ and $g \sim 1.0$, in very reasonable agreement with the values deduced on purely observational grounds. Subsequent calculations for the line profiles, however, indicate that although the weaker C I lines can be reproduced fairly accurately with the assumed abundances, the metallic lines cannot; even the very weakest of those observed occur under such low pressure conditions, that LTE calculations are virtually meaningless. Work is now in progress to study variations in abundances, and to investigate the departures from LTE in both the continuum and the lines of hydrogen deficient carbon stars. This necessitates firstly a consideration of the rate coefficients for the processes involved in determining the statistical equilibrium for, in particular, C^-, and secondly, reasonable estimates of the collisional excitation cross sections for the first few levels of, for example C I, Mg I, Si I, Fe I and Fe II.

An equivalent-width analysis of five cooler non-variable carbon stars has been presented by Warner [332]. With the possible exception of sodium, all the observable elements heavier than oxygen appear to have normal abundances relative to iron. Carbon is overabundant by factors of three to ten, and hydrogen is deficient, by more than a factor of 10^5 in four of the five stars considered. It is suggested that the unusual abundances in these objects result from the loss of an outer envelope, by stars of about a solar mass, at an evolutionary stage shortly after the onset of helium burning, and that mixing from the helium → carbon reaction causes the excess of carbon.

Among the hotter carbon stars, BD+10° 2179 has been analyzed in detail using a grid of LTE model atmospheres by Hunger and Klinglesmith [336]. The sources of opacity included were H, H^-, He, He^+, He^-, C^+ and, by comparison of the computed HeI profiles and the equivalent widths of twenty CII lines with the corresponding observed quantities, the suggested parameters are $T_e \sim 16000\,°K$, $\log g \sim 2.8$. The star appears to be extremely hydrogen deficient and carbon rich, with $N(C)/N(He) \sim 0.4$; there is also evidence for slight overabundances of nitrogen, aluminium, phosphorus and iron.

Many of the cooler carbon stars have recently been observed in the infrared; all the spectra show complex molecular absorption features [286]. Bands of C_2, $^{12}C\,^{16}O$, $^{13}C\,^{16}O$ and the red system of CN have been identified in W Ori, 19 Psc, U Hya and X Cnc [337, 338], while some of the results presented by Gaustad and his collaborators [339] indicate the possibility of either HCN or C_2H_2 absorption at 3μ. R CrB has been observed in the infrared [340], the data being taken when the star was just below maximum intensity in the visual. The results indicated an infrared excess, which was attributed as due to black-body radiation from a circumstellar cloud of particles at a temperature of about 940 °K; the deficiency in the visual flux then balanced well with the amount observed in the infrared excess. Observations of RY Sag in both the visual and infrared have also been presented [341, 342]; over a period of eight months the visual intensity of the star appeared to increase, while the infrared intensity decreased. These results tend to support a general model for R CrB stars in which the infrared flux is ascribed to black body radiation from a circumstellar cloud of particles, probably graphite grains, ejected at the time of deep minimum. Qualitative correlations of observations indicate that stars which exhibit low excitation emission lines are likely to emit excess infrared radiation [343] which suggests that all such stars have circumstellar envelopes in which rapid grain formation is occurring. Quantitative results await the development of detailed dynamical models, in which such envelopes are produced by small-scale mass loss.

§ 5-7. Problems requiring further study

The most abundant elements, other than hydrogen and helium, in stars of normal composition are carbon, nitrogen, oxygen, sodium, magnesium, aluminium, silicon, sulphur, calcium and iron (see Table 5-6-1). In order to estimate more precisely the relative abundances of these elements, particularly in the atmospheres of unusual stars, much more atomic data is needed.

Photoionization from the ground configurations of the above elements contributes strongly to the continuous opacity in the solar ultraviolet (see Fig. 5-6-3). While the general formula for bound–free transitions discussed in § 5-3B is probably fairly accurate for the excited states, photoionization from the ground states, especially those containing equivalent electrons, requires much more careful treatment. The best calculations so far are those involving close coupling functions and results using this method are readily available for He [42], N, O [44] and Al, Si, S [45]. These are in good agreement with experimental data [43, 46, 47, 48], but further experimental work on photoionization cross-sections for atoms with ground configurations $(np)^q$, $q = 1, 2, \ldots, 6$; $n = 2, 3$ is needed for comparison with the results of the general program [49]. In the hotter stars, where carbon, nitrogen and oxygen are primarily in the form CII, CIII, NII, NIII and OII, OIII, it is particularly desirable to obtain experimental data on the photoionization cross-sections of these positive ions for comparison with theoretical work [40]. For some unusual stars, much more reliable results are needed for photoionization from excited states of, for example, CI in R CrB stars, CII in BD+10° 2179 and Si in Ap stars (see § 5-6E). Measurements of the continuous absorption by these elements where only excited states can contribute (such as those of Boldt [36] for N and O), would be a very useful check of the present predictions which are based on the general formula. Theoretically, the problem of the validity of the dipole length versus the dipole velocity form of the matrix element when approximate wave functions are used needs clarification, particularly in the case of photodetachment from negative ions such as C^-, which are important in the cooler hydrogen deficient stars (see § 5-3C and § 5-6E). For the late type stars much more detailed information on molecular absorption is required in order to determine the relative importance of, for example, the opacity due to H^- compared with that provided by H_2O (see § 5-6E).

Rather similar comments apply to the determination of oscillator strengths, where much more reliable results are needed especially for excited states of the abundant elements C, N, O, Na, Mg, Al, Si, Ca and Fe. The available data for metallic ions is also sketchy and its accuracy is difficult to estimate, while Wiess et al. [87] quote no experimental information for CIII, CIV, NIII and NIV. Recent theoretical work [96, 100, 101, 102], which takes into account many states in configuration interaction calculations, and new techniques such as beam foil spectroscopy [97] may gradually remedy the situation. The continuity rule (5-4-9) between bound–bound and bound–free transitions can be a useful check on the consistency of sufficiently detailed experimental or theoretical data.

The broadening of spectral lines by electron impact is often calculated using an impact approximation (see § 5-4C). For these problems data on the scattering of electrons by atoms or ions in excited states is required and, at present, a full quantum mechanical treatment can be applied only for a few special cases. Any experimental information on collision cross-sections for transitions between excited states, or elastic scattering by atoms such as C, N, O, Mg, Si, Fe in excited states would be extremely welcome, but perhaps impossible to obtain. Measurements of differential scattering cross-sections would provide a much more sensitive test of theory than is currently available, and would be particularly useful for those systems already studied in the close-coupling approximation (see e.g., [49]). Theoretically, more might be done to apply the results of existing close-coupling calculations to line broadening, for example, the best results on the electron–hydrogen problem [178] could be used to obtain the profile of Lyα emitted by a hydrogen plasma, and possibly also that of Hα. As discussed in § 5-4C the agreement between the present results using the semi-classical impact approximation for the electrons [134, 136] and experimental data [130–133, 135], for the broadening of the Balmer lines is reasonable (see Fig. 5-4-2), but it is nowhere near so satisfactory for the Lyman lines (see Fig. 5-4-3). Even larger discrepancies occur between the theoretical results [145, 146] and the experimental profiles [144] for the HeI line (2p) ^{3}P–(4d) ^{3}D at 4471 Å and its forbidden components (2p) ^{3}P–(4f) ^{3}F at 4470 Å, and consequently there are many difficulties to be resolved before such strong helium lines can be used in abundance determinations or to characterise NLTE effects [263] in stellar atmospheres. Among the heavier atoms, the resonance lines of MgII and CaII have been studied with a full quantum mechanical treatment [154–156], but the agreement with experiment [149, 152, 157] is again unsatisfactory. Much more work, both theoretical and experimental is needed to resolve these discrepancies. Due to the above difficulties, it seems that semi-classical methods based on perturbation theory will still be used in the majority of line broadening calculations for the foreseeable future. It should be realized, however, that the results may not be very accurate, and the collisional excitation and ionization cross-sections involved should be checked for consistency with oscillator strengths and photoionization cross-sections using the Bethe approximation (5-5-21) and (5-5-25).

All these problems are each involved in several aspects of the interpretation of stellar spectra. Collision rates affect the populations of the atomic levels and the shape of the line profiles, while the bound–bound and bound–free transitions contribute not only to the calculation of the level populations and line profiles but also determine the continuous opacity. In recent years the importance of NLTE in stellar atmospheres has been increasingly realized, and in order to solve the coupled equations (5-2-10) and (5-2-66) even with only a few levels allowed to depart from LTE, a knowledge of many individual cross-sections is needed. It must be emphasized again that the calculation of such cross-sections, and the line broadening, is a continually developing art, and consequently careful attention must be given to the accuracy of the atomic data before elaborate predictions are made on the basis of NLTE studies. Much more information is needed, particularly on cross-sections for metallic atoms and ions and the effects of turbulence, in order to understand through detailed NLTE calculations, whether (and why) any LTE results for the abundances of the elements heavier than hydrogen in stellar atmospheres are meaningful.

We hope that this article has gone some way towards clarifying some of the difficulties and helping both astrophysicists and atomic physicists each to appreciate better the problems of the other.

Acknowledgments

We should like to thank Professor M. J. Seaton and Dr. J. A. Tully for their helpful comments and Mrs. Lynn Parry for her careful typing of the manuscript.

References

1. L. H. ALLER, The Atmospheres of the Sun and Stars (Ronald, New York, 2nd ed., 1963).
2. C. W. ALLEN, Astrophysical Quantities (Athlone, London, 2nd ed., 1963).
3. S. CHANDRASEKHAR, Radiative Transfer (Oxford, 1950).
4. C. MARK, Phys. Rev. 72 (1947) 558.

5. L. G. HENYEY, Astrophys. J. **148** (1967) 207.
6. E. H. AVRETT, Smithsonian Astrophys. Obs. Rept. **167** (1964) 83.
7. B. E. J. PAGEL, Proc. Roy. Soc. **A306** (1968) 91.
8. D. L. LAMBERT and B. E. J. PAGEL, Mon. Not. R. Astr. Soc. **141** (1968) 299.
9. R. WILDT, Astrophys. J. **89** (1939) 295.
10. D. F. CARBON and O. GINGERICH, in: Theory and Observation of Normal Stellar Atmospheres, ed. O. Gingerich (MIT, 1969) p. 377 seq.
11. R. KURUCZ, in: Theory and Observation of Normal Stellar Atmospheres, ed. O. Gingerich (MIT, 1969) p. 375.
12. H. ZIRIN, The Solar Atmosphere (Blaisdell, 1966).
13. L. GOLDBERG and A. PIERCE, The Photosphere of the Sun, Handbuch der Physik **52** (1959) 1.
14. D. HUMMER, J. Quant. Spectr. Radiative Transfer **8** (1968) 193.
15. D. HUMMER, Mon. Not. R. Astr. Soc. **145** (1969) 95.
16. R. N. THOMAS, Some Aspects of Non-Equilibrium Thermodynamics in the Presence of a Radiation Field (University of Colorado, 1965).
17. M. J. SEATON, Advances in Atomic and Molecular Physics, Vol. 4 (Academic Press N.Y., 1968) p. 331.
18. M. BROCKLEHURST, Mon. Not. R. Astr. Soc. **148** (1970) 147.
19. A. BURGESS and H. P. SUMMERS, Astrophys. J. **157** (1969) 1007.
20. M. J. SEATON, Plan. and Space Sci. **12** (1964) 55.
21. A. BURGESS, Astrophys. J. **139** (1964) 776.
22. L. H. AUER and D. MIHALAS, Astrophys. J. **160** (1970) 233.
23. D. HUMMER, Proc. 2nd Harvard Smithsonian Conference on Stellar Atmospheres (1965) p. 143.
24. A. BURGESS and M. J. SEATON, Mon. Not. R. Astr. Soc. **120** (1960) 121.
25. D. H. MENZEL and C. L. PEKERIS, Mon. Not. R. Astr. Soc. **96** (1935) 77.
26. A. BURGESS, Mem. R.A.S. **69** (1964) 1.
27. L. M. BIBERMAN and G. E. NORMAN, Soviet Phys. Usp. **10** (1967) 52.
28. R. D. HUDSON and L. J. KIEFFER, Bibliography of Photoabsorption Cross Section Data, J.I.L.A. Information Center Report no. 11 (Boulder, Colorado, 1970).
29. G. PEACH, Mem. R.A.S. **71** (1967) 13.
30. G. PEACH, Mem. R.A.S. **71** (1967) 29.
31. G. PEACH, Mem. R.A.S. **73** (1970) 1.
32. M. J. SEATON, Mon. Not. R. Astr. Soc. **118** (1958) 504.
33. G. PEACH, Mem. R.A.S. **71** (1967) 1.
34. G. PEACH, Mon. Not. R. Astr. Soc. **130** (1965) 361.
35. L. D. TRAVIS and S. MATSUSHIMA, Astrophys. J. **154** (1968) 689.
36. G. BOLDT, Z. Physik **154** (1959) 330.
37. E. CLEMENTI, Tables of Atomic Wave Functions (I.B.M., San Jose, 1965).
38. G. PEACH, unpublished.
39. E. J. MCGUIRE, Phys. Rev. **175** (1968) 20; Atomic Subshell Photoionization Cross Sections for $2 \leq Z \leq 54$, Report No. SC-RR-70-721.
40. D. L. MOORES, J. Phys. B, to be published.
41. W. EISSNER and H. NUSSBAUMER, J. Phys. B **2** (1969) 1028.
42. D. W. NORCROSS, J. Phys. B **4** (1971) 652.
43. J. F. LOWRY, D. H. TOMBOULIAN and D. L. EDERER, Phys. Rev. **137** (1965) A1054.
44. R. J. W. HENRY and R. E. WILLIAMS, Publ. Astron. Soc. Pacific **80** (1968) 669.
45. M. J. CONNEELY, K. SMITH and L. LIPSKY, J. Phys. B **3** (1970) 493.
46. J. C. RICH, Astrophys. J. **148** (1967) 275.
47. F. J. COMES and A. ELZER, Z. Naturforsch. **23A** (1968) 133.
48. F. J. COMES, F. SPEIER and A. ELZER, Z. Naturforsch. **23A** (1968) 25.

49. M. J. CONNEELY, L. LIPSKY, K. SMITH, P. G. BURKE and R. J. W. HENRY, Comp. Phys. Commun. **1** (1970) 306.
50. L. M. BRANSCOMB, in: Atomic and Molecular Processes, ed. Bates (Academic Press, N.Y., 1962) Ch. 4, p. 100.
51. T. F. O'MALLEY, L. SPRUCH and L. ROSENBERG, J. Math. Phys. **2** (1961) 491.
52. S. CHANDRASEKHAR, Astrophys. J. **102** (1945) 223.
53. S. CHANDRASEKHAR, Astrophys. J. **102** (1945) 395.
54. N. A. DOUGHTY, P. A. FRASER and R. P. MCEACHRAN, Mon. Not. R. Astr. Soc. **132** (1966) 255.
55. K. L. BELL and A. E. KINGSTON, Proc. Phys. Soc. **90** (1967) 895.
56. N. A. DOUGHTY and P. A. FRASER, Mon. Not. R. Astr. Soc. **132** (1966) 267.
57. J. L. STILLEY and J. CALLAWAY, Astrophys. J. **160** (1970) 245.
58. S. J. SMITH and D. S. BURCH, Phys. Rev. Letters **2** (1959) 165.
59. S. J. SMITH and D. S. BURCH, Phys. Rev. **116** (1959) 1125.
60. A. BOHM and L. REHDER, Z. Naturforsch. **20A** (1965) 114.
61. L. M. BRANSCOMB and B. E. J. PAGEL, Mon. Not. R. Astr. Soc. **118** (1958) 258.
62. M. S. VARDYA, Mem. R.A.S. **71** (1967) 249.
63. M. L. SEMAN and L. M. BRANSCOMB, Phys. Rev. **125** (1962) 1602.
64. L. M. BRANSCOMB, S. J. SMITH and G. TISONE, J. Chem. Phys. **43** (1965) 2906.
65. R. S. BERRY and C. W. REIMANN, J. Chem. Phys. **38** (1963) 1540.
66. H. P. POPP, Z. Naturforsch. **22A** (1967) 254.
67. R. S. BERRY, C. W. REIMANN and G. N. SPOKES, J. Chem. Phys. **37** (1962) 2278.
68. G. MUCK and H. P. POPP, Z. Naturforsch. **23A** (1968) 1213.
69. D. E. ROTHE, Phys. Rev. **177** (1969) 93.
70. J. W. COOPER and J. B. MARTIN, Phys. Rev. **126** (1962) 1482.
71. E. J. ROBINSON and S. GELTMAN, Phys. Rev. **153** (1967) 4.
72. V. P. MYERSCOUGH and M. R. C. MCDOWELL, Mon. Not. R. Astr. Soc. **128** (1964) 287.
73. V. P. MYERSCOUGH and M. R. C. MCDOWELL, Mon. Not. R. Astr. Soc. **132** (1966) 457.
74. R. J. W. HENRY, J. Chem. Phys. **44** (1966) 4357.
75. R. J. W. HENRY, Phys. Rev. **162** (1967) 56.
76. R. J. W. HENRY, Phys. Rev. **172** (1968) 99.
77. A. TEMKIN, Phys. Rev. **107** (1957) 1004.
78. B. SCHNEIDER, M. WEINBERG, J. TULLY and R. S. BERRY, Phys. Rev. **182** (1969) 133.
79. W. R. GARRETT and H. T. JACKSON Jr., Phys. Rev. **153** (1967) 28.
80. YU V. MOSKVIN, Opt. Spectr. (USSR) **17** (1964) 270; Soviet Phys. High Temp. **3** (1965) 765.
81. G. PEACH, unpublished.
82. W. B. SOMERVILLE, Astrophys. J. **141** (1965) 811.
83. T. L. JOHN, Mon. Not. R. Astr. Soc. **138** (1968) 137.
84. M. R. C. MCDOWELL, J. H. WILLIAMSON and V. P. MYERSCOUGH, Astrophys. J. **144** (1966) 827.
85. K. B. GEBBIE, Mon. Not. R. Astr. Soc. **135** (1967) 181.
86. S. P. TARAFDAR and M. S. VARDYA, Mon. Not. R. Astr. Soc. **145** (1969) 171.
87. W. L. WIESE, M. W. SMITH and B. M. GLENNON, Atomic Transition Probabilities, NSRDS-NBS4 Vol. I (Washington D.C., 1966).
88. W. L. WIESE, M. W. SMITH and B. M. MILES, Atomic Transition Probabilities, NSRDS-NBS4 Vol. II (Washington D.C., 1969).
89. R. W. NICHOLLS and A. L. STEWART, in: Atomic and Molecular Processes, ed. Bates (Academic Press, N.Y., 1962) Ch. 2, p. 47.
90. R. H. GARSTANG, in: Atomic and Molecular Processes, ed. Bates (Academic Press, N.Y., 1962) Ch. 1, p. 1.

91. L. C. Green, R. P. Rush and C. D. Chandler, Astrophys. J. Suppl. **3** (1957) 37.
92. H. C. Goldwire, Astrophys. J. Suppl. **17** (1968) 445.
93. D. H. Menzel, Astrophys. J. Suppl. **18** (1969) 221.
94. D. R. Bates and A. Damgaard, Phil. Trans. Roy. Soc. **A242** (1949) 101.
95. H. R. Griem, Plasma Spectroscopy (McGraw-Hill, N.Y., 1964).
96. C. Froese and A. Underhill, Astrophys. J. **146** (1966) 301.
97. S. Bashkin, Beam Foil Spectroscopy (Gordon and Breach, N.Y., 1968).
98. W. Ervens and H. F. Berg, Z. Physik **222** (1969) 180.
99. W. Hofmann, Z. Naturforsch. **24a** (1969) 990.
100. B. Warner, Mon. Not. R. Astr. Soc. **139** (1968) 1.
101. H. Nussbaumer, Mon. Not. R. Astr. Soc. **145** (1969) 141.
102. A. W. Weiss, Phys. Rev. **188** (1969) 119;
 P. S. Bagus and C. M. Moser, J. Phys. B **2** (1969) 1214.
103. H. E. Saraph, Thesis (University of London, 1970).
104. P. Westhaus and O. Sinanoğlu, Phys. Rev. **183** (1969) 56.
105. O. Sinanoğlu, Comments on Atomic and Molecular Physics **2** (1970) 73.
106. A. L. Stewart, J. Phys. B **1** (1968) 844.
107. R. H. Garstang, Mem. Soc. Roy. Sci. Liege **17** (1969) 35.
108. D. G. Hummer, Mem. R.A.S. **70** (1965) 1.
109. E. Lindholm, Ark. Mat. Astron. Fys. **28A** (1942) No. 3.
110. E. Lindholm, Ark. Mat. Astron. Fys. **32A** (1946) No. 17.
111. H. M. Foley, Phys. Rev. **69** (1946) 616.
112. G. Traving, Über die Theorie der Druckverbreiterung von Spektrallinien (G. Braun, Karlsruhe, 1960).
113. M. Baranger, Phys. Rev. **111** (1958) 481.
114. M. Baranger, Phys. Rev. **111** (1958) 494.
115. M. Baranger, Phys. Rev. **112** (1958) 855.
116. S. Sahal-Bréchot, Astron. and Astrophys. **1** (1969) 91.
117. J. Holtsmark, Ann. Physik **58** (1919) 577.
118. R. G. Breene, The Shift and Shape of Spectral Lines (Pergamon, N.Y., 1961).
119. M. Margenau, Phys. Rev. **48** (1935) 755.
120. E. W. Smith, C. R. Vidal and J. Cooper, Phys. Rev. **185** (1969) 140;
 C. R. Vidal, J. Cooper and E. W. Smith, J. Quant. Spectr. Radiative Transfer **10** (1970) 1011.
121. D. Voslamber, Z. Naturforsch. **24a** (1969) 1458.
122. H. R. Griem, Comments on Atomic and Molecular Physics **2** (1970) 19.
123. G. Peach, to be published.
124. G. Peach, to be published.
125. M. Baranger and B. Mozer, Phys. Rev. **115** (1959) 521.
126. B. Mozer and M. Baranger, Phys. Rev. **118** (1960) 626.
127. C. F. Hooper, Phys. Rev. **149** (1966) 77.
128. C. F. Hooper, Phys. Rev. **165** (1963) 215.
129. A. Ben-Reuven, Advances in Atomic and Molecular Physics, Vol. 5 (Academic Press, N.Y., 1969) p. 201.
130. R. A. Hill and J. B. Gerardo, Phys. Rev. **162** (1967) 45.
131. J. C. Morris and R. U. Krey, Phys. Rev. Letters **21** (1968) 1043.
132. J. B. Shumaker and C. H. Popenoe, Phys. Rev. Letters **21** (1968) 1046.
133. R. D. Bengtson, M. H. Miller, W. D. Davis and J. R. Greig, Astrophys. J. **157** (1969) 957.
134. P. Kepple and H. R. Griem, Phys. Rev. **173** (1968) 317.
135. R. D. Bengtson, J. D. Tannich and P. Kepple, Phys. Rev. **A1** (1970) 532.
136. G. Peach, to be published;

B. E. J. PAGEL, J. Phys. B **4** (1971) 279.
137. G. BOLDT and W. S. COOPER, Z. Naturforsch. **19a** (1964) 968.
138. R. C. ELTON and H. R. GRIEM, Phys. Rev. **135A** (1964) 1550.
139. H. R. GRIEM, Phys. Rev. **140A** (1965) 1140; **144** (1966) 366.
140. H. R. GRIEM, Astrophys. J. **148** (1967) 547.
141. M. BROCKLEHURST and S. LEEMAN, Astrophys. Letters, in press;
G. PEACH, to be published.
142. F. N. EDMONDS, H. SCHLÜTER and D. C. WELLS, Mem. R.A.S. **71** (1967) 271.
143. A. EBERHAGEN and R. WUNDERLICH, Z. Physik **232** (1970) 1.
144. D. D. BURGESS and C. J. CAIRNS, J. Phys. B **3** (1970) L67.
145. H. R. GRIEM, Astrophys. J. **154** (1968) 1111.
146. A. J. BARNARD, J. COOPER and L. J. SHAMEY, Astron. and Astrophys. **1** (1969) 28.
147. H. A. GIESKE and H. R. GRIEM, Astrophys. J. **157** (1969) 963.
148. J. D. E. FORTUNA, NRL Report No. 6950 (NRL, Washington D.C., 1969).
149. J. R. ROBERTS and K. L. ECKERLE, Phys. Rev. **159** (1967) 104.
150. J. MEYER and R. J. BECK, Astron. and Astrophys. **8** (1970) 93.
151. S. SAHAL-BRÉCHOT, Astron. and Astrophys. **2** (1969) 322.
152. J. CHAPELLE and S. SAHAL-BRÉCHOT, Astron. and Astrophys. **6** (1970) 415.
153. D. E. ROBERTS, Astron. and Astrophys. **6** (1970) 1.
154. K. S. BARNES and G. PEACH, J. Phys. B **3** (1970) 350.
155. O. BELY and H. R. GRIEM, Phys. Rev. **A1** (1970) 97.
156. K. S. BARNES, J. Phys. B, to be published.
157. M. YAMAMOTO, Phys. Rev. **146** (1966) 137.
158. W. R. HINDMARSH, A. N. DU PLESSIS and J. M. FARR, J. Phys. B **3** (1970) L5.
159. P. W. ANDERSON, Phys. Rev. **76** (1949) 647.
160. K. HUNGER, Z. Astrophys. **49** (1960) 129.
161. B. WARNER, Observatory **89** (1969) 11.
162. C. DE JAGER and L. NEVEN, Solar Phys. **11** (1970) 3.
163. E. ROUEFF, Astron. and Astrophys. **7** (1970) 4.
164. D. D. BURGESS and J. E. GRINDLEY, Astrophys. J. **161** (1970) 343.
165. Y. CUNY, Theory and Observation of Normal Stellar Atmospheres, ed. O. Gingerich (M.I.T., 1969) p. 173.
166. K. SANDS, R. O. DOYLE and A. DALGARNO, Astrophys. J. **157** (1969) L143.
167. F. H. MIES, J. Chem. Phys. **48** (1968) 482.
168. M. J. SEATON, in: Atomic and Molecular Processes, ed. Bates (Academic Press, N.Y., 1962) p. 374.
169. R. PETERKOP and V. VELDRE, Advances in Atomic and Molecular Physics **2** (1966) 264.
170. P. G. BURKE, Intern. Conf. on Atomic Physics (Plenum, 1968) p. 265.
171. B. L. MOISEIWITSCH and S. J. SMITH, Rev. Mod. Phys. **40** (1968) 238.
172. W. L. FITE, in: Atomic and Molecular Processes, ed. Bates (Academic Press, N.Y., 1962) p. 421.
173. D. W. O. HEDDLE and R. G. W. KEESING, Advances in Atomic and Molecular Physics **4** (1968) 267.
174. M. J. SEATON, Intern. Conf. on Atomic Physics (Plenum, 1968) p. 295.
175. L. J. KIEFFER and G. H. DUNN, Rev. Mod. Phys. **38** (1966) 1.
176. M. R. H. RUDGE, Rev. Mod. Phys. **40** (1968) 564.
177. O. BELY and H. VAN REGEMORTER, Ann. Rev. Astron. and Astrophys. **8** (1970) 329; H. VAN REGEMORTER, Astrophys. J. **136** (1962) 906.
178. P. G. BURKE, S. ORMONDE and W. WHITAKER, Proc. Phys. Soc. **92** (1967) 319.
179. P. G. BURKE, D. F. GALLAHER and S. GELTMAN, J. Phys. B **2** (1969) 1142.
180. S. GELTMAN and P. G. BURKE, J. Phys. B **3** (1970) 1062.

181. P. G. Burke and T. G. Webb, J. Phys. B **3** (1970) L131.
182. M. J. Seaton, to be published.
183. I. M. Cheshire, D. F. Gallaher and A. J. Taylor, J. Phys. B **3** (1970) 813.
184. D. L. Moores and H. Nussbaumer, J. Phys. B **3** (1970) 161;
     D. G. Economides and M. R. C. McDowell, J. Phys. B **2** (1969) 1323.
185. G. Peach, J. Phys. B **1** (1968) 1088.
186. G. Peach, J. Phys. B **3** (1970) 349.
187. G. Peach, J. Phys. B, to be published.
188. D. H. Sampson and L. B. Golden, Astrophys. J. **161** (1970) 321.
189. A. Burgess and I. C. Percival, Advances in Atomic and Molecular Physics **4** (1968) 109.
190. M. J. Seaton, Proc. Phys. Soc. **79** (1962) 1105.
191. A. Burgess, Culham Conf. on Atomic Collisions, AERE Rept. No. 4818 (1964) 63.
192. K. Alder, A. Bohr, T. Huus, B. Mottelson and A. Winther, Rev. Mod. Phys. **28** (1956) 432.
193. M. Abramowitz and I. A. Stegun, Handbook of Mathematical Functions (Dover, N.Y., 1965).
194. R. M. Pengelly and M. J. Seaton, Mon. Not. R. Astr. Soc. **127** (1964) 165.
195. A. D. Stauffer and M. R. C. McDowell, Proc. Phys. Soc. **85** (1965) 61.
196. B. L. Moiseiwitsch, Proc. Phys. Soc. **87** (1966) 885.
197. I. C. Percival and D. Richards, J. Phys. B **3** (1970) 315.
198. H. E. Saraph, Proc. Phys. Soc. **83** (1964) 763.
199. I. C. Percival and D. Richards, Astrophys. Letters **4** (1970) 235;
     I. C. Percival and D. Richards, J. Phys. B **4** (1971) 918, 932.
200. J. W. B. Hughes, Proc. Phys. Soc. **91** (1967) 810.
201. R. Cayrel and G. Cayrel de Strobel, Ann. Rev. Astron. and Astrophys. **4** (1966) 1.
202. M. Minnaert and G. F. Mulders, Z. Astrophys. **1** (1930) 192; **2** (1931) 165.
203. D. H. Menzel, Astrophys. J. **84** (1936) 462.
204. A. Unsöld, Physik der Sternatmosphären (Springer-Verlag, 1955).
205. M. Wrubel, Astrophys. J. **109** (1949) 66; **111** (1950) 157.
206. L. Goldberg, E. A. Müller and L. H. Aller, Astrophys. J. Suppl. **5** (1960) 1.
207. L. H. Aller and A. K. Pierce, Astrophys. J. **116** (1952) 176.
208. C. Jordan, Mon. Not. R. Astr. Soc. **132** (1966) 463, 515.
209. S. R. Pottasch, Bull. Astron. Inst. Neth. **19** (1967) 113.
210. D. L. Lambert, Mon. Not. R. Astr. Soc. **138** (1968) 143.
211. D. L. Lambert and B. Warner, Mon. Not. R. Astr. Soc. **138** (1968) 181, 213.
212. B. Warner, Mon. Not. R. Astr. Soc. **138** (1968) 229.
213. J. Richter, Z. Physik **151** (1958) 114.
214. J. Richter, Z. Astrophys. **51** (1961) 177.
215. J. Solarski and W. L. Wiese, Phys. Rev. **135A** (1964) 1236.
216. C. H. Corliss and W. R. Bozmann, Nat. Bur. Std. Monograph **53** (1962).
217. C. H. Corliss and B. Warner, Astrophys. J. Suppl. **8** (1964) 395.
218. B. Warner, Mem. R.A.S. **70** (1967) 165.
219. T. Garz and M. Kock, Astron. and Astrophys. **2** (1969) 274.
220. T. Garz, H. Holweger, M. Kock and J. Richter, Astron. and Astrophys. **2** (1969) 446.
221. B. Baschek, T. Garz, H. Holweger and J. Richter, Astron. and Astrophys. **4** (1970) 229.
222. R. B. King and A. S. King, Astrophys. J. **82** (1935) 381; **87** (1938) 24.
223. J. M. Bridges and W. L. Wiese, Astrophys. J. **161** (1970) L71.
224. H. Holweger, Z. Astrophys. **65** (1967) 365.
225. A. Unsöld, Joint discussion of Commissions 29 and 36, I.A.U. Sussex, 1970.

REFERENCES

226. R. BONNET, Ann. Astrophys. **31** (1968) 597.
227. W. H. PARKINSON and E. M. REEVES, Solar Phys. **10** (1969) 342.
228. O. GINGERICH and C. DE JAGER, Solar Phys. **3** (1968) 5.
229. O. GINGERICH, R. NOYES, W. KALKOFEN and Y. CUNY, Solar Phys. (1971), in press.
230. W. KALKOFEN, Repts. Commission 36, I.A.U. Sussex, 1970.
231. V. P. MYERSCOUGH, in: Theory and Observation of Normal Stellar Atmospheres, ed. O. Gingerich (M.I.T., 1969) p. 153.
232. O. GINGERICH, I.A.U. Symposium No. 36: Ultraviolet Spectra and Related Ground-based observations, eds. L. Houziaux and H. E. Butler (1970) p. 140.
233. R. CAYREL, Smithsonian Astrophys. Obs. Rept. **167** (1964) 169.
234. J. L. LINSKY, Smithsonian Astrophys. Obs. Rept. **274** (1968) 1.
235. J. T. JEFFERIES, Astrophys. J. **132** (1960) 775.
236. J. ZIRKER, Solar Phys. **3** (1968) 164.
237. A. BURGESS and M. J. SEATON, Mon. Not. R. Astr. Soc. **127** (1964) 355.
238. D. L. FLOWER, J. Phys. B **4** (1971) 697.
239. A. B. UNDERHILL, Publ. Dominion Astrophys. Obs. **11** (1962) 467.
240. O. GINGERICH, Astrophys. J. **138** (1963) 576.
241. S. E. STROM and E. AVRETT, Astrophys. J. Suppl. **12** (1965) 1.
242. D. MIHALAS, Astrophys. J. Suppl. **9** (1965) 321; **11** (1965) 184.
243. O. GINGERICH, Smithsonian Astrophys. Obs. Rept. **167** (1964) 17.
244. D. MIHALAS and D. C. MORTON, Astrophys. J. **142** (1965) 253.
245. T. F. ADAMS and D. C. MORTON, Astrophys. J. **152** (1968) 195.
246. D. C. MORTON, in: Theory and Observation of Normal Stellar Atmospheres, ed. O. Gingerich (M.I.T., 1969) p. 253.
247. R. C. BLESS, A. D. CODE and T. E. HOUCK, Astrophys. J. **153** (1968) 561.
248. R. WILSON and A. BOKSENBERG, Ann. Rev. Astron. and Astrophys. **7** (1969) 421.
249. S. E. STROM, in: Theory and Observation of Normal Stellar Atmospheres, ed. O. Gingerich (M.I.T., 1969) p. 99.
250. G. W. VAN CITTERS and D. C. MORTON, Astrophys. J. **161** (1970) 695.
251. W. KALKOFEN, Smithsonian Astrophys. Obs. Rept. **167** (1964) 175.
252. W. KALKOFEN, J. Quant. Spectr. Radiative Transfer **6** (1966) 633.
253. D. PETERSON and S. E. STROM, in: Theory and Observation of Normal Stellar Atmospheres, ed. O. Gingerich (M.I.T., 1969) p. 279.
254. D. PETERSON and S. E. STROM, Astrophys. J. **157** (1969) 1341.
255. D. MIHALAS, Astrophys. J. **149** (1967) 169.
256. H. R. GRIEM, Astrophys. J. **147** (1967) 1092.
257. L. H. AUER and D. MIHALAS, Astrophys. J. **156** (1969) 157.
258. D. MIHALAS and L. H. AUER, Astrophys. J. **161** (1970) 1129.
259. J. NORRIS, Thesis, Mount Stromlo and Siding Springs Observatory, Australian National University (1970).
260. H. JOHNSON and A. POLAND, J. Quant. Spectr. Radiative Transfer **9** (1969) 1151.
261. A. POLAND, Astrophys. J. **160** (1970) 609.
262. H. L. SHIPMAN and S. E. STROM, Astrophys. J. **159** (1970) 183.
263. A. G. HEARN, Mon. Not. R. Astr. Soc. **150** (1970) 227.
264. M. A. J. SNIJDERS and A. B. UNDERHILL, Mon.Not. R. Astr. Soc. **151** (1971) 215.
265. H. ABT, A. MEINEL, W. MORGAN and J. TAPSCOTT, An Atlas of Low Dispersion Grating Stellar Spectra (Kitt Peak Nat. Obs., Steward Obs., Yerkes Obs., 1968).
266. D. MIHALAS and L. H. AUER, Astrophys. J. **160** (1970) 1161.
267. L. H. AUER, Repts. Commission 36, I.A.U., Sussex, 1970.
268. K. B. GEBBIE and M. J. SEATON, Nature **199** (1963) 580.
269. K.-H. BÖHM and W. DEINZER, Z. Astrophys. **61** (1965) 1; **63** (1966) 177.
270. K.-H. BÖHM, Astron. and Astrophys. **2** (1969) 180.

271. D. Hummer and D. Mihalas, Mon. Not. R. Astr. Soc. **147** (1970) 339.
272. L. H. Aller, Gaseous Nebulae (Wiley, N.Y., 1956) Chs. 4 and 5.
273. M. J. Seaton, Repts. Prog. Physics **23** (1960) 313.
274. M. J. Seaton and D. E. Osterbrock, Astrophys. J. **125** (1957) 66.
275. H. E. Saraph and M. J. Seaton, Mon. Not. R. Astr. Soc. **148** (1970) 367.
276. P. de A. P. Martins and M. J. Seaton, J. Phys. B **2** (1969) 333.
277. S. J. Czyzak, T. K. Krueger, P. de A. P. Martins, H. E. Saraph and M. J. Seaton, Mon. Not. R. Astr. Soc. **148** (1970) 361.
278. B. Höglund and P. Mezger, Science **150** (1965) 339.
279. L. Goldberg, Astrophys. J. **144** (1966) 1225.
280. A. Duprée and L. Goldberg, Ann. Rev. Astr. and Astrophys. **8** (1970) 231.
281. R. J. Harman and M. J. Seaton, Mon. Not. R. Astr. Soc. **132** (1966) 15.
282. J. B. Kaler, Astrophys. J. **160** (1970) 887.
283. M. J. Seaton, Joint discussion B on Helium abundance, I.A.U., Sussex (1970).
284. M. Peimbert and H. Spinrad, Astrophys. J. **159** (1970) 809.
285. P. Palmer, B. Zuckermann, H. Penfield and A. Lilley, Astrophys. J. **156** (1969) 887.
286. H. Spinrad and R. F. Wing, Ann. Rev. Astron. and Astrophys. **7** (1969) 249.
287. B. E. J. Pagel, Quart. J. Roy. Astr. Soc. **11** (1970) 172.
288. J. Auman, Astrophys. J. Suppl. **14** (1967) 171.
289. V. Kunde, Astrophys. J. **153** (1968) 435.
290. H. L. Johnson and I. Marenin, Proc. Amer. Astron. Soc., Boulder 1970, paper 27.03.
291. O. Gingerich and S. S. Kumar, Astron. J. **69** (1964) 139.
292. S. S. Kumar, Mem. Soc. Roy. Sci. Liege **9** (1964) 476.
293. O. Gingerich, D. W. Latham, J. L. Linsky and S. S. Kumar, Colloquium on Late-type Stars, ed. M. Hack (Trieste, Osservatorio Astronomico di Trieste, 1966).
294. T. Tsuji, Colloquium on Late-type Stars, ed. M. Hack (Trieste, Osservatorio Astronomico di Trieste, 1966).
295. O. Gingerich, Smithsonian Astrophys. Obs. Rept. **240** (1967) 1.
296. J. Auman, Astrophys. J. **157** (1969) 799.
297. G. Goon and J. Auman, Astrophys. J. **161** (1970) 533.
298. M. Spinrad and M. S. Vardya, Astrophys. J. **149** (1966) 399.
299. H. L. Johnson, I. Coleman, R. I. Mitchell and D. L. Steinmetz, Comm. Lunar and Plan. Obs. **7** (1968) 83.
300. H. Spinrad and B. J. Taylor, Astrophys. J. **157** (1969) 1279.
301. F. Hoyle and N. C. Wickramasinghe, Mon. Not. R. Astr. Soc. **124** (1962) 417.
302. B. Donn, N. C. Wickramasinghe, J. P. Hudson and T. P. Stecher, Astrophys. J. **153** (1968) 451.
303. K. S. Krishna Swamy, Astrophys. J. **162** (1970) 259.
304. G. B. Field and J. Hitchcock, Phys. Rev. Letters **16** (1966) 817.
305. P. Thaddeus and J. Clauser, Phys. Rev. Letters **16** (1966) 819.
306. W. M. Goss, Astrophys. J. Suppl. **15** (1968) 131.
307. P. Palmer and B. Zuckermann, Astrophys. J. **148** (1967) 727.
308. A. C. Cheung, D. M. Rank, C. H. Townes, D. D. Thornton and W. J. Welch, Phys. Rev. Letters **21** (1968) 1701; Nature **221** (1969) 626.
309. L. Snyder, D. Buhl, B. Zuckermann and P. Palmer, Phys. Rev. Letters **22** (1969) 679.
310. B. Zuckermann, D. Buhl, P. Palmer and L. Snyder, Astrophys. J. **160** (1970) 485.
311. C. E. Heiles, Astrophys. J. **151** (1968) 919.
312. P. Palmer, B. Zuckermann, D. Buhl and L. Snyder, Astrophys. J. **156** (1969) L147.
313. W. B. Somerville, Repts. Prog. Phys., in press 1971.

314. W. L. W. SARGENT and L. SEARLE, Astrophys. J. **139** (1964) 793.
315. W. L. W. SARGENT and P. STRITTMATTER, Astrophys. J. **145** (1966) 938.
316. W. L. W. SARGENT and L. SEARLE, Magnetic and Related Stars, ed. R. Cameron (Mono. Book Co., 1967) p. 209.
317. S. E. STROM, O. GINGERICH and K. M. STROM, Astrophys. J. **146** (1966) 880.
318. S. E. STROM and K. M. STROM, Astrophys. J. **155** (1969) 17.
319. L. H. AUER, D. MIHALAS, L. H. ALLER and J. ROSS, Astrophys. J. **145** (1966) 153.
320. K. M. STROM, Astron. and Astrophys. **2** (1969) 182.
321. L. H. ALLER and W. BIDELMAN, Astrophys. J. **139** (1964) 171.
322. G. MICHAUD, Astrophys. J. **160** (1970) 641.
323. D. PETERSON, Astrophys. J. **161** (1970) 685.
324. W. D. WATSON, Astrophys. J. **162** (1970) L45.
325. V. WEIDEMANN, Ann. Rev. Astron. and Astrophys. **6** (1968) 351.
326. V. WEIDEMANN, Astrophys. J. **131** (1960) 638.
327. P. A. STRITTMATTER and D. T. WICKRAMASINGHE, Mon. Not. R. Astr. Soc. **152** (1971) 47.
328. P. HILL, Mon. Not. R. Astr. Soc. **129** (1965) 137.
329. G. HERBIG, Astrophys. J. **110** (1949) 143.
330. L. SEARLE, Astrophys. J. **133** (1961) 531.
331. P. C. KEENAN and J. L. GREENSTEIN, Mt. Wilson and Palomar Obs. reprint (1963).
332. B. WARNER, Mon. Not. R. Astr. Soc. **137** (1967) 119.
333. I. DANZIGER, Mon. Not. R. Astr. Soc. **130** (1965) 199.
334. V. P. MYERSCOUGH, Astrophys. J. **153** (1968) 421.
335. H. SPINRAD and B. TAYLOR, private communication, 1967.
336. K. HUNGER and D. KLINGLESMITH, Astrophys. J. **157** (1969) 721.
337. R. THOMPSON, H. SCHNOPPER, R. MITCHELL and H. L. JOHNSON, Astrophys. J. **158** (1969) L55.
338. R. THOMPSON and H. SCHNOPPER, Astrophys. J. **160** (1970) L97.
339. J. GAUSTAD, F. GILLETT, R. KNACKE and W. STEIN, Astrophys. J. **158** (1969) 613.
340. W. STEIN, J. GAUSTAD, F. GILLETT and R. KNACKE, Astrophys. J. **155** (1969) L3.
341. T. LEE and M. FEAST, Astrophys. J. **157** (1969) L173.
342. M. FEAST, Repts. Commission 29, IAU Sussex (1970).
343. S. GIESEL, Astrophys. J. **161** (1970) L105.

CHAPTER 6

POLARIZED ORBITAL APPROXIMATIONS

BY

R. J. DRACHMAN and A. TEMKIN

Goddard Space Flight Center,
National Aeronautics and Space Administration, Greenbelt, Maryland 20771,
U.S.A.

Contents

	Page
6-1. Introduction	401
6-2. The basic method and notation	402
6-3. The polarized target wave function	404
A. The adiabatic approximation	404
B. Perturbation theory	405
C. The one-electron target	406
D. Variational-perturbation methods	414
E. The exact static solution for the one-electron target	417
F. The many-electron target; Sternheimer's approximation	421
6-4. The total polarized orbital wave function and the scattering problem	423
A. Introductory remarks	423
B. The total wave function and non-variational methods	424
C. Variational and variationally motivated methods	426
6-5. Applications to electron scattering and reactions	433
A. Scattering from one-electron targets	433
1. Elastic scattering	433
2. Inelastic scattering	440
B. Scattering from helium	445
C. Scattering from other (non-highly polarizable) atoms	448
D. Scattering from highly polarizable atomic systems	451
E. Electron–molecule scattering	454
F. Photoionization, autoionization and ionization	456
6-6. Applications to positron scattering and annihilation	462
A. Positron–hydrogen scattering	463
B. Positron–helium scattering and annihilation	469
References	477

§ 6-1. Introduction

In this article we propose to examine a class of approximations which have proven to be quite useful in the solution of electron and positron scattering problems. Collectively called polarized-orbital approximations, they combine effectively a considerable degree of accuracy with simplicity, and, as an added virtue, are understandable in physical terms. It is this feature which lends these methods their special usefulness and justifies their detailed study.

Even for the case of the simplest atoms, hydrogen and helium, the Schrödinger equation of electron–atom scattering is of too high dimension to be solved directly using present electronic computers. The goal of approximation theory at low energies (where the Born and related approximations are unreliable) is usually to expand the scattering wave function in some appropriate set of basis functions, which after truncation yields a finite set of coupled equations. These may be ordinary differential or integro-differential equations as in the close-coupling method [1], lower-dimensional partial differential equations as in the non-adiabatic method [2] or the algebraic (matrix) equations resulting from the application of variational principles [3]. In all these methods, a reasonably rapid convergence rate is essential to their practicality.

By contrast, the polarized orbital methods begin with an attempt to incorporate the essential physics of the problem in the form of the wave function. It is recognized at the outset, that the incoming electron will perturb the atomic system in such a way as to induce in it electric multipole moments, which in turn accelerate the electron inward towards the atom. If the atom is sufficiently polarizable, this induced attraction can be a dominant contributor to the effective scattering potential acting on the electron. It is especially important in the case of positron scattering where the undistorted charge density of the atomic target generates a repulsive potential, that the attractive polarization potential be represented as accurately as possible, since a considerable degree of cancellation can occur.

But the polarized orbital approximations do not simply produce an effective polarization potential in which the scattered particle is to move, as does the so-called adiabatic approximation. In addition, they offer a prescription for the wave function of the system. This has several important consequences: For electron scattering, it enables one to take into account the effect of the Pauli principle and to calculate photoionization; for positron scattering, it makes possible the computation of annihilation rates and angular correlations. By suitable use of the polarized-orbital wave function, one can also obtain bounds on the scattering length and phase shifts which compare favorably with those obtained in other ways.

Since its introduction in 1957 [4] the basic idea of polarized orbitals has undergone a fairly wide series of modifications. One of our principal aims in this article is to define clearly the various modified polarized orbital approximations, contrasting them with one another and with the original method of polarized orbitals, since much confusion and some criticism has appeared in the literature.

§ 6-2. The basic method and notation

We are concerned with the problem of low-energy scattering of electrons (which will be understood to include positrons) from atoms, molecules and ions. In most cases the nuclei will be assumed infinitely heavy. The non-relativistic Schrödinger equation is to be solved approximately, and only Coulomb forces act between the various particles. If we are considering electron scattering from an atomic system (of nuclear charge Z) then the complete Hamiltonian is

$$H = -\sum_{j=1}^{N+1} \nabla_j^2 - \sum_{j=1}^{N+1} \frac{2Z}{r_j} + \sum_{i>j=1}^{N+1} \frac{2}{r_{ij}}, \tag{6-2-1a}$$

where r_j is measured from the nucleus and $r_{ij} = |r_i - r_j|$. Energies are measured in rydberg (13.6 eV), and lengths are in units of the Bohr radius $a_0 = \hbar^2/me^2$. If the incident particle is a positron, eq. (6-2-1a) is modified to a somewhat more unsymmetric notation to emphasize the fact that the incident particle differs from the orbital electrons. In that case we write

$$H = -\sum_{j=1}^{N} \nabla_j^2 - \nabla_x^2 - \sum_{j=1}^{N} \frac{2Z}{r_j} + \frac{2Z}{x} + \sum_{i>j=1}^{N} \frac{2}{r_{ij}} - \sum_{j=1}^{N} \frac{2}{|x-r_j|}. \tag{6-2-1b}$$

In this case the positron is described by the unsymmetric co-ordinate x to emphasize its nonidentical nature to the orbital electrons r_j. Furthermore, we shall more often use unsymmetrical notation to emphasize the fact, discussed in detail in the next section, that in the static problem the rôle of the incident particle is treated quite unsymmetrically even if it is an electron.

Basic to the general class of polarized orbital approximations [4] is the division of the scattering problem into two parts which can be treated sequentially: the calculation of the distorted target wave function and the calculation of the scattering cross-section. In order to make this convenient division possible, we use the adiabatic approximation: the idea that the electrostatic potential produced by the incoming electron is assumed to vary so slowly that the target electrons can smoothly adjust. Semi-classically, this means that the speed of the incoming electron must be considerably less than the orbital speed of the atomic electrons. (A more quantitative discussion of the non-adiabatic corrections will be given later.) We then must obtain a wave function for the target in which the co-ordinate of the incoming electron is treated as a parameter.

Given such a target function ϕ, which must be normalizable with respect to the coordinates of the target electrons, we assume that the scattering wave function has the form

$$\Psi = \chi(x)\phi(r_1, r_2, \ldots, r_N, x), \tag{6-2-2}$$

where x is the coordinate of the incoming electron, and χ has the proper asymptotic form for pure elastic scattering:

$$\chi \sim e^{i\mathbf{k}\cdot\mathbf{x}} + f(\theta)\frac{e^{ikx}}{x}, \tag{6-2-3}$$

while ϕ approaches the unperturbed target wave function ϕ_0 for large x. If the incoming particle is a positron, eq. (6-2-2) is complete; if it is an electron it is necessary to antisymmetrize Ψ.

The basic method for determining the function χ and, from it, the scattering cross-section, imposes the following simple condition [5]:

$$\int dr_1 \ldots dr_N \phi_0(r_1, \ldots, r_N)[H-E]\Psi(r_1, \ldots, r_N, x) = 0. \tag{6-2-4}$$

This is a necessary condition for Ψ to satisfy the Schrödinger equation, but it is obviously not sufficient. Only if the projection of $(H-E)\Psi$ onto *all* the states of the target vanished would we have an exact solution. Nevertheless, eq. (6-2-4) is often surprisingly good, *provided* the perturbed target function ϕ is good enough.

To demonstrate the derivation of the basic polarized orbital differential equation without exchange, we re-write the Hamiltonian [eq. (6-2-1b)] as follows:

$$H = H_0(r_1, \ldots, r_N) - \nabla_x^2 + V(r_1, \ldots, r_N, x), \tag{6-2-1c}$$

where

$$V(r_1, \ldots, r_N, x) = \frac{2Z}{x} - \sum_{j=1}^{N}\frac{2}{|r_j - x|}, \tag{6-2-5}$$

and H_0 contains those parts of H which are independent of the co-ordinate x. The total energy is $E = E_0 + k^2$, where

$$(H_0 - E_0)\phi_0(r_1, \ldots, r_N) = 0. \tag{6-2-6}$$

With this notation and making use of eq. (6-2-6) we can re-write eq. (6-2-4) in the simple form

$$-[\nabla_x^2 + k^2]g\chi + \langle\phi_0|V|\phi\rangle\chi = 0, \tag{6-2-7}$$

where $g(x) \equiv \langle\phi_0|\phi\rangle$, and the bracket notation implies integration over all coordinates except x. To put this equation into the closest analogy with a one-particle Schrödinger equation, let us define a new function $\bar{\chi}$ by

$$\bar{\chi}(x) \equiv g(x)\chi(x). \tag{6-2-8}$$

Then eq. (6-2-7) becomes

$$-[\nabla_x^2 + k^2]\bar{\chi} + V_{\mathrm{AD}}(x)\bar{\chi} = 0, \tag{6-2-9}$$

with the further definition of the adiabatic potential

$$V_{\mathrm{AD}}(x) \equiv \frac{\langle\phi_0|V|\phi\rangle}{\langle\phi_0|\phi\rangle}. \tag{6-2-10}$$

[The analog of eq. (6-2-9) when Ψ must be antisymmetrized will be discussed in § 6-3.] Since $g(x) \to 1$ for large x, it follows that $\bar{\chi} \to \chi$, and the phase shifts obtained from eqs. (6-2-7) and (6-2-9) are identical. Eq. (6-2-10) is a well-known result in the context of perturbed stationary states [6].

§ 6-3. The polarized target wave function

A. THE ADIABATIC APPROXIMATION

Let us now examine methods for obtaining the distorted (or polarized) atomic wave function. The adiabatic approximation is mathematically defined as the problem which arises when the kinetic energy of the projectile is neglected; therefore the adiabatic Hamiltonian is [cf. eq. (6-2-1c)]:

$$H_{\mathrm{AD}} \equiv H_0 + V.$$

The resultant adiabatic Schrödinger equation, $H_{\mathrm{AD}}\phi = E(x)\phi$, is considerably simpler than the full Schrödinger equation, but it too cannot generally be solved exactly. In the following we shall discuss perturbative approximations but it is well to keep in mind that the adequacy of such approximations

B. Perturbation theory

We consider the adiabatic Schrödinger equation

$$(H_0 + V)\phi = E(x)\phi. \tag{6-3-1}$$

Multiplying on the left by ϕ_0 and integrating over all coordinates but x gives

$$V_{AD}(x) = E(x) - E_0 \tag{6-3-2}$$

when the definition of eq. (6-2-10) is used. We would like to obtain approximate solutions for ϕ and approximate eigenvalues $E(x)$.

Let us introduce a parameter λ in the usual way to keep track of the various orders of approximation. Let

$$\phi = \sum_{j=0}^{\infty} \lambda^j G_j \phi_0, \quad E(x) = \sum_{j=0}^{\infty} \lambda^j E_j(x), \quad V \to \lambda V, \tag{6-3-3}$$

substitute in eq. (6-3-1) and equate terms of like powers of λ. The first three equations are

$$H_0 \phi_0 = E_0 \phi_0, \tag{6-3-4a}$$

$$(H_0 G_1 + V)\phi_0 = (E_0 G_1 + E_1)\phi_0, \tag{6-3-4b}$$

$$(H_0 G_2 + V G_1)\phi_0 = (E_0 G_2 + E_1 G_1 + E_2)\phi_0. \tag{6-3-4c}$$

Without loss of generality we can assume $\langle \phi_0 | G_n | \phi_0 \rangle \equiv 0$ for $n > 0$, and by multiplying eq. (6-3-4) on the left by ϕ_0 and integrating we find, as usual

$$E_0 = \langle H_0 \rangle \tag{6-3-5a}$$

$$E_1 = \langle V \rangle \tag{6-3-5b}$$

$$E_2 = \langle V G_1 \rangle. \tag{6-3-5c}$$

From now on we let $\langle \rangle$ represent expectation values with respect to ϕ_0; thus these quantities are functions of x only. In addition we shall use Dirac bra and ket notation for wavefunctions, when convenient. Thus we can make a first-order approximation to $|\phi\rangle$ in the following form

$$|\phi\rangle = [1 + G]|\phi_0\rangle, \tag{6-3-6}$$

where we have let $\lambda = 1$ and defined $G \equiv G_1$.

Eq. (6-3-4b) is equivalent to [7]

$$[H_0, G]|\phi_0\rangle = (\langle V\rangle - V)|\phi_0\rangle. \tag{6-3-7}$$

The square bracket in (6-3-7) means the commutator, and use has also been made of (6-3-5a). Eq. (6-3-7) is the fundamental equation for G in first-order perturbation theory. To second order in perturbation theory the adiabatic potential is given by

$$V_{AD}^{(1+2)}(x) = \langle V\rangle + \langle VG\rangle, \tag{6-3-8}$$

[see (6-3-2), (6-3-3) and (6-3-5) or (6-2-10) and (6-3-6)] and we define the two contribution to $V_{AD}^{(1+2)}$ as

$$V_1(x) \equiv \langle V\rangle$$

and

$$V_2(x) \equiv \langle VG\rangle. \tag{6-3-9}$$

It is of interest to relate the above to the formulation in terms of intermediate states. In this familiar method we have an expression for the second-order adiabatic potential

$$V_2(x) = \sum_{n \neq 0} \frac{\langle \phi_0|V|\phi_n\rangle\langle \phi_n|V|\phi_0\rangle}{E_0 - E_n}. \tag{6-3-10}$$

From eq. (6-3-7) we obtain

$$\langle \phi_n|[H_0, G]|\phi_0\rangle = (E_n - E_0)\langle \phi_n|G|\phi_0\rangle = -\langle \phi_n|V|\phi_0\rangle. \tag{6-3-11}$$

When this expression is inserted in eq. (6-3-10), we obtain

$$V_2(x) = \sum_{n \neq 0} \langle \phi_0|V|\phi_n\rangle\langle \phi_n|G|\phi_0\rangle. \tag{6-3-12}$$

Use of the closure relation together with the requirement $\langle G\rangle = 0$ enables us to recover the expression (6-3-9). It is, therefore, formally immaterial which method is used to obtain the perturbed target function and adiabatic potential. Unless, however, a very small number of discrete states dominates the sum in eq. (6-3-10) one does better to work with the differential equation.

C. THE ONE-ELECTRON TARGET

To examine the perturbation method in more detail, let us specialize the problem. Consider a hydrogen atom fixed at the origin of co-ordinates and perturbed by a (positive) charge at position x. Then we have [from eq. (6-2-5)]

$$V(r, x) = \frac{2}{x} - \frac{2}{|r-x|}, \tag{6-3-13}$$

and $\phi_0(r) = \pi^{-\frac{1}{2}} e^{-r}$. The first-order perturbation equation (6-3-7) takes the form

$$\nabla_r^2 G + 2\phi_0^{-1} \nabla_r G \cdot \nabla_r \phi_0 = V(r, x) - V_1(x), \tag{6-3-14}$$

and, after inserting the form of $\phi_0(r)$ and carrying out the integration indicated by the brackets, we obtain

$$\nabla_r^2 G - 2\frac{\partial G}{\partial r} = \frac{2}{x}[1 - e^{-2x}(x+1)] - \frac{2}{|r-x|}. \tag{6-3-15}$$

This is the equation whose solutions, approximate and exact, will concern us for most of this section. We note in passing that the function $G(r, x)$ is determined only to within an additive function of x, so we will always be able to satisfy the condition $\langle G \rangle = 0$.

A solution of eq. (6-3-15) can be obtained as an expansion in Legendre polynomials, since the last term in the equation has a simple multipole expansion, and the operators ∇^2 and d/dr are diagonal in an angular momentum representation. Let

$$G(r, x) = \sum_{l=0}^{\infty} g_l(r, x) P_l(\gamma)$$

and

$$V(r, x) = \sum_{l=0}^{\infty} v_l(r, x) P_l(\gamma), \tag{6-3-16}$$

where $\gamma \equiv \cos \theta_{rx}$. The multipole components satisfy the equation

$$\frac{\partial^2 g_l}{\partial r^2} + 2\left(\frac{1}{r} - 1\right)\frac{\partial g_l}{\partial r} - \frac{l(l+1)}{r^2} g_l = v_l - 2e^{-2x}\left(1 + \frac{1}{x}\right)\delta_{l0}. \tag{6-3-17}$$

The adiabatic potential is then

$$V_2(x) = \sum_{l=0}^{\infty} \frac{1}{2l+1} \int dr \, \phi_0^2 g_l v_l.$$

For $l = 0$: $v_0 = 0$ for $x \geq r$ and $v_0 = [x^{-1} - r^{-1}]$ for $x \leq r$. For $l \neq 0$:

$v_l = -2r^l x^{-(l+1)}$ for $x \geqq r$ and $v_l = -2x^l \, r^{-(l+1)}$ for $x \leqq r$. Particular solutions are easily obtained for $l \neq 0$. They are

$$\bar{g}_{l>} = r^l x^{-(l+1)} \left(\frac{1}{l} + \frac{r}{l+1} \right), \quad \text{for } x > r$$

and

$$\bar{g}_{l<} = -x^l r^{-(l+1)} \left(\frac{1}{l+1} + \frac{r}{l} \right), \quad \text{for } x < r. \tag{6-3-18}$$

To these solutions one adds the general solution of the corresponding homogeneous equation and adjusts the arbitrary functions of x to satisfy boundary and continuity conditions. Before carrying out this process, however, we will now examine an approximate form of adiabatic potential derived from the particular solution $g_{l>}$.

It has been argued [8] that only the asymptotic region $(x \gg r)$ satisfies the adiabatic conditions sufficiently well and thus one should not take seriously those parts of the potential V_{AD} which come from the region $x < r$. The Callaway–Temkin potential [4, 9, 10] is derived by assuming that $g_{l<} = 0$, retaining the form of $g_{l>}$ given by eq. (6-3-18) and omitting $l = 0$

TABLE 6-3-1

Adiabatic potential for the e^+–H system. Columns (a), (b), (c), (d), are the same as for the e^-–H case. In all cases the first-order potential has been removed and the negative of the potential is listed

x	(a)	(b)	(c)	(d)	(e)	(f)
0	0	0	0	1	$\frac{2}{3}$	1
0.2	—	—	—	0.9220	0.616	—
0.4	—	—	—	0.7731	—	0.7123
0.5	0.144	0.199	0.122	0.6965	0.485	—
1.0	0.147	0.207	0.170	0.399	0.326	0.313
1.5	—	0.159	—	0.234	0.235	—
2.0	0.0799	0.109	0.0983	0.141	0.171	0.0724
3.0	0.0348	0.0452	0.0395	0.0514	0.0795	0.0228
4.0	—	0.0183	—	0.0195	0.0295	0.0108
5.0	0.0068	0.0080	0.0070	0.0082	0.0109	0.0061

(a) Callaway–Temkin ($l = 1$), eq. (7-3-19).
(b) Callaway–Temkin (all l), eq. (7-3-23).
(c) Bethe–Reeh ($l = 1$), eq. (7-3-29).
(d) Dalgarno–Lynn, eq. (7-3-37).
(e) Variation-perturbation, eq. (7-3-43) and Fig. 7-3-3.
(f) "Exact", eq. (7-3-52) with $\alpha = \beta$; see Table 7-3-2.

terms entirely; that is, $g_l = \varepsilon(x, r) \bar{g}_{l>}(r, x)$ where ε is the unit step function. The leading term ($l = 1$) of the Callaway–Temkin potential is (see Table 6-3-1)

$$V_{\text{CT},l=1} = -\frac{8}{3x^4} \int_0^x dr\, e^{-2r}(r^4 + \tfrac{1}{2}r^5)$$

$$= -\frac{9}{2x^4}[1 - e^{-2x}(1 + 2x + 2x^2 + \tfrac{4}{3}x^3 + \tfrac{2}{3}x^4 + \tfrac{4}{27}x^5)]. \quad (6\text{-}3\text{-}19)$$

This goes over, as it should, to the famous x^{-4} long-range polarization potential; the coefficient $\tfrac{9}{2}$ is the exact polarizability of hydrogen.

This dipole-only Callaway–Temkin potential has been used with great success in the electron–atom problem (see § 6-5A). For positron–atom scattering, as we shall see, it is not adequate to include only one multipole, and it is interesting to sum over all $l \geq 1$. We write

$$V_{\text{CT}}(x) = -\sum_{l=1}^{\infty} \frac{4\pi}{x^{2l+2}} \int_0^x dr\, r^2 \frac{e^{-2r}}{\pi}\left(\frac{r^{2l}}{l} + \frac{r^{2l+1}}{l+1}\right)\int_{-1}^{1} d\gamma\, P_l^2(\gamma)$$

$$= -8x \int_0^1 dp\, p^2 e^{-2xp}[S_1(p) + xS_2(p)], \quad (6\text{-}3\text{-}20)$$

where we have interchanged the order of summation and integration, set $r = px$, and have made the definitions

$$S_1(p) = \sum_{l=1}^{\infty} \frac{p^{2l}}{l(2l+1)}$$

and

$$S_2(p) = \sum_{l=1}^{\infty} \frac{p^{2l+1}}{(l+1)(2l+1)}. \quad (6\text{-}3\text{-}21)$$

These infinite series can be summed and give

$$S_1(p) = 2 - \left(\frac{1}{p}+1\right)\ln(1+p) + \left(\frac{1}{p}-1\right)\ln(1-p)$$

and

$$S_2(p) = -p + \left(\frac{1}{p}+1\right)\ln(1+p) + \left(\frac{1}{p}-1\right)\ln(1-p). \quad (6\text{-}3\text{-}22)$$

We can evaluate the integral in eq. (6-3-20) using these expressions:

$$V_{CT}(x, = \frac{5}{x^2} - e^{-2x}\left(\frac{5}{x^2} + \frac{2}{x} - 2 + 4x\right)$$

$$+ 2e^{-2x}\left(1 + \frac{1}{x}\right)^2 [\ln(2\gamma x) - \text{Ei}(2x)]$$

$$+ 2e^{2x}\left(1 - \frac{1}{x}\right)^2 [\text{Ei}(-4x) - \text{Ei}(-2x)]$$

$$- 2e^{-2x}\left(\frac{1}{x^2} + \frac{2}{x} + 1 - 4x\right)\ln 2, \qquad (6\text{-}3\text{-}23)$$

where $\text{Ei}(-z) = -\int_z^\infty dy\, e^{-y}/y$ and Euler's constant $\ln \gamma = 0.577\,215\,7\ldots$. We may note from eq. (6-3-20) that $V_{CT}(x)$ vanishes near the origin linearly with x; in fact, $V_{CT}(x) \to \frac{8}{9}x\,(6\ln 2 - 5)$ in this limit. For large x, we can neglect all exponentially decreasing terms and find the limiting form

$$V_{CT}(x) \underset{x \to \infty}{=} \frac{5}{x^2} - 2\left(1 + \frac{1}{x}\right)^2 e^{-2x} \text{Ei}(2x) - 2\left(1 - \frac{1}{x}\right)^2 e^{2x} \text{Ei}(-2x).$$

$$(6\text{-}3\text{-}24)$$

Using the asymptotic expansion for the exponential integral, we can obtain the asymptotic series [7]

$$V_{CT}(x) \sim -2 \sum_{l=1}^{\infty} \frac{(2l+2)!(l+2)}{l(l+1)(2x)^{2l+2}}, \qquad (6\text{-}3\text{-}25)$$

which also is derivable from eq. (6-3-20) and exhibits the long-range behavior of each term in the multipole expansion.

Let us return now and consider the exact solutions of eq. (6-3-17). For $l > 0$ we have already derived particular solutions in both regions $x > r$ and $x < r$, and we may note that they are well-behaved for large and small x respectively. At the point $x = r$, the functions g_l and dg_l/dr must be continuous; these particular solutions are not continuous although by accident their derivatives are. It is thus necessary to solve the homogeneous equation obtained from (6-3-17) by setting the right-hand side equal to zero. It is easy to find two linearly independent solutions having the forms

$$g_{l1}(r) = r^{-(l+1)} \sum_{n=0}^{l+1} c_n r^n,$$

and

$$g_{12}(r) = e^{2r} r^{-(l+1)} \sum_{n=0}^{l-1} d_n r^n, \qquad (6\text{-}3\text{-}26)$$

with coefficients satisfying the recursion relations

$$c_{n+1}/c_n = 2(n-l-1)(n+1)^{-1}(n-2l)^{-1}$$

and

$$d_{n+1}/d_n = -2(n-l+1)(n+1)^{-1}(n-2l)^{-1}. \qquad (6\text{-}3\text{-}27)$$

For large r, g_{11} is well-behaved, while the rising exponential rules out g_{12}. For small r, a regular solution is constructed by setting $c_0 = d_0$ and using the difference $g_{11} - g_{12}$, which goes to zero as r^l. One solution of the homogeneous equation is thus applicable in each region, $x \geq r$ and $x \leq r$, and each can be multiplied by an arbitrary function of x, which is then determined by imposing the continuity conditions at $x = r$.* The leading (dipole) term yields the following expressions:

$$g_{1>} = \frac{1}{x^2}(r + \tfrac{1}{2}r^2) + \tfrac{3}{2}e^{-2x}\left(1 + \frac{1}{x}\right)^2\left(1 + \frac{1}{r} + \frac{1}{2r^2} - \frac{e^{2r}}{2r^2}\right),$$

$$g_{1<} = -x\left(\frac{1}{r} + \frac{1}{2r^2}\right) + \frac{3}{2}\left[1 - \frac{1}{x^2} + e^{-2x}\left(1 + \frac{1}{x^2}\right)\right]\left(1 + \frac{1}{r} + \frac{1}{2r^2}\right), \qquad (6\text{-}3\text{-}28)$$

and the resulting Bethe–Reeh potential is [11, 12]

$$V_{BR, l=1} = -\frac{9}{2x^4}\left[1 - \tfrac{1}{3}e^{-2x}(1 + 2x + 6x^2 + \tfrac{20}{3}x^3 + \tfrac{4}{3}x^4) - \tfrac{2}{3}e^{-4x}(1+x)^4\right], \qquad (6\text{-}3\text{-}29)$$

and is given in Table 7-3-1. This potential vanishes near the origin like x^2, in contrast to the behavior of the Callaway–Temkin potential. Higher multipoles have been treated by Reeh [12].

It is also simple to solve eq. (6-3-17) directly for the case of $l = 0$, since the g_0 term is missing. Regularity requirements fix the integration constants, and one obtains

---

* It is also possible to construct the Green's function from these two solutions of the homogeneous equation. The complete solution of the inhomogeneous equation (6-3-17) is then obtained by quadrature.

$$g_{0>} = e^{-2x}\left(1+\frac{1}{x}\right)\left[r-\frac{1}{2r}+\ln(2\gamma r)-\text{Ei}(2r)+\frac{e^{2r}}{2r}+\ln(2\gamma x)-\frac{e^{2x}}{2x}\right]$$

$$+\left(1-\frac{1}{x}\right)\text{Ei}(-2x)+\frac{1}{2x}\left(1+\frac{1}{x}\right)+e^{-2x}\left(x-\tfrac{3}{2}-\frac{1}{x}\right),$$

$$g_{0<} = e^{-2x}\left(1+\frac{1}{x}\right)\left[r-\frac{1}{2r}+\ln(2\gamma r)+\ln(2\gamma x)-\text{Ei}(2x)\right]$$

$$+\left(1-\frac{1}{x}\right)\left[r-\frac{1}{2r}+\ln\left(\frac{r}{x}\right)+\text{Ei}(-2x)\right]-r+1+\frac{1}{x}$$

$$+e^{-2x}\left(x-\tfrac{3}{2}-\frac{1}{x}\right). \tag{6-3-30}$$

Appropriate functions of x have been included here to insure continuity and also to satisfy the condition $\langle g_0 \rangle = 0$. The corresponding monopole potential can be obtained analytically [12], but it is rather complicated and will not be given here; it is found to decrease exponentially for large x and is negligible in the region where the dipole term goes as x^{-4}. For small x however, it is the dominant part of the adiabatic potential and it equals -1 at $x = 0$.

Eq. (6-3-15) can also be separated in elliptical coordinates and solved in closed form [13]. The integrals needed to evaluate $V_2(x)$ can also be carried out analytically. To some extent this is just a mathematical tour-de-force, and some workers prefer to use the Bethe–Reeh multipole expansion, since in practice only about three or four terms are needed to get a good approximation to the sum. The result obtained by Dalgarno and Lynn [13], however, is fairly simple and elegant, and will be useful for us further on, so we will discuss it here.

Since we keep the nucleus and the positron fixed and solve eq. (6-3-15) only for the electron wave function, it is convenient to let the polar axis pass through the two fixed particles. We will be interested only in solutions independent of the azimuthal angle ϕ. The two remaining spherical coordinates r and γ are written as

$$r = \tfrac{1}{2}x(\lambda+\mu)$$
$$\gamma \equiv \cos\theta_{xr} = (1+\lambda\mu)(\lambda+\mu)^{-1}. \tag{6-3-31}$$

(The situation is shown in Fig. 6-3-1.) Since $0 \le r < \infty$ and $-1 \le \gamma \le 1$, it follows that the ranges of the elliptical coordinates λ and μ are $1 \le \lambda < \infty$

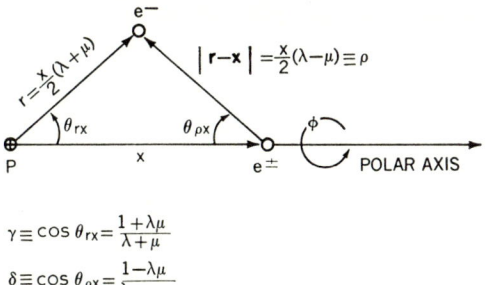

Fig. 6-3-1. Coordinate system in which the adiabatic Schrödinger equation (6-3-1) and first-order perturbation equation (6-3-15) are separable. Only solutions independent of the azimuthal angle ϕ are considered here; the variables λ and μ define the elliptical coordinate system.

and $-1 \leq \mu \leq 1$. The three-dimensional volume element is transformed as follows:

$$d^3r = \tfrac{1}{8}x^3(\lambda^2 - \mu^2) d\lambda \, d\mu \, d\phi. \tag{6-3-32}$$

Other quantities relevant to the problem are

$$\phi_0 = \pi^{-\frac{1}{2}} e^{-\frac{1}{2}x(\lambda + \mu)},$$

and

$$V = \frac{2}{x}(\lambda - \mu - 2)(\lambda - \mu)^{-1}. \tag{6-3-33}$$

With these definitions, eq. (6-3-15) in elliptical coordinates can be written as

$$\frac{\partial}{\partial \lambda}(\lambda^2 - 1)\frac{\partial G}{\partial \lambda} + \frac{\partial}{\partial \mu}(1 - \mu^2)\frac{\partial G}{\partial \mu} - x(\lambda^2 - 1)\frac{\partial G}{\partial \lambda} - x(1 - \mu^2)\frac{\partial G}{\partial \mu} =$$
$$= A(x)(\mu^2 - \lambda^2) - x(\lambda + \mu), \tag{6-3-34}$$

where $A(x) = \tfrac{1}{2}x[-1 + (1 + x)e^{-2x}]$. This equation is additively separable in λ and μ, and is an inhomogeneous equation of second order with G itself missing. Eq. (6-3-34) thus reduces to two ordinary differential equations for $L_x(\lambda)$ and $M_x(\mu)$, where $G = L + M$:

$$\left(\frac{d}{d\lambda} - x\right)(\lambda^2 - 1)\frac{dL}{d\lambda} = -A\lambda^2 - x\lambda + B(x),$$
$$\left(\frac{d}{d\mu} - x\right)(\mu^2 - 1)\frac{dM}{d\mu} = -A\mu^2 + x\mu + B(x), \tag{6-3-35}$$

where B is a separation constant. Because the unknowns L and M do not appear explicitly, these equations can be solved directly, and by demanding that G be regular everywhere one can fix $B(x)$ and the four integration constants which appear. By further requiring that $\langle G \rangle = 0$ another additive function of x is determined. The final result is

$$G_x(\lambda, \mu) = Ax^{-1}(\lambda+\mu)+(1+2Ax^{-2})\ln(1+\lambda)$$
$$+(1+x^{-1})\{\text{Ei}\,[x(\mu-1)]-\ln(1-\mu)-e^{-2x}[\text{Ei}\,(x[\mu+1])-\ln(1+\mu)]\}$$
$$+2x^{-1}[\ln 2-\text{Ei}\,(-2x)]+e^{-2x}[2(1+x^{-1})\ln\gamma x+x-\tfrac{1}{2}]. \quad (6\text{-}3\text{-}36)$$

Making use of the expressions (6-3-32) and (6-3-33) one can express the complete second-order adiabatic potential as

$$V_2(x) \equiv \langle VG \rangle = x^{-2}\{5-(4x^2+8x+10)e^{-2x}+(4x^3+7x^2+8x+5)e^{-4x}$$
$$-2(x+1)^2(e^{-2x}+e^{-4x})(\text{Ei}\,[2x]-2\ln[2\gamma x])$$
$$-2\,\text{Ei}\,[-2x]([x-1]^2e^{2x}+[x^2+2x-3]+4[x+1]e^{-2x})\}.$$
$$(6\text{-}3\text{-}37)$$

This expression goes as $-1+\tfrac{8}{3}x^2$ near $x=0$, and reduces to (6-3-24) for large x; selected values are given in Table 6-3-1.

D. Variational-Perturbation Methods

We emphasize that even the Dalgarno–Lynn functions are exact solutions of only the first order wavefunction and second order energy associated with (6-3-1). It is natural to attempt to improve upon a perturbation calculation by using the form of wavefunction suggested by perturbation theory in conjunction with a Rayleigh-Ritz type of variational principle. For the electron–atom scattering problems of interest here these methods have not been particularly valuable. They have a close connection, however, with the rigorous bound methods to be discussed later (§ 6-4C) and so we will outline them here.

Suppose we make a simple generalization of the first-order perturbed wave function of eq. (6-3-6) as follows:

$$|\phi\rangle = [1+T(x)\,G]|\phi_0\rangle. \quad (6\text{-}3\text{-}38)$$

The function G still satisfies the perturbation equation (6-3-7), but another function $T(x)$ has been added in an attempt to improve the wave function.

We wish then to minimize the expectation value of the adiabatic Hamiltonian using the trial function ϕ:

$$\overline{V}(x) = \frac{\langle \phi_0 | [1+TG][H_0+V][1+TG] | \phi_0 \rangle}{\langle \phi_0 | [1+TG]^2 | \phi_0 \rangle} - E_0. \qquad (6\text{-}3\text{-}39)$$

Let us define some new functions ("potentials") which will be of interest again in § 6-4C:

$$V_3(x) \equiv \langle G^2 V \rangle,$$
$$N(x) \equiv \langle G^2 \rangle, \qquad (6\text{-}3\text{-}40)$$

and

$$C(x) \equiv \langle G[H_0, G] \rangle.$$

Then we can write

$$\overline{V}(x) = [V_1 + 2TV_2 + T^2(V_3 + C)][1 + T^2 N]^{-1}, \qquad (6\text{-}3\text{-}41)$$

and the condition for an extremum of $\overline{V}$ is $d\overline{V}/dT = 0$. This leads to the quadratic equation

$$NV_2 T^2 + (NV_1 - C - V_3)T - V_2 = 0, \qquad (6\text{-}3\text{-}42)$$

whose solution is

$$T(x) = \frac{(C+V_3-NV_1)}{2NV_2}\left[1 - \left(1 + \frac{4NV_2^2}{[C+V_3-NV_1]^2}\right)^{\frac{1}{2}}\right], \qquad (6\text{-}3\text{-}43)$$

where the negative sign before the square root is chosen to give a minimum value for $\overline{V}$. It is now easy to show that the potential used in the basic method, V_{AD} [as defined in eq. (6-2-10)] turns out to be identical to $\overline{V}$ when this optimum T is used: $V_{AD} = V_1 + TV_2$.

To get some feeling for the changes brought about by this modification of the adiabatic method, we will apply it to the complete perturbation method of Dalgarno and Lynn for the case of hydrogen. From eq. (6-3-7) we find that the function $C(x) = -V_2(x)$, since $\langle G \rangle = 0$. In the Appendix of [14] appears the evaluation of the various potentials of eq. (6-3-40) in terms of the exact G in elliptical co-ordinates; we will use here certain limiting values. For large x, N and V_2 decrease like x^{-4} while V_3 decreases like x^{-7} and V_1 vanishes exponentially. The square root in eq. (6-3-43) thus can be expanded, and $T(x) \to 1$. For $x \to 0$, $V_3 = NV_1$, $V_2 = -1$, $N = \frac{3}{4}$, and therefore $T(0) = \frac{2}{3}$. The adiabatic potential is thus less attractive at the origin but retains its asymptotic value at large x. Since odd powers of V appear in eq. (6-3-43), it is not surprising that T differs considerably when the sign of the incoming charge is reversed. For an incoming positive charge, T reaches a maximum of about 1.6 at $x \approx 3.5$; for a negative charge, T is always less than 1 and has a small local maximum at $x = 1$ and a broad minimum near $x = 3$. These functions are plotted in Fig. 6-3-2, and the corresponding potential, TV_2, is given in Table 6-3-1 for an incoming positive charge and in Fig. 6-3-3 for a negative charge.

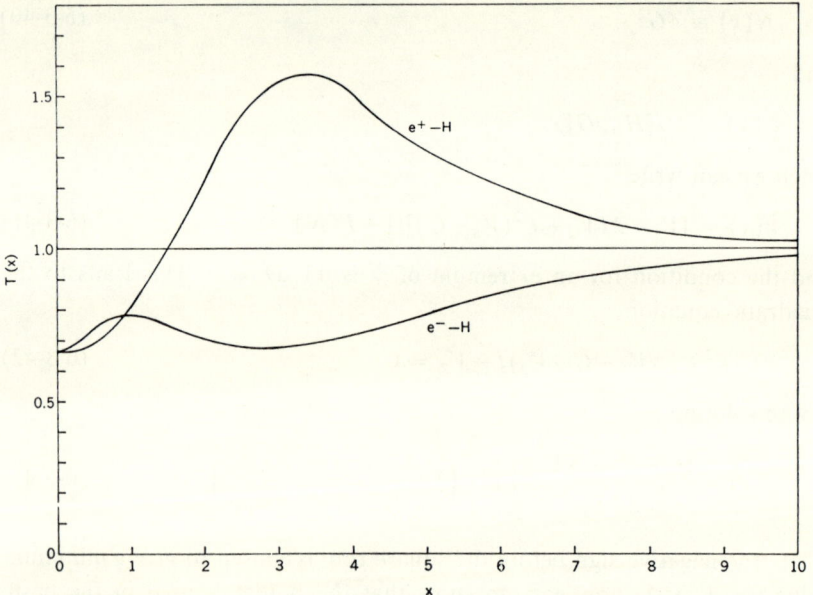

Fig. 6-3-2. Variational-perturbation function $T(x)$ for e^+–H and e^-–H scattering. The function $T(x)$ modifies the polarized orbital wave function as shown in eq. (6-3-38) and is computed from eq. (6-3-43) using the Dalgarno–Lynn form for G given in eq. (6-3-36).

Stone [15] tried to simplify the use of the polarized target function in a quite similar way. He replaced G in eq. (6-3-38) by a dipole form as in the multipole expansion of eq. (6-3-16). Instead of using the exact solution or the Callaway–Temkin form he took the simple asymptotic form (see eqs. (6-3-16) and (6-3-18))

$$G = (r + \tfrac{1}{2}r^2) \cos \theta_{rx} \qquad (6\text{-}3\text{-}44)$$

which describes the wave function of the perturbed atom for very large x. The dependence on x then comes entirely from the variational function T.

To see what happens, we can use the same expression for T (eq. (6-3-43)) with new forms for the potentials.

Doing this we find the following expressions:

$$N = \tfrac{43}{8}, \quad C = \tfrac{9}{2},$$

$$V_2 = -\frac{9}{2x^2}[1-e^{-2x}(1+2x+2x^2+\tfrac{10}{9}x^3+\tfrac{2}{9}x^4)],$$

$$V_3 = -\frac{48}{x^3}[1-e^{-2x}(1+2x+\tfrac{427}{192}x^2+\tfrac{325}{192}x^3+\tfrac{15}{16}x^4+\tfrac{3}{8}x^5+\tfrac{3}{32}x^6+\tfrac{1}{96}x^7)].$$

(6-3-45)

For large x, eq. (6-3-43) tells us that $T \to x^{-2}$, so the correct long-range behavior is recovered. For small x, $V_2 = -x$, $V_3 - NV_1 = \tfrac{13}{2}$ and $T \to \tfrac{1}{11}x$, so $V_{AD}^{(2)}$ vanishes quadratically near $x = 0$. Stone's method has not been used much to date, but the idea of using the simple, factorable form for G (eq. (6-3-44)) will reappear in § 6-4C.

E. THE EXACT STATIC SOLUTION FOR THE ONE-ELECTRON TARGET

We now come to a discussion of the exact solution of the adiabatic equation (6-3-1) obtainable for the hydrogen target. Since all orders of the perturbation occur, the solutions are quite different when the sign of the incoming charge is reversed. We will examine both cases in turn.

As in the case of the exact perturbation solution of Dalgarno and Lynn, the adiabatic equation proves to be separable in elliptical coordinates. Eq. (6-3-1) can be written in terms of these coordinates in the following form:

$$\frac{\partial}{\partial \lambda}(\lambda^2-1)\frac{\partial \phi}{\partial \lambda} + \frac{\partial}{\partial \mu}(1-\mu^2)\frac{\partial \phi}{\partial \mu}$$
$$+ \left\{\tfrac{1}{4}x^2\left(E(x)-\frac{2}{x}\right)(\lambda^2-\mu^2)+x[\lambda(q+1)+\mu(q-1)]\right\}\phi = 0, \quad (6\text{-}3\text{-}46)$$

where q is the charge of the incoming particle (± 1). The variables are separated by letting $\phi = M(\mu)L(\lambda)$, and one obtains the two equations

$$\frac{d}{d\mu}(\mu^2-1)\frac{dM}{d\mu} + [A-p^2\mu^2+x\mu(1-q)]M = 0$$

$$\frac{d}{d\lambda}(\lambda^2-1)\frac{dL}{d\lambda} + [A-p^2\lambda^2+x\lambda(1+q)]L = 0 \quad (6\text{-}3\text{-}47)$$

where A is a separation constant and $p^2 = \tfrac{1}{4}x^2[2/x - E(x)]$.

These equations have been solved in terms of expansions, recursion relations and continued fractions [16]. There is an important physical difference between the solutions for $q = \pm 1$, which we will have to consider, before applying the exact solutions to the scattering problem. For $q = +1$, $M(\mu)$ is either an even or odd function of μ, while for $q = -1$, no such symmetry holds, since the differential equation is not invariant under the substitution $\mu \rightarrow -\mu$. The meaning of this symmetry is easy to see when we examine the definitions in Fig. 6-3-1: interchange of μ and $-\mu$ is equivalent to an interchange of r and $r-x$. Only when both of the fixed charges have the same sign does this symmetry hold.

For the case of $q = +1$, we will find it difficult to construct solutions of (6-3-47) which will go over to the correct asymptotic form at large x; this should be $\phi = \exp\{-\tfrac{1}{2}x(\lambda+\mu)\}$, which is clearly not symmetric in μ. This problem will be examined later.

For $q = -1$, the ground state has the correct asymptotic form, and from the work of Wightman [17] and Bates [18] we can obtain numerical values for $V_{\mathrm{AD}}(x)$, which are given in Fig. 6-3-3, compared with the exact second-order perturbation results of Dalgarno and Lynn ([11] and § 6-3D). At a certain critical radius x_c, the total energy of the system equals the energy of a rearranged system consisting of an electron at ∞ and a pair of fixed $(+ \text{ and } -)$ charges at a separation x_c. That is, when $V_{\mathrm{AD}} - 1 \geq -2/x$, it becomes energetically favorable for the atomic electron to escape. For values of $x < x_c$, the adiabatic potential is not realistic, although exchange effects are large in this region and may mask details of the potentials. The exact value of $x_c = 0.639$, while perturbation theory gives $x_c = 1.25$. Beyond this latter point there is not a very large difference between exact and perturbation results.

The symmetry that exists for $q = +1$ separates all solutions into *gerade* (g) and *ungerade* (u), respectively even and odd under the substitution $\mu \rightarrow -\mu$. The two lowest lying levels are called $1s\,\sigma_g$ and $2p\,\sigma_u$, since they correspond to 1s and 2p states of the He$^+$ ion when $x = 0$, have no angular momentum component along the internuclear axis (σ), and are even (g) or odd (u). For large x they have the limiting forms

$$\phi_{\substack{g\\u}} \sim e^{-\tfrac{1}{2}x\lambda}(e^{-\tfrac{1}{2}x\mu} \pm e^{\tfrac{1}{2}x\mu}). \tag{6-3-48}$$

Because of the degeneracy at large x, a linear combination can be constructed in the form

$$\phi = \tfrac{1}{2}(\phi_g + \phi_u) = e^{-\tfrac{1}{2}x(\lambda+\mu)} = e^{-r}, \tag{6-3-49}$$

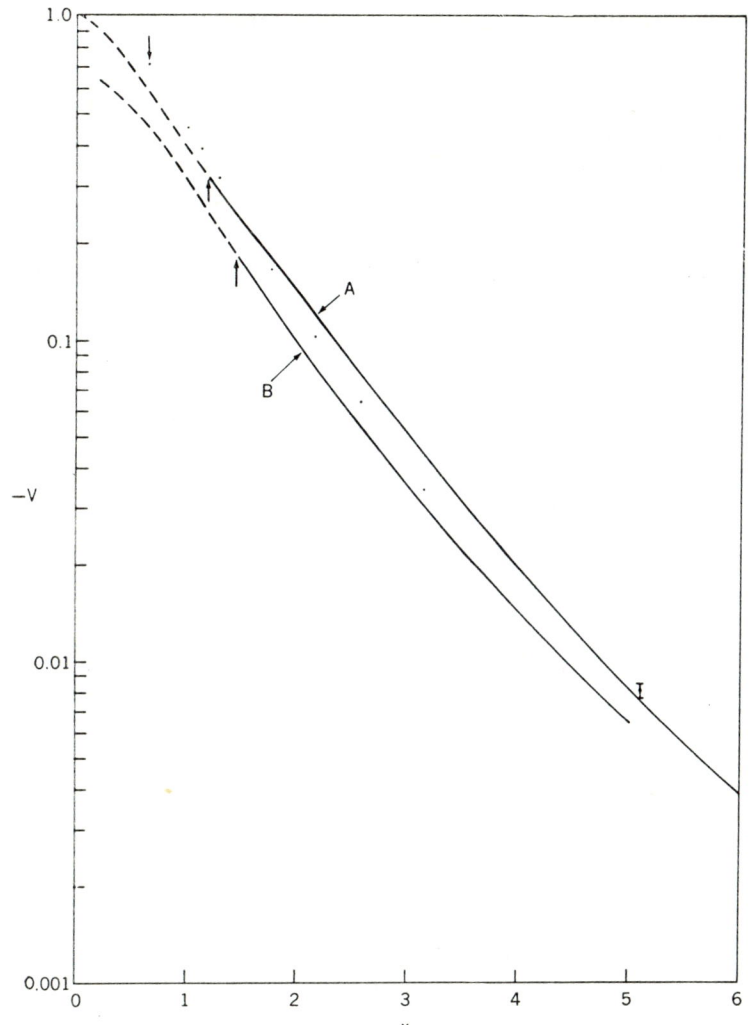

Fig. 6-3-3. Adiabatic potentials for e^-–H scattering. Curve A is Dalgarno and Lynn's result V_2 as given in eq. (6-3-37). Curve B is the variational-perturbation result TV_2, using the values of $T(x)$ shown in Fig. 6-3-2 and V_2 from curve A. The points are from exact solutions of eq. (6-3-46) [18], with $V_1(x)$ subtracted for comparison purposes. In each case, an arrow represents the critical separation x_c, below which the atomic electron can escape.

satisfying the asymptotic condition required for a polarized orbital target function. Unfortunately, this linear combination is not correct at the origin, where only the spherically symmetric state $1s\,\sigma_g$ can remain. Therefore, one might propose the following polarized orbital target function:

$$\phi = \phi_g + \alpha(x)\phi_u. \qquad (6\text{-}3\text{-}50)$$

Here $\alpha(x)$ varies smoothly between 0 and 1 as x goes from 0 to ∞.

To obtain the adiabatic potential for this form of target function we use eq. (6-2-10) and note that $(H_0 + V)\phi_{g,u} = E_{g,u}(x)\phi_{g,u}$. We obtain

$$V_{\text{AD}}(x) = \frac{\langle\phi_0|[(H_0+V)-H_0]|\phi\rangle}{\langle\phi_0|\phi\rangle} = \frac{V_g N_g + \alpha V_u N_u}{N_g + \alpha N_u}, \qquad (6\text{-}3\text{-}51)$$

where

$$V_{g,u} \equiv E_{g,u}(x) - E_0, \quad N_{g,u} \equiv \langle\phi_0|\phi_{g,u}\rangle.$$

If we further define $\beta(x) \equiv N_u/N_g$, we can write

$$V_{\text{AD}}(x) = (V_g + \alpha\beta V_u)(1 + \alpha\beta)^{-1}. \qquad (6\text{-}3\text{-}52)$$

Using the wave functions in elliptical coordinates as given by Bates et al. [16], we evaluated $\beta(x)$ and list its value at a number of points in Table 6-3-2.

TABLE 6-3-2

The functions $\beta(x)$, $V_g(x)$ and $V_u(x)$ as defined in eq. (6-3-51), to be used in deriving the "exact" e^+–H adiabatic potential appearing in Table 6-3-1

x	$\beta(x)$ (a)	$V_g(x)$ [16]	$V_u(x)$ [16]
0	0	$2/x-3$	$2/x$
0.4	0.1163	2.398	5.022
1.0	0.2842	0.096	1.870
2.0	0.5246	-0.20525	0.66493
3.0	0.7045	-0.15511	0.26382
4.0	0.8320	-0.09216	0.10890
5.0	0.9122	-0.04884	0.04542
7.0	0.9791	-0.0112	0.0075

(a) Derived using the 1s σ_g and 2p σ_u wave function parameters of [16].

There is no convincing way of determining $\alpha(x)$, short of a full variational lower bound calculation (see § 6-4C), but since $\beta(x)$ has the same limiting behavior as $\alpha(x)$ we have made the assumption that $\alpha = \beta$. With this assumption we obtain the form of $V_{\text{AD}}(x)$ shown in Table 6-3-1, compared with some of the other approximate adiabatic potentials discussed in this section. [If, instead, $\alpha = \beta^2$, a result identical (within 5%) to that of Dalgarno and Lynn is obtained.]

F. THE MANY-ELECTRON TARGET; STERNHEIMER'S APPROXIMATION

The starting point for constructing the polarized target wave function is with the first order perturbation expression (6-3-7). Although the complete adiabatic equation (6-3-1) might still be considered for the many-electron target, it would be far too complicated to be solved in its entirety in non-hydrogenic cases, nor is it clear, as we shall discuss, whether it would be advantageous to use it even if one could evaluate it. Even (6-3-7) is too difficult to solve in the general case, and we shall for practical purposes be further restricted to Sternheimer's approximation [19]. That approximation is in turn closely tied to the Hartree–Fock approximation for the unperturbed target system, so that ultimately one is restricted even in the form of target wave function one may practically use. (The use of an approximate target wave function gives rise to other difficulties which we shall point out as they come along.)

Sternheimer's approximation, which was originally introduced to calculate polarizabilities of atomic systems, was rather heuristically derived and only subsequently was it shown by Allen [20] also Dalgarno [21] how it approximately corresponds to perturbation theory. The reduction has also been given in [4], and it will only briefly reviewed here.

Assume the target state is described by a single determinant and that the static perturbation is described in the dipole approximation (for $x > r_j$; $j = 1, 2, \ldots, N$; $Z = N$):

$$-\frac{2N}{x} + \sum_{j=1}^{N} \frac{2}{|r_j - x|} \approx \sum_{j=1}^{N} \frac{2r_j}{x^2} \cos \theta_{jx} \qquad (6\text{-}3\text{-}53)$$

$$\Phi_0 = \frac{1}{(N!)^{\frac{1}{2}}} \det(\varphi_1, \varphi_2, \ldots, \varphi_N) \qquad (6\text{-}3\text{-}54)$$

where φ_i is the ith orbital of a target electron:

$$\varphi_i = [u_{n_i l_i}(r)/r] Y_{l_i m_i}(\Omega) \chi(\tfrac{1}{2} m_s) \qquad (6\text{-}3\text{-}55)$$

and χ is a spin (one-half) function.

The dipole approximation, over and above the adiabatic approximation (which simply assumes x is fixed), assumes $x \gg r_j$; so that one may associate the quantity $1/x^2$ with the perturbation theory expansion parameter λ which can then be considered a small, fixed quantity compared to the rest of the perturbation. Under these circumstances the unperturbed target function acquires a polarized increment

$$\Phi_0 \to \Phi_0 + \lambda \sum_{j=1}^{N} \Phi'_j \qquad (6\text{-}3\text{-}56)$$

where

$$\Phi'_j = \frac{1}{(N!)^{\frac{1}{2}}} \det(\varphi_1, \ldots, \varphi_{j-1}, \varphi'_j, \varphi_{j+1}, \ldots, \varphi_N) \qquad (6\text{-}3\text{-}57)$$

with the subsidiary equations

$$\varphi' = \sum_{l'm'} \frac{u_{nl \to l'}(r)}{r} C_{ll'}^{(mm')} Y_{l'm'}(\Omega) Y_{1\, m-m'}(\Omega_x) \chi(\tfrac{1}{2}m_s) \qquad (6\text{-}3\text{-}58)$$

where

$$C_{ll'}^{(mm')} = (-1)^{m-m'+1} \left[\frac{16\pi}{3} \left(\frac{2l+1}{2l'+1} \right) \right]^{\frac{1}{2}} (1l00|l'0)(1l\, m'-m\, m|l'm'). \qquad (6\text{-}3\text{-}59)$$

(The latter two parenthetical expressions are Clebsch–Gordan coefficients in the notation of Condon and Shortley [147].) In Sternheimer's approximation the functions $u_{nl \to l'}(r)$ satisfy the equation

$$\left[-\frac{d^2}{dr^2} + V_{nl \to l'}(r) \right] u_{nl \to l'}(r) = r u_{nl}(r), \qquad (6\text{-}3\text{-}60)$$

where

$$V_{nl \to l'}(r) = \frac{1}{u_{nl}(r)} \left[\frac{d^2}{dr^2} u_{nl}(r) \right] + \frac{l'(l'+1) - l(l+1)}{r^2}. \qquad (6\text{-}3\text{-}61)$$

The first term of eq. (6-3-61) will contain singularities for $n-l > 1$, however these singularities are artifacts of Sternheimer's approximation. In general they are narrow tangent-like singularities and they can be simply smoothed over in numerically integrating (6-3-60).

The final polarized part of the wave function is the second part of (6-3-56):

$$\Phi^{(\text{pol})}(1, 2, \ldots, N; x) \equiv \lambda \sum_{j=1}^{N} \Phi'_j \qquad (6\text{-}3\text{-}62)$$

$$= \frac{1}{x^2} \frac{1}{(N!)^{\frac{1}{2}}} \sum_{n_j l_j l'_j} C_{l_j l'_j}^{(m_j m'_j)} Y_{1\, m_j - m'_j}(\Omega_x)$$

$$\times \det \left(\varphi_1, \ldots, \varphi_{j-1}, \frac{\varepsilon(x, r)}{r} u_{nl_j \to l'_j}(r) Y_{l'_j m'_j}(\Omega) \chi(\tfrac{1}{2}m_s), \ldots, \varphi_N \right). \qquad (6\text{-}3\text{-}63)$$

(The polarized orbitals have also been multiplied by a step-function, $\varepsilon(x, r)$, which will be justified in the next section.) Verbally this means that each determinant in the original wave function is replaced by a sum of determinants in which each orbital is replaced by a subsum over all of its polarized orbitals (hence the origin of the name). If the complete set of suborbitals is retained and if the original function Φ_0 is an eigenfunction of angular momentum, then $\Phi^{(\text{pol})}$ will also be an eigenfunction of angular momentum. If the target state is a sum of determinants belonging to a given configuration, then each determinant is handled the same way; and if the polarized orbitals, for a given n, l, l' are the same in all determinants then again the above eigenfunction character will not be lost.

§ 6-4. The total polarized orbital wave function and the scattering problem

A. Introductory Remarks

The foregoing sections have been concerned primarily with the solution of the static (or adiabatic) Schrödinger equation (6-3-1). Although many approximations have been discussed, the problem is still a relatively straightforward one, and the success of the various methods, in so far as they approximate exact solutions of (6-3-1), can easily be assessed. In the present sections we shall be dealing with taking these approximate solutions and constructing from them total wave functions and associated methods for describing the whole (scattering) problem; i.e., finding good approximations of the complete Schrödinger equation:

$$H\Psi = E\Psi. \tag{6-4-1}$$

The manner in which one has confronted this task has given rise to the greatest controversy concerning the method of polarized orbitals. This stems from the fact that the success of various total wave functions and scattering methods based on approximate Φ's of eq. (6-3-1) is not in a one to one relation with how accurately Φ solves the adiabatic problem. That this is so is not hard to see because as the projectile approaches the target the scattering particle and orbital electrons interact strongly, nonadiabatic effects become important and eq. (6-3-1) becomes an increasingly artificial problem.

Our philosophy therefore has been to keep our attention on the important thing, eq. (6-4-1). In particular the object is to devise wave functions and methods such that the scattering parameters associated with our approximations are as close to the exact ones as possible. At the same time the idea has been to construct a method: i.e., a prescription which in principle is well-defined and applicable to all atomic targets. We shall find that two basic approximations for Φ have been most useful and used in this context: Callaway–Temkin dipole function (6-3-19) and the full Dalgarno–Lynn function or its relative partial wave expansion (Bethe–Reeh) form (6-3-16, 28, 29).

Given the total wave function (§ 6-3B) one can subdivide the methods for deriving scattering equations into nonvariational (§ 6-3C) and variational (§ 6-3D).

B. The total wave function and non-variational methods

The total wave function and non-variational method have in essence been described in § 6-2. Our purpose here is to specialize this somewhat and to clarify the special validity of non-variational method in electron scattering.

First, for the purposes of most calculations one is restricted to a partial wave expansion of the scattered amplitude and correspondingly to a specific total angular momentum component of (6-2-2). This will still be referred to as the total wave function Ψ_l:

$$\Psi_l = \frac{u_l(x)}{x} Y_{lm}(\Omega_x) \Phi(1, 2, \ldots, N; x). \tag{6-4-2}$$

For $\Phi(1, 2, \ldots, N; x)$ one can now in principle use any approximation of solution of the static problem (6-3-1) and substitute it into the basic non-variational expression (6-2-4). In practice the static problem is approximated by first order perturbation theory and further reduced to the dipole approximation and Sternheimer's approximations for many-electron targets. All of these approximations physically imply that the static coordinate x be large compared to those of the orbital electrons, however in Sternheimer's approximation the static parameter $\lambda = 1/x^2$ diverges in a ridiculous fashion if the above restriction is not imposed explicitly in Φ. The manner in which such a restriction has been imposed is to augment the polarized orbital with a step function factor which cuts off when $x < r$. This accounts for the fact that the polarized orbitals in (6-3-63) are multiplied by

$$\varepsilon(x, r) \equiv \begin{cases} 1 & x > r \\ 0 & x < r. \end{cases} \tag{6-4-3}$$

Some controversy has already attended the incorporation of the step-function because it is not a continuous function – seemingly a necessary condition for a quantum mechanical wave function [23]. In point of fact, however, (6-4-3) gives a well-defined finite expression when used in conjunction with (6-2-4). The reason for this is that if one envisions $\varepsilon(x, r)$ as the limit of a smooth cut-off factor, then the leading order contribution that such a smooth cut-off would give is in fact independent of the shape of the cut-off – a very important result!

The significance of this result, however, takes on added weight when it is combined with the requirement of an antisymmetry of the total wave function in the case that the incident particle is an electron. Mathematically the prescription for the total wave function in that case is

$$\Psi_l^A = \sum_{P_c} (-1)^{P_c} \frac{u_l(r_{N+1})}{r_{N+1}} Y_{lm}(\Omega_{N+1}) \chi(\tfrac{1}{2} m_s)$$
$$\times [\Phi_0(1, 2, \ldots, N) + \Phi^{(\text{pol})}(1, 2, \ldots, N; r_{N+1})]. \quad (6\text{-}4\text{-}4)$$

Here we have written the perturbed wave function explicitly as a sum of its unperturbed and polarized parts [cf. (6-3-56)]. In addition we now use the coordinates r_{N+1} rather than x to emphasize that the scattered particle is identical to the orbital electrons. P_c is a cyclic permutation of the indices $1, 2, \ldots, N, N+1$. The sum over P_c is sufficient to make Ψ_l^A completely antisymmetric, since the Φ's are themselves antisymmetric in $1, \ldots, N$. Finally the orbital angular momentum of the incident particle must be vector coupled to the orbital angular momentum of the target [4] in order that the scattering parameters be independent of the magnetic quantum numbers. A similar statement holds for the spins.

The efficacy of (6-4-4) as compared to its unsymmetrized counterpart (6-4-2) stems from the following circumstance. If Ψ_l^A were an exact solution of the Schrödinger equation the symmetry of the wave function would not yield any additional dynamical information that could not be recovered from a suitable linear combination of asymmetric solutions. However since Ψ_l^A is not exact, by imposing the correct symmetry we force the approximate wave function to partake correctly of this property and thereby impose additional dynamical constraints ultimately on the scattering function $u_l(r)$ in (6-4-4). These additional terms turn out to be exchange polarization terms, so-named because they are associated with the exchange parts of $\Phi^{(\text{pol})}$ in (6-4-4). The direct terms of $\Phi^{(\text{pol})}$ give rise to the classical α/r^4 polarizability potential and we have argued above that they should be insensitive in lowest order to the specific form of cut-off. With the inclusion of the appropriate exchange polarization terms, one can be sure that this insensitivity becomes greatly enhanced; thus even before various forms of cut-off had actually been tried, one could confidently say "it is not the exact form of direct polarization potential for small values of r ... but rather its accuracy for larger values of *r and the consistent incorporation of the exchange polarization terms* which are important ..." [10]. Recently Duxler, Poe and LaBahn [24] (to be discussed more in § 6-5B) have completed calculations in which a Bethe–Reeh polarization function was used in addition to the usual step-function in both e–H and e–He scattering. Their results completely corroborate this expectation (cf. Fig. 6-5-2).

When the exchange polarization terms are neglected, then the resulting approximation, which is still non-variational in nature, has been called the "exchange-adiabatic" or "adiabatic-exchange" approximation. If additionally the direct polarization terms are also dropped, then the approximation reduces to "exchange approximation" of Morse and Allis [25]. There are two additional non-variational methods called "dynamic polarization" [26] and "extended-polarization" [27] in which additional terms beyond direct polarization are included to attempt to incorporate non-adiabatic effect. Because these additional terms can best be understood from the variational point of view they will be described in the next section § 6-4C. It is to be emphasized, however, because of the peculiar power of the total symmetry argument, that no method, variational or non-variational, depending on only *one* undetermined function has been as generally successful in quantitative calculations of electron–atom scattering as the original polarized orbital method associated with (6-4-4) and (6-2-4).

C. VARIATIONAL AND VARIATIONALLY MOTIVATED METHODS

We next discuss variational and variationally motivated methods. Strictly variational methods of the kind we consider give lower bounds on phase shifts, and we shall concentrate on two relatively new methods in that category. One was introduced as a modification of the close-coupling approximation designed to account exactly for the polarizability [28], while the other was originally an extension of the polarized orbital methods to account for short-range and non-adiabatic effects [14]. They can be given a unified treatment, following Matese [29], and we shall introduce them in that way. Along with another, intermediate method introduced by Perkins [30] they fall into the category of pseudo-state expansions. In addition a group of other approximations can be treated in this same framework. (In what follows, we will restrict the formalism to the case of positron–hydrogen scattering.)

To introduce the derivation, we review the optical potential formalism of Feshbach [31]. We define an "open channel" projection operator P

which must satisfy the usual condition of idempotency $P^2 = P$. Its only other defining property requires that it satisfy the asymptotic condition $P\Psi_{x\to\infty} = \Psi$. For most purposes it is convenient to construct P out of the state vector $|\phi_0\rangle$ of the target atom, as follows:

$$P = |\phi_0\rangle\langle\phi_0|, \tag{6-4-5}$$

as long as elastic scattering is the only energetically allowed process. A "closed channel" projection operator is also defined as $Q = 1-P$, and one sees that $Q^2 = Q$ and $QP = 0$.

Feshbach's derivation of the optical potential equation begins with the Schrödinger equation for the scattering wave function Ψ:

$$(H-E)\Psi = (H-E)(P+Q)\Psi = 0, \tag{6-4-6}$$

where the condition $P+Q = 1$ has been used. Equation (6-4-6) is equivalent to the projected equations

$$P(H-E)(P+Q)\Psi = 0 \tag{6-4-7}$$

and

$$Q(H-E)(P+Q)\Psi = 0, \tag{6-4-8}$$

where (6-4-7) is projected onto the space of open-channel functions and (6-4-8) onto the remainder of the space. Solving (6-4-8) for $Q\Psi$ and recalling the idempotency of P and Q we obtain

$$Q\Psi = \frac{1}{Q(E-H)Q} \cdot QHP \cdot P\Psi, \tag{6-4-9}$$

where $QEP = EQP$ is zero. Using (6-4-9), we can write (6-4-7) in the form

$$(\mathcal{H}-E)P\Psi = 0, \tag{6-4-10}$$

where the effective Hamiltonian is

$$\mathcal{H} = PHP + PHQ \cdot \frac{1}{Q(E-H)Q} \cdot QHP, \tag{6-4-11}$$

and the second term in eq. (6-4-11) is called the optical potential. So far, everything has been exact; (6-4-10) represents simply another way to obtain the scattering [31].

Gailitis [32] has proven an important theorem which leads to a whole class of bounded approximations of the optical potential. He introduces still another projection operator $R = R^2$, and constructs a modified optical potential by replacing Q everywhere by $\hat{Q}$, where $P + \hat{Q} = R$ and $\hat{Q}^2 = \hat{Q}$. As long as the denominator in the optical potential is negative definite, the phase shifts $\hat{\eta}$ obtained with $\hat{Q}$ replacing Q satisfy the relation $\hat{\eta} \leqq \eta$, and the equality holds only if $R \equiv 1$. Any $\hat{Q}$ which can be explicitly written can be used, in principle, to obtain a lower bound on the phase shifts. Of two different forms of $\hat{Q}$ the better one yields higher phase shifts.

To determine the form of $\hat{Q}$ which corresponds to the polarized orbital type of wave function, we recall that the latter can be written:

$$\Psi = \chi[1+G]|\phi_0\rangle. \tag{6-3-6}$$

From (6-4-5) we find that

$$P\Psi = \chi[1+\langle G\rangle]|\phi_0\rangle, \tag{6-4-12}$$

but since $\langle G \rangle = 0$ [see text following eq. (6-3-4)], it follows that $G|\phi_0\rangle$ represents the space projected by the operator $\hat{Q}$. We can construct this approximate projector as

$$\hat{Q} = |\omega\rangle\langle\omega| \tag{6-4-12}$$

where $|\omega\rangle = N^{-\frac{1}{2}}G|\phi_0\rangle$, and N is defined in eq. (6-3-40). The various quantities which enter the approximate form of eq. (6-4-11) are:

$$PHP \equiv |\phi_0\rangle\langle\phi_0|[H_0 - \nabla_x^2 + V]|\phi_0\rangle\langle\phi_0| = |\phi_0\rangle[-\nabla_x^2 + E_0 + V_1]\langle\phi_0|,$$

$$PH\hat{Q} \equiv |\phi_0\rangle\langle\phi_0|[H_0 - \nabla_x^2 + V]G|\phi_0\rangle\langle\phi_0|GN^{-1} = V_2 N^{-\frac{1}{2}}|\phi_0\rangle\langle\omega|,$$

$$\hat{Q}HP = V_2 N^{-\frac{1}{2}}|\omega\rangle\langle\phi_0|,$$

$$\hat{Q}(E-H)\hat{Q} \equiv N^{-\frac{1}{2}}|\omega\rangle\langle\phi_0|G(\nabla_x^2 - H_0 - V + E_0 + k^2)G|\phi_0\rangle\langle\omega|N^{-\frac{1}{2}}$$

$$= N^{-\frac{1}{2}}|\omega\rangle L\langle\omega|N^{-\frac{1}{2}},$$

$$L \equiv N(\nabla_x^2 + k^2) - C - V_3 - W + \nabla_x N \cdot \nabla_x, \tag{6-4-13}$$

where most of the terms are defined in (6-3-40), while $W(x) = -\langle G\nabla_x^2 G\rangle$. The scattering equation (6-4-10) takes the form

$$\left\{-(\nabla^2 + k^2) + V_1 + V_2 \frac{1}{L} V_2\right\}\chi = 0, \tag{6-4-14}$$

where $\chi \equiv \langle\phi_0|P\Psi\rangle$ is a one-particle function containing all the scattering information. Before analyzing eq. (6-4-14) itself, we will discuss some further approximations.

Let us first specify that the particular G to be used in the projection operator $\hat{Q}$ [eq. (6-4-12)] satisfy the first-order perturbation equation (6-3-7) so that $C = -V_2$ defined in (6-3-40). In the expression for L, several terms are "smaller" than the leading term, V_2, although no rigorous ordering can be carried out. The third-order term, V_3, and the non-adiabatic term W are of shorter range than V_2 and hence may be neglected beyond some radius. The other terms involve the gradient operator or k^2, both of which are, in some sense, velocity dependent; they may perhaps be neglected for low enough energies. If, then, we let $L \approx V_2$, eq. (6-4-14) reduces exactly to the second order adiabatic equation ((6-2-9) and (6-3-8)) which is seen to be a particular approximation to the optical potential equation.

A higher-order approximation can also be obtained by continuing the expansion of L^{-1} to first order in the "small" terms discussed above:

$$L^{-1} \approx V_2^{-1} - V_2^{-1}[N(\nabla^2 + k^2) - V_3 - W + \nabla N \cdot \nabla]V_2^{-1}. \qquad (6\text{-}4\text{-}15)$$

Then eq. (6-4-14) becomes [14]

$$-(\nabla^2 + k^2)\chi + (1+N)^{-1}(V_1 + V_2 + V_3 + W - \nabla N \cdot \nabla)\chi = 0. \qquad (6\text{-}4\text{-}16)$$

This is precisely the equation obtained by the simplest variational treatment of the scattering using the original polarized orbital wave function, eq. (6-2-2) with a first order perturbed target (cf. [23]).

Two further simplifications of eq. (6-4-16) have been introduced by Callaway's group. The first, called "dynamical polarization" [26] focuses attention on the fact that the potentials in eq. (6-4-16) are of mixed (or "inconsistent") order, in the perturbing potential V. Although it is variationally correct, as mentioned above, the authors of [26] preferred to keep only those terms of first and second order in V. The resulting equation

$$-(\nabla^2 + k^2)\chi + (V_1 + V_2 + W - \nabla N \cdot \nabla)\chi = 0 \qquad (6\text{-}4\text{-}17)$$

is the basis of the dynamical polarization method. (For electron scattering, the static-exchange terms are added to (6-4-17).)

The velocity-dependent gradient term in eq. (6-4-17) is often inconvenient to employ, and it can easily be transformed to a simpler form. By defining a new scattering function

$$\bar{\chi} = (1 + \tfrac{1}{2}N)\chi \qquad (6\text{-}4\text{-}18)$$

which has the same phase shifts as χ since N vanishes for large x, one can transform eq. (6-4-17) into the form

$$-(\nabla^2+k^2)\bar{\chi}+(V_1+V_2+W)\bar{\chi}+(2-N)^{-1}[(\nabla N)^2 \\ +\nabla^2 N+N\nabla N\cdot\mathbf{V}]\,\bar{\chi} = 0. \quad (6\text{-}4\text{-}19)$$

Once again retaining only those terms of first and second order in V, which in the above expression is equivalent to retaining terms linear in N, and using the identity

$$W+\tfrac{1}{2}\nabla^2 N = \langle \nabla G \cdot \nabla G \rangle \equiv V_D \quad (6\text{-}4\text{-}20)$$

one obtains the "extended polarization" equation [27]

$$-(\nabla^2+k^2)\bar{\chi}+(V_1+V_2+V_D)\bar{\chi} = 0, \quad (6\text{-}4\text{-}21)$$

where V_D is the "distortion potential". This modification of eq. (6-4-16) has the attractive feature that $V_2+V_D = 0$ at $x = 0$, supplying the desired decrease in the strength of the adiabatic potential (see § 6-6 and [8]). As $x \to \infty$, V_D vanishes like x^{-6}. For intermediate values of x neglect of higher order terms is, however, not well justified.

Returning to the original equation (6-4-14) we note that it is really of fourth order, since the denominator contains a second-order differential operator. It can thus be decomposed into the following two second-order equations:

$$[-(\nabla^2+k^2)+V_1]\chi = -V_2 F \quad (6\text{-}4\text{-}22a)$$

$$LF = V_2\chi. \quad (6\text{-}4\text{-}22b)$$

This set of equations thus yields lower bounds to the scattering phase shifts and represents the natural generalization, through the optical potential formalism, of the more restricted class of polarized orbital approximations discussed above. It is quite easy to show that eq. (6-4-22) can be derived from a variational scattering principle, if the trial function has the form

$$|\Psi\rangle = [\chi+FG]|\phi_0\rangle, \quad (6\text{-}4\text{-}23)$$

where χ and F depend only on $\mathbf{x}$. [In this form the method has obvious connections with the variational-perturbation method, eq. (6-3-38) et seq.]

If we use the form of G which satisfies eq. (6-3-7), then eqs. (6-4-22) and (6-4-23) are identical to those introduced by Drachman [14]. This method extended the polarized orbital idea, making it variationally correct and introducing the new function F to adjust the correlation function G in the region of small x, where non-adiabatic effects were expected to be important. For large x, $F \to \chi$ and eq. (6-4-22a) once again reduces to the adiabatic equation. We will discuss this method in detail in § 6-5 and § 6-6.

Equation (6-4-23) was also reached from a different direction by Damburg and Karule [28]. The standard close-coupling wave function (restricted to include 1s and 2p states only) has the form (for s-states):

$$\Psi_{1s\,2p} = \chi(x)\phi_{1s}(r) + F(x)\phi_{2p}(r)\cos\theta_{xr}, \quad (6\text{-}4\text{-}24)$$

where $\phi_{2p}(r) = (32\pi)^{-\frac{1}{2}} r e^{-\frac{1}{2}r}$. This has the general form of eq. (6-4-23), where $G|\phi_0\rangle = \phi_{2p}(r)\cos\theta_{xr}$ and $\langle G \rangle = 0$ due to the orthogonality of the angular functions. For large x, the leading term in the denominator of eq. (6-4-14) is

$$-C \equiv -\langle G[H_0, G]\rangle = (E_{1s} - E_{2p})\langle G^2 \rangle = -\tfrac{3}{4}. \quad (6\text{-}4\text{-}25)$$

The asymptotic form of V_2 for large x is

$$\lim_{x\to\infty} V_2 = -2x^{-2}\int dr\,\phi_{2p}(r)\,r\,\phi_{1s}(r)\cos^2\theta_{rx} = -8\sqrt{2}(\tfrac{2}{3})^5 x^{-2}. \quad (6\text{-}4\text{-}26)$$

Then the asymptotic potential appearing in eq. (6-4-14) is

$$-\tfrac{4}{3}V_2^2 = -2.9596 x^{-4}; \quad (6\text{-}4\text{-}27)$$

this has the x-dependence of the exact long-range polarization potential but the coefficient is only 65.8 % of the exact value, $\tfrac{9}{2}$.

The fact that a finite length close coupling expansion does not yield the correct polarizability is well-known. It will be recalled from the discussion of § 6-3C that the complete dipole polarizability can be obtained from the function $\bar{g}_{1>}$ of eq. (6-3-18). With somewhat different notation this function was introduced into the scattering problem via the method of the (dipole) polarization potential. In order to motivate the next method to be discussed as well as its many-body generalization, we shall revert to the original notation:

$$2x^2 e^{-r}\bar{g}_{1>} \equiv r^{-1} u_{1s\to p}(r) = 2e^{-r}(r + \tfrac{1}{2}r^2). \quad (6\text{-}4\text{-}28)$$

Now Damburg and Karule, knowing these facts, made the simple but important observation that one could modify the close-coupling expansion by substituting the polarized orbital for the excited eigenstate in the close-coupling ansatz:

$$\Psi_{\text{CCPO}} = \chi(x)\phi_{1s}(r) + F(x)r^{-1}u_{1s\to p}(r)\cos\theta_{xr}, \quad (6\text{-}4\text{-}29)$$

and that in the radial equation associated with χ one would obtain a polarization term with the correct polarizability. From the point of view of the polarized orbital method wherein the total wave function has the analogous form

$$\Psi_{PO} = \chi(x)\phi_{1s}(r) + \chi(x)x^{-2}\varepsilon(x,r)r^{-1}u_{1s\to p}(r)\cos\theta_{xr}, \quad (6\text{-}4\text{-}30)$$

eq. (7-4-29) has the advantage that the cut-off function $\varepsilon(x,r)$ is removed (thus allowing a completely variational calculation to be carried out) with the additional freedom of an extra function, $F(x)$, to describe short range correlations.

One noteworthy aspect of the use of pseudo-states is the emergence of pseudo-thresholds. Specifically, if one derives the longest range part of the equation acting on the "closed-channel" function $F(x)$ in (6-4-29), one finds that it satisfies the equation

$$\lim_{x\to\infty}\left(\frac{d^2}{dx^2} + k^2 - \tfrac{36}{43}\right)F(x) = -\frac{36}{43x^2}\chi(x). \quad (6\text{-}4\text{-}31)$$

Since χ is a bounded function; $F(x)$ *can not* have vanishing boundary conditions unless

$$k^2 < \tfrac{36}{43}$$

corresponding to a total energy

$$E = k^2 - 1 < -0.163 \text{ Ry}. \quad (6\text{-}4\text{-}32)$$

Thus (6-4-32) provides a threshold beyond which these "closely-coupled polarized orbital" calculations cannot be rigorously carried out for the specific polarized orbital $u_{1s\to p}$. That energy however is above the $n = 2$ threshold at $E = -0.25$ Ry and the positronium threshold $E = -0.5$ Ry for positron scattering. We shall confront pseudo-thresholds again in § 6-5A.2 when we deal with $n = 2$ polarized orbitals and applications.

In this method one can also introduce higher multipole distortions, each multiplied by its own undetermined function. Finally within Sternheimer's approximation the method is defined and can be carried out even for many-electron targets, and when fully antisymmetrized for electron collisions, the total wave function contains all the additional advantages discussed above for ordinary polarized orbitals which come from the short-range dynamical effects implicit in the exclusion principle. Motivated by the comparison of (6-4-24), (6-4-29) and (6-4-30) we call this method "closely coupled polarized orbitals". It is destined to have a very important place in low-energy electron–atom scattering theory.

We note that the closely-coupled polarized orbital function (6-4-29) is identical in form to that used by Stone if F is proportional to T, but the methods of determining the respective functions are not the same. From (6-4-22) one can show in (6-4-29) that $F_{x\to\infty} \to x^{-2}\chi$ in accord with the fact that (6-4-30) is indeed correct for large x.

Beginning also with the close-coupling trial function (6-4-24) Perkins [30] made a different modification. He replaced ϕ_{2p} by

$$\phi_{2p'}(r) = (\delta^5/\pi)^{\frac{1}{2}} r e^{-\delta r}, \tag{6-4-33}$$

and treated δ as a variational parameter, to be determined by maximizing the phase shifts obtained from the solution of eq. (6-4-22). We will discuss his results for positron–hydrogen scattering in § 6-6, but will here simply make some observations on the long-range potential resulting from eq. (6-4-33). Following the technique leading to eq. (6-4-27), we find the asymptotic potential to have the form

$$\lim_{x\to\infty} V(\delta, x) = -\frac{4096\delta^5}{(1+\delta)^{10}(1-\delta+\delta^2)}\frac{1}{x^4}, \tag{6-4-34}$$

which has its minimum value for $\delta = 0.797$, when $V = -4.475 x^{-4}$, differing by only about one-half percent from the exact value. Perkins' 2p' function is thus almost as good as the $u_{1s\to p}$ function in describing the long-range polarizability of the hydrogen atom. (The two functions are actually quite similar; the overlap integral of the normalized $u_{1s\to p}$ and 2p' functions equals 0.9987.)

§ 6-5. Applications to electron scattering and reactions

A. SCATTERING FROM ONE-ELECTRON TARGETS

1. *Elastic scattering*

Although we shall be mainly interested in complete polarized orbital calculations, it should be realized that such calculations only exist in a restricted number of cases. In many other cases only the subsidiary exchange adiabatic or adiabatic exchange approximation (AE) have been calculated although often called polarized orbital approximation. As mentioned above the AE approximation drops all exchange polarization terms. It can equally well be viewed as an augmentation of the basic exchange approximation by the addition of a direct-polarization potential. Historically in fact, this is how the approximation was introduced [33], and the necessary cut-off at the origin was supplied using a Buckingham potential

$$-V_p(r) = \begin{cases} \alpha/(r^2+d^2)^2 & \text{(6-5-1a)} \\ \alpha/(r^4+D). & \text{(6-5-1b)} \end{cases}$$

[Strictly speaking (6-5-1a) is of the Buckingham form, but some of the calculations we shall refer to use (6-5-1b). Differences should be minimal.] When viewed from the polarized orbital approach the polarization potential comes naturally to have the Callaway–Temkin form (6-3-19). Replacing the scattering coordinate x by r in this whole section, § 6-5, we see that the latter provides a much sharper cut-off (going to zero like r) than (6-5-1). In Fig. 6-5-1 we present ^1P (top most curves) e–H elastic scattering phase shifts based

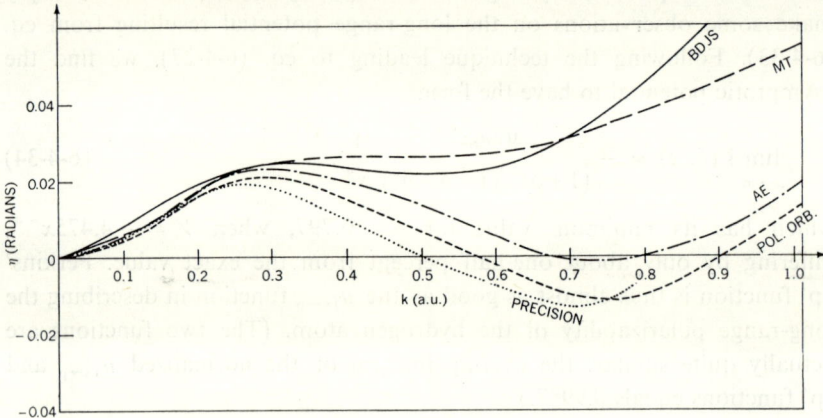

Fig. 6-5-1. ^1P electron–hydrogen phase shifts in various approximations: BDJS = Bransden et al. [149]; MT = Malik and Trefftz [150]; AE = Temkin and Lamkin [5]; POL. ORB. = Sloan [35]; Precision = Armstead [124].

on the two forms of potential in the (AE) approximation. This partial wave is of interest because it exhibits *three* changes of sign in the region below $k^2 = 0.75$ Ry. This behavior was first obtained in the adiabatic exchange calculation of Temkin and Lamkin [5], and it demonstrates that the sharper cut-off implicit in (6-3-19) is advantageous over the Buckingham form. Although this conclusion has not often been directly tested, we believe it is generally true in electron scattering from atomic systems.

The basic although not the first application of the full polarized orbital approximation was to electron–hydrogen scattering. In Table 6-5-1 we present phase shifts for this case in a variety of approximations in addition to full polarized orbitals. The triplet results are seen to lie close together and to exhibit smooth behavior. This is because they are all dominated by the spatial antisymmetry of the wave function which keeps the electrons apart, and deemphasizes the short-range correlations. A much more stringent test of the methods is provided by the singlet phase shifts. The three alterations of sign of the ^1P phase shifts mentioned above are seen to be even more closely approximated in full polarized orbitals as can also be seen in Fig. 6-5-1. On the other hand the extended polarization approximation (with static exchange terms), discussed in § 6-4, does not give the last change in sign. Taken together with its decidedly lesser accuracy in the ^1S case this shows very definitely, in our opinion, that it is not as accurate a method as polarized orbitals even though it was introduced many years later. It should also be noted that in the *non-resonant* energy range ordinary 3-state close coupling is also, by and large, not as accurate as polarized orbitals whereas closely coupled polarized orbitals is generally superior to both.

The P wave phase shifts of Table 6-5-1 are those of Sloan [35] and not those of Temkin and Lamkin. This is because in carrying the basic prescription (6-2-4) with the appropriately antisymmetrized version of (6-4-30) an

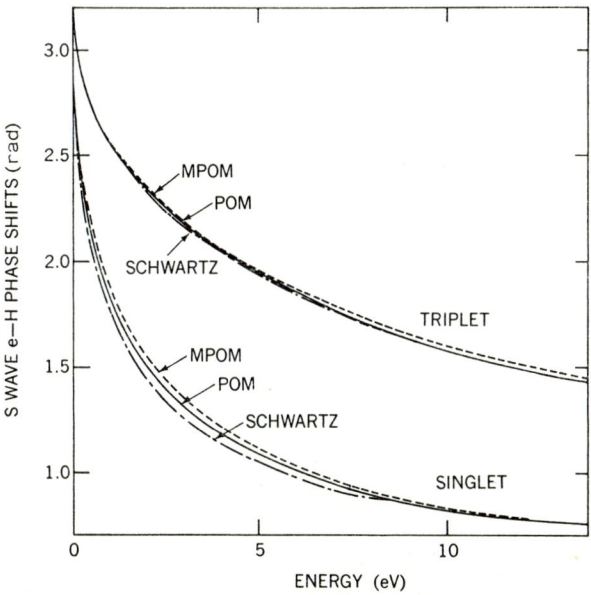

Fig. 6-5-2. Comparison of step function cut-off (POM) and Bethe–Reeh (MPOM) dipole functions in a full polarized orbital calculation for S-wave e–H phase shifts. Of note is the similarity of the results. In the case e–He phase shifts, the differences are much smaller. Calculations and curves from [24].

Electron–hydro

Partial wave	k(a.u.)	Extended polarization[a]	Polarized orbital	Closely coupled polarized orbital[c]	Close-coupling[d] (1s-2s-2p)	Precision
			Singlet			
S	0	7.419	5.6[b]	—	—	5.965[e]
	0.1	2.4537	2.583	2.529	2.492	2.553
	0.3	1.5371	1.750	1.657	1.596	1.696
	0.5	1.0423	1.251	1.155	1.093	1.202
	0.7	—	0.947	0.875	0.817	0.930
	0.8	—	0.854	0.823	0.773	0.886
	1.0	0.5258	0.758	—	—	—
P	0.1	0.0063	0.0067[g]	0.006	0.003	0.007[e]
	0.3	0.0160	0.0210	0.014	0.004	0.017
	0.5	−0.0068	0.0064	−0.007	−0.029	−0.0007
	0.7	—	−0.010	−0.026	−0.059	−0.013
	0.75	−0.0404	—	—	—	—
	0.8	—	−0.0095	−0.023	−0.059	−0.004
	1.0	−0.0424	0.0135	—	—	—
D	0.1	0.0013	—	0.001	0.001	—
	0.3	0.0105	0.0113[b]	0.011	0.008	—
	0.5	0.0215	0.0266	0.027	0.021	0.027[e]
	0.7	—	—	0.050	0.041	0.051
	0.75	0.0290	0.0456	—	—	—
	0.8	—	—	0.070	0.061	0.073
	1.0	0.0316	0.0627	—	—	—

[a] Callaway et al. [27].
[b] Temkin and Lamkin [5].
[c] Burke et al. [137].
[d] Burke and Schey [138].
[e] Schwartz [122].

extra set of terms *for P-waves only* arises. They were first derived by Sloan; for e–H they are

$$\mp \tfrac{4}{3}\delta_{l1} e^{-r_1} \left\{ (\tfrac{3}{2}r_1^2 + \tfrac{1}{2}r_1 - 3)u_l(r_1) - (\tfrac{1}{2}r_1^2 + r_1)\left[\frac{d}{dr}u_l(r_1)\right] \right\} \qquad (6\text{-}5\text{-}2)$$

to be added to the right-hand side of eq. (2.11) of Temkin and Lamkin [5]. It can be seen the eq. (6-5-2) contains a derivative (i.e. velocity dependent) term. Thus it represents nonadiabatic effects and the fact that it improves phase shifts particularly in the tricky ^1P case illustrates the arguments above (§ 6-4B), that the known symmetry of the approximate wave function

6-5 APPLICATIONS TO ELECTRON SCATTERING

phase shifts (radians)

Partial wave	k(a.u.)	Extended polarization[a]	Polarized orbital	Closely coupled polarized orbital[c]	Close-coupling[d] (1s-2s-2p)	Precision
			Triplet			
S	0	1.676	1.7[j]	—	—	1.768[e,f]
	0.1	2.9477	2.945[b]	2.937	2.936	2.939
	0.3	2.5254	2.519	2.498	2.497	2.500
	0.5	2.1454	2.133	2.102	2.097	2.105
	0.7	—	1.815	1.777	1.767	1.780
	0.8	—	1.682	1.641	1.633	1.644
	1.0	1.4854	1.460	—	—	—
P	0.1	0.0102	0.0109[g]	0.010	0.007	0.012[h]
	0.3	0.1047	0.1151	0.103	0.090	0.106
	0.5	0.2607	0.2867	0.262	0.236	0.271
	0.7	—	0.4063	0.383	0.353	0.393
	0.75	0.3871	—	—	—	—
	0.8	—	0.4351	0.418	0.391	0.428
	1.0	0.4119	0.4520	—	—	—
D	0.1	0.0013	—	0.001	0.001	—
	0.3	0.0119	0.0118[b]	0.011	0.009	—
	0.5	0.0334	0.0350	0.029	0.024	0.030[i]
	0.7	—	—	0.053	0.044	0.055
	0.75	0.0713	0.0746	—	—	—
	0.8	—	—	0.066	0.055	0.068
	1.0	0.1074	0.112	—	—	—

Temkin and Sullivan [139].
Sloan [35].
Armstead [124].

[i] Gailitis [140].
[j] Corrected in [141].

gives extra control when used with the basic scattering prescription (6-2-4). A final demonstration of this control is afforded in an important and careful calculation of Duxler, Poe and LaBahn [24] wherein when the function $\varepsilon(r_1, r_2)u_{1s\to p}(r_2)$ in (6-4-30) was replaced by the Bethe–Reeh function $g_{1>}$ of (6-3-28), the function antisymmetrized and used in (6-2-4), the resultant phase shifts differed very little from the original ones. Figure 6-5-2 displays the differences for S-waves, for e–H scattering. In the case of e–He scattering the analogous calculations yielded differences too small to be observed in a comparable graph.

Finally we note that the finiteness of the Sloan term even for a step-function cut-off can be understood from the fact that even though the kinetic energy gives a singular second derivative of $\varepsilon(r_1, r_2)$ the integration back over r in (6-2-4) effectively reduces that to a first derivative of $\varepsilon(r_1, r_2)$ which is equal to a well-defined delta function.

Sloan himself was interested in electron scattering from He^+. This is the other basic one electron target case. Results are summarized in Table 6-5-2 where they are compared with close coupling results and virtually exact

TABLE 6-5-2

S-wave e–He^+ phase shifts (relative to a pure Coulomb) in various approximations

k^2(Ryd)	Close coupling[a] (1s-2s-2p)	Polarized orbital[b]	Close coupling[c] +19 Hylleraas
		^1S	
0.4	0.375	0.4301	0.4092
0.8	0.355	0.3986	0.3912
1.0	0.350	0.3823	0.3850
1.4	0.342	0.3579	0.3780
1.8	0.343	0.3457	0.3779
		^3S	
0.4	0.887	0.9235	0.8873
0.8	0.838	0.8723	0.8462
1.0	0.820	0.8371	0.8253
1.4	0.789	0.7591	0.7895
1.8	0.758	0.6832	0.7593

[a] Burke and McVicar [36]. Numbers in this table obtained from graph in [37].
[b] Sloan [35].
[c] Burke and Taylor [37].

results wherein 19 Hylleraas terms were added to the close-coupling expansion. Here again we see that the triplet phase shifts are dominated by exchange; the close coupling results are superior whereas in the more demanding singlet case the polarized orbitals are significantly better. In all cases, discussed in this section, we have avoided inelastic thresholds which are dominated by resonances which the non-variational polarized orbital cannot reproduce, but which have produced the most striking successes of close coupling.

One further calculation should be added in the category of fundamental elastic scattering results – calculations of the electron affinity of H^-. Although this is a bound state result, dynamically speaking the outer electron is quite far removed from the inner electron and is therefore governed by the same dynamics as the low energy scattering problem. This is evidenced by the small electron affinity (only the ^1S state of H^- is bound) and the fact that all the methods can be well applied here with appropriate modification of boundary conditions. Oberoi and Callaway [38] undertook a series of calculations for this quantity to test various approximations and in Table 6-5-3 we list their results. Unfortunately both their nomenclature and their historical account of the development of the polarized orbital method is misleading. Because Bethe was the first to derive the exact form of $g_1(r,x)$, eq. (6-3-28), they attribute the polarized orbital method to him. This ignores

TABLE 6-5-3
Electron affinity of H^- (^1S) in various approximations[a]

Approximation	Electron affinity (Ry)
Adiabatic exchange (AE)	−0.0513
Extended polarization	−0.0295
Mittleman–Peacher (variational)	−0.0273
Polarized orbital	−0.0642
Drachman	−0.0544
Precision (Pekeris)	−0.0555

[a] Results in this table from [38]. See text for explanation of difference in nomenclature of approximations.

the tremendous importance of Sternheimer's work [19] for treating many-electron targets but most of all it identifies the polarized orbital method with the static problem whereas, in fact, the method is primarily one for doing the scattering problem. (The name "method of polarized orbitals" was coined in [10].) The approximation labelled "Mittleman–Peacher" in Table 6-5-3 is a full variational calculation using the Bethe–Reeh dipole correlation function, eq. (6-3-28), in the polarized orbital wave function. The name derives from a calculation of Mittleman and Peacher [23] where such a wave function was used variationally with varying smooth cut-offs multiplying $u_{1s \to p}(r)$. That paper is unfortunately even more misleading than Oberoi–Callaway. It defines the variational calculation as the method of polarized orbitals and concludes that the method is not very good. What it really shows is that a naive variational calculation is not very good. Good variational methods require two functions. Of these the one using Drachman's method in Table 6-5-3 is the only result thus far calculated using his method for electrons (i.e. total Ψ appropriately antisymmetrized). The result is seen to be very good! The adiabatic exchange result was obtained using the Bethe–Reeh potential; it is seen to be slightly better than the polarized orbital result; this corresponds to the fact that at low energies the adiabatic exchange phase shifts (Table 6-5-1) are slightly superior to the full polarized orbital results. However by $k = 0.5$ the situation reverses itself; out of a total of twelve cases (singlet and triplet S,P,D, waves for e–H and e–He$^+$ scattering) where reliable comparisons are available, the e–H ^{1}S is practically the only case in which adiabatic exchange does better than polarized orbitals. This is another point at which the Oberoi–Callaway calculation is misleading – it tests only one partial wave and that effectively only in a narrow energy range.

2. *Inelastic scattering*

There has been one further fundamental development in closely coupled polarized orbitals which we should like to discuss. And that is the nontrivial extension to the $n = 2$ states of hydrogen. This problem is complicated by the degeneracy of these target states: if one were naively to write down Sternheimer's equation for, say, the 2s state one would write [in an obvious notation from (6-3-60)]:

$$\left[-\frac{d^2}{dr^2} + \frac{2}{r^2} - \frac{2}{r} - \frac{1}{4} \right] u_{2s \to p}(r) = 2r u_{2s}(r).$$

But that equation is inconsistent as can readily be seen by multiplying it from the left by $u_{2p}(r)$ and integrating over r. The operator in brackets can then be transferred onto u_{2p} by hermiticity, and it is then seen to be zero because it is precisely the equation which the hydrogen orbital u_{2p} satisfies. On the other hand the right-hand side, $2\langle u_{2p}|r|u_{2s}\rangle$ is not zero. To make the above equation consistent one must therefore subtract the non-orthogonal part of u_{2p} from the right-hand side of the above equation. Using this analysis one can readily write the set of three polarized orbital equations associated with the $n = 2$ states:

$$\left(-\frac{d^2}{dr^2} + \frac{2}{r^2} - \frac{2}{r} - \frac{1}{4} \right) u_{2 \to p}(r) = 2\{r u_{2s} - \langle u_{2p}|r|u_{2s}\rangle u_{2p}\}$$

$$\left(-\frac{d^2}{dr^2} - \frac{2}{r} - \frac{1}{4} \right) u_{2 \to s}(r) = 2\{r u_{2p} - \langle u_{1s}|r|u_{2p}\rangle u_{1s} - \langle u_{2s}|r|u_{2p}\rangle u_{2s}\}$$

$$\left(-\frac{d^2}{dr^2} + \frac{6}{r^2} - \frac{2}{r} - \frac{1}{4} \right) u_{2 \to d}(r) = 2r u_{2p}. \tag{6-5-2}$$

6-5 APPLICATIONS TO ELECTRON SCATTERING

The complete closely coupled polarized orbital wave function, including exchange, is in this case

$$\Psi_L^{(CCPO)} = \sum_n \left\{ \sum_{l_1 l_2 = 0}^{n-1} \frac{F_{nl_1}(r_1)}{r_1} \frac{u_{nl_2}(r_2)}{r_2} \mathscr{Y}(l_1 l_2 L; \Omega_1, \Omega_2) \right.$$
$$\left. + \sum_{\lambda_1} \sum_{\lambda_2 = |l_2 - 1|}^{l_2 + 1} \frac{v_{n\lambda_1}^{(L)}(r_1)}{r_1} \frac{u_{n \to \lambda_2}(r_2)}{r_2} \cos\theta_{12} \mathscr{Y}(\lambda_1 \lambda_2 L; \Omega_1 \Omega_2) \right\} \pm (1 \rightleftarrows 2).$$
(6-5-3)

In (6-5-3) the $\mathscr{Y}$'s are the appropriately vector coupled linear combinations of spherical harmonics to form total angular momentum L. As with the polarized orbital it will be found that the $v_{n\lambda}^{(L)}$ are limited by pseudo-thresholds which are three in number for $n = 2$ and are found to be (in Ry):

$$E_s = \frac{2[3 + \frac{1}{9}(\frac{4}{3})^9]}{57 - \frac{2}{9}(\frac{4}{3})^9} - \frac{1}{8} = -0.0979 \ldots$$
$$E_p = -\frac{5}{52} = -0.09616 \ldots$$
$$E_d = -\frac{9}{88} = -0.10228 \ldots$$
(6-5-4)

They all lie *above* the $n = 3$ threshold

$$E_3 = -\frac{1}{9} = -0.111 \ldots,$$

thus integrations with rigorously required vanishing boundary conditions for the $v_{2\lambda}^{(L)}$ can be carried out to this point, in complete analogy with (6-4-32). It is interesting to observe, however, that solutions above these energies can be carried out if "pseudo-current" is allowed to flow into the pseudo-states. Such current is of course fictitious, but it can be a measure of the real current that will flow into inelastic channels that are not included. We therefore suggest that incorporation of polarized orbital pseudo-states above pseudo-thresholds provides an attractive alternative to the usual technique of a complex potential. (*Added note*: Burke and Webb [148] have carried out a type of such calculation with encouraging results at intermediate energies.)

Before showing some results, we give the normalized $n = 2$ polarized orbitals

$$u_{2 \to s} = \sqrt{\frac{\frac{3}{2}}{57 - \frac{2}{9}(\frac{4}{3})^9}} \, r \{ e^{-\frac{1}{2}r}(1 - \frac{1}{2}r - \frac{1}{2}r^2 + \frac{1}{12}r^3) + \frac{1}{2}(\frac{4}{3})^6 e^{-r} \}$$
$$u_{2 \to p} = \frac{1}{2}\sqrt{\frac{5}{26}} \, e^{-\frac{1}{2}r}(1 - \frac{1}{30}r^2)r^2$$
$$u_{2 \to d} = \frac{1}{8} \frac{1}{\sqrt{55}} \, e^{-\frac{1}{2}r} r^3 (1 + \frac{1}{6}r).$$
(6-5-5)

Damburg and Geltman [39], who are responsible for the above analysis, suggested a 6 state calculation including the hydrogenic 1s, 2s, 2p states plus the three polarized orbitals $u_{2\to s}$, $u_{2\to p1}$, $u_{2\to d}$. This provides most of the correct polarizabilities associated with the $n = 2$ channels which are only partially realized in ordinary close-coupling even including all $n = 3$ states. (One would obtain all of the correct polarizabilities if one would also include the part of $u_{1s\to p}$ orthogonal to u_{2p} [40]

$$\bar{u}_{1s\to p}(r) = 0.34r^2 e^{-\frac{1}{2}r} - 0.966 r^2 e^{-r}(1+\tfrac{1}{2}r); \; E_{\bar{p}} = +0.07806 \tag{6-5-6}$$

in the closely-coupled polarized orbital expansion.)

The calculation suggested by Damburg and Geltman was carried out by Geltman and Burke [40]. The fact that the pseudo-threshold energies are almost but not quite degenerate leads to tremendous difficulties in numerically integrating the scattering equations. The difficulty was overcome by assuming the energies exactly degenerate (with some mean value) in the asymptotic region and then integrating in. The resultant R matrices were found to retain their symmetry; thus we believe with the authors that their results can be trusted. In Fig. 6-5-3 we display the 1s–2s and 1s–2p excitation cross sections in the three best calculations compared with the experimental results of Kauppila, Ott and Fite [41] and McGowan, Williams and Curley [42] respectively. Overall there seems now to be some credence in the belief that the experimental curves may still need to be renormalized to the calculated values [40], although ratios of experimental and theoretical values are still in remarkable agreement with each other. It is gratifying also to note that of the theoretical calculations the closely coupled polarized orbital is by and large the best judged by the fact that the sum of its eigenphases is larger than the other two methods in 75 % of the cases calculated [40].

The above completes our survey of fundamental electron scattering applications. We shall now discuss additional scattering calculations which are not as fundamental because (a) they do not obtain scattering parameters from radial equations derived directly from an ansatz for the total wave function and/or (b) they involve more than one electron targets and hence the initial state wave function cannot be given exactly. In § 6-5F we shall discuss applications in which polarized orbital wave functions are used in integral expressions for other physical observables (for example photoionization, autoionization). These applications have much in common with the first application we shall discuss, Lloyd and McDowell [43], wherein an integral expression was used to calculate inelastic transition amplitudes

$$T_{if} = \langle \phi_f V_f \Psi_i^+ \rangle. \tag{6-5-7}$$

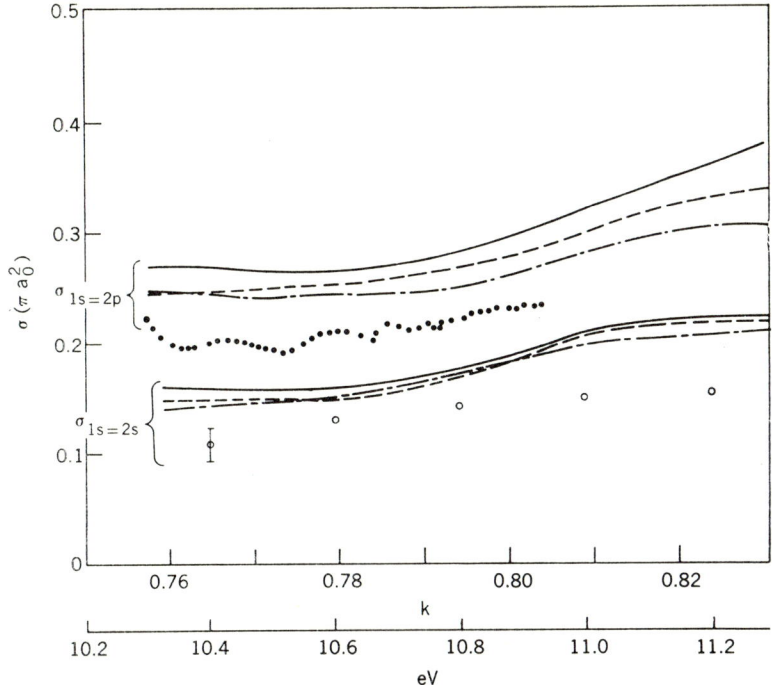

Fig. 6-5-3. Inelastic 1s–2s and 1s–2p cross sections near threshold. Open circles and closed circles refer to the respective experiments [41, 42]. Calculations are: solid line = 6 state close coupling, Burke et al. [151]. Dot–dash line = 3 state close coupling plus 16 Hyll. correlation terms [37]. Dashed line = closely coupled polarized orbital ($n = 2$ p.o.) [40]. Note that the smaller oscillations in the McGowan et al. points [42] do not come out of any of these calculations.

Eq. (6-5-7) is an identity if Ψ_i^+ is exact; the point however is to solve for $\Psi_i^{(+)}$ with only elastic scattering boundary conditions

$$\lim_{r_1 \to \infty} \Psi_i^+(r_1, r_2) = \left(e^{i k_0 \cdot r_1} + f(\theta_1) \frac{e^{i k_0 r_1}}{r_1} \right) \varphi_0(r_2). \tag{6-5-8}$$

In practice a partial wave expansion of $\Psi_i^{(+)}$ was solved for in the extended polarization approximation with exchange and using a multipole Bethe–Reeh form (6-3-28) for the polarized orbital. (There are other slight modifications – the reader is referred to Lloyd and McDowell for details.) In Figs. 6-5-4 and 6-5-5 we present results of their calculation together with the Born approximation and the recently much-discussed Glauber approximation [44]. The latter is seen to be very good in the 1s–2p cross section however at the lower energies the agreement would appear to be fortuitous as judged

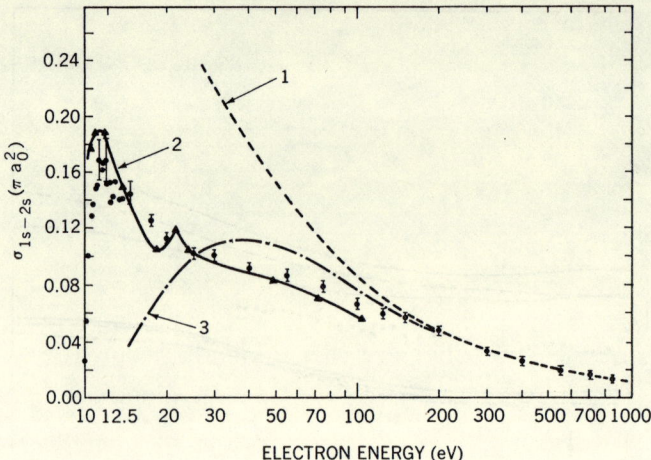

Fig. 6-5-4. Inelastic 1s–2s cross sections. Solid circles are experiment [41]. Dashed curve = Born approximation. Solid curve is calculation using polarized-orbital type elastic scattering wave function in integral for inelastic scattering amplitude [43]. Dot–dash curve = Glauber approximation [44].

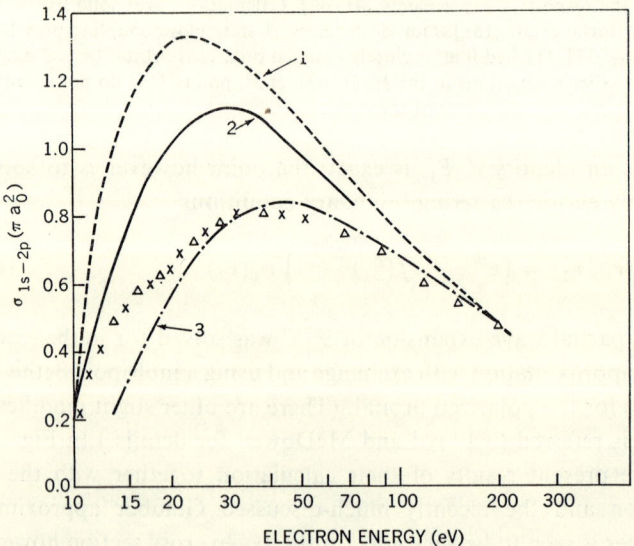

Fig. 6-5-5. Same as Fig. 6-5-4. Here experiments are denoted by × [42], and △ [152].

by the 1s–2s results. There the Lloyd and McDowell calculation is at its best, the secondary maximum at 27 eV is due to primarily a large ^3D phase shift in the extended polarization approximation. There is some hint of a secondary maximum in the experimental data at somewhat higher energies; it would be of interest to see what a complete polarized orbital calculation would give, but the general idea seems good.

B. Scattering from Helium

With helium we come to the first example of the general case wherein the target wave function is not known exactly. This means at the outset that one replaces the exact target function φ_0 by an approximate one $\bar{\varphi}_0$ and correspondingly the exact ground state energy E_0 by the approximate energy

$$\bar{E}_0 = \langle \bar{\varphi}_0 H_0 \bar{\varphi}_0 \rangle. \tag{6-5-9}$$

$\bar{\varphi}_0$ is assumed normalized, and in accord with eq. (6-2-1c) H_0 is the Hamiltonian of the target system. For impacting energy k^2 (in Ry) the total energy must then be taken to be:

$$E_{\text{total}} = \bar{E}_0 + k^2. \tag{6-5-10}$$

If there is no exchange, the ambiguity of deriving radial equations for the scattering function can be minimized by adopting a model for $H_0 (\to \bar{H}_0)$ for which $\bar{\varphi}_0$ becomes the exact Hamiltonian [45]. This will be explained in greater detail in the positron applications below. However for electron scattering this device will not work because one will encounter exchange terms of the form for example

$$I_{\text{ex}}(i, k) = \langle \bar{\varphi}_0(jk) | H_0(ij) | \bar{\varphi}_0(ij) \rangle, \tag{6-5-11}$$

where the "bra-ket" notation indicates integration over co-ordinates with repeated indices in the $\bar{\varphi}_0$'s. If one adopts a model or simply assumes

$$H_0(ij) \bar{\varphi}_0(ij) \approx \bar{E}_0 \bar{\varphi}_0(ij),$$

then one obtains

$$I_{\text{ex}}(i, k) = \bar{E}_0 \langle \bar{\varphi}_0(jk) | \bar{\varphi}_0(ij) \rangle. \tag{6-5-12}$$

On the other hand if one directly evaluates (6-5-11) one will in general obtain something different. This has had the effect of causing differences in calculations not only from different $\bar{\varphi}_0$ used, but also in the way the various $\bar{\varphi}_0$ are manipulated. For that reason one must exercise some caution in judging the *a priori* quality of methods (by which we mean the quality of

TABLE 6-5-4
Electron–helium phase shifts

k	AE (V_p = Callaway–Temkin)[a]	AE (V_p = Bethe–Reeh)[b]	Extended polarization[c]	Polarized orbital[a]
		S-Wave		
0	1.121	1.132	1.152	1.110
0.01	—	3.1306	3.12995	—
0.05	—	3.0822	3.0812	—
0.1	3.0192	3.0186	3.0170	3.0203
0.25	2.8193	2.8189	2.8156	2.8216
0.5	2.4918	2.4934	2.4878	2.4942
0.75	2.2011	2.2050	2.1979	2.2012
1	1.9571	1.9632	1.9550	1.9530
1.25	1.7572	1.7651	1.7560	1.7483
1.5	1.5943	1.6034	1.5932	1.5811
1.75	1.4610	1.4704	1.4584	1.4445
2.0	1.3507	1.3601	1.3450	1.3321
		P-Wave		
0.01	—	0.29×10^{-4}	0.29×10^{-4}	—
0.05	—	0.80×10^{-3}	0.75×10^{-3}	—
0.1	0.00335	0.0033	0.00306	0.00332
0.25	0.0229	0.0228	0.0204	0.0224
0.5	0.0971	0.1006	0.0845	0.0926
0.75	0.2015	0.2142	0.1729	0.1891
1	0.2959	0.3176	0.2508	0.2749
1.25	0.3600	0.3862	0.3014	0.3332
1.5	0.3967	0.4235	0.3277	0.3671
1.75	0.4152	0.4405	0.3383	0.3852
2.0	0.4232	0.4460	0.3399	0.3941
		D-Wave		
0.1	0.00044	0.00044	0.00035	0.00044
0.25	0.00281	0.0027	0.00245	0.00280
0.5	0.0116	0.0111	0.0104	0.0115
0.75	0.0270	0.0267	0.0239	0.0262
1	0.0477	0.0491	0.0419	0.0458
1.25	0.0712	0.0753	0.0620	0.0676
1.5	0.0943	0.1017	0.0813	0.0890
1.75	0.1152	0.1256	0.0980	0.1086
2.0	0.1333	0.1460	0.1118	0.1256

[a] Duxler, Poe and LaBahn [24].
[b] LaBahn and Callaway [47].
[c] Callaway, LaBahn, Poe and Duxler [27].

the phase shifts that they would produce if one could use the exact target state) by the agreement of phase shifts calculated using approximate target states with experimentally determined ones. (Cf. Peterkop [46] for some calculations testing the sensitivity to $\bar{\varphi}_0$.)

Three adiabatic exchange calculations have been carried out for e–He elastic scattering. LaBahn and Callaway [47] and Williamson and McDowell [48] have both used a Bethe–Reeh type polarization potential but differing ground state wave functions – an analytic Hartree–Fock and open-shell variational function respectively. Their results are very similar. They are given under AE (B-R) in Table 6-5-4 where they are compared with a recent adiabatic exchange calculation of Duxler, Poe and LaBahn [24] wherein the direct polarization potential was derived using a step-function cut-off in the polarized orbital according to the original prescription of Temkin; that V_p therefore is the analogue of the analytic Callaway–Temkin potential for the one electron target. It vanishes like r near the origin (as opposed to r^2 for the Bethe–Reeh potential) but is generally less attractive in the intermediate and major range of r. Comparison of the two AE results with the extended polarization [27] and recently completed full polarized orbital results [24] suggests again that the overall lesser attraction of the step-function cut-off is superior. With regard to the comparison of the latter two methods (both using the same Hartree–Fock ground state) there are no more definitive calculations by which they may be compared. The analogous e–H results would suggest the polarized orbital should be better, however, the polarizability being smaller ($1.3\ a_0^3$ versus $4.5\ a_0^3$), such a simplistic analogy may not be adequate, particularly in view of the approximate target wave functions. The phase shift analysis of Bransden and McDowell [49] as modified by McDowell [50], giving a least square fit to a whole host of experimental and theoretical consistency information, suggests that the adiabatic exchange P-wave results are too large and that the D-wave extended polarization phase shifts are too small. On the other hand individual swarm and scattering data suggest a scattering length $a = 1.17 \pm 0.05\ a_0$ [51], whereas the polarized orbital result, Table 6-5-4, is 1.10. The difference is comparable to the (singlet) e–H polarized orbital and precision results, Table 6-5-1. It would appear that phase shifts derived from the least square analysis should be considered only interim in character.

Adiabatic exchange and extended polarization calculations of e–He differential cross-sections have been carried out at higher energies. Although the original calculations of Khare and Moiseiwitsch [52] appeared too low at small angles, it has been shown [53] that inclusion of all necessary partial waves greatly improves results, so that the adiabatic exchange results are comparable to extended polarization results of LaBahn and Callaway [54 a]. Experimental results range from 100–400 eV [54b] and 500 eV [54c] and they are very accurate. Calculated results in forward directions show very good agreement with experiment when polarization is included and not so when it is neglected. However we would advise some caution in highly quantitative comparisons between these calculations and experiment, particularly as a method of judging various methods, because all calculations have in no way included the effect of inelastic channels which could very well be quite important at these energies. In fact a recent dispersion theory fit [49b] of the experimental data shows that the extended polarization calculation does not give an accurate shape to the scattered intensity in precisely the forward direction as a function of the energy.

C. Scattering from other (non-highly polarizable) atoms

Only a few calculations in this category have been performed in the full polarized orbital approximation. In the original e–O calculation of the polarized orbital method, Temkin [4] only included $u_{2p \to d}$ orbitals whereas Henry [55] included $u_{2s \to p}$ and $u_{2p \to s}$ orbitals as well. He also corrected a mistake in one of Temkin's exchange polarization terms although this did not effect Temkin's conclusions of the general size and direction of the exchange polarization terms. The effect of $(2p \to s)$ and $(2s \to p)$ orbitals is to raise the polarizability from 3.7 a_0^3 to 5.15 a_0^3, the latter to be compared

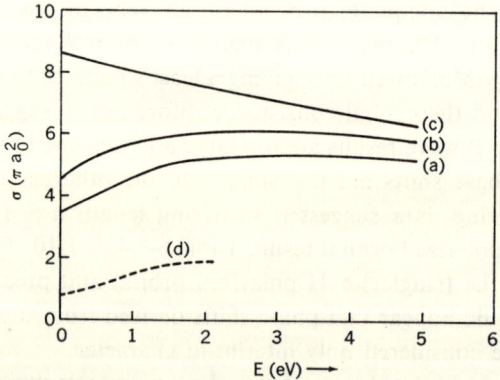

Fig. 6-5-6. Electron–oxygen S-wave cross-section in exchange approximation (c), adiabatic-exchange (a), and polarized orbital (b), the latter two using only the (2p–d) contribution to the polarizability. The lower (dashed) curve is an adiabatic-exchange S-wave result using the whole polarizability [33]. This figure is taken from [4].

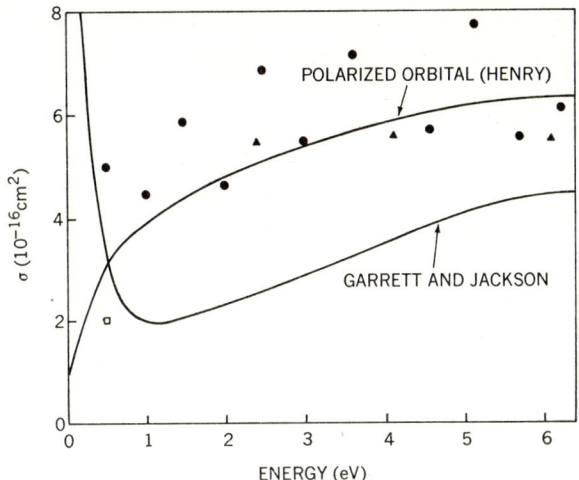

Fig. 6-5-7. Electron–oxygen elastic cross-section. Solid curves are calculations discussed in text. Experimental results are: solid circles = Sunshine et al. [153]; solid triangle = Neynaber et al. [154]; open square = Lin and Kivel [155].

with an experimental value of 5.19 a_0^3 [56]. The effect of various terms on S-wave cross-sections is given in Fig. 6-5-6 in the (2p → d) approximations [4]. The results are confirmed by Henry and both adiabatic exchange and polarized orbital results are lowered in the direction of the Bates–Massey result [33] when all the $n = 2$ polarized orbitals are included. Figure 6-5-7 gives Henry's final results (including a Born estimate of $l > 2$ contributions) compared to some experimental results and the calculation of Garrett and Jackson [56] wherein exchange was included via the Slater [57] approximation; this accounts, in our opinion, for its spurious shape at low energies.

Henry [55] has also calculated the photodetachment of O^- using his e–O scattering wave function for the final state; we shall discuss his results, with some comments, in § 6-5F. Finally we note that Henry, using Lagrange multipliers, forced his scattering function to be orthogonal to all the bound orbitals. Such orthogonality simplifies the radial scattering equations and can be shown to hold rigorously in triplet S-wave e–H scattering and in S-wave e–He scattering when a Hartree–Fock closed shell target state wave function is used. In general, however, we cannot understand the justification of such orthogonality, except as an approximation. Indeed it appears that if a low-energy shape resonance occurs, it will manifest itself principally in a lack of orthogonality with an *unperturbed* target orbital. Such orthogonality was originally imposed in close coupling calculation [58], but when it was relaxed [59], it gave rise to low energy shape resonances in e–C and e–N systems. We therefore believe that Henry's [60] polarized orbital calculations of e–C and e–N scattering (and related photodetachment calculations) require relaxation of the orthogonality constraint.

Remaining calculations on non-highly polarizable atoms have been carried out only in the adiabatic exchange approximation. Thompson [61] has calculated electron scattering from the noble gases neon and argon. In constructing the direct polarizability potential he has followed the prescription of Temkin [4] very closely. For neon, only the (2p → d) polarized orbital was included; it gives a polarizability of 2.2 a_0^3 as compared to an accurate value of 2.7 a_0^3 [62]. A typical result is the total cross-section as compared to experimental results principally of Salop and Nakano [63] is shown in Fig. 6-5-8. The deviation of Thompson's curve from Salop and Nakano can almost completely be accounted for by his 20 % underestimate of the polarizability. In the case of argon, only the (3p → d) polarized orbitals were included; the Sternheimer approximation for them, 14.2 a_0^3, overshoots an accurate (3p → d) value of 13.9 a_0^3 [64]. The polarization potential was renormalized to that value. Angular distributions in this

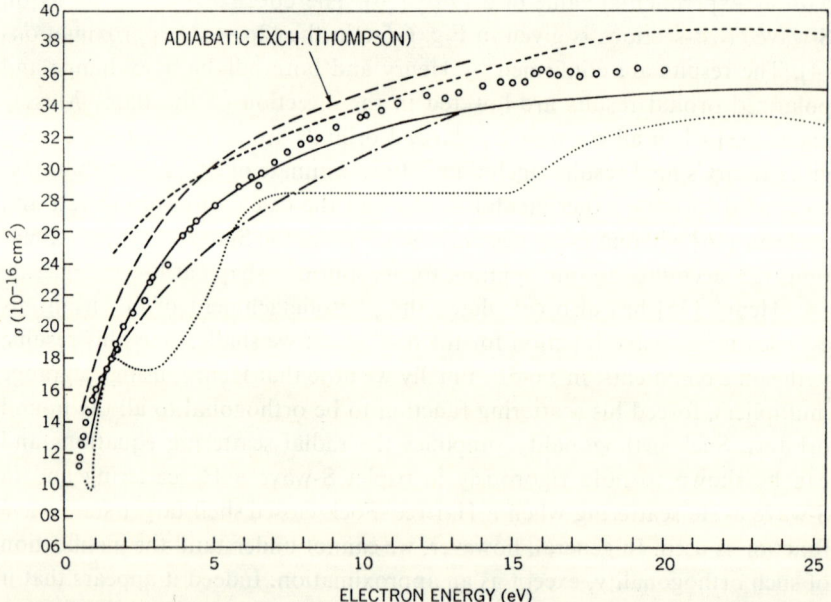

Fig. 6-5-8. Elastic scattering of electrons from neon. Adiabatic exchange calculation of Thompson [61] is longer dash curve and circles are the experimental results of Salop and Nakano [63]. Remaining curves are other experimental results cited in [63].

approximation also appear to improve agreement with experiment, but they are by no means perfect.*

Walker [65] has carried out calculations for the scattering of electrons from mercury in the adiabatic exchange approximation modified to be appropriate to the Dirac equation. Only the 6s electrons were included and in this case one gets two polarized orbitals (6s → p+ and 6s → p−). The polarization potential was renormalized to give the experimental values $\alpha = 34.5 \pm 0.1\, a_0^3$. Differential, spin polarization, and total cross-sections were calculated from 3.5 eV to 100 eV. The results do not show any marked improvement coming from the inclusion of direct polarization over ordinary exchange when compared to experiment [66]; however, in an added note Walker states that new spin polarization experiments do agree better with adiabatic-exchange results.

D. Scattering from Highly Polarizable Atomic Systems

Highly polarizable targets present a very demanding test for polarized orbital approximations. On the one hand the large polarizability renders first order perturbation theory in general and the dipole approximation in particular very uncertain as a means of constructing the polarized wave function. On the other hand the interaction between the scattered and outer atomic electron is very strong, so that non-adiabatic effects can also be expected to increase in importance. For this reason the closely coupled polarized orbital method presents a natural and intrinsically more satisfactory alternative. In fact since the polarizability is supplied chiefly by the first excited state in alkali targets, the first excited state itself is an excellent approximation of the polarized part of the wave function; this means the ordinary two-state close coupling expansion is almost equivalent to a closely coupled polarized orbital expansion and it likewise should provide a good approximation. (This similarity would not be true for inclusion of quadrupole distortion, however.) Such calculations have been performed by Karule and Karule and Peterkop [67], Marriott and Rotenberg [68] and Burke and Taylor [69].

A somewhat more general approach has been given by Mittleman [70] wherein he treats the outer two, valence and scattered, electrons exactly and replaces the remaining core by an effective non-local potential. The approach is in a sense a generalization of Temkin's non-adiabatic theory [2], and it has

---

* M. R. C. McDowell informs us [private communication] that the experimental angular distributions do not seem completely compatible with other experimental data if one tries to fit them all with a unique set of phase shifts.

considerable theoretical appeal; unfortunately no calculations have been reported.

With respect to non-variational polarized orbital approximations – it is our opinion that if it succeeds at all it will be due to the extra control afforded by the known symmetry of the total wave function as discussed in § 6-4A. Thus we tend to be suspicious of the adiabatic-exchange approximation in these cases. Garrett [71] has performed in essence an adiabatic-exchange calculation, but he has concentrated on evaluating polarized orbitals which are correctly matched at $r_1 = r_i$, where r_i is the coordinate of a bound electron. This has led to a considerably more difficult calculation for the scattering problem. The results appeared to be similar to experimental results of Perel, Englander and Bederson [72]. Those results however were normalized relative to electron–potassium results of Brode [73]. The latter results have now been shown to be in error [74], and when the Perel et. al. curve is normalized relative to the new e–K results Fig. 6-5-9 results [75] for e–Li scattering. We have included other calculated results as well. That Garrett's phase shifts cannot be correct at the lowest energies follows from the fact his zero energy absolute phase shifts [76] approach $3\pi, 2\pi, \pi$ radians for s, p and d partial waves respectively whereas from Swan's "theorem" [77] assuming one bound state of Li^- [78], one would expect s-waves to approach 2π and all remaining ones to approach 0. Stone's [15] somewhat different type of adiabatic-exchange calculation appears to give 2π for the s-wave phase shifts and π for the ^3P phase shift, with remaining phase shifts approaching 0; this immediately shows the superiority of his phase shifts over Garrett's, but it appears that his ^3P phase shift is still incorrect.

As we complete this article, we have just received a preprint of a full polarized orbital calculation by Vo Ky Lan [79] in which all the above suspicions are evinced. The exchange polarization terms make the ^3P phase shift approach zero radians, but they rise rapidly to give an effective "shape resonance" just above zero energy (see inset in Fig. 6-5-9). This accounts for the almost step function rise and steep descent of the cross-section characteristic also of the close coupling results. In fact the only difference with close coupling is a slight overall reduction of the cross-section which fits the renormalized experimental results even better. The situation can only be described as an unexpected success of the full polarized orbital method. Nevertheless further experiments are desirable. We believe that precisely in the field of electron–alkali scattering will experiments with polarized electrons and/or polarized target atoms afford the most definite information about phase shifts (cf. Raith [80]).

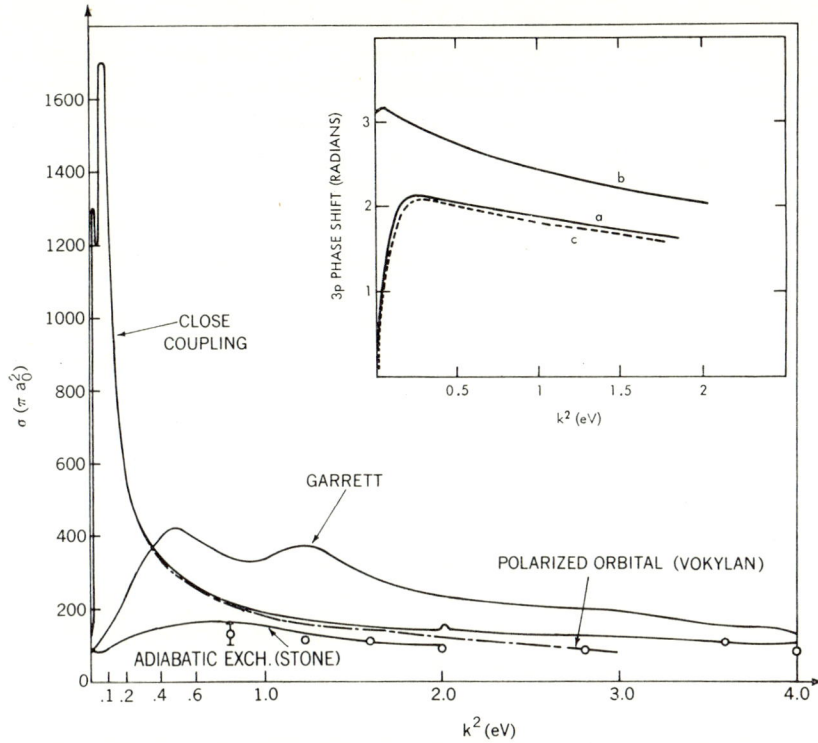

Fig. 6-5-9. Scattering of electrons lithium. Circles are renormalized experimental results [75]; these results go up sharply near threshold but the errors are correspondingly larger [B. Bederson, private communication]. The theoretical calculations are close coupling [67, 68, 69]; adiabatic exchange [15], Garrett's substantially altered AE calculation [71] and polarized orbital [79]. In the inset the ^3P phase shift is shown in various approximations: c = close coupling, a = polarized orbital, b = adiabatic exchange. Only in the former two does this phase shift correctly go to zero radians at zero energy, and this accounts for the huge cross-section at low energies. An analogous situation occurs in the other alkalis.

The other class of atomic targets with large polarizabilities are the negative ions. They are similar to alkali atoms in that the outermost electron is far outside the remaining electrons and is weakly bound; thus they give rise to essentially the same difficulties as the alkalis. A calculation of elastic scattering of electrons of H^- has been carried out by McDowell [81] in the adiabatic-exchange approximation using both dipole and quadrupole Bethe–Reeh polarization potentials. It is significant that the quadrupole potential changes the dipole results minimally in view of the fact that the quadrupole polarizability α_q $(= 1300\ a_0^5)$ is much larger than α $(= 203$

a_0^3); the reason is because where $\alpha_q(r)$ achieves its asymptotic value it is much more strongly cut-down by the stronger r^{-6} dependence of its denominator. At smaller distances the contribution is small because it turns out that $\alpha_q(r)$ decreases even more rapidly than $\alpha(r)$. A very similar situation occurs in a full polarized calculation of Abiodun [82]. This is a necessary condition for the meaningfulness of any such method. Unfortunately it is not a sufficient condition, nor does it prove that adiabatic-exchange results will be close to those of polarized orbital; the latter values are not available.

E. Electron–Molecule Scattering

The idea of polarization has been introduced into this field from several directions. The rotational close coupling theory of Arthurs and Dalgarno [83] has been augmented to include exchange [84] and polarization [85]; calculations on e–H_2 scattering [86] appear reasonably satisfactory.

It turns out, however, that the most natural way of introducing polarized orbitals is in the context of the fixed-nuclei theory [87]. Assuming the nuclei fixed introduces the dependence of the angles of the internuclear axis with respect to the incoming beam direction $\boldsymbol{\beta}_0$ into the scattered amplitude. The important point is that this dependence can be factored and the $\boldsymbol{\beta}_0$ part given analytically as follows [88]:

$$f(\boldsymbol{\beta}_0, \Omega') = \sum_{m'} \sum_{l_i l_j} f_{l_i l_j m'}(\boldsymbol{\beta}_0) Y_{l_i m'}(\Omega'), \tag{6-5-13}$$

where Ω' is the scattered angle in the lab system and

$$f_{l_i l_j m'}(\boldsymbol{\beta}_0) = \sum_m a_{l_i l_j m} \mathscr{D}_{m'm}^{(l_i)}(\boldsymbol{\beta}_0) \mathscr{D}_{0m}^{(l_j)*}(\boldsymbol{\beta}_0). \tag{6-5-14}$$

Only the scattering parameters $a_{l_i l_j m}$ are determined from solutions of scattering equations, and these can be derived from assumptions of the total wave function in the body-fixed frame (i.e., $\boldsymbol{\beta}_0$ is absent).

For example, in the scattering of electrons from H_2^+ in the uncoupled approximation [87] the polarized orbital wave function is taken to be:

$$\Psi_{lm}^{(N)} = \frac{u_{lm}(r_1)}{r_1} Y_{lm}(\Omega_1) \left[\Phi_0^{(N)}(r_2) - \frac{\varepsilon(r_1, r_2)}{r_1^2} \frac{\varphi_0^{(pol)}(r_2)}{r_2} \frac{\cos\theta_{12}}{\sqrt{4\pi}} \right] \pm (1 \rightleftarrows 2). \tag{6-5-15}$$

$\Psi_{lm}^{(N)}$ is a single-center wave function; thus the scattered and orbital electrons, r_1 and r_2, are measured from the midpoint between the two molecular nuclei. This renders a very strong analogy with electron–atom scattering and motivates a whole host of methods that can be used for molecules also. The radial functions $u_{lm}(r)$ and similarly the phase shifts η_{lm} depend on the magnetic number m by virtue of the non-spherical symmetry of the Hamiltonian. This lack of sphericity also applies to the target state itself wherein we write (for the ground state)

$$\Phi_0^{(N)}(r_2) = \sum_{n=0}^{N}{}'' \frac{\varphi_n^{(N)}(r_2)}{r_2} P_n(\cos\theta_2). \tag{6-5-16}$$

In principle the sum in (6-5-16) contains an infinite number of terms, but in practice it must be truncated: $\frac{1}{2}N$ is the number of terms retained and thus it measures the order of approximation in that expansion. Calculations for e–H_2^+ scattering in zeroth $(N = 0)$ and first order $(\frac{1}{2}N = 1)$ have been performed in the uncoupled [87] and coupled [88] approximation and in those cases the contribution of the permanent distortion ($N = 2$ term in $\Phi_0^{(N)}$) are very small. Single-center calculations of e–H_2 scattering are under way [89]. Very elaborate single-center fixed-nuclei e–N_2 calculations have been performed by Burke and Sinfailam [90] with qualitative rather then quantitative success. This can in part be attributed to the non-inclusion of polarization.

Fixed-nuclei calculations in electron–molecule scattering are by no means restricted to single-center expansions. The most successful fixed-nuclei calculation of e–H_2 scattering was done in spheroidal coordinates by Hara [91] in essentially an adiabatic-exchange approximation using a spheroidal polarization potential [92]. It is likely in our opinion that two-center coordinates will be essential for high accuracy.

Fixed-nuclei calculations carry an extra bonus in that once one knows the scattering parameters, one can calculate rotational excitation with them to high accuracy. The pertinent "adiabatic" theory was introduced by Chase [93] in the context of nuclear physics. Oksyuk [94] was the first to introduce it into molecular rotational excitation using something substantially better than Born type calculations for the fixed-nuclei scattered amplitude with impressive qualitative results. The realization that the true quantitative potentialities of the theory was come only recently [95] as a result of the developments in the fixed-nuclei theory discussed above, as we shall explain in the next paragraph. The essential point is that if one can calculate the fixed-nuclei amplitude accurately, then rotational excitation can be calculated from it by a simple process of quadrature.

It turns out that much that has been done in rotational excitation has explicitly or implicitly been confined to Σ-states of the target molecule. Only recently has the adiabatic theory been worked out explicitly for electronic states of arbitrary Λ symmetry [96]. The basic formula can be reduced to the form

$$f_{\Gamma'\Gamma}(\Omega') = C_{j'j} \int \mathscr{D}_{\Lambda,m}^{(j')*}(\boldsymbol{\beta}_0) f(\boldsymbol{\beta}_0, \Omega') \mathscr{D}_{\Lambda,m}^{(j)}(\boldsymbol{\beta}_0) \mathrm{d}\boldsymbol{\beta}_0 + \mathrm{O}\left(\frac{m}{M}\right). \quad (6\text{-}5\text{-}17)$$

$C_{j'j}$ is a normalization constant depending on the initial (j) and final (j') rotational quantum numbers, and $\Gamma(\Gamma')$ stands for the whole set of final (initial) quantum numbers. The important points of (6-5-17) is that the error is of order of the mass of the electron to that of the nuclei and therefore enormously small, except near threshold [97]. Secondly, using the fixed nuclei amplitude (6-5-13) in the above allows the right-hand side to be analytically evaluated. For example, the total cross-section turns out to be [96]:

$$\sigma_{\Gamma'\Gamma} = \frac{k_{\Gamma'}}{k_\Gamma} \sum \frac{a_{l\lambda m} a_{l\lambda\mu}^*}{(2\lambda+1)} (-1)^{m+\mu} (l\lambda m - m|J0)(l\lambda\mu - \mu|J0)(jJ\Lambda - \Lambda|j0)^2.$$

$$(6\text{-}5\text{-}18)$$

From (6-5-18) one learns specifically from the Clebsch–Gordan coefficient $(jJ\Lambda - \Lambda|j'0)^2$ that $|j'-j|$ = even integer only if $\Lambda = 0$ (Σ states). For $\Lambda \neq 0$, $|j'-j|$ can be anything contrary to what many people have thought.

The most successful adiabatic calculation of rotational excitation has been done by Hara [98] using his previously discussed e–H_2 fixed nuclei scattering parameters. The theory has also been used for e–N_2 by Burke and Sinfailam [90], and it promises to be an indispensible adjunct of the fixed-nuclei theory.

The discussion of this section has necessarily been very abbreviated. For somewhat more complete reviews the reader is referred to Takayanagi [99] and Golden, Lane, Temkin and Gerjuoy [100].

F. PHOTOIONIZATION, AUTOIONIZATION AND IONIZATION

This section deals, again briefly, with applications of polarized orbital scattering wave functions to other physical quantities, specifically the ones mentioned in the title (including photodetachment). In principle the idea is the same as Lloyd and McDowell [43] wherein the wave function, calculated in the usual way, is substituted in an integral expression for the quantity in question. A prime example is the photodetachment of H^-, because the matrix element is exact in principle, so that deviations from the experiment are directly related to inadequacies of the wave functions. Two principal forms of this cross-section (or atomic absorption coefficient) have been used:

$$K_L = 6.812 \times 10^{-20} k(k+I) \left| \int \Psi_F^*(z_1+z_2)\Psi_B \, d\tau \right|^2 \text{cm}^2$$

$$K_V = 2.725 \times 10^{-19} \frac{k}{k+I} \left| \int \Psi_F^* \left(\frac{\partial}{\partial z_1} + \frac{\partial}{\partial z_2} \right) \Psi_B \, d\tau \right|^2 \text{cm}^2. \quad (6\text{-}5\text{-}19)$$

K_L and K_V are equivalent if exact initial (bound Ψ_B) and final (free Ψ_F) state wave functions are used (k = wave number of emerging electron and I = ionization energy in rydberg). For H^- since Ψ_B is a ^1S state, only ^1P continuum functions contribute. Bell and Kingston [101] have used the appropriate polarized orbital wave functions for Ψ_F and a 70 parameter Hylleraas function for Ψ_B. Results are given in Fig. 6-5-10 wherein length and velocity forms are compared together with close coupling results and experimental data. In Table 6-5-5 we give a comparison of sum rules

$$S_n = \left[\frac{\frac{1}{137}}{2\pi a_0} \right]^{n+2} \frac{137^2}{\pi a_0} \int_0^{\lambda_E} K\lambda^n \, d\lambda, \quad (6\text{-}5\text{-}20)$$

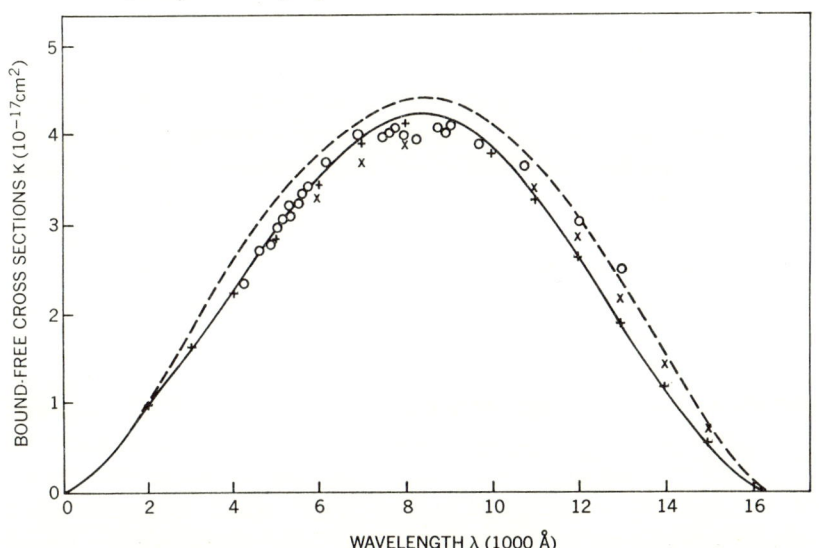

Fig. 6-5-10. The bound–free absorption cross-section of H^-. Open circles are experiment, Smith and Burch [156]. Solid (dashed) curves are obtained using polarized orbital free wave functions in the length (velocity) matrix element together with a 70 parameter bound function [101]. Plusses and crosses refer to the same type calculations using 3-state close coupling free wave functions [142].

TABLE 6-5-5

Sum rules associated with photodetachment of H$^-$ for 70-parameter Ψ_B and various Ψ_F (from [101])

Ψ_F	$S_1(\times 10^{-3})$		$S_0(\times 10^{-2})$		$S_{-1}(\times 10^{-1})$	
	L	V	L	V	L	V
Close coupling[a]	3.63	3.72	2.01	2.02	1.44	1.42
Polarized orbital[b]	3.70	4.15	2.05	2.27	1.47	1.60
Exact	4.00±0.30[c]		2.06±0.03[d]		1.50[e]	

[a] Doughty, Fraser and McEachran [142].
[b] Bell and Kingston [101].
[c] C. Schwartz (unpublished).
[d] Chung and Hurst [143].
[e] Pekeris [144].

where λ is the wave length of the incident radiation (λ_E is the threshold wave length)

$$\lambda = 911.267/(k^2+I) \quad \text{Å}, \qquad (6\text{-}5\text{-}21)$$

and K is either K_L or K_V.

Table 6-5-5 shows rather more clearly than Figure 6-5-10 that although close coupling results give somewhat better consistency between length and velocity forms of the matrix element, the length form of polarized orbital results is more accurate than both. The p.o. results should be most accurate at long wave lengths and the discrepancy with experiment there has caused Bell and Kingston to question the accuracy of the experimental results there [102].

Bell and Kingston originally used the same technique to calculate the photoionization of He [103]. These results and Li$^+$ [103] are similar in quality but perhaps slightly less accurate compared to close coupling calculations and experiment, as expected from their larger Z.

Matese and LaBahn [104] have calculated the photoionization of Li using the adiabatic exchange (they use the abbreviation AED, D implying dipole) as well as extended polarization continuum functions. Their AE results are very close to the (corrected) Brueckner–Goldstone (BG) many-body theory results of Chang and McDowell [105] but the EP results appear closer to experimental results near threshold [106]. McDowell and Chang have also found that they could well approximate various integrals of their many-body calculation using AE wave functions and they have used this device to extend their calculations to members of the lithium isoelectronic sequence [107].

The BG method is of importance because it corrects systematically for the Hartree–Fock approximation in *both* the initial (bound) and final (free) states of the target system. For simple systems one can use brute force variational calculations to get accurate wave functions for the bound state. However for many-body systems, most especially if they are negative ions, this is no longer the case. This brings us to the photodetachment calculations of Henry [55, 60] because he used ordinary Hartree–Fock wave functions for the negative ion. However as discussed in § 6-5A the polarized orbital method itself can be used to calculate energies and wave functions for the negative ions which should be superior to ordinary exchange (i.e. essentially Hartree–Fock) calculations. One can then confidently predict that the use of such wave functions, which may clearly be obtained as an adjunct of one's scattering calculation, should improve the accuracy of a photoionization calculation. Also since bound and continuum wave functions are calculated from the same effective Hamiltonian, they should automatically lead to satisfaction of the sum rules. Finally the fact that Hartree–Fock bound state wave function was used by Henry [55] for O^- photodetachment may account for the fact that his velocity form was more accurate than length. For the length form weights the outer portions of the wave functions more heavily, and it is there that $\Psi_F^{(pol.\ orb.)}$ is most accurate providing its accuracy is not offset by a poor Ψ_B in the same region. This means that Ψ_B should contain a polarized part beyond the perturbed bound state orbital as done in [104]. Even that alteration, however, was found to provide a significant improvement [104].

The extension of polarized orbital to electron–molecule scattering being new, there have not been many applications to photoionization of molecules, however Khare [108] has used the single-center polarized orbital e–H_2^+ wave functions of Temkin and Vasavada [87] (modified to the equilibrium radius of H_2) to calculate the photoionization of H_2 with auspicious results.

We next turn to autoionization. It will be seen that the explicit formulae are given in terms of P and Q operators which were introduced in § 6-4. It is to be emphasized that since we are dealing with electron applications here, the basic formulae for the operators, even for one electron targets, must be augmented to include exchange; they are [109]

$$P = P_1 + P_2 - P_1 P_2, \qquad (6\text{-}4\text{-}5b)$$

where P_i is the operator given in eq. (6-4-5) and i represents the coordinates of the ith electron ($i = 1, 2$) in the ground state wave function φ_0:

$$P_i = \varphi_0(r_i)\rangle\langle\varphi_0(r_i); \tag{6-4-5a}$$

Q is then defined as $1 - P$.

The two required integrals are the width [110]

$$\Gamma_n(E) = 2k|\langle P\Upsilon(E)|H|Q\Phi_n\rangle|^2, \tag{6-5-22}$$

where $\Gamma_n \equiv \Gamma_n(\mathscr{E}_n)$ is the width at resonance, and the shift [111]

$$\Delta_n = \mathscr{P}\int_0^\infty \frac{\Gamma_n(E')dE'}{\mathscr{E}_n - E'}. \tag{6-5-23}$$

In (6-5-23) $\mathscr{P}$ means the principal value integral. $Q\Phi_n$ is the nth eigenfunction of a projected Schrödinger equation with eigenvalue $\mathscr{E}_n$:

$$QHQ\Phi_n = \mathscr{E}_n\Phi_n. \tag{6-5-24}$$

We shall not discuss the $Q\Phi_n$ calculation further except to state that the energy of a "Feshbach" resonance is related to $\mathscr{E}_n$ by

$$E_n = \mathscr{E}_n + \Delta_n \tag{6-5-25}$$

and that for one-electron targets [112] $Q\Phi_n$ can be extremely accurately calculated. Of concern here is the calculation of the width and shift which requires a non-resonant scattering function $P\Upsilon(E)$ which is formally the solution of the complete Schrödinger equation less the resonant term [110]

$$\left[PHP + \sum_{j\neq n} \frac{PHQ\Phi_j\rangle\langle\Phi_j QHP}{E - \mathscr{E}_j} - E\right] P\Upsilon(E) = 0. \tag{6-5-26}$$

The optical potential in this equation differs from the complete optical potential in eq. (6-4-11) in that the resonant term, $j = n$, is absent here. The solution of PHP is the exchange approximation which is certainly the natural first approximation to $P\Upsilon(E)$. Beyond that, however, the discrete terms in the non-resonant optical potential [$\sum_{j\neq n}$ terms in (6-5-26)] make only a small contribution (cf. [113]). The major contribution comes from the continuum which sets in for E greater than the first excited state energy. As usual this is very difficult to incorporate outright; Bhatia and Temkin [113] have identified the continuum with polarization effects and used polarized orbital wave functions for $\Upsilon(E)$ (and operated with P to get $P\Upsilon$). Table 6-5-6 is a precis of some of their results for the ^1P ($n = 1$) autoionization state of He which has been accurately enough measured [114]

TABLE 6-5-6

^1P ($n = 1$) autoionization state parameters of He

	ε_n (eV)	Γ_n (eV)	Δ_n (eV)	E_n (eV)	q
Calculation (Bhatia and Temkin [113])	60.150	0.0374 $(0.0365)^{(a)}$ $(0.0350)^{(b)}$	−0.007	60.143	−2.275
Experiment (Madden and Codling [114])		0.038±0.004		60.130±0.015	−2.80±0.25

(a) Using exchange approximation for $P\Upsilon$.

(b) Using adiabatic exchange approximation for $P\Upsilon$.

by vacuum ultraviolet techniques to afford a stringent test of the theory. We see that the polarized orbital wave function gives a width closest to the central value of the experiment, although both exchange and adiabatic exchange approximations are within the experimental error. The shift is seen to be small but nevertheless essential to get within the experimental error for E_n. (In fact this is the only calculation to yield a value of E_n within the experimental uncertainty [113].) The number q is Fano's [111] shape parameter which occurs in his resonance profile formula

$$\sigma(E) = (q+\varepsilon)^2/(1+\varepsilon^2) \tag{6-5-27}$$

where

$$\varepsilon = E - E_n/(\tfrac{1}{2}\Gamma_n). \tag{6-5-28}$$

An integral expression for q can be derived [111] which, it turns out, is much more sensitive to the interior of the continuum than the shift. The fact that q in Table 7-5-6 is outside of the experimental error shows that even polarized orbital wave functions do not have requisite accuracy in the interior region. This plus the fact that E_n is barely within the experimental error has motivated a closely-coupled polarized approach. The idea is to use the closely coupled polarized orbital function (6-5-3) less the hydrogenic functions φ_{2s} and φ_{2p}, which are chiefly responsible for the resonance. One will still retain two undetermined functions which should greatly improve the description of the interior region [115].

The final topic we shall discuss in this section of electron applications concerns electron impact ionization. Here the idea has been to use a form of the matrix element from the unperturbed initial system (plane wave on the target state) to a final state which is approximated by a plane wave for the outer (scattered) electron and an elastic scattering wave function for the slower (ejected) electron scattered by the residual ion. Sloan [116] has used polarized orbital e–He$^+$ wave functions for the impact ionization of He and a slightly different version of e–He$^+$ scattering equations was used by McDowell and Economides [117] with essentially identical results; the latter have also extended the calculation to e–Li$^+$ ionization retaining plane waves for the outer electron in that case also. Comparison with experiment [118, 119] is very good in both cases in spite of the fact that this kind of treatment involves an additional order of approximation – neglect of the interaction between the outer electron and the residual system – beyond the polarized orbital approximation. (The accuracy of this approximation can be expected to improve at high energies, and the results bear out this expectation.)

§ 6-6. Applications to positron scattering and annihilation

In the course of our previous discussions of both target distortions (§ 6-3) and bound calculations (§ 6-4C), we used the example of the positron–hydrogen system for simplicity, since indistinguishability of the projectile and target was not at issue. For the same reason, however, positron scattering is a severe test of the Polarized Orbital Approximations, since it probes the inner reaches of the electronic cloud. In addition, annihilation takes place during collisions at a rate dependent upon details of the scattering wave function. In this section we will apply our previous developments to e$^+$–H and e$^+$–He systems in several approximations. Some comparison with experiment and other theories will also be given, although two very complete reviews [120] may be consulted for details.

We have already seen (in § 6-5) that very good results are obtained for electron–atom scattering using only the dipole term in the distorted wave function of the atom. This is emphatically not the case for positron scattering, so in what follows we will use only those approximations which include the higher multipole contributions; the Dalgarno–Lynn method will prove particularly useful. We will discuss the scattering of positrons first from hydrogen and then from helium; for the latter case we will also consider the annihilation rate and γ-ray angular correlation.

A. POSITRON–HYDROGEN SCATTERING

The direct application of the basic method ((6-2-9), (6-2-10)) to this problem is very straightforward, depending on the solution of a standard scattering type of equation. The only interesting question concerns the precise form of adiabatic potential to use. Contrary to the electron–hydrogen case (§ 6-5A) it is entirely inadequate to use only the dipole term of equation (6-3-16), in either the Callaway–Temkin form (6-3-19), the Bethe-Reeh form (6-3-29) or the variational form of eq. (6-3-44). The complete Dalgarno–Lynn potential, eq. (6-3-37), was applied in order to include all the multipole terms [121]. For s-waves it gave phase shifts in qualitative agreement with the best variational results [122, 123] but significantly too high at all energies. [The scattering length found was $a = -2.54$ compared with Schwartz's value [122] of -2.10.] The so-called modified adiabatic method [121] took the potential in the more flexible form

$$V^{(2)}_{MAD} = V^{(2)}_{AD} + (\alpha - 1)V^{(2)}_{AD, l=0}, \qquad (6\text{-}6\text{-}1)$$

where α is the monopole suppression factor measuring how much of the monopole term is to be retained in G(eq. (6-3-16)). The idea was that the least adiabatic situation occurs when the positron is near the nucleus, where complicated correlations may occur; since the monopole ($l = 0$) term is purely short-range it should be treated separately from the $l > 0$ terms. It was found that α must be nearly zero ($\alpha = 0.1$) to give good agreement with Schwartz's [122] scattering length; the s-wave phase shifts up to the positronium threshold are then well represented by the same value of α. This monopole suppression gives a small value of the potential at $x = 0$, in agreement with the expectation of Ob'edkov [8]. In fact, the Callaway–Temkin method summed over all $l \neq 0$ (eq. (6-3-20)) has properties similar to the modified adiabatic method; it yields a high but respectable value of the scattering length ($a = -1.83$) and correspondingly low phase shifts.

For higher partial waves, even the full adiabatic potential gives phase shifts that fall below the best values [124, 125] at the higher energies, due presumably to the lack of sufficient virtual positronium in the function G. There seems to be no simple modification of G which can reliably predict partial wave phase shifts for $L > 0$. These phase shifts are shown in Figs. 6-6-1 and 6-6-2.

The situation is thus clearly worse for positron scattering than for electron scattering. We have seen that the original form of the polarized orbital method including exchange, cutoff and dipole terms only, is a

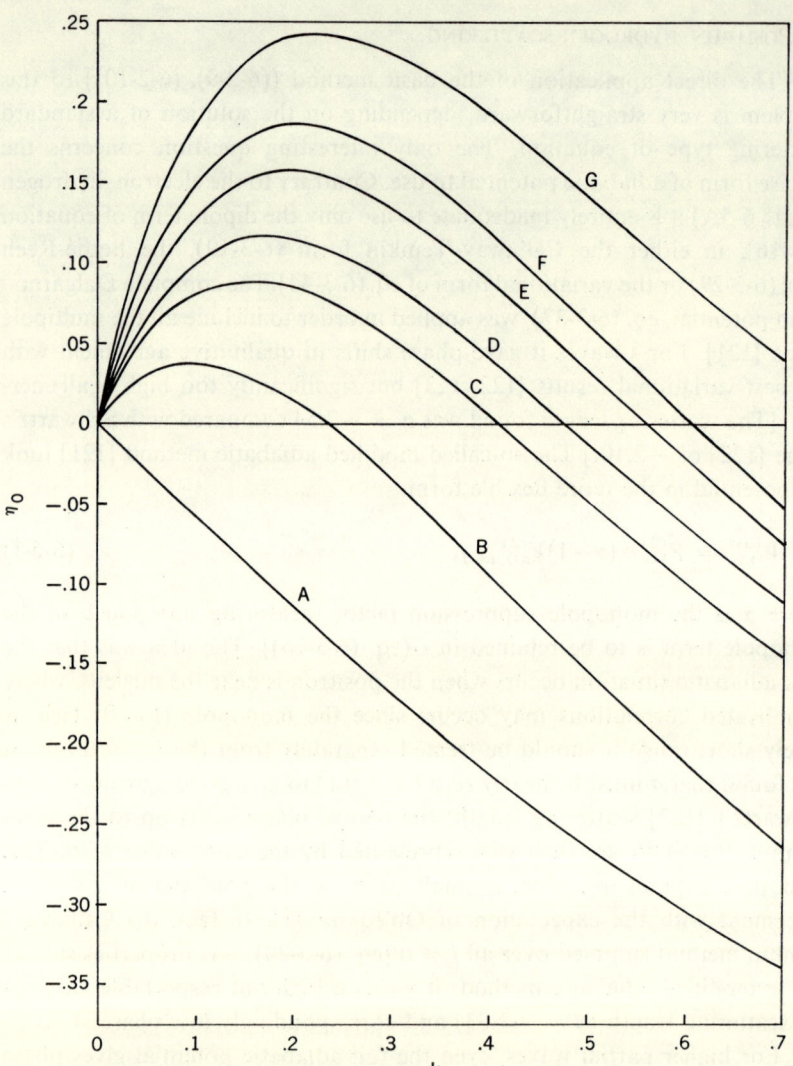

Fig. 6-6-1. S-wave positron–hydrogen phase shift η_0 in various approximations: Curve A – static or Hartree; curve B – extended polarization [27]; curve C – Callaway–Temkin (dipole only) [145]; curve D – Bethe–Reeh (dipole only) [146]; curve E – Drachman [14], eq. (6-4-22); curve F – Bhatia et al. [123]; curve G – non-variational Dalgarno–Lynn [12]. Curves A and E are lower bounds to the true phase shift, and curve F is essentially exact.

predictive method for electron scattering. For positrons monopole suppression is good but arbitrary, and it seems worthwhile to make use of a variational principle to minimize the errors due to the assumed form of scattering

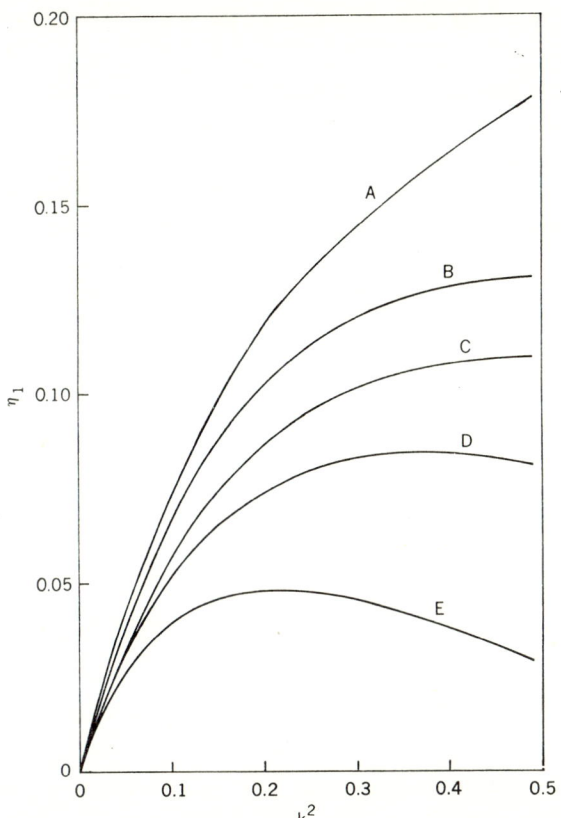

Fig. 6-6-2. P-wave positron–hydrogen phase shift η_1 in various approximations: Curve A – Armstead [124]; curve B – non-variational Dalgarno–Lynn [121]; curve C – Drachman [14]; curve D – Bethe–Reeh (dipole only) [146]; curve E – extended polarization [27]. Curve C is a lower bound to the true phase shift while curve A is essentially exact.

function. The variational equations (6-4-22) and (6-4-23) form a suitable basis for calculation. We have already seen in § 6-5 how the anti-symmetrized form of equation (6-4-23) can be used in a Rayleigh–Ritz calculation for the energy of the H^- ion. Here the numerical solution of equations (6-4-22) will be described, for the case of elastic positron–hydrogen scattering, using the correlation function G of Dalgarno and Lynn [13].

The numerical integration of eqs. (6-4-22) in partial wave decomposition would be straightforward were it not for difficulties connected with the asymptotic form of the closed-channel part of the trial function. The problem is very similar to those arising in the close-coupling method with closed channels, and our treatment follows the review by Burke and Smith [1] fairly closely.

Since we are dealing here with a set of two coupled second-order differential equations, four linearly independent solutions can be found, with each solution consisting of a pair of functions (χ, F). We will see later that only two of these solutions are sufficiently regular at $x = 0$ to be admissable. The general solution of eqs. (6-4-22) is thus a linear combination of these two solutions, and our problem is to determine such a combination which is sufficiently regular as $x \to \infty$. Making partial wave expansions as follows

$$\chi(x) = \sum_{L=0}^{\infty} x^{-1} u_L(x) P_L(\cos\theta)$$
$$F(x) = \sum_{L=0}^{\infty} x^{-1} g_L(x) P_L(\cos\theta),$$
(6-6-2)

one can write the partial wave equivalent of eqs. (6-4-22):

$$u_L'' + [k^2 - L(L+1)x^{-2} - V_1]u_L = V_2 g_L \qquad (6\text{-}6\text{-}3\text{a})$$

$$g_L'' + \left[k^2 - L(L+1)x^{-2} + N^{-1}\left(V_2 - V_3 - W - \frac{dN}{dx}x^{-1}\right)\right] g_L$$

$$+ N^{-1}\frac{dN}{dx} g_L' = N^{-1} V_2 u_L. \qquad (7\text{-}6\text{-}3\text{b})$$

In the Appendix of [14] the dominant terms of the potentials near $x = 0$ were obtained. Representing each linearly independent solution as a vector $\psi = (u, g)$, we find two linearly independent regular solutions near $x = 0$:

$$\psi_L^{(a)}(x) = (t_L, 0)$$
$$\psi_L^{(b)}(x) = (0, t_L),$$
(6-6-4)

where $t_L \approx x^{L+1}[1 + x/(L+1)]$, plus higher powers of x. Eq. (6-6-3) is numerically integrated, using these starting forms, up to a point $x = \bar{x}$, and the general solution there is

$$\psi_L(\bar{x})_< = A\psi_L^{(a)}(\bar{x}) + B\psi_L^{(b)}(\bar{x}), \qquad (6\text{-}6\text{-}5)$$

where A and B are constants to be determined.

The asymptotic form of eqs. (6-6-3) is

$$u_L'' + k^2 u_L = 0 \qquad (6\text{-}6\text{-}6\text{a})$$

$$g_L'' + (k^2 - \tfrac{36}{43})g_L = -\tfrac{36}{43} u_L, \qquad (6\text{-}6\text{-}6\text{b}) \quad \text{cf. (6-4-31)}$$

although for numerical purposes one should retain some x^{-1} and x^{-2} terms which vanish very slowly (see [14]). From eqs. (6-6-6) one sees that $g_L = u_L + C\exp(-\gamma x)$, where $\gamma = [\frac{36}{43} - k^2]^{\frac{1}{2}}$. This means that the original polarized orbital function of eq. (6-3-6) is recovered at large x and that the deviation from that function decreases as if there were an excited state [see § 6-4C] ("pseudo-state") of the target with an energy $E = \frac{36}{43}$, measured from the ground state. Since $k^2 \leq \frac{1}{2}$, the pseudo-state threshold is not reached in these calculations. (Note that the rising exponential solution is excluded by normalizability requirements.)

Using the three linearly independent large-x solutions

$$\psi_L^{(c)} = (0, e^{-\gamma x})$$
$$\psi_L^{(d)} = (xj_L, p_{Lj}) \tag{6-6-7}$$
$$\psi_L^{(e)} = (xn_L, p_{Ln})$$

(where j_L, n_L are spherical Bessel functions and p_{Lj}, p_{Ln} can be expressed in terms of asymptotic expansions in x^{-1}, with leading terms xj_L, xn_L respectively), one integrates eqs. (6-6-3) inward from some asymptotic point x_0 to the matching point, $\bar{x}$. At that point, the solution is

$$\psi_L(\bar{x})_> = C\psi_L^{(c)}(\bar{x}) + \psi_L^{(d)}(\bar{x}) - \tan \eta_L \psi_L^{(e)}(\bar{x}). \tag{6-6-8}$$

The requirement of continuity of u_L, g_L and their first derivatives gives four linear equation in A, B, C, and $\tan \eta_L$, which should be relatively insensitive to $\bar{x}$ and x_0.

For $k = 0$, it is well known that the long-range polarization potential significantly affects the open-channel part of the wave function. This is taken into account by modifying two of the asymptotic solutions in eq. (6-6-7) to read

$$\psi_0^{(d)} = \left(x - \frac{9}{4x}, x - \frac{253}{36x}\right)$$
$$\psi_0^{(e)} = \left(1 - \frac{3}{4x^2}, 1 - \frac{3}{4x^2}\right). \tag{6-6-9}$$

The solution sketched here was carried out [14] for s-waves and the results are included in Fig. 6-6-1, along with the non-variational results [121], those of Bhatia et al. [123] and others. To assess the accuracy of a given approximate calculation, which depends on how well the closed channel is represented, a quality factor $Q(k)$ has been defined [14, 30]:

$$Q(k) = \frac{\eta(k) - \eta_H(k)}{\eta_{ex}(k) - \eta_H(k)}. \tag{6-6-10}$$

Here η is the approximate phase shift being tested, η_{ex} is the "exact" phase shift [123] and η_H is the "Hartree" or undistorted-atom value obtained by retaining only V_1 in the basic scattering eqs. ((6-2-9), (6-3-8)). This is a useful measure, since no error is associated with the calculation of η_H. Clearly, for a bound calculation $Q(k) \leqq 1$; the variational calculation outlined above gives an almost constant $Q(k) = 0.90$–0.92. [For comparison, the Dalgarno–Lynn potential used non-variationally gives values of $Q(k)$ varying smoothly with energy from 1.16 (at $k = 0$) to 1.31 (at $k = 0.7$).] Table 6-6-1 gives the zero-energy values $Q(0)$ obtained in different approximations.

TABLE 6-6-1

$Q(k)$ values at zero energy for e^+–H scattering. $Q(0) = (a - a_H)/(a_{ex} - a_H)$, where a_H and a_{ex} are the Hartree and Schwartz values respectively

Hartree (static)	0 (defined)
Extended polarization [27]	0.51
Non-variational, Callaway–Temkin ($l = 1$) [145]	0.69
Perkins [30]	0.7
Non-variational, Callaway–Temkin (all l)	0.90
Variational, Dalgarno–Lynn [14]	0.92
Schwartz [122]	1 (defined)
Non-variational, Dalgarno–Lynn [121]	1.16

It is interesting to examine the behavior of the second function $F(x)$ obtained from the numerical solution of eqs. (6-6-3). (The zero-energy case is simplest to discuss, since χ and F do not oscillate as $x \to \infty$). It is found that $\chi > F$ for small values of x; this is the variational counterpart of the "monopole suppression" imposed *ad hoc* in the modified adiabatic method of eq. (6-6-1). It was not originally anticipated, however, that $F > \chi$ for $x > 4$, although a similar enhancement was subsequently observed when the binding energy of H^- was calculated by the same method (see § 6-5, and ref. [38]). In fact, one can rewrite eq. (6-4-23) to make it look more like the original polarized orbital wave function:

$$|\psi\rangle = \chi(x)[1 + T(x)\, G(x, r)]|\phi_0\rangle, \tag{6-6-11}$$

where $T = F/\chi$. It would be possible to calculate a new adiabatic potential from eq. (6-6-11) as

$$V_{\text{ADT}} = V_1 + TV_2, \tag{6-6-12}$$

which has obvious affinities with Stone's method, eq. (6-3-38), but with the T obtained from the method of eqs. (6-4-22). The result is a greater suppression of the potential at small x ($T(0) \approx \frac{1}{3}$ rather than $\frac{2}{3}$) and a reduced maximum at $x \approx 5$ ($T_{\max} \approx 1.1$ rather than 1.6).

Using Perkins's method [30] including 1s and 2p' states (see eq. (6-4-7)), one finds $Q(k)$ values near 0.5. With the addition of a second pseudo-state (3d'), Perkins can improve $Q(k)$ to range between about 0.7 and 0.6. No calculations of this type have yet been done using the closely-coupled polarized orbital method [28], but its similarity to Perkins' method implies that it will not be much better unless many multipoles are included. It seems fair to say that the Dalgarno–Lynn correlation function used with eqs. (6-4-22) gives about the best positron–hydrogen s-wave phase shifts obtainable with only one pseudo-state. Additional monopole suppression would probably be an improvement, but would complicate the evaluation of the potentials.

The s-wave phase shifts obtained from either the dynamical or extended polarization approximations (eqs. (6-4-17) and (6-4-21)) are similar and quite poor. For the latter method, Callaway et al. [27] have found a scattering length of only -0.783 (to be compared with Schwartz's -2.10), and their phase shifts are also plotted in Figs. 6-6-1 and 6-6-2. The values of $Q(k)$ in this case vary from 0.51 at $k = 0$ to 0.26 at $k = 0.7$. These results show clearly that the distortion potential V_D is much too repulsive for intermediate values of x, although correct as $x \to \infty$; as the energy increases, the positron penetrates closer to the nucleus and experiences more of this repulsion. The same argument indicates that for higher partial waves the results should improve as the centrifugal barrier forces the positron away from the nucleus.

B. POSITRON–HELIUM SCATTERING AND ANNIHILATION

Before describing the results of several calculations involving the application of polarized orbital techniques to the positron–helium system, it is necessary to treat the problem of the inexactness of the atomic target wave functions. Most of the previous discussion (with the exception of §§ 6-3F, 6-5) has assumed that the asymptotic wave function is known exactly (see eqs. (6-2-1) and (6-2-2)):

$$\lim_{x \to \infty} \Psi = e^{i\mathbf{k} \cdot \mathbf{x}} \phi_0(\mathbf{r}_1, \ldots, \mathbf{r}_N), \tag{6-6-13}$$

where ϕ_0 is an eigenfunction of the atomic Hamiltonian H_0 (eq. (7-2-6)). Except for the positron–hydrogen case, ϕ_0 is in fact only known approximately; usually a Rayleigh–Ritz variational technique is employed to obtain a reasonable form for this function. In what follows, a certain point of view will be employed consistently, in order to make it possible to apply previously derived results even when ϕ_0 is not known exactly.

Suppose that a certain approximate target function $\bar{\phi}_0$ is given, and that it satisfies the model eigenvalue equation

$$(\bar{H}_0 - \bar{E}_0)\bar{\phi}_0 = 0, \tag{6-6-14}$$

where $\bar{H}_0$ differs from H_0 only in the potential energy term. For the two-electron helium case this means

$$\bar{H}_0 = -\nabla_1^2 - \nabla_2^2 + \mathscr{V}(r_1, r_2), \tag{6-6-15}$$

where $\mathscr{V}$ is a local potential which vanishes as r_1, r_2, and $r_{12} \equiv |r_1 - r_2|$ approach ∞. It is then straightforward to find $\bar{E}_0$ and $\mathscr{V}$ if desired:

$$\mathscr{V} - \bar{E}_0 = \bar{\phi}_0^{-1}(\nabla_1^2 + \nabla_2^2)\bar{\phi}_0, \tag{6-6-16}$$

and

$$\bar{E}_0 = -\lim_{r_1 r_2 r_{12} \to \infty} (\mathscr{V} - \bar{E}_0). \tag{6-6-17}$$

(For some forms of $\bar{\phi}_0$, the $\mathscr{V}$ may be singular or vanish at unphysical points, because of the form of (6-6-16).) For our purposes, it is most important to remember that

$$[H_0, F] = [\bar{H}_0, F] \tag{6-6-18}$$

for any function $F(r_1, r_2)$ since the potentials appearing in H_0 and $\bar{H}_0$ commute with F. The model scattering system is then completely characterized by the Hamiltonian

$$\bar{H} = \bar{H}_0 - \nabla_x^2 + V, \tag{6-6-19}$$

where V is the true potential involving the positron. If the approximate wave function $\bar{\phi}_0$ is good enough, the model phase shifts and scattering function should approximate the true values quite well. In fact, one can imagine a sequence of models with scattering parameters eventually converging to the exact ones. At every stage in this sequence one has a well-defined scattering problem to solve and can thus obtain rigorous bounds by the techniques of § 6-4C; these may be called "quasi-bounds" to the scattering parameters of the original problem defined by the Hamiltonian H.

In order to use the Dalgarno–Lynn results (§ 6-3C) for helium, an extremely simple form must be chosen for $\bar{\phi}_0$. First, a Hartree–Fock function is assumed, as discussed in § 6-3F. For helium, this takes the form

$$\bar{\phi}_0(r_1, r_2) = \theta(r_1)\,\theta(r_2)\,S(12), \tag{6-6-20}$$

where $S(12)$ is an antisymmetric (singlet) spin function, and can be suppressed from now on. Then the polarized orbital scattering function is

$$|\Psi\rangle = \chi(x)[1 + G(r_1, r_2, x)]|\bar{\phi}_0\rangle. \tag{6-6-21}$$

The first-order perturbation equation (6-3-7), with the model Hamiltonian $\bar{H}_0$, and the approximate form of ϕ_0 becomes

$$[(\nabla_1^2 + \nabla_2^2), G]|\bar{\phi}_0\rangle = (\langle V \rangle - V)|\bar{\phi}_0\rangle. \tag{6-6-22}$$

Since $V = \sum_{j=1}^{2} \bar{V}(r_j, x)$ where

$$\bar{V}(r, x) = \frac{2}{x} - \frac{2}{|x - r|}, \tag{6-6-23}$$

and $\bar{\phi}_0$ is factorable (eq. (6-6-20)), it is easy to see that (6-6-22) separates and

$$G(r_1, r_2, x) = \bar{G}(r_1, x) + \bar{G}(r_2, x). \tag{6-6-24}$$

(Equation (6-6-24) is equivalent to Sternheimer's approximation (6-3-56).) The functions $\bar{G}$ then satisfy an equation analogous to (6-3-14):

$$\nabla_r^2 \bar{G}(r, x) + 2\theta^{-1}(r)\nabla_r \bar{G}(r, x) \cdot \nabla_r \theta(r) = \bar{V}(r, x) - \langle \bar{V} \rangle, \tag{6-6-25}$$

where the bracket notation now represents the expectation value with respect to the one-particle function θ. This equation could be solved, after multipole expansion, for any $\theta(r)$; if the shielded hydrogenic function is used, however, the solution of Dalgarno and Lynn [13] can be easily adapted and employed. That is, if

$$\theta(r) = Z^{3/2}\pi^{-1/2}e^{-Zr} \tag{6-6-26}$$

then eq. (6-6-25) becomes

$$\nabla_r^2 \bar{G} - 2Z\frac{\partial \bar{G}}{\partial r} = \frac{2}{x} - 2e^{-2Zx}\left(\frac{1}{x} + Z\right) - \frac{2}{|r - x|}. \tag{6-6-27}$$

By making the substitutions $y = Zr$, $v = Zx$, one obtains an equation just like the original form for hydrogen, eq. (6-3-15),

$$Z\left(\nabla_y^2 \bar{G} - 2\frac{\partial \bar{G}}{\partial y}\right) = \frac{2}{v}[1 - e^{-2v}(v+1)] - \frac{2}{|y-v|}. \quad (6\text{-}6\text{-}28)$$

From this equation one sees that the helium correlation function can be obtained as

$$\bar{G}_{He}(r, x) = Z^{-1} G_H(Zr, Zx), \quad (6\text{-}6\text{-}29)$$

and a set of helium potentials can be defined by analogy with those of eqs. (6-3-9), (6-3-40) and (6-4-13):

$$[V_p(x)]_{He} = Z^{2-p}[V_p(Zx)]_H, \quad p = 1, 2, 3$$
$$[N(x)]_{He} = Z^{-2}[N(Zx)]_H, \quad (6\text{-}6\text{-}30)$$
$$[W(x)]_{He} = [W(Zx)]_H.$$

All the quantities appearing in the hydrogen scattering problem (using the Dalgarno–Lynn solution) now go over to *twice* their scaled values (e.g., $V_2 \to 2[V_2(zx)]_H$) except for the third-order term:

$$V_3 \to 2\{Z^{-1}[V_3(Zx)]_H + Z^{-1}[N(Zx)]_H[V_1(Zx)]_H\}. \quad (6\text{-}6\text{-}31)$$

With these correspondences, one can calculate variational and non-variational e^+–He phase shifts, using the techniques of § 6-6A.

The scale parameter Z of (6-6-26) so far remains unspecified. Each proposed value of Z simply defines a distinct model, in the sense of eqs. (6-6-14), (6-6-15). The best Rayleigh–Ritz variational energy is obtained for $Z = \frac{27}{16} = 1.6875$; this gives quite a poor value of the polarizability of the helium atom, however. To see this, one recalls that the second-order potential for helium $= 2[V_2(Zx)]_H \sim -9(Zx)^{-4}$, so the polarizability is $9Z^{-4}$. For $Z = \frac{27}{16}$, this gives a value of 1.110, rather than the correct value 1.376 [126]. We prefer the value $Z = 1.5992$, which gives the correct value for the polarizability; the corresponding variational energy is only slightly worse.

Having now defined the helium model to be considered, which should give good low-energy phase shifts (since the polarizability is correct), we can examine the helium spectrum corresponding to H_0. Using (6-6-16), and (6-6-17), one finds

$$\bar{E}_0 = -2Z^2 = -5.1149$$
$$\mathscr{V} = -2Z(r_1^{-1} + r_2^{-1}). \quad (6\text{-}6\text{-}32)$$

The first excitation potential is $\frac{3}{4}Z^2$ and the first ionization potential is Z^2. As usual, the positronium formation threshold lies $\frac{1}{2}$ Ry below the ionization level, or $E_{ps} = Z^2 - \frac{1}{2}$. From these numbers one can see that the lowest inelastic threshold of the model is not that of positronium (as in the actual system) but excitation to the 1s 2s atomic level. This occurs at a wave number of $k_{ex} = 1.385$, as compared with the real positronium threshold at $k_{ps} =$

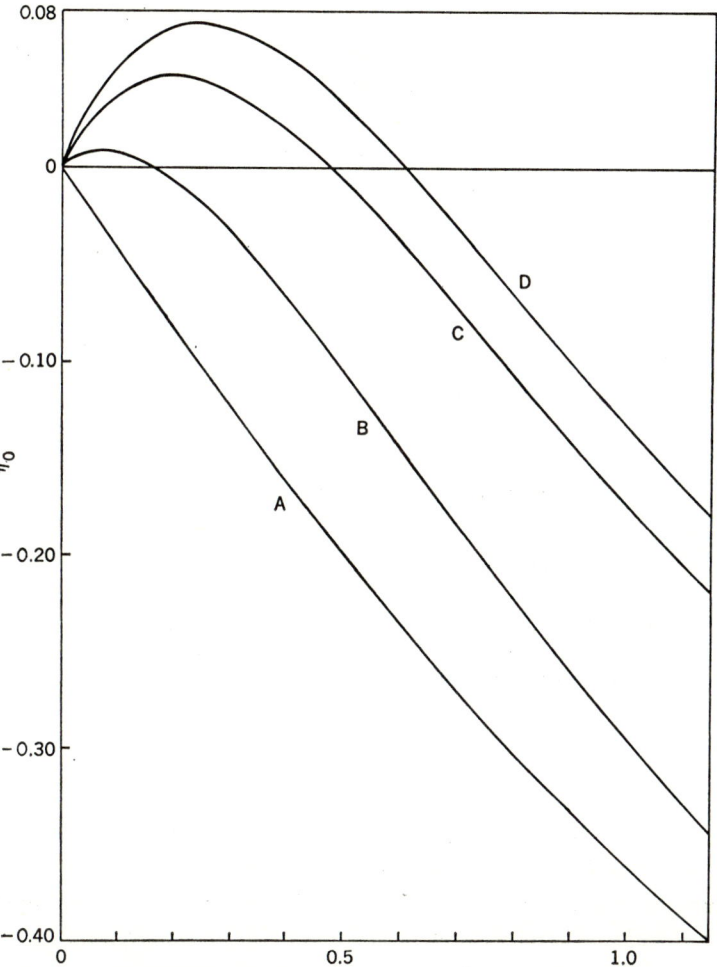

Fig. 6-6-3. S-wave positron–helium phase shift η_0 in various approximations: Curve A – static or Hartree; curve B – extended polarization [27]; curve C – Drachman [14]; curve D – non-variational ($l > 0$) [127]. Curve C is a lower bound to the true *model* phase shift as discussed in § 6-6B.

1.145. The model thus has no spurious thresholds in the elastic scattering region, but might not represent well the region near the real positronium threshold.

In Fig. 6-6-3 various approximations to the s-wave e^+–He phase shift are plotted. These include the unpolarized "Hartree", the non-variational method with full monopole suppression (see eq. (6-6-1)), the variational method of (6-4-22), (6-4-23) and the extended polarization method (6-4-21). (These are all based on the model described above, except for the last: in that case monopole, dipole and quadrupole terms are retained, with different values of Z used for each one: $Z_0 = 1.6875$, $Z_1 = 1.594$, $Z_2 = 1.531$.) At zero energy, the scattering lengths obtained are listed in Table 6-6-1. There is now evidence [45] from a variational calculation using Hylleraas correlation terms that, for this model, the non-variational phase shifts (even for $\alpha = 0$) are too high, while the variational results of (6-4-22), (6-4-23) are almost correct (see Table 6-6-2). Reference [127] gives the non-variational results for s, p, d waves as well as total and momentum-transfer cross-sections; these latter are in fair agreement with experiment [128, 129], although they may be too low near threshold [130]. In any case, better experiments are needed before quantitative agreement can be expected.

TABLE 6-6-2
Positron–helium scattering length

Approximation	Scattering length
Hartree (static)	0.420
Extended polarization [27]	−0.202
Variational, Dalgarno–Lynn [14]	−0.511
Variational, Hylleraas [45]	−0.524
Non-variational ($l > 0$) [14]	−0.659

There are two quantities capable of more accurate measurement than the cross-sections themselves: the total annihilation rate of positrons in helium, and the angular correlation between the two gamma rays produced by the annihilation. Ferrell [131] and Neamtan [132] have given the general formulation of the problem based on the annihilation process $e^+ + e^- \to 2\gamma$, which can occur only while the electron and positron are within a distance of about a Compton wavelength ($\lambda_c = \hbar/mc$) of each other; for atomic problems, this can be considered a point interaction since $\lambda_c/a_0 = \frac{1}{137}$.

The annihilation cross-section (in units of πa_0^2) is

$$\sigma_i(q) = \alpha^4(c/v)Z_i(q), \tag{6-6-33}$$

where i describes the state in which the residual He^+ ion is following annihilation, q is the vector sum of the momenta of the two outgoing γ-rays, $\alpha = e^2/\hbar c$ is the fine-structure constant, c and v are the speed of light and the asymptotic speed of the incident positron respectively. All the atomic physics is in the quantity $Z_i(q)$ which we would like to calculate from the definition

$$Z_i(q) \equiv (4\pi^3)^{-1} \left| \int d\mathbf{x} \int d\mathbf{r}\, \Psi(\mathbf{r}, \mathbf{x}, \mathbf{x}) \phi_i(\mathbf{r}) e^{i\mathbf{q}\cdot\mathbf{x}} \right|^2, \tag{6-6-34}$$

where $\Psi(\mathbf{r}, \mathbf{x}, \mathbf{x})$ is the scattering wave function (e.g. eq. (6-6-21)) evaluated when the positron and one electron are at the same point, and $\phi_i(\mathbf{r})$ is the ith state of the residual ion. It is difficult to measure the partial cross-sections σ_i, so we will carry out the summation over all states of the residual ion

$$Z(q) \equiv \sum_i Z_i(q) = (4\pi^3)^{-1} \int d\mathbf{r} \left| \int d\mathbf{x}\, \Psi(\mathbf{r}, \mathbf{x}, \mathbf{x}) e^{i\mathbf{q}\cdot\mathbf{x}} \right|^2, \tag{6-6-35}$$

where use has been made of the completeness relation

$$\sum_i \phi_i(\mathbf{r}) \phi_i^*(\mathbf{r}') = \delta(\mathbf{r}-\mathbf{r}'). \tag{6-6-36}$$

If, in addition, no measurement of the γ-ray momenta is made, one obtains the effective electron number Z_{eff}:

$$Z_{\text{eff}} \equiv \int d\mathbf{q}\, Z(q) = 2 \iint d\mathbf{r}\, d\mathbf{x} |\Psi(\mathbf{r}, \mathbf{x}, \mathbf{x})|^2, \tag{6-6-37}$$

where the identity

$$\int d\mathbf{q}\, e^{i\mathbf{q}\cdot(\mathbf{x}-\mathbf{x}')} = (2\pi)^3 \delta(\mathbf{x}-\mathbf{x}') \tag{6-6-38}$$

has been used. Then the total annihilation cross-section is

$$\sigma_a = \alpha^4(c/v) Z_{\text{eff}}, \tag{6-6-39}$$

and if Ψ is given its plane-wave approximation

$$\Psi = e^{i\mathbf{k}\cdot\mathbf{x}} \phi_0(\mathbf{r}_1, \mathbf{r}_2), \tag{6-6-40}$$

then $Z_{\text{eff}} = 2$, and the annihilation rate is that for positrons in a non-interacting electron gas; both electrons in the helium atom are counted.

The basic experimental parameter in positron annihilation physics is Z_{eff}, which depends on positron energy through the wave function Ψ. Using the polarized orbital wave function (6-6-21) with the separable wave function (6-6-20) and correlation function (6-6-24), one finds [133]:

$$Z_{\text{eff}} = 2\int d\mathbf{x}\, \chi^2(x)\theta^2(x)[(1+\bar{G}(\mathbf{x},\mathbf{x}))^2 + N(x)]. \qquad (6\text{-}6\text{-}41)$$

For the variational method this expression is modified in an obvious way [14]:

$$Z_{\text{eff}} = 2\int d\mathbf{x}\, \theta^2(x)[|\chi(\mathbf{x})+F(\mathbf{x})\bar{G}(\mathbf{x},\mathbf{x})|^2 + |F|^2 N(x)]. \qquad (6\text{-}6\text{-}42)$$

(In the Appendix of [14] the evaluation of $\bar{G}(\mathbf{x},\mathbf{x})$ is outlined.) Values of Z_{eff} as a function of positron momentum were obtained in these two approximations [14, 133] and we will limit the present discussion to $k = 0$, where only s-waves contribute. The results are: for the non-variational calculation with full monopole suppression, $Z_{\text{eff}} = 6.32$; for the variational calculation, $Z_{\text{eff}} = 3.66$. The quite significant discrepancy is resolved in favor of the lower, variational result when one makes comparison with the experimental values at near zero energy. The best value appears to be $Z_{\text{eff}} = 3.677 \pm 0.025$ [128], while other measurements [129] cluster nearby. Matese and LaBahn [104, 134] have emphasized the fact that the positron current is not strictly conserved in the non-variational method, and a normalization correction can be made which reduces the value of Z_{eff}, although not by a significant amount in the present case. The correction is, in any case, unnecessary in the variational calculation. A recent recalculation [45], using a Kohn variational approximation, yields a value of Z_{eff} about 15 % higher, and hence in poorer agreement with experiment. More accurate atomic helium models are evidently needed.

The final experimental test measures the angular correlation between the two outgoing γ-rays that result from the annihilation of thermalized ($k \approx 0$) positrons in helium. This can be obtained easily from the momentum distribution $Z(\mathbf{q})$ of (6-6-35), but since the experiment measures only a certain projection of the correlation angle it is necessary to eliminate two components of the vector $\mathbf{q}$. The result is

$$P(q_3) = \int_{-\infty}^{\infty} dq_1 \int_{-\infty}^{\infty} dq_2\, Z(\mathbf{q}), \qquad (6\text{-}6\text{-}43)$$

and the relation with the correlation angle is $\theta = \hbar q_3/mc$. For the variational calculation this can be put in the form [135]

$$P(q_3) = \int_{q_3}^{\infty} dq\, q\, S(q),$$

where

$$S(q) = (2\pi^2)^{-1} \left\{ \left| \int d\mathbf{x}\, e^{i\mathbf{q}\cdot\mathbf{x}} \theta(x)[\chi(x) + F(x)\bar{G}(\mathbf{x}, x)] \right|^2 + \langle |T(\mathbf{r}, \mathbf{q})|^2 \rangle \right\},$$

(6-6-44)

and

$$T(\mathbf{r}, \mathbf{q}) = \int d\mathbf{x}\, e^{i\mathbf{q}\cdot\mathbf{x}} \theta(x) F(x) \bar{G}(\mathbf{r}, \mathbf{x}),$$

since for q_3 held constant $dq_1\, dq_2 = 2\pi q\, dq\, d\phi$ with $q \geq q_3$ and $\phi = \tan^{-1}(q_2/q_1)$. The non-variational result is obtained when $F \equiv \chi$ and the static or Hartree result when $F = 0$; in each case the corresponding numerical values of χ and F are used in a numerical integration to obtain the angular correlation function P. (The difficult term involving $T(\mathbf{r}, \mathbf{q})$ is very small and has not been evaluated.) The "Born approximation" corresponds to setting $\chi = \exp(i\mathbf{k}\cdot\mathbf{x}) \to 1$ for low energies and $F = 0$; in this case the integration can be easily done analytically and the result depends only on the momentum distribution of the atomic electrons. In Table 6-6-3 the full width at half maximum of the angular correlation distribution is given for experiment [136] and for each of these approximations. Again, the variational method proves best.

TABLE 6-6-3

Two-photon angular correlation width Γ and annihilation rate parameter Z_{eff} in zero-energy positron–helium annihilation

Method	Γ in milliradians		Z_{eff}	
Experiment	9.2±0.2	[136]	3.677±0.025	[128]
Variational	9.47	[135]	3.66	[14]
Non-variational	10.07	[135]	6.32	[133]
Hartree (static)	10.37		0.71	
Born	11.90	[135]	2	(eq. (6-6-40))

References

1. H. S. W. Massey and C. B. O. Mohr, Proc. Roy. Soc. (London) **A136** (1932) 289; cf. also P. G. Burke and K. Smith, Rev. Mod. Phys. **34** (1962) 458.
2. A. Temkin, Phys. Rev. **126** (1962) 130.

3. W. Kohn, Phys. Rev. **74** (1948) 1763;
   L. Hulthén, Kgl. Fysiograf. Sällskab. Lund, Forh. **14**, No. 21 (1944).
4. A. Temkin, Phys. Rev. **107** (1957) 1004.
5. A. Temkin and J. C. Lamkin, Phys. Rev. **121** (1961) 788.
6. N. F. Mott and H. S. W. Massey, Theory of Atomic Collisions (Clarendon Press, Oxford, third edition, 1965).
7. A. Dalgarno and J. T. Lewis, Proc. Roy. Soc. (London) **A233** (1955) 70;
   A. Dalgarno and A. L. Stewart, Proc. Roy. Soc. (London) **A238** (1956) 276.
8. V. D. Ob'edkov, Opt. i Spektroskopiya **17** (1964) 189 (English transl.: Opt. Spectry. (USSR) **17** (1964) 101).
9. J. Callaway, Phys. Rev. **106** (1957) 868.
10. A. Temkin, Phys. Rev. **116** (1959) 358.
11. H. A. Bethe, Handbuch der Physik (Edward Brothers Inc., Ann Arbor, 1943) Vol. 24, Part 1, p. 339.
12. H. Reeh, Z. Naturforsch. **15a** (1960) 377.
13. A. Dalgarno and N. Lynn, Proc. Phys. Soc. (London) **A70** (1957) 223.
14. R. J. Drachman, Phys. Rev. **173** (1968) 190.
15. P. M. Stone, Phys. Rev. **141** (1966) 137.
16. D. R. Bates, K. Ledsham and A. L. Stewart, Proc. Phil. Trans. Roy. Soc. (London) **A246** (1953) 215.
17. A. S. Wightman, Phys. Rev. **77** (1950) 516.
18. D. R. Bates, private communication.
19. R. M. Sternheimer, Phys. Rev. **96** (1954) 951.
20. L. Allen, Phys. Rev. **118** (1960) 167.
21. A. Dalgarno, Proc. Roy. Soc. (London) **A251** (1959) 292.
22. A. Temkin, Ph.D. thesis M.I.T., 1956, unpublished.
23. M. Mittleman and J. L. Peacher, Phys. Rev. **173** (1968) 160.
24. W. Duxler, R. J. Poe and R. LaBahn, Phys. Rev. A **4** (1971).
25. P. M. Morse and W. Allis, Phys. Rev. **44** (1933) 269.
26. R. W. LaBahn and J. Callaway, Phys. Rev. **147** (1966) 28.
27. J. Callaway, R. W. LaBahn, R. J. Poe and W. Duxler, Phys. Rev. **168** (1968) 12.
28. R. Damburg and E. Karule, Proc. Phys. Soc. (London) **90** (1967) 637.
29. J. J. Matese and A. C. Fung, Phys. Rev. A **3** (1971) 928.
30. J. R. Perkins, Phys. Rev. **173** (1968) 164.
31. H. Feshbach, Ann. Phys. (N.Y.) **19** (1962) 287.
32. M. Gailitis, Zh. Eksperim. i Teor. Fiz. **47** (1964) 160 (English transl.: Soviet Phys.-JETP **20** (1965) 107).
33. D. R. Bates and H. S. W. Massey, Proc. Roy. Soc. (London) **A192** (1947) 1.
34. R. A. Buckingham, Proc. Roy. Soc. (London) **A160** (1937) 94.
35. I. A. Sloan, Proc. Roy. Soc. (London) **A281** (1964) 181.
36. P. G. Burke and D. D. McVicar, Proc. Phys. Soc. (London) **86** (1965) 989.
37. P. G. Burke and A. J. Taylor, Proc. Phys. Soc. (London) **88** (1966) 549.
38. R. S. Oberoi and J. Callaway, Phys. Rev. A **1** (1970) 45.
39. R. Damburg and S. Geltman, Phys. Rev. Letters **20** (1968) 485.
40. S. Geltman and P. G. Burke, J. Phys. B **3** (1970) 1062.
41. H. E. Kauppila, W. R. Ott and W. L. Fite, Phys. Rev. A **1** (1970) 1099, 1089.
42. J. W. McGowan, J. F. Williams and E. Curley, Phys. Rev. **180** (1969) 132.
43. M. D. Lloyd and M. R. C. McDowell, J. Phys. B **2** (1969) 1313.
44. H. Tai, R. Bassel, E. Gerjuoy and V. Franco, Phys. Rev., in press.
45 S. K. Houston and R. J. Drachman, Phys. Rev. A **3** (1971) 1335.
46. R. K. Peterkop, Zh. Eksperim. i Teor. Fiz. **54** (1968) 1581 (English transl.: Soviet Phys.-JETP **27** (1968) 836).

47. R. W. LaBahn and J. Callaway, Phys. Rev. **135** (1964) A1539.
48. J. H. Williamson and M. R. C. McDowell, Proc. Phys. Soc. (London) **85** (1964) 719.
49. B. H. Bransden and M. R. C. McDowell, J. Phys. B **2** (1969) 1187.
49b. B. H. Bransden and M. R. C. McDowell, J. Phys. B **3** (1970) 29.
50. M. R. C. McDowell, in: Proc. 2nd Intern. Conf. on Atomic Physics (Plenum Press, N.Y.) to be published.
51. R. W. Crompton, M. J. Elford and R. L. Jory, Aust. J. Phys. **20** (1967) 369;
  D. Golden and H. W. Bandel, Phys. Rev. **138** (1965) A14.
52. S. P. Khare and B. Moiseiwitsch, Proc. Phys. Soc. (London) **85** (1965) 821.
53. S. P. Khare and P. Shobha, VI I.C.P.E.A.C. Abstracts (M.I.T. Press, Cambridge, Mass., 1969) p. 844.
54a. R. W. LaBahn and J. Callaway, Phys. Rev. **180** (1969) 91; **188** (1969) 520.
54b. L. Vriens, C. Kuyatt and S. Mielczarek, Phys. Rev. **170** (1968) 163.
54c. J. Bromberg, J. Chem. Phys. **50** (1969) 3906.
55. R. J. W. Henry, Phys. Rev. **162** (1967) 56.
56. W. R. Garrett and H. T. Jackson Jr., Phys. Rev. **153** (1967) 28.
57. J. C. Slater, Phys. Rev. **81** (1951) 385.
58. K. Smith, R. Henry and P. Burke, Phys. Rev. **147** (1966) 21.
59. K. Smith, M. Connelly and L. A. Morgan, Phys. Rev. **177** (1969) 196;
  P. G. Burke and A. L. Sinfailam, Phys. Rev. **178** (1969) 218.
60. R. J. W. Henry, Phys. Rev. **172** (1967) 99.
61. D. G. Thompson, Proc. Roy. Soc. (London) **A294** (1966) 160.
62. A. Dalgarno and D. Parkinson, Proc. Roy. Soc. (London) **A250** (1959) 242.
63. A. Salop and H. H. Nakano, Phys. Rev. A **2** (1970) 127.
64. S. Kaneko, J. Phys. Soc. Japan **14** (1959) 1600.
65. D. W. Walker, J. Phys. B **3** (1970) 788.
66. H. Deichsel, H. Reichert and H. Steidl, Z. Physik **189** (1966) 212.
67. E. Karule, Abstracts IV I.C.P.E.A.C. (Science Bookcrafters, Inc., Hastings on Hudson, 1965) p. 139;
  E. Karule and R. K. Peterkop, ibid., p. 134.
68. R. Marriott and M. Rotenberg, Abstracts V I.C.P.E.A.C. (Nauka, Leningrad, 1967) p. 379.
69. P. G. Burke and A. J. Taylor, J. Phys. B **2** (1969) 869.
70. M. Mittleman, Phys. Rev. **147** (1966) 69.
71. W. R. Garrett, Phys. Rev. **140** (1965) A705.
72. J. Perel, P. Englander and B. Bederson, Phys. Rev. **128** (1962) 1148.
73. R. Brode, Phys. Rev. **34** (1929) 673.
74. B. Bederson, Phys. Rev., to be published.
75. R. E. Collins, B. Bederson and M. Goldstein, Phys. Rev., to be published.
76. A. Temkin, J. Math. Phys. **2** (1961) 336.
77. P. Swan, Proc. Roy. Soc. (London) **A228** (1955) 10.
78. A. W. Weiss, Phys. Rev. **166** (1968) 70.
79. Vo Ky Lan, J. Phys. B, submitted for publication.
80. W. Raith, in: Atomic Physics, eds. B. Bederson, V. W. Hughes and F. Pechanik (Plenum Press, N.Y., 1969) p. 389.
81. M. R. C. McDowell, Phys. Rev. **175** (1968) 189.
82. R. Abiodun, University College, London, Ph.D. Thesis 1968 (unpublished).
83. A. M. Arthurs and A. Dalgarno, Proc. Roy. Soc. (London) **A256** (1960) 540.
84. R. Ardill and W. Davison, Proc. Roy. Soc. (London) **A304** (1968) 465.
85. N. Lane and R. J. W. Henry, Phys. Rev. **173** (1968) 183.
86. R. J. W. Henry and N. Lane, Phys. Rev. **183** (1969) 221;
  N. Lane and S. Geltman, Phys. Rev. **160** (1967) 53.

87. A. TEMKIN and K. V. VASAVADA, Phys. Rev. **160** (1967) 109.
88. A. TEMKIN, K. V. VASAVADA, E. S. CHANG and A. SILVER, Phys. Rev. **186** (1969) 59.
89. A. K. BHATIA and A. TEMKIN, work in progress.
90. P. G. BURKE and A. L. SINFAILAM, J. Phys. B **3** (1970) 641.
91. S. HARA, J. Phys. Soc. Japan **27** (1969) 1009.
92. S. HARA, J. Phys. Soc. Japan **27** (1969) 1262.
93. D. M. CHASE, Phys. Rev. **104** (1956) 838.
94. YU. D. OKSYUK, Zh. Eksperim. i Teor. Fiz. **49** (1965) 1261 (English transl.: Soviet Phys.-JETP **22** (1966) 873).
95. E. S. CHANG and A. TEMKIN, Phys. Rev. Letters **23** (1969) 399.
96. A. TEMKIN and F. H. M. FAISAL, Phys. Rev. A **3** (1971) 520.
97. E. GERJUOY and S. STEIN, Phys. Rev. **97** (1955) 1671;
    E. S. CHANG and A. TEMKIN, J. Phys. Soc. Japan **29** (1970) 172.
98. S. HARA, J. Phys. Soc. Japan **27** (1969) 1592.
99. K. TAKAYANAGI, Prog. Theoret. Physics, Suppl. No. 40 (1967).
100. D. E. GOLDEN, N. F. LANE, A. TEMKIN and E. GERJUOY, Rev. Mod. Phys. **43** (1971).
101. K. L. BELL and A. E. KINGSTON, Proc. Phys. Soc. **90** (1967) 895.
102. K. L. BELL and A. E. KINGSTON, Proc. Phys. Soc. **90** (1967) 901.
103. K. L. BELL and A. E. KINGSTON, Proc. Phys. Soc. **90** (1967) 31; 337.
104. J. MATESE and R. W. LABAHN, Phys. Rev. **188** (1969) 17.
105. E. S. CHANG and M. R. C. MCDOWELL, Phys. Rev. **176** (1968) 126.
106. R. HUDSON and V. I. CARTER, J. Opt. Soc. Am. **57** (1967) 651.
107. M. R. C. MCDOWELL and E. S. CHANG, Mon. Not. R. Astr. Soc. **142** (1969) 455.
108. S. P. KHARE, Phys. Rev. **173** (1968) 43.
109. Y. HAHN, T. F. O'MALLEY and L. SPRUCH, Phys. Rev. **128** (1962) 932.
110. T. F. O'MALLEY and S. GELTMAN, Phys. Rev. **137** (1965) A1344.
111. U. FANO, Phys. Rev. **124** (1961) 1866.
112. A. K. BHATIA, A. TEMKIN and J. F. PERKINS, Phys. Rev. **153** (1967) 177.
113. A. K. BHATIA and A. TEMKIN, Phys. Rev. **182** (1969) 15.
114. R. P. MADDEN and K. CODLING, Astrophys. J. **141** (1965) 364.
115. A. K. BHATIA, D. F. GALLAHER and A. TEMKIN, work in progress.
116. I. H. SLOAN, Proc. Phys. Soc. **85** (1966) 435.
117. M. R. C. MCDOWELL and D. G. ECONOMIDES, J. Phys. B **2** (1969) 1323;
    cf. also M. R. C. MCDOWELL, in: Case Studies in Atomic Collision Physics, Vol. I (North-Holland, Amsterdam, London, 1969) Ch. 2.
118. D. RAPP and P. ENGLANDER (Golden), J. Chem. Phys. **43** (1965) 1464.
119. B. PEART and K. T. DOLDER, J. Phys. B **1** (1969) 872; **2** (1969) 1169.
120. B. H. BRANSDEN, in: Case Studies in Atomic Collision Physics, Vol. I (North-Holland, Amsterdam, London, 1969) Ch. 4;
    P. A. FRASER, in: Advan. in Atomic and Molecular Physics, Vol. 4 (Academic Press, New York, London, 1968) p. 63.
121. R. J. DRACHMAN, Phys. Rev. **138** (1965) A1582.
122. C. SCHWARTZ, Phys. Rev. **124** (1961) 1468.
123. A. K. BHATIA, A. TEMKIN, R. J. DRACHMAN and H. EISERIKE, Phys. Rev. A **3** (1971) 1328.
124. R. L. ARMSTEAD, Phys. Rev. **171** (1968) 91.
125. C. J. KLEINMAN, Y. HAHN and L. SPRUCH, Phys. Rev. **140** (1965) A413.
126. K. F. HERZFELD and K. L. WOLF, Ann. Physik **76** (1925) 71 and 567;
    A. DALGARNO and A. E. KINGSTON, Proc. Roy. Soc. (London) **A259** (1960) 424;
    D. R. JOHNSTON, G. J. OUDEMANS and R. H. COLE, J. Chem. Phys. **33** (1960) 1310.
127. R. J. DRACHMAN, Phys. Rev. **144** (1966) 25.
128. C. Y. LEUNG and D. A. L. PAUL, J. Phys. B **2** (1969) 1278.

129. G. F. Lee, P. H. R. Orth and G. Jones, Phys. Letters **28A** (1969) 674; S. J. Tao and T. M. Kelly, Phys. Rev. **185** (1969) 135.
130. D. E. Groce, D. G. Costello, J. W. McGowan and D. F. Herring, Abstracts VI I.C.P.E.A.C. (M.I.T. Press, Cambridge, Mass., 1969) p. 757 and private communications.
131. R. Ferrell, Rev. Mod. Phys. **28** (1956) 308.
132. S. M. Neamtan, G. Darewych and G. Oczkowski, Phys. Rev. **126** (1962) 193.
133. R. J. Drachman, Phys. Rev. **150** (1966) 10.
134. R. W. LaBahn, private communication (1968).
135. R. J. Drachman, Phys. Rev. **179** (1969) 237.
136. C. V. Briscoe, S.-I. Choi and A. T. Stewart, Phys. Rev. Letters **20** (1968) 493.
137. P. G. Burke, D. F. Gallaher and S. Geltman, J. Phys. B **2** (1969) 1142.
138. P. G. Burke and A. M. Schey, Phys. Rev. **126** (1962) 147.
139. A. Temkin and E. Sullivan, Phys. Rev. **129** (1963) 1250.
140. M. Gailitis, Abstracts IV I.C.P.E.A.C. (M.I.T. Press, Cambridge, Mass., 1969) p. 10.
141. A. Temkin, Phys. Rev. Letters **6** (1961) 355.
142. N. A. Doughty, P. A. Fraser and R. P. McEachran, Mon. Not. R. Astr. Soc. **132** (1966) 255.
143. K. Chung and T. Hurst, Phys. Rev. **152** (1966) 35.
144. C. L. Pekeris, Phys. Rev. **126** (1962) 1470.
145. W. J. Cody, J. Lawson, H. S. W. Massey and K. Smith, Proc. Roy. Soc. **A278** (1964) 479.
146. B. H. Bransden and Z. Jundi, Proc. Phys. Soc. (London) **89** (1966) 7.
147. E. U. Condon and G. H. Shortley, The Theory of Atomic Spectra (Cambridge University Press, Cambridge, 1951).
148. P. G. Burke and T. G. Webb, J. Phys. B **3** (1970) L131.
149. B. H. Bransden, A. Dalgarno, I. John and M. J. Seaton, Proc. Phys. Soc. **71** (1958) 877.
150. F. B. Malik and E. Trefftz, Z. Astrophysik **50** (1960) 96.
151. P. G. Burke, S. Ormonde and W. Whittaker, Proc. Phys. Soc. **92** (1967) 319.
152. G. E. Chamberlain, S. J. Smith and D. W. O. Heddle, Phys. Rev. Letters **12** (1964) 647.
153. G. Sunshine, B. Aubrey and B. Bederson, Phys. Rev. **154** (1967) 1.
154. R. H. Neynaber, L. Marino, E. W. Rothe and S. Irujillo, Phys. Rev. **123** (1961) 148.
155. S. C. Lin and B. Kivel, Phys. Rev. **114** (1959) 1026.
156. K. Smith and D. S. Burch, Phys. Rev. **116** (1959) 1125.

REFERENCES

129. G. F. Litz, R. H. Garth and G. Jones, Phys. Letters 28A (1969) 615.
130. L. Tao and T. M. Kelly, Phys. Rev. 185 (1969) 135.
131. D. Bloor, O. G. Coetzee, L. W. McGowan and D. F. Hopkins, Abstracts of I.C.P.E.A.C. (M.I.T. Press, Cambridge, Mass. 1969), p. 337 and previous publications.
132. R. L. Liboff, Rev. Mod. Phys. 38 (1966) 98.
133. M. Inokuti, G. Dupeyrat and C. Dupeyrat, Phys. Rev. 126 (1962) 194.
134. R. J. Damburg, Russ. Rev. 150 (1966) 10.
135. R. W. LaBahn, private communication (1968).
136. R. J. Damburg, Phys. Rev. 179 (1969) 231.
137. C. J. Kleinman, S. L. Cunn and A. F. Starace, Phys. Rev. Letters 20 (1968) 993.
138. R. G. Boyd, D. E. Gillespie and S. Geltman, J. Phys. B 2 (1969) 1147.
139. R. C. Bobb and A. M. Smith, Phys. Rev. 176 (1968) 147.
140. A. Temkin and R. Sullivan, Phys. Rev. 129 (1963) 1250.
141. M. Gailitis, Abstracts IV I.C.P.E.A.C. (M.I.T. Press, Cambridge, Mass., 1966) p. 10.
142. T. Temkin, Phys. Rev. Letters 6 (1961) 354.
143. R. A. Oberoi, V. P. A. Lonkar and R. P. McEachran, Abstr. New R. astr. Soc. 143 (1969) 355.
144. K. Takayanagi, J. Theor. Phys. Res. 24 (1967) 1060-85.
145. C. L. Gerjuoy, Phys. Rev. 126 (1962) 1470.
146. V. J. Franco, J. Fawcett, H. S. W. Massey and K. Smith, Proc. Roy. Soc. A297 (1968) 479.
147. R. H. Bransden and Z. Jundi, Proc. Phys. Soc. (London) 92 (1968) 880.
148. L. D. Coxton and L. H. Spruch, in: The Theory of Atomic Structure (Cambridge University Press, Cambridge, 1933).
149. Y. G. Borek and T. M. Wang, J. Chem. Phys. (1970) 5131.
150. R. H. Bransden, A. Daigarno, J. Jones and M. J. Seaton, Proc. Phys. Soc. 71 (1958) 877.
151. T. B. Massey and P. Takeda, Z. Astrophys. 50 (1960) 12.
152. P. O. Burke, S. Ormond and W. Whittaker, Proc. Phys. Soc. 92 (1967) 319.
153. D. F. Crothers, V. S. A. Smith and R. W. B. Ardill, Phys. Rev. 180 (1969) 54.
154. O. Sinanoglu, J. Chem. Phys. and R. Stromberg, Phys. Rev. 174 (1968) 1517.
155. R. B. Bernstein, K. Murray, J. A. Kohls and S. Geltman, J. Chem. Phys. (1970) 244.
156. S. C. Liu and R. Kelly, Phys. Rev. 178 (1969) 1055.
157. W. R. Smith and D. S. Burge, Phys. Rev. 116 (1959) 1215.

CHAPTER 7

PHOTODETACHMENT: CROSS SECTIONS AND ELECTRON AFFINITIES

BY

BRUCE STEINER

*National Bureau of Standards,
Washington, D.C., U.S.A.*

Contents

		Page
7-1.	Introduction	485
7-2.	Photodetachment cross sections	487
	A. Introduction	487
	B. Theory	490
	C. Experiment	491
	D. Double photon detachment	492
7-3.	Photodetachment thresholds	493
	A. Experimental energy determination	493
	B. Threshold behavior theory	494
	1. Atoms	494
	2. Molecules	495
	C. Comparison of experiment with theory	496
7-4.	Experimental approaches	497
	A. Plasma	497
	B. Beams	498
	1. Introduction	498
	2. Classical experimental approach	498
	3. Detachment using tunable dye laser	512
	4. Detachment using fixed laser and electron energy analysis	515
	5. Double quantum detachment	517
7-5.	Electron affinities	517
	A. Techniques	517
	1. Theory	517
	2. Empirical fitting	518
	3. Experiment	520
	B. Case study: S^-	526
	C. Selected list of electron affinities	528
Acknowledgment		539
References		539

§ 7-1. Introduction

Negative ions have proved to be exceptionally elusive atomic and molecular particles. The energy levels of atoms and atomic positive ions fill Charlotte Moore's classic three volumes [1] with six digit numbers, but no negative analogues were considered reliable enough for inclusion with such august company. Because of the propensity of negative ions for evading close study, they are substantially less well characterized than are either their neutral or positive counterparts.

There are a number of reasons for the difficulty, and therefore the challenge, encountered with negative ions. Perhaps the dominant factor is the typically weak nature of the electronic binding involved. The minimum energy required for removal of an electron from a ground state negative ion is referred to as the electron affinity. This binding is seldom higher than 3 eV; the highest affinity known with near certainty is that of CN, 3.82 eV [2]. Related to such low binding energies are relatively large collisional detachment cross sections, typically of gas-kinetic magnitude [3–7]. These cross sections arise partially from the weak binding involved and the relatively small interaction energy required to raise a level in the collision complex into the continuum. Perhaps more essential, however, is the relatively diffuse nature of the outer orbital of negative ions; mutual repulsion of the electrons spreads them out.

A third challenge in negative ion work is the primitive one of the initial formation of the ions in the laboratory. Since the process is necessarily exothermic, electron attachment must take place by: (1) a radiative process, (2) a dissociative process, (3) a three body process, or (4) charge transfer. Since the cross sections for radiative process are small, since three-body collisions imply a number of constraints in addition to gas kinetic cross sections, and since charge transfer involves double or consecutive beam formation, dissociative processes typically lead to negative ions much more readily (or less infrequently) than do the other processes. The difficulty of formation of negative ions thus limits the range of ions accessible to study and the ease of work with most of these.

Similarly, the same basic characteristics, small binding energy and diffuse wavefunctions, lead to theoretical challenges which parallel the experimental ones. But as is true in experimental work, the challenge is being picked up increasingly by theoreticians. The table of electron affinities is very much fuller now than it was eight years ago [8], for which theory can claim much of the credit.

These characteristics of negative ions ultimately determine the techniques that are required to investigate the ions either experimentally or theoretically. From the experimental point of view, one must be prepared with a few notable exceptions to work either with the rapid techniques and quasinormal particle densities of shock waves, or with beams and ultrahigh vacua, and even then, with unusually intense photon beams and sensitive detection such as lock-in amplification or gating. These are our constraints; they determine what follows.

A number of summaries of work on stable negative ions have appeared in recent years [8–16]. It is already widely recognized that negative ions have a severely restricted number of bound states, if indeed one exists at all. In fact, firm evidence for an excited bound state is only now coming to light [17–19] indicating the existence in C^- not only of the ^4S ground state, but also a ^4D state. Similarly, the multiplet splitting of negative ions is only recently beginning to be explored. In both cases studied so far, S^- [20] and O^- [21], the results are difficult to interpret in detail. The challenge in negative ion work continues even after one has succeeded in getting data.

Section 2 presents a comprehensive table of cross section determinations, while the last section of this chapter presents a compilation of the energy level determinations judged to be the most reliable. The emphasis in the rest of the chapter, in keeping with the rest of this volume, will be on *techniques* employed in the determination of cross sections and electron affinities.

Work with "microsecond" metastable ions will be included where the data warrant. The best known example is the ^4P state of He^-, whose immediate decay into the ^1S helium atom and an electron is delayed by the time required for one of the electrons to "flip its spin", 345 ± 90 and 11 ± 5 microseconds for the $J = \frac{5}{2}$, and $\frac{3}{2}$ and $\frac{1}{2}$, respectively [22]. A second type of metastability is represented by SF_6^-, [23] for which the lifetime of ~ 20 microseconds [24, 25] is very likely generated by the "loss" of vibrational energy of formation among the vibrational and rotational degrees of freedom [26]. In this case, the ions in their ground vibrational state are presumably stable indefinitely. Other aspects of the importance of geometrical considerations in work with molecular ions have been emphasized by Ferguson, Fehsenfeld and Schmeltekopf [27].

These above metastable ions are exceptionally long lived resonances, which have been the subject of an enormous amount of work in recent years and generally are outside of the scope of the present chapter. More typically, the lifetimes of such states are so short that photodetachment probabilities are vanishingly small. Resonances are generally categorized [28] into two groups: open channel, and therefore very short lived (broad) shape resonances, and the closed channel, and therefore longer lived (narrow) Feshbach resonances. It is these latter resonances that live long enough for occasional observation by "conventional" negative ion techniques such as photodetachment.

Future work on stable negative ions will probably center on the exciting new photodetachment techniques described in the following sections. In addition, however, two other areas of work will undoubtedly receive emphasis: work on cluster ions and the use of photodetachment for the production of well-characterized neutral beams. Work with water cluster ions is surely the beginning of a large field of study [29–33]. In a similar way, the production of excited neutral oxygen atoms by photodetachment [34] suggests an exciting new field that is already being exploited for the study of definite and distinct states of atoms.

§ 7-2. Photodetachment cross sections

A. INTRODUCTION

In spite of the importance of photodetachment cross sections in determination of the physics of late-type stars, and the potential importance of these cross sections in understanding planetary atmospheres, relatively few cross sections have been determined (Tables 7-2-1 and 7-2-2). These cross sections are not only of intrinsic interest but also of indirect importance, since they are frequently the most accessible route to cross sections for the reverse process, radiative attachment. The two processes are of course related by detailed balance:

$$\sigma_{att} = \left(\frac{h\nu}{mc\nu}\right)^2 \left(\frac{g_-}{g_0}\right) \sigma_{det}$$

where $h\nu$ is the photon energy, $m\nu$ the electron momentum, and g_- and g_0 are the negative ion and neutral statistical weights respectively. Detachment cross sections typically fall in the decade below one atomic unit, that is roughly 10^{-17} to 10^{-16} cm^2; but attachment cross sections are orders of magnitude lower, as can be observed from the fact that the ratio $h\nu/mc\nu$ typically falls in the range 10^{-2}–10^{-3}.

TABLE 7-2-1
Photodeteachment cross sections: atomic ions

Ion	First channel[a] Maximum $\times 10^{18}$ cm^2	Method	Reference
H$^-$	43±4	Beam experiment	35–36
	~45	Plasma experiment	37
	39	Self consistent field (SCF)	38
	39	Variation	39
	42	Polarized orbital	40
	40	Exchange approximation (EA)	41
	41	Numerical Eigenfunction	42
	Resonance only	Close coupling	43
Li$^-$	210	SCF; exchange; planewave	44
	72	Analytic coulomb; EA = 0.62 eV	45
	78	Analytic coulomb; EA = 0.80 eV	46
	82	Analytic coulomb; EA = 0.38 eV	47
C$^-$	14+0.5	Beam experiment	19
	10	Beam experiment	48
	15	Analytic SCF	49
	21	Polarized orbital	50
	17	Central field	51
	17	Central field	52
	14	Central field, exchange	53
	>10	Central field, correlation	54
N$^-$	9.6	Polarized SCF	50
	1.1	Analytic SCF	49
O$^-$	6.3	Beam experiment	55
	(6.3)	Beam experiment	34
	7.0	Beam experiment	56
	6.4	Polarized SCF and exchange	57
	6	Analytic SCF	49
	7.4	Polarized SCF	58
	9.5	Central field	52
	8.6	Central field	51
	9.5	Born	59
	>9	Numerical	60
	2.1	Pseudopotential	61
F$^-$	6.0±1.5	Plasma experiment	62
	3.3±2.4	Plasma experiment	63
	2.7±0.8	Plasma experiment	64
	12.2	Central field	52
	12.8	Central field	51
	12.5	SCF, exchange; plane wave	44
	30	SCF; plane wave	49
Na$^-$	220	SCF, exchange; plane wave	44

TABLE 7-2-1 (continued)

Ion	First channel[a] Maximum $\times 10^{18}$ cm^2	Method	Reference
Si$^-$	54	Central field	51
S$^-$	54	Central field	51
Cl$^-$	25	Plasma emission experiment	65
	12.5	Plasma emission experiment	66
	15^{+12}_{-5} ($^2P_{\frac{3}{2}}$)	Plasma emission experiment	67
	39	SCF exchange; plane wave	44
	50	Central field	52
	55	Central field; plane wave	51
	615	Impulse	68
K$^-$	450	SCF exchange; plane wave	44
Br$^-$	45	Plasma emission experiment	65
	37 ($^2P_{\frac{3}{2}}$)	Plasma emission experiment	69
	12 ($^2P_{\frac{3}{2}}$)	Plasma emission experiment	67
	65	Central field	51
I$^-$	$>31.\pm5.$ ($^2P_{\frac{3}{2}}$)	Beam experiment	70
	22. ($^2P_{\frac{3}{2}}$)	Plasma emission experiment	65
	15. ($^2P_{\frac{3}{2}}$)	Plasma experiment	67
	21 ± 11 ($^2P_{\frac{3}{2}}$)	Beam experiment	71
	98	Central field	51

[a] The "first channel" here denotes the first state accessible experimentally. In the case of a multiplet, therefore unless otherwise indicated, all components are included.

TABLE 7-2-2

Photodetachment cross sections: molecular ions

Ion	First channel[a] Maximum $\times 10^{18}$ cm^2	Method	Reference
O_2^-	2.4	Beam experiment	72
CH	8	Beam experiment	48
OH$^-$	11	Beam experiment	73
SH$^-$	19 ± 4	Beam experiment	74
SO$^-$	12	Beam experiment	48
NO_2^-	1.4	Beam experiment	75
SO_2^-	2.7	Beam experiment	48

[a] "First channel" is taken to mean the first pleateau experimentally observable.

B. Theory

The calculation of photodetachment cross sections is straightforward in principle but complex in practice. It is useful by suitable normalization of the continuum wave function to resolve the cross section into the product of transition probability into a single continuum state,

$$|\langle\psi_b|p|\psi_c\rangle|^2,$$

and the density of continuum states, which is proportional to the root of the electron energy:

$$\sigma \propto |\langle\psi_b|p|\psi_c\rangle|^2 E_e^{\frac{1}{2}}.$$

The matrix element for the dipole transition probability is commonly formulated in one of three ways, conventionally identified as dipole length, dipole velocity, and dipole acceleration. These give identical results if the wave functions are exact solutions to the Schrödinger equation. The "acceleration" operator contains the inverse cube of the distance of the electrons from the nucleus and thus gives a large weight to the regions of the wave function close to the nucleus. The "velocity" operator weights the electron–nucleus distance less heavily than does the "length" operator. Thus the three formulations are sensitive to the accuracy of the wave functions in different regions of space, and agreement of calculations using the various operators suggests that the wave functions are sufficiently accurate for the purpose and increases confidence in the results.

Detailed ab initio calculations in which the convergence of these forms can be demonstrated have been carried out only for H^-. The dipole acceleration formulation, with its emphasis on the interior of the ion, is particularly sensitive to the proper inclusion of electron correlation. Because this is difficult, the agreement of "acceleration" result with the others is generally poor. Only recently [38] has this agreement come within 25%, and here only for H^-. Doughty, Fraser and McEachran employed a 70 parameter bound state wave function and six state self-consistent field continuum function.

Confidence is gained also in another way. In addition to the convergence of the results of the three formulations, using the same wave functions, one can also observe agreement of the "velocity" form results with different wave functions. The earlier Geltman "velocity" results [39] using the same Schwartz bound states but with a variational continuum agree with the Doughty, Fraser and McEachran velocity formulation to well within 1%.

These in turn agree with the earlier work of John [41], who employed a twenty parameter bound state of Hart and Herzberg and exchange approximation continuum functions. One is tempted to conclude that almost any function with enough adjustable parameters can be made to represent the wave function properly over a limited region of space, the intermediate "velocity" operator region.

For ions other than H^-, wave functions of comparable accuracy are not available for either initial or final states. Computational techniques rely on using an effective central potential for both initial and final states, with the effects of correlation included either semiempirically in the central potential [51, 52] or alternatively using Hartree–Fock approximations modified to include the important correlations via the method of adiabatic polarizability [50, 53, 54, 67]. Both methods appear capable of predicting the correct order of magnitude of the cross section and its spectral shape.

For the larger ions, however, the polarizability becomes much more important, and indeed the entire shape of the calculated cross section depends strongly on the inclusion of polarization.

C. Experiment

Detachment cross sections may in principle be determined from observation of: (1) photon absorption, (2) negative ion disappearance, (3) electron appearance, or (4) neutral appearance. Each of these types of experiments has been performed.

Photon absorption (1) is observable only with the high densities obtainable in a plasma. For a one percent absorption in a process with typical cross section of 10^{-18} cm^2, one would need a particle density of 10^{16} ions/cm^3 (1 mmHg) for an absorption path length of one centimeter.

Observation of negative ion disappearance (2) also involves looking for a small signal in a very much larger one. It has been attempted only recently with the very interesting employment of ion cyclotron resonance techniques [76]. It is probably more promising for threshold energy determination than for the measurement of absolute cross sections.

The most widely used method for cross sections involves the observation of the detached electron (3) in a crossed beam experiment. Until recently, the total production of electrons in all directions was observed. Electrons are also produced by collisional detachment, but these can be discriminated against by chopping the photon beam and synchronously detecting the electron signal. The resolution utilized here is then limited by the noise in the collisional detachment background at the chopping frequency. This experiment measures the cross section in a manner simple in concept but difficult in practice. It is examined in detail in § 7-4B.

The recent mating of a high resolution Kuyatt–Simpson electron energy analyser with a negative ion beam and laser provides another dimension to the information, the coupling between polarization and the differential cross section [18, 77]. This technique permits relatively high energy resolution since the collisional background is heavily discriminated against by the energy resolution and to a lesser extent by the small acceptance angle of the analyser. The information contained in the differential cross section and its variation with energy provides far more information on the nature of the transitions involved than does the total cross section. In addition, the total cross section can of course be obtained in principle by integrating over all angles. The opportunity for error in total cross sections with this procedure is perhaps somewhat greater.

Finally, monitoring of the neutral atoms and molecules formed (4) is another interesting path to cross sections. It has been used for the first time recently in dye laser photodetachment [20], where the quantity of primary interest is the energy dependence of the cross section rather than its absolute value. The neutrals were monitored by measuring the current secondary electrons in an electron multiplier. The uncertainty in the neutral/electron conversion efficiency for various neutrals is a hurdle to be overcome if the method is to be applied to absolute cross sections.

D. Double photon detachment

An even more challenging experiment (and theory) is double photon detachment. Indeed it has been made possible only with the employment of the photon densities possible with pulsed lasers. The double detachment of I^- is the only such experiment of this type performed [78].

The transition probability involved is approximately proportional to the single photon cross section and to the square of the photon flux density [79]. The probability observed experimentally was a factor of five higher than that predicted originally by Geltman using a plane wave continuum state and an early imprecise experimental cross section [71]. (The published result is corrected here for the incorrect application of a coherence function.) The more recent calculation with a central field model [57] brings the calculation up to the edge of the 40 % error bars. Application of the recent more precise experimental cross section [70] in turn brings the theory and experiment into fortuitously close agreement.

§ 7-3. Photodetachment thresholds

A. Experimental energy determination

The threshold energy can be determined by any of the cross section approaches listed above. These fall into two general classes: 1) those which depend on varying the energy of the photon beam and for which the threshold energy determination depends on knowledge of the cross section behavior as a function of energy and 2) those which involve simultaneous measurement of two energies: that of the detaching photon and that of the detached electron.

The experiment involving the simultaneous measurement of both photon and electron has several decided advantages and disadvantages compared with the more classical experiment. In the first place, determination of the electron energy involves location of a peak rather than identification of a threshold; peak location is, of course, typically more precise than location of a threshold. Furthermore, as already noted the energy analysis of the electrons discriminates strongly against the large background of collisionally detached electrons. This discrimination permits one to work with a weaker, more highly resolved photon beam and thus, also, to identify the threshold more precisely.

But the increased precision of the photo-electron energy resolution experiment comes at the expense of decreased accuracy: an inherently accurate photon energy measurement (wavelength) must be combined with a somewhat more complex electron energy measurement. Contact potentials in the photodetachment region can be observed to change even during an experiment; but this disadvantage has been partially offset under certain circumstances by simultaneous measurement of two detachment processes, one known and one unknown. The electron energy measurement in this case becomes an energy difference measurement. A second moderate disadvantage is the increased complexity of locating and understanding the trajectories of electrons in a complex apparatus, typically one with a large, magnetic mass analyser.

On the other hand, the classical experimental determination of threshold has its own strengths and weaknesses. The measurement involves the energy measurement only of photons. But threshold energy identification by following the cross section in the vicinity of a threshold demands some knowledge of cross section behavior, because the resolution of the instrument is finite. Cross section behavior is a subject in which both theoretical and experimental activity has continued for many years. Definitive understanding will continue to be an exciting challenge for the forseeable future. A gigantic step in this direction has been taken with the dye laser [20]. This work initially has employed a resolution of the order of one ångstrom, or typically less than 1 meV or a few wave numbers.

B. Threshold Behavior Theory

1. Atoms

Nevertheless, at any resolution, threshold behavior continues to be a subject of fascination both to experimentalist and theoreticians. Photodetachment comes within the range of cases covered by Wigner in his classic treatment of threshold behavior [80]. The essential idea in this approach is that the behavior is governed by the forces between the separating particles. Photodetachment will thus behave in a manner very different from that in photoionization, with which it is frequently compared. Photoionization is dominated by the Coulomb attraction between the electron and positive ion, which obscures the influence of the detailed structure of the positive ion. Threshold behavior is here independent of energy above threshold. For photodetachment, on the other hand, Wigner indicated that the cross section will go as k^{2l+1}, where k is proportional to the square root of energy above threshold and l is the angular momentum of the departing electron. Thus, for H^-, initially in a ^1S state, detachment occurs via a P state, with a corresponding k^3 threshold law or $E^{\frac{3}{2}}$, where E is the energy above threshold. In the more general case of an outer p electron, detachment into an S final state can occur yielding $E^{\frac{1}{2}}$, or parabolic, behavior. Close to threshold then, detachment into an S continuum state will be more evident than detachment into a P continuum.

Geltman has shown [56] however, that for short range forces the Wigner expression is actually the first order term of an expansion in odd powers of k. Unfortunately, the relative size of the higher order terms was not derived explicitly.

More recently, O'Malley [81] has included more realistic long range multipole forces and shown that they introduce other terms. The size of these terms depends both on the symmetry of the long range forces between the atom and electron and on the polarizability of the atom in question. For quadrupole forces, a term containing $k^2 \ln k$ and the polarizability appears in the cross section expression. The form of the polarizability contribution is a function of the presence or absence of symmetry in the r^{-4} potential. Where dipole forces are involved, very interesting behavior is predicted. A weak dipole produces the Wignerian k^{2l+1}. As the strength of the dipole force constant is increased to $(l+\tfrac{1}{2})^2$ (in atomic units) the exponent of k decreases monotonically from $2l+1$ to zero. Thus for this value, a photoionization-type step function is predicted. Above this value, oscillations are predicted to occur that are constant in amplitude for a given dipole moment but increase in frequency as the threshold is approached. As the dipole moment is increased further the frequency of the oscillations a given distance above threshold increases but their magnitude decreases.

2. Molecules

Geltman has shown [82] that, for detachment from molecular negative ions, the behavior of the cross section depends on molecular symmetry, even upon consideration only of short range forces. He derives the expression $\sigma = v k^m (1 + a_1 k^2 + a_2 k^4 + \ldots)$, where v is the photon frequency and m is given in Table 7-3-1.

TABLE 7-3-1

Threshold exponent m

$|\lambda_0|$*		Heteronuclear	Homonuclear g	Homonuclear u
0		1	3	1
$|\lambda_0| \geq 1$	odd	$2|\lambda_0|-1$	$2|\lambda_0|+1$	$2|\lambda_0|-1$
	even	$2|\lambda_0|-1$	$2|\lambda_0|-1$	$2|\lambda_0|+1$

* Axial angular momentum quantum number.

The analysis of O'Malley for atoms with dipole potentials is of obvious relevance to detachment from heteronuclear molecular negative ions, but this extension has not yet been made. The problem is complicated by the fact that the dipole force felt by the electron will be a function of the orientation of the molecule. The departure time of the electron is small compared to the periods of vibration and rotation of molecule: a 0.1 eV electron moves an atomic unit, half an ångstrom, in less than 10^{-15} seconds; but vibrational and rotational periods are much longer, $\approx 10^{-13}$ and $\approx 10^{-10}$ respectively, so that time averaging will probably not be necessary although orientation averaging over the large number of events will. Presumably, however, the presence of dipole forces in heteronuclear molecules will drastically affect the threshold behavior predicted by Geltman. There is some preliminary evidence in the next section that this is in fact so.

C. Comparison of Experiment with Theory

The earliest crossed beam photodetachment work [35, 36, 56] was performed using broad band filters. The threshold region either was not explored, as in the case of H^-, or was explored with "low" resolution, as in the case of O^-; and then the parameters in a two-term Geltman-type expansion were fitted to the data. For O^-, a satisfactory fit to the experimental data was obtained with the first two terms in the series: $E^{\frac{1}{2}}$ and $E^{\frac{3}{2}}$. Similar filter data for I^- [71] could be fitted with the assumption that the rising portion of the cross section was so small that a step function was a satisfactory approximation, consistent with early plasma results of Berry and colleagues. This treatment gives an upper limit to the threshold energy.

The introduction of a monochromator has permitted one to study threshold behavior with a resolution of 20 Å (15 meV) rather than to assume the theory to be complete. Data taken with this resolution could be very satisfactorily fitted with the first two terms for the entire rising portion of the curve, a range of 165 Å, 0.12 eV, or 0.01 rydberg. At an energy of 0.01 rydberg above threshold, the energy terms in the series will thus decrease two orders of magnitude per term. The coefficients of higher terms would have to be very large to have an influence over the rising part of the I^- threshold [83].

Sulfur seemed to offer a more critical test of threshold theory [84] than iodine because its smaller size and consequently smaller polarizability would lead to a smaller second term and hence to a longer region of rising cross section. Unfortunately, the ground state negative ion is a doublet; and the final neutral state is a triplet, so that six transitions are involved within 0.2 eV. The individual thresholds are thus not resolved with the available resolution of 50 Å, or 25 meV. Therefore the quality of the fit for the sum of six transitions was used to check the relative suitability of four threshold forms: $k(1+k^2)$; $k(1+bk^2 \ln k + ck^2)$, $k(1+k)$, and (bk^c). In addition to the electron affinity, the spin–orbit splitting and the statistical weights of the various final states were adjusted for best fit. The threshold form $k^{0.34}$ gave a significantly beter fit than the others. These results have been thrown open to question by the recent work of Lineberger [20], who has observed unexpected breaks in the cross section with the higher resolution possible with dye lasers. The detailed behavior of this cross section thus still awaits explicit explanation, although resonance with the continuum may play a role.

In the early O_2^- work [72] the threshold region was also inaccessible, but the lowest part of the observed region was consistent with the slowly rising behavior predicted by Geltman for a homonuclear molecule.

The work on OH⁻ [73] was interpreted in terms of a series of step functions, one for each rotational level in each member of the doublet. This behavior is consistent with that predicted by O'Malley's atomic model for relatively large electric dipoles.

The SH⁻ cross section [74] also involves a number of unresolved rotational levels in an electronic doublet neutral. A fitting process similar to that for S⁻ was more successful for the form $E^{\frac{1}{2}}+aE$ than for the forms $E^{\frac{1}{2}}$, E^{b}, or $E^{\frac{1}{2}}+E^{\frac{3}{2}}$. This was interpreted to mean that the best average form of the cross section behavior for permanent dipoles was closer to this empirical form than to one of the forms derived using either short range forces or an ensemble of identical weak dipoles.

§ 7-4. Experimental approaches

A. PLASMA

Plasmas, and particularly shock waves, present the opportunity to acquire relatively dense concentrations of transient species, with partial pressures approaching one mm Hg. With these pressures, and with suitably rapid optics one can then observe certain negative ions with the high resolution of otherwise conventional spectroscopic techniques.

The basic approach is to take a tube, long compared with its width, divided by a thin scored metallic membrane into two compartments, an evacuated one, and one filled with a "driver" gas. When the membrane is burst, either by a large pressure on one side of the membrane or by mechanical means, a shock wave runs the length of the tube. In the 3000 K plasma formed, readily ionizable components of the driver gas or strategically placed salts are ionized and vaporized. A light beam is made to traverse the shock wave several times at the appropriate moment, yielding upon resolution in a spectrograph the absorption or emission spectrum of the shock wave.

To date negative ion plasma spectroscopy has involved principally work with the atomic halides and oxygen negative ions [21, 62–67, 69, 85]. Indeed, where other less readily examined ions are studied, the results are more difficult to interpret (Table 7-2-1, ref. [37]). However, the elegant inversion of the absorption work by Berry and colleagues [21] and more recently by Rothe [65] to observe radiative attachment suggests that one may not be constrained to work with the more readily formed ions.

Balancing the magnificent resolution obtainable with these techniques is the need for care in two areas. The determination of the concentration of the species being studied has been made by several techniques which do not always yield the same answers [9, 67, 69, 86]. Moreover, the ions if present in sufficient concentration can apparently influence the shape of the observed continua [9, 67, 68, 86] and perhaps even lead to problems in threshold identification. For example, Berry and Reimann [9, 63] observe fluorine doublet thresholds at 3595 ± 5 Å and 3542 ± 2 Å, while Popp [64, 85] observes thresholds at 3646 ± 2 Å and 3595 ± 2 Å. One threshold is thus common to both experiments. The disagreement in the location of the other lends excitement to an area one would have thought so elegant as to be uncomplicated.

B. Beams

1. *Introduction*

Photodetachment experiments over the past twenty years have led to some of the more precise and less questionable values for electron affinities. Photodetachment cross section determinations, however, illustrate many of the problems which plague all atomic cross section measurements. Results which future research will not change by more than 10% are probably more a dream than a reality at this point. Recent work with the dye laser and with the electron energy analyser provides some elegant new solutions to problems of photodetachment threshold energy determination. In keeping with the purpose of this book to examine specific areas in considerable detail, we shall look closely at classical photodetachment experiments as developed at the National Bureau of Standards in Washington. Similar experiments have been performed in Mainz and in Bedford, Massachusetts. Following this examination we shall look more briefly at recent developments.

2. *Classical experimental approach*

a. Negative ion sources. As noted earlier, the production of negative ions constitutes a severe challenge to the experimenter, particularly if he wishes to form a wider variety of ions than the handful of those more commonly observed. Since the physical process furnishing the most abundant supply of negative ions is dissociative attachment, this process is most frequently utilized, typically in plasma sources. But the plasma produces a wide variety of interactions, a situation not conducive to complete understanding of individual ion formation processes. The ions actually observed frequently bear little resemblance to those one intended to produce.

Because stability of the plasma is of considerable importance and most readily obtainable with hot cathode arcs, they have been used in most recent work. Ammonia provides the most stable medium for a discharge; to it are added substances to provide the desired negative ions. Ammonia is useful in addition because it provides H^- ions, which are compared with the desired ions to provide their absolute cross section. Since H^- is relatively easily studied both theoretically and experimentally, its cross section is far more accurately known than any of the others. It thus provides the surest route to other cross sections. The configuration used (Fig. 7-4-1) utilizes a tungsten hairpin cathode and both electrostatically floating plates and magnetic confinement of the arc [74]. Ion beams of more than 10^{-7} A have been obtained with such sources, but currents of many ions are several orders of magnitude less than this. Reengineering of the earlier source [19] has reduced the currents of impurity ions but has not substantially increased the total currents available.

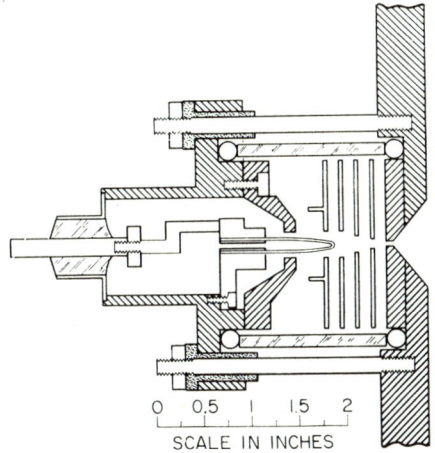

Fig. 7-4-1. Axially symmetric plasma, hot-cathode ion source. Magnetically permeable elements shown crosshatched. Permanent magnets around circumference not shown.

Recently the duoplasmatrons, which had been optimized as negative ion sources for Van de Gaaffs [87], have been used for low energy collision work [88]. The duoplasmatron had long been used for positive ion production, but the discovery that negative ions were formed perferentially at the edges of the arc, while the electrons remained on the center axis, led to the practical utilization of the source for direct extraction of negative ions. These sources have not yet been applied to detachment research, although an exploratory source has been studied [89].

A second promising negative ion source that has not been applied to detachment work is the mercury pool source developed by Dawton [90, 91]. He has been able to obtain currents of about 1 μA for a number of ions whose photodetachment spectra have not been studied: B^-, Al^-, Ga^-, In^-, Tl^-, P^-, As^-, Sb^- and Bi^-.

A rotating disc ion source [92] has also provided similar ion currents by secondary emission of Cu^-, Ag^-, Au^-, Ta^-, Cr^-, Fe^- and C^-.

The cleanliness of the photodetachment source has been a dominant factor in the nature of the ions produced. For example, months after we had stopped work with I^- produced from iodine vapor and had cleaned the entire ion source, I^- was still the dominant ion produced by about two orders of magnitude. Even after: (1) complete dissembly of the photodetachment apparatus, (2) reconstruction after a move of 20 miles, and (3) introduction of a largely new gas handling system for the source, I^- was still observed, although very weakly.

On the other hand, extreme attention to cleanliness in the new source and gas handling system led to the identification and study of an unexpected cluster ion: $H_3O_2^-$ [31]. In the reconstruction, a new metal manifold with ultra-high vacuum copper gaskets and all-metal ultra-high vacuum valves were used; only a few small pieces of interconnecting glass of an earlier system remain. The new system is pumped with sputter pumps. No water was deliberately introduced into the system and great attention was paid to the cleanliness of the reconstructed new source itself. The dominant ion in the ammonia discharge was, as before the original introduction of iodine, OH^-. This ion presumably came from the residual water surviving cleaning with alcohol and acetone; the very small amount of CH_3^- observed suggested that the OH^- itself probably did not stem from cleaning alcohols. Further indication that water must be involved was the observation of ions of mass 35 without the associated ions of mass 37. Thus chloride was not a major component. High resolution work identified most of the mass 35 ions, and a very much smaller current of mass 37 ions, as primarily $H_3O_2^-$. A miniscule chloride component was also identified, however, both by isotope ratio and mass splitting (Fig. 7-4-2). The complexity of the ion formation process is illustrated by the fact that the production of $H_3O_2^-$, although substantially increased by the deliberate introduction of water, was not increased to dominance over other ions. Thus the formation of ions in an arc source remains largely a mystery. This problem has perhaps held up negative ion research more severely than any other single factor.

A second and related aspect of the complicated nature of the ion production process is the possibility of the production of ions in more than one state. Early work on C^- [19] suggested that this ion was formed in two states, one detached with 1.27 eV photons, and an excited state so close to the continuum that a threshold for it could not be observed. The existence

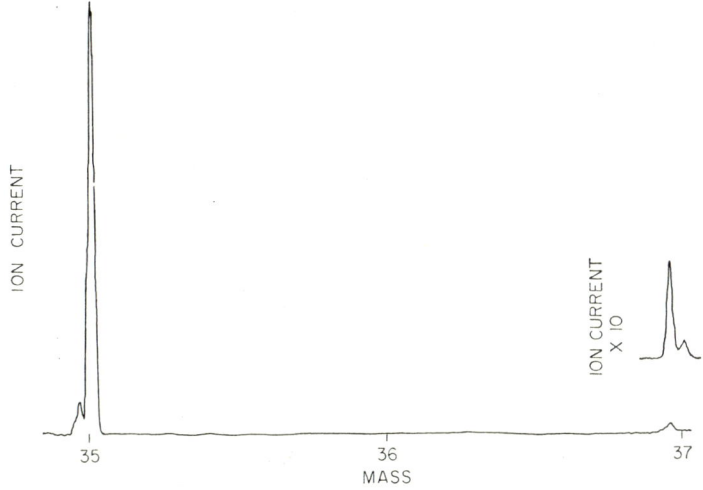

Fig. 7-4-2. Spectrum of ^{35}Cl, $^1H_3^{16}O_2$, ^{37}Cl, and $^1H_3^{16}O^{18}O$.

of the second state has been confirmed [17, 18]. The complexity of the arc processes intensifies the problem of ambiguous identification of the negative ion states populated under given conditions. Five ameliorating factors reduce the severity of this problem, however, and in fact permit useful work to be performed with complex ion sources. In the first place, most work can be interpreted in terms of the assumption that only the ground state negative ion is present. Second, other work performed by completely unrelated techniques, although sometimes imprecise, gives values for electron affinities close enough to the photodetachment values that it is unlikely that one is dealing with more than one electronic state. Third, for atomic ions the multiplet structure, although seldom resolved until recently, is consistent with assignment of most ions formed to the ground electronic state. For molecular ions, the vibrational states can also be identified. Fourth, the flight time in beam experiments is typically of the order of a microsecond; ions with substantially shorter lifetimes would not be observed. And fifth, perturbing the arc conditions in a beam apparatus whose operation is otherwise understood does not change the cross sections observed. That is, data taken on different days, with different arc parameters, is observed to be under statistical control. It is unlikely that the relative populations would remain the same under different arc conditions. Nevertheless, the lack of detailed knowledge of the processes taking place in ion sources is disturbing and presents an interesting challenge for the future.

A second method of formation must be employed for those ions that cannot be formed by dissociative attachment and which in addition are so fragile that the higher densities and temperatures required for encouraging three body attachment destroy those which might have been formed. A primary example is He^- ions, which have been formed most abundantly by means of alkali atom charge exchange [93]. The formation of these ions is made difficult by the necessity to obtain simultaneously the successful operation of the positive ion optics with that of an alkali metal oven. The alkali itself creates severe electrical insulation problems. The mere observation of an He^- beam is thus a considerable achievement; the continued operation of such a source represents a major accomplishment.

b. Initial ion focusing. The ion source is separated from the rest of the apparatus by a large ceramic-to-metal ultra-high vacuum seal that provides insulation for 20 kV (Fig. 7-4-3, cross-hatched area near source). To date an accelerating potential of 2.5 kV has been employed. This provides a suitable compromise between ion residence time in the photon beam and ion beam deflection in the electron collection field. The exit hole in the anode of the ion source is imaged onto the entrance slit of the mass analyser by a Septier lens [94], which is designed for minimum spherical abberation. Somewhat higher efficiency in extraction of the ions from the source can be provided by operating the first element of the Septier lens not at ground, but at a small positive potential.

Immediately following the 3rd Septier element, a split ring provides the opportunity for steering the beam in the horizontal direction onto the entrance slit.

c. Mass analysis. The nature of the ion source determines special requirements for the mass analyser. First, the considerable energy spread of the ion beam from an arc source will limit the resolution. Second, because of the difficulties in predicting relative ion abundances and because of the role of substances not deliberately added ("impurities"), high resolution in the mass analyser is essential for positive identification of negative ions. Third, the relatively weak beams from negative ion sources present the necessity for care in the ion optics of the analyser so that transmission is as high as possible.

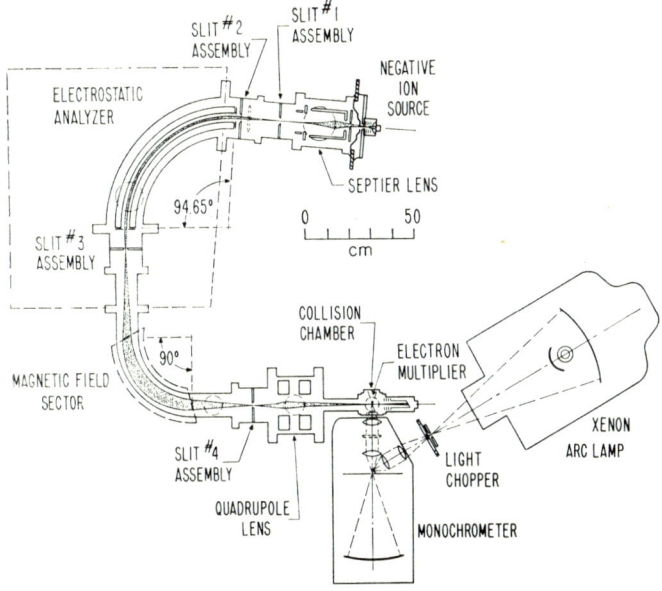

Fig. 7-4-3. Washington NBS photodetachment apparatus.

These three requirements indicate the utility of a mass analyser that is double focusing in the mass spectrometer sense: focusing both in angular spread and energy spread. Double focusing in the ion optical sense, that is in both directions perpendicular to the ionic beam direction, would also be a useful feature. But inasmuch as focusing in energy spread is important, and since stigmatic focusing can be obtained with separate lenses, complete second order, double focusing is employed [95]. The configuration selected involves a $\frac{1}{2}\pi$ magnetic sector preceded by a nearly $\frac{1}{2}\pi$ (94.65°) cylindrical electrostatic sector (Fig. 7-4-3).

Externally adjustable slits at each of the three foci and an additional one for angular definition behind the entrance slit are mounted on sapphire balls and connected to feed-throughs so that ion currents to each can be monitored. External adjustment of the slits permits one to increase temporarily the resolution for precise mass identification without either disassembly or modification of the arc. Thus the ionic composition of a given arc can be monitored *"in situ"*. The adjustable feature also permits one to explore the beam profile. Rapid optical adjustment is greatly enhanced by electrical isolation of the individual slits. By monitoring the current to each of the individual slits, one can identify and adjust the ion optics much more easily than would otherwise be the case. Indeed, without some way to monitor the beam at the various slits adjustment of the system may be nearly impossible.

Since the tandem fields focus only in the horizontal plane, albeit to second order, stigmatic focusing is provided by additional lenses in the vertical plane. These are parallel bars located immediately before and after the electrostatic sector. Each bar of the pairs is controlled independently so that vertical steering of the beam can also be accomplished. The fact that the vertical lenses are located at different distances from the entrance and exit mass analyser slits respectively means, of course, that the beam is vertically magnified. In the horizontal plane the magnetic field then demagnifies the beam. The entrance and exit slits are thus not the same size and shape for a triangular slit function.

d. Final ion focusing. A quadrupole doublet between the mass analyser exit slit and the ion–photon interaction region is designed in such a way that its magnification is the inverse of the magnification of the mass analyser. Thus a final ion image roughly similar in size and shape to the entrance slit is formed in the interaction region. The quadrupole is also advantageous in that it provides steering in both directions, in the horizontal direction to compensate for the slight fringe field of the mass analyser magnet, and in the vertical direction to permit partial compensation of the upward deflection of the ion beam in the electron collecting field of the interaction region.

e. Ion monitoring. The primary requirement for ion collection is naturally that all ions are measured. This requirement is not severe, but the deflection of the ion beam in the electron collection region prevents the problem from being completely trivial.

It is well known that a Faraday cup must be sufficiently deep that the aperture as seen from the base subtends a relatively small solid angle [96]. The specular component is reduced by tilting the cup base with respect to the axis (Fig. 7-4-4). A series of discs at the entrance to the cup is negatively biased so that secondary electrons directed toward the mouth of the cup are driven to one of the walls.

The effectiveness of such a system can be tested in several ways [70]. First, scanning the repelling voltage will surely affect the fraction of electrons or ions escaping. Inelastically reflected particles can thus be monitored. Elastically reflected particles may or may not be affected depending on their detailed trajectories.

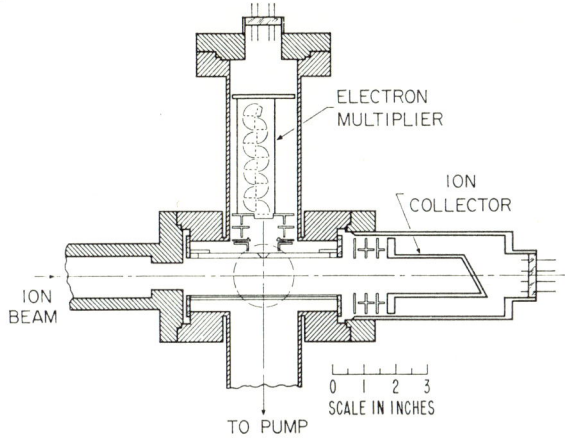

Fig. 7-4-4. Collision chamber.

Second, elastically reflected ions could be detected by the application of magnetic fields in the vicinity of the Faraday cup. Fields strong enough to deflect an H^- beam from the entrance of the cup do not measurably increase the measured ion currents. A third method is to scan the ion beam electrostatically across the base of the cup. This scanning is accomplished either by application of a voltage between the two halves of the entrance plates or by varying the preceding electron collection field. Such scanning procedures would be expected to affect the quantity of ions "lost" from the Faraday cup. No such effect was observed in the measured beams.

The splitting of the Faraday cup entrance plates serves an additional function. The detached electrons are collected by means of purely electrostatic field in the present NBS apparatus. The field also affects the ion trajectory. Ion deflection is reversed on approach to the Faraday cup by application of a reverse voltage to the halves of the first entrance plates. These reverse voltages then permit a stronger electron collection field in the interaction region than would otherwise be possible.

f. Electron monitoring. Measurement of the electron signals in classical photodetachment can be resolved into two distinct problems: (1) complete focusing of all photodetached electrons onto the cathode of an electron multiplier, and (2) relatively uniform amplification of this electron current by the multiplier.

Although the possible imposition of a magnetic field makes the solution to the first problem easier, it complicates greatly the solution to the second problem. Moreover the use of a magnetic field for electron collection would present two other problems. On the one hand it would complicate the quantitative analysis of the electron lenses involved to the point that the entire trajectory of electrons from the detachment region to the photomultiplier could not be predicted with certainty. On the other hand, the magnetic field so imposed will have a different effect on ions of different mass. Both ion and electron trajectories will then be mass dependent, severely complicating the problem of assuring uniform electron gain for all ions. Therefore, for the present NBS instrument electrostatic electron optics have been developed (Figures 7-4-4 and 7-4-5). The initial handling of the electrons takes place in as uniform an electrostatic field as possible. The beam thus formed serves as the source for an Einzel lens which focuses the beam onto a small spot on the photomultiplier cathode.

At this point it is important to observe that the primary noise source in classical photodetachment above 4000 Å is the collisional detachment of the background gas. Below 4000 Å, wall photodetachment (the photoelectric effect) also becomes a problem. Photodetachment under the conditions of the classical equipment occurs with a probability less than 10^{-7}, while collisional detachment typically occurs with a probability greater than 10^{-7}.

The power available in the photon beam corresponds to less than 100 mW/cm^2; each photon corresponds to 2×10^{-19} J; thus approximately 5×10^{17} photons per second per cm^2 are available. Since the cross sections are of the order of 10^{-17} cm^2 and since the ions spend 10^{-9} seconds in the beam, the photodetachment probability per ion is less than 10^{-8}. On the other hand, the background pressure of 3×10^{-8} mm Hg presents approximately 10^{-9} background particles/cm^3 to the beam. Since detachment cross sections are of gas kinetic size, 10^{-16} cm^2, collisional detachment occurs with a probability of 10^{-7} over a path of 1 centimeter.

Thus, in addition to the necessity for chopping and phase sensitive detection, the vacuum is of the greatest importance. Apart from the general employment of ultra high vacuum techniques, attention must be paid to the differential pumping. Each ion slit has been designed so that its gas conductance is dominated by the slit aperture itself. The Septier lens chamber is pumped by a turbine pump, which also serves for initial pump down of the entire system. The electrostatic sector, magnetic sector, quadrupole chamber, and interaction region, are all pumped with separate sputter ion pumps. The pump in the interaction region is baffled to prevent direct flow of uncharged, energetic particles (presumably metastable molecules and/or photons) which otherwise create a substantial background signal at the multiplier.

Virtually all of the charged particles from the pump have low energy (≤ 10 eV) and are negatively charged. One presumes that they are electrons. The lower of the pair of plates which forms the field in which photodetachment take place (Fig. 7-4-4) serves to repel the charged particles back toward the pump. Because of the necessity for rapid pumping through this lower plate, it is made of stretched high transmission mesh.

Conversely, because of the desirability of discrimination between photodetached and collisionally detached electrons, and since electrons are collisionally detached along the entire ion beam while electrons are photodetached along only a relatively small section of the path (~ 4 mm), the upper plate of the pair producing the electron collecting field is solid and contains an externally adjustable slit.

Because a major purpose of the NBS apparatus is to measure cross sections reliably and since the measurement of the photodetached electron current contributes directly to the cross section reliability, care must be taken not to "lose" electrons. In view of the problems in a magnetic lens already noted, this requirement dictates a relatively high electrostatic field for electron collection. The higher the field the smaller the divergence of the collected electron beam and the narrower can be the slit, or alternatively the higher above the threshold energy one can go without losing electrons. Since one wants to measure as much of the cross section as possible, one wants the electrostatic field as high as possible. But since the ions are also deflected in the same field, although to a much less extent, one must reach a compromise. For the Washington NBS, this compromise for 2.5 keV ions and a 7 mm slit permits work on cross sections within one half of an electron volt above the threshold.

If the electrostatic electron collection field were not continued above the slit for some distance, the slit would act as a lens with considerable astigmatism, which in turn would cause the electron beam image on the photomultiplier cathode to be much larger than is desirable, as we shall see. Therefore, the field is continued above the slit by introducing a series of closely spaced plates at proper potentials (Fig. 7-4-5).

At the upper end of the plates two operations must be performed: (1) focusing the electrons onto a small spot on the dynode and (2) sweeping

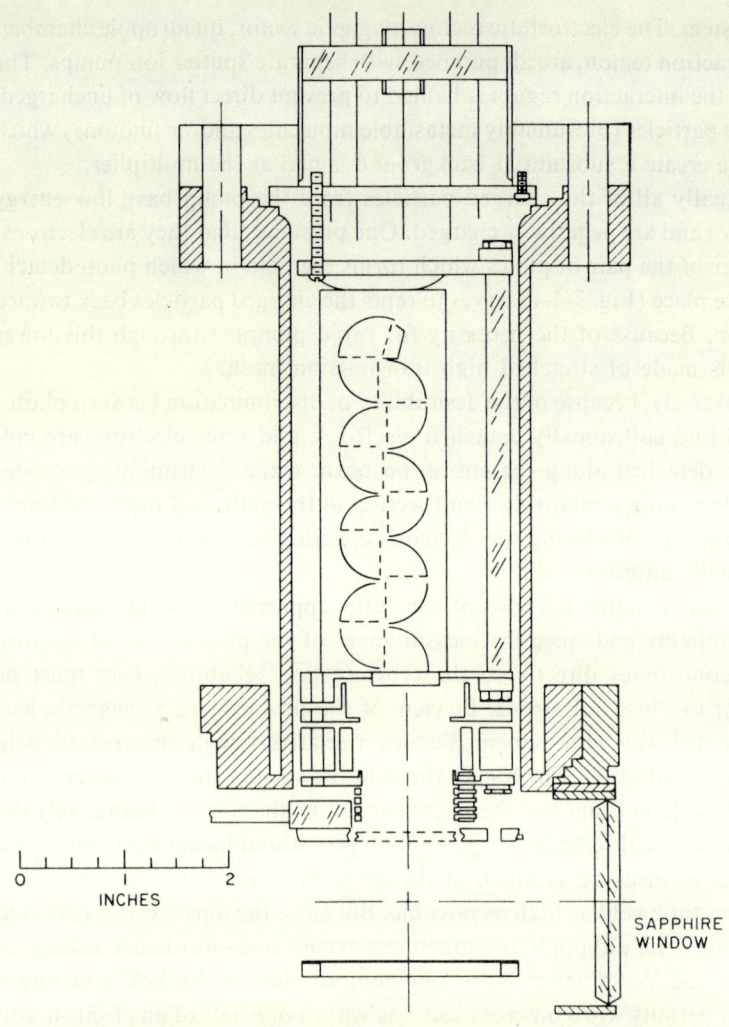

Fig. 7-4-5. Electron focusing plates and multiplier mount.

the spot across the dynode to explore the multiplier gain as a function of the location of the impact of the original electron. Both functions are achieved by means of an Einzel lens element consisting of a cylindrical section split longitudinally into four sections. Although such a deflection system is far from ideal and involves some change in focal length with substantial deflections, the compactness of the lens insures a smaller image (demagnification) than would otherwise be the case.

The importance both of careful focusing and the ability to sweep the image across the multiplier cathode are illustrated in Figures 7-4-6–7-4-8. Since the cathode is a 15 mm square box, the beam is clearly being focused within one mm when near the center of the dynode. The most rapid attenuation is associated with the adjacent dynode: apparently at a certain point the field formed by the adjacent dynode attracts the entire beam into this dynode, so that it misses the first dynode completely. The more gradual attenuation in the opposite direction is probably associated with less efficient collection by the second dynode of secondary electrons from the first. The

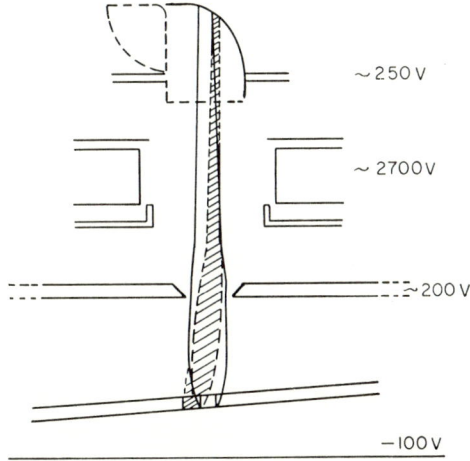

Fig. 7-4-6. Electron trajectories from I^- (clear) and H^- (crosshatched).

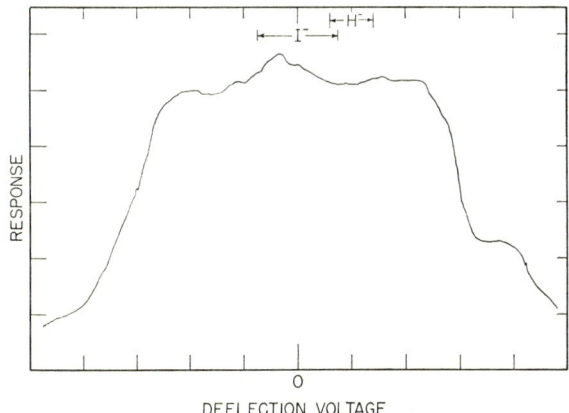

Fig. 7-4-7. Electron multiplier response versus impact position parallel to cathode curvature. Impact position of electrons from two different ions indicated.

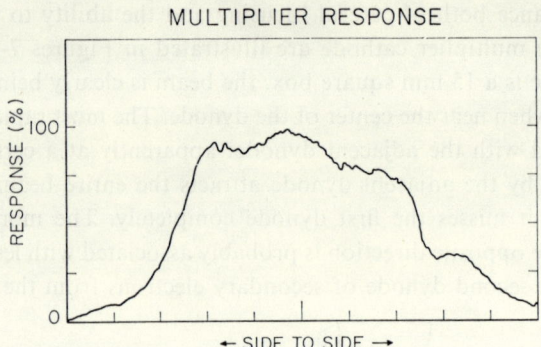

Fig. 7-4-8. Later example of electron multiplier response versus position parallel to cathode curvature.

structure in the transverse direction is less pronounced and less obviously correlated with the gross features of the multiplier electron optics.

One major aspect of these observations on the electron optics pertains directly to the reliability of cross section measurements. Each electron carries with it a certain fraction of the kinetic energy of the ion from which it is detached. This energy is directed along the direction of ion motion, of course, while the electron velocities associated with the energy above the threshold for the photodetachment process have a wide distribution in angle.

The trajectories of electrons from a light ion, say H^-, will be substantially displaced from those of a heavy ion, say I^-, by several millimeters for the parameters used at NBS. The electrons from the two different ion beams will traverse the slit at different points if the light beam is not appropriately displaced. Even with displacement of the photodetachment region

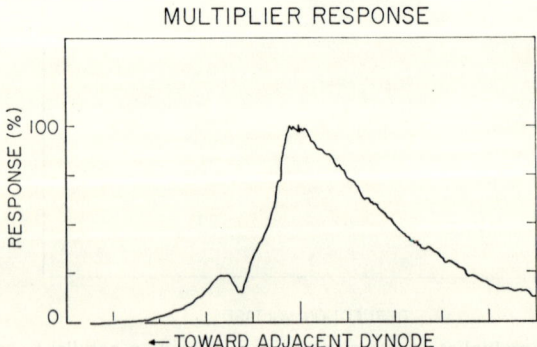

Fig. 7-4-9. Electron multiplier response versus position around curvature of cathode.

so that electrons from each beam in turn successfully negotiate the electron slit (Fig. 7-4-6), the electron beams from different ions will impinge on somewhat different regions of the cathode with measurably different gains associated with each. Only by exploring these effects can appropriate corrections be made in the resulting apparent cross sections (Fig. 7-4-9).

g. Photon optics. Until recently, the most intense and relatively stable source of visible photons was a commercial high pressure xenon arc. The 2.5 kW model is about 40 % efficient, producing about 1 kW over approximately 10 000 Å, or 100 mW per ångstrom. The commercial source has ellipsoidal and spherical mirrors and fills an aperture of f/2, so that roughly 5 mW of power per ångstrom, or ~ 250 mW/50 Å is available over a few square millimeters at the monochromator entrance slit. A quasi-Ebert monochromator with f/1.2 optics (f/1.5 in effective aperture) is mated by collimating optics to this source. The exit optical system is driven so that the image remains fixed in space with respect to the monochromator face (and the ion beam). Also included in the exit optics is a beam splitter and a second imaging lens so that $\approx 10\%$ of the photon beam can be monitored during the experiment. A specially constructed disc thermopile is normally located at the monitor image. A duplicate thermopile can be located at the main image for calibration purposes. Of course, the calibration constant so determined is a strong function of wavelength, presumably because of various well-known polarization effects connected with gratings. But since great care has been exercised, as noted, to collect all electrons independent of initial direction, the polarization of the light, although interesting, will not affect total cross section measurements.

Auxiliary slits are located at the image between light source and monochromator for two reasons: (1) the light beam height is thereby limited so that it is not larger than the ion beam height. This is useful both for focusing the light beam onto the ion beam and for improving the correlation between the light beam that the ions "see" and that which the detector "sees" (2). Horizontal slits at the same place reduce the heat loading on the monochromator entrance slit.

It has already been noted that chopping reduces the collisional background noise, which is several times that of the photodetachment. This noise was determined to be relatively independent of frequency above 200 Hz, except for harmonics of 60 Hz. The frequency of 2160 Hz is safely above the prominent harmonics, but still usable with synchronous motors. The chopper is located adjacent to the auxiliary slits described above.

h. Beam overlap. One major challenge in all crossed beam research is the determination of the beam overlap factor, OF. Typically the two beams are measured independently, while the overlap integral is in fact the parameter of interest:

$$\text{OF} = \frac{\int A(x)B(x)\,dx}{\int A(x)\,dx \int B(x)\,dx}.$$

This factor approaches unity as one of the beams becomes uniform. In practice the light beam is made as uniform as possible and this factor calculated to determine the uncertainty in the cross section due to beam nonuniformity.

The problem is complicated if the quantities are varying rapidly as functions of time. The first goal, then, is to stabilize the beams as much as possible. For both beams this requirement is most easily fulfilled with a source that is as stable as possible with respect to time and position. The

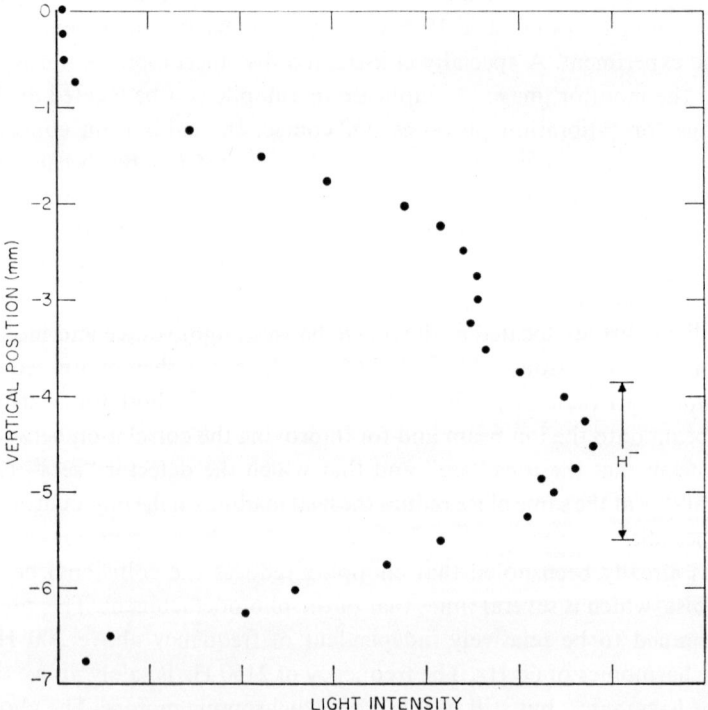

Fig. 7-4-10. Vertical photon distribution at 9930 Å. Position of H⁻ beam indicated.

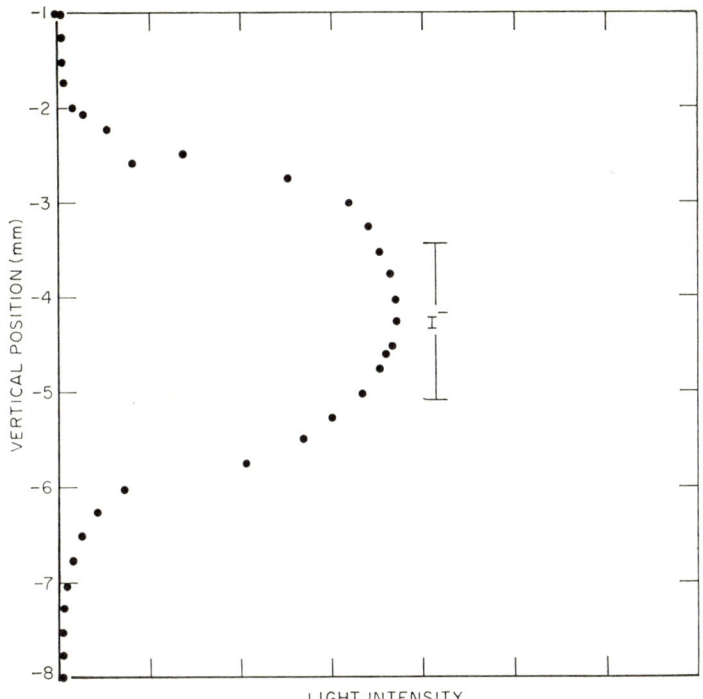

Fig. 7-4-11. Vertical photon distribution of 3470 Å. Position of I⁻ beam indicated.

commercial 2.5 kW xenon arc at NBS fulfills this criterion relatively well. That is its stability, in contrast to that of the earlier carbon arcs, is not a limiting factor in the experiment. Monitoring is directly assisted by reducing the light beam so that it is slightly larger than the ion beam in height (Figures 7-4-10 and 7-4-11). The size of the photon beam, a few millimeters, is then sufficiently small and the light collection optics sufficiently large that the photon density is quite uniform over the desired area. This fact is verified by movement of a slitted photomultiplier over the photon image in the vertical direction (Figures 7-4-10 and 7-4-11). Since the ion beam moves in the horizontal direction, knowledge of the horizontal distribution of the photon beam is not required.

Photon detection uniformity is checked in a similar manner. For this purpose the slits between the light source and monochromator are adjusted to reduce the photon image. Both the monitor photon detector and the calibration detector temporarily located at the main image are scanned systematically across the beams. The stability of the entire photon optical system, the alignment of the photon beam with respect to the ion and electron slits, and the execution of the diagnostic tests are all facilitated by the mounting of the light source, monochromator, intermediate slits, and chopper on a large pair of plates separated by balls in kinematic grooves. The upper plate moves in a direction orthogonal both to the lower plate and to the ion beam.

i. *Instrumental alignment.* The above photon optical base plates as well as all of the ion optics are mounted in turn on a large aluminum table in the surface of which the ion optical path is scribed. Alignment is then performed with the aid of a height gauge and a theodolite mounted temporarily on the tables. Each flange of the vacuum housing is centered on the optical axis and the slit micrometers adjusted visually for slit closure. In each case, the slit closure determined in this manner agreed within 0.003" with that from ion transmission measurements during operation.

3. *Detachment using tunable dye laser*

Essentially the same experiment as that just described can now be performed with the much higher resolution and photon power of a dye laser [20]. A rhodamine 6 G laser provides, in a band width of approximately one ångstrom, the radiant flux of the 50 Å resolved xenon arc and can be scanned continuously over a 500 Å region in a manner analogous to the scanning of a monochromator. Indeed, the laser cavity contains an echelle grating to provide wavelength sensitive gain as an essential element. The photon beam intensity is monitored with a beam splitter and external thermopile as in the classical experiment.

The ion beam is monitored, as in the classical approach, by a Faraday cup, but in this case after electrostatic separation from the neutrals formed. It is the neutrals formed, as monitored with a multiplier, rather than the electrons detached, which provides the measure of the cross section.

The threshold information obtained in the one experiment published so far [20] is considerably in excess of that from a recent classical experiment [84] (Fig. 7-4-12). In this experiment $S^-(^2P)$ is detached to $S(^3P)$. Six transitions are expected, with predictable separation within each pair of triplets. The separation of the two pairs of triplets will then yield the ^2P spin–orbit splitting of the negative ion. Four of these transitions are in fact resolved for the first time in the dye laser experiments. Although the absence of the remaining two is not particularly disturbing, because they are the weakest and the exact strengths of the transitions cannot yet be predicted, the appearance of two other breaks in the cross section is very puzzling indeed, although the general explanation may lie in interaction with the continuum.

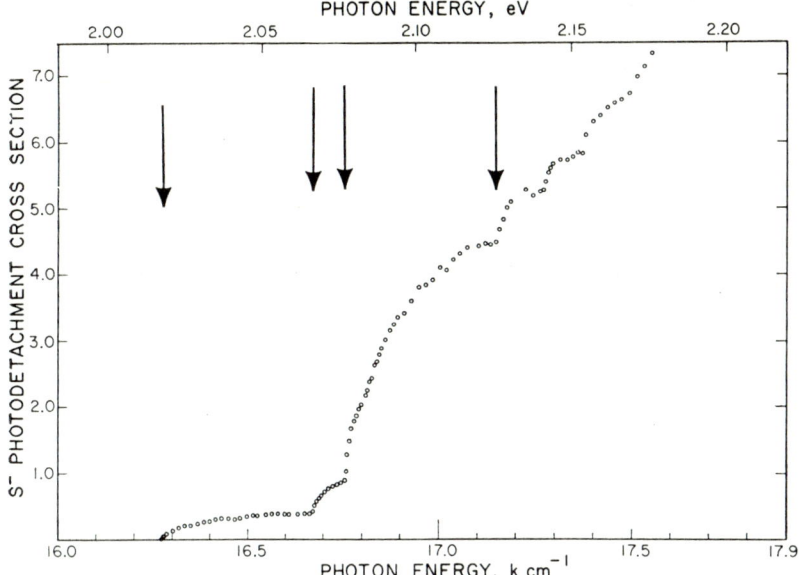

Fig. 7-4-12. Threshold behavior of S⁻ photodetachment cross section obtained with a dye laser, courtesy of Carl Lineberger.

The technique is so new that it is difficult now to say exactly what the limitations will be. The unexpected features of the S⁻ cross section present one real challenge, both to the theoretician and the experimentalist. The reconciliation of the two detachment experiments, whose electron affinity determinations do not quite overlap, $(2.0772 \pm 0.0005$ eV and 2.095 ± 0.015 eV) will depend on the detailed understanding of cross section behavior.

The determination of absolute cross sections with dye lasers presents two problems, the absolute measurement of the neutral particle intensity and the measurement of the beam overlap factors of pulsed beams.

At the moment, the spectral range accessible with available dyes is limited, but this is surely a temporary problem. The enormous increase in resolution available is very exciting.

4. *Detachment using fixed laser and electron energy analysis*

An alternative detachment experiment that uses photons of a relatively narrow fixed energy (gas laser) and analysis of the energy of the electrons detached is also very promising. Some of the advantages and limitations of this approach have already been desdribed earlier in the chapter.

The primary requirement is that a fixed, preferably continuous, laser be available with a wavelength shorter than the detachment threshold. The ability to lock the laser on more than one transition is very useful in identifying the appropriate energy scale of the electron monochromator; a dispersive element is then included to permit the selection of one transition. An argon ion laser has been used for previous work [93, 97]. The polarization of the laser influences the angular distribution of the detached electrons in such a way as to give information on the electronic states involved.

The directional character of the electrons emitted by the polarized beam, and the momentum imparted to the electron by the ion, coupled with the small spatial and angular acceptance of the electron energy analyzer complicate the problem of electron collection. The detachment takes place inside a hollow carbon block so that the stray electrostatic fields are reduced as much as can conveniently be done. This block then acts as the source for a deflection analyzer of the Kuyatt–Simpson type capable of a resolution of a few milli-electron volts.

The importance of surface potentials is demonstrated by an observed electron energy shift with time after the composition of the ion beam is changed in a bare copper chamber. No shifts are observed when carbon is used [98]. The effects of surface potential shifts are minimized by working simultaneously with two beams, a reference beam and one whose detachment threshold is to be determined. This is accomplished by means of a Wien filter, whose mass dispersion can be controlled by simultaneously varying both electrostatic and magnetic fields. Since the thresholds for detachment of different ions will usually differ, no ambiguity need result from the elimination of ion mass analysis once the ions in the beam have been identified. In some cases the ion beam is alternated rapidly, although this presents a challenge to the ion optics.

A detailed understanding of the electron analyzer performance is very important in the analysis of the molecular detachment data. In such a case, the identification and the characterization of the vibrational levels involved is of great importance in the determination of electron affinities. This identification follows from one of several different approaches: (1) isotope substitution; (2) comparison of the relative vibrational levels with Frank–Condon factor calculations assuming a Morse potential; (3) observation of anharmonicities, that is the energy differences between the vibrational levels, and (4) search for a break in the intensities, which presumably would indicate a 0-0 transition. The latter three techniques imply detailed understanding of the electron energy analyzer behavior. In addition, the identification of characteristic energies will also depend on the nature of the unresolved rotational transitions, as is also the case for classical detachment. The proper energy is therefore associated with the peak energy but perhaps not identical to it. Thus maximum utilization of electron energy analysis will require detailed understanding both of the electron optics and of the molecular physics of the transition.

The information implicit in the cross section dependence on polarization comes at the expense of difficulty in total cross section determination. This experiment is thus complementary to the others.

5. *Double quantum detachment*

Finally, an interesting avenue of photodetachment research in which no current activity is apparent is double-photon detachment [78], or the use of a negative ion as a sort of frequency doubler. The probability of two subthreshold photons being simultaneously in the vicinity of a negative ion so that detachment can occur goes as the square of the photon flux [79]. Such an event happens with observable frequency only with a coherent source. The detachment of I^- has been observed with a Q-switched ruby laser and was found to agree with the Geltman predictions within the 50 % experimental uncertainty inherent in such a difficult experiment.

The electron and ion detection are similar to the early classical beam experiments [19]. In this case, however, the electron multiplier was cleverly calibrated by completely detaching an H^- beam with the laser. The photon beam was monitored with a high speed photodiode and a travelling wave oscilloscope. The photodiode had been calibrated by comparison with a liquid cell calorimeter. The coherence coefficient of $\frac{3}{2}$ used in the initial work [19] has since been withdrawn [51].

§ 7-5. Electron affinities

A. TECHNIQUES

Electron affinity determinations fall into three general categories, (1) theory, (2) empirical fitting, (3) experiment, and into a number of distinct sub groups.

1. *Theory*

Because of the small size of the $(Z+1)^{\text{th}}$ electron binding energies in negative ions in comparison with the total sum for all of the electrons, the accurate calculation of electron affinities has remained a difficult and challenging occupation.

For H⁻, the problem, although involving three bodies and thus not susceptible to analytic solution, is sufficiently simple to yield to the elaborate variational techniques of Pekeris [99, 100]. His value of the affinity, 0.75421 eV is thought to be several orders of magnitude more accurate than any other *a priori* electron affinities. The disadvantage of this approach, of course, is that it is not extensible to heavy ions.

For the rest of the Periodic Table Hartree–Fock (HF) techniques have yielded some very useful results in recent years. Although relativistic corrections are negligible for the lighter atoms, correlation effects are not. Indeed, the correlation effects are frequently as large as the entire affinity: the Hartree–Fock energy may not predict a bound ion at all. In their pioneering "Chicago school" calculations, Clementi and colleagues have used isoelectronic extrapolation of ionization potentials and experimental affinities to predict correlation corrections to their *a priori* HF calculations [101–103].

Very recently, considerable effort has been made to calculate correlation effects and to apply them to negative ions. For example, Weiss has employed a superposition of configuration technique [104, 105]. Sinanoglu and Öksüz have investigated in detail different types of correlation [106, 107]. Schaefer and Harris have used a configuration interaction term for part of the correlation [108, 109].

For the diatomic molecules OH⁻ and SH⁻, Cade has employed an isoelectronic estimation technique analogous to Clementi's atomic ion calculations to obtain affinities from Hartree–Fock calculations [110, 111].

These various approaches to correlation energy seem to have reduced the uncertainty in such calculations to the order of 0.1 eV. Greater accuracy in *a priori* calculation of affinities will probably be very difficult to obtain, but present accuracy far surpasses the level of agreement of many experiments.

2. *Empirical fitting*

In common with other areas of spectroscopy, the difficulty of *a priori* calculation of energy levels has led to empirical attempts to estimate energy levels of negative ions. These rely on observed regularities in the Periodic Table of the elements with respect to various aspects of electron configuration. But partly because of the absence of a sufficiently detailed and universally obvious theoretical foundation, partly because of the paucity of completely reliable information on the levels of multiply ionized atoms, and partly because of the scarcity of completely reliable experimental electron affinities with which to evaluate the reliability of calculations, a wide variety of such empirical approaches has been attempted. These tend to fall into six distinct types, although some hybrid forms have also appeared.

a. Ionization energy. Inasmuch as an electron affinity is analogous to the ionization energy of neutral atoms and positive ions (all analogies have their weaknesses and limitations), the most elementary form of extrapolation utilizes ionization energies of isoelectronic species. Glockler [112] pioneered this approach in a comprehensive fashion with the energy relationship:

$$E_0 = 3E_1 - 3E_2 + E_3$$

where the subscript refers to the charge of the *final* state involved in the various successive states of ionization of isoelectronic ions. These ionization energies, E, are in turn calculated by determination of empirical constants a, b and c in the equation $E = a + bZ + cZ^2$, where Z is the nuclear charge. These calculations were repeated by Bates [113] and Pritchard [15] with newer spectroscopic data and extended by Charkin and Dyatkina [114]. Geltman [115] derived lower limits by not restricting himself to starting with a neutral state in the isoelectronic series. More elaborate series forms with negative powers of the ionic charge have been used by Johnson and Rohrlich [116], Edlén [117] and Kaufman [118].

b. First excitation energy. An attempt in a similar direction was made by Bates and Moiseiwitsch [119], who employed a formalism similar to that used with isoelectronic ionization potential extrapolation, but worked with the first excited term energies rather than ionization potentials. The assumption is that the negative ion continuum is adjacent to the first excited state. The method has the advantage that first excitation energies are generally more precisely known than are ionization energies.

c. Binding energy per p electron. Another modification to the basic concept of isoelectronic extrapolation of ionization energies is to normalize the energies involved by the number of p electrons. First developed by Hellmann and Mamotenko [120] the method has been utilized also by Margrave [121, 122] and Pritchard and Skinner [123].

d. Configuration energy. An essentially different approach has been developed by Bacher and Goudsmit [124]. Instead of working with an isoelectric extrapolation, they developed a series formulation involving the successive states of ionization of a given atom. The method has been employed by Ta-You Wu [125, 126] for several light atoms.

e. *Electron affinity horizontal analysis.* With the availability of some reliable experimental affinities, much of the most recent activity in the area of empirical fitting has been generated by yet another method. Since successive experimental ionization energies vary in a similar manner as a function of atomic number within a given row of the Periodic Table, the behavior of electron affinities, one speculates, may also follow a similar path. If so, a single reliable experimental affinity will permit one to generate others for atoms in the same row. This technique of "horizontal analysis" has been applied to negatively ions apparently independently by Ginsberg and Miller [127] and Edie and Rohrlich [128]. More recently, Crossley [129] and Zollweg [130] have applied the newest experimental affinities.

f. *Radius–vertical analysis.* A completely different approach has been used recently to estimate affinities of atoms lower in the Periodic Table from those of atoms in the first two rows. Polizer has observed [131] that a log-plot of the difference between the first ionization potential and the electron affinity against the atomic radius for atoms within each period fals on a straight line. The two best known affinities can thus be used with the ionization potential and atomic radii to obtain the other affinities. Although the assumptions seem questionable, the halide affinities certainly do fall on a straight line. Polizer has used this approach to estimate atomic affinities not previously attempted.

g. *Correlation energy.* Finally, the absolute calculations of the affinities have until recently, not accounted *a priori* for electron correlation. This problem was faced by some of the earlier workers [124, 132] and a correction made by extrapolating the differences between experimental and *a priori* calculations in an isoelectronic sequence. Clementi and co-workers [101–103] have utilized this technique with great succes. Crossley's approach noted above [129] actually incorporates this approach in a hybrid with horizontal analysis.

3. *Experiment*

There are at least eight relatively simple physical processes, the measurement of which leads to electron affinities. Each one is subject to certain problems as well as presenting certain advantages over others. Because of the serious nature of the assumptions involved in each, experimental values can be considered firm only when confirmed by another technique. It is essential, then, to evaluate the major techniques employed.

In addition to those cited below, an even wider variety of methods has been applied to larger molecules. Since these are not so precise as to provide a basis for discrimination and since there are so many of them, they fall outside of the scope of this chapter.

a. Photodetachment and radiative attachment. Conceptually the simplest way to measure the energy of the interaction:

$$Y^- + h\nu = Y + e^-$$

is to measure in a beam experiment either the photon energy for the process releasing zero energy electrons (threshold) [20, 56, 70] or to measure both photon and electron energy simultaneously in photoelectron spectroscopy [93]. In a plasma experiment the threshold can be measured either for absorbtion (detachment) [63, 67] or emission (attachment) [21].

The advantages and problems in each method have been analyzed earlier in this chapter. The chief general advantage is the spectroscopic precision available. In addition, the beam experiment, and to a slightly less extent, the plasma experiment offer freedom from interaction with other systems.

The chief disadvantage is the frequent lack of definition of the states involved, a problem shared with other experimental techniques. Virtually all work is interpretable in terms of the absence of excited states, but unambiguous proof of the identity of the transitions involved is generally lacking. A practical disadvantage is the experimental complexity. For this reason, relatively few detachment experiments have been performed. For most affinities then, one must turn to other experiments or to theory. And even where detachment experiments have been performed, other techniques are necessary to confirm the interpretation of the transitions involved.

An "experiment" closely related to laboratory plasma spectroscopy is implicit in the spectrum of the star α-Tauri [133]. Berry has used this spectrum to derive the most precise experimental value for the affinity of atomic hydrogen, 0.756 ± 0.013 eV [9].

b. Photoionization. Almost as simple conceptually is the measurement of the photon energy in the process:

$$AB + h\nu = A^+ + B^-.$$

This is a resonant process because conservation of momentum leaves little freedom in energy. In addition to determination of the resonant energy, the bond dissociation enery and the requisite ionization energy must be known. However, the interaction

$$AB + h\nu \rightarrow A^+ + B + e^-$$

can generally also be observed, so that the affinity is derived from the difference between the resonant energy for the former process and the threshold for the latter. Because the threshold laws are unambiguous, one can thus determine electron affinities by observing only the positive ion, without ever observing directly the negative ion in question.

Dibeler, Walker and McCulloh [134] have thus supported $(3.47 \pm 0.03$ eV) Berry's atomic fluorine affinity, 3.448 ± 0.005 eV [63], which differs from the other plasma affinity value of Popp, 3.400 ± 0.002 eV [85]. Recently also, Berkowitz, Chupka and Walter have thus determined the affinity of CN, 3.82 ± 0.02 eV [2]. Unfortunately, the experiment of Elder, Villarejo and Inghram [135], which yields a value of 1.461 ± 0.024 eV, is not precise enough to distinguish between the atomic oxygen affinities of Branscomb, Burch, Smith and Geltman, 1.465 ± 0.005 eV [56], and of Berry, Mackie, Taylor and Lynch, 1.478 ± 0.002 eV [21]. Earlier work of Morrison, Hurzeler and Inghram [136] had given the first strong indication that the then accepted affinities of the halogen atoms might be in error by 0.1–0.2 eV. This question was later confirmed, first by plasma and then by beam photodetachment.

The observation of the photoionization continuum of atoms and molecules sometimes indicates ionization below the continuum, probably due to the processes:

$$Y^* + Y \rightarrow Y^+ + Y^-$$

where Y^* is an excited state. Such an interaction is of course second order in pressure and can thus be distinguished from the first order absorption. The affinity thus determined is, of course, a lower limit. Lee and Mahan have determined lower limits of alkali atom affinities [137] and Samson and Cairns [138] have obtained a value $(0.20 \pm 0.01$ eV) for O_2, which is significantly lower than the most recent work.

c. *Dissociative electron attachment.* The reaction

$$XY + e^- \rightarrow X + Y^-$$

has long been used to determine electron affinities by combining it with external information on the dissociation energy of the molecule. Thus, although the reaction is conceptually analogous to photoionization, the absence of a positive ion to monitor means that in this case one must actually use auxiliary bond energy data.

The necessity for such auxiliary data reduces the applicability of the method to those cases for which bond energy data of sufficient reliability is available. But recently a number of more basic problems have come to light which severely complicate the interpretation and suggest several sources of the pronounced discrepancies with determinations by other techniques.

The first source of complication to be identified involves the role of thermal translational energy. Chantry and Schulz have shown [139] by elementary mechanics that near threshold the thermal energy broadens the ion energy distribution, so that even work with monoenergetic electrons would not produce a clean threshold. The effect is sufficiently large that data originally thought to yield an electron affinity of 2 eV for atomic oxygen has been shown consistent with an affinity of 1.5 eV. Newer experiments [140] confirm the interpretation and the lower affinity. More recently, Schulz and Spence [141] and Henderson, Fite and Brackman [142] have pointed out that the molecular internal energy, vibrational and rotational, must be taken into account as well. They indicate that results must be extrapolated to 0 K to give meaningful electron affinities. These effects are associated with the very rapid change of Frank–Condon factors with vibrational level for noncongruent energy surfaces.

The nature of the upper level is therefore important and Chantry has alluded to this factor [143]. Dorman has also shown that the product fragments may be internally excited [144]. For metastable negative ions, the distribution of energy in the product ion may be very complex indeed [24, 145].

d. Thermochemistry. In principle, the energy of the transition

$$Y + e^- = Y^- + h\nu$$

can also be determined on a macroscopic scale by traditional thermodynamic techniques. That is, the change in the ratio of Y^- to other species can be followed as a function of temperature. A number of different approaches have been taken.

Honig pioneered the mass spectrometric observation of negative self-ionization in a determination of the affinity of carbon [146].

Recently Scheer and Fine have determined the affinities of a number of atoms by a similar technique [147–151]. By the application of the Saha–Langmuir equation the ratio of positive ion and negative ion currents, X^+/X^-, can be used to derive the electron affinity,

$$X^+/X^- = (w_+/w_-) \exp\left[(2\Phi - I - EA)/kT\right]$$

where the w's are statistical weights, Φ is the work function, I is the ionization potential, and EA the electron affinity. Thus the work function must be known and assumed to be the same for both positive and negative surface ionization. Since the surface very likely displays regions of high and low work function simultaneously, the positive and negative ions may come from different regions of the surface [152]. Another additional assumption is that the optics is the same for both ions when the applied potentials are reversed, a goal difficult to achieve in the presence of contact potentials.

The same data may be used with the heat of sublimation of the neutral species, l_0, to obtain another value for the affinity:

$$EA = l_0 - l_- + \phi.$$

Here the uncertainties associated with the two external pieces of data, the l_0 and ϕ, are a factor.

Bailey [152] employed a similar technique for work with salts rather than filament self-ionization. He determined the work function in situ, although the problem of uniformity of positive and negative ion emission is present. In addition, reaction of the salt with the filament may play a role. The work function problem has been minimized in some cases by determination of electron affinity differences [152–155]. Pyatigorskii has pointed out that one must be careful to work at sufficiently high temperatures and low salt vapor pressures [156].

Another experimental approach has been exploited by Page and colleagues for work with molecules that react with the surfaces. They measured [157] both electron and ion currents, so that the work function disappears, but its place is taken by the frequently complex chemistry of the gas-surface interaction. Bond dissociation energies and intermediate states play a complex role. The lack of mass analysis is thus a particularly serious drawback [158], although, not an unavoidable one.

Ya'akobi, in an interesting variation on these experiments, has used the power balance in exploding lithium wires and the temperature during an anomaly in his balance to speculate on the vaporisation of Li^- and thus the affinity [159].

Finally, the application of Second and Third Law thermodynamics to flames has led to estimates of oxide affinities [160–162]. The reactions involved are quite complex.

e. Born–Haber cycle. The earliest measure of electron affinities came from application of a Born–Haber cycle to ionic crystals [163, 164].

For one such calculation for example, one must know the energy of formation of the salt with respect to the elements, the lattice energy, the ionization energy of the alkali, the sublimation energy of the alkali, and the dissociation energy of the molecular electronegative molecule. The precise determination of all of these quantities is a challenge. Nevertheless, the method is still being used occasionally, most recently for OH^- [165] and O^{2-} [166].

The cycle is now used more frequently in reverse; that is, precise affinities are now being utilized to give information on lattice energies [167].

f. Charge exchange. The interaction

$$X + Y^- \rightarrow X^- + Y$$

can be used to give information on the energy involved in such a transfer. Thus if the exchange is observed for low energy ions, then the reaction is exothermic and the affinity of X is greater than that of Y. One must be careful, of course, that the product ion is the one expected [168]. In addition, many of the considerations surveyed in the section on dissociative attachment apply. The method has been used extensively to give lower limits to affinities by Ferguson and Fehsenfeld [168, 169], and by Paulson [170].

g. Solid spectroscopy. The abundant spectroscopic information on solids has furnished a tempting field for information on electron affinities. The basic problem is the extent and nature of the interactions of the ions with the surrounding matrix. Even when the interactions are too strong to observe a significant threshold, one can sometimes use other spectroscopic information. For example, vibrations ascribed to ozonide are observed [171] to be higher than those for ozone, indicating stronger chemical bonding in the negative ion. Since the two species can be assumed to dissociate to $O_2 + O$ and $O_2 + O^-$ respectively, we can thus speculate that the affinity for ozone is greater than 1.5 eV.

An exceedingly promising advance in such work has been made recently by Milligan and Jacox [172]. They have succeeded in isolating a number of species in an argon matrix at cryogenic temperatures. Bombardment of the material with electrons during deposition of the matrix yields characteristics spectra which are ascribed to a negative ion. Because of the very low temperature and inert nature of the surrounding material, interaction between an ion and the matrix is thought to be small. An upper limit to the affinity of NO_2 has thus been determined.

h. Field detachment. Ions with a small affinity are susceptible to autodetachment if subjected to a relatively strong electric field. Such an approach is restricted to ions with a very small affinity. The interpretation of the results requires a detailed knowledge of the negative ion structure. Such an analysis for $He^-(^4P)$ gives an affinity of 0.060 ± 0.005 eV [173], a value which is in moderate though not serious disagreement with the photodetachment value of 0.080 ± 0.008 eV.

i. Electron scattering. An approach that has not yet yielded an *a priori* electron affinity, but has been used to check for consistency and with refinement promises to provide independent affinities is resonant scattering of electrons, both elastic [174] and total [175]. The resonances in the total scattering on O_2 may be identified with vibrational levels both of the negative ion and of the neutral molecule. Extrapolation of the negative ion energy levels to the accepted affinity yields a consistent identification of the negative ion vibrational levels. One can imagine that with development of the techniques one will be able to perform the extrapolation independent of auxiliary data.

B. Case study: S^-

An example of the application of a variety of techniques to one species is perhaps instructive in order to permit the reader to arrive at his own evaluation of the state of the art of electron affinity determinations. The affinity of atomic sulfur has been determined probably as many times as that of any molecule with the possible exception of O^- and O_2^-. A wide variety of techniques has been used. It has even been the subject of three separate photodetachment experiments.

Table 7-5-1 summarizes the results and permits one to make some generalizations which seem to stand up to close scrutiny:

First, although the Born–Haber cycle is still being used to estimate electron affinities, it is subject to considerable uncertainty.

Second, dissociative electron attachment also yields affinities subject to substantial uncertainty if detailed analysis is not performed.

Electron affinity of atomic sulfur (eV)

Year	Experiment					Empirical fitting				Theory	Ref.
	Born–Haber cycle	Dissociative electron attachment	Photo-detachment	Surface ionization		Isoelectric extrapolation			Horizontal analysis		
				Mass spect.	Magnetron	Ionization	Excitation	Energy/p electron			
1926	6.46										176
1932						≈2.82					177
1934						2.06					112
1938						2.13					178
1947						1.5					113
1953		1.78±1.09				2.17±1.00; 3.3					179
						2.4					180
1954		≈3.5									181
		≥2.2									182
				2							155
1955							2.5$_5$				119
1956			2.07±0.07								183
1957				2.13±0.05		>2.1					115
1958						2.07					184
1959				2.11±0.05		2.79					185
								2.40	2.24		127
1960						2.15					116
											153–154
											117
1961										1.25	186
1962					2.09$_0$±0.06$_5$	1.92					187
											188
1964									2.08		128
											102
1965						2.03±0.10			2.00	2.12	129
1969						1.79			2.04		10
											130
1970			2.095±0.015								84
			2.0772±0.0005								20

Third, although photodetachment yields relatively precise results, the interpretation of the threshold region must be performed with great care.

Fourth, surface ionization can yield accurate, if not the most precise, affinities if performed with proper ion identification and reaction analysis.

Fifth, isoelectronic extrapolation of ionization potentials is a hazardous business. The proximity of the first such attempt by Glockler to the most probable value is unfortunately fortuitous. Other values, for example the atomic oxygen affinity, are a factor of two from the value now considered most probable.

Sixth, the Bates extrapolation of first excitation potentials is not notably better than extrapolation of ionization potentials.

Seventh, the Hellmann–Mamotenko extrapolation of energy per p electron is also not substantially more reliable than other isoelectronic techniques.

Eight, "horizontal analysis", which is pinned to a reliable affinity, seems to yield relatively reliable results, particularly when it is noted that the early Ginsberg–Miller value was tied to halogen affinities now thought to be too high by about 0.2 eV.

And, finally, the Hartree–Fock approach with suitable correlation energy correction seems to yield relatively reliable results.

C. A SELECTED LIST OF ELECTRON AFFINITIES

Since there are few, if any, definitive determinations of electron affinities, the list of most probable values is a continually changing one. Each author thus properly feels called upon to exercise his own critical judgement, and to continue a winnowing of the field. This process is inherently an imperfect one. Without it, however, the number of values to be considered would proliferate to such an extent as to be subject only to statistical analysis. Such an approach would certainly be analogous to the monetary situation of the "bad" money driving out the "good".

In the following tables several criteria for inclusion have been used. Atoms are listed according to atomic number. Molecules listed first are homonuclear and then heteronuclear, with increasing number of atoms. Within a given group items are listed by increasing atomic number. Within a given state, values are listed according to preference, the most preferred first. Although there is a rough correlation of preference with accuracy, stated accuracies are estimates of the respective authors and thus are not located on a consistent base and are not strictly comparable. Prejudicial preference is given to the least questionable experiments and to comprehensive *a priori* theory. More recent values have been looked at less critically than earlier ones, partly because they are frequently based on more reliable external data and partly because they have not yet been examined comprehensively in comparison with other techniques.

The labeling of techniques described in the text is hopefully self-explanatory. The application of theoretical *variational calculations* to the lighter atoms has been described. The observation in *stellar* spectra of *detachment* continua has been noted. *Beam* photodetachment experiments are of course the main subject of this chapter. Strong electric *field detachment* has also been mentioned. Self-consistent field calculations of the type first performed by *H*artree and *F*ock and extended by Roothan and his Chicago school have also been mentioned. The various *a priori* and empirical approaches to correlation energy are not distinguished, but those familiar with the field will know that Weiss has approached correlation energy in an *a priori* manner, while the others have used various empirical or semi-empirical techniques.

Horizontal analysis is the empirical technique employing one affinity in each row of the Periodic Table to determine the others in the row. In retrospect, *ionization* potential *extrapolation* seems to have been applied consistently only by Edlén. The *exploding wire* has been described. To be observed in a *mass spectrum* an ion must be stable at least for about a microsecond. All such ions thus fall within the scope of this chapter. Although the *M*any *E*lectron *T*heory results seem to be significantly different from the others, any *a priori* theory which can be extended is in principle promising. A variety of *thermochem*ical techniques have been described. They all involve study of the variation of reactions with temperature. *Charge exchange* has been described. *Plasma* photodetachment and attachment includes both shock wave and stationary techniques. The problems of electron *attachment* have been described. The observation of pair production in *photoionization* has been noted. The empirical correlation of the difference between the affinity and ionization energy with atomic radius in a period of the table has been referred to as *vertical analysis*.

Because of the greater complexity of molecular negative ion structure, less precise techniques have sometimes been employed here than for atomic ions. Original references should be consulted, since the approximations involved may influence the results strongly. The *R*ydberg-*K*lein-*R*ees method generates potential energy curves from spectroscopic data. It has been applied to the data on neutral molecules. Various configuration *i*nteraction techniques have been used for molecular ion states. Although use of the *Born–Haber* cycle has been completely superseded for atomic ions, for

TABLE 7-5-2

Atomic electron affinities (eV)

Atomic no.	Ion	State	Affinity	Technique	References
1	H⁻	$1s^2$ ^1S	0.7542	Variational calc.	Pekeris [99]
			0.756±0.013	Stellar detachment	Berry [9]
			0.776±0.020	Beam	Feldman [48]
		$2p^2$ ^3P	>0.0095	Variational calc.	Drake [189]
2	He⁻	1s 2s 2p ^4P (metastable)	0.080±0.008	Beam	Brehm [93]
			0.060±0.005	Field detachment	Smirnov [173]
			≧0.033	Variational calc.	Holøien [190]
3	Li⁻	$1s^2$ $2s^2$ ^1S	0.62±0.10	H. F.	Weiss [105]
			0.58±0.05	H. F.	Clementi [101]
			0.85±0.20	Thermochem.	Scheer [148]
			0.59	Horizontal anal.	Crossley [129]
			0.64	Ionization extrap.	Edlén [191]
			0.82	Ionization extrap.	Edlén [117]
			∼0.6	Exploding wire	Ya'akobi [159]
4	Be⁻	2s $2p^2$ ^4P (metastable)	0.24±0.10	H. F.	Weiss [105]
		$2s^2$ 3s $^2S_{1/2}$	0.38±0.20	Horizontal anal.	Zollweg [130]
			>0	Mass spectrum	Bethge [192]
		$2s^2$ 2p $^2P^o_{1/2}$ (metastable)	−0.68	Horizontal anal.	Crossley [129]
			−0.38	Horizontal anal.	Edie [128]
			−0.19	Ionization extrap.	Edlén [117]
5	B⁻	$2p^2$ $^3P_0$	0.187±0.100	H. F.	Schaefer [108]
			0.30±0.05	H. F.	Clementi [101]
			0.18±0.20	Horizontal anal.	Zollweg [130]
			0.16	Horizontal anal.	Crossley [129]
			0.33	Horizontal anal.	Edie [128]
			0.33	Ionization extrap.	Edlén [117]
			>0	Mass spectrum	Branscomb [183]
6	C⁻	$2s^2$ $2p^3$ $^4S^o_{3/2}$	1.270±0.010	Beam	Hall [18]
			1.25±0.03	Beam	Seman [19]
			1.24±0.10	H. F.	Schaefer [108]
			1.17±0.06	H. F.	Clementi [101]
			1.17±>0.10	MET	Öksüz [106]
			1.29±0.20	Horizontal anal.	Zollweg [130]
			1.33	Horizontal anal.	Crossley [129]
			1.38	Horizontal anal.	Edie [128]
			1.24	Ionization extrap.	Edlén [117]
			1.2	Thermochem.	Honig [146]
		$2s^2$ $2p^3$ $^2D^o$	0.050±0.025	Beam	Hall [193]
			0.062	Beam	Hall [18]
			>0	Beam	Seman [19]

TABLE 7-5-2 (continued)

Atomic no.	Ion	State	Affinity	Technique	References
			<0.5	Beam	Feldman [48]
			−0.105±0.100	H. F.	Schaefer [108]
			−0.08±0.05	H. F.	Clementi [101]
			−0.13±>0.10	MET	Öksüz [106]
7	N⁻	$2p^4\ ^2P_2$	−0.213	H. F.	Schaefer [108]
			0.05	Ionization extrap.	Edlén [117]
			>0	Charge exchange	Fogel [194]
8	O⁻	$2p^5\ ^2P_{3/2}$	1.465±0.005	Beam	Branscomb [56]
			1.478±0.002	Plasma	Berry [21]
			1.461±0.024	Photoionization	Elder [135]
			1.5±0.1	e attachment	Chantry [140]
			1.46±0.10	H. F.	Schaefer [108]
			1.31±0.15	e attachment	Henderson [142]
			1.22±0.14	H. F.	Clementi [101]
			1.24±>0.10	MET	Öksüz [106]
			1.39	Horizontal anal.	Crossley [129]
			1.47	Ionization extrap.	Edlén [117]
9	F⁻	$2p^6\ ^1S_0$	3.448±0.005	Plasma	Berry [63]
			3.47±0.03	Photoionization	Dibeler [134]
			3.400±0.002	Plasma	Popp [64, 85]
			3.47	H. F.	Weiss [104]
			3.45±0.10	H. F.	Schaefer [108]
			3.37±0.08	H. F.	Clementi [101]
			3.23±>0.10	MET	Öksüz [106]
			3.50±0.20	Horizontal anal.	Zollweg [130]
			3.47	Horizontal anal.	Edie [128]
			3.50	Ionization extrap.	Edlén [117]
11	Na	$2s^2\ ^1S$	0.54±0.10	H. F.	Weiss [105]
			0.78±0.06	H. F.	Clementi [102]
			$0.41^{+0.06}_{-0.02}$	Charge exchange	Bydin [195]
			0.22	Horizontal anal.	Crossley [129]
			0.47	Ionization extrap.	Edlén [117]
12	Mg	$2s\ 2p^2\ ^4P^o$	0.32±0.10	H. F.	Weiss [105]
		(metastable)	>0	Mass spectrum	Bethge [192]
		$4s\ ^2S_{1/2}$	−0.22	Horizontal anal.	Zollweg [130]
		$3p\ ^2P^o_{1/2}$	−0.52	Horizontal anal.	Zollweg [130]
			−0.69	Horizontal anal.	Crossley [129]
			−0.61	Horizontal anal.	Edie [128]
			−0.32	Ionization extrap.	Edlén [117]
13	Al	$3p^2\ ^3P_0$	0.52	H. F.	Clementi [102]
			0.20±0.20	Horizontal anal.	Zollweg [130]
			0.27	Horizontal anal.	Crossley [129]

TABLE 7-5-2 (continued)

Atomic no.	Ion	State	Affinity	Technique	References
			0.28	Horizontal anal.	Edie [128]
			0.52	Ionization extrap.	Edlén [117]
			>0	Mass spectrum	Khvostenko [196]
			>0	Mass spectrum	Branscomb [183]
		^1D	0.23	H. F.	Clementi [102]
14	Si	$3p^3\ ^4S^0_{3/2}$	1.39	H. F.	Clementi [102]
			1.36±0.20	Horizontal anal.	Zollweg [130]
			1.40	Horizontal anal.	Crossley [129]
			1.36	Horizontal anal.	Edie [128]
			1.46	Ionization extrap.	Edlén [117]
			>0	Mass spectrum	Dukelskii [197]
		^2D	0.58	H. F.	Clementi [102]
		^2P	0.08	H. F.	Clementi [102]
15	P	$3p^4\ ^3P_2$	0.78	H. F.	Clementi [102]
			0.71±0.20	Horizontal anal.	Zollweg [130]
			0.62	Horizontal anal.	Crossley [129]
			0.72	Horizontal anal.	Edie [128]
			0.77	Ionization extrap.	Edlén [117]
			>0	Mass spectrum	Branscomb [183]
		^1D	0.01	H. F.	Clementi [102]
16	S	$3p^5\ ^2P^0_{3/2}$	2.0772±0.0005	Beam	Lineberger [20]
			2.095±0.015	Beam	Steiner [84]
			2.07±0.07	Beam	Branscomb [183]
			2.090±0.065	Thermochem.	Ansdell [188]
			2.12	H. F.	Clementi [102]
			2.04±0.20	Horizontal anal.	Zollweg [130]
			2.03	Horizontal anal.	Crossley [129]
			2.08	Horizontal anal.	Edie [128]
			2.15	Ionization extrap.	Edlén [117]
17	Cl	$3p^6\ ^1S_0$	3.613±0.003	Plasma	Berry [63]
			3.616±0.003	Plasma	Mück [66]
			3.628±0.005	Plasma	Berry [67]
			3.56	H. F.	Clementi [102]
			3.70	Ionization extrap.	Edlén [117]
19	K	$4s^2\ ^1S_0$	0.47±0.10	H. F.	Weiss [105]
			0.90±0.05	H. F.	Clementi [103]
			0.40±0.20	Horizontal anal.	Zollweg [130]
			$0.22^{+0.08}_{-0.06}$	Charge exchange	Bydin [195]
			0.5	Ionization extrap.	Charkin [114]
20	Ca	$3d\ 4s^2\ ^2D$	−1.88±0.20	Horizontal anal.	Zollweg [130]
			−1.6	Ionization extrap.	Charkin [114]
21	Sc	$3d^2\ 4s^2\ ^3F$	−0.14±0.10	H. F.	Clementi [103]

TABLE 7-5-2 (continued)

Atomic no.	Ion	State	Affinity	Technique	References
			-0.70 ± 0.20	Horizontal anal.	Zollweg [130]
			-0.4	Ionization extrap.	Charkin [114]
22	Ti	$3d^3\,4s^2\,{}^4F$	0.39 ± 0.20	H. F.	Clementi [103]
			-0.03 ± 0.20	Horizontal anal.	Zollweg [130]
			0.15	Ionization extrap.	Charkin [114]
23	V	$3d^4\,4s^2\,{}^5D$	0.94 ± 0.25	H. F.	Clementi [103]
			0.56 ± 0.20	Horizontal anal.	Zollweg [130]
			0.65	Ionization extrap.	Charkin [114]
24	Cr	$3d^5\,4s^2\,{}^6S$	0.98 ± 0.35	H. F.	Clementi [103]
			0.85 ± 0.20	Horizontal anal.	Zollweg [130]
			0.85	Ionization extrap.	Charkin [114]
25	Mn	$3d^6\,4s^2\,{}^5D$	-1.07 ± 0.20	H. F.	Clementi [103]
			-1.07 ± 0.20	Horizontal anal.	Zollweg [130]
			-1.2	Ionization extrap.	Charkin [114]
26	Fe	$3d^7\,4s^2\,{}^4F$	0.58 ± 0.20	H. F.	Clementi [103]
			0.40 ± 0.20	Horizontal anal.	Zollweg [130]
			0.1	Ionization extrap.	Charkin [114]
27	Co	$3d^8\,4s^2\,{}^3F$	0.94 ± 0.15	H. F.	Clementi [103]
			1.03 ± 0.20	Horizontal anal.	Zollweg [130]
			0.7	Ionization extrap.	Charkin [114]
28	Ni	$3d^9\,4s^2\,{}^2D$	1.28 ± 0.20	H. F.	Clementi [103]
			1.62 ± 0.20	Horizontal anal.	Zollweg [130]
			1.1	Ionization extrap.	Charkin [114]
29	Cu	$3d^{10}\,4s^2\,{}^1S$	1.80 ± 0.10	H. F.	Clementi [103]
			1.80 ± 0.20	Horizontal anal.	Zollweg [130]
			1.5 ± 0.5	Thermochem.	Bakulina [198]
			1.4	Ionization extrap.	Charkin [114]
30	Zn	$5s\,{}^2S_{\frac{1}{2}}$ (metastable)	0.09 ± 0.20	Horizontal anal.	Zollweg [130]
31	Ga	$4p^2\,{}^3P_0$	0.37 ± 0.20	Horizontal anal.	Zollweg [130]
			0.50	Radius extrap.	Politzer [131]
			>0	Mass spectrum	Khvostenko [196]
32	Ge	$4p^3\,{}^4S^o_{\frac{3}{2}}$	1.44 ± 0.20	Horizontal anal.	Zollweg [130]
			1.37	Vertical anal.	Politzer [131]
			>0	Mass spectrum	Dukel'skii [197]
33	As	$4p^4\,{}^3P_2$	1.07 ± 0.20	Horizontal anal.	Zollweg [130]
			0.74	Vertical anal.	Politzer [131]
34	Se	$4p^5\,{}^2P^o_{\frac{3}{2}}$	2.12 ± 0.20	Horizontal anal.	Zollweg [130]

TABLE 7-5-2 (continued)

Atomic no.	Ion	State	Affinity	Technique	References
			2.11	Vertical anal.	Politzer [131]
			5.4 ± 2.0	e attachment	Prichard [15]
35	Br	$4p^6\ ^1S_0$	3.363 ± 0.003	Plasma	Berry [63]
			3.378 ± 0.005	Plasma	Berry [67]
			3.53 ± 0.12	Photoionization	Morrison [136]
37	Rb	$5s^2\ ^1S$	0.40 ± 0.20	Horizontal anal.	Zollweg [130]
			0.16 ± 0.06	Charge exchange	Bydin [195]
			>0.20	Photoionization	Lee [137]
			0.6	Ionization extrap.	Charkin [114]
38	Sr	$4d\ 5s^2\ ^2D$	-1.33 ± 0.20	Horizontal anal.	Zollweg [130]
			-0.5	Ionization extrap.	Charkin [114]
39	Y	$4d^2\ 5s^2\ ^3F$	-0.15 ± 0.20	Horizontal anal.	Zollweg [130]
			0.3	Ionization extrap.	Charkin [114]
40	Zr	$4d^3\ 5s^2\ ^4F$	0.62 ± 0.20	Horizontal anal.	Zollweg [130]
			1.0	Ionization extrap.	Charkin [114]
41	Nb	$4d^4\ 5s^2\ ^5D$	1.23 ± 0.20	Horizontal anal.	Zollweg [130]
			1.3	Ionization extrap.	Charkin [114]
42	Mo	$4d^5\ 5s^2\ ^6S$	1.0 ± 0.2	Thermochem.	Scheer [147, 149]
			1.15	Horizontal anal.	Zollweg [130]
			1.3	Ionization extrap.	Charkin [114]
43	Tc	$4d^6\ 5s^2\ ^5D$	0.94 ± 0.20	Horizontal anal.	Zollweg [130]
			1.0	Ionization extrap.	Charkin [114]
44	Ru	$4d^7\ 5s^2\ ^4F$	1.49 ± 0.20	Horizontal anal.	Zollweg [130]
			1.45	Ionization extrap.	Charkin [114]
45	Rh	$4d^8\ 5s^2\ ^3F$	1.68 ± 0.20	Horizontal anal.	Zollweg [130]
			1.35	Ionization extrap.	Charkin [114]
46	Pd	$4d^9\ 5s^2\ ^2D$	1.03 ± 0.20	Horizontal anal.	Zollweg [130]
			1.4	Ionization extrap.	Charkin [114]
47	Ag	$4d^{10}\ 5s^2\ ^1S$	2.0 ± 0.2	Thermochem.	Bakulina [198]
			2.00 ± 0.20	Horizontal anal.	Zollweg [130]
			1.5	Ionization anal.	Charkin [114]
48	Cd	$5s^2\ 5p\ ^2P^o_{1/2}$	-0.78 ± 0.20	Horizontal anal.	Zollweg [130]
		$5s^2\ 6s\ ^2S_{1/2}$	-0.27 ± 0.20	Horizontal anal.	Zollweg [130]
49	In	$5p^2\ ^3P_0$	0.20 ± 0.20	Horizontal anal.	Zollweg [130]
			0.72	Vertical anal.	Politzer [131]
			>0	Mass spectrum	Khvostenko [196]
50	Sn	$5p^3\ ^4S^o_{3/2}$	1.03 ± 0.20	Horizontal anal.	Zollweg [130]
			1.47	Vertical anal.	Politzer [131]
			>0	Mass spectrum	Dukel'skii [197]

ELECTRON AFFINITIES

TABLE 7-5-2 (continued)

Atomic no.	Ion	State	Affinity	Technique	References
51	Sb	$5p^4\ ^3P_2$	0.94 ± 0.20	Horizontal anal.	Zollweg [130]
			0.61	Vertical anal.	Politzer [131]
			>0	Mass spectrum	Dukel'skii [199]
52	Te	$5p^5\ ^2P^o_{3/2}$	1.96 ± 0.20	Horizontal anal.	Zollweg [130]
			2.21	Vertical anal.	Politzer [131]
			>0	Mass spectrum	Dukel'skii [199]
53	I	$5p^6\ ^1S_0$	3.059 ± 0.002	Beam	Steiner [83]
			3.063 ± 0.003	Plasma	Berry [63]
			3.078 ± 0.005	Plasma	Berry [67]
			$\leq 3.076 \pm 0.005$	Beam	Steiner [71]
			3.13 ± 0.12	Photoionization	Morrison [136]
55	Cs	$6s^2\ ^1S$	0.40 ± 0.30	Hroizontal anal.	Zollweg [130]
			$0.13^{+0.07}_{-0.06}$	Charge exchange	Bydin [195]
			>0.19	Photoionization	Lee [137]
56	Ba	$5d\ 6s^2\ ^2D$	-0.48 ± 0.30	Horizontal anal.	Zollweg [130]
57	La	$5d^2\ 6s^2\ ^3F$	0.55 ± 0.30	Horizontal anal.	Zollweg [130]
72	Hf	$5d^3\ 6s^2\ ^4F$	-0.63 ± 0.30	Horizontal anal.	Zollweg [130]
73	Ta	$5d^4\ 6s^2\ ^5D$	0.8 ± 0.3	Surface ionization	Scheer [147]
			0.15 ± 0.30	Horizontal anal.	Zollweg [130]
74	W	$5d^5\ 6s^2\ ^6S$	0.5 ± 0.3	Surface ionization	Scheer [147, 150, 151]
			1.23 ± 0.30	Horizontal anal.	Zollweg [130]
75	Re	$5d^6\ 6s^2\ ^5D$	0.15 ± 0.10	Surface ionization	Scheer [147, 150]
			0.38 ± 0.30	Horizontal anal.	Zollweg [130]
76	Os	$5d^7\ 6s^2\ ^4F$	1.44 ± 0.30	Horizontal anal.	Zollweg [130]
77	Ir	$5d^8\ 6s^2\ ^3F$	1.97 ± 0.30	Horizontal anal.	Zollweg [130]
78	Pt	$5d^9\ 6s^2\ ^2D$	2.56 ± 0.30	Horizontal anal.	Zollweg [130]
79	Au	$5d^{10}\ 6s^2\ ^1S$	2.8 ± 0.1	Surface ionization	Bakulina [198]
			2.80 ± 0.30	Horizontal anal.	Zollweg [130]
80	Hg	$6s^2\ 6p\ ^2P^o_{1/2}$	-0.67 ± 0.30	Horizontal anal.	Zollweg [130]
		$6s^2\ 7s\ ^2S_{1/2}$	-0.19 ± 0.30	Horizontal anal.	Zollweg [130]
81	Tl	$6s^2\ 6p^2\ ^3P_0$	0.5 ± 0.1	e attachment	Khvostenko [200]
			0.32 ± 0.30	Horizontal anal.	Zollweg [130]
			1.21	Vertical anal.	Politzer [131]
			>0	Mass spectrum	Khvostenko [196]
82	Pb	$6p^3\ ^4S^o_{3/2}$	1.03 ± 0.30	Horizontal anal.	Zollweg [130]
			1.79	Vertical anal.	Politzer [131]
			>0	Mass spectrum	Dukel'skii [197]
83	Bi	$6p^4\ ^3P_2$	0.95 ± 0.30	Horizontal anal.	Zollweg [130]
			−0.34	Vertical anal.	Politzer [131]
			>0	Mass spectrum	Dukel'skii [199]
84	Po	$6p^5\ ^2P^o_{3/2}$	1.32 ± 0.30	Horizontal anal.	Zollweg [131]
			1.97	Vertical anal.	Politzer [131]

TABLE 7-5-3

Molecular electron affinities

Ion	State	Affinity	Technique	References
H_2^-	$^2\Sigma_u^+$	−2.5	RKR–CI	Sharp [201]
		<0	CI	Taylor [202]
		<0	CI	Somerville [203]
		<0	CI	Taylor [204]
		<0	CI	Dalgarno [205]
		>0	Mass spectrum	Carter [206]
		>0	Mass spectrum	Khvostenko [207]
C_2^-	$^2\Sigma^+$	≦3.54±0.05	Beam	Feldman [48]
		4.0	Thermochem.	Honig [146]
C_3^-		2.5	Thermochem.	Honig [146]
N_2^-	Π_g	−1.6±1.0	e scattering	Gilmore [208]
N_3^-		∽3.5±0.2	Spectrum	Gray [167]
		∽2.3		Buchner [209]
O_2^-	$^3\Sigma_g^-$	0.430±0.030	Beam	Celotta [97]
		0.43±0.02	Thermochem.	Pack [210]
		0.15±0.05	Beam	Burch [72]
		≧0.21	Photoionization	Samson [138]
O_3^-		1.92±0.13	Plasma	Wong [211]
		1.91±0.43		Wood [212]
		∽>1.5	Spectrum	Herman [171]
O_4^-		>0	Mass spectrum	Conway [213]
S_2^-		≈>2.0	e attachment	Jäger [214]
Cl_2^-		<1.7	e attachment	Baker [215]
		≈2.54	Dissociation energy	Geiger [216]
Se_2^-		>0	Mass spectrum	Dukel'skii [199]
Se_3^-		>0	Mass spectrum	Dukel'skii [199]
Se_4^-		>0	Mass spectrum	
Br_2^-		>0	Mass spectrum	Blewett [217]
		≈2.62	Dissociation energy	Geiger [216]
Sb_2^-		>0	Mass spectrum	Dukel'skii [199]
Sb_3^-		>0	Mass spectrum	Dukel'skii [199]
Te_2^-		>0	Mass spectrum	Dukel'skii [199]
I_2^-		>0	Mass spectrum	Hogness [218]
		≈2.42	Dissociation energy	Geiger [216]
I_3		>0	Mass spectrum	Hogness [218]
Bi_2^-		>0	Mass spectrum	Dukel'skii [199]
Bi_3^-		>0	Mass spectrum	
BeH^-		∽1.0_0	Atom affinity	Gaines [219]
BH^-		∽0.09	Atom affinity	Gaines [219]
CH^-	$^3\Sigma^-$	0.74±0.05	Beam	Feldman [48]
		1.61±0.20	H. F.	Cade [110]
NH^-	$^2\Pi$	0.22±0.20	H. F.	Cade [110]
OH^-	$^1\Sigma^+$	1.83±0.04	Beam	Branscomb [73]
		1.91±0.10	H. F.	Cade [111]
MgH^-		∽1.08	Atom affinity	Gaines [219]
AlH^-		∽0.04	Atom affinity	Gaines [219]

7-5 ELECTRON AFFINITIES

TABLE 7-5-3 (continued)

Ion	State	Affinity	Technique	References
SiH^-	$^3\Sigma^-$	1.46 ± 0.20	H. F.	Cade [110]
PH^-	$^2\Pi$	0.93 ± 0.20	H. F.	Cade [110]
SH^-	$^1\Sigma^+$	2.319 ± 0.010	Beam	Steiner [74]
		2.28 ± 0.15	Resonance	Brauman [76]
		2.25 ± 0.10	H. F.	Cade [111]
		2.30 ± 0.04	Thermochem.	Ansdell [188]
SeH^-	$^1\Sigma^+$	>0	Mass spectrum	Neuert [220]
		~2.0	Born–Haber	Pritchard [15]
BO^-		≥2.5	Thermochem.	Jensen [161]
		~2.12	Atom affinity	Gaines [219]
CN^-	$^1\Sigma^+$	3.82 ± 0.02	Photoionization	Berkowitz [2]
		$\sim3.6\pm0.3$	e attachment	Berkowitz [2]
		3.17 ± 0.04	Thermochem.	Page [221]
CO^-		$<-1.8\pm0.1$	e scattering	Rempt [222]
NO^-	$^3\Sigma_g^-$	0.70 ± 0.30	Beam	Siegel [223]
		≤0.430	Charge exchange	Fehsenfeld [169]
		~0.15	Isoelectronic model	Gilmore [208]
		0.030 ± 0.030	Beam	Celotta [224]
		0.070 ± 0.030	Beam	Siegel [225]
		0.89 ± 0.11	Thermochem.	Farragher [226]
		0.85	e impact	Williams [227]
AlO^-		~2.60	Atom affinity	Gaines [219]
PO^-	$^3\Sigma^-$	<1.13	H. F.	Boyd [228]
SO^-		$\leq1.09\pm0.05$	Beam	Feldman [48]
SF^-	$^1\Sigma$	$\leq2.5\pm0.5$	H. F.	O'Hare [229]
SeF^-	$^1\Sigma$	$\leq2.8\pm0.5$	H. F.	O'Hare [229]
BH_2^-		~1.39	Atom affinity	Gaines [219]
NH_2^-		0.6	e attachment	Dorman [230]
		1.21	Thermochem.	Page [231]
OH_2^-		~0.9	Spectrum	Gray [167]
AlH_2^-		~2.12	Atom affinity	Gaines [219]
SiH_2^-		~3.47	Atom affinity	Gaines [219]
PH_2		>0	Mass spectrum	Neuert [220]
		~1.60	Atom affinity	Gaines [219]
SH_2^-		1.11 ± 0.09	Thermochem.	Ansdell [188]
C_2H^-		$\leq3.73\pm0.05$	Beam	Feldman [48]
HF_2^-		~3	H. F.	Noble [232]
HO_2^-		~4.6	Born–Haber	Weiss [233]
HSi_2		~4.08	Atom affinity	Gaines [219]
BO_2		4.07 ± 0.21	Thermochem.	Jensen [162]
CNS^-		1.99	Thermochem.	Page [231]
CO_2^-		>-0.9	e scattering	Bardsley [234]
N_2O^-		<1.5	Charge exchange	Paulson [170]
		>0.5	e attachment	Ferguson [235]
		$\sim>0$	Charge exchange	Ferguson [27]
		≈0	e attachment	Chantry [143]
		≈0	Charge exchange	Chantry [236]
		≈0	e attachment	Bardsley [234]

TABLE 7-5-3 (continued)

Ion	State	Affinity	Technique	References
NO_2^-		3.10±0.05	Beam	Warneck [75]
		>3.6	Charge exchange	Curran [237]
		3.99±0.16	Thermochem.	Farragher [226]
$SiCl_2^-$		≥2.6	e attachment	Vought [238]
SO_2^-		<1.0±0.1	Beam	Feldman [48]
BeH_3^-		3.8	H. F.	Joshi [239]
CH_3^-		~1.8	Dipole moment extrap.	Sklar [240]
SiH_3		~2.73	Atom affinity	Gaines [219]
CCl_3^-		≧2.10±0.35	e attachment	Curran [241]
		1.6±0.2	e attachment	Reese [242]
		1.44±0.05	Thermochem.	Gaines [243]
CO_3^-		>0	Mass spectrum	Moruzzi [244]
WO_3^-		≈3.6	Thermochem.	Jensen [160]
$Si_2H_3^-$		~3.17	Atom affinity	Gaines [219]
$CHCl_3^-$		1.76±0.05	Thermochem.	Gaines [243]
CO_4^-		1.22±0.07	Thermochem.	Pack [245]
		>0	Mass spectrum	Moruzzi [244]
CCl_4^-		2.12±0.10	Thermochem.	Gaines [243]
HWO_4^-		4.4	Thermochem.	Jensen [160]
SF_5^-		3.66±0.04	Thermochem.	Kay [246]
SF_6^-		1.49±0.22	Thermochem.	Kay [246]
$O(H_2O)^-$		>0	Mass spectrum	Moruzzi [244]
$OH(H_2O)^-$		2.95±0.15	Beam	Golub [31]
		3.35±0.15	Collision	De Paz [30]
$O_2(H_2O)^-$		>0	Mass spectrum	Moruzzi [244]

molecular ions it is probably no less reliable than most other techniques used to obtain comparable results.

Knowledge of atomic electron affinities has been used with assumptions concerning the chemical binding energy to generate estimates of polyatomic electron affinities. Thus, Mulliken's assumption that the ionic chemical *dissociation energy* is approximately equal to one-half of that for the neutral molecule and knowledge of atomic electron affinities has led Geiger [216] to estimates of molecular halogen affinities. If these results are to be employed, they should be corrected for more recent values of the data used. In contrast to this approach, Gaines and Page assume that the dissociation energies of ion and neutral species are identical and determine the molecular affinity directly from the *atomic affinity* defined for the appropriate electronic state. Photodetachment with ion cyclotron *resonance* has been mentioned earlier in the chapter. Electron *scattering* has also been described. Gilmore has used an *isoelectronic model* for information on dissociation energies to derive molecular affinities. Williams and Hamill [210] believed that they observed pair production under electron *impact*. Sklar [240] has observed regularity in the dipole moment of a diatomic molecular bond with respect to the affinity of various subsituent members of the atom pair and extrapolated to zero *dipole moment*. Heavy particle *collisions* have also been used to determine bond energies, and these in turn can be used to determine the molecular affinity if the affinity of one of the separated species is known.

In summary, the agreement among selected methods of atomic electron affinity determinations is now substantially better than agreement among molecular affinity determinations. In many cases, there is broad agreement within 0.1 eV. There are still a number of atoms lower in Table 7-5-2, however, with 0.5 eV lack of agreement.

The situation with regard to molecular affinities is still very unsatisfactory. The general level of agreement is poor. Greater accuracy now will require improvement not only in experimental and computational techniques but also in understanding of the molecular dynamics, in particular the role of rotations and vibrations.

Acknowledgment

I am grateful to Alex Mandl, Robert Celotta, and Jan Hall for communication of data and experimental details prior to publication.

References

1. C. E. MOORE, Atomic Energy Levels, U.S. NBS Circular 467 (1952–1958).
2. J. BERKOWITZ, W. A. CHUPKA and T. A. WALTER, J. Chem. Phys. **50** (1969) 1497.
3. R. N. COMPTON and T. L. BAILEY, J. Chem. Phys. **53** (1970) 454.
4. T. L. BAILEY and P. MAHADEVAN, J. Chem. Phys. **52** (1970) 179.
5. M. J. WYNN, J. D. MARTIN and T. L. BAILEY, J. Chem. Phys. **52** (1970) 191.
6. J. B. HASTED, in: Atomic and Molecular Processes, ed. D. R. Bates (Academic Press, New York 1962) Ch. 18.
7. L. FROMMHOLD, Fortschritte der Physik **12** (1964) 597.
8. L. M. BRANSCOMB, in: Atomic and Molecular Processes, ed. D. R. Bates (Academic Press, New York 1962) Ch. 4.
9. R. S. BERRY, Chem. Revs. **69** (1969) 533.
10. B. L. MOISEIWITSCH, Advan. Atom. Molec. Phys. **1** (1965) 61.
11. B. M. SMIRNOV, Teplofiz. Vysok. Temp. **3** (1965) 775 [Engl. transl.: High Temperature **3** (1965) 716].
12. L. M. BRANSCOMB, Ann. Geophys. **20** (1964) 88.
13. V. I. VENDENEYEV, L. V. GURVICH, V. N. KONDRAT'YEV, V. A. MEDVEDEV and YE. L. FRANKEVICH, Energiya razryva khimicheskikh svyazei: Potentsialy ionizatsii i srodstvo k elektronu (U.S.S.R. Acad. Sci.) [Engl. transl.: Bond energies, ionization potentials and electron affinities (St. Martin's Press, New York 1966)].

14. N. S. BUCHEL'NIKOVA, Usp. Fiz. Nauk **65** (1958) 351 [Engl. transl.: p. 243].
15. H. O. PRITCHARD, Chem. Revs. **52** (1953) 529.
16. H. S. W. MASSEY, Negative Ions (Cambridge University Press, 1950).
17. J. F. PAULSON, J. Chem. Phys. **52** (1970) 5491.
18. J. L. HALL and M. W. SIEGEL, J. Chem. Phys. **48** (1968) 943.
19. M. L. SEMAN and L. M. BRANSCOMB, Phys. Rev. **125** (1962) 1602.
20. W. E. LINEBERGER and B. W. WOODWARD, Phys. Rev. Leters **25** (1970) 424.
21. R. S. BERRY, J. C. MACKIE, R. L. TAYLOR and R. LYNCH, J. Chem. Phys. **43** (1965) 3067.
22. L. M. BLAU, R. NOVICK and D. WEINFLASH, Phys. Rev. Letters **24** (1970) 1268.
23. J. A. D. STOCKDALE, R. N. COMPTON and H. C. SCHWEINLER, J. Chem. Phys. **53** (1970) 1502.
24. R. N. COMPTON, L. G. CHRISTOPHOROU, G. S. HURST and P. W. REINHARDT, J. Chem. Phys. **45** (1966) 4634.
25. D. EDELSON, J. E. GRIFFITHS and K. B. MCAFEE JR., J. Chem. Phys. **37** (1962) 917.
26. W. M. HICKAM and R. E. FOX, J. Chem. Phys. **25** (1956) 642.
27. E. E. FERGUSON, F. C. FEHSENFELD and A. L. SCHMELTEKOPF, J. Chem. Phys. **47** (1967) 3085.
28. J. N. BEARDSLEY and F. MANDL, Repts. Prog. Phys. **31** (1968) 471.
29. M. DEPAZ, S. EHRENSON and L. FRIEDMAN, J. Chem. Phys. **52** (1970) 3362.
30. M. DEPAZ, A. G. GIARDINI and L. FRIEDMAN, J. Chem. Phys. **52** (1970) 687.
31. S. GOLUB and B. W. STEINER, J. Chem. Phys. **49** (1968) 5191.
32. P. KEBARLE, A. ARSHADI and J. SCARBOROUGH, J. Chem. Phys. **49** (1968) 817.
33. J. L. MORUZZI and A. V. PHELPS, J. Chem. Phys. **45** (1966) 4617.
34. L. M. BRANSCOMB, S. J. SMITH and G. TISONE, J. Chem. Phys. **43** (1965) 2906.
35. S. J. SMITH and D. S. BURCH, Phys. Rev. **116** (1959) 1125.
36. S. J. SMITH and D. S. BURCH, Phys. Rev. Letters **2** (1959) 165.
37. A. BÖHM and L. REHDER, Z. Naturforsch. **20.** (1965) 114.
38. N. A. DOUGHTY, P. A. FRASER and R. P. MCEACRHAN, Mon. Notices Roy. Astron. Soc. **132** (1966) 255.
39. S. GELTMAN, Astrophys. J. **136** (1962) 935.
40. K. L. BELL and A. E. KINGSTON, Proc. Phys. Soc. **90** (1967) 895.
41. T. L. JOHN, Mon. Notices Roy. Aston. Soc. **121** (1960) 41.
42. W. S. KROGDAHL and J. E. MILLER, Astrophys. J. **150** (1967) 273.
43. J. MACEK, Proc. Phys. Soc. **92** (1967) 365.
44. YU. V. MOSKVIN, Teplofiz. Vysokikh Temperatur **3** (1965) 821 [Engl. transl.: High Temp. **3**, 765].
45. B. YA'AKOBI, Phys. Rev. **184** (1969) 246.
46. V. A. ZHIRNOV, Zh. Eksp. Teoret. Fiz. **42** (1962) 1097 [Engl. transl.: Soviet Phys. JETP **15** (1962) 758].
47. S. GELTMAN, Phys. Rev. **104** (1956) 346.
48. D. FELDMAN, Z. Naturforsch. **25a** (1970) 621.
49. YU. V. MOSKVIN, Opt. i. Spektroskopiya **17** (1964) 499 [Engl. transl.: Opt. Spectry (USSR) **17**, 270].
50. R. J. W. HENRY, Phys. Rev. **172** (1968) 99.
51. E. J. ROBINSON and S. GELTMAN, Phys. Rev. **153** (1967) 4.
52. J. W. COOPER and J. B. MARTIN, Phys. Rev. **126** (1962) 1482.
53. V. P. MYERSCOUGH and M. R. C. MCDOWELL, Mon. Notices Roy. Astron. Soc. **128** (1964) 287.
54. V. P. MYERSCOUGH, Proc. Phys. Soc. **85** (1965) 33.
55. S. J. SMITH, Proc. 4th Intern. Conf. Ioniz. Phen. Gases, Uppsala, 1959, IC 219 (1960).
56. L. M. BRANSCOMB, D. S. BURCH, S. J. SMITH and S. GELTMAN, Phys. Rev. **111** (1958) 504.

57. R. J. W. HENRY, Phys. Rev. **162** (1967) 56.
58. W. R. GARRETT and H. T. JACKSON JR., Phys. Rev. **153** (1967) 28.
59. J. GILLESPIE, Phys. Rev. **135** (1964) A75.
60. R. G. BREENE JR., J. Quant. Spectr. Radiative Transfer **5** (1965) 449.
61. B. SCHNEIDER, M. WEINBERG, J. TULLY and R. S. BERRY, Phys. Rev. **182** (1969) 133.
62. A. MANDL, Phys. Rev., in press.
63. R. S. BERRY and C. W. REIMANN, J. Chem. Phys. **38** (1963) 1540.
64. H. P. POPP, Z. Naturforsch **22a** (1967) 254.
65. D. E. ROTHE, Phys. Rev. **177** (1969) 93.
66. G. MÜCK and H. P. POPP, Z. Naturforsch. **23**a (1968) 1213.
67. R. S. BERRY, C. W. REIMANN and G. N. SPOKES, J. Chem. Phys. **37** (1962) 2278.
68. B. B. ROBINSON, Phys. Rev. **140** (1965) A764.
69. R. S. BERRY, T. CERNOCH, M. COPLAN and J. J. EWING, J. Chem. Phys. **49** (1968) 127.
70. B. W. STEINER, Phys. Rev. **173** (1968) 136.
71. B. W. STEINER, M. L. SEMAN and L. M. BRANSCOMB, J. Chem. Phys. **37** (1962) 1200.
72. D. S. BURCH, S. J. SMITH and L. M. BRANSCOMB, Phys. Rev. **112** (1958) 171.
73. L. M. BRANSCOMB, Phys. Rev. **148** (1966) 11.
74. B. W. STEINER, J. Chem. Phys. **49** (1968) 5097.
75. P. WARNECK, Chem. Phys. Letters **3** (1969) 532.
76. J. I. BRAUMAN and K. C. SMYTH, J. Am. Chem. Soc. **91** (1969) 7778.
77. J. COOPER and R. N. ZARE, J. Chem. Phys. **48** (1968) 942.
78. J. L. HALL, E. J. ROBINSON and L. M. BRANSCOMB, Phys. Rev. Letters **14** (1965) 1013.
79. S. GELTMAN, Phys. Letters **4** (1963) 168.
80. E. P. WIGNER, Phys. Rev. **73** (1948) 1002.
81. T. F. O'MALLEY, Phys. Rev. **137** (1965) A1668.
82. S. GELTMAN, Phys. Rev. **112** (1958) 176.
83. B. W. STEINER, M. L. SEMAN and L. M. BRANSCOMB, in: Atomic Collision Processes, ed. M. R. C. McDowell (Amsterdam, North-Holland, 1964) p. 537.
84. B. W. STEINER, Proc. 6th Intern. Conf. Physics Electronic and Atomic Collisions (Cambridge, MIT Press, 1969) p. 535.
85. H. P. POPP, Z. Naturforsch **20a** (1965) 642.
86. R. S. BERRY, C. W. DAVID and J. C. MACKIE, J. Chem. Phys. **42** (1965) 1541.
87. G. P. LAWRENCE, R. K. BEAUCHAMP and J. L. MCKIBBEN, Nucl. Inst. Methods **32** (1965) 357.
88. W. ABERTH and J. R. PETERSON, Rev. Sci. Instr. **38** (1967) 745.
89. B. W. STEINER, unpublished work.
90. R. H. V. M. DAWTON, Nucl. Inst. Methods **67** (1969) 341.
91. R. H. V. M. DAWTON, Nucl. Inst. Methods **24** (1963) 285.
92. G. HORTIG, P. MOKLER and M. MÜLLER, Z. Physik **210** (1968) 312.
93. B. BREHM, M. A. GUSINOW and J. L. HALL, Phys. Rev. Letters **19** (1967) 737.
94. A. SEPTIER, CERN Report No. 60-39 (1960).
95. H. HINTENBERGER and L. A. KÖNIG, Z. Naturforsch. **12a** (1957) 773.
96. C. E. KUYATT, Methods of Experimental Phys. **7A** (Academic Press, New York, 1968) p. 1.
97. R. CELOTTA, R. BENNETT, J. HALL, J. LEVINE and M. W. SIEGEL, Proc. 23rd Gas. Elect. Conf. (Hartford, 1970) p. 80.
98. J. A. HALL, private communication.
99. C. L. PEKERIS, Phys. Rev. **126** (1962) 1470.
100. C. L. PEKERIS, Phys. Rev. **112** (1958) 1649.
101. E. CLEMENTI and A. D. MCLEAN, Phys. Rev. **133** (1964) A419.
102. E. CLEMENTI, A. D. MCLEAN, D. L. RAIMONDI and M. YOSHIMINE, Phys. Rev. **133** (1964) A1274.

103. E. CLEMENTI, Phys. Rev. **135** (1964) A980.
104. A. W. WEISS, Phys. Rev. **A2** (1971) 126.
105. A. W. WEISS, Phys. Rev. **166** (1968) 70.
106. I. ÖKSÜZ and O. SINANOĞLU, Phys. Rev. **181** (1969) 54.
107. O. SINANOĞLU and I. ÖKSÜZ, Phys. Rev. Letters **21** (1968) 507.
108. H. F. SCHAEFER, R. A. KLEMM and F. E. HARRIS, J. Chem. Phys. **51** (1969) 4643.
109. H. F. SCHAEFER and F. E. HARRIS, Phys. Rev. **170** (1968) 108.
110. P. E. CADE, Proc. Phys. Soc. (London) **91** (1967) 842.
111. P. E. CADE, J. Chem. Phys. **47** (1967) 2390.
112. G. GLOCKLER, Phys. Rev. **46** (1934) 111.
113. D. R. BATES, Proc. Roy. Irish Acad. **A51** (1947) 151.
114. O. P. CHARKIN and M. E. DYATKINA, Zh. Strukt. Khim. **6** (1965) 422 [Engl. transl.: J. Struct. Chem. (USSR) **6** (1966) 397].
115. S. GELTMAN, J. Chem. Phys. **25** (1956) 782.
116. H. R. JOHNSON and F. ROHRLICH, J. Chem. Phys. **30** (1959) 1608.
117. B. EDLÉN, J. Chem. Phys. **33** (1960) 98.
118. M. KAUFMAN, Astrophys. J. **147** (1963) 1296.
119. D. R. BATES and B. L. MOISEIWITSCH, Proc. Phys. Soc. (London) **A68** (1955) 540.
120. H. HELLMANN and M. MAMOTENKO, Acta Physicochem. URSS **7** (1937) 127.
121. J. L. MARGRAVE, J. Chem. Phys. **22** (1954) 636.
122. J. L. MARGRAVE, J. Chem. Phys. **22** (1954) 1937.
123. H. O. PRITCHARD and H. A. SKINNER, J. Chem. Phys. **22** (1954) 1936.
124. R. F. BACHER and S. GOUDSMIT, Phys. Rev. **46** (1934) 948.
125. TA-YOU WU, Phil. Mag. **22** (1936) 837.
126. TA-YOU WU, Phys. Rev. **100** (1955) 1195.
127. A. P. GINSBERG and J. M. MILLER, J. Inorg. Nucl. Chem. **7** (1958) 351.
128. J. W. EDIE and F. ROHRLICH, J. Chem. Phys. **36** (1962) 623; erratum: **37** (1962) 1151.
129. R. J. S. CROSSLEY, Proc. Phys. Soc. (London) **83** (1964) 375.
130. R. J. ZOLLWEG, J. Chem. Phys. **50** (1969) 4251.
131. P. POLITZER, Trans. Faraday Soc. **64** (1968) 2241.
132. B. L. MOISEIWITSCH, Proc. Phys. Soc. **A67** (1954) 25.
133. M. J. WOLF, M. SCHWARZSCHILD and W. K. ROSE, Astrophys. J. **140** (1964) 833.
134. V. H. DIBELER, J. A. WALKER and K. E. MCCULLOH, J. Chem. Phys. **51** (1969) 4230.
135. F. A. ELDER, D. VILLAREJO and M. G. INGHRAM, J. Chem. Phys. **43** (1965) 758.
136. J. D. MORRISON, H. HURZELER and M. G. INGHRAM, J. Chem. Phys. **33** (1960) 821.
137. Y. LEE and B. H. MAHAN, J. Chem. Phys. **42** (1965) 2893.
138. J. A. R. SAMSON and R. B. CAIRNS, J. Opt. Soc. Am. **56** (1966) 769.
139. P. J. CHANTRY and G. J. SCHULZ, Phys. Rev. Letters **12** (1964) 449.
140. P. J. CHANTRY and G. J. SCHULZ, Phys. Rev. **156** (1967) 134.
141. G. J. SCHULZ and D. SPENCE, Phys. Rev. Letters **22** (1969) 47.
142. W. R. HENDERSON, W. L. FITE and R. T. BRACKMANN, Phys. Rev. **183** (1969) 157.
143. P. J. CHANTRY, J. Chem. Phys. **51** (1969) 3369.
144. F. H. DORMAN, J. Chem. Phys. **44** (1966) 3856.
145. C. E. KLOTZ, J. Chem. Phys. **46** (1967) 1197.
146. R. E. HONIG, J. Chem. Phys. **22** (1954) 126.
147. D. SCHEER, J. Res. Natl. Bur. of Stads. **74A** (1970) 37.
148. M. D. SCHEER and J. FINE, J. Chem. Phys. **50** (1969) 4343.
149. J. FINE and M. D. SCHEER, J. Chem. Phys. **47** (1967) 4267.
150. M. D. SCHEER and J. FINE, J. Chem. Phys. **46** (1967) 3998.
151. M. D. SCHEER and J. FINE, Phys. Rev. Letters **17** (1966) 283.
152. T. L. BAILEY, J. Chem. Phys. **28** (1958) 792.

153. I. N. BAKULINA and N. I. IONOV, Zh. Fiz. Khim. **39** (1965) 157 [Engl. transl.: p. 78].
154. I. N. BAKULINA and N. I. IONOV, Zh. Fiz. Khim. **33** (1959) 2063 [Engl. transl.: p. 286].
155. I. N. BAKULINA and N. I. IONOV, Dokl. Akad. Nauk. SSSR **99** (1954) 1023.
156. G. M. PYATIGORSKII, Zh. Tekh. Fiz. **35** (1965) 1127 [Engl. transl.: Soviet Phys. - Tech. Phys. **10** (1965) 867].
157. F. M. PAGE and G. C. GOODE, Negative Ions and the Magnetron (Wiley, London 1969).
158. J. T. HERRON, H. M. ROSENSTOCK and W. R. SHIELDS, Nature **206** (1965) 611.
159. B. YA'AKOBI, Phys. Letters **23** (1966) 655.
160. D. E. JENSEN and W. J. MILLER, J. Chem. Phys. **53** (1970) 3287.
161. D. E. JENSEN, J. Chem. Phys. **52** (1970) 3305.
162. D. E. JENSEN, Trans. Faraday Soc. **65** (1969) 2123.
163. M. BORN, Problems of Atom Dynamics (1960) p. 168.
164. C. D. WEST, J. Phys. Chem. **39** (1935) 493.
165. I. D. CAMPBELL and C. K. COOGAN, J. Chem. Phys. **42** (1965) 2738.
166. E. S. GAFNEY and T. J. AHRENS, J. Chem. Phys. **51** (1969) 1088.
167. P. GRAY and T. C. WADDINGTON, Proc. Roy. Soc. **A235** (1956) 481.
168. D. B. DUNKIN, F. C. FEHSENFELD and E. E. FERGUSON, J. Chem. Phys. **53** (1970) 987.
169. F. C. FEHSENFELD, E. E. FERGUSON and A. L. SCHMELTEKOPF, J. Chem. Phys. **45** (1966) 1844.
170. J. F. PAULSON, Adv. in Chem. **58** (1966) 28.
171. Bond data from K. HERMAN and P. A. GIGUÈRE, Can. J. Chem. **43** (1965) 1746.
172. P. E. MILLIGAN, M. E. JACOX and W. A. GUILLORY, J. Chem. Phys. **52** (1970) 3864.
173. B. M. SMIRNOV and M. I. CHIBISOV, Zh. Eksperim. i. Theor. Fiz. **49** (1965) 841 [Engl. transl.: Soviet Phys. JETP **22** (1966) 585].
174. M. J. W. BONESS and G. J. SCHULZ, Phys. Rev. **A2** (1970) 2182.
175. D. SPENCE and G. J. SCHULZ, Phys. Rev. **A2** (1970) 1802.
176. H. SENFTLEBEN, Z. Physik **37** (1926) 539.
177. "Eine ganz rohe Abschätzung": J. E. MEYER and M. M. MALTBIE, Z. Physik **75** (1932) 748.
178. E. LISITZIN, Soc. Sci. Fenn. Comm. Physic. Math. **10** (1938) 4.
179. Ref. 15; dissociative attachment data used from H. Neuert and H. Clasen, Z. Naturforsch. **7a** (1952) 410.
180. H. A. SKINNER and H. O. PRITCHARD, Trans. Faraday Soc. **49** (1953) 1254.
181. H. NEUERT, Z. Naturforsch. **8a** (1953) 459.
182. O. ROSENBAUM and H. NEUERT, Z. Naturforsch. **9a** (1954) 990.
183. L. M. BRANSCOMB and S. J. SMITH, J. Chem. Phys. **25** (1956) 598.
184. I. N. BAKULINA and N. I. IONOV, Dokl. Akad. Nauk. SSSR **1** (1957) 41 [Engl. transl.: **2** (1957) 423. Published value corrected for best bromine affinity, ref. 63].
185. L. M. BRANSCOMB, Advan. Electron. Electron Phys. **9** (1957) 43.
186. P. GOMBÁS and K. LADÁNYI, Z. Physik **158** (1960) 261.
187. E. C. BAUGHAN, Trans. Faraday Soc. **57** (1961) 1863.
188. D. A. ANSDELL and F. M. PAGE, Trans. Faraday Soc. **58** (1962) 1084.
189. G. W. F. DRAKE, Phys. Rev. Letters **24** (1970) 126.
190. E. HOLØIEN and S. GELTMAN, Phys. Rev. **153** (1967) 81.
191. B. EDLEN, in press.
192. K. BETHGE, E. HEINICKE and H. BAUMANN, Phys. Letters **23** (1966) 542.
193. J. HALL, private communication.
194. YA. M. FOGEL, V. F. KOZLOV and A. A. KALMYKOV, Zh. Eksperim. i Teor. Fiz. **36** (1959) 1354 [Engl. transl.: Soviet Phys. JETP **36** (1959) 963].
195. YU. F. BYDIN, Zh. Eksperim. i Teor. Fiz. **46** (1964) 1612 [Engl. transl. Soviet Phys. - JETP **19** (1964) 1091].

196. V. I. Khvostenko and A. Sh. Sultanov, Zh. Eksperim. i Teor. Fiz. **46** (1964) 1605 [Engl. transl.: Soviet Phys. JETP **19** (1964) 1086].
197. V. M. Dukelskii and V. M. Sokolov, Zh. Ekserim. i Teor. Fiz. **32** (1957) 394 [Engl. transl.: Soviet Phys. JETP **5** (1957) 306].
198. I. N. Bakulina and N. I. Ionov, Dokl. Akad. Nauk SSSR **155** (1964) 309 [Engl. transl.: Soviet Phys. – Doklady **9** (1964) 217].
199. V. M. Dukelskii and N. I. Ionov, Doklady Akad. Nauk SSSR **81** (1951) 767.
200. V. I. Khvostenko and A. Sh. Sultanov, Zh. Fiz. Khim. **39** (1965) 475 [Engl. transl.: Russ. J. Phys. Chem. **39** (1965) 252].
201. T. E. Sharp, Lockhead Report LMSC 5-10-69-9.
202. H. S. Taylor, Proc. Phys. Soc. **90** (1967) 877.
203. W. B. Somerville, Proc. Phys. Soc. **89** (1966) 185.
204. H. S. Taylor and F. E. Harris, J. Chem. Phys. **39** (1963) 1012.
205. A. Dalgarno and M. R. C. McDowell, Proc. Phys. Soc. **A69** (1956) 615.
206. E. B. Carter and R. H. Davis, Rev. Sci. Inst. **34** (1963) 93.
207. V. I. Khvostenko and V. M. Dukels'kii, Zh. Eksperim. i Teor. Fiz. **34** (1958) 1026 [Engl. transl.: Soviet Phys. JETP **7** (1958) 709].
208. F. R. Gilmore, J. Quant. Spectry. Radiative Transfer **5** (1965) 369.
209. E. H. Buchner, Rec. Trav. Chim. Pays-Bas **69** (1950) 329.
210. J. L. Pack and A. V. Phelps, J. Chem. Phys. **44** (1966) 1870.
211. S. F. Wong, T. V. Vorburger and S. B. Woo, Proc. 23rd Gas. Elect. Conf. Hartford (1970) p. 82.
212. R. H. Wood and L. A. Dorzaio, J. Phys. Chem. **69** (1965) 2562.
213. D. C. Conway and L. E. Nesbitt, J. Chem. Phys. **48** (1968) 509.
214. K. Jäger and A. Hengelein, Z. Naturforsch. **21a** (1966) 1251.
215. R. F. Baker and J. T. Tate, Phys. Rev. **53** (1938) 683.
216. W. Geiger, Z. Physik **140** (1955) 608.
217. J. P. Blewett, Phys. Rev. **49** (1936) 900.
218. T. R. Hogness and R. W. Harkness, Phys. Rev. **32** (1928) 784.
219. A. F. Gaines and F. M. Page, Trans. Faraday Soc. **62** (1966) 3086.
220. H. Neuert and H. Clasen, Z. Naturforsch. 7a (1952) 410.
221. F. M. Page, J. Chem. Phys. **49** (1968) 2466.
222. R. D. Rempt, Phys. Rev. Letters **22** (1969) 1034.
223. M. W. Siegel, thesis, U. Colorado, 1970.
224. R. Celotta, private communication.
225. M. W. Siegel, R. Celotta, J. Levine and J. L. Hall, Bull. Am. Phys. Soc. **15** (1970) 326; corrected for misprint.
226. A. L. Farragher, F. M. Page and R. C. Wheeler, Disc. Faraday Soc. **37** (1964) 203.
227. J. M. Williams and W. H. Hamill, J. Chem. Phys. **49** (1968) 4467.
228. D. B. Boyd and W. N. Lipscomb, J. Chem. Phys. **46** (1967) 910.
229. P. A. G. O'Hara and A. C. Wahl, J. Chem. Phys. **53** (1970) 2834.
230. F. H. Dorman, J. Chem. Phys. **44** (1966) 3856.
231. F. M. Page, Adv. Chem. **36** (1962) 68.
232. P. N. Noble and R. N. Kortzeborn, J. Chem. Phys. **52** (1970) 5375.
233. J. Weiss, Trans. Faraday Soc. **31** (1935) 966.
234. J. N. Bardsley, J. Chem. Phys. **51** (1969) 3384.
235. E. E. Ferguson, Bull. Am. Phys. Soc. **14** (1969) 266.
236. P. J. Chantry, J. Chem. Phys. **51** (1969) 3380.
237. R. K. Curran, Phys. Rev. **125** (1962) 910, corrected for new chlorine affinity.
238. R. H. Vought, Phys. Rev. **71** (1947) 93.
239. B. D. Joshi, J. Chem. Phys. **46** (1967) 875.
240. A. L. Sklar, J. Chem. Phys. **7** (1939) 984.

241. R. K. Curran, J. Chem. Phys. **34** (1961) 2007.
242. Reese, V. H. Dibeler and F. Mohler, J. Res. NBS **57** (1956) 367.
243. A. F. Gaines, J. Kay and F. M. Page, Trans. Faraday Soc. **62** (1966) 874.
244. J. L. Moruzzi and A. V. Phelps, J. Chem. Phys. **45** (1966) 4617.
245. J. L. Pack and A. V. Phelps, J. Chem. Phys. **45** (1966) 4316.
246. J. Kay and F. M. Page, Trans. Faraday Soc. **60** (1964) 1042.

CHAPTER 8

THE ROLE OF METASTABLE PARTICLES IN COLLISION PROCESSES

BY

R. D. RUNDEL and R. F. STEBBINGS

Rice University, Houston, Texas, USA

Contents

	Page
8-1. Introduction	549
8-2. Production of metastables	551
A. Thermal metastables	551
1. Recoil effects	552
2. Production cross sections	555
B. Fast metastables	559
8-3. Detection and identification	561
A. Collisions with surfaces	561
B. Collisions with gas atoms or molecules	568
C. Collisions with photons	569
D. Collisions with electrons	570
E. Interaction with d.c. electric and magnetic fields	571
F. Magnetic resonance techniques	572
8-4. Chemiionization	574
A. General principles of thermal energy metastable collision experiments	576
B. Absolute measurements	577
1. The afterglow technique	578
2. The beam technique	583
3. Results	588
C. Measurements of singlet–triplet cross section ratios	593
D. Models of chemiionization reactions	597
E. Theoretical treatments of chemiionization	608
F. Chemiionization involving highly excited atoms	612
8-5. Ion beam studies	613
References	624

§ 8-1. Introduction

The existence of atoms and molecules in long-lived excited states was recognized many years ago, but it has been only quite recently that the crucial role of such excited particles in collision processes has been properly appreciated. In addition, the emergence over the past ten years of a number of highly sophisticated experimental techniques has made possible experiments which were not formerly feasible, and data obtained from the use of these techniques have stimulated new efforts toward theoretical understanding of collision processes.

A metastable excited state is often defined as one whose natural lifetime is greater than 10^{-6} seconds. A more practical definition for the discussion of collision phenomena might be a state whose natural lifetime is comparable with the mean collision time of the particle in some specified environment. For example, the $2\,^1S$ state of helium, whose natural lifetime is 38 msec, is metastable in a typical helium discharge where its mean collision time is of the order of microseconds, but it is not considered a metastable in the environment of a planetary nebula, where the mean collision time is about 3 hours. The $2\,^3S$ state of helium, however, whose natural lifetime is about 10^6 seconds, is metastable in both environments.

Metastable states may be subdivided into three different types. The first is comprised of those states for which the electric dipole matrix elements for transitions to all lower states are zero (or very small, as in the case of hydrogen atoms in the 2s state). This is the most important class of metastables. Collision processes involving particles in such states are of great importance in understanding the behavior of complicated systems such as planetary and stellar atmospheres, gas discharges, and flames. The metastability of such states also has permitted the development of important techniques such as optical pumping, which can be used to produce spin-polarized particles, and of several methods of laser population inversion. The energies and lifetimes of various dipole-forbidden states are given in Table 8-1-1.

The second type of metastables consists of an atom or molecule whose outermost electron is in a Rydberg state with large principal quantum number. Although electric dipole transitions are allowed, the coupling of the electron and nucleus is so weak that the probability of decay is small. Comparatively little work has been done on such states.

Vibrational excitation of molecules provides the third type of metastability. For homonuclear diatomic molecules, electric dipole transitions

TABLE 8-1-1

Characteristics of some interesting metastable species

Species	Ground State	Metastable State	Excitation Energy (eV)	Lifetime (sec.)	Ref.
H	$1\,^2S_{\frac{1}{2}}$	$2\,^2S_{\frac{1}{2}}$	10.20	1/7	1
He	$1\,^1S_0$	$2\,^3S_1$	19.82	6×10^5	2
		$2\,^1S_0$	20.61	3.8×10^{-2}	3
				2×10^{-2}	4
				1.95×10^{-2}	5
N	^4S	$^2D_{\frac{5}{2}}$	2.38	6.3×10^4	6
		$^2D_{\frac{3}{2}}$		1.4×10^5	6
		$^2P_{\frac{3}{2}}$	3.58	13	6
		$^2P_{\frac{1}{2}}$		13	6
N^+	^3P	^1D	1.89	250	6
		^1S	4.04	9.2×10^{-1}	6
O	^3P	^1D	1.96	110	6
		^1S	4.17	7.8×10^{-1}	6
		^5S	9.13	Long	
O^+	^4S	$^2D_{\frac{3}{2}}$	3.31	7.7×10^3	6
		$^2D_{\frac{5}{2}}$		2.4×10^4	6
		$^2P_{\frac{1}{2}}$	5.00	6.0	6
		$^2P_{\frac{3}{2}}$		5.9	6
Ne	$^1S_0$	$^3P_2$	16.62	Long	
		$^3P_0$	16.72	Long	
Ar	$^1S_0$	$^3P_2$	11.55	Long	
		$^3P_0$	11.72	Long	
Kr	$^1S_0$	$^3P_2$	9.92	Long	
		$^3P_0$	10.56	Long	
Xe	$^1S_0$	$^3P_2$	8.32	Long	
		$^3P_0$	9.45	Long	
Hg	$^1S_0$	$^3P_0$	4.66	Long	
		$^3P_2$	5.43	Long	
H_2	$X\,^1\Sigma_g^+$	$C\,^3\Pi_u$	11.86	Long	
N_2	$X\,^1\Sigma_g^+$	$A\,^3\Sigma_u^+$	6.17	12.6	7
		$a'\,^1\Sigma_u^-$	8.40	0.7	8
		$a\,^1\Pi_g$	8.55	1.7×10^{-4}	9
		$E\,^3\Sigma_g^+$	11.87	2.7×10^2	10
N_2^+	$X\,^2\Sigma_g^+$	$A\,^2\Pi_u$	1.1	3×10^{-6}	6
		$^4\Sigma_u^+$	~6	Long	
NO	$X\,^2\Pi$	$a\,^4\Pi$	4.7	$\geq 1.6 \times 10^{-1}$	6
		$A\,^2\Sigma^+$	5.5	2×10^{-6}	6
		$B\,^2\Pi$	5.6	1.2×10^{-5}	6
NO^+	$X\,^1\Sigma^+$	$a\,^3\Sigma^+$	~5	Long	
		$^3\Delta$	~7	Long	
O_2	$X\,^3\Sigma_g^-$	$a\,^1\Delta_g$	0.98	2.7×10^3	11
		$b\,^1\Sigma_g^+$	1.63	7	6
		$C\,^3\Delta_u$	~4.2	Long	
		$A\,^3\Sigma_u^+$	4.43	~10^3	12
		$c\,^1\Sigma_u^-$	4.5	Long	
O_2^+	$X\,^2\Pi_u$	$a\,^4\Pi_u$	4.0	Long	
		$A\,^2\Pi_u$	~4.8	7×10^{-7}	6
		$b\,^4\Sigma_g^-$	~2.1	1.1×10^{-6}	6

between different vibrational levels of the same electronic state are forbidden, and for other molecules the transition probability is usually small. It is difficult to assess the collisional effects of vibrational excitation in more than a semi-quantitative way experimentally, since the distribution of particles among the various vibrational levels is difficult to determine.

A great deal of work, particularly experimental, has been devoted to the study of the effects of metastable states on collision processes. The subject is as broad as atomic collision physics itself, and we shall make no attempt here to cover the entire field. In particular, we shall not discuss in detail the important areas of elastic scattering of metastables and excitation transfer, which have been the subjects of many recent experiments, and of electron and photon ionization of metastables, for which little data are presently available. We shall instead concentrate our efforts on those areas in which there is a strong current interest, and which, we feel, have not thus far been adequately reviewed. The areas are experimental techniques for the production and detection of metastable atoms, molecules, and ions, both fast and slow; chemiionization; and ion–neutral reactions. The latter two areas demonstrate many examples of the general principle that internal energy is much more efficient than translational energy in causing a given reaction to occur.

§ 8-2. Production of metastables

Perhaps the most extensive studies involving metastable atoms have been concerned with their elastic and inelastic scattering at thermal energies. It is often convenient to conduct such studies by passing a beam of thermal metastable atoms through a target gas which may be contained in a scattering chamber, or which may itself be in the form of a beam. In this section we shall review some of the techniques employed in the production of such metastable beams.

A. THERMAL METASTABLES

Thermal metastable particles are most readily produced through electron impact excitation of the parent atom or molecule, although excitation could be achieved by heavy particle impact [13]. Excitation is most commonly effected in a gas discharge or by use of an electron beam.

In the first method excited particles produced in the discharge are allowed to effuse through a small orifice in the wall of the discharge tube and are then collimated by a series of apertures. High intensities of excited atoms may be produced in this way but several excited species may be present in the beam, together with energetic photons whose effects may be difficult to distinguish from those of the metastable particles.

In the second method a beam of electrons is passed through the gas, which may itself be collimated into a beam. In this event it is important to recognize that, particularly for the lighter atoms, considerable momentum transfer may accompany excitation, and the resulting excited atoms may suffer appreciable scattering. This effect and its consequences have often been overlooked and we shall therefore now examine it in some detail.

1. *Recoil effects*

We consider the case of a beam of ground state particles effusing through a small aperture into a region where it is crossed by an electron beam which may be gated on for periods of the order of a few microseconds. The time of flight of the resulting metastable particles in their passage to a detector may thus be determined.

The velocity distribution in the incident beam of ground state particles is the modified Maxwell–Boltzmann distribution given by

$$N(v) = Cv^3 \exp\left[\frac{-v^2}{\alpha^2} - \frac{DPF(v/\alpha)}{T}\right] \qquad (8\text{-}2\text{-}1)$$

where

$$\alpha = (2kT/M)^{\frac{1}{2}}. \qquad (8\text{-}2\text{-}2)$$

The second exponential factor allows for collisional attenuation of slow particles in the neighborhood of the slit. Here P is the source pressure, T is the temperature, D is a constant which contains the elastic scattering cross section, and $F(v/\alpha)$ is a function derived by Estermann et al. [14].

The excitation of this beam by electron bombardment leads to a metastable beam whose velocity distribution is further modified from that described by (8-2-1). Thus for continuous electron bombardment the velocity distribution formula should be divided by v since slower particles spend more time in the bombardment region and hence have a correspondingly higher probability of being excited. When pulsed bombardment is used this correction applies only to the fastest molecules, since those with velocities $v_c < L/t_p$, where L is the length of the collision path and t_p is the duration of the electron pulse, are all bombarded for the same length of time, i.e. t_p.

The effect of momentum transfer from the exciting electron to the heavy particle will alter its velocity, and may cause slow particles to recoil sufficiently to miss the detector. Table 8-2-1 gives the recoil at threshold of

TABLE 8-2-1

Threshold recoil of atoms and molecules excited by electron impact, at normal incidence, to their lowest metastable state

Gas	Atomic weight	Most probable velocity v ($\times 10^4$ cm/sec)	Excitation energy (eV)	Recoil velocity v_{rec} ($\times 10^4$ cm/sec)	v_{rec}/v
He	4.00	12.35	19.81	3.62	0.293
Ne	20.2	5.50	16.62	0.66	0.119
Ar	39.9	3.91	11.55	0.28	0.071
Kr	82.9	2.71	9.92	0.12	0.046
Xe	130	2.17	8.32	0.07	0.033
N_2	28.0	4.67	6.169	0.29	0.062
H_2	2.0	17.35	11.86	0.55	0.318
CO	28.0	4.67	6.01	0.29	0.061

several atoms and molecules excited to their lowest metastable state by electrons at normal incidence. The influence of this recoil on the observed velocity distribution may be approximated by multiplying eq. (8-2-1) by v^g where g is an adjustable parameter. Values of $g > 0$ give added weight to higher velocities and thus indicate the removal of slow particles. The value of g depends upon the detailed dimensions of the apparatus, the source temperature, and the mass of the neutral targets. In the limit of very heavy molecules, their velocity is not affected by the collision, and the value of g is zero.

Following Freund and Klemperer [15] we define

$$n = g+3 \quad \text{for} \quad v < v_c \tag{8-2-3a}$$

$$n = g+2 \quad \text{for} \quad v > v_c \tag{8-2-3b}$$

and the velocity distribution in the metastable beam is then given by

$$N(v) = Cv^n \exp\left[-\frac{v^2}{\alpha^2} - \frac{DPF(v/\alpha)}{T}\right] dv. \tag{8-2-4}$$

This may be converted into a time-of-flight distribution function by the substitution $v = L/t$ where L is the distance from the bombarder to detector and t is the flight time to travel this distance.

Then

$$N(t) = C \frac{L^{n+1}}{t^{n+2}} \exp\left[-\frac{L^2}{\alpha^2 t^2} - \frac{DPF(L/\alpha t)}{T}\right] dt. \tag{8-2-5}$$

It was pointed out by Briglia [16] that if the radiative lifetime τ of the metastable state is less than about thirty times the most probable time of flight, then a significant excess of slow metastables will radiate during flight. To include this effect equation (8-2-5) is modified to

$$N(t) = C \frac{L^{n+1}}{t^{n+2}} \exp\left[-\frac{L^2}{\alpha^2 t^2} - \frac{DPF(L/\alpha t)}{T} - \frac{t}{\tau}\right] dt. \tag{8-2-6}$$

Taking advantage of the fact that the molecular mass appears in this function through the parameter α, Freund and Klemperer [15] used time of flight measurements for mass identification of neutral molecules excited to metastable states. They do not claim high accuracy or resolution because the time-of-flight distribution is so broad; nonetheless they point out that such measurements are valuable in identifying observed metastable states. Discrimination against photons is of course also achieved by time of flight measurements. A summary of their preliminary results on a number of neutral species is shown in Table 8-2-2. It can be seen that the mass measurements are good to about 5%.

Of considerable interest are the tabulated values of n, which are considered preliminary but which nevertheless indicate an increasing departure from the modified Maxwell–Boltzmann distribution with decreasing mass.

TABLE 8-2-2

Time-of-flight mass spectroscopy on neutral metastables

Molecule	Mass	Mass measured by TOF	n (= Power of v)
Helium	4.0	3.6	5.5
Neon	20.2	20.0	5.0
Carbon monoxide	28.0	28.7	4.6
Nitrogen	28.0	28.9	3.7
Argon	39.9	40.5	4.5
Benzene	78.1	78.3	4.0
Krypton	83.8	89.4	3.6
Toluene	92.1	91.2	3.8
Xenon	131.3	138.0	3.1

A somewhat similar study was reported by Robiscoe and Shyn [17] for the case of hydrogen atom 1s–2s excitation by electron impact. They crossed an H-atom beam with a pulsed electron beam and detected the resulting 2s atoms with a channeltron electron multiplier positioned to give maximum signal. The output from the multiplier was fed to a single channel scaler which was gated open for 5 μsec at various delay times following the electron pulse. The time of flight spectrum obtained in this manner was found to be similar to that appropriate to a velocity distribution characterized by $n = 4$.

The observed deficiency of slow 2s atoms is very marked. Robiscoe and Shyn point out that such deficiencies have been noticed in other experiments with light metastable atoms, and are usually attributed to poor definition of oven temperature or self scattering in the beam. Robiscoe and Shyn, however, contend that the kinematic effects are a major cause in distorting metastable velocity distributions.

Van Dyck et al. [4], in their time-of-flight measurement of the lifetime of $He(2\,{}^1S)$, observed that in producing metastables by electron impact at normal incidence they obtained a metastable beam whose velocity distribution varied across the transverse dimension of the beam. This effect produced serious errors in their lifetime measurements until it was eliminated by using electrons directed antiparallel to the helium beam.

Locke and French [18, 19] have made a detailed theoretical and experimental study of the feasibility of using metastable time-of-flight as a diagnostic tool for studying gas flow velocity distributions. They have obtained experimental time-of-flight distributions for a number of different metastables, but have not attempted to analyze their data in terms of eq. (8-2-6).

It is evident that a metastable beam excited by electron impact may have a velocity distribution significantly different from that of the parent ground state beam. However when the momentum transfer accompanying excitation is small compared with the original molecular momentum, the velocity distribution of the metastables will, apart from the $1/v$ factor discussed earlier, be essentially the same as that of the beam from which they are created. In this event the metastable time-of-flight technique may be used to determine the velocity distribution of the ground state beam.

2. Production cross sections

In planning an experiment involving the use of metastables, it is important to know the relevant metastable production cross sections. We consider first the case of H(2s $^2S_{\frac{1}{2}}$), whose production by electron impact has been investigated experimentally by a number of workers [20–22]. For some time it was not possible to reconcile the various experimental results either among themselves or with the various theoretical predictions. The situation has apparently been clarified through recent measurements of Kauppila et al. [23]. A summary of the pertinent data is shown in Fig. 8-2-1. The cross

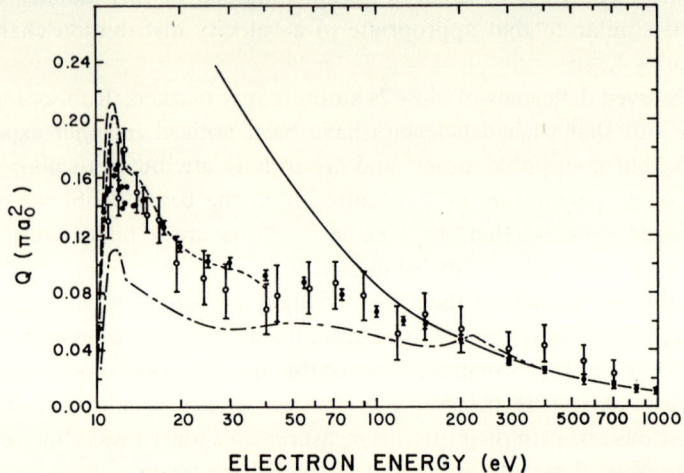

Fig. 8-2-1. Total cross section for production of H(2s) by electron impact. ● – Kauppila et al.; ○ – Stebbings et al. (using correct polarization, see text); – – – Hils et al. (normalized to Born approximation at 500 eV); ---- Lichten and Schultz (normalized to Kauppila et al. at 25 eV); – · – close coupling approximation; —— Born approximation.

section values published originally by Stebbings et al. [21] were derived from their experimental data assuming a value of unity for the polarization fraction of the Lyman-α radiation following quenching of the 2s atoms in an electric field. Recent measurements [24] have shown this assumption to be incorrect. Use of the correct polarization value leads to the cross section values plotted in Fig. 8-2-1. Satisfactory agreement is then observed with the close coupling calculation and, except at the highest energies, with the results of Kauppila et al. The results of Hils et al. [22] now appear definitely to be too low. The angular distribution of the 2s atoms was determined by Stebbings et al. who noted that for the production of well-defined beams of H(2s) atoms there is considerable merit in using high energy electrons rather than working near threshold where the cross section is at a maximum. Larger electron currents may be used at higher energies without appreciably quenching the 2s atoms by space charge fields, and in addition the angular divergence of the resulting 2s atom beam due to recoil is quite small at these energies.

The production of beams of excited helium atoms is complicated by the presence of two closely adjacent metastable levels, $2\,^1S$ and $2\,^3S$. In a gas discharge the $2\,^1S$ atoms are rapidly converted to $2\,^3S$ atoms in superelastic collisions with slow electrons, and beams containing only $1\,^1S$ and $2\,^3S$ atoms can be obtained in this way. However when excited by electron impact under single collision conditions both states will be present. (It is of course possible to excite only the $2\,^3S$ state by careful control of the energy of the exciting electrons, but to avoid exciting $2\,^1S$ atoms one must work within about 0.8 eV of the $2\,^3S$ threshold where the cross section is still small.) It is of interest therefore to know the ratio $R = N_s/N_t$ of the singlet and triplet metastables in a mixed beam as a function of electron energy. A number of experimenters have measured this ratio [25–29] and the published data are given in Fig. 8-2-2. Substantial discrepancies exist between the different measurements and are discussed by Hotop et al. [28]. They point out that R may not in general be equated to the ratio of the electron impact excitation cross sections for the singlet and triplet states because subsequent effects such as elastic scattering, deactivation, and superelastic collisions with slow electrons will not in general act equally upon both components. Hotop et al. further point out that kinematic effects of the type de-

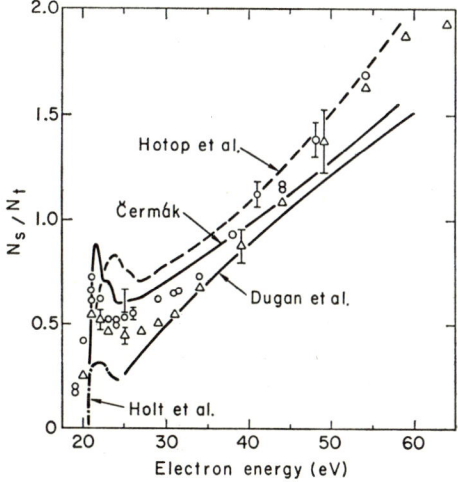

Fig. 8-2-2. Measured ratios of the numbers of He($2\,^1S$) to He($2\,^3S$) produced in a thermal beam by electron impact. Circles and triangles represent the results of Dunning and Smith, using two different methods of measurement.

scribed in § 8-2A-1 may be even more influential in governing R, since only if the apparatus geometry is such that all the metastables produced are collected can R have any well-defined relationship to the cross section ratio. Nonetheless in view of all the possible sources of disagreement the published data are in reasonable accord.

It is evident that even if R were reliably known the presence of two metastable components in a beam is still an inconvenience. A number of techniques have therefore been developed to ascertain the separate effects due to the singlet and triplet species. Richards and Muschlitz [30], and Rothe et al. [31] used an inhomogeneous magnetic field to spatially separate the triplet metastables into three beams corresponding to magnetic quantum numbers $m_s = \pm 1, 0$. The $m_s = 0$ was undeflected, as were the $2\,^1S$ atoms, so only a partial separation of the metastable atoms was effected. However Rothe et al. demonstrated that in their experiment $2\,^1S$ atoms were present in very small numbers and were able to show that the cross section for scattering of $2\,^3S_1$ helium in the rare gases is independent of the magnetic sublevel within their experimental error of 10 %.

An advance on their technique was made by Holt and Krotkov [27] who, in addition to spatially separating the three magnetic sublevels, also quenched the singlet atoms in a strong electric field. However, due to the large separation (0.6 eV) of the $2\,^1S$ level from the closest radiating state ($2\,^1P$), application of a field of 2×10^5 V/cm was required to achieve 90 % quenching. More recently a method has been developed [28, 32] by which beams of metastable helium containing less than 1 % $2\,^1S$ atoms can be produced. The method involves irradiation of a helium beam containing metastable $2\,^3S$ and $2\,^1S$ atoms with radiation from a helium discharge. The $2\,^1S$–$2\,^1P$ 2μ radiation emitted by this discharge is strongly absorbed by the $2\,^1S$ atoms, raising them to the $2\,^1P$ state, from which they decay preferentially to the ground state with emission of 584 Å radiation. The $2\,^3S$ atoms will also absorb radiation from the discharge but subsequent reradiation will result in their return to the $2\,^3S$ state, since it is the lowest state of the triplet system. A modification of this technique utilizes a cold cathode discharge in helium contained within a glass spiral wound around the beam axis. Again essentially complete quenching of $2\,^1S$ is achieved and a beam containing only $2\,^3S$ and $1\,^1S$ atoms results.

It is not in fact always necessary to remove one metastable component from a mixed beam in order to ascertain separately the effects due to each metastable species. In measurements of Penning ionization, for example, determination of the energy of the ejected electron serves to identify the particular metastable participating in the reaction (see eq. (8-2-4)) and thus measurements employing mixed $2\,^1S$ and $2\,^3S$ beams may be readily interpreted.

Metastable species have been produced by electron excitation of many other atoms and molecules including Ar, Kr, Xe, N, O, Hg, H_2, N_2, O_2, CO, N_2O, and CO_2 [33] (see Table 8-1-1). In several instances more than one metastable state may be excited and while the technique of selective optical absorption in principle could often be applied to quench one or more components, this has apparently not been done. Nonetheless attempts have been made to experimentally identify the metastable states produced in complex cases and are described in § 8-3.

B. Fast metastables

The techniques described above are appropriate to the production of thermal energy metastable particles. They are however quite unsuited to the generation of fast metastable beams, which are most conveniently produced through the mechanism of charge transfer. The cross sections σ for collisions involving symmetric resonance charge transfer are normally characterised by an expression of the form

$$\sigma^{\frac{1}{2}} = a - b \log v \tag{8-2-7}$$

where a and b are constants and v is the collision velocity. The cross section thus increases with decreasing velocity. On the other hand non-symmetric reactions of the type

$$A^+ + B \rightarrow A + B^+ + \Delta E \tag{8-2-8}$$

are generally characterized by a cross section which is small at low and at high velocities and passes through a maximum value at an intermediate velocity [34, 35]. ΔE is called the energy defect and makes allowance for the transfer between potential and kinetic energy. The velocity at which this maximum cross section occurs tends to decrease as ΔE decreases and for very small values of ΔE the variation of the cross section with velocity is often quite similar to that for symmetric charge transfer [36, 37]. Small values of ΔE may occur in cases of electron capture into excited states, for example in

$$A^+ + B \rightarrow A^* + B^+ + (\Delta E \sim 0) \tag{8-2-9}$$

which leads to production of excited A^* atoms. Several of these cases of accidental or asymmetric resonance have been used to produce beams of fast metastable particles. Donnally et al. [38] reported measurements of the cross section for

$$H^+ + Cs \rightarrow H(2s) + Cs^+ + 0.49 \text{ eV}. \tag{8-2-10}$$

They found a large cross section having essentially resonant form with a value at 200 eV of $\sim 4 \times 10^{-15}$ cm^2.

Lorents and Peterson [39] have since studied capture by He$^+$ and Ar$^+$ in Rb and Cs in the energy range from 10–1500 eV.

For He$^+$ (1s) the energy defects for electron capture into various neutral atom final states are

He final state	ΔE(Rb)	ΔE(Cs)
1 ^{1}S	19.31	19.6
2 ^{3}S	0.60	0.89
2 ^{1}S	−0.21	0.08
2 ^{3}P	−0.56	−0.27
2 ^{1}P	−0.81	−0.52

Results for total electron capture by He$^+$ ions in Rb and Cs are shown in Fig. 8-2-3. The authors note that the cross sections are larger than 5×10^{-15} cm^2 except at the lowest energies and point out that these large values denote a long-range interaction of the order of $(7-10) \times 10^{-8}$ cm. Capture into excited states is therefore predominant since the ground state of a noble gas atom cannot interact strongly at such large distances because of its closed shell configuration.

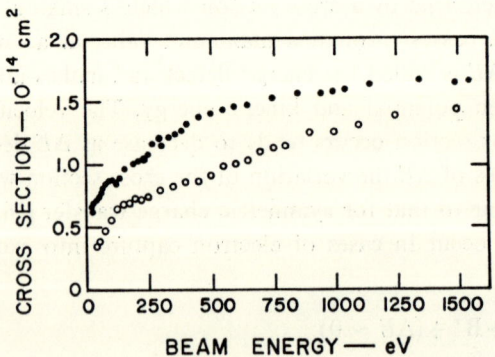

Fig. 8-2-3. Charge transfer cross sections for He$^+$ incident on Cs (closed circles) and Rb (open circles).

It is apparent that charge transfer into a number of excited states of helium can occur with energy defects below 1 eV. The structure in the curve of Fig. 8-2-3 may indeed be an indication that the cross sections for individual excited states come into prominence as the collision velocity changes, although other interpretations are possible. Unfortunately the charge transfer cross section itself cannot be used to determine the intensity of the resulting metastable beam since some of the states which are initially excited may not be metastable but may decay radiatively to the ground state. An interesting technique described in § 8-3 has however been developed by Hollstein et al. [64] to provide data on the $2\,{}^1S$ and $2\,{}^3S$ components of such a mixed beam. It seems reasonable to suppose that their technique will find general applicability in this context.

§ 8-3. Detection and identification

A problem that is central to many studies involving metastable particles is that of their detection. It is sometimes sufficient to make relative measurements of the density of metastable particles in a discharge or in a beam, although frequently absolute measurements are needed. This section will be concerned with an evaluation of the numerous techniques which are applicable to metastable particle detection and identification.

It is convenient to classify these techniques according to the nature of the collision or interaction experienced by the metastable particle. The following cases may be identified:

A. Collisions with surfaces
B. Collisions with gas atoms or molecules
C. Collisions with photons
D. Collisions with electrons
E. Interaction with d.c. electric and magnetic fields
F. Magnetic resonance techniques.

In principle all of these procedures may be used to detect metastable atoms and indeed most of them have been utilized to some extent. We shall now examine each method in turn and attempt to indicate its usefulness and general applicability.

A. Collisions with surfaces

When a metastable particle impacts on a surface a wide variety of possible events may occur [40]. However by far the greatest emphasis has been accorded to the phenomenon of secondary electron ejection and we turn our attention now to this effect, which was first observed in 1924 by Webb [41] for mercury atoms incident on a nickel surface. Since that time many experimenters have investigated secondary electron ejection by metastable particles and it is at first glance surprising that so little reliable data exist concerning the ejection mechanism. The recent emergence of new techniques, however, gives promise of a rapid advance in understanding of this phenomenon. Clearly one vital piece of information is a knowledge of the absolute ejection efficiencies for various metastable species incident on well specified and reproducible surfaces. Data of this kind are now becoming available.

Perhaps the most significant work in recent years has been that of MacLennan and Dunning and their respective colleagues [42, 43]. In each case their experimental method was based upon that developed earlier by Stebbings [44], which we shall now briefly describe with reference to Fig. 8-3-1. We consider a beam of metastable particles, of type X*, flowing from a source S through a chamber C and impinging upon the surface D for which the secondary emission coefficient γ is to be determined.

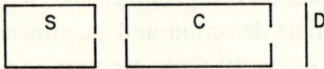

Fig. 8-3-1. Schematic view of apparatus for determining the secondary electron emission coefficient for metastables striking a surface.

With the collision chamber evacuated let the electron current ejected from the detector surface be I_1, due to N_1 metastables incident upon it per second. Then

$$I_1 = \gamma N_1 e \tag{8-3-1}$$

where e is the electronic charge. Let a gas Y, whose ionization potential is less than the excitation energy of X* be introduced into C. Penning ionization will occur

$$X^* + Y \rightarrow X + Y^+ + e \tag{8-3-2}$$

and the detector current will accordingly be reduced to

$$I_2 = \gamma N_2 e \tag{8-3-3}$$

where N_2 is the flux, now smaller, of metastable atoms striking D.

If the ions produced in Penning collisions are collected the resulting current is given by

$$I_3 = (N_1 - N_2)e \qquad (8\text{-}3\text{-}4)$$

since each metastable atom lost from the beam results in the formation of a Y^+ ion.

Combining equations (8-3-1), (8-3-3), and (8-3-4) gives

$$\gamma = (I_1 - I_2)/I_3 \qquad (8\text{-}3\text{-}5)$$

from which γ can be determined.

This simple treatment assumes that all metastables which fail to reach D when gas is introduced into the chamber C are de-excited in ionizing collisions with Y atoms, and it is essential therefore that proper account be taken of elastic scattering collisions, which would reduce I_2 without a corresponding increase in I_3. The studies reported to date compensate for elastic scattering in different but equally valid ways.

The major distinction to be drawn between the early work of Stebbings and that of MacLennan et al. and Dunning et al. is that the latter measurements have provided yield data for atomically clean surfaces while the earlier work, which demonstrated the feasibility of the method, provided data solely for a contaminated surface. The work of Dunning et al. is also different in two other respects. Firstly the surfaces under investigation could be produced by the continuous deposition of fresh material from an oven, and secondly the collision chamber could be replaced by a crossed atom beam. A number of advantages accrue from these features, of which perhaps the greatest is the suitability of this arrangement for the study of low lying metastable levels. The earlier methods utilized stable gases of relatively high (> 10 eV) ionization potentials in the collision chamber, together with high work function surfaces, and were thus restricted to metastable particles with excitation energies greater than 10 eV. The use of crossed beam and surface deposition techniques using alkalis, with low ionization potentials and low work functions, extends greatly the range of metastable particles which may conceivably be studied with this technique. Nonetheless to date Dunning et al. have concentrated on the rare gas metastables, and have determined the yields for a number of atomically clean and contaminated surfaces.

Before the discussion of these and other data it is perhaps worthwhile here to comment briefly on the production of atomically clean surfaces and their subsequent contamination by gas adsorption.

A number of techniques have been employed for the preparation of atomically clean surfaces and are described by Farnsworth [45]. All the studies pertinent to this discussion utilized the technique of heating in high vacuum which has however the following limitations.

1. Surface contamination in the form of a stable chemical compound may not necessarily be removed by heating below the melting point.

2. Bulk impurities may diffuse to the surface and may not be removed by heating below the melting point.

3. Diffusion of surface impurities into a thin underlying layer may occur.

4. Heat treatment and the rate of subsequent cooling may affect the surface structure.

All of these are not necessarily applicable to the case of tungsten surfaces; nevertheless it is evident that the mere heating of a metal in vacuum will not inevitably lead to a reproducible atomically clean surface and that care must be exercised in comparing results for different "atomically clean" surfaces.

The mechanism of gas adsorption by a surface exposed to a given pressure of gas is not well understood. It is generally viewed, following Langmuir [46], as a dynamic process whereby gas molecules striking the surface may rebound, or may be adsorbed and remain on the surface until they acquire enough energy to leave. At equilibrium the rate of adsorption will equal to the rate of desorption.

The essence of the Langmuir theory, in which the basic assumption is that adsorption is limited to formation of a unimolecular layer, is that the fraction (Θ) of the surface covered with adsorbed molecules is related to the pressure p by

$$\Theta = bp/(1+bp) \tag{8-3-6}$$

where

$$b = \frac{\alpha_0 \, e^{q/kT}}{k_0 (2\pi mkT)^{\frac{1}{2}}}. \tag{8-3-7}$$

α_0 is the ratio (close to unity) of the inelastic collisions which result in adsorption to the total number of collisions of gas molecules with the bare surface, q is the binding energy and k_0 is a term that depends upon the entropy of adsorption.

This theory has certain limitations but nonetheless it serves to emphasize the strong temperature dependence of Θ at low pressures. In addition, because of the dynamic nature of the adsorption process, the customary description of the extent of contamination as being proportional to the product of the ambient pressure and the exposure time may not be valid.

The time necessary to reach an equilibrium situation depends upon the nature of the adsorbent. For the adsorption of argon, krypton, and xenon on zirconium films an essentially instantaneous equilibrium was observed by Hansen [47] while equilibrium times of several hours have been reported [48] for the adsorption of nitrogen on various porous adsorbents.

In Table 8-3-1 are given the results for secondary emission from atomically clean tungsten surfaces obtained by heating at elevated temperature (~ 2300 °K) in a high vacuum. It is interesting to note that the values for metastable atoms are in substantial agreement and are closely similar to those obtained by extrapolating, to zero energy, the data of Hagstrum [49] for helium and neon ions. This lends credence to Hagstrum's hypothesis that as the metastable atoms approach the surface they are first ionized in an energy resonant process whereby the excited electron "tunnels" into a vacant energy level in the metal. The electron ejection is thus a consequence of ion impact with the surface regardless of whether the original particle was an ion or a metastable particle.

TABLE 8-3-1

Measured values of γ for atomically clean surfaces

Surface	Tungsten	Tungsten	Cadmium	Sodium
Impacting species				
He($2\ ^1S$)	0.28 ± 0.06 [42]		0.41 ± 0.06 [43]	
He($2\ ^3S$)	0.32 ± 0.03 [42]		0.42 ± 0.06 [43]	
H($2\ ^1S, 2\ ^3S$)	0.31 ± 0.03 [42]	0.295 ± 0.06 [43]	0.38 ± 0.07 [43]	
He$^+$		0.29 [49]		
Ne($^3P_{0,2}$)	0.215 ± 0.02 [42]	0.205 ± 0.04 [43]	0.23 ± 0.06 [43]	
Ne$^+$		0.213 [49]		
Ar($^3P_{0,2}$)		0.085 ± 0.16 [43]	0.33 ± 0.03 [43]	0.14 ± 0.06 [43]
Ar$^+$		0.095 [49]		

We conclude from these data that, notwithstanding the earlier remarks concerning the potential dangers in production of atomically clean surfaces by heating in vacuum, such surfaces do in fact exhibit reproducible secondary emission characteristics. It should be noted, however, that for an atomically clean surface the yield will in general be influenced by:

1. The structure of the surface – crystalline or amorphous;
2. If crystalline, the yields from different crystal planes may differ;
3. If amorphous, the yield will depend upon the degree of surface roughness.

It is to be anticipated that the secondary emission coefficient for an initially atomically clean surface will change as it becomes contaminated. It is well known, for example, that the presence of a small amount of surface contaminant can cause appreciable changes in the work function for the surface. In addition, since some of the ejected electrons originate from beneath the surface, the presence of sub-surface contaminants may also influence the yield.

It is in fact generally observed that the yields for various surfaces are modified by gas contamination. However the results in these circumstances are conflicting. For example MacLennan observed that exposure of his tungsten surface always resulted in a reduction of the yield for incident helium metastables and values as low as $\gamma = 0.06$ were observed. Delchar et al. [42] also observed a change in yield which was however less dramatic. For nitrogen contamination the yields for the (111) and (110) planes of tungsten were observed to remain essentially unaltered while for an essentially (100) oriented polycrystalline tungsten surface the yield dropped by about 42%. Contamination with hydrogen and carbon monoxide also resulted in varying degrees of yield reduction.

In marked contrast to these results are those of Dunning et al. which showed an increase in yield to a value ~ 0.5 as their tungsten surface was contaminated with air.

There is no reason to doubt the correctness of any of these observations since the determination of changes in yield as a consequence of surface contamination is less susceptible to error than the determination of the absolute yield for a flashed surface, for which, as we have seen, there is good agreement. We therefore take the view that these various measurements refer to surfaces which differ in some, as yet unidentified, way in their contamination.

It is perhaps useful to comment briefly upon the suggestions advanced to account for the various observations. Delchar et al. suggest that gas adsorption occurs via a co-valent bonding mechanism whereby electrons are shared between the conductive band of the metal and the adsorbed gas atom. These electrons are then viewed as participants in the ejection process which will then be influenced not only by the nature of the adsorbed atom but also by its position in the surface and the surface density.

An alternative view which could qualitatively explain the results of Dunning et al. is that collision of incident metastables with adsorbed gas molecules could result in their Penning ionization. The resulting electrons will probably escape, while the ions will be neutralized at the surface, possibly with the ejection of a second electron. With the duality of processes occurring an increased secondary electron yield over the atomically clean surface value is to be anticipated. It is interesting to note that in a measurement of the photoelectron yield of atomically clean tungsten, Waclawski et al. [50] observed an increase in electron yield after exposure of the surface to oxygen, which they attributed to photoionization of the adsorbed oxygen atoms.

The observation by Hagstrum that, for He^+, γ drops as the tungsten surface is contaminated is consistent with this picture since charge transfer may now occur between the He^+ and adsorbed gas. This will however merely produce a slower ion of the same species, or one having a lower ionization potential, for which γ would probably be smaller. The principal effect of the adsorbed gas for incident ions is then to inhibit the escape of secondary electrons and thus to reduce γ.

In summary then it appears that when a surface becomes contaminated the yield may increase or decrease according to the nature and abundances of the adsorbed gases and vapor. In view of these observations it is our view that comparison between the results for different contaminated surfaces is permissable only if their initial atomically clean state and their subsequent history were identical. In this latter regard the surface temperatures and the nature and partial pressures of the various gases and vapors to which they were exposed should be similar. Because of these considerations we do not consider it profitable to tabulate values of γ obtained for contaminated surfaces. We merely note as an example that for $He(2\,^3S)$ atoms incident on contaminated gold surfaces values of 0.29 ± 0.03 [44] and 0.67 ± 0.1 [43] have been obtained while for $Ne(^3P_{0,2})$ incident on contaminated gold γ's of 0.27 ± 0.14 [51] and 0.55 ± 0.06 [43] have been observed.

It should also be noted that Dunning et al. observed that for chemically cleaned surfaces, metastable atoms of helium, neon, and argon appeared consistently to have a yield near unity. This result may well be a consequence, however, of their cleaning procedure and residual gas background.

It is evident that much more work is necessary before a detailed understanding of metastable-surface interactions can be gained. The application of techniques such as LEED [45, 53]* (Low Energy Electron Diffraction)

---

* For a discussion of relevant techniques see ref. [52].

would be useful in providing more precise details of the surface conditions, while the determination of the energy distribution of the ejected electrons would provide further insight into the precise nature of the ejection mechanism.

B. COLLISIONS WITH GAS ATOMS OR MOLECULES

Penning ionization will be dealt with at length in § 8-4. Here we merely wish to draw attention to those aspects of this process which bear upon metastable particle detection and identification. Clearly the mere occurrence of Penning ionization is indicative of the presence of excited neutral particles. Two refinements introduced by Cermák [54] permit insight into the identity of the metastable species in question and may be readily understood as follows.

Penning ionization such as

$$A^* + BC \rightarrow A + BC^+ + e \tag{8-3-6}$$

may occur only when the excitation energy of A^* exceeds the ionization potential of BC. For a given metastable A^* a series of different targets BC with slowly increasing ionization potentials are employed. The targets may then be separated into two groups, those which are ionized in collision with A^* and those which are not. The excitation energy of A^* is then greater than the highest ionization potential of the particles in the first group and lower than the lowest ionization potential of particles in the second group. There are certain complications to this simple interpretation such as the possible occurrence of associative ionization,

$$A^* + BC \rightarrow ABC^+ + e \tag{8-3-7}$$

which can however be resolved by the use of mass spectrometric detection. Using this approach, Cermák identified some previously unknown metastable states of N_2 and CO.

Referring again to reaction (8-3-6) it is clear that, if the ionization potential of the target molecule BC is known, an energy analysis of the ejected electrons enables the excitation energy of A^* to be determined and in general this will allow its identification. This technique, which is analogous to photoelectron spectroscopy is sometimes referred to as Penning electron spectroscopy and has been used by a number of workers [55–62].

The phenomenon of excitation transfer has received a great deal of attention in the past few years, in part due to its relevance to many laser systems. We mention it here solely in the context of its usefulness as a technique of metastable detection, and we cite as an example the recent work of Schmeltekopf and Fehsenfeld [63], who detected helium metastables in their flowing afterglow by the addition of neon, which has excited states in close energy resonance with He($2\,^3S$) and He($2\,^1S$). Excitation transfer may then occur, followed by the emission of neon radiation (see § 8-4).

The detection schemes described so far have been appropriate to metastable particles having thermal energies. However if when a metastable particle collides with an atomic or molecular target the collision velocity is sufficiently great, other processes may occur which serve to detect and identify the metastable species in question.

Thus at energies in the range 150–2200 eV Hollstein et al. [64] observed that in collision with normal helium atoms, metastable atoms of helium are excited with high efficiency to upper levels from which they decay with emission of characteristic radiation. They were thus able to monitor the $2\,^1S$ and $2\,^3S$ populations in their fast helium beam by observation of the singlet ($3\,^1D$–$2\,^1P$ 6678 Å) and triplet ($3\,^3D$–$2\,^3P$ 5867 Å) radiation. This technique appears suitable for more general application.

At still higher energies in the range 75–250 keV Gilbody et al. [65] determined the metastable atom fractions in fast helium atom beams through the mechanism of electron loss collisions. The procedure is analogous to that used to determine the excited state fractions in ion beams and described in § 8-5, and is dependent upon the fact that for the metastable atoms the electron loss (stripping) cross section greatly exceeds that for the ground state atoms.

C. COLLISIONS WITH PHOTONS

Metastable atoms may be detected by the method of selective optical absorption. This involves passing through the region containing the metastable atoms a beam of photons which are strongly absorbed by these atoms. The absorption is then a measure of the number of metastable atoms present.

Let $\Delta I/I$ be the fractional reduction of intensity of the radiation in the absorption line due to the excited atoms. Then [66]

$$\frac{\Delta I}{I} = \frac{L \int k_v \, dv}{\Delta v} \qquad (8\text{-}3\text{-}8)$$

where k_v is the absorption coefficient of the radiation in the frequency range between v and $v+dv$, L the path length of the photons through the region containing the excited atoms, and Δv is the width of the adsorption line. Also

$$\int k_v \, dv = \pi e^2 N_j f_{kj}/mc \qquad (8\text{-}3\text{-}9)$$

where f_{kj} is the oscillator strength associated with the transition giving rise to the absorption and N_j the number of excited atoms per unit volume. It follows that

$$N_j = \left(\frac{mc}{\pi e^2}\right)\left(\frac{\Delta v}{Lf_{kj}}\right)\left(\frac{\Delta I}{I}\right) \qquad (8\text{-}3\text{-}10)$$

enabling N_j to be determined.

This technique has been utilized by a number of people, most recently by Ellis and Twiddy [67] in studies of excited atom concentrations in argon afterglows.

Another optical technique which shows great promise for the detection and identification of molecular metastable species is that of Raman spectroscopy. In the Raman effect when a molecule is irradiated by a monochromatic source part of the scattered light suffers a wavelength change. The differences in frequency between the incident and scattered radiation are called the Raman displacements and are related to the frequencies of rotational and vibrational motion within the molecule. In that these frequencies differ from one electronic state to another, their determination enables the electronic state of the scattering molecule to be ascertained.

Although Raman scattering was first observed in 1928 [68] its use as a detection tool has been hindered because of the low intensity of the scattered radiation. However the availability of high power laser sources will undoubtedly lead to a growing use of this technique. The first published work known to the authors is that of Braunlich and Lambropoulous [69] who used laser Raman spectroscopy to detect metastable deuterium atoms.

D. COLLISIONS WITH ELECTRONS

Collisions with electrons may be used to detect and identify metastable particles. Both superelastic collisions and those resulting in further excitation or ionization are pertinent. Mercury metastable atoms in the $6\,^3P_0$ state were detected by Latyscheff and Leipunsky [70] using the technique of superelastic collisions, although little subsequent use appears to have been made of this method.

Metastable particles may also be detected by virtue of the fact that their ionization by electron impact can occur at an electron energy lower than the normal ionization potential by an amount equal to their excitation energy. This approach has been used by Foner and Hudson [71] in studies of oxygen and nitrogen. They detected $O_2(^1\Delta_g)$ in the presence of ground state $O_2(^3\Sigma_g^-)$ and also, by use of mass spectrometric techniques, they demonstrated the existence of $N(^2D)$, $N(^2P)$, and $N_2(A\ ^3\Sigma_u^+)$ in nitrogen and helium–nitrogen discharges. The method is not however without its limitations which may be indicated with reference to the case of oxygen atoms. The lowest ionization potential for $O(^3P)$ atoms is 12.08 eV leading to $O^+(^4S)$. Metastable (^1D) and (^1S) atoms can only be ionized to $O^+(^4S)$ however by a collision involving electron exchange. The cross sections for such collisions are expected to be very small and the presence of $O(^1D)$ and $O(^1S)$ may well go undetected by this method. Ionization of these metastable atoms will in fact proceed primarily by simple electron removal leading to excited $O^+(^2D)$ and $O^+(^2P)$ with threshold energies greater than 12.08 eV.

E. Interaction with d.c. Electric and Magnetic Fields

A variety of techniques employing direct and oscillatory fields may be used to detect metastable atoms. The techniques, however, are not universally applicable, and are in general appropriate only in certain special circumstances.

Thus metastable hydrogen atoms may be detected by subjecting them to a weak electric field, which causes the $2\ ^2S_{\frac{1}{2}}$ state to become mixed with the $2\ ^2P_{\frac{1}{2}}$ state, and to a lesser extent the $2\ ^2P_{\frac{3}{2}}$ state, as a result of the Stark effect. An optically allowed transition to $1\ ^2S_{\frac{1}{2}}$ ground state can then occur with the emission of a Lyman-α photon [72]. The technique, which has been widely used for the detection of metastable H-atoms is not generally applicable to metastable particle detection because the field strengths necessary to cause mixing with the closest radiating state are in general exorbitantly large. However $He(2\ ^1S)$ atoms have been detected this way [27] by coupling the $2\ ^1S$ state to the nearby (0.6 eV) $2\ ^1P$ state with fields of the order of 2×10^5 V/cm. Metastable He^+ ions may also be quenched in an electric field of order 2 kV/cm which mixes the $2\ ^2S_{\frac{1}{2}}$ and $2\ ^2P_{\frac{1}{2}}$ states. An optically allowed transition to the $1\ ^2S_{\frac{1}{2}}$ ground state may then occur with emission of a 304 Å photon [73].

Field ionization of highly excited metastable species has also been utilized as a detection mechanism and may be understood as follows. An electron in an atomic system is bound by the nuclear Coulomb attraction. For an isolated atom the binding energy represents the energy required to remove the electron to infinity. In the presence of an electric field, however, the bound electron is separated from a region at lower potential by a potential barrier, which in the case of an extremely intense field is quite narrow. Tunnelling of the electron through this barrier may occur, leading to ionization of the atom. The effect was first observed by Rausch von Traubenberg et al. [74] during the study of the Stark effect in the Balmer spectrum.

More recently it has been utilized by Riviere [75] to observe highly excited hydrogen atoms formed by electron capture. These atoms are passed through an electrode system where they are subjected to an electric field of up to 1.2×10^5 V/cm. Such a field is sufficiently strong to ionize atoms in levels with $n > 9$ in 10^{-10} sec. To distinguish the protons so formed from those resulting from stripping in the residual gas, the quench voltage was modulated by a small a.c. potential and the a.c. component of the ion current was determined by phase sensitive detection. The a.c. signal was effectively the differential of the ionization of the atom beam with respect to field strength. The observed modulation of the proton signal is shown in Fig. 8-3-2 as a function of the electric field strength, and the ionization of successive levels of the atom is clearly revealed by the series of maxima.

Magnetic fields have also found application in such studies since, if the hydrogen atoms are travelling with sufficient velocity v through a magnetic field B, they experience an electric field $\mathbf{E} = \mathbf{v} \times \mathbf{B}$ in their own frame which can lead to ionization.

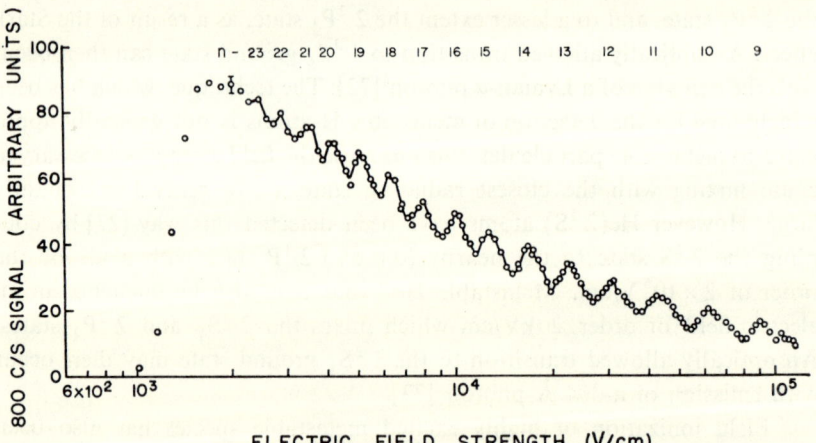

Fig. 8-3-2. Differential signal versus electric field strength for field ionization of highly excited H atoms.

F. MAGNETIC RESONANCE TECHNIQUES

If a paramagnetic particle, of magnetic moment μ, is placed in a magnetic field B it acquires an increment of energy W given by

$$W = -\boldsymbol{\mu} \cdot \boldsymbol{B} \qquad (8\text{-}3\text{-}11a)$$

$$= -\mu_{\text{eff}} B \qquad (8\text{-}3\text{-}11b)$$

where μ_{eff} is the effective magnetic moment of the particle in the magnetic field. For sufficiently weak fields, and if there is Russell–Saunders coupling,

$$\mu_{\text{eff}} = m_J g \mu_0 \qquad (8\text{-}3\text{-}12)$$

where μ_0 is the Bohr magneton, m_J is the magnetic quantum number, which may take $2J+1$ values, and g is the Landé g-factor. Thus a given level will be split by an external field into $2J+1$ levels, and transitions between these levels may in certain circumstances occur. The frequencies of such transitions are given by

$$v = \frac{\mu_0 g B}{h} \Delta m_J. \qquad (8\text{-}3\text{-}13)$$

Measurement of these transition frequencies together with knowledge of the magnitude of the magnetic field enables the g-factor to be determined. The g-factor is however a property of the state of the particle being observed and may be used therefore to identify the state.

Two experimental techniques for the determination of g have been used to study metastable species. Each involves the determination of the field and frequency at which a transition within the Zeeman multiplet takes place.

In the electron spin resonance method the particles are confined within a region throughout which is maintained a uniform magnetic field. Radio frequency power is then beamed through this region and adjustments made to either the field or the frequency until a resonance is achieved as demonstrated by the absorption of electromagnetic energy. This technique has been applied by Evenson et al. [76] and by Falick et al. [77] who observed the presence of $O_2(^1\Delta_g)$ in a flowing afterglow.

An alternative experimental arrangement is that of molecular beam magnetic resonance which was first used by Rabi et al. [78]. In this method a beam of particles is first passed through an inhomogeneous magnetic field which both polarizes them and deflects those for which $m_J \neq 0$. They then pass through a second, uniform, magnetic field region into a third field which is again inhomogeneous. The atoms therefore suffer further deflection and by suitable employment of stops and by proper choice of field strengths and directions it is possible to arrange that only one magnetic subcomponent of the beam arrives at a suitably placed detector. If radio frequency power is now applied in the uniform field region molecular transitions will occur, as described above, when the oscillator frequency equals one of the transition frequencies. Such transitions cause a change in the orientation of the particles in the magnetic field and they will therefore follow a different trajectory when they enter the second inhomogeneous field. Transitions may then be identified through the occurrence of intensity changes in the detector current. This technique has been employed by Brink [79] to examine the neutral species emerging from a microwave discharge in oxygen and helium–oxygen mixtures.

In both experimental arrangements the observed transitions take place within the Zeeman multiplet of a given atomic or molecular state at magnetic fields of the order of tens of gauss. Therefore only the linear Zeeman effect is involved and the states may be identified by means of their weak field g-factors. For the case of atoms these are usually known from theory to sufficient accuracy to permit state identification. For molecules the g-factors can usually be obtained from a simple vector-model calculation. However they depend upon the rotational state as well as on the electronic state and it is therefore possible to observe many resonances in a given molecular electronic state.

The technique is of course limited to paramagnetic species and is thus not suitable for such interesting particles as $O_2(^1\Sigma_g^+)$, $O(^1S)$ and $He(2\ ^1S)$. However a large number of metastable species are amenable to study in this manner. Brink [79] has reported measurements of $O_2(a\ ^1\Delta_g)$ and $O(^5S)$.

§ 8-4. Chemiionization

In this section we shall consider several closely related reaction processes which can be conveniently referred to as chemiionization reactions. Consider a collision between an excited atom A* and a molecule BC. We shall assume (1) that the excitation energy of A* is greater than the ionization potential of BC, and (2) that the collision takes place at thermal energy. By thermal energy we mean that the particles can be characterized by some effective temperature low enough that $kT/E_i \ll 1$ where E_i is the ionization potential. The possible reaction channels are

$$A^* + BC \rightarrow A^* + BC \qquad (8\text{-}4\text{-}1)$$

$$A + BC + h\nu \qquad (8\text{-}4\text{-}2)$$

$$A + B + C \qquad (8\text{-}4\text{-}3)$$

$$A + BC^+ + e \qquad (8\text{-}4\text{-}4)$$

$$A + B^+ + C + e \tag{8-4-5}$$

$$AB^+ + C + e \tag{8-4-6}$$

$$ABC^+ + e. \tag{8-4-7}$$

Reactions (8-4-4)–(8-4-7), in which internal energy is transferred to an electron causing it to make a transition from a bound to a free state, are called chemiionization reactions.

Reaction (8-4-7) is usually termed associative ionization. Its existence was originally suggested to explain the formation of diatomic molecular ions (such as He_2^+) in monatomic gases [80]. Hornbeck and Molnar [81], in a study of the appearance potentials of He_2^+, Ne_2^+, and Ar_2^+, demonstrated that these ions are formed by a collision of an excited atom with a ground state atom, rather than by an ion–atom collision. More recently, the work of Cermák and Herman [58, 82–84] has indicated that associative ionization is possible for almost all reactant species.

Reactions (8-4-4) and (8-4-5), in which the products include a ground state A atom and an ion, are usually termed Penning ionization (although this term has occasionally been used to refer to all chemiionization reactions). Penning ionization was first observed by Kruithoff and Penning [85], and has since been extensively studied by a number of investigators.

Reaction (8-4-6), which might be termed rearrangement ionization, was first observed quite recently by Kuprianov [86].

We shall not consider here reactions (8-4-1)–(8-4-3), which do not lead to ionization, nor shall we consider excitation transfer,

$$A^* + BC \rightarrow A + BC^* \tag{8-4-8}$$

which can occur when $E(A^*) < E_i(BC)$. This latter reaction is a very important one, and has been the subject of other recent reviews [87, 88].

Since the lifetime of excited states which can decay via electric dipole radiation is usually much shorter than the mean collision time, most of the experimental work on chemiionization has involved metastable excited states. It has been qualitatively demonstrated, however, that short-lived excited states can be important under some circumstances [89, 90], and in fact the associative ionization reaction

$$He^* + He \rightarrow He_2^+ + e \tag{8-4-9}$$

can only occur (at room temperature) if the He* is in a state higher than $3\,^3S$. Two rather different types of metastable states have been studied: excited states with small principal quantum number which are not optically connected with the ground state, and high-lying Rydberg levels with large principal quantum number which can make optically allowed transitions but with a small transition probability (the lifetime of the Rydberg states with principal quantum number n, averaged over all values of the angular momentum quantum number, increases as $n^{4.5}$ [72], so that states with $n \gtrsim 10$ are observable in a typical beam experiment). As will be seen below, these two types of metastables have rather different collisional properties.

Theoretical studies, on the other hand, are easier to carry out for excited states which are not metastable, and most of the available calculations are for this case. Recently an increasing amount of experimental data seems to have generated an awakening of theoretical interest in chemiionization involving metastable states, and careful quantum mechanical treatments of this case, which previously were non-existent, have begun to appear.

Many of the theoretical aspects of chemiionization have been reviewed recently by Berry [91], who has pointed out the close relationship of chemiionization reactions to autoionization. Experimental work on chemiionization and other reactions involving metastables has been reviewed by Muschlitz [92].

A. General Principles of Thermal Energy Metastable Collision Experiments

Although several quite different techniques have been used to study Penning ionization and associative ionization processes, the essential elements of any of these techniques are the same. Any experimental apparatus must consist of three basic parts: a Metastable Production Region, an Interaction Region, and a Detection Region. In the Metastable Production Region, ground state atoms or molecules are excited to a metastable state. The techniques for this have already been discussed in some detail in § 8-2A. We must emphasize, however, the importance of recoil effects in the excitation process, and the need to consider such effects in determining the effective metastable temperature.

In the Interaction Region, metastables of one species are allowed to react with ground state neutrals of some different species. Ideally, only the particular reaction under study should be allowed to occur. If, in fact, other reactions occur simultaneously, the Detection Region of the apparatus must be capable of discriminating against the products of these other reactions. In addition, if absolute cross sections are to be determined, the reactant concentration and the distance over which reactions occur must be known.

In the Detection Region, some effect resulting from reactions is detected, and provides a quantitative measure of what is occurring in the interaction region. There are two basic types of detection measurement which can be made; as the density of the reactant species is increased, one can measure either the decrease in the metastable beam flux transmitted through the Interaction Region, or the increase in one of the products of the reaction. In the former case, one measures a "total destruction cross section" for metastables colliding with the reactant species, while in the latter case one is measuring some partial production cross section for one of the reaction products. In general, the attenuation method has been used to measure absolute cross sections, while the product detection method has been used to determine relative efficiencies for different specific reaction channels. The attenuation method is based on the fact that a beam of particles traversing a region where scattering can occur is attenuated such that the ratio of the particle current I_0 entering the region to the current I leaving the region is given by

$$I/I_0 = \exp(-NLQ) \tag{8-4-10}$$

where N is the particle density of reactant species, L is the path length through the scattering region, and Q is the sum of the cross sections for all reactions which remove particles from the beam. If the particles in the beam are metastables, the measured cross section Q includes all the reactions (8-4-1) to (8-4-7). However, it is often assumed that Penning ionization is the dominant reaction, and Q is presented as a cross section for Penning ionization.

B. ABSOLUTE MEASUREMENTS

Most of the absolute measurements of cross sections for metastable destruction or chemiionization which have been measured to date are for helium metastables incident on various target species. The excitation energies of the metastable states of helium, $2\,^3S$ (19.82 eV) and $2\,^1S$ (20.61 eV), exceed the ionization potential of all other species except neon (21.56 eV) so that ionization can readily occur in such collisions. Two quite different experimental techniques have been used to investigate these reactions, the afterglow method and the beam method. Each of these methods has its individual strengths and weaknesses, especially in coping with the problem of measuring the $2\,^3S$ and $2\,^1S$ cross sections separately. Since, as will be seen below, the two methods give rather different results for the $2\,^1S$ reactions while agreeing well for the $2\,^3S$ reactions, it is worthwhile looking at both of these techniques in some detail.

1. *The afterglow technique*

In a gas discharge electrons, ions, photons, and metastable neutrals are continually created and lost through a wide variety of collision processes. If the discharge is turned off, however, energy is no longer being pumped into the gas, and only exothermic or resonant reactions can occur. There is a period of time, called the afterglow, during which the gas relaxes back to its equilibrium state. Although the situation is complicated, it is still much easier to analyze than the discharge itself, and by monitoring the concentration of various species in the afterglow a number of reaction rates may be inferred. In particular, metastables can no longer be created by direct excitation in the afterglow, but are lost, primarily through diffusion to the walls or through ionizing collisions with other neutrals. Benton et al. [93] used this technique to measure destruction cross sections for metastable helium. They monitored the helium $2\,^1S$ and $2\,^3S$ densities in the afterglow using selective absorption of radiation at 3889 Å ($2\,^3S-3\,^3P$) and 5016 Å ($2\,^1S-3\,^1P$). By measuring these densities in the afterglow as a function of time, first in a pure helium afterglow, and second with a small amount of some impurity gas added, they could determine the cross section for the destruction of helium metastables in collisions with the impurity species. However, due to an incomplete analysis of the other reactions occurring in the afterglow, at least some of their cross sections are probably incorrect.

There are two important difficulties connected with the time-dependent afterglow method. The afterglow period only lasts for some milliseconds, so that fast response time detection must be used. Also, when the afterglow is extinguished, the discharge must be restarted and then turned off to create another afterglow, so that the useful detection time is only a small fraction of the total experiment time. The second problem is that the impurity gas which is added may itself be affected during the discharge – in particular, if the impurity is a molecule it may become dissociated, so that the metastable destruction cross section which is measured is for some unknown mixture of impurity atoms and molecules.

As a consequence of this, the time-dependent afterglow method has been superceded, to a large extent, by the so-called flowing afterglow method, in which these problems are avoided. By flowing gas rapidly down a long tube, a time-dependent afterglow is converted to a spatially dependent afterglow, and reactant species may be added after the discharge region. A typical flowing afterglow apparatus is shown in Fig. 8-4-1. Helium gas flows through the tube at a velocity of about 10^4 cm/sec, at a pressure of about 1 torr, and is ionized and excited by either a hot cathode or cold cathode discharge. Since the afterglow persists for about 10 milliseconds in time, at the above flow velocity it will persist for about 1 meter in space. Reactant gas is added on axis through an entry nozzle in the tube, and the increase or decrease of various species is monitored either by observing the reaction zone optically or by use of the mass spectrometer M.

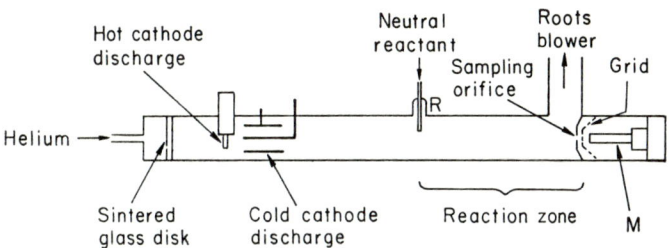

Fig. 8-4-1. Typical flowing afterglow apparatus.

An accurate mathematical analysis of this system is complicated, and has already been discussed in some detail [94–96]. It can be shown that to a good approximation the axial variation of the primary species (say helium metastables) is given by

$$I = I_0 \exp\left[-\left(D + \frac{Bk\dot{N}}{\pi a^2 v_{av}}\right)\frac{L}{v_{av}}\right] \qquad (8\text{-}4\text{-}11)$$

where I is the primary species density at some distance L down the tube, I_0 is the density at $L = 0$, D is a term which is due to loss of primaries through diffusion to the walls, $\dot{N}$ is the rate at which reactant gas is added, k is the reaction rate for the destruction of the primaries in collisions with the reactant, a is the tube radius, v_{av} is the average flow velocity of the helium, and B is a (pressure-dependent) constant, near unity, which allows for the fact that the flow velocity is not constant but depends on both axial and radial position in the tube. The behavior of eq. (8-4-11) at various positions in the afterglow tube is shown in Fig. 8-4-2, where the log of the density of the primary species is plotted as a function of distance along the tube. In region I primaries are being lost only due to diffusion. In region II, eq. (8-4-11) does not hold, since the concentration of reactant is not uniform. In region III, the log of the primary density is again decreasing uniformly, but at a faster rate due to the additional loss process of collisions with the reactant.

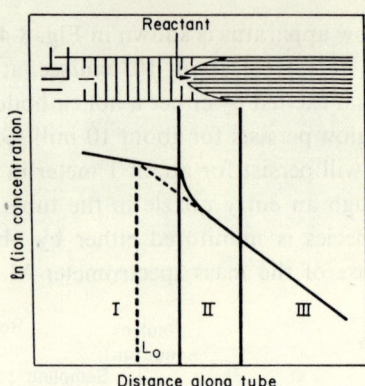

Fig. 8-4-2. Axial variation of primary species concentration.

If a plot is made of the primary species density at some distance L from the point where the reactant is introduced, versus the rate at which reactant is added, the result will be a straight line of the form

$$\log I = C - \left(\frac{BkL}{\pi a^2 v_{av}^2}\right) \dot{N} \tag{8-4-12}$$

where C is a constant. Thus the reaction rate k can be obtained from the slope of this line. This equation is completely analogous to eq. (8-4-10), since $\dot{N}/\pi a^2 v_{av}$ is the reactant number density, k/v_{rel} is the cross section Q (v_{rel} is the average relative velocity between primary and reactant) and Lv_{rel}/v_{av} is the interaction path length in the reference frame where the reactant is at rest. Fig. 8-4-2 shows that the effective position L_0 of reactant gas entry does not coincide with the actual physical position, so that to make absolute measurements it is important that this be taken into consideration.

The two principal measurements of helium metastable destruction cross sections using the flowing afterglow technique are those of Bolden et al. [97] and Schmeltekopf and Fehsenfeld [63]. Bolden et al. measured only the He($2\,^3$S) reactions, since they found that the reaction

$$e + He(2\,^1S) \rightarrow e + He(2\,^3S) + 0.79 \text{ eV} \tag{8-4-13}$$

which for thermal electrons has a cross section of about 10^{-14} cm^2 [98] converted effectively all of the helium singlet metastables to the $2\,^3$S state in the region immediately following the discharge. Their apparatus is shown in Fig. 8-4-3. The discharge was excited by a hot cathode which was pulsed on and off at a frequency of 7 Hz by the pulse generator P. The purpose of this pulse system was to permit sensitive measurement of the metastable

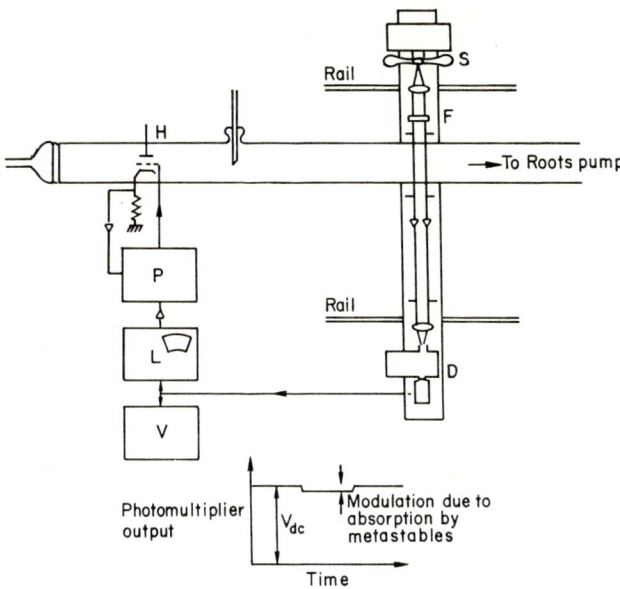

Fig. 8-4-3. Apparatus of Bolden et al.

helium density using selective absorption (see § 8-3D) and a lock-in amplifier. Radiation at 3889 Å was created in the helium discharge S, selected by the filter F, and the radiation transmitted through the afterglow tube was detected with the photomultiplier D. The lock-in amplifier L and voltmeter V were used to detect the a.c. component of the photomultiplier signal at 7 Hz, and thus gave a relative measure of the He ($2\,^3S$) density in the afterglow tube. Since the entire selective absorption system was mounted on rails, it could be moved along the afterglow tube, and by measuring the He ($2\,^3S$) density as a function of position the effective entry position L_0 of the reactant gas (see Fig. 8-4-2) could be determined experimentally. The metastable destruction cross sections obtained in this experiment will be discussed in § 8-4B-3.

The experiment of Schmeltekopf and Fehsenfeld [63] measured cross sections for the destruction of both $2\,^3S$ and $2\,^1S$ metastables. Their apparatus is shown in Fig. 8-4-4. They used an electron gun instead of a discharge to excite (and ionize) the helium and create an afterglow. Two alternative methods were employed to prevent He($2\,^1S$) destruction through reaction (8-4-13). One method was to set their electron gun at an accelerating voltage less than the ionization potential of helium, so that the creation of slow

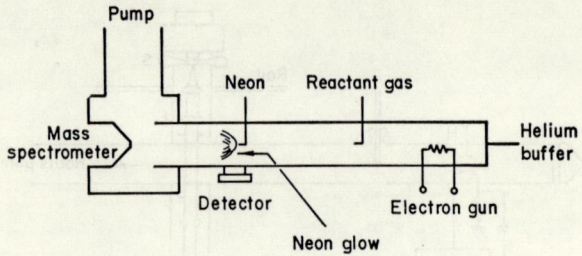

Fig. 8-4-4. Apparatus of Schmeltekopf and Fehsenfeld.

electrons was minimized. The other method was to add a small amount of SF_6 to the helium. The reaction

$$e + SF_6 \rightarrow SF_6^- \tag{8-4-14}$$

has a very large cross section for thermal electrons [99], in fact large enough so that most of the thermal electrons react with SF_6 rather than with $He(2\,^1S)$. Although the SF_6 also reacts with helium metastables, the reaction with thermal electrons is much faster so that the thermal electrons are almost completely removed without significant attenuation of the metastables. In almost all the $2\,^1S$ measurements which were made, the first method was used.

The method of monitoring the helium metastable density used by Schmeltekopf and Fehsenfeld was quite different from that used by Bolden et al. As shown in Fig. 8-4-4, a second reactant inlet was inserted in the afterglow tube downstream from the first, and neon was added through this inlet. If a sufficient concentration of neon is introduced, all the helium metastables which have survived to this point will react with the neon via excitation transfer, producing excited neon in radiating states. The experiment was run in two modes. In mode 1, the electron gun accelerating voltage was kept high (and no SF_6 was added) so that only $2\,^3S$ metastables were present. There are four neon states of the $2p^5\,4s$ configuration (with energies of 19.66, 19.69, 19.76 and 19.78 eV) which are nearly resonant with $He(2\,^3S)$. These states decay either to the ground state or to the $2p^5\,3p$ configuration. Since the former transitions are in the far ultraviolet and the latter in the infrared, neither is convenient for observation. Thus the transition which was actually observed was one from the $2p^5\,3p$ configuration to the $2p^5\,3s$ ($2p_{10}$–$1s_5$ in the Paschen notation) at 7032.4 Å. Since no $2\,^1S$ metastables were present when this line was observed, this transition provided an unambiguous measure of the $2\,^3S$ density.

In mode 2, the electron gun accelerating voltage was reduced (or SF_6 added) so that both $He(2\,^1S)$ and $He(2\,^3S)$ were present. The $2\,^1S$ state of helium is nearly resonant with a state at 20.56 eV of the neon $2p^5\,5s$ configuration ($3s_5$ in the Paschen notation), and this state cannot be excited by $He(2\,^3S)$. Thus the transition from $2p^5\,5s$ ($3s_5$) to $2p^5\,3p$ ($2p_{10}$) at 5689.8 Å was used to monitor $He(2\,^1S)$.

However, there is a serious problem which must be overcome before the flowing afterglow technique can be used to measure $He(2\,^1S)$ destruction cross sections. As the $2\,^1S$ metastables encounter the reactant gas Penning ionization occurs as the most probable reaction channel. This process creates electrons which are quickly thermalized by elastic collisions with ground state helium. These thermal electrons can then react with the $2\,^1S$ metastables via reaction (8-4-13), thus producing additional attenuation. A quantitative estimate of this effect can be obtained by considering $k\dot{N}$ in eq. (8-4-11) to consist of two separate contributions,

$$k\dot{N} = k_{Pen}\dot{R} + k_{ST}\dot{e} \qquad (8\text{-}4\text{-}15)$$

where k_{Pen} is the Penning ionization rate, $\dot{R}$ is the rate at which reactant is added, k_{ST} is the singlet to triplet conversion rate for reaction (8-4-13), and $\dot{e}$ is the rate at which electrons are formed by Penning ionization. Since

$$\dot{e} = k_{Pen}\dot{R}ML/v_{rel} \qquad (8\text{-}4\text{-}16)$$

where M is the metastable density, eq. (8-4-15) becomes

$$k\dot{N} = k_{Pen}\dot{R}(1 + k_{ST}ML/v_{rel}). \qquad (8\text{-}4\text{-}17)$$

The rate for singlet to triplet conversion is $k_{ST} = 3.5 \times 10^{-7}$ cm³/sec [93], and putting in typical values for the other parameters, $M = 3 \times 10^9/\text{cm}^3$, $L = 40$ cm, $v_{rel} = 10^5$ cm/sec, the result is that more than 30% of the singlet metastables are lost due to reaction (8-4-13). Thus the metastable $2\,^1S$ destruction cross section would be overestimated by a factor of 1.5. Schmeltekopf (private communication) has indicated that this problem was avoided by working with extremely low metastable densities ($\sim 10^6$ cm⁻³) so that the additional term in eq. (8-4-17) would be negligible. It is conceivable, however, that at such low primary density other spurious effects, such as those due to resonance photons created in the discharge region, could influence the measurement.

2. The beam technique

Only one beam experiment, that of Sholette and Muschlitz [100], has measured absolute chemiionization cross sections for helium metastable–neutral collisions (although several others have measured relative cross sections – see § 8-4C). The problem in a beam experiment is that if one uses the attenuation method, so that absolute metastable intensities are not required, then one must somehow avoid beam attenuation due to elastic scattering, the cross section for which is typically 10 times larger than that for ionization. The afterglow technique avoids this problem by using a very high background pressure of helium so that the "elastic scattering loss" of helium metastables is via diffusion in the background helium gas and is not a function of reactant gas density. A beam experiment, however, is inherently a low pressure experiment, and elastic scattering must be taken into account by some other means. If, on the other hand, a beam experiment uses the collision product detection method, then in order to derive absolute cross sections an absolute measurement of the metastable intensity must be made. This is a difficult measurement to make accurately, as is discussed in § 8-3.

The experiment of Sholette and Muschlitz basically uses the collision product detection method, although the apparatus geometry is unusual and contains elements of both methods. Their apparatus is shown in Fig. 8-4-5. The entire region to the left of the aperture H_1 is filled with helium at a pressure of 0.02 to 0.1 torr. The plate P is the mounting for an electron gun of variable energy which is used to excite some of the helium to the metastable levels. A beam of helium atoms, containing a small frac-

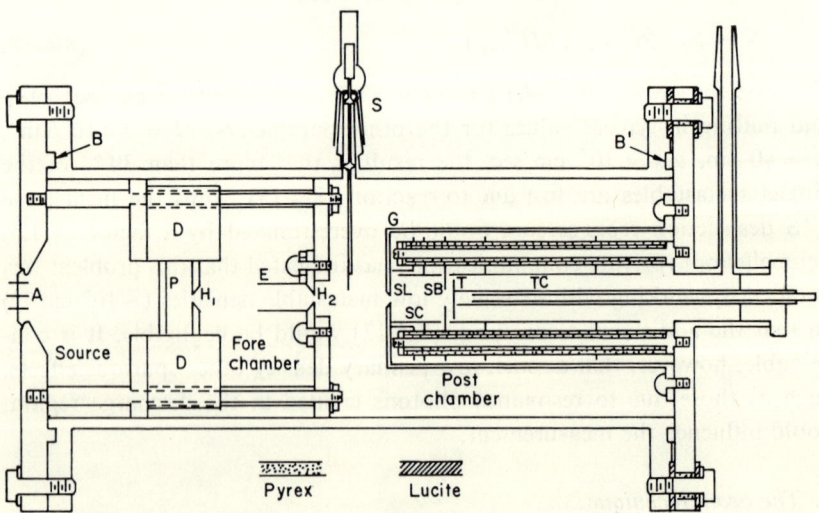

Fig. 8-4-5. Apparatus of Scholette and Muschlitz.

tion of metastables, ions, and electrons, emerges from the aperture H_1. Charged particles are removed by a potential applied across the plates E, and the beam is then collimated by the aperture H_2. The temperature of the helium gas in the source is measured by a thermocouple located immediately behind the electron gun collector. Since there are thermal gradients in the region of the electron gun this method may not give a very reliable beam temperature, and in fact the measured temperature is probably higher than the mean gas temperature. The velocity distribution of the beam is a Maxwellian multiplied by v, since atoms with a higher velocity have a larger probability per unit time of emerging from H_1. For the metastables, however, this is counteracted by the fact that the probability of excitation per unit time is proportional to $1/v$, so that if metastables make few collisions before emerging from H_1 the two effects tend to cancel. If the exciting electrons are forward scattered the metastables are heated by the excitation process but since the location of the thermocouple tends to make the measured temperature too high these two effects may also cancel, and the assumption that the metastables have a Maxwellian distribution of velocities characterized by the temperature read by the thermocouple may be a reasonable approximation.

The beam passes through the fore chamber and post chamber, in each of which the pressure is about 10^{-5} torr, and enters the scattering chamber which contains reactant gas at a pressure of the order of 6×10^{-4} torr. The scattering chamber consists of four electrically independent electrodes SL, SC, SB, and T, shielded from the rest of the apparatus by the cylindrical can G. Metastables which suffer no ionizing or elastic scattering collision will reach electrode T, where they will liberate a secondary electron with some probability γ. Elastically scattered metastables will generally fail to reach T and will be collected on SB or SC, also liberating secondary electrons. If SB and SC are made sufficiently negative, while SL and T are at ground, secondary electrons produced at T will not be able to escape because of the field between T and SB. However, secondary electrons produced at SB and SC will be attracted to SL or to T, and SL will also collect the electrons produced in metastable-neutral ionizing collisions. Thus the sum of the (negative) currents read at SL and T is

$$-I_1 = I_{\text{Ion}} + \bar{\gamma} I_{\text{ScM}} \qquad (8\text{-}4\text{-}18)$$

where I_{Ion} is the current of electrons produced by ionizing reactions and I_{ScM} is current of elastically scattered metastables. Now if SC and SB are made positive (with SL and T still at ground), secondary electrons produced at T will be attracted to SB, and a positive current I_2 will be read at T. Also, SL will now collect a current I_3 due to the positive ions produced in ionizing reactions.

$$I_2 = \bar{\gamma} I_{\text{UnscM}} \tag{8-4-19}$$

$$I_3 = I_{\text{Ion}} \tag{8-4-20}$$

where I_{UnscM} is the current of metastables which suffer no collisions. The current of metastables entering the collision region is therefore

$$I_0 = I_3 + (I_2 - I_1 - I_3)/\bar{\gamma} \tag{8-4-21}$$

and the current of positive ions produced is

$$I_+ = I_3. \tag{8-4-22}$$

From eq. (8-4-10)

$$I_+ = I_0(1 - \exp(-\bar{Q}_i NL)) \tag{8-4-23}$$

so that

$$\bar{Q}_i = (1/NL)[\ln I_0/(I_0 - I_+)] \tag{8-4-24}$$

and the cross section can thus be determined absolutely. In contrast to the afterglow experiments, the experiment of Sholette and Muschlitz measures the total ionization cross section rather than the total metastable destruction cross section.

The above analysis, however, does not take into account the fact that there are two metastable states of helium, and this complicates the experiment a great deal. It means that

$$\bar{\gamma} = (\gamma_T I_T + \gamma_S I_S)/(I_T + I_S) \tag{8-4-25}$$

where the subscripts S and T refer respectively to singlet and triplet metastable states. Thus unless $\gamma_T = \gamma_S$ one must know not only the individual γ's but also the ratio R of singlet to triplet metastable intensity in the beam, in order to derive $\bar{Q}_i$. In the same way,

$$\bar{Q}_i = (Q_T I_T + Q_S I_S)/(I_T + I_S) = (Q_T + R Q_S)/(R + 1) \tag{8-4-26}$$

($R = I_S/I_T$), and if R is known as a function of electron energy in the metastable source then a plot of $(R+1)\bar{Q}_i$ versus R will give a straight line whose slope is Q_S and whose intercept on the ordinate gives Q_T.

The most serious potential error in the experiment of Sholette and Muschlitz is in their assumption about the secondary emission coefficient for metastable helium striking a gas-covered surface. They took the value of Stebbings [49] of 0.29 for $2\,^3S$ metastables, and assumed it to be the same for $2\,^1S$ metastables as well. More recent work (see § 8-3A) indicates that the secondary emission coefficient is a strong function of the amount of gas and other impurities which are adsorbed on the surface, and that the actual coefficient appropriate to the Sholette and Muschlitz apparatus could have been quite different. Dunning et al. [47, 48] and MacLennan [44] have observed γ's differing by as much as an order of magnitude for contaminated surfaces. Also, the assumption that $\gamma_S/\gamma_T = 1$ is probably not accurate; however, the measured cross sections are not very sensitive to this ratio, so that comparatively little error is introduced by this assumption. Other possible sources of error which were considered by Sholette and Muschlitz are the effect of helium metastables elastically scattered in the backward direction ejecting secondary electrons from SL, and the ejection of secondary electrons by the positive ions which are collected. Sholette and Muschlitz estimated that neither of these effects could produce more than a 10 % error in the measured cross sections. Other spurious effects may have been present as well, but the possible error due to these is probably much less than that introduced by their assumptions for γ.

The above effects concern the measurement of the quantity $\bar{Q}_i$. There are also other potential errors in the extraction of separate cross sections Q_T and Q_S from $\bar{Q}_i$. If the same $\bar{Q}_i$ is observed when the electron gun accelerating voltage is changed, then $\bar{Q}_i = Q_T = Q_S$, and this was observed by Sholette and Muschlitz for all target gases except H_2. However, $\bar{Q}_i$ is rather insensitive to changes in R, as is shown in eq. (8-4-26). It is difficult to estimate R as a function of electron energy, since it depends not only on the electron excitation functions for singlet and triplet metastables, but also on the effects of beam scattering and the imprisonment of resonance radiation (see § 8-2A-2). From the various measurements which have been made [25–29] the maximum possible variation in R as the electron energy is varied from 30 to 60 eV is from 0.5 to 1.5. If Q_S/Q_T were 1.5, the variation in $\bar{Q}_i$ would only be 10 %. Thus this is inherently not a very accurate way of extracting Q_T and Q_S and leads to large uncertainties unless the scatter in the data is extremely small. Also, for the case of H_2, in which some variation of $\bar{Q}_i$ with electron energy was observed, Sholette and Muschlitz used values of R from unpublished work of Frost and Phelps to find Q_S and Q_T. This is undoubtedly a hazardous procedure, since R is generally apparatus-dependent, but must be known quite accurately in order to derive the separate cross sections.

3. Results

Before giving the various results obtained for absolute cross sections, it is useful to define more explicitly what these cross sections represent. In general, any cross section varies with the energy of the particles, and a cross section is well-defined only when it is measured at some definite, known interaction energy. When measurements are carried out on some ensemble of particles having a distribution of velocities, the cross section is averaged over this velocity distribution. This true average cross section is generally not the same as the apparent cross section derived, say, from eq. (8-4-10), and in fact the true average cross section cannot be found unless its functional dependence on energy (and hence velocity) is known. Thus approximations must be resorted to, and the simplest one is to assume the cross section is constant with energy and to carry out the proper velocity averaging procedure, keeping the cross section as a constant. If the cross section does not vary too rapidly with velocity over the region in which the velocity distribution function is large this will be a fairly good approximation.

In the afterglow experiments, the quantity which comes most naturally from the experimental data is the reaction rate, k. In terms of the cross section,

$$k = \int v Q(v) f(v) \, dv \qquad (8\text{-}4\text{-}27)$$

where v is the relative velocity of the colliding particles, and $f(v)$ is the distribution function of these relative velocities. If we replace $Q(v)$ by a constant $\bar{Q}$, then the average relative velocity can be calculated [101], with the result that

$$k = \bar{Q}(\bar{v}_1^2 + \bar{v}_2^2)^{\frac{1}{2}} \qquad (8\text{-}4\text{-}28)$$

where $\bar{v}$ is the average velocity of the particle,

$$\bar{v} = (8 k_B T / \pi m)^{\frac{1}{2}} \qquad (8\text{-}4\text{-}29)$$

T being the temperature, m the mass, and k_B the Boltzman constant. This simple result is due partly to the fact that both species of particles are described accurately by a Maxwellian distribution of velocities.

For a beam traversing a region containing gas at some temperature T_g, the situation is somewhat more complicated. As a simple illustration, assume that the beam is monoenergetic, each beam atom having velocity v_A, and that the target gas particles each have velocity v_G randomly distributed in direction. The time that a beam atom spends in the target gas region is

$$t = L/v_A \tag{8-4-30}$$

where L is the path length of the beam atoms through the target gas. If $\bar{v}_r$ is the mean relative velocity of beam and target atoms, then the effective collision path length is

$$X = L(\bar{v}_r/v_A) \tag{8-4-31}$$

and the effective collision path length differs from the physical path length unless $v_A \gg v_G$. The "true" cross section Q (i.e., for target particles at rest) is then related to the measured cross section Q_m by

$$Q = Q_m(v_A/\bar{v}_r). \tag{8-4-32}$$

The relative velocity for the assumptions made above is given by

$$v_r = (v_A^2 + v_G^2 - 2v_A v_G \cos\theta)^{\frac{1}{2}} \tag{8-4-33}$$

and averaging over angle,

$$\bar{v}_r = v_A + (v_G^2/3v_A) \quad \text{if} \quad v_A > v_G \tag{8-4-34a}$$

$$\bar{v}_r = v_G + (v_A^2/3v_G) \quad \text{if} \quad v_A < v_G. \tag{8-4-34b}$$

From this,

$$Q = Q_m(1+\tfrac{1}{3}\alpha)^{-1} \quad \text{if} \quad \alpha < 1 \tag{8-4-35a}$$

$$Q = Q_m\left[\alpha^{\frac{1}{2}}\left(1+\frac{1}{3\alpha}\right)\right]^{-1} \quad \text{if} \quad \alpha > 1 \tag{8-4-35b}$$

where $\alpha = v_G^2/v_A^2$.

If the cross section itself is velocity-dependent, there is another correction to be made. The calculated "true" cross section Q is $Q(\bar{v}_r)$, and v_r depends on v_G. In order to correct for this, one must know (or assume) the velocity-dependence of the cross section

$$Q(v) = Q_0 g(v) \tag{8-4-36}$$

and then

$$Q(v_A) = Q(\bar{v}_r)\frac{g(v_A)}{g(\bar{v}_r)} \tag{8-4-37}$$

where $Q(\bar{v}_r)$ is calculated from eq. (8-4-35).

If now we allow the target gas to have a thermal distribution of velocities, the measured cross section will be related to the "true" cross section by

$$Q_{\mathrm{m}}(v_{\mathrm{A}}, T_{\mathrm{G}}) = \iint Q(v_{\mathrm{r}}) \frac{v_{\mathrm{r}}}{v_{\mathrm{A}}} f(v_{\mathrm{G}}) dv_{\mathrm{G}} d\omega \tag{8-4-38}$$

where ω is the angle between v_{A} and v_{G}, and $f(v_{\mathrm{G}})$ is the velocity distribution function for the gas. If the beam particles also are distributed in velocity, then

$$Q_{\mathrm{m}}(T_{\mathrm{a}}, T_{\mathrm{G}}) = \iiint Q(v_{\mathrm{r}}) \frac{v_{\mathrm{r}}}{v_{\mathrm{A}}} f'(v_{\mathrm{A}}) f(v_{\mathrm{G}}) dv_{\mathrm{A}} dv_{\mathrm{G}} d\omega. \tag{8-4-39}$$

Again, the velocity-dependence of the cross section must be known in order to extract the "true" cross section from the measured cross section. Rosin and Rabi [102] have calculated correction factors, assuming a velocity-independent cross section, for the case

$$f'(v_{\mathrm{A}}) = 2(h_{\mathrm{A}} M_{\mathrm{A}})^2 v_{\mathrm{A}}^3 \exp(-h_{\mathrm{A}} M_{\mathrm{A}} v_{\mathrm{A}}^2) \tag{8-4-40}$$

$$f(v_{\mathrm{G}}) = (4h_{\mathrm{G}}^{3/2}/\pi^{1/2}) v_{\mathrm{G}}^2 \exp(-h_{\mathrm{G}} v_{\mathrm{G}}^2) \tag{8-4-41}$$

where $h = 1/k_{\mathrm{B}} T$ and k_{B} is the Boltzmann constant. They express their correction term as

$$Q = 2\pi^{1/2} Q_{\mathrm{m}} \Phi(\alpha) \tag{8-4-42}$$

where $\alpha = \bar{v}_{\mathrm{G}}^2/\bar{v}_{\mathrm{A}}^2 = T_{\mathrm{G}} M_{\mathrm{A}}/T_{\mathrm{A}} M_{\mathrm{G}}$. Values of $\Phi(\alpha)$ are given in Table 8-4-1. It is interesting to note that the correction factors calculated from eq. (8-4-35) are very close to those calculated by Rosin and Rabi, indicating that the correction is not very sensitive to the exact form of the velocity distribution functions.

TABLE 8-4-1

Values of $\Phi(\alpha)$ for a number of values of α

α	$\Phi(\alpha)$	α	$\Phi(\alpha)$	α	$\Phi(\alpha)$	α	$\Phi(\alpha)$
0	$1/2\sqrt{\pi}$	0.545	0.236	3.0	0.155	10	0.094
0.05	0.275	0.8	0.221	4.0	0.140	15	0.0813
0.1	0.265	0.9	0.216	5.28	0.127	21	0.0714
0.2	0.256	1.0	0.211	7.0	0.113	25	0.0643
0.3	0.248	2.0	0.173	8.0	0.107	30	0.0589
						35	0.0548

More recently, Berkling et al. [103] have calculated correction factors using several different assumptions about the velocity distribution functions and the velocity-dependence of the cross section.

TABLE 8-4-2

$He(2\ ^3S)$ destruction cross sections in $Å^2$ (SA: Stationary Afterglow, FA: Flowing Afterglow, B: Beam)

Target species	Method	SA	FA	FA	B	FA	SA	B	SA	SA	FA	B
	Error	20–50%	20%	30%	10–25%	30%	9%	30%	10%	25%	30%	+100% −50%
	Ref.	[93]	[97]	[63]	[100][a]	[104]	[105]	[44][a]	[106]	[107]	[108]	[109]
Ne		0.28		0.28								
Ar		6.6	7	5.3	7.6		7.7	9	0.97			
Kr		10.3	8	7.7	9							
Xe		13.9	11	9.8	12							
H_2		6.0	1.5	1.5	1.4[b]						2.5	
D_2				1.5								
N_2		6.4	5	5.2	7	10.5						
CO			8	7.3	7	18.7						
NO			16	18.2								
O_2			15	16.2	14	38.5						
CO_2			43	44.2		76.9						
N_2O				33.3								
NH_3			53	60.0								
SF_6				20.6								
CH_4			12	9.8								
C_2H_6				18.8								
C_3H_8				24.4								
C_4H_{10}				32.3								
Hg									140			
H										22		
Na												14

[a] cross section for production of ions.
[b] corrected for H_2 velocity.

Finally, the results of the various absolute measurements which have been made are presented in Tables 8-4-2–8-4-4. The errors quoted are the authors' estimates of maximum possible systematic error. In a number of cases, however, these are probably underestimates. It is interesting to note that, with a few exceptions, there is good agreement between the various experiments for $He(2\ ^3S)$ cross sections, but that there is marked disagreement for the $He(2\ ^1S)$ cross sections. Thus there is good justification for performing an experiment in which the ratio $\bar{Q}_S/\bar{Q}_T$ is measured. Recently two such experiments have been reported, which are described below. Discussion of possible causes of the disagreement between various measurements of $He(2\ ^1S)$ cross sections will be deferred until the results of the ratio measurements have been presented.

TABLE 8-4-3

He(2 ^{1}S) destruction cross sections in Å^2 (SA: Stationary Afterglow, FA: Flowing Afterglow, B: Beam)

Target species	Method Error Ref.	SA 20–50% [93]	FA 30% [63]	B 10–25% [100][a]	B 30% [44][a]	B +100% −50% [109][a]
Ne		4.1	4.7			
Ar		55	16.4	7.6	9	
Kr		64	28.2	9		
Xe		103	36.4	12		
H_2			2.3	0.94[b]		
D_2			2.4			
N_2			12.5	7		
CO			23.7	7		
NO			32.6			
O_2			44.6	14		
CO_2			84.6			
N_2O			70.6			
NH_3			95.0			
SF_6			52.6			
CH_4			26.8			
C_2H_6			42.6			
C_3H_8			53.8			
C_4H_{10}			69.8			
Na						17

[a] cross section for production of ions.
[b] corrected for H_2 velocity.

TABLE 8-4-4

Miscellaneous reactions

Reaction	Reference	$Q(10^{-16} cm^2)$
Ne* + H_2 → Ne	[107]	7
Ne* + D_2 → Ne		11
Ne* + CH_4 → Ne		25
Ne* + NH_3 → Ne		80
Ne* + Ar → Ar$^+$	[106]	2.6
	[53]	4.8*
Ar* + Ar → Ar$_2^+$	[110]	310
Hg* + Hg* → Hg$^+$	[111]	500
He(2 ^{3}S) + He(2 ^{3}S) → He$^+$	[112]	120
	[113]	100
He(3 ^{3}D) + He → He$^+$	[114]	14
He(3 ^{3}P) + He → He$^+$		2.1
He(3 ^{1}D) + He → He$^+$		4.6
He(3 ^{1}P) + He → He$^+$		1.9

* at $v = 5.5 \times 10^4$ cm/sec.

C. Measurements of Singlet–Triplet Cross Section Ratios

The ratio of cross sections for ionization by He($2\,^1S$) and He($2\,^3S$) has been measured recently by Dunning and Smith [115]. Their apparatus is similar to that shown in Fig. 8-3-1. Briefly, it utilizes a beam of metastable helium traversing the reactant gas, with the ions produced in metastable–neutral collisions being detected. Singlet metastables in the beam can be selectively removed by irradiating the beam prior to the interaction region with light from a helium discharge lamp. Thus when all the singlet metastables are removed a pure triplet metastable cross section is measured; when the discharge lamp is turned off, the increase in metastable current is a measure of the singlet fraction of the beam, and the increase in positive ion current is due to singlet reactions. The apparatus was previously used to make accurate absolute measurements of the secondary electron yields for both singlet and triplet metastables, so that no a priori assumptions had to be made about these quantities. The electron gun used by Dunning and Smith to produce helium metastables was unusual in that it produced an electron beam colinear with, but oppositely directed to, the helium beam. Application of momentum and energy conservation shows that the effective temperature of the metastables is reduced, and assuming an original beam temperature of 315 °K and an electron energy of 30 eV the effective metastable temperature is about 225 °K, if the electrons are scattered primarily in the forward direction.

Ratio measurements have also been carried out by Hotop et al. [28]. Their apparatus is shown in Fig. 8-4-6. A beam of helium atoms from the multichannel source M is partially excited by a crossed electron beam,

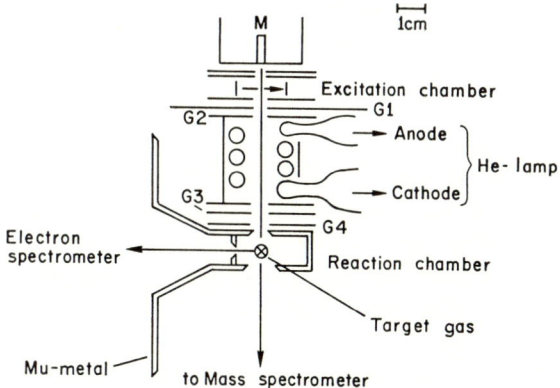

Fig. 8-4-6. Apparatus of Hotop and Niehaus.

traverses a region in which it can be irradiated by light from a helium discharge lamp in order to quench singlet metastables, and then enters the interaction region containing a target gas beam. Ions formed from ionizing collisions may be extracted into a mass spectrometer. The electron spectrometer is capable of separating electrons produced in singlet metastable collisions from those produced in triplet metastable collisions, and in this experiment was used to ensure that the radiation used to remove singlet metastables from the beam did not alter the flux of triplet metastables. The current of ions of some given mass was measured first with the helium discharge lamp on and then with it off, and the data analyzed using

$$(I_{\text{off}} - I_{\text{on}})/I_{\text{on}} = N_S Q_S / N_T Q_T \qquad (8\text{-}4\text{-}43)$$

where N_S and N_T are respectively the singlet and triplet metastable beam fluxes. Since Hotop et al. had no direct knowledge of the appropriate secondary emission coefficients for singlet and triplet metastable helium, they chose to determine the ratio N_S/N_T by assuming that Q_S/Q_T for N_2 is 1.0, as was found by Sholette and Muschlitz [100]. Added support for this procedure was provided by the later experiment of Dunning and Smith, who found Q_S/Q_T for N_2 to be 1.02 ± 0.30. This normalization procedure is further strengthened in that after making the above assumption, Hotop et al. could then derive the ratio N_S/N_T as a function of electron energy and compare the results with other measurements of the same quantity. These results have already been shown in Fig. 8-2-2; although this ratio is apparatus-dependent, the agreement is fairly good, and on this basis it is unlikely that the assumption made for N_2 is in error by more than a factor of two.

Hotop et al. also made measurements of Q_S/Q_T for rare gas targets using two different helium beam temperatures, 320 °K and 90 °K. In connection with these measurements, it is interesting to look at the effect of the excitation process on the metastable beam temperature. The electrons, with an energy of 50 eV, collided with ground state helium atoms at an angle of 90°. For a helium atom having a velocity corresponding to $\bar{v}$ for 320 °K, the minimum scattering angle of the helium is $5\frac{1}{4}°$ for triplet excitation and $5\frac{1}{2}°$ for singlet excitation. However, according to their apparatus drawing the maximum acceptance angle of the apparatus was about 5°. Thus the only metastables which can reach the collision region are either those which emerge from the multichannel source travelling off-axis and are scattered toward the beam axis in the excitation process, or those which are travelling on-axis but have a velocity somewhat greater than $\bar{v}$, and thus are scattered through a smaller angle in the excitation process. The effect of this certainly results in the metastable tamperature exceeding that of the ground state atoms, although the amount of the increase depends on the angular distribution of helium atoms emerging from the multichannel source. Also, since the differential cross section for electron excitation of the singlet metastable level falls off much faster with increasing scattering angle than that for the triplet level [116], the average scattering angle of the triplets will be greater than for the singlets, so that the ratio N_S/N_T will be affected as well. At a helium temperature of 90 °K the situation is even worse, since the minimum helium scattering angle is then more than 10°, and the effect on the ratio N_S/N_T may be much greater. Thus their quoted temperatures (which were appropriate for the ground-state atoms) may bear little resemblance to the actual effective metastable temperatures in the beam.

The results of various measurements of Q_S/Q_T for several reactant species are given in Table 8-4-5. The error given is the authors' estimate, and is included only for cases in which an error estimate was made specifically for the cross secton ratio.

TABLE 8-4-5

Ratio of destruction cross sections for He(2 ^{1}S) and He(2 ^{3}S)
(SA: Stationary Afterglow, FA: Flowing Afterglow, B: Beam)

Ref.	Meth.	Error	Q_S/Q_T								
			Ar	Kr	Xe	O_2	N_2	CO	NO	H_2	Na
[93]	SA		8.3	6.2	7.4						
[63]	FA	10%	3.1	3.6	3.7	2.7	2.4	3.1	1.8	1.5	
[100][a]	B		1	1	1	1	1	1		0.67	
[115][a]	B	25%	1.1	1.3	1.3	1.3	1.0	0.93	0.93	0.67	
[44][a]	B		1								
[28][a]	B		1.2	1.5	1.5	0.75[b]	1	1.1	0.69	0.57[c]	
[109][a]	B										1.2

[a] cross section for production of ions.
[b] production of O_2^+.
[c] production of H_2^+.

A clear discrepancy is apparent between the results of afterglow experiments and those of beam experiments. The experiment of Benton et al. [93] almost certainly obtained singlet cross sections which are too large, due to not taking account of the effect of thermal electrons in their afterglow. However, this still leaves a discrepancy of a factor of about $2\frac{1}{2}$ between the flowing afterglow measurement of Schmeltekopf and Fehsenfeld and the beam measurements. Even though several of the beam measurements required a priori assumptions, it is difficult to believe that they could be in error by more than 50 % for the cross section ratio, since taking a ratio tends to eliminate many of the sources of systematic error.

One possible cause for this disagreement could lie in the fact that afterglow experiments measure the cross section for destruction of metastables, while the beam experiments measure the cross section for the production of ionization. Since the $2\,^1S$–$1\,^1S$ transition is less strongly forbidden than the $2\,^3S$–$1\,^1S$, it is possible (although unlikely) that in reactions of singlet metastables there is a significant probability of radiative quenching, without production of an ion, but that this probability is insignificant for triplet metastables. However, the cross section for quenching of singlet metastables in collision with ground state helium atoms has been measured by Phelps [117] to be 3×10^{-20} cm^2, and calculated by Burhop and Marriott [118] to be less than 5×10^{-22} cm^2. Even though theory and experiment disagree, they are unanimous in that the cross section for this process is exceedingly small; if this holds true for other targets as well, as seems likely, then this process must be an insignificant metastable loss mechanism.

If the flowing afterglow measurements are correct, then a number of separate beam experiments have given invalid results. It is difficult to conceive of any physical effect which would invalidate the beam method for determining these cross section ratios. On the other hand, the flowing afterglow data are the result of only one experiment which was performed under difficult conditions, and it would seem that a possible explanation for the discrepancy may lie in some spurious effect in the use of the flowing afterglow technique at low primary density.

D. Models of Chemiionization Reactions

In addition to the experiments discussed in the preceding section, which attempted to measure absolute cross sections or reaction rates, a great deal of work has been done on other aspects of metastable–neutral collisions. There are many interesting aspects which can be studied, such as the probability of reactions proceeding via the various channels indicated by reations (8-4-1) to (8-4-7), the way in which these probabilities and the total reaction cross section depend on temperature (or energy), the distribution of excited states in the product ions, and the energy spectrum of product electrons. Information gained from various experiments has led to the construction of models for these chemiionization reactions. The word model is used here rather than theory, in that the principal effort has been to elucidate the essential physics of such reactions and, sometimes, to provide semiquantitative estimates of cross sections, rather than to do a rigorous quantum-mechanical calculation. In this section we shall discuss a number of these models and their validity in light of experimental data.

One of the earliest models put forth for chemiionization reactions is that of the orbiting collision. If the interaction potential between two particles can be represented by

$$V(R) = -CR^{-S} \tag{8-4-44}$$

then a simple classical analysis shows that if the impact parameter is less than some critical value

$$\rho_0 = \left(\frac{C(S-2)}{2E}\right)^{1/S}\left(\frac{S}{S-2}\right)^{\frac{1}{2}} \tag{8-4-45}$$

then the time derivative of R is always negative – i.e., the particles spiral inwards toward each other until $R = 0$, or until a repulsive part of the interaction potential is reached. Thus the cross section for orbiting collisions is given by

$$Q_0 = \pi\rho_0^2 = \frac{\pi S}{S-2}\left(\frac{C(S-2)}{2E}\right)^{2/S} \tag{8-4-46}$$

and, averaging this over a Maxwellian velocity distribution, the corresponding rate coefficient is

$$k_0 = \frac{2\pi^{\frac{1}{2}}}{\mu^{\frac{1}{2}}}\frac{SC^{2/S}\Gamma(2-2/S)(2k_B T)^{(\frac{1}{2}-2/S)}}{(S-2)^{1-2/S}} \tag{8-4-47}$$

where μ is the reduced mass and T is the temperature. Due to this spiraling the interaction time will be comparatively long, and thus if a reaction is possible, one might expect that it will occur with high probability. This model has been used with some success for ion–molecule reactions, where it is often found that the reaction cross section is approximately equal to the orbiting cross section.

This model has been applied to metastable helium ractions by Ferguson [119], Bates et al. [120], and Bell et al. [121]. The long-range interaction between two neutral particles is a Van der Waals interaction, for which $S = 6$, so that

$$Q_0 = 1.66 \times 10^{-16}(C/E)^{\frac{1}{3}} \quad \text{cm}^2 \tag{8-4-48}$$

and

$$k_0 = 6.35 \times 10^{-9} C^{\frac{1}{3}} T^{\frac{1}{3}}/\mu^{\frac{1}{2}} \quad \text{cm}^3/\text{sec.} \tag{8-4-49}$$

When these expressions are evaluated, using appropriate values for C, it is found that, although there is some disagreement between the orbiting cross sections calculated by Ferguson and those calculated by Bates et al. and Bell et al., the measured cross sections are generally smaller than the orbiting cross sections by at least a factor of 10. Thus most of the predicted orbiting collisions do not produce ionization, but only elastic scattering. Since an orbiting collision produces large angular deflection of the particles, one would expect to see a large amount of large-angle elastic scattering of metastables. There is some experimental evidence, however, that this is not in fact the case (Smith and Muschlitz [122], Richards and Muschlitz [30], Smith and Dunning, private communication).

Other experimental work pertinent to this model concerns the distribution among vibrational states of the molecular ions produced in metastable–molecule collisions. This has been studied both spectroscopically [123–125] and by observation of the energy distribution of the Penning electrons [55, 126, 127]. It is found (with one or two exceptions) that the dstribution of vibrational states is very close to that for a Franck–Condon transition from the ground state of the neutral molecule to one of the states of the molecular ion. This indicates that ionization occurs in such a way that the nuclear part of the molecular wave function is not significantly perturbed by the metastable. If an orbiting complex were formed, one would expect more perturbation to be observed.

The orbiting model has also been questioned by Jones and Robertson [128], who have pointed out the importance of the repulsive part of the potential curve. The He($2\,^3$S)–Ar interaction potential is known experi-

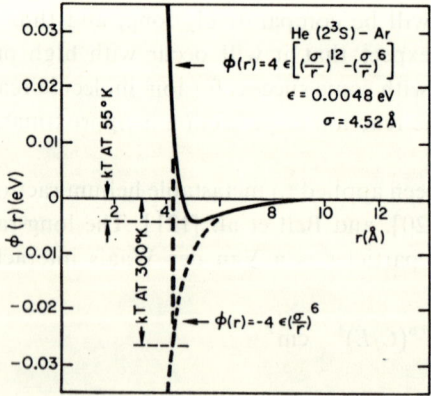

Fig. 8-4-7. Lennard–Jones He($2\,^3$S)–Ar interaction potential.

mentally from the total scattering measurements of Rothe et al. [31], and is shown in Fig. 8-4-7. It is characterized by an extremely shallow potential well at rather large internuclear separation, and a strong repulsive core. If the critical impact parameter for orbiting given by eq. (8-4-45) is less than the radius of the repulsive core, orbiting cannot occur. The critical impact parameter is energy-dependent, so that one can calculate a maximum energy for orbiting reactions to occur; for He(2 ^{3}S)–Ar this corresponds to a temperature of 35 °K! Thus for this case the collision at room temperature is better described as a hard sphere collision rather than as an orbiting complex. It is expected that similar considerations apply for other target species. In the absence of curve-crossing, the polarizability of the target must be exceedingly large in order that orbiting may occur at room temperature interaction energies.

Jones and Robertson [129] have also attempted to provide experimental evidence for the influence of the repulsive core by measuring the temperature dependence of the cross section for destruction of He(2 ^{3}S) in collision with Ar in a stationary afterglow. Their results, covering a temperature range from 200 °K to 500 °K, are compared in Fig. 8-4-8 with cross section temperature dependences calculated using (A) a Lennard–Jones (6, 12) potential using the parameters given in Fig. 8-4-7, (B) an attractive potential $V(R) =$

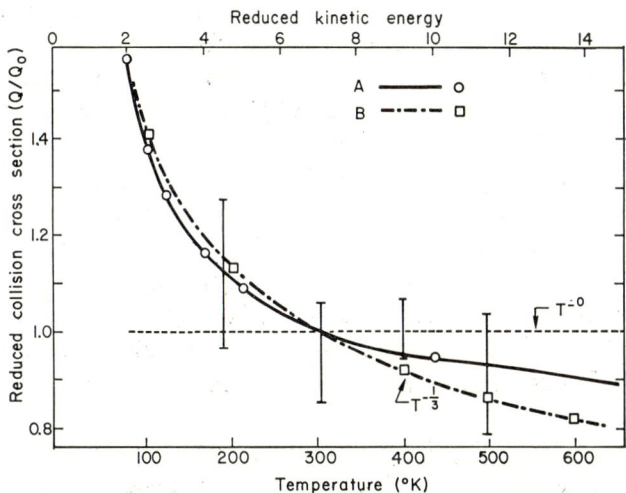

Fig. 8-4-8. Temperature variation of the He(2 ^{3}S)–Ar collision cross section for (A) the Lennard–Jones (6,12) potential, (B) the attractive potential $U = ar^{-6}$, and the hard sphere interaction $Q = Q_0 T^{-0}$. Also shown is the experimental data of Jones and Robertson.

CR^{-6}, and a hard sphere potential, each normalized to $Q = 1$ at 300 °K. Unfortunately, the error bars are too large to indicate any definite conclusions. Also, such a treatment neglects the fact that the probability of ionization occurring during a scattering event may be (and in fact, probably is) dependent on temperature as well.

Another model which has been utilized by a number of authors is the excitation transfer model. In this model, ionization is visualized as a two step process:

$$A^* + BC \rightarrow A + BC^* \qquad (8\text{-}4\text{-}50)$$

$$BC^* \rightarrow BC^+ + e \qquad (8\text{-}4\text{-}51a)$$

$$\rightarrow B + C. \qquad (8\text{-}4\text{-}51b)$$

Here the excitation transfer is considered as a resonant process, so that BC^* is a "superexcited" state of BC which overlaps the continuum, and a competition exists between preionization and predissociation of BC^*. (Of course, if the target is an atom, only reaction (8-4-51a) will occur.) This model was first developed by Platzman [130, 131] in order to explain certain effects in the experimental results of Jesse and Platzman [132] and Jesse [133]. Jesse and his coworkers [132–137] had been studying the ionization produced in a cell containing a noble gas when a collimated beam of alpha particles or electrons was directed through the cell. They found that the presence of a small concentration of impurity atoms or molecules (of the order of 0.1 %) increased the ionization by as much as 50 %. Some typical curves of ionization increase as a function of impurity concentration are shown in Fig. 8-4-9. The ionization increases rapidly as the impurity is first added, but reaches a plateau region over which the ionization is no longer a strong function of impurity concentration. This effect was attributed to the ionization of the impurity by metastable noble gas atoms excited by the alpha or beta rays. If no impurity is present, metastables are lost radiatively or diffuse to the walls; as the impurity is added, metastable–impurity ionization reactions start to occur, until enough impurity is present so that all metastables produce ionization.

As can be seen in Fig. 8-4-9, it was found that in most cases deuterated impurities produced more ionization than undeuterated impurities. The excitation transfer model is capable of explaining this isotope effect in that in the competition between preionization and predissociation of the "superexcited" molecule, the deuterated species will have a larger preionization

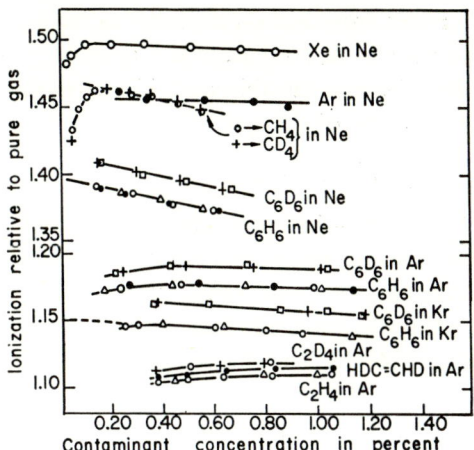

Fig. 8-4-9. Ionization as a function of impurity concentration for electron bombardment of gas mixtures.

probability because the predissociation time increases with the mass of the separating fragments.

However, the experiment was performed under high pressure conditions, the pressure in the gas cell being of the order of one atmosphere. Under these conditions, there is severe trapping of resonance radiation, and the effective lifetime of optically allowed excited states is increased to the point where it is comparable with the mean time for noble gas–impurity collisions (10^{-7}–10^{-8} sec). Thus, since the cross section for alpha particle or electron excitation of optically allowed states is much greater than that for metastable states, most of the ionization is probably due to collisions of noble gas atoms in optically allowed states with impurities, rather than from metastable–impurity collisions, and the experiment then tells us nothing about the validity of the model for metastable–neutral collisions.

This hypothesis has recently been confirmed by the work of Kubota [138], who studied the effect of adding a neon impurity in an apparatus similar to that of Jesse. As previously noted, neon cannot be ionized by helium metastables, but it can be ionized by helium in any excited state for which $n \geq 3$. Kubota observed that the amount of ionization increased sharply as neon was added, indicating that non-metastable excited states of helium are indeed important in this situation.

The excitation transfer model has also been used by several authors [139–143] for quantitative cross section calculations. Using the impact parameter method, and making certain simplifying assumptions, it is found that the excitation transfer cross section depends on the dipole matrix element for the transition $A^* \to A$, so that, in this approximation, excitation transfer does not occur for a metastable state. The cross sections obtained from this theory for reactions of the type

$$\text{He}(2\,^1\text{P}) + X \to \text{He}(1\,^1\text{S}) + X^+ + e \qquad (8\text{-}4\text{-}52)$$

are of the order of 10^{-14} cm^2 – i.e., more than a factor of ten greater than the measured cross sections for metastable ionization processes.

Isotope effects have also been observed in experiments of Muschlitz and his coworkers [144, 145], who have advanced the excitation transfer model as an explanation of their results. They measured ratios of cross sections for ionization of H_2, HD, and D_2 by a mixed beam of $2\,^1\text{S}$ and $2\,^3\text{S}$ helium metastables. The interpretation of their data is complicated by the fact that the apparent cross section ratios must be corrected for the velocity of the target molecules, as discussed in § 8-4B-3. Muschlitz et al. used the correction factors of Berkling et al. [103] assuming both beam atoms and target molecules had Maxwellian velocity distributions and equal temperatures. Their results are shown in Table 8-4-6. The data have been corrected in two different ways, one assuming Q to be velocity-independent (hard sphere model) and the other assuming Q to vary as $v^{-2/5}$ (Van der Waals R^{-6} potential). In the latter case the cross section ratios are taken at a constant velocity – i.e., the correction factor expressed by eq. (8-4-37) has been taken into account. Muschlitz et al. find a cross section which seems to increase with the mass of the target, and they explain this on the basis of an excitation transfer and subsequent competition between preionization and predissociation.

TABLE 8-4-6

Cross section ratios for Penning ionization of the molecular hydrogen isotopes

	Uncorrected	Corrected $Q = Q_0$	Corrected $Q = Q_0 V^{-2/5}$	Corrected $Q = Q_0 V^{-1}$	Corrected $Q = Q_0 V^{-7/5}$
$Q(\text{HD})/Q(\text{H}_2)$	1.03±0.05[a]	1.18[a] 1.18[b]	1.12[a] 1.12[b]	1.03[b]	0.97[b]
$Q(\text{D}_2)/Q(\text{H}_2)$	1.15±0.05[a]	1.43[a] 1.41[b]	1.31[a] 1.29[b]	1.13[b]	1.04[b]

[a] From [145].
[b] Calculated from eqs. (8-4-35) and (8-4-37).

However, there is an alternative explanation of these results. When Penning ionization is compared with the total scattering cross sections calculated using a Lennard–Jones or hard sphere potential [128] it can be seen that ionization occurs only about once in every ten collisions. The ionization probability must depend on the collision time, and since the probability of a reaction occurring during a collision is low, it is reasonable to assume that the relationship is linear, and hence the ionization probability is inversely proportional to the collision velocity. Including this effect, then, the cross section is proportional to v^{-1} for a hard sphere model, or $v^{-7/5}$ for a Van der Waals potential model. Correction factors analogous to those of Berkling et al. have not been calculated for these cases, but we have approximated these corrections using eqs. (8-4-35) and (8-4-37). This approximation seems to be justified by the agreement with the Berkling corrections for the cases $Q = Q_0$ and $Q = Q_0 v^{-2/5}$. The results of this calculation are also shown in Table 8-4-6. It can be seen that the apparent isotope effect tends to disappear when the stronger cross section velocity-dependences are used. Further information on this problem will have to await actual measurement of the cross section velocity-dependence.

Herce et al. [146] have studied the ionization of CH_4 and CD_4 by both $He(2\,^3S)$ and $He(2\,^1S)$ metastables and found no significant isotope effect.

Hotop and Niehaus [61, 147] have noted tht the fact that $He(2\,^3S)$ and $He(2\,^1S)$ metastables have nearly the same ionization cross section even though the transition probability to the ground state differs greatly suggests that simple energy exchange is not a valid model of the reaction. They propose, by analogy with the Auger ejection of electrons from a metal surface, that the reaction proceeds via electron exchange,

$$A^*(1) + B(2) \to A(2) + B^+ + e(1). \tag{8-4-53}$$

This process is indicated schematically in Fig. 8-4-10. Using the impact parameter method, the cross section for this process is given by

$$Q = 2\pi \int_0^\infty P(b, v) b \, db \tag{8-4-54}$$

where

$$P(b, v) = 1 - \exp\left(-2 \int_{R_{\min}}^\infty [W(R)/v(R)] dR\right) \tag{8-4-55}$$

and P is the ionization probability, b the impact parameter, $R_{\min}$ the classical turning point, $W(R)$ the transition frequency, and $v(R)$ the radial

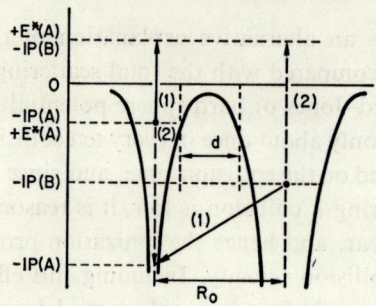

Fig. 8-4-10. Schematic illustration of two possible models of Penning ionization, (1) the exchange mechanism, (2) the direct mechanism.

velocity. Since, as discussed above, the ionization probability is small,

$$P(b, v) \approx 2 \int_{R_{\min}}^{\infty} [W(R)/v(R)] dR. \qquad (8\text{-}4\text{-}56)$$

The transition frequency can be calculated using a tunnelling model in which the probability of electron 2 tunnelling through the potential barrier is given by

$$T(R) = \exp\left[-\frac{2}{\hbar}\int_a^b p\, dx\right] \qquad (8\text{-}4\text{-}57)$$

which can be approximated by

$$T(R) = \exp\left[-\frac{2}{\hbar}(R - R_0 + d)\sqrt{(2M_e E_i)}\right] \qquad (8\text{-}4\text{-}58)$$

where d is the effective width of the potential barrier of height E_i at an internuclear distance R_0. In order to obtain the transition probability per second, $T(R)$ must be multiplied by the orbital frequency of electron 2, $v = E_i/h$. However, there is also a "steric factor" to be considered, which arises due to the fact that the potential barrier does not surround atom B uniformly. Hotop and Niehaus evaluate this steric factor f by assuming the R-dependence of $F(R)$ to be weak compared with that of $W(R)$, so that

$$\bar{f}(R) \approx f(R_0) \qquad (8\text{-}4\text{-}59)$$

and noting that from eq. (8-4-58) the effective tunnelling distance (i.e., the distance over which $T(R)$ decreases by a factor of e) is $\hbar/\sqrt{(8M_e E_i)}$, so that $\hbar^2/8M_e E_i$ is a measure of the "active surface" over which tunnelling can occur. Thus

$$f \approx \hbar^2/8M_e r_B^2 E_i \qquad (8\text{-}4\text{-}60)$$

and

$$W(R) = \frac{n\hbar}{16\pi M_e^{\frac{3}{2}} r_B^2} \exp\left[-\frac{2}{\hbar}(R - R_0 + d)\sqrt{(2M_e E_i)}\right] \quad (8\text{-}4\text{-}61)$$

where n is the number of outer shell electrons on atom B. Then

$$Q = \frac{n\hbar}{4M_e^{\frac{3}{2}} r_B^2} \exp\left[-\frac{2}{\hbar}(d - R_0)\sqrt{E_i}\right] \int_0^\infty \int_{R_{\min}}^\infty \frac{\exp\left[-(2/\hbar) R \sqrt{E_i}\right]}{v(R)}$$
$$\times b\, dR\, db. \quad (8\text{-}4\text{-}62)$$

Assuming $v(R) = v$ and taking a hard sphere model of the collision,

$$Q = \pi R_0^2 \frac{10^7 n \exp(-d\sqrt{E_i})}{\sqrt{E_i}\, v r_B^2} \left[1 + \frac{2(\sqrt{E_i}\, R_0 + 1)}{E_i R_0^2}\right] \quad (8\text{-}4\text{-}63)$$

where the constants have been evaluated and E_i is expressed in eV, v in cm/sec, r_B and R_0 in Å, and Q in Å². The second term in the brackets is small and, certainly to the accuracy of this approximation, can be neglected.

Aside from the many approximations made in this derivation, the parameter d is still not known. Hotop and Niehaus calculate d for the case of He* + Ar using the measured cross section, and then assume that d will be the same for other cases. They find good agreement with experiment for cross sections calculated in this manner for He* + Kr, Xe, H_2. We feel that the agreement is perhaps fortuitous, in that if the calculation is extended to other target species, the results differ from the measured cross sections by as much as a factor of 3. Also, we feel that the estimate by Hotop and Niehaus of the "steric factor" is not in accord with a realistic model of the effect. Taking a model of two touching spheres, each, for simplicity, of radius r_B, a simple geometric calculation of the fractional area of one sphere which is closer than a distance $\hbar/\sqrt{(8M_e E_i)}$ to the other sphere gives

$$f = \frac{1}{2\pi} \left(\frac{\hbar}{r_B \sqrt{(8M_e E_i)}}\right)^{\frac{1}{2}}. \quad (8\text{-}4\text{-}64)$$

This gives a much weaker dependence on E_i and r_B than the result of Hotop and Niehaus, and alters the calculated cross sections by as much as 50%.

It seems unlikely that the parameter d will be the same for all target species, and, since d appears in an exponential, Q is extremely sensitive to small changes in d. Also, the calculation indicates that most reactions take place in the immediate vicinity of the classical turning point, which is just the region in which such a semiclassical calculation is invalid.

Perhaps the most useful way of thinking about chemiionization reactions lies in considering the potential curves of the molecule A*B which exists for a short time during the collision. Fig. 8-4-11a shows schematically such potential curves. The particles enter along the curve A*+B, with some kinetic energy E_k. This potential curve is embedded in a set of continuum states $A+B^+ +e$, and there is some probability that a transition will occur.

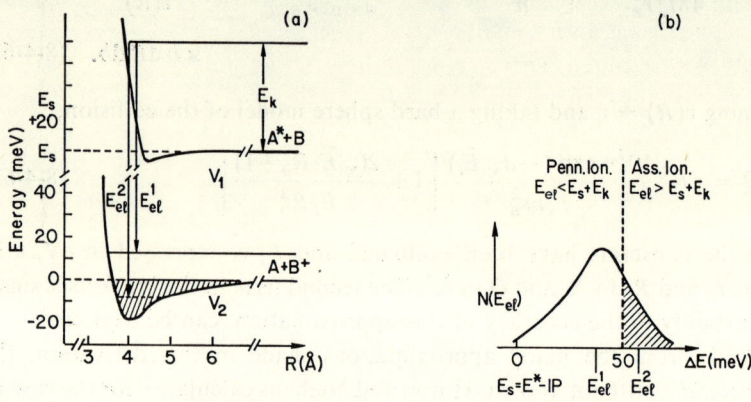

Fig. 8-4-11. (a) Schematic diagram of potential curves involved in chemiionization, (b) energy distribution of the ejected electrons.

(This process is obviously closely related to autoionization.) Using the Born–Oppenheimer approximation, one assumes that transitions must preserve the state of motion of the nuclei (i.e., position and velocity), and that the transition probability depends only on the electronic part of the wave function. The validity of this approximation is supported by the observations, previously described, of the relative vibrational state populations of molecular ions produced in Penning ionization.

The electrons produced in chemiionization reactions will have a distribution of energies, and this distribution will depend on the particular potential curves and on the variation of the transition probability with internuclear separation. In general, as indicated in Fig. 8-4-11b, the electron energy may be either greater or less than E_s+E_k, where E_s is the difference between the excitation potential of A* and the ionization potential of B. If $E_{el} > E_s+E_k$, the AB^+ molecule is bound, and the result is associative ionization; if $E_{el} < E_s+E_k$, then simple Penning ionization occurs, the additional energy being given to the nuclei. Such a correlation between the ratio of associative to Penning ionization and the energy of the ejected electrons has been observed by Hotop and Niehaus [61].

Hotop and Niehaus [109], in a study of the system He*–Na, have noted that, under special conditions, the depth of the potential well for any system He*X can be obtained from the width of the electron energy distribution. The conditions are that the potential well be much deeper for He*X than for HeX$^+$, and that the minimum in the potential curve for He*X occurs at a larger internuclear separation than that for HeX$^+$; these conditions hold for He*–Na, because He* and Na both have large polarizability, while He has a very small polarizability. Hotop and Niehaus also showed that their measured electron energy distributions were consistent with this model. They took the probability of measuring an electron with energy E_{el}, $P(E_{el})$ to be

$$P(E_{el}) = \int_0^\infty \frac{W_{el}(R)}{v_r(R,b)} \left| \frac{dE_{el}}{dR} \right|^{-1} b \, db \qquad (8\text{-}4\text{-}65)$$

where $v_r(R, b)$ is the relative collision velocity, given by

$$v_r(R,b) = \left(\frac{2}{\mu}\right)^{\frac{1}{2}} \left(-V^*(R) + E_k \left\{1 - \frac{b^2}{R^2}\right\}\right)^{\frac{1}{2}}, \qquad (8\text{-}4\text{-}66)$$

E_{el} is given by the difference between the potential curves,

$$E_{el} = V^*(R) - V^+(R), \qquad (8\text{-}4\text{-}67)$$

R is the internuclear separation, b is the impact parameter, $V^*(R)$ is the potential curve for the incoming channel, and $V^+(R)$ is the potential curve for the outgoing channel.* By estimating the shape of the potential curves, and taking the transition probability $W_{el}(R)$ to be of the form $\exp(-\alpha R/R_0)$ with α of the order of 6, they found that eq. (8-4-65) gave energy distributions which agreed well with their measurements.

This temporary molecular state model also identifies possible causes for a difference in chemiionization cross sections for He(2 ^{1}S), and He(2 ^{3}S). For He* reacting with any atom in a singlet state, say argon, the He*Ar will be in either a singlet or triplet molecular state, depending on the He*. The final HeAr$^+$ will be a doublet, and combining this with the continuum electron we get a singlet or a triplet. Thus Penning ionization by both He(2 ^{1}S) and He(2 ^{3}S) occurs through an allowed transition. Differences in the cross sections will therefore result mainly from differences in the He*Ar potential curves.

---

* In general, there may be more than one value of R for a given E_{el}. If this is the case, eq. (8-4-65) must be summed over these values.

The situation is different, however, for He* reacting with an atom in a doublet state, say sodium. For He($2\,^3S$)+Na, the initial state can be either a doublet or a quartet while the final state (with continuum electron) is a doublet, so that only $\frac{1}{3}$ of the collisions can lead to Penning ionization. For He($2\,^1S$)+Na, the initial state can only be a doublet, so all collisions can lead to ionization. If the potential curves were the same in the two cases, we would expect $Q_s = 3Q_t$; for Na, however the He($2\,^3S$)Na potential well is much deeper than that of He($2\,^1S$)Na, so that the cross sections are approximately equal.

E. Theoretical Treatments of Chemiionization

Quite recently, several theoretical treatments [148–150] of the reaction

$$\text{He}^* + \text{H} \rightarrow \text{He} + \text{H}^+ + e \tag{8-4-68}$$

have appeared. These calculations represent the first attempts at a careful quantum mechanical treatment of a chemiionization reaction. Chemiionization is in many aspects similar to autoionization and dissociative attachment, in that the reaction process depends upon the configuration interaction between a discrete molecular state and the continuum of molecular states in which this discrete state is embedded. Nakamura [151] has derived a rigorous formalism for this problem, in which the non-adiabatic effects due to the relative motion of the particles are properly included. However, the use of this formalism involves very complicated and lengthy calculations, and it has not been applied.

Since chemiionization is an exact resonance process, i.e. transitions can occur even adiabatically, the adiabatic approximation, which neglects the influence of the relative motion, can be expected to be good (at least at low energies). Nakamura [151] has used the method of Fano [152] to derive a formalism for chemiionization in the adiabatic approximation. The total electronic wave function Φ_ε is expanded as

$$\Phi_\varepsilon(r, R) = a(\varepsilon)\varphi_d + \int b(\varepsilon, \varepsilon')\varphi_{\varepsilon'}\,d\varepsilon' \tag{8-4-69}$$

where φ_d and φ_ε are the zeroth order discrete and continuum states. From the results derived by Fano,

$$|a(\varepsilon)|^2 = \frac{\Gamma(R)/\pi}{|\varepsilon(R) - W_d(R)|^2} \tag{8-4-70}$$

and from this, we can define a local complex adiabatic potential,

$$W_d(R) = \varepsilon(R) - i\Gamma(R) \tag{8-4-71}$$

where

$$\varepsilon(R) = \langle \varphi_d | H_{el} | \varphi_d \rangle + \int d\varepsilon' \frac{\langle \varphi_d | H_{el} | \varphi_{\varepsilon(R)} \rangle}{\varepsilon(R) - \varepsilon'} \tag{8-4-72}$$

and

$$\Gamma(R) = \pi \langle \varphi_d | H_{el} | \varphi_{\varepsilon(R)} \rangle. \tag{8-4-73}$$

One can now make a partial wave expansion of the Schroedinger equation using this potential, and the cross section for chemiionization can be calculated from the imaginary part of the phase shift (the real part of the phase shift gives the elastic scattering cross section). A formalism in the adiabatic approximation has also been derived in a similar way by Miller [153].

Both Fujii et al. [148] and Miller and Schaefer [149] have started out using the adiabatic approximation in their calculations of reaction (8-4-68). Each has calculated the real part of the potential for the He* + H system, and their results are shown in Fig. 8-4-12. (The $^4\Sigma$ state resulting from He($2\,^3$S) + H is not shown since transitions from it to the He + H$^+$ curve cannot occur.) Fujii et al. made a comparatively simple valence bond calculation, using properly antisymmetrized atomic orbitals, whereas Miller and Schaefer performed a sophisticated configuration interaction calculation, and their results are quite different. There are apparently several important avoided crossings which the calculation of Fujii et al. did not take into account. The He ($2\,^3$S) + H curve has an avoided crossing with the He$^+$ + H$^-$ ionic curve (the attractive part of the ionic curve is indicated in Fig. 8-4-12; by analogy with the LiH molecule, this curve will turn back up at about $3a_0$) which makes a deeper minimum at larger internuclear separation. More important, the He($2\,^1$S) + H curve, which is repulsive in the zero order approximation, has an avoided crossing with the attractive He($2\,^3$P) + H curve, which makes the He($2\,^1$S) + H potential attractive.

Fujii et al. go on to calculate the imaginary part of the potential, and use the JWKB approximation to obtain the phase shifts, and hence the cross sections. However, their results may be in error due to the approximations they used in calculating the potential curves. In particular, changing the He($2\,^1$S) + H curve from a repulsive to an attractive potential should have a large effect on the cross section.

Unfortunately, Miller and Schaefer were apparently not able to extract the width of the resonance (i.e., the imaginary part of the potential) from their calculations, so that instead of a quantum mechanical calculation of

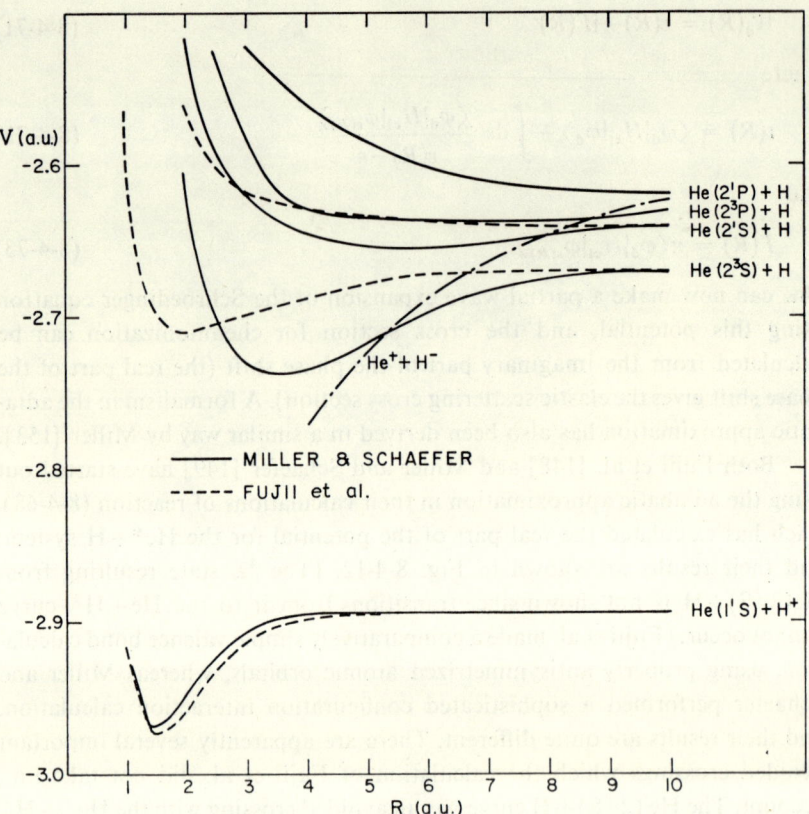

Fig. 8-4-12. Relevant potential curves for He*+H → He+H$^+$+e.

the cross section, they used essentially the orbiting approximation, applied to the potential curves which they had already calculated. It is difficult to estimate how accurate this method is liable to be. However, they predict a cross section for He($2\,^3$S)+H which is more than a factor of two above the experimental value of Shaw et al. [108], and they predict that the product will be mostly HeH$^+$, in contradiction also with the experiment.

Bell [150] has followed a somewhat different approach in his calculation of reaction (8-4-68) for the case of He($2\,^3$S). He uses an impact parameter formulation [139–141] in which the collision probability is taken as

$$P(b, E) = 1 - \exp\left[\frac{54.16}{E^{\frac{1}{2}}} \int_{R_C}^{\infty} \gamma(R) \left(1 - \frac{b^2}{R^2} - \frac{U(R)}{E}\right)^{-\frac{1}{2}} dR\right] \quad (8\text{-}4\text{-}74)$$

where $\gamma(R)$ is the probability per unit time for an electronic transition to a continuum state to occur at internuclear separation R, $U(R)$ is the interatomic potential, and R_c is the classical turning point defined by

$$1 - \frac{b^2}{R_c^2} - \frac{U(R_c)}{E} = 0. \tag{8-4-75}$$

Bell then calculates the transition probability γ in terms of the matrix element of the electronic Hamiltonian between the states $He(2\,^3S)+H$ and $He(1\,^1S)+H^+ + e$ using wave functions which have been properly antisymmetrized, and calculates $U(R)$ using the method of Heitler and London.

The result of Bell gives a cross section at 300 °K of 35 Å^2. This compares well with the result of Fujii et al. of 33 Å^2, although both are higher than the experimental result by about 50 %. The agreement between the two calculations is perhaps not surprising since, although the methods are different, both use semiclassical approximations, and both arrive at essentially the same potential curve. Also, Bell has observed that his result is insensitive to the location of the repulsive part of $U(R)$, and this is the part which is most difficult to calculate accurately. It is difficult to say at present whether the disagreement between theory and experiment is due to experimental error, or to the various approximations, particularly those of a semiclassical nature, in the calculations. For the case of $He(2\,^1S)+H$, there are as yet no experimental results, but the theoretical result of Fujii et al., $Q = 0.36$ Å^2 at 300 °K, is almost certainly too low, due to not taking account of the effect of avoided crossings on the $He(2\,^1S)+H$ potential curve. Further theoretical and experimental effort on these reactions would be of great value in understanding the basic mechanism of Penning ionization.

The only other quantum mechanical calculation of a chemiionization reaction which has been made is that of Matsuzawa and Katsuura [154] for the reaction

$$He(2\,^3S) + He(2\,^3S) \rightarrow He(1\,^1S) + He^+ + e. \tag{8-4-76}$$

Using the impact parameter formulation, they calculate a cross section which is a little less than half of the experimental cross section. In this case, however, the cross section apparently is sensitive to the location of the repulsive core of the interatomic potential. In a note added in proof, Matsuzawa and Katsuura point out that a new calculation by Klein [155] of the $He(2\,^3S)-He(2\,^3S)$ potential curve gives a smaller dimension for the repulsive core, tending to increase the calculated cross section.

F. CHEMIIONIZATION INVOLVING HIGHLY EXCITED ATOMS

The existence of atoms in very highly excited, long-lived states was observed experimentally in 1964 by Cermák [156], and in 1965 Kuprianov [86] reported the production of HeH_2^+ and ArH^+ in collisions of highly excited He and Ar with H_2. Quantitative studies of such reactions have recently been carried out by Hotop and Niehaus [157, 158], showing that the collisional properties of such atoms are quite different from those of atoms in low-lying optically forbidden states.

If one of the outer electrons of some atom is excited to one of the Rydberg levels near the continuum, shielding of the nuclear potential by the remaining electrons is virtually complete, and the electron moves in a hydrogen-like potential. For an electron with, say, a principal quantum number $n = 30$, the mean lifetime (averaged over all angular momentum substates) is about 0.3 msec, the ionization potential is 0.015 eV, and the "orbital radius" is 476 Å.

Hotop and Niehaus [157] found that for highly excited noble gas atoms incident on polyatomic molecules, the collisional ionization process

$$A^{**} + B \rightarrow A^+ + e + B \qquad (8\text{-}4\text{-}76)$$

was highly efficient, with cross sections of the order of 10^{-12} cm^2. Their measured cross sections are shown in Table 8-4-7. In one case, they also observed negative ion formation via the reaction

$$Ar^{**} + SF_6 \rightarrow Ar^+ + SF_6^- . \qquad (8\text{-}4\text{-}77)$$

TABLE 8-4-7

Cross sections ($\times 10^{12}$ cm^2) for reactions (8-4-76), (8-4-77)

A^{**}	Cross section derived from A^+ measurement					From SF_6^- measurement
	H_2O	NH_2	SO_2	C_2H_5OH	SF_6	
He**	0.29±0.05	0.15±0.04	0.11±0.05	0.27±0.08	0.76±0.08	
Ne**	0.78±0.06		0.27±0.07	0.50±0.09	1.4 ±0.09	
Ar**	1.2 ±0.08	0.74±0.07	0.40±0.09	0.80±0.11	1.5 ±0.11	1.7 ±0.11

They did not, however, find reaction (8-4-76) occurring for the target molecules H_2, N_2, O_2, NO, and CH_4. Since it has often been observed that it is easier to supply the energy needed in an endothermic reaction from internal energy of one of the colliding particles rather than from kinetic energy, it may be that reaction (8-4-76) proceeds efficiently only when the target molecule is such that it may be vibrationally excited at room temperature.

The actual numbers given in Table 8-4-7 should probably be considered only as estimates of the order of magnitude of these reactions, since the specific excited state composition of the metastable beam depends on the flight time between excitation region and collision region, and on the presence of electric fields (an atom with its outermost electron having $n = 30$ can be ionized by an electric field of the order of a few hundred volts per centimeter).

In a later experiment, Hotop and Niehaus [158] studied collisions of H_2 and HD with He*, Ne*, Ar**, and Kr**. They found that while the metastables He* and Ne* produced mostly H_2^+ (or HD^+), the highly excited atoms Ar** and Kr** reacted almost entirely via rearrangement ionization, producing ArH^+ and KrH^+ (and ArD^+ and KrD^+ in the case of HD). The cross sections for these reactions were estimated to be of the order of 10^{-14} cm². Such a result might be expected if one views a highly excited atom as essentially a positive ion, with an electron somewhere in the vicinity, since the reactions

$$Ar^+ + H_2 \rightarrow ArH^+ + H \qquad (8\text{-}4\text{-}78)$$

$$Kr^+ + H_2 \rightarrow KrH^+ + H \qquad (8\text{-}4\text{-}79)$$

are known to occur with high efficiency [159, 160]. This model of the collision is strengthened by the observation that the production ratios ArH^+/ArD^+ and KrH^+/KrD^+ for collisions with HD are in accord with those observed for the corresponding ion–molecule reactions [161].

It is intriguing to speculate that further experimental results may indicate that thermal energy ion–molecule reactions could be studied using highly excited atoms instead of ions.

§ 8-5. Ion beam studies

The collision properties of positive ions have been extensively investigated during the past 50 years, so it is perhaps surprising that only in the past few years has the importance of long lived excited ions been generally appreciated. Prior to this their presence was usually ignored or unrecognized, and it was customarily assumed that all ions were in the ground state.

Among the first indications that this represented a gross oversimplification were observations of Lindholm and coworkers [162] that mass selected beams of O^+ ions behaved differently in collision with various gases depending on whether they were produced through dissociative ionization of CO, N_2O, or CO_2. In collision with N_2 the charge transfer cross sections were observed to differ by a factor greater than 4, clearly demonstrating the varying fraction of metastable excited states in the ion beam. Others, including Gilbody and Hasted [163], interpreted structure in their observed cross sections as arising from the influence of long lived excited ions.

A considerable advance in the understanding of the role of excited ions came with the use of ion sources utilizing electrons of controlled energy. With such ion sources it was observed that the collision properties of the resulting ion beam were often sensitive to the energy of the source electrons. This is a consequence of the fact that at sufficiently low electron energy, ions are formed only in their ground state whereas with increasing electron energy the production of excited ions becomes energetically possible. Thus, for example, measurement of the cross section for charge transfer at fixed ion energy as a function of the electron energy in the ion source provides qualitative information on the charge transfer cross sections for the excited ions. This technique has been exploited by a number of workers and an illustrative set of data obtained by Stebbings et al. [164] for the reaction $O^+ + N_2 \rightarrow O + N_2^+$ is shown in Fig. 8-5-1.

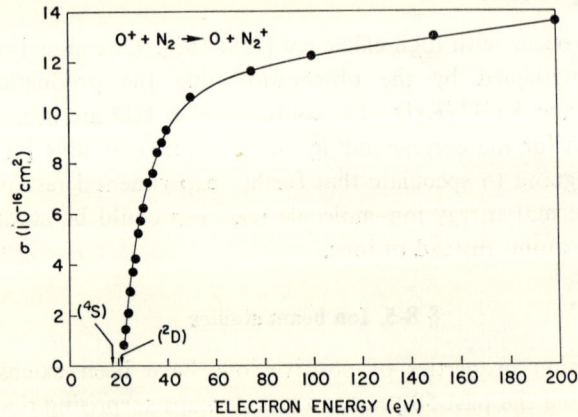

Fig. 8-5-1. Dependence of the cross section of N_2^+ production in O^+–N_2 collisions on the energy of the electrons used in the ion source to produce O^+ from O_2.

This reaction is endothermic for ground state $O^+(^4S)$ ions but exothermic for excited $O^+(^2D)$ and $O^+(^2P)$ ions, which have natural lifetimes of 1.3×10^4 sec and 6 sec respectively. The observed cross sections for N_2^+ production at a fixed ion energy of 100 eV are shown as a function of the energy of the electrons producing O^+ from O_2 in the ion source. The uncertainty in the energy scale is at least 1 eV; nevertheless the observed threshold corresponds closely with that at 20.6 eV for $O^+(^2D)$ production via

$$e + O_2(X^3\Sigma_g^-) \rightarrow O^+(^2D) + O^-(^2P) + e. \tag{8-5-2}$$

The absence of reaction for the ground state ^4S ions, which have a threshold at 17.3 eV, is well demonstrated and shows that the energy required to overcome the energy defect is not readily provided from translational energy at this ion energy.

The curve in Fig. 8-5-1 may be understood as follows. At electron energies below 20.6 eV the O^+ beam is entirely composed of ^4S ions which do not react with N_2 to give N_2^+. At higher electron energies ^2D and possibly ^2P ions are produced and undergo charge transfer with N_2 with high efficiency. The variation of the observed cross section for N_2^+ production therefore merely reflects the increasing fraction of these excited ions in the beam as the electron energy is increased. It is to be noted that for the first 20 eV or so above threshold the cross section, and therefore the excited state fraction, rises rapidly. Above about 100 eV the composition of the beam, as inferred from constancy of the apparent cross section for charge transfer, is relatively insensitive to the electron energy.

The observed cross section σ may be expressed in the form

$$\sigma = \sum_n \sigma_n f_n \tag{8-5-3}$$

where σ_n is the cross section for those ions in state n which are present in fractional abundance f_n. It is evident that without knowledge of the fractional abundances of the various constituents of the ion beam the individual cross sections σ_n cannot be determined. These measurements therefore merely demonstrate that the ground state cross section is very small or zero, and the cross section for some excited state or states is large. Quantitative interpretation of the data shown in Fig. 8-5-1 is contingent upon the determination of the ion beam composition. Stebbings et al. [164] reported initial measurements of this type for an O^+ beam, and subsequently Turner et al. [165] extended this work. Their procedure may be understood as follows.

Let a beam of ions all in state i pass through a chamber of length L filled with gas at a number density N with which the ions react in a two body process. Ions which do not react reach a collector and give an ion current

$I_{i,N}$. The ions that do react as well as any product ions, are assumed not to reach the collector. $I_{i,N}$ is then related to $I_{i,0}$, the ion current at zero number density of gas, by

$$I_{i,N} = I_{i,0} \exp(-NQ_i L) \tag{8-5-4}$$

where Q_i represents the sum of the cross sections for all processes by which ions in state i are lost. A semilogarithmic plot of $I_{i,N}/I_{i,0}$ versus gas pressure thus gives a straight line such as A in Fig. 8-5-2. For ions in a different state, a similar equation would hold,

$$I_{j,N} = I_{j,0} \exp(-NQ_j L) \tag{8-5-5}$$

where Q_j represents the corresponding sum of the cross sections for ions in state j. For $Q_j > Q_i$ a plot of $I_{j,N}/I_{j,0}$ gives a line such as B in Fig. 8-5-2.

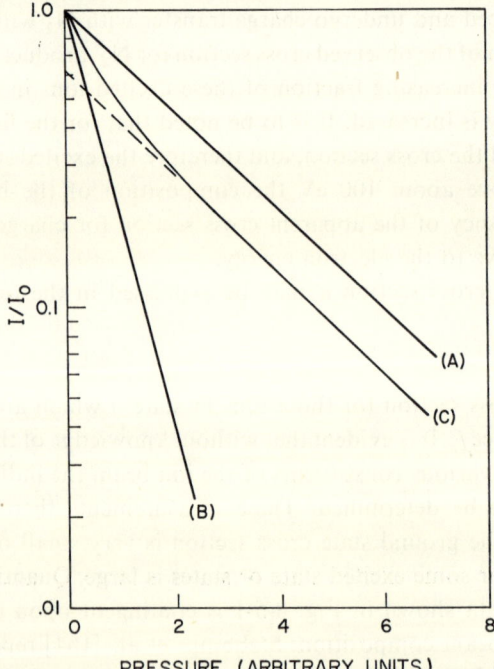

Fig. 8-5-2. Attenuation curves for ions traversing a region of reactant gas. (A) type i only; (B) type j only; (C) mixture of types i and j.

If now a beam of ions containing a fraction f of ions in state i, the remainder in state j, passes through the same chamber the detector current is given by

$$I = (1-f)I_0 \exp(-NQ_jL) + fI_0 \exp(-NQ_iL). \tag{8-5-6}$$

A plot of I/I_0, where I_0 is the ion current at zero number density, would then have the shape of curve C. Thus a curve of this type indicates the presence of (at least) two states in an ion beam, and may in principle yield the fractional abundances of these states. This is achieved by extrapolating to zero pressure the straight portion of curve C obtained at high pressures. The intercept is then the fraction of ground state ions i. The excited state fraction may then immediately be determined. The apparatus used by Turner et al. [165] is shown in Fig. 8-5-3. The ion source is of the electron bombardment type and is identical to that used to obtain the data shown in Fig. 8-5-1. The ions are extracted from the ion source and mass analyzed prior to entry into the reaction chamber, into which gas may be introduced. Within the reaction chamber, which is 10 cm in length, are three Faraday cages. One, located near the entrance to the chamber, may be inserted into the ion beam path to determine the ion current entering the chamber. When this cage is removed the ions pass through the chamber toward two Faraday cages marked 2 and 3. Two grids placed before these cages are used to prevent slow ions produced in the chamber from reaching these cages and to prevent secondary electron emission from the cages. In general, measurements were made with cages 2 and 3 connected together, although separate measurements of their currents provided information as to the extent of elastic scattering.

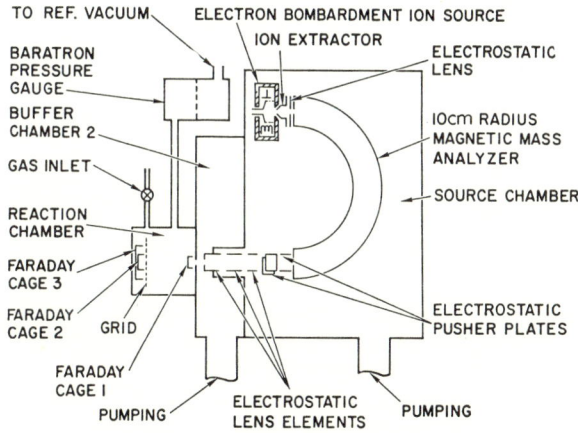

Fig. 8-5-3. Apparatus of Turner et al.

A typical set of data is shown in Fig. 8-5-4, where I/I_0 is plotted semilogarithmically versus pressure for O^+ ions passing through N_2. Data are given for three different values of the energy of the electrons used in the ion source to produce O^+ from O_2. For $E_s = 20$ eV, all ions must be $O^+(^4S)$, and the experimental data fall on a straight line A. Curves B and C, for which E_s is 50 eV and 100 eV respectively, show an initial rapid fall followed by a straight section approximately parallel to A. Turner et al. analyzed these data as described above. Curve D is obtained as the difference between the experimental data points of curve B and the straight line extrapolation of the high pressure part of B to zero pressure. It clearly represents the attenuation of the excited state component of the O^+ beam in passing through the chamber.

Fig. 8-5-4 is therefore explained as follows: The beam of pure ground state ions produced at $E_s = 20$ eV is attenuated by elastic scattering (and

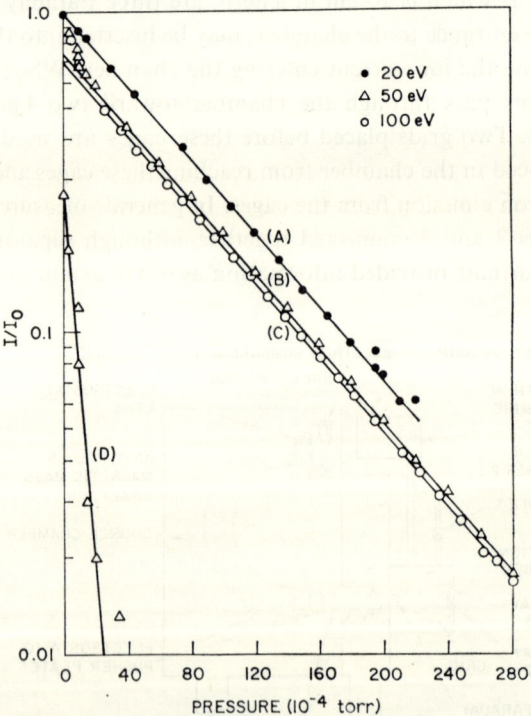

Fig. 8-5-4. Measured attenuation curves of O^+ in N_2. (A) $E_s = 20$ eV; (B) $E_s = 50$ eV; (C) $E_s = 100$ eV. Curve (D) represents the decay of excited ions alone and is derived from (B) as described in the text.

conceivably by inelastic processes other than charge transfer, for which the cross section has already been shown to be zero). At $E_s = 50$ eV the excited ions are rapidly removed from the beam by charge transfer, and the remaining ground state ions are then slowly removed by elastic scattering as the pressure is increased.

The fact that curve D, which gives the fraction and attenuation of the excited ions, is linear implies that only one excited species is present in the beam, although it is of course possible that two excited species having identical scattering behavior in N_2 could be present. The case for a single excited component is greatly strengthened, however, by the observation that substitution of argon for N_2 in the collision chamber again leads to a linear form for the excited component.

A summary of the results of Turner et al. [165] for oxygen ions is shown in Fig. 8-5-5. They show that when electrons with energy well above threshold are used to ionize O_2 to form O^+ and O_2^+, substantial fractions of these

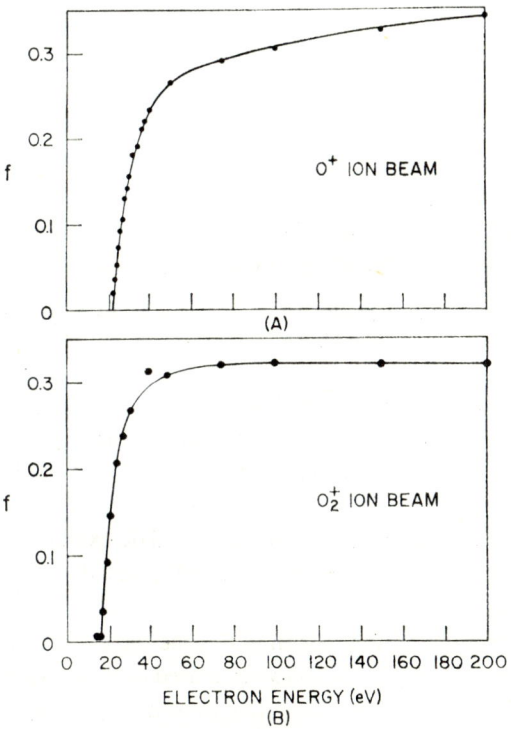

Fig. 8-5-5. Fraction f of metastable excited states in beams of O^+ and O_2^+ as a function of electron energy in the ion source.

ions are formed in excited states. In each case the results indicate that only one excited species is present. For the case of O^+ the excited state is identified as ^2D. Primary evidence for this comes from the threshold behavior in Fig. 8-5-5. The reaction with N_2 is attributed by Stebbings et al. [166] to

$$O^+(^2D) + N_2(X\ ^1\Sigma_g)_{v=0} \to O(^3P) + N_2^+(A\ ^2\Pi_u)_{v=1} \qquad (8\text{-}5\text{-}7)$$

which is near resonant. Data in Fig. 8-5-6 can then be obtained by using the measured apparent cross sections [166] together with the known composition of the ion beam.

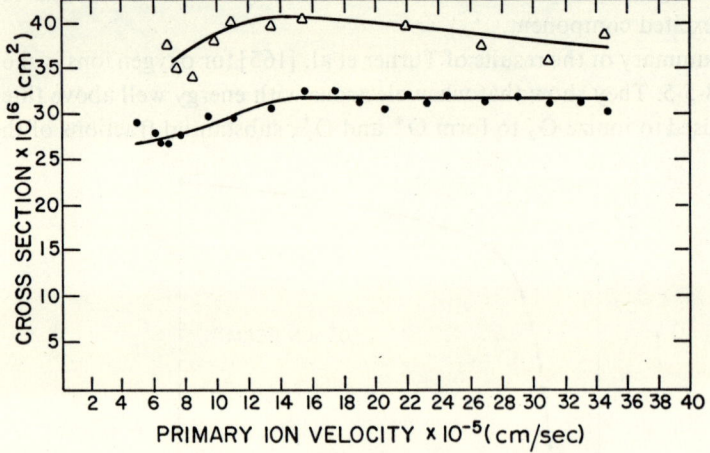

Fig. 8-5-6. Cross sections for N_2^+ production in collisions between N_2 and $O^+(^2D)$ – △ and $O_2^+(a^4\Pi_u)$ – ●.

For O_2^+ the excited state is identified as the $a\ ^4\Pi_u$ state. The cross section for charge transfer with N_2 is observed to be small (perhaps zero) for ground state $O_2^+(X\ ^2\Pi_u)$ while the cross section for the excited ion is shown in Fig. 8-5-6.

Certain complications attributed to collisional de-excitation of the metastable ions are discussed by Turner et al. [165] but do not substantially alter the discussion given above.

It must be remembered that for the O^+ ions the excited state fractions are strictly appropriate only to the ion source used in these experiments. This is because for dissociative ionization the ions are typically formed with some kinetic energy, which is different for each O^+ state. The relative numbers of extracted ions in given states may then be strongly influenced by the nature of the extraction field within the source. However, the source used in this work is of fairly typical design and the results are probably representative of many electron impact sources. This complication does not of course arise for the molecular ions, which acquire negligible kinetic energy during formation, and hence all states will be extracted with the same efficiency.

This work has been extended to the case of NO^+ by Mathis et al. [167] using the same apparatus. In this case the fraction of ions in an excited state, presumably a $^3\Sigma^+$, was found to be as high as 46% for 100 eV electrons.

One interesting point not discussed by Turner et al. relates to vibrational excitation of $O_2^+(X\,^2\Pi_g)$. As the energy of the ionizing electrons in the ion source is raised above the $X\,^2\Pi_g$ threshold at 12.08 eV vibrational excitation of this state becomes energetically possible. Examination of the form of the relevant potential curves indicates however that direct excitation of those levels with $v > 3$ is unlikely, and indeed the Franck Condon factors tabulated by Hallmann and Lanlicht [168] confirm this. However, as the electron energy is raised still further, production of higher electronically excited states of O_2^+ become possible, and some of these may then decay radiatively to $O_2^+(X\,^2\Pi_g)$. It is evident that if the radiative lifetime of these upper states is comparable with or longer than a vibrational period, population of upper vibrational levels of O_2^+ $(X\,^2\Pi_g)$ will result. In the measurements of Turner et al. the presence of such vibrationally excited ions would not be apparent since their recombination energies in Franck Condon transitions are generally insufficient for them to undergo efficient charge transfer with N_2 or Ar at these collision energies. They are thus indistinguishable in this work from O_2^+ $(X\,^2\Pi_g)_{v=0}$. In other experiments however, this may well not be the case and it must be recognized that the "ground state" fractions tabulated by Turner et al. almost certainly include vibrationally excited O_2^+ $(X\,^2\Pi_g)$ ions. The particular vibrational levels which are populated and their corresponding populations will be functions of the energy of the source electrons.

Evidence suggesting the influence of vibrational excitation on charge transfer cross sections has been presented by Amme and Hayden [169] who determined the charge transfer cross section for N_2^+–Ar collisions as a function of the ion source electron energy at a number of ion energies. Analysis of their data was hindered by lack of information regarding the composition of the N_2^+ beam in terms of the excited states. However earlier measurements by Amme and Utterback [170] and by McGowan and Kerwin [171] showed that the N_2^+–N_2 cross section at 1 keV decreased by about 15% as the ionizing electron energy was increased, thus indicating the presence of an excited state of N_2^+. Amme and Hayden analyzed their N_2^+–Ar data assuming that their beam contained 15 % of excited ions. Their measurements show that in this event these excited ions would have a cross section for charge transfer with argon which exhibits a resonance-like behavior between 50–900 eV. This would be a somewhat unlikely result if the excited ions were $N_2^+(A\ ^2\Pi_u)$ since the energy defect would then be at least 1.3 eV. However, if the ions were excited vibrationally rather than electronically, much closer energy resonance in $N_2^+(v)$–Ar charge transfer is possible, and the observed resonance-like behavior of the excited ions becomes more plausible. Recent measurements of Hollstein et al. [172] show that the lifetimes of $N_2^+(A\ ^2\Pi_u)_{v=3,4,5}$ are close to 12 μsec, and such ions are therefore almost certainly present in the N_2^+ beam of Amme and Hayden.

Other studies which are somewhat easier to interpret have been carried out with H_2^+. The H_2^+ ion is particularly suited for investigating effects of vibrational excitation, because its higher vibrational levels can be populated when it is created by electron impact ionization of H_2. In addition, electronically excited H_2^+ ions are unlikely to be present in significant amounts because these states are predominantly dissociating. Amme and Hayden [170] show that the cross section for H_2^+–Ar charge transfer falls as the ionizing electron energy is increased from 16 eV to 17 eV, indicating that the vibrational-state population changes over this interval. The relative constancy of the H_2^+–H_2 cross section over this same range of electron energy however indicates that the different vibrational states involved have comparable charge transfer efficiencies in H_2. McGowan and Kerwin [173] have investigated the effects of vibrational excitation of H_2^+ on its collision-induced dissociation. They observed a marked increase in the cross section as the energy of the ionizing electrons was raised above the threshold for ground state H_2^+ production.

The prime difficulty in the interpretation of data of this type is the absence of detailed information regarding the relative populations of the various vibrational levels. It seems reasonable to conjecture that progress in this area may result from the application of ion sources utilizing photoionization, since the state composition of the resulting ions is then known. Using the technique of photoelectron spectroscopy Turner [174] obtained the data for ionization of O_2 by helium resonance radiation shown in Fig. 8-5-7. The 0–0 component of the $O_2^+(X\ ^2\Pi_g)$ band occurs at about 12.07 eV and four additional components with a mean spacing of 0.22 eV are seen. The vibrational structures of the $A\ ^4\Pi_u$ and the $b\ ^4\Sigma_g^-$ states, with respective thresholds at 16.12 eV and 18.17 eV, are again well resolved. The state as-

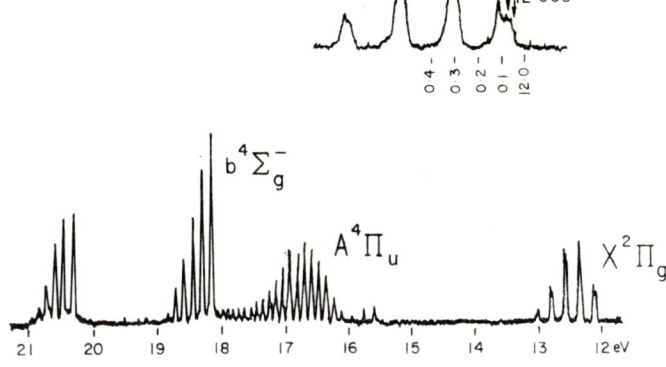

Fig. 8-5-7. Photoelectron energy spectrum of oxygen ionized by helium resonance radiation. The upper curve gives the X $^2\Pi_g$ data more clearly.

sociated with the structure having an onset at 20.29 eV has not been positively identified although it has been tentatively designated both as $^4\Sigma_u$ and as $^2\Sigma_g^-$.

It requires no great stretch of the imagination to believe that ingenious operation of an ion source utilizing photoionization could significantly advance our understanding of the role of excitation, particularly vibrational excitation, in ion–neutral collisions.

In summary, it may be said that metastable ions of certain species have been shown to exhibit collision properties markedly different from the respective ground state ions. We may extrapolate from this and assert that in general excited ionic species are likely to have collision properties as dissimilar from the respective ground state ions as from quite different ionic species. The description of a given experiment involving ions is thus clearly incomplete without a precise statement regarding the involvement of long lived excited ions.

Note added in proof: Since this article was written, there have been several new results pertaining to the collision of helium metastables with hydrogen atoms. Cohen and Lane [175] have calculated cross sections for these reactions using the interaction potentials of Miller and Schaefer [149] and the autoionization lifetimes of Fujii et al. [148]. They find that, at 0.03 eV, $Q_T = 41$ Å^2, $Q_S = 7.5$ Å^2, and that, while Q_T is relatively insensitive to changes in the assumed lifetime, Q_S is strongly dependent on this quantity. They have also checked the validity of using the JWKB method (see § 8-4E) in this problem by explicitly solving the Schroedinger equation containing the local complex potential $V(R) - i\Gamma(R)$, with the result that the use of the JWKB method results in overestimation of the thermal energy cross sections by only 7–10 %.

Miller et al. [176] have extended the calculations of Miller and Schaefer [149] and have obtained the autoionization lifetime (as a function of internuclear separation) for the $He(2\,^3S) + H(1\,^2S)$ system. Using this lifetime, they calculate a cross section at 0.03 eV of 33 $Å^2$. Their calculations also include the ratio of associative to Penning ionization, and the energy distribution of the ejected electrons.

Cohen and Lane (private communication) have pointed out that there is reason to believe that the autoionization lifetimes should be similar for both $He(2\,^3S) + H$ and $He(2\,^1S) + H$, and find that if the lifetime calculated by Miller et al. [176] is used for the $He(2\,^1S) + H$ system, then $Q_S = 74$ $Å^2$ at 0.03 eV.

Hotop et al. [177] have measured the energy distribution of the ejected electrons. From this, they can infer the ratio Q_S/Q_T (see § 8-4C), and they obtain a ratio of 1.5 ± 50 %. They also obtain from the measured distribution the well-depth of the potential curves. For $He(2\,^3S) + H$ they obtain the value 2.44 ± 0.2 eV, as compared with a calculated well-depth [149] of 1.91eV. However, Hotop et al. had to correct their measured distributions due to the presence of undissociated H_2, so that the possible error in their value of the well-depth may be larger than their estimate. This could result in there being no significant disagreement with the theory.

In summary, there seems to be a significant disagreement in the value of Q_T between the measurement of Shaw et al. [108] of 22 ± 6 $Å^2$ and the various calculated cross sections. Calculations of Q_S are very sensitive to the assumed autoionization lifetime, so that it is difficult to compare theoretical values of Q_S/Q_T with the experimental value of Hotop et al. Clearly, it would be extremely desirable to have a more accurate calculation of the autoionization lifetime for $He(2\,^1S) + H$, and further absolute experimental measurements of Q_T and Q_S.

References

1. G. BREIT and E. TELLER, Astrophys. J. **91** (1940) 215.
2. H. R. GRIEM, Astrophys. J. **156** (1969) L103.
3. A. S. PEARL, Phys. Rev. Letters **24** (1970) 703.
4. R. S. VAN DYCK JR., C. E. JOHNSON and H. A. SHUGART, Phys. Rev. Letters **25** (1970) 1403.
5. G. W. F. DRAKE, G. A. VICTOR and A. DALGARNO, Phys. Rev. **180** (1969) 25.

6. R. W. NICHOLLS, Ann. Geophys. **20** (1964) 144.
7. T. WENTINK JR. and L. ISAACSON, J. Chem. Phys. **46** (1967) 822.
8. S. G. TILFORD, P. G. WILKINSON and J. T. VANDERSLICE, Astrophys. J. **141** (1965) 427.
9. W. LICHTEN, J. Chem. Phys. **26** (1957) 306.
10. R. S. FREUND, J. Chem. Phys. **50** (1969) 3734.
11. R. M. BADGER, A. C. WRIGHT and R. F. WHITLOCK, J. Chem. Phys. **43** (1965) 4191.
12. R. A. YOUNG and G. BLACK, J. Chem. Phys. **44** (1966) 3741.
13. A. DALGARNO and R. H. G. REID, Mem. Soc. Roy. Sci. Liége **17** (1958) 57.
14. I. ESTERMANN, O. C. SIMPSON and O. STERN, Phys. Rev. **71** (1947) 238.
15. R. S. FREUND and W. KLEMPERER, J. Chem. Phys. **47** (1967) 2897.
16. D. D. BRIGLIA, Ph. D. Thesis, University of California at Los Angeles, 1963, unpublished.
17. R. T. ROBISCOE and T. W. SHYN, Phys. Rev. Letters **24** (1970) 559.
18. J. W. LOCKE and J. B. FRENCH, Vacuum Sci. and Techn. **7**, No. 1 (1969).
19. J. W. LOCKE, UTIAS Report No. 143, Institute for Aerospace Studies, University of Toronto, Canada (1969).
20. W. LICHTEN and S. SCHULTZ, Phys. Rev. **116** (1959) 1132.
21. R. F. STEBBINGS, W. L. FITE, D. G. HUMMER and R. T. BRACKMANN, Phys. Rev. **119** (1960) 1939.
22. D. HILS, H. KLEINPOPPEN and H. KOSCHMIEDER, Proc. Phys. Soc. **89** (1966) 35.
23. H. E. KAUPPILA, W. R. OTT and W. L. FITE, Phys. Rev. A **1** (1970) 1099.
24. W. R. OTT, W. E. KAUPPILA and W. L. FITE, Phys. Rev. A **1** (1970) 1089.
25. V. CERMÁK, J. Chem. Phys. **44** (1966) 3774.
26. J. L. G. DUGAN, H. L. RICHARDS and E. E. MUSCHLITZ JR., J. Chem. Phys. **46** (1966) 346.
27. H. HOLT and R. KROTKOV, Phys. Rev. **144** (1966) 82.
28. H. HOTOP, A. NIEHAUS and A. L. SCHMELTEKOPF, Z. Physik **229** (1969) 1.
29. F. B. DUNNING and A. C. H. SMITH, Phys. Letters **32A** (1970) 287.
30. H. L. RICHARDS and E. E. MUSCHLITZ JR., J. Chem. Phys. **41** (1964) 559.
31. E. W. ROTHE, R. H. NEYNABER and S. M. TRUJILLO, J. Chem. Phys. **42** (1965) 3310.
32. E. S. FRY and W. L WILLIAMS, Rev. Sci. Inst. **40** (1969) 1141.
33. R. CLAMPITT and A. S. NEWTON, J. Chem. Phys. **50** (1969) 1997.
34. D. RAPP and W. E. FRANCIS, J. Chem. Phys. **37** (1962) 2631.
35. W. LICHTEN, J. Chem. Phys. **47** (1967) 2897.
36. R. F. STEBBINGS and J. A. RUTHERFORD, J. Geophys. Res. **73** (1968) 1035.
37. R. F. STEBBINGS, A. C. H. SMITH and H. EHRHARDT, J. Geophys. Res. **69** (1964) 2349.
38. B. L. DONNALLY, T. CLAPP, W. SAWYER and M. SCHULTZ, Phys. Rev. Letters **12** (1964) 502.
39. D. C. LORENTS and J. R. PETERSON, Stanford Research Inst., Final Report, Project PAU-5027 (1965).
40. M. KAMINSKY, Atomic and Ionic Impact Phenomena on Metal Surfaces (Academic Press Inc., New York, 1965).
41. H. W. WEBB, Phys. Rev. **24** (1924) 113.
42. D. A. MACLENNAN, Phys. Rev. **148** (1966) 218;
T. A. DELCHAR, D. A. MACLENNAN and A. M. LANDERS, J. Chem. Phys. **50** (1969) 1779;
D. A. MACLENNAN and T. A. DELCHAR, J. Chem. Phys. **50** (1969) 1772.
43. F. B. DUNNING, A. C. H. SMITH and R. F. STEBBINGS, J. Phys. B **4** (1971) 1683;
F. B. DUNNING and A. C. H. SMITH, J. Phys. B **4** (1971) 1696.
44. R. F. STEBBINGS, Proc. Roy. Soc. **A241** (1957) 270.
45. E. H. FLOOD, editor, The Solid Gas Interface, Vol. I (Marcel Dekker, Inc., New York, 1966) p. 431.

46. I. Langmuir, J. Am. Chem. Soc. 30 (1918) 1361.
47. N. Hansen, Vakuum Tech. 3 (1962) 70.
48. M. G. Kagener, Proc. Acad. Sci. USSR (1958) 663.
49. H. D. Hagstrum, Phys. Rev. 96 (1954) 336; 96 (1954) 325; 91 (1953) 543; 89 (1953) 244.
50. J. Waclawski, L. R. Hughey and R. P. Madden, Appl. Phys. Letters 10 (1967) 305.
51. S.-Y. Tang, Ph.D. Thesis, University of Florida, 1970, unpublished;
    S. Y. Tang, A. B. Marcus and E. E. Muschlitz Jr., J. Chem. Phys. (1971) to be published.
52. E. H. Flood, editor, The Solid Gas Interface, Vol. II (Marcel Dekker Inc., New York, 1966).
53. E. G. McRae, J. Chem. Phys. 45 (1966) 3258.
54. V. Cermák, J. Chem. Phys. 44 (1966) 1318.
55. V. Cermák, J. Chem. Phys. 44 (1966) 3781.
56. V. Cermák, Coll. Czech. Chem. Commun. 33 (1968) 2739.
57. Z. Herman and V. Cermák, Coll. Czech. Chem. Comm. 33 (1968) 468.
58. Z. Herman and V. Cermák, Coll. Czech. Chem. Comm. 31 (1966) 649.
59. V. Cermák and Z Herman, Chem. Phys. Letters 2 (1968) 359.
60. V. Fuchs and A. Niehaus, Phys. Rev. Letters 21 (1968) 1136.
61. H. Hotop and A. Niehaus, Z. Physik 228 (1969) 68.
62. H. Hotop and A. Niehaus, J. Mass Spectr. Ion Phys. 5 (1970) 415.
63. A. L. Schmeltekopf and F. C. Fehsenfeld, J. Chem. Phys. 53 (1970) 3173.
64. M. Hollstein, J. R. Sheridan, J. R. Peterson and D. C. Lorents, Phys. Rev. 187 (1969) 118.
65. H. B. Gilbody, R. Browning, G. Levy, A. I. McIntosh and K. F. Dunn, J. Phys. B 1 (1968) 863.
66. A. C. G. Mitchell and M. W. Zemansky, Resonance Radiation and Excited Atoms (McMillan Co., 1934).
67. E. Ellis and N. D. Twiddy, J. Phys. B 2 (1969) 1366.
68. C. V. Raman, Indian J. Phys. 2 (1928) 387.
69. P. Braunlich and P. Lambropoulous, Phys. Rev. Letters 25 (1970) 986.
70. K. Latyscheff and B. Leipunsky, Z. Physik 65 (1930) 111.
71. S. N. Foner and R. L. Hudson, J. Chem. Phys. 37 (1962) 1662; 25 (1956) 601.
72. H. A. Bethe and E. E. Salpeter, Quantum Mechanics of One- and Two-electron Atoms (Springer-Verlag, Berlin, 1957).
73. M. F. A. Harrison, D. F. Dance, K. T. Dolder and A. C. H. Smith, Rev. Sci. Instr. 36 (1965) 1443.
74. H. Rausch von Traubenberg, R. Gebauer and G. Lewin, Naturwissenschaften 18 (1930) 417.
75. A. C. Riviere, in: Methods of Experimental Physics, Vol. 7A, eds. B. Bederson and W. L. Fite (Academic Press, 1968).
76. K. M. Evenson and D. S. Burch, J. Chem. Phys. 45 (1966) 2450.
77. A. M. Falick, B. H. Mahan and R. J. Meyers, J. Chem. Phys. 42 (1965) 1837.
78. I. I. Rabi, J. R. Zacharias, S. Millman and P. Kusch, Phys. Rev. 53 (1938) 318.
79. G. O. Brink, Cornell Aeronautical Labs, Report No. RM-2156-P-1, 1966; RM-2156-P-2, 1968; J. Chem. Phys. 46 (1967) 4531.
80. F. L. Mohler, P. D. Foote and R. L. Chenault, Phys. Rev. 27 (1926) 37;
    F. L. Mohler and C. Boeckner, J. Res. Nat. Bur. Stand. 5 (1930) 51, 399, 831;
    F. L. Arnot and M. B. McEwen, Proc. Roy. Soc. A166 (1938) 543; A171 (1939) 106.
81. J. A. Hornbeck and J. P. Molnar, Phys. Rev. 84 (1951) 621.
82. Z. Herman and V. Cermák, Coll. Czech. Chem. Comm. 28 (1963) 799.
83. V. Cermák, J. Chem. Phys. 43 (1965) 4527.

84. Z. HERMAN and V. CERMÁK, Nature **199** (1963) 588.
85. A. A. KRUITHOFF and F. M. PENNING, Physica **4** (1937) 430; F. M. PENNING, Z. Phys. **46** (1928) 335.
86. S. E. KUPRIANOV, Sov. Phys. JETP **21** (1965) 311; **24** (1967) 674.
87. E. C. ZIPF, Can. J. Chem. **47** (1969) 1863.
88. R. J. DONOVAN and D. HUSAIN, Chem. Rev. **70** (1970) 489.
89. W. KAUL, P. SEYFRIED and R. TAUBERT, Z. Naturforsch. **18a** (1963) 432, 884.
90. P. M. BECKER and F. W. LAMPE, J. Chem. Phys. **42** (1965) 3857.
91. R. S. BERRY, in: Molecular Beams and Reaction Kinetics, ed. Ch. Schlier, Course 44 of Proc. Intern. School of Physics "Enrico Fermi" (Academic Press, 1970).
92. E. E. MUSCHLITZ JR., in: Advances in Chemical Physics, Vol. 10: Molecular Beams, ed. J. Ross (Interscience Pyblishers, 1966); Science **159** (1968) 599.
93. E. E. BENTON, E. E. FERGUSON, F. A. MATSEN and W. W. ROBERTSON, Phys. Rev. **128** (1962) 206.
94. R. W. HUGGINS and J. H. CAHN, J. Appl. Phys. **38** (1967) 180.
95. R. C. BOLDEN, R. S. HEMSWORTH, M. J. SHAW and N. D. TWIDDY, J. Phys. B **3** (1970) 45.
96. E. E. FERGUSON, F. C. FEHSENFELD and A. L. SCHMELTEKOPF, Adv. Atomic Molec. Phys. **5** (1969) 1.
97. R. C. BOLDEN, R. S. HEMSWORTH, M. J. SHAW and N. D. TWIDDY, J. Phys. B **3** (1970) 61.
98. A. V. PHELPS, Phys. Rev. **99** (1955) 1307.
99. D. RAPP and D. D. BRIGLIA, J. Chem. Phys. **43**, (1965) 1480.
100. W. P. SHOLETTE and E. E. MUSCHLITZ, J. Chem. Phys. **36** (1962) 3368.
101. E. H. KENNARD, Kinetic Theory of Gases (McGraw-Hill, 1938).
102. S. ROSIN and I. I. RABI, Phys. Rev. **48** (1935) 373.
103. K. BERKLING, R. HELBING, K. KRAMER, H. PAULY, CH. SCHLIER and P. TOSCHEK, Z. Phys. **166** (1962) 406.
104. M. CHER and C. S. HOLLINGSWORTH, Can. J. Chem. **47** (1969) 1937.
105. C. R. JONES and W. W. ROBERTSON, J. Chem. Phys. **49** (1968) 4240.
106. M. A. BIONDI, Phys. Rev. **88** (1952) 660.
107. T. MARSHALL, J. Appl. Phys. **36** (1965) 712.
108. M. J. SHAW, R. C. BOLDEN, R. S. HEMSWORTH and N. D. TWIDDY, Chem. Phys. Letters **8** (1971) 148.
109. H. HOTOP and A. NIEHAUS, Z. Physik **238** (1970) 452.
110. P. M. BECKER and F. W. LAMPE, J. Chem. Phys. **42** (1965) 3857.
111. K. L. TAN and A. VON ENGEL, J. Phys. D **1** (1968) 258.
112. W. B. HURT, J. Chem. Phys. **45** (1966) 2713.
113. A. V. PHELPS and J. P. MOLNAR, Phys. Rev. **89** (1953) 1020.
114. M. P. TETER, F. E. NILES and W. W. ROBERTSON, J. Chem. Phys. **44** (1966) 3018.
115. F. B. DUNNING and A. C. H. SMITH, J. Phys. B **3** (1970) L60.
116. J. A. SIMPSON, M. G. MENENDEZ and S. R. MIELCZAREK, Phys. Rev. **150** (1966) 76.
117. A. V. PHELPS, Phys. Rev. **99** (1955) 1307.
118. E. H. S. BURHOP and R. MARRIOTT, Proc. Phys. Soc. **A69**.
119. E. E. FERGUSON, Phys. Rev. **128** (1962) 210.
120. D. R. BATES, K. L. BELL and A. E. KINGSTON, Proc. Phys. Soc. **91** (1967) 288.
121. K. L. BELL, A. DALGARNO and A. E. KINGSTON, J. Phys. B **1** (1968) 18.
122. G. M. SMITH and E. E. MUSCHLITZ JR., J. Chem. Phys. **33** (1960) 1819.
123. W. W. ROBERTSON, IV Int. Conf. Phys. Elec. Atom. Coll., Quebec, 1965.
124. W. W. ROBERTSON, J. Chem. Phys. **44** (1966) 2456.
125. A. L. SCHMELTEKOPF, E. E. FERGUSON and F. C. FEHSENFELD, J. Chem. Phys. **48** (1968) 2966.

126. H. Hotop and A. Niehaus, Chem. Phys. Letters **3** (1969) 687.
127. H. Hotop and A. Niehaus, J. Mass. Spec. Ion Phys. **5** (1970) 415.
128. C. R. Jones and W. W. Robertson, J. Chem. Phys. **49** (1968) 4241.
129. C. R. Jones and W. W. Robertson, J. Chem. Phys. **49** (1968) 4240.
130. R. L. Platzman, J. Phys. Radium **21** (1960) 853.
131. R. L. Platzman, Vortex **23**, No. 8 (1962).
132. W. P. Jesse and R. L. Platzman, Nature **195** (1962) 790.
133. W. P. Jesse, J. Chem. Phys. **41** (1964) 2060.
134. W. P. Jesse and J. Sadauskis, Phys. Rev. **88** (1952) 417.
135. W. P. Jesse and J. Sadauskis, Phys. Rev. **94** (1954) 764.
136. W. P. Jesse and J. Sadauskis, Phys. Rev. **97** (1955) 1668.
137. W. P. Jesse and J. Sadauskis, Phys. Rev. **100** (1955) 1755.
138. S. Kubota, J. Phys. Soc. Japan **29** (1970) 1017.
139. K. Katsuura, J. Chem. Phys. **42** (1965) 3771.
140. B. M. Smirnov annd O. B. Firsov, Sov. Phys. JETP Letters **2** (1965) 297.
141. M. Mori, J. Phys. Soc. Japan **21** (1966) 979.
142. M. Mori and H. Fujita, J. Phys. Soc. Japan **20** (1965) 432.
143. T. Watanabe and K. Katsuura, J. Chem. Phys. **47** (1967) 800.
144. E. E. Muschlitz Jr. and J. R. Penton, Fifth Int. Conf. Phys. Elec. Atom Coll., Leningrad, 1967.
145. J. R. Penton and E. E. Muschlitz Jr., J. Chem. Phys. **49** (1968) 5083.
146. J. S. Herce, J. R. Penton, R. J. Cross and E. E. Muschlitz Jr., J. Chem. Phys. **49** (1968) 958.
147. H. Hotop and A. Niehaus, Sixth Int. Conf. Phys. Elec. Atom. Coll., Boston, 1969
148. H. Fujii, H. Nakamura and M. Mori, J. Phys. Soc. Japan **20** (1970) 1030.
149. W. H. Miller and H. F. Schaefer, J. Chem. Phys. **53** (1970) 1421.
150. K. L. Bell, J. Phys. B **3** (1970) 1308.
151. H. Nakamura, J. Phys. Soc. Japan **26** (1969) 1473.
152. U. Fano, Phys. Rev. **124** (1961) 1866.
153. W. H. Miller, J. Chem. Phys. **52** (1970) 3563.
154. M. Matsuzawa and K. Katsuura, J. Chem. Phys. **52** (1970) 3001.
155. J. Klein, J. Chem. Phys. **50** (1969) 5151.
156. V. Cermák and Z. Herman, Coll. Czech. Chem. Comm. **29** (1964) 953.
157. H. Hotop and A. Niehaus, J. Chem. Phys. **47** (1967) 2507.
158. H. Hotop and A. Niehaus, Z. Physik **215** (1968) 395.
159. F. C. Fehsenfeld, A. L. Schmeltekopf and E. E. Ferguson, J. Chem. Phys. **46** (1967) 2802.
160. G. Gioumousis and D. P. Stevenson, J. Chem. Phys. **29** (1958) 294.
161. F. S. Klein and L. Friedman J. Chem. Phys. **41** (1964) 1789.
162. E. Gustafsson and E. Lindholm, Ark. Fys. **18** (1960) 219;
    E. Petterson and E. Lindholm, ibid. **24** (1963) 49;
    E. Lindholm, ibid. **8** (1954) 433.
163. H. R. Gilbody and J. R. Hasted, Proc. Roy. Soc. **A238** (1956) 334; **A240** (1957) 382.
164. R. F. Stebbings, B. R. Turner and J. A. Rutherford, J. Geophys. Res. **71** (1966) 771.
165. B. R. Turner, J. A. Rutherford and D. M. J. Compton, J. Chem. Phys. **48** (1968) 1602.
166. R. F. Stebbings, M. A. Fineman, J. W. McGowan and B. R. Turner, Technical Summary Report, General Atomic GA 6699 (1965).
167. R. F. Mathis, B. R. Turner and J. A. Rutherford, J. Chem. Phys. **49** (1968) 2051.
168. M. Hallmann and I. Lanlicht, J. Chem. Phys. **43** (1965) 1503.
169. R. C. Amme and H. C. Hayden, J. Chem. Phys. **42** (1965) 2011.

170. R. C. Amme and N. G. Utterback, Proc. Third Intern. Conf. on the Physics of Electric and Atomic Collisions (London, 1963) p. 847.
171. J. W. McGowan and L. Kerwin, Can. J. Phys. **42** (1964) 2086.
172. M. Hollstein, D. C. Lorents, J. R. Peterson and J. R. Sheridan, Can. J. Chem. **47** (1969) 1858.
173. J. W. McGowan and L. Kerwin, Can. J. Phys. **42** (1964) 972.
174. D. W. Turner, Proc. Roy. Soc. **A307** (1968) 15.
175. J. S. Cohen and N. F. Lane, Chem. Phys. Letters **10** (1971) 623.
176. W. H. Miller, C. A. Slocomb and H. F. Schaeffer, J. Chem. Phys. (1971) to be published.
177. H. Hotop, E. Illenberger, H. Morgner and A. Niehaus, Chem. Phys. Letters **10** (1971) 493.

AUTHOR INDEX

ABERTH, W., 499, *541*
ABIODUN, R., 454, *479*
ABRAMOWITZ, M., 362, *394*
ABT, H., 378, *395*
ACCAD, Y., 268, *290*
ADAMS, T. F., 374, *395*
AHRENS, T. J., 525, *543*
ALDER, K., 363, *394*
ALLEN, C. W., 234, *289*, 295, *389*
ALLEN, J. S., 108, 136, *157*
ALLEN, L., 421, *478*
ALLER, L. H., 295, 303, 305, 328, 335, 368–371, 379, 382, *389, 394, 396, 397*
ALLIS, W., 426, *478*
ALLIS, W. P., 98, 100, 102, 103, *156*
AMALDI, U., 164, 187, 196, *206*
AMME, R. C., 621, 622, *628, 629*
ANDERSON, P. W., 353, *393*
ANSDELL, D. A., 527, 532, 537, *543*
ARDILL, R., 454, *479*
ARMSTEAD, R. L., 434, 437, 463, 465, *480, 481*
ARNOT, F. L., 575, *626*
ARSHADI, A., 487, *540*
ARTHURS, A. M., 454, *479*
ASUNDI, R. K., 182, *207*
AUBREY, B., 448, 449, *481*
AUER, L. H., 311, 374–376, 378, 382, *390, 395, 397*
AUMAN, J., 380, *396*
AVRETT, E., 374, *395*
AVRETT, E. H., 301, 303, *390*

BACHER, R. F., 519, 520, *542*
BADGER, R. M., 550, *625*
BAGOT, C. H., 132, *158*
BAGUS, P. S., 332, *392*
BAILEY, T. L., 485, 524, *539, 542*
BAILEY, V. A., 154, *158*

BAKER, R. F., 536, *544*
BAKULINA, I. N., 524, 527, 533–535, *543, 544*
BANDEL, H. W., 447, *479*
BARANGER, M., 340, 341, 346, 350, *392*
BARDSLEY, J. N., 537, *544*
BARNARD, A. J., 351, 377, 388, *393*
BARNES, K. S., 352, 358, 388, *393*
BASCHEK, B., 369, *394*
BASHKIN, S., 332, 388, *392*
BASSEL, R., 443, 444, *478*
BATES, D. R., 6, 9, 10, 19, 22, 25, 27–29, 36, 66, 73, 75, 78, 80, 81, 84, 88–90, 331, 368, *392*, 418–420, 433, 448, 449, *478*, 519, 527 *542*, 597, *627*
BAUGHAN, E. C., 527, *543*
BAUMANN, H., 530, *543*
BEARDSLEY, J. N., 487, *540*
BEAUCHAMP, R. K., 499, *541*
BECK, R. J., 352, *393*
BECKER, P. M., 575, 592, *627*
BEDERSON, B., 448, 449, 452, *479, 481*
BEHRING, W., 231, *289*
BELL, K. L., 324, 325, *391*, 457, 458, *480*, 488, *540*, 597, 608, 610, *627, 628*
BELY, O., 218, 219, 232–234, 237, 239, 245, 249, 269, 270, 272, 279, 280, 288–291, 352, 354, 360, 388, *393*
BENGSTON, R. D., 346, 348, 350, 388, *392*
BEN-REUVEN, A., 346, *392*
BENTON, E. E., 578, 583, 591, 592, 595, *627*
BERG, H. F., 332, *392*
BERKLING, K., 590, 602, *627*
BERKOWITZ, J., 485, 522, 537, *539*
BERRY, R. S., 325–327, *391*, 486, 488, 489, 491, 497, 498, 521, 522, 530–534, *539–541*, 576, *627*
BETHE, H. A., 411, 418, *478*, 570, 576, *626*
BETHGE, K., 530, *543*

BHATIA, A. K., 455, 460, 461, 463, 464, 467, 468, *480*
BIBERMAN, L. M., 316, 323, *390*
BIDELMAN, W., 382, *397*
BIONDI, M. A., 151, *158*, 591, 592, *627*
BIRD, R. B., 64, *90*
BLACK, G., 550, *625*
BLAHA, M., 219, 245, 249, 287, *288, 290*
BLAU, L. M., 486, *540*
BLESS, R. C., 374, *395*
BLEWETT, J. P., 536, *544*
BOCKASTEN, K., 231, 243, *289, 290*
BOECKNER, C., 575, *626*
BÖHM, A., 324, 325, *391*, 488, 497, *540*
BÖHM, K. H., 378, *395*
BOHR, A., 363, *393*
BOKSENBERG, A., 374, *395*
BOLAND, B. C., 226, 236, 268, *288, 290*
BOLDEN, R. C., 579, 580, 591, 610, 624, *627*
BOLDT, G., 319, 349, 387, *390, 393*
BONESS, M. J. W., 526, *543*
BONNET, R., 369, 371, 372, *395*
BONSEN, T. P. M., 164, 187, 191, 192, 195, 199, *206*
BOOZ, J., 11, *89*
BORN, M., 525, *543*
BORTNER, T. E., 108, 132, 136, *157*
BOWE, J. C., 107, 135, 136, *157*
BOYD, D. B., 537, *544*
BOYD, T. J. M., 27, *89*
BOZMANN, W. R., 368, *394*
BRACKMANN, R. T., 182, *207*, 225, *288*, 523, 531, 542, 556, *625*
BRADBURY, N. E., 109, 132, 135, *157, 158*
BRANSCOMB, L. M., 322, 325–327, *391*, 486–489, 492, 494, 496, 497, 501, 517, 521, 522, 527, 530–532, 535, 536, *539–541, 543*
BRANSDEN, B. H., 434, 447, 448, 462, 464, 465, *478–481*
BRAUMAN, J. I., 491, 537, *541*
BRAUNLICH, P., 570, *626*
BREENE, R. G., 344, *392*, 488, *541*
BREHM, B., 502, 516, 521, 530, *541*
BREIT, G., 550, *624*
BRIDGES, J. M., 369, *394*
BRIGLIA, D. D., 554, 582, *625, 627*
BRINK, G. O., 574, *626*
BRION, C. E., 169, 202, 204, *207, 208*
BRISCOE, C. V., 477, *481*
BROCKLEHURST, M., 310, 350, 379, *390, 393*
BRODE, R., 452, *479*

BROMANDER, J., 243, *290*
BROMBERG, J., 448, *479*
BROSE, H. L., 132, *158*
BROWN, M., 268, *290*
BROWNING, R., 569, *626*
BRUECKNER, K. A., 8, 28, 56, 62, *89*
BUCHEL'NIKOVA, N. S., 486, *540*
BUCHNER, E. H., 536, *544*
BUHL, D., 381, *396*
BURCH, D. S., 124, *157*, 324, 325, *391*, 457, *481*, 488, 489, 494, 496, 521, 522, 531, 536, *540, 541*, 573, *626*
BURGESS, A., 218–220, 232, 233, 238, 246, 249, 270, 271, *288*, 310, 315–317, 319, 361, 363, 373, *390, 394, 395*
BURGESS, D. D., 351, 353, 377, *393*
BURHOP, E. H. S., 61, *90*, 183, 184, *207*, 596, *627*
BURKE, P., 449, *479*
BURKE, P. G., 232, 233, 236, 237, 239, 268, *289, 290*, 322, 326, 354, 356, 387, 388, *391, 393, 394*, 436, 438, 441–443, 450, 451, 453, 455, 456, 466, *478–481*
BURTON, W. M., 243, 255, 262, 263, *290*
BYDIN, Yu. F., 531, *543*

CADE, P. E., 518, 536, 537, *542*
CAHN, J. H., 579, *627*
CAIRNS, C. J., 351, 377, 388, *393*
CAIRNS, R. B., 522, 536, *542*
CALLAWAY, J., 324, 325, *391*, 408, 426, 429, 430, 436, 439, 446–448, 464, 465, 468, 469, 473, 474, *478, 479*
CAMPBELL, I. D., 525, *543*
CARBON, D. F., 303, *390*
CARLTON, T. S., 13, *89*
CARRIER, G., 6, 10, 65, 66, *88*
CARTER, E. B., 536, *544*
CARTER, V. I., 458, *480*
CAVALLERI, G., 99, 137, 152, *156, 158*
CAYREL DE STROBEL, G., 368, *394*
CAYREL, R., 368, 372, *394, 395*
CELOTTA, R., 516, 536, 537, *541, 544*
ČERMÁK, V., 557, 568, 575, 587, 598, 612, *625–628*
CERNOCH, T., 489, 497, 498, *541*
CHAMBERLAIN, G. E., 184, 185, *207*, 444, *481*
CHAN, Y. M. C., 268, *290*
CHANDLER, C. D., 330, *392*
CHANDRASEKHAR, S., 65, *90*, 296, 300, 301, 324, *389, 391*
CHANG, E. S., 455, 456, 458, *480*

AUTHOR INDEX

CHANIN, L. M., 151, *158*
CHANTRY, P. J., 523, 531, 537, *542*, *544*
CHAPELLE, J., 352, 388, *393*
CHARKIN, O. P., 519, 532–534, *542*
CHASE, D. M., 455, *480*
CHENAULT, R. L., 575, *626*
CHER, M., 591, *627*
CHESHIRE, I. M., 356, *394*
CHEUNG, A. C., 381, *396*
CHIBISOV, M. L., 526, 530, *543*
CHOI, S.-I., 477, *481*
CHRISTODOULIDES, A. A., 109, *157*
CHRISTOPHOROU, L. G., 108, 109, *157*, 486, 523, *540*
CHUBB, T. A., 231, *289*
CHUNG, K., 458, *481*
CHUPKA, W. A., 485, 522, 537, *539*
CLAMPITT, R., 559, *625*
CLAPP, T., 560, *625*
CLARKE, E. M., 202, *208*
CLASEN, H., 527, 537, *543*, *544*
CLAUSER, J., 381, *396*
CLEMENTI, E., 320, 326, *390*, 518, 520, 527, 530–533, *541*, *542*
COCHRAN, L. W., 132, 154, *158*
CODE, A. D., 374, *395*
CODLING, K., 460, 461, *480*
CODY, W. J., 464, 468, *481*
COHEN, J. S., 623, *629*
COHEN, L., 231, 268, *289*, *290*
COHEN, M., 237, 239, *289*
COLE, R. H., 472, *480*
COLEMAN, L., 380, *396*
COLLI, L., 109, 136, *157*
COLLINS, R. E., 452, 453, *479*
COMES, F. J., 321, 387, *390*
COMMETTI, A., 136, *158*
COMPTON, D. M. J., 615, 617, 619, 620, *628*
COMPTON, R. N., 108, *157*, 185, 204, *207*, 485, 486, 523, *539*, *540*
CONDON, E. U., 422, *481*
CONNEELY, M. J., 321, 322, 326, 327, 387, 388, *390*, *391*
CONNELLY, M., 450, *479*
CONWAY, D., 62, *90*, 536, *544*
COOGAN, C. K., 525, *543*
COOPER, J., 345, 350, 351, 365, 377, 388, *392*, *393*, 492, *541*
COOPER, J. W., 326, 391, 488, 489, 491, *540*
COOPER, W. S., 349, *393*
COPLAN, M., 489, 497, 498, *541*
CORLISS, C. H., 368, *394*

COSTELLO, D. G., 474, *481*
CRAGGS, J. D., 182, *207*
CRAWFORD, O. H., 81, *90*
CREASER, R. P., 103, 139, *156*
CROMPTON, R. W., 94, 103–107, 110–117, 123, 125–132, 134, 135, 140, 141, 143, 145–152, 154, 155, *156–158*, 447, *479*
CROSS, R. J., 603, *628*
CROSSLEY, R. J. S., 520, 527, 530–532, *542*
CUNY, Y., 353, 370, 372, *393*, *395*
CURLEY, E., 442–444, *478*
CURRAN, R. K., 538, *544*, *545*
CURTISS, C. H., 64, *90*
CZYZAK, S. J., 379, *396*

DALGARNO, A., 28, 63, 64, 81, *89*, *90*, 231, 268, 269, *289*, *290*, 353, *393*, 406, 410, 412, 421, 434, 450, 454, 465, 471, 472, *478–481*, 536, *544*, 550, 551, 597, *624*, *625*, *627*
DAMGAARD, A., 331, 368, *392*
DAMBURG, R., 426, 431, 442, 469, *478*
DAMON, K. R., 255, *290*
DANCE, D. F., 225, *288*, 571, *626*
DANZIGER, I., 384, *397*
DAREWYCH, G., 474, *481*
DAVID, C. W., 498, *541*
DAVIDSON, P. M., 99, *156*
DAVIS, F. J., 109, 135–137, *157*, *158*
DAVIS, R. H., 536, *544*
DAVIS, W. D., 346, 388, *392*
DAVISON, W., 454, *479*
DAVYDOV, B., 97, *156*
DAWTON, R. H. V. M., 500, *541*
DE A. P. MARTINS, P., 379, *396*
DE HEER, F. J., 182, 185, *207*
DEICHSEL, H., 451, *479*
DEINZER, W., 378, *395*
DE JAGER, C., 353, 369, *393*, *395*
DELCHAR, T. A., 562, 565, 566, *625*
DEPAZ, M., 487, 538, *540*
DESI, S., 108, 136, *157*
DESILVA, A. W., 228, *288*
DESLOGE, E. A., 141, *158*
DETWILER, C. R., 231, *289*
DIBELER, V. H., 522, 531, 538, *542*, *545*
DOEHRING, A., 132, *158*
DOLDER, K. T., 462, *480*, 571, *626*
DONN, B., 381, *396*
DONNALLY, B., 268, *290*
DONNALLY, B. L., 560, *625*
DONOVAN, R. J., 575, *627*

Dorman, F. H., 523, 537, *542*, 544
Dorzaio, A., 536, *544*
Doschek, G. A., 231, 277, *289*, *291*
Doughty, N. A., 324, 325, *391*, 457, 458, *481*, 488, 490, *540*
Doyle, R. O., 353, *393*
Drachman, R. J., 415, 426, 429, 430, 445, 463–468, 473, 474, 476, 477, *478*, *480*, *481*
Drake, G. W. F., 268–270, 273, *290*, *291*, 530, *543*, 550, *624*
Dugan, J. L. G., 557, 587, *625*
Dukelskii, V. M., 532–536, *544*
Dunkin, D. B., 525, *543*
Dunn, G. H., 177, *207*, 354, *393*
Dunn, K. F., 569, *626*
Dunning, F. B., 557, 562, 565, 567, 587, 593, *625*, *627*
Du Plessis, A. N., 353, *393*
Dupree, A., 379, *396*
Dutton, J., 63, 65, *90*
Duxler, W., 426, 430, 435–437, 446, 447, 464, 465, 468, 469, 473, 474, *478*
Dyatkina, M. E., 519, 532–534, *542*

Eberhagen, A., 350, 351, *393*
Ebert, H. G., 11, *89*
Eckerle, K. L., 352, 388, *393*
Economides, D. G., 183, *207*, 360, *394*, 462, *480*
Edelson, D., 486, *540*
Ederer, D. L., 321, 387, *390*
Edie, J. W., 520, 527, 530–532, *542*
Edlén, B., 231, 243, 268, *289*, *290*, 519, 527, 530–532, *542*
Edmonds, F. N., 350, 374, *393*
Egidi, A., 164, 170, 187, 196, *206*, *207*
Ehrenson, S., 487, *540*
Ehrhardt, H., 167, 177, 186, 187, 199, *207*, 559, *625*
Eiserike, H., 463, 464, 467, 468, *480*
Eissner, W., 234, 244–247, 253, 254, *289*, *290*, 321, 355, *390*
Elder, F. A., 522, 531, *542*
Elford, M. T., 103–106, 110–117, 122, 123, 125, 127, 129–132, 134, 135, 141, 143, 145, 146, 148–152, 154, *156–158*
Ellis, E., 570, *626*
El-Menshawy, M. F., 268, 280, *290*
El-Sherbini, Th. M., 185, *207*
Elton, R. C., 268, 269, 280, 281, 283, *290*, *291*, 349, *393*
Elwert, G., 213, *288*

Elzer, A., 321, 387, *390*
Engelhardt, A. G., 107, 123, 130, 131, *157*
Engelhardt, W., 287, *291*
Englander, P., 452, 462, *479*, *480*
Englander-Golden, P., 182, *207*
Errett, D., 132, *158*
Ervens, W., 332, *392*
Estermann, I., 552, *625*
Evans, D. E., 226, *288*
Evenson, K. M., 573, *626*
Ewing, J. J., 489, 497, 498, *541*
Eyring, H., 8, *89*

Facchini, U., 108, 136, *157*
Faisal, F. H. M., 456, *480*
Falick, A. M., 573, *626*
Fano, U., 460, 461, *480*, 608, *628*
Farr, J. M., 353, *393*
Farragher, A. L., 537, 538, *544*
Faucher, P., 249, 269, *290*
Fawcett, B. C., 231, 243, 268, 283, *289–291*
Feast, M., 386, *397*
Fehsenfeld, F. C., 486, 525, 537, *540*, *543*, 569, 579–581, 591, 592, 598, 613, *626*, *627*, *628*
Feibelman, P. J., 9, 62, 65, *89*
Feldman, D., 488, 489, 530, 531, 536–538, *540*
Feldman, U., 231, 268, *289*, *290*
Ferguson, E. E., 486, 525, 537, *540*, *543*, *544*, 578, 579, 583, 591, 592, 595, 597, 598, 613, *627*, *628*
Ferrell, R., 474, *481*
Feshbach, H., 426, 427, *478*
Field, G. B., 381, *396*
Fine, J., 524, 530, 534, 535, *542*
Fineman, M. A., 620, *628*
Firsov, O. B., 601, 610, *628*
Fisher, D. L., 227, *288*
Fisk, G. A., 13, *89*
Fite, W. L., 182, *207*, 225, *288*, 354, *393*, 442–444, *478*, 523, 531, 542, 556, *625*
Flannery, M. R., 6, 28, 29, 42, 49, 62, 66, 75, 78, 81, 83, 84, *89*, *90*
Flood, E. H., 564, 567, *625*, *626*
Flower, D. L., 373, *395*
Flower, D. R., 234, *289*
Fogel, Ya. M., 531, *543*
Foley, H. M., 336, *392*
Foner, S. N., 570, *626*
Foote, P. D., 575, *626*
Forrester, D. W., 132, 154, *158*

FORTUNA, J. D. E., 352, *393*
FOX, R. E., 202, *207*, 486, *540*
FRANCEY, J. L. A., 141, *158*
FRANCIS, W. E., 559, *625*
FRANCO, V., 443, 444, *478*
FRANKEVICH, Ye. L., 486, *539*
FRASER, P. A., 324, 325, *391*, 357, 458, *481*, 488, 490, *540*
FREEMAN, F. F., 255, 269, 270, 275–277, *290*, *291*
FRENCH, J. B., 555, *625*
FREUND, R. S., 550, 553, 554, *625*
FRIEDMAN, H., 231, 270, *289*, *291*
FRIEDMAN, L., 487, 538, *540*, 613, *628*
FRIEDRICH, H., 244, *290*
FRITZ, G., 270, *291*
FROESE, C., 331, 332, 388, *392*
FROMMHOLD, L., 125, 132, *157*, *158*, 485, *539*
FROST, D. C., 202, *208*
FROST, L. S., 97, 98, 105, 110, 130, 131, 135, *156*, *157*
FRY, E. S., 558, *625*
FUCHS, V., 568, *626*
FUENO, T., 8, *89*
FUJII, H., 608, 609, 623, *628*
FUJITA, H., 601, *628*
FUNG, A. C., 426, *478*

GABRIEL, A. H., 221, 226–229, 231, 241, 242, 270, 274, 275–281, 283, 286, 287, *288*, *289*, *291*
GAFNEY, E. S., 525, *543*
GAILITIS, M., 244, *290*, 427, 437, *478*, *481*
GAINES, A. F., 536–538, *544*, *545*
GALLAHER, D. F., 356, *393*, *394*, 436, 461, *480*, *481*
GARDNER, M. E., 5, 11, *88*
GARRETT, W. R., 185, 204, *207*, 326, *391*, 449, 452, 453, *479*, 488, *541*
GARSTANG, R. H., 244, 265, 268–270, *290*, *291*, 329, 333, *391*, *392*
GARZ, T., 369, *394*
GASCOIGNE, J., 112, 146, 152, *157*
GAUSTAD, J., 386, *397*
GEBAUR, R., 572, *626*
GEBBIE, K. B., 328, 378, *391*, *395*
GEIGER, W., 536, 538, *544*
GELTMAN, S., 202, *207*, 326, 327, 356, *391*, *393*, 436, 442, 454, 460, *478–481*, 488–492, 494–496, 517, 519, 521, 522, 527, 530, 531, *540–543*

GERARDO, J. B., 346, 388, *392*
GERJUOY, E., 443, 444, 456, *478*, *480*
GIARDINI, A. G., 487, 538, *540*
GIBSON, D. K., 94, 98, 106, 107, 123, 130, 131, *156*, *157*
GIESEL, S., 386, *397*
GIESKE, H. A., 351, 377, *393*
GIGUERE, P. A., 525, 536, *543*
GILBODY, H. B., 569, *626*
GILBODY, H. R., 614, *628*
GILLESPIE, J., 488, *541*
GILLETT, F., 386, *397*
GILMORE, F. R., 536, 537, *544*
GINGERICH, O., 303, 369–374, 380, 382, *390*, *395–397*
GINSBERG, A. P., 520, 527, *542*
GIOUMOUSIS, G., 44, *89*, 613, *628*
GLASSGOLD, A. E., 167, 192, 197, *207*
GLENNON, B. M., 231, 244, 268, *289*, 329, 331, 332, 388, *391*
GLOCKLER, G., 519, 527, *542*
GOLDBERG, L., 304, 368–371, 379, *390*, *394*, *396*
GOLDEN, D., 447, *479*
GOLDEN, D. E., 106, *157*, 456, *480*
GOLDEN, L. B., 360, *394*
GOLDSMITH, S., 268, *290*
GOLDSTEIN, H., 42, *89*
GOLDSTEIN, M., 452, *479*
GOLDWIRE, H. C., 331, *392*
GOLUB, S., 487, 500, 538, *540*
GOMBÁS, P., 527, *543*
GOODRICH, M., 183, *207*
GOON, G., 380, *396*
GOSS, W. M., 381, *396*
GOUDSMIT, S., 519, *542*
GRANT, I. P., 270, *291*
GRAY, P., 525, 536, 537, *543*
GREEN, L. C., 330, *392*
GREEN, T. S., 227, *288*
GREENSTEIN, J. L., 384, *397*
GREIG, J. R., 346, 388, *392*
GRIEM, H. R., 269, 270, 278–281, 283, 286, *288*, *290*, *291*, 331, 340, 341, 346, 348–352, 374, 377, 388, *392*, *393*, *395*, 550, *624*
GRIFFIN, W. G., 227, 231, 278, 283, 287, *288*, *289*, *291*
GRIFFITHS, J. E., 486, *540*
GRINDLEY, J. E., 353, *393*
GRISSOM, J. T., 185, 204, *207*
GROCE, D. E., 474, *481*
GROVE, A., 191, *207*

GROVE, D. J., 202, *207*
GRÜNBERG, R., 125, 126, *157*
GUILLORY, W. A., 525, *543*
GURVICH, L. V., 486, *539*
GUSINOW, M. A., 502, 516, 521, 530, *541*
GUSTAFSSON, E., 614, *628*

HADJIANTONIOU, A., 109, *157*
HAGSTRUM, H. D., 565, 587, *626*
HAHN, Y., 459, 463, *480*
HAKE Jr., R. D., 107, 132, *157*
HALL, C. A., 255, *290*
HALL, J. L., 487, 492, 501, 502, 516, 517, 521, 530, 537, *540, 541, 543, 544*
HALLIN, R., 231, 243, *289, 290*
HALLMANN, M., 621, *628*
HAMILL, W. H., 537, *544*
HANSEN, N., 565, 586, *626*
HARA, S., 455, 456, *480*
HARDCASTLE, R. A., 231, 243, *289*
HARKNESS, R. W., 536, *544*
HARMAN, R. J., 379, *396*
HARPER, W. R., 5, *88*
HARRIS, F. E., 518, 536, *542, 544*
HARRISON, M. F. A., 225, *288*, 571, *626*
HASKELL, R. E., 67, *90*
HASSE, H. R., 30, 44, *89*
HASTED, J. B., 485, *539*, 614, *628*
HASTED, J. R., 614, *628*
HAYDEN, H. C., 621, *628*
HAYS, P. B., 10, 22, 28, 36, 81, *89, 90*
HEALEY, R. H., 103, 132, *156*
HEARN, A. G., 377, 388, *395*
HEDDLE, D. W. O., 225, *288*, 354, *393*, 444, *481*
HEILES, C. E., 381, *396*
HEINICKE, E., 530, *543*
HELBING, R., 590, 602, *627*
HELLMANN, H., 519, *542*
HEMSWORTH, R. S., 579, 580, 610, 624, *627*
HENDERSON, W. R., 523, 531, *542*
HENGELEIN, A., 536, *544*
HENRY, R. J. W., 321, 322, 326, 327, 387, 388, *390, 391*, 448–450, 454, 459, *479*, 488, 491, 492, *540, 541*
HENYEY, L. G., 300, *390*
HERBIG, G., 384, *397*
HERCE, J. S., 603, *628*
HERMAN, K., 525, 536, *543*
HERMAN, Z., 568, 575, 612, *626–628*
HEROUX, L., 235, 237, 239, *289*
HERRENG, P., 107, *157*

HERRING, D. F., 474, *481*
HERRON, J. T., 524, *543*
HERZFELD, K. F., 472, *480*
HERZOG, R., 168, *207*
HESSELBACHER, K. H., 167, 177, 186, 187, *207*
HICKAM, W. H., 202, *207*, 486, *540*
HICKAM, W. M., 202, *207*
HILL, P., 384, *397*
HILL, R. A., 346, 388, *392*
HILS, D., 556, *625*
HINDMARSH, W. R., 353, *393*
HINTENBERGER, H., 503, *541*
HINTEREGGER, H. E., 255, *290*
HIRSCHFELDER, J. O., 64, *90*
HITCHCOCK, J., 381, *396*
HOFMANN, W., 332, *392*
HÖGLUND, B., 379, *396*
HOGNESS, T. R., 536, *544*
HOLLINGSWORTH, C. S., 591, *627*
HOLLSTEIN, M., 561, 569, 622, *626, 629*
HOLØIEN, E., 530, *543*
HOLSTEIN, T., 96, 97, 99, *156*
HOLT, E. H., 67, *90*
HOLT, H., 557, 558, 571, 587, *625*
HOLTSMARK, J., 343, 344, *392*
HOLWEGER, H., 369, *394*
HONIG, R. E., 523, 530, 536, *542*
HOOPER, C. F., 346, 348, *392*
HORNBECK, J. A., 107, *157*, 575, *626*
HORTIG, G., 500, *541*
HOTOP, H., 557, 558, 568, 587, 591–593, 598, 603, 606, 607, 612, 613, *624–629*
HOUCK, T. E., 374, *395*
HOUSE, L. L., 231, *289*
HOUSTON, S. K., 445, 474, 476, *478*
HOWELLS, P., 63, 65, *90*
HOYLE, F., 381, *396*
HUBER, P., 136, *158*
HUDSON, J. P., 381, *396*
HUDSON, R., 458, *480*
HUDSON, R. D., 316, 323, 324, *390*
HUDSON, R. L., 570, *626*
HUGGINS, R. W., 579, *627*
HUGHES, A. L., 183, *207*
HUGHES, J. W. B., 365, *394*
HUGHES, T. P., 231, *289*
HUGHEY, L. R., 567, *626*
HULTHÉN, L., 401, *478*
HUMMER, D., 305, 311, 312, 378, *390, 396*
HUMMER, D. G., 219, 270, 271, *288*, 336, *392*, 556, *625*

HUNGER, K., 353, 386, *393*, *397*
HURST, C. A., 139, 140, 144, *158*
HURST, G. S., 108, 109, 132, 135, 136, *157*, 486, 523, *540*
HURST, T., 458, *481*, 486, *540*
HURT, W. B., 592, *627*
HURZELER, H., 522, 534, 535, *542*
HUSAIN, D., 575, *627*
HUUS, T., 363, *394*
HUXLEY, L. G. H., 99, 102, 104, 132, 138–140, *156*, *158*

IALONGO, G., 167, 192, 197, *207*
ILLENBERGER, E., 624, *629*
INGHRAM, M. G., 522, 531, 534, 535, *542*
IONOV, N. I., 524, 527, 533–536, *543*, *544*
IRONS, F. E., 268, *290*
IRUJILLO, S., 449, *481*
ISAACSON, L., 550, *625*

JACKSON Jr., H. T., 326, *391*, 449, *479*, 488, *541*
JACOX, M. E., 525, *543*
JAGER, K., 536, *544*
JAHODA, F. C., 226, 236, *288*
JEFFERIES, J. T., 372, *395*
JENSEN, D. E., 525, 537, 538, *543*
JESSE, W. P., 600, *628*
JOHN, J., 434, *481*
JOHN, T. L., 328, *391*, 488, 491, *540*
JOHNSON, C. E., 550, 555, *624*
JOHNSON, H., 377, *395*
JOHNSON, H. L., 380, 386, *396*, *397*
JOHNSON, H. R., 519, 527, *542*
JOHNSTON, D. R., 472, *480*
JOHNSTON, W. D., 219, 228, 236, 237, 246, 254, *288*
JONES, B. B., 231, 255, 269, 270, 275–277, 283, *289*, *291*
JONES, C. R., 591, 598, 603, *627*, *628*
JONES, G., 474, 476, *481*
JONES, T. J. L., 226, 236, *288*
JORDAN, C., 221, 231, 238, 240, 241, 254, 256, 262, 263, 270, 272, 274–277, *288*, *289*, *291*, 368, *394*
JORY, R. L., 103, 105, 110, 122, 130, 134, 135, 139, 140, 146, 148, 150, 152, 154, *156*, *158*, 447, *479*
JOSHI, B. D., 538, *544*
JUNDI, Z., 10, 66, 73, 75, *89*, 464, 465, *481*
JUNG, K., 167, 177, 186, 187, *207*

KAGENER, M. G., 565, 587, *626*
KALER, J. B., 379, *396*
KALKOFEN, W., 370, 372, 374, *395*
KALMYKOV, A. A., 531, *543*
KAMINSKY, M., 561, *625*
KANEKO, S., 450, *479*
KARULE, E., 426, 431, 451, 453, 469, *478*, *479*
KATSUURA, K., 601, 610, 611, *628*
KATZENSTEIN, J., 226, *288*
KAUFMAN, M., 519, *542*
KAUFMAN, S., 286, *291*
KAUL, W., 575, *627*
KAUPPILA, H. E., 442–444, *478*, 556, *625*
KAY, J., 538, *545*
KEBARLE, P., 487, *540*
KECK, J. C., 6, 10, 64–66, 69, 71, *88*
KEENAN, P. C., 384, *397*
KEESING, R. G. W., 225, *288*, 354, *393*
KELLY, R. L., 239, 243, 267, *289*
KELLY, T. M., 474, 476, *481*
KENNARD, E. H., 588, *627*
KEPPLE, P., 346, 348, 350, 388, *392*
KERVIN, L., 168, 169, *207*
KERWIN, L., 621, 622, *629*
KHARE, S. P., 447, 459, *479*, *480*
KHVOSTENKO, V. I., 532, 534, 535, 536, *544*
KIEFFER, L. J., 177, *207*, 316, 323, 324, 354, *390*, *393*
KING, A. S., 369, *394*
KING, R. B., 368, *394*
KINGSTON, A. E., 10, 19, 73, 80, *89*, *90*, 324, 325, *391*, 457, 458, 472, *480*, 488, *540*, 597, *627*
KIRSCHNER, E. J. M., 108, *157*
KIRSHNER, R. P., 263, *290*
KISTEMAKER, J., 182, *207*
KIVEL, B., 449, *481*
KJELDAAS, T., 202, *207*
KLEIN, F. S., 613, *628*
KLEIN, J., 611, *628*
KLEINMAN, C. J., 463, *480*
KLEINPOPPEN, H., 556, *625*
KLEMA, E. D., 108, 136, *157*
KLEMM, R. A., 518, 530, *542*
KLEMPERER, W., 553, 554, *625*
KLINGLESMITH, D., 386, *397*
KLOTZ, C. E., 523, *542*
KNACKE, R., 386, *397*
KOCK, M., 369, *394*
KOEPP, R., 11, *89*

KOHN, W., 401, *478*
KONDRAT'YEV, V. N., 486, *539*
KONIG, L. A., 503, *541*
KOPAL, Z., 43, *89*
KÖPENDORFER, W. W., 280, 281, 287, *291*
KORTZEBORN, R. N., 537, *544*
KOSCHMIEDER, H., 556, *625*
KOZLOV, F., 531, *543*
KRAMER, K., 590, 602, *627*
KREPLIN, R. W., 231, 270, *289, 291*
KREY, R. U., 346, 388, *392*
KRISHNA SWAMY, K. S., 381, *396*
KROGDAHL, W. S., 488, *540*
KROTKOV, R., 557, 558, 571, 587, *625*
KRUEGER, T. K., 379, *396*
KRUGER, P. G., 268, *290*
KRUITHOFF, A. A., 575, *627*
KUBOTA, S., 601, *628*
KUMAR, S. S., 380, *396*
KUNDE, V., 380, *396*
KUNZE, H. J., 219, 226, 228, 229, 236, 237, 246, 254, 278–281, 286, 287, *288, 291*
KUPRIANOV, S. E., 575, 612, *627, 628*
KUREPA, M. V., 182, *207*
KURUCZ, R., 303, 374, *390*
KUSCH, P., 572, *626*
KUYATT, C. E., 448, *479*, 504, *541*

LaBAHN, R., 426, 435–437, 446, *478*
LaBAHN, R. W., 426, 429, 430, 436, 446–448, 458, 459, 464, 465, 468, 469, 473, 474, 476, *478–480*
LADANYI, K., 527, *543*
LAMBERT, D. L., 302, 368, 369, *390, 394*
LAMBROPOULOUS, P., 570, *626*
LAMKIN, J. C., 436, *478*
LAMPE, F. W., 575, 592, *627*
LAN, Vo Ky, 452, 453, *479*
LANDAU, L. D., 16, *89*
LANDERS, A. M., 562, 565, 566, *625*
LANDON, S. A., 6, 10, 64–66, 69, 71, *88*
LANE, N., 454, *479*
LANE, N. F., 456, *480*, 623, *629*
LANGEMUIR, I., 564, *626*
LANGEVIN, P., 4, *88*
LANLICHT, I., 621, *628*
LASSETTRE, E. N., 184, 185, *207*
LATHAM, D. W., 380, *396*
LATYSCHEFF, K., 570, *626*
LAWRENCE, G. P., 499, *541*
LAWSON, J., 464, 468, *481*
LAWSON, P. A., 139, *158*

LEDSHAM, K., 418, 420, *478*
LEE, G. F., 447, 476, *481*
LEE, T., 386, *397*
LEE, Y., 522, 534, 535, *542*
LEEMAN, S., 350, *393*
LEGLER, W., 126, *157*
LEIPUNSKY, B., 570, *626*
LEUNG, C. Y., 474, 476, 477, *480*
LEVINE, J., 537, *544*
LEVINE, N. E., 132, *158*
LEVY, G., 569, *626*
LEWIN, G., 572, *626*
LEWIS, B. A. 231, 233, 236, 237, 239, *289*
LEWIS, J. T., 406, 410, *478*
LICHTEN, W., 550, 556, 559, *625*
LIFSHITZ, E. M., 16, *89*
LILEY, B. S., 139, 140, 151, *158*
LILLEY, A., 379, *396*
LIN, S. C., 449, *481*
LINDHOLM, E., 336, *392*, 614, *628*
LINEBERGER, W. E., 486, 492, 496, 514, 521, 527, 532, *540*
LINSKY, J. L., 372, 380, *395, 396*
LIPSCOMB, W. N., 537, *544*
LIPSKY, L., 321, 322, 326, 327, 387, 388, *390, 391*
LISITZIN, E., 527, *543*
LLOYD, M. D., 442, 444, 456, *478*
LOCKE, J. W., 555, *625*
LOEB, L. B., 7, 8, 11, 28, 65, 87, *89*, 107, *157*
LÖFSTRAND, B., 268, *290*
LORENTS, D. C., 560, 561, 569, 622, *625, 626, 629*
LOWKE, J. J., 99, 100, 105, 110, 117, 118, 120, 124, 125, 131, 134–136, *156, 157*
LOWRY, J. F., 321, 387, *390*
LUCAS, J., 139, *158*
LYNCH, R., 486, 497, 521, 522, 531, *540*
LYNN, N., 412, 465, 471, *478*

McAFFEE Jr., K. B., 486, *540*
McCULLOH, K. E., 522, 531, *542*
McDANIEL, E. W., 6, 11, 13, 44, 55, 62, 63, 65, 78, 81, 84, 86, *89*, 100, 107, *156, 157*
McDOWELL, C. A., 202, *208*
McDOWELL, M. R. C., 183, *207*, 326–328, 360, 364, *391, 394*, 442, 444, 447, 448, 453, 456, 458, 462, *478–480*, 488, 491, 536, *540, 544*
McEACHRAN, R. P., 324, 325, *390, 391*, 457, 458, *481*, 488, *540*
McEWEN, M. B., 575, *626*

AUTHOR INDEX

McGowan, J. W., 202, *208*, 442–444, 474, *478*, *481*, 620–622, *628*, *629*
McGowan, S., 11, 64, 65, 83, *89*
McGuire, E. J., 320, *390*
McIntosh, A. I., 106, 107, 111, 114, 125, 128, 130, 131, 141, 143, 145, 147–149, 151, 152, 154, *157*, *158*, 569, *626*
McKibben, J. L., 499, *541*
Mackie, J. C., 486, 497, 498, 521, 522, 531, *540*, *541*
McKnight, L. G., 63, *90*
McLean, A. D., 518, 520, 527, 530, 531, 532, *541*
MacLennan, D. A., 562, 565, 566, *625*
McMillen, J. H., 183, *207*
McRae, E. G., 567, 592, *626*
McVicar, D. D., 438, *478*
McWhirter, R. W. P., 10. 19, *89*, 211, 226, 227, 229, 236, 246, 252, 268, 273, 278, 287, *288–290*
Macek, J., 488, *540*
Mächler, W., 84, *90*
Madden, R. P., 460, 461, *480*, 567, *626*
Mahadevan, J., 485, *539*
Mahan, B. H., 6, 9, 10, 12, 64, 71, *88*, *89*, 522, 534, 535, *542*, 573, *626*
Malik, F. B., 434, *481*
Maltbie, M. M., 527, *543*
Mamotenko, M., 519, *542*
Mandl, A., 488, *541*
Mandl, P., 66, *90*, 487, 497, *540*
Marchand, P., 202, *208*
Marconero, R., 164, 170, 196, *206*, *207*
Marcus, A. B., 567, *626*
Marenin, I., 380, *396*
Margenau, H., 97, *156*
Margenau, M., 345, *392*
Margrave, J. L., 519, *542*
Marino, L., 449, *481*
Mark, C., 300, *389*
Marmet, P., 168, 169, 202, *207*, *208*
Marriott, R., 451, 453, *479*, 596, *627*
Marrus, R., 269, 270, *290*, *291*
Marshall, L. C., 8, 11, 28, *89*
Marshall, T., 591, 592, *627*
Martin, D. W., 13, 62, 63, 65, 84, *89*
Martin, J. B., 326, *391*, 488, 489, *540*
Martin, J. D., 485, 491, *539*, *540*
Mason, E. A., 42, *89*
Massey, H. S. W., 61, *90*, 183, 184, *207*, 401, 404, 433, 449, 464, 466, 468, *477*, *478*, *481*, 486, *540*

Matese, J., 458, 459, 476, *480*
Matese, J. J., 426, *478*
Mathis, J., 269, *290*
Mathis, R. F., 621, *628*
Matsen, F. A., 578, 583, 591, 592, 595, *627*
Matsushima, S., 319, 370, 371, *390*
Matsuzawa, J., 611, *628*
Medvedev, V. A., 486, *539*
Meekins, J. F., 231, 270, 277, *289*, *291*
Meinel, A., 378, *395*
Menendez, M. G., 595, *627*
Menzel, D. H., 315, 330, 331, 365, 368, *390*, *392*, *394*
Meyer, J., 352, *393*
Meyer, J. E., 527, *543*
Meyers, R. J., 573, *626*
Mezger, P., 379, *396*
Michaud, G., 382, *397*
Mielczarek, S., 448, *479*, 595, *627*
Mielczarek, S. R., 184, 185, *207*
Mies, F. H., 353, *393*
Mihalas, D., 311, 374–376, 378, 382, *390*, *395–397*
Miles, B. M., 231, 244, *289*, 329, 331, *391*
Miller, J. E., 488, *540*, *543*
Miller, J. M., 520, 527, *542*
Miller, M. H., 346, 388, *392*
Miller, W. H., 608, 609, 623, 624, *628*, *629*
Miller, W. J., 525, 538, *543*
Milligan, P. E., 525, *543*
Millman, S., 573, *626*
Minnaert, M., 368, *394*
Mitchell, A. C. G., 569, *626*
Mitchell, R., 386, *397*
Mitchell, R. D., 141, *158*
Mitchell, R. I., 380, *396*
Mittleman, M., 424, 429, 439, 451, *478*
Moffett, R. J., 6, 19, 25, *88*
Mohler, F., 538, *545*
Mohler, F. L., 575, *626*
Mohr, C. B. O., 183, *207*, 401, 466, *477*
Moiseiwitsch, B. L., 218, 225, *288*, 354, 364, *393*, *394*, 447, *479*, 486, 519, 520, 527, *539*, *542*
Mokler, P., 500, *541*
Molnar, J. P., 575, 592, *626*, *627*
Moore, C. E., 231, 243, 267, *289*, 485, *539*
Moores, D. L., 273, *291*, 321, 360, 387, *390*, *394*
Morgan, F. J., 227, *288*

MORGAN, L. A., 450, *479*
MORGAN, W., 378, *395*
MORGNER, H., 624, *629*
MORI, M., 601, 608–610, 623, *628*
MORRIS, J. C., 346, 388, *392*
MORRISON, J. D., 522, 534, 535, *542*
MORSE, P. M., 426, *478*
MORTON, D. C., 374, *395*
MORUZZI, J. L., 487, 538, *540*, *545*
MOSER, C. M., 332, *392*
MOSKVIN, Yu. V., 326, *391*, 488, 489, *540*
MOTT, N. F., *207*, 404, *478*
MOTTELSON, B., 363, *394*
MOZER, B., 346, 350, *392*
MÜCK, G., 325, 327, *391*, 489, 497, 532, *541*
MULDERS, G. F., 368, *394*
MÜLLER, E. A., 368–371, *394*
MÜLLER, M., 500, *541*
MUSCHLITZ Jr., E. E., 557, 558, 567, 576, 583, 587, 591, 592, 594, 598, 602, 603, *625–628*
MYERSCOUGH, V. P., 326–328, 370, 384, 385, *391*, *395*, *397*, 488, 491, *540*

NAFF, W. T., 185, 204, *207*
NAGY, L., 108, 136, *157*
NAGY, T., 108, 136, *157*
NAKAMURA, H., 608, 609, 623, *628*
NAKANO, H. H., 450, *479*
NAQVI, A. M., 244, *290*
NATANSON, G. L., 5, 6, 8, 28, 62, 78, *88*
NEAMTAN, S. M., 474, *481*
NELSON, D. R., 137, *158*
NESBITT, L. E., 536, *544*
NEUERT, H., 527, 437, *543*, *544*
NEUPERT, W. M., 231, 277, *289*
NEVEN, L., 353, *393*
NEWELL, W. R., 191, *207*
NEWTON, A. A., 227, *288*
NEWTON, A. S., 559, *625*
NEYNABER, R. H., 449, *481*, 558, 599, *625*
NICHOLLS, R. W., 329, *391*, 550, *624*
NICOLL, F. H., 183, *207*
NIEHAUSS, A., 557, 558, 568, 587, 591–593, 598, 603, 606, 607, 612, 613, *624–629*
NIELSEN, R. A., 109, 132, 135, 136, *157*, *158*
NILES, F. E., 592, *627*
NOBLE, P. N., 537, *544*
NORCROSS, D. W., 321, 355, 387, *390*
NORMAN, G. E., 316, 323, *390*
NORRIS, J., 377, 383, *395*
NOVICK, R., 486, *540*

NOYES, R., 370, 372, *395*
NOYES, R. W., 263, *290*
NUSSBAUMER, H., 234, 244, 265, *289*, *290*, 321, 332, 355, 360, 388, *390*, *392*, *394*

OB'EDKOV, V. D., 408, 430, 463, *478*
OBEROI, R. S., 439, 468, *478*
OCZKOWSKI, G., 474, *481*
O'HARA, H., 43, *89*
O'HARA, P. A. G., 537, *544*
O'KELLY, L. B., 109, *157*
OKSUZ, I., 518, 530, 531, *542*
OKSYUK, YU. D., 455, *480*
OLSEN, L. A., 204, *207*
O'MALLEY, T. F., 106, *156*, 323, *391*, 459, 460, *480*, 494, *541*
ORMONDE, S., 356, 388, *393*, 443, *481*
ORTH, P. H. R., 474, 476, *480*
OSTERBROCK, D. E., 244, 246, 254, 264, 265, *290*, 379, *396*
OTT, W. R., 442–444, 556, *625*
OUDEMANS, G. J., 472, *480*

PACK, J. L., 62, *90*, 110, 112, 130, 132, 135, 136, *157*, *158*, 536, 538, 539, *544*, *545*
PAGE, F. M., 524, 527, 532, 536–538, *543–545*
PAGEL, B. E. J., 302, 303, 325, 326, 348–350, 365, 369, 380, *390*, *391*, *396*
PAGET, T. M., 221, 229, 241–243, 280, 287, *288–290*
PALENIUS, H. P., 243, *290*
PALMER, P., 379, 381, *396*
PALMER, R. R., 183, *207*
PALUMBO, L. J., 269, 283, *290*
PAQUET, C., 202, *208*
PARKER Jr., J. H., 99, 100, 132, 141, 148, 152, 154, *156*, *158*
PARKER, R. A. R., 264, *290*
PARKINSON, D., 450, *479*
PARKINSON, W. H., 369, 371, 372, *395*
PARKS, E. K., 8, 13, 61, *89*
PARKS, J. E., 109, 135, *157*
PAUL, D. A. L., 474, 476, 477, *480*
PAULSON, J. F., 486, 501, 525, 537, *540*, *543*
PAULY, H., 590, 602, *627*
PEACH, G., 183, 185, 186, *207*, *208*, 317–322, 326, 327, 346, 348–350, 352, 358, 360, 365, 388, *390–394*
PEACHER, J. L., 424, 429, 439, *478*
PEACOCK, N. J., 231, *289*
PEARL, A. S., 550, *624*

PEREL, J., 452, *479*
PEART, B., 462, *480*
PEIMBERT, M., 379, *396*
PEKERIS, C. L., 268, *290*, 315, 330, *390*, 458, *481*, 518, 530, *541*
PENFIELD, H., 379, *396*
PENGELLY, R. M., 364, *394*
PENNING, F. M., 575, *627*
PENTON, J. R., 602, 603, *628*
PERCIVAL, I. C., 361, 364, 365, 379, *394*
PERKINS, J. F., 460, *481*
PERKINS, J. R., 426, 433, 468, 469, *479*
PERROTT, R. H., 268, *290*
PERSON, J. C., 9, 13, 64, *89*
PETERKOP, R., 354, *393*, 447, 451, *479*
PETERKOP, R. K., 478, *479*
PETERKOP, R. P., 199, 202, *207*
PETERSON, D., 374, 383, *395*, *397*
PETERSON, J. R., 499, *541*, 560, 561, 569, 622, *625*, *626*, *629*
PETRINI, D., 234, *289*
PETTERSON, E., 614, *628*
PHELPS, A. V., 62, *90*, 94, 97, 98, 105, 107, 110, 112, 123, 130–132, 135, 136, 151, *156–158*, 487, 536, 538, 539, *540*, *544*, *545*, 580, 592, 596, *627*
PIERCE, A., 304, *390*
PIERCE, A. K., 368, *394*
PITAEVSKII, L. P., 6, 10, 65, 66, 69, *88*
PIZELLA, G., 164, 170, 196, *206*, *207*
PLATZMAN, R. L., 600, *628*
PLIMPTON, S. G., 11, *89*
POE, R. J., 426, 430, 435–437, 446, 447, 464, 465, 468, 469, 473, 474, *478*
POLAND, A., 376, 377, *395*
POLITZER, P., 520, 533–535, *542*
POPENOE, C. H., 346, 388, *392*
POPP, H. P., 325, 327, *391*, 488, 489, 497, 498, 522, 531, 532, *541*
POTTASCH, S. R., 239, *289*, 368, *394*
PRASAD, A. N., 268, 280, *290*
PRESNYAKOV, L., 202, *207*
PRITCHARD, H. O., 486, 519, 527, 534, 537, *540*, *542*, *543*
PURCELL, J. D., 231, *289*
PYATIGORSKII, G. M., 524, *543*

RABI, I. I., 573, 590, *626*, *627*
RAIMONDI, D. L., 518, 520, 527, 531, 532, *541*
RAITH, W., 452, *479*
RAMAN, C. V., 570, *626*
RANK, D. M., 381, *396*
RAPP, D., 182, *207*, 462, *480*, 559, 582, *625*, *627*
RAUSCH VON TRAUBENBERG, H., 572, *626*
REE, T., 9, *89*
REED, J. W., 103, 132, *156*
REEH, H., 411, 412, 464, *478*
REES, J. A., 132, *158*
REESE, 538, *545*
REEVES, E. M., 369, 371, 372, *395*
REHDER, L., 324, 325, *391*, 488, 497, *540*
REICHERT, H., 451, *479*
REID, R. H. G., 551, *625*
REIMANN, C. W., 325, 327, *391*, 488, 489, 491, 497, 498, 521, 522, 531, 532, 534, *541*
REINHARDT, P. W., 108, *157*, 486, 523, *540*
REMPT, R. D., 537, *544*
RICH, J. C., 321, 387, *390*
RICHARDS, D., 364, 365, 379, *394*
RICHARDS, H. L., 557, 558, 587, 598, *625*
RICHTER, J., 368, 369, *394*
RIDGELEY, A., 243, 255, 262, 263, *290*
RIEMANN, W., 132, *158*
RISK, C. G., 131, *158*
RIVIERE, A. C., 572, *626*
ROBERTS, D. E., 352, *393*
ROBERTS, J. R., 352, 388, *393*
ROBERTSON, A. G., 94, 104–106, 111–113, 115–117, 120, 122, 123, 125–131, 134, 135, 152, *156*, *158*
ROBERTSON, W. W., 578, 583, 591, 592, 595, 598, 603, *627*, *628*
ROBINSON, B. B., 489, 498, *541*
ROBINSON, E. J., 326, 327, *391*, 488, 489, 491, 492, 517, *540*, *541*
ROBISCOE, R. T., 555, *625*
ROHRLICH, F., 519, 520, 527, 530–532, *542*
ROSE, W. K., 521, *542*
ROSENBAUM, O., 527, *543*
ROSENBERG, L., 323, *391*
ROSENSTOCK, H. M., 524, *543*
ROSIN, S., 590, *627*
ROSS, J., 382, *397*
ROTENBERG, M., 451, 453, *479*
ROTHE, D. E., 325, 327, *391*, 489, 497, *541*
ROTHE, E. W., 449, *481*, 558, 599, *625*
ROUEFF, E., 353, *393*
RUDGE, M. R. H., 177, 185, 196, 197, 199, *207*, 354, *393*
RUGGE, H. R., 231, 242, 268, 270, 277, 286, *289*, *291*
RÜMELIN, G., 11, *89*

Rush, R. P., 330, *392*
Rutherford, J. A., 559, 614, 515, 617, 619–621, *625, 628*

Sadauskis, J., 600, *628*
Sahal-Brechot, S., 340, 341, 352, 361, 388 *392, 393*
Salop, A., 450, *479*
Salpeter, E. E., 570, 576, *626*
Sampson, D. H., 360, *394*
Samson, J. A. R., 522, 536, *542*
Sands, K., 353, *393*
Saraph, H. E., 234, 244, *289*, 332, 365, 379, *392, 394, 396*
Sargent, W. L. W., 382, *397*
Sawyer, G. A., 231, *289*
Sawyers, W., 560, *625*
Sayers, J., 5, 11, *88, 89*
Scarborough, J., 487, *540*
Schaefer, H. F., 518, 530, 531, *542*, 608, 609, 623, 624, *628, 629*
Scheer, D., 524, *542*
Schey, A. M., 436, *481*
Schiff, B., 268, *290*
Schlier, Ch., 590, 602, *627*
Schlumbohm, H., 132, *158*
Schlüter, H., 350, 374, *393*
Schmeltekopf, A. L., 486, 525, 537, *540, 543*, 557, 558, 569, 579–581, 587, 591–593 598, 613, *625–628*
Schmieder, R. W., 269, 270, *290, 291*
Schneider, B., 326, *391*, 488, *541*
Schnopper, H., 386, *397*
Schram, B. L., 182, *207*
Schultz, M., 560, *625*
Schultz, S., 556, *625*
Schulz, G. J., 169, *207*, 523, 526, 531, *542, 543*
Schulz, M., 167, 177, 186, 187, 199, *207*
Schummers, J. H., 13, 62, 63, 65, 84, *89*
Schwartz, C., 436, 463, 468, *480, 481*
Schwarzchild, M., 521, *542*
Schweinler, H. C., 486, *540*
Searle, L., 382, 384, 385, *397*
Seaton, M. J., 218, 234, 244, *288, 289*, 310, 315, 317, 319, 354, 356, 362–364, 373, 378, 379, *390, 393–396*, 434, *481*
Sellin, I. A., 268, *290*
Seman, M. L., 325, 327, *391*, 486, 488, 489, 492, 496, 501, 517, 530, 535, *540, 541*
Senftleben, H., 527, *543*

Septier, A., 502, *541*
Sesta, G., 99, *156*
Seyfried, P., 575, *627*
Shamey, L. J., 244, 265, *290*, 351, 377, 388, *393*
Sharp, T. E., 536, *544*
Shaw, M. J., 579, 580, 591, 610, 624, *627*
Sheer, D., 534, 535, *542*
Sheer, M. D., 524, 530, 534, 535, *542*
Shemming, J., 234, 244, *289*
Sheridan, J. R., 561, 569, 622, *626, 629*
Shields, W. R., 524, *543*
Shipman, H. L., 377, *395*
Shobha, P., 447, *479*
Sholette, W. P., 583, 591, 592, 594, *627*
Shortley, G. H., 422, *481*
Shugart, H. A., 550, 555, *624*
Shumaker, J. B., 346, 388, *392*
Shyn, T. W., 555, *625*
Siegel, M. W., 486, 492, 501, 530, 537, *540, 544*
Silvermann, S. M., 184, 185, *207*
Simpson, J. A., 184, 185, *207*, 595, *627*
Simpson, O. C., 552, *625*
Sinanoglu, O., 332, 333, *392*, 518, 530, *542*
Sinfailam, A. L., 450, 455, 456, *479, 480*
Skinker, M. F., 132, *158*
Skinner, H. A., 519, 527, *542, 543*
Sklar, A. L., 538, 539, *544*
Slater, J. C., 449, *479*
Sloan, I. A., 434, 435, 437, 438, *478*
Sloan, I. H., 177, *207*, 462, *480*
Slocomb, C. A., 624, *629*
Smirnov, B. M., 486, 526, 530, 539, *543*, 601, 610, *628*
Smith, A. C., H., 225, *288*, 557, 559, 562, 565, 567, 571, 587, 593, *625–627*
Smith, E. W., 345, 350, 365, *392*
Smith, F. J., 43, *89*
Smith, G. M., 598, *627*
Smith, K., 321, 322, 326, 327, 387, 388, *390, 391*, 449, 450, 457, 464, 466, 468, *479, 481*
Smith, M. W., 231, 244, 268, *289*, 329, 331, 332, 388, *391*
Smith, S. J., 218, 225, *288*, 324, 325, 354, *391, 393*, 444, *481*, 487–489, 494, 496, 521, 522, 527, 530–532, 536, *540, 541, 543*
Smith, W. W., 268, *290*
Smyth, K. C., 491, 537, *541*
Snijders, M. A. J., 377, *395*
Snuggs, R. M., 13, 62, 63, 65, 84, *89*
Snyder, L., 381, *396*

Sobel'man, J., 202, *207*
Sokolov, V. M., 532–534, *544*
Solarski, J., 368, *394*
Somerville, W. B., 328, 381, *391*, *396*, 536, *544*
Sommer, J., 287, *291*
Speier, F., 321, 387, *390*
Spence, D., 523, 526, *542*, *543*
Sperli, F., 170, *207*
Spinrad, H., 379, 380, 384, 386, *396*, *397*
Spinrad, M., 380, *396*
Spokes, G. N., 325, 327, *391*, 489, 491, 497, 498, 521, 532, 534, *541*
Sprevak, D., 10, 22, 28, 36, *89*
Spruch, L., 323, *391*, 459, 463, *480*
Stauffer, A. D., 364, *394*
Stebbings, R. F., 556, 559, 562, 565, 567, 587, 591, 592, 614, 615, 620, *625*, *628*
Stecher, T. P., 381, *396*
Stegun, I. A., 363, *394*
Steidl, H., 451, *479*
Stein, S., 456, *480*
Stein, W., 386, *397*
Steiner, B. W., 487, 489, 492, 496–498, 500, 505, 514, 521, 527, 532, 535, 537, 538, *540*, *541*
Steinmetz, D. L., 380, *396*
Stern, O., 552, *625*
Sternheimer, R. M., 421, 439, *478*
Stevenson, A., 108, *157*
Stevenson, D. P., 44, *89*, 613, *628*
Stewart, A. L., 268, *290*, 329, 333, *391*, *392*, 406, 410, 418, 420, *478*
Stewart, A. T., 477, *481*
Stilley, J. L., 324, 325, *391*
Stockdale, J. A., 109, *157*, 185, 204, *207*, 486, *540*
Stone, P. M., 416, 452, 453, *478*
Stone, W. G., 108, 132, 136, *157*
Strittmatter, P., 382, *397*
Strittmatter, P. A., 383, *397*
Strom, K. M., 382, *397*
Strom, S. E., 374, 377, 382, *395*, *397*
Sullivan, E., 437, *481*
Sultanov, A. Sh., 532, 534, 535, *544*
Summers, H. P., 310, *390*
Sunshine, G., 448, 449, *481*
Sutton, D. S., 140, 146, 154, *158*
Swan, P., 452, *479*
Swartz, M., 231, 277, *289*

Tai, H., 443, 444, *478*

Tait, J. H., 232, 233, 236, 237, 239, *289*
Takayanagi, K., 456, *480*
Tan, K. L., 592, *627*
Tang, S. Y., 567, *626*
Tannich, J. D., 348, 350, *392*
Tao, S. J., 474, 476, *481*
Tapscott, J., 378, *395*
Tarafdar, S. P., 329, *391*
Tate, J. T., 183, *207*, 536, *544*
Taubert, R., 575, *627*
Taylor, A. J., 356, *394*, 438, 443, 451, 453, *478*, *479*
Taylor, B., 384, *397*
Taylor, B. J., 380, *396*
Taylor, H. J., 236, 237, *289*
Taylor, H. S., 536, *544*
Taylor, R. L., 486, 497, 521, 522, 531, *540*
Tekaat, T., 167, 177, 186, 187, 199, *207*
Teller, E., 550, *624*
Temkin, A., 326, *391*, 401–403, 408, 421, 425, 426, 434, 436, 437, 439, 448–452, 454–456, 459–461, 463, 464, *477–481*
Teter, M. P., 592, *627*
Thaddeus, P., 381, *396*
Thomas, G. E., 202, *208*
Thomas, R. N., 305, *390*
Thompson, D. G., 450, *479*
Thompson, R., 386, *397*
Thomson, G. P., 11, *89*
Thomson, J. J., 5, 6, 10, 11, 65, 85, 87, *88*, *89*
Thornton, D. D., 381, *396*
Tilford, S. G., 550, *625*
Tisone, G., 325, *391*, 487, 488, *540*
Tizard, H. T., 102, 137, 138, *156*
Toffolo, D. S., 108, *157*
Tomboulian, D. H., 321, 387, *390*
Tondello, G., 229, 231, 243, 246, 252, *289*, *290*
Toschek, P., 590, 602, *627*
Tousey, R., 231, *289*
Townes, C. H., 381, *396*
Townsend, J. S., 102, 137, 138, 154, *156*, *158*
Tozer, B. A., 182, *207*
Traving, G., 336, *392*
Travis, L. D., 319, 370, 371, *390*
Trefftz, E., 244, *290*, 434, *481*
Trujillo, S. M., 558, 599, *625*
Tsuji, T., 380, *396*
Tully, J., 326, *391*, 488, *541*
Tully, J. A., 219, 271, *288*, *291*
Turner, B. R., 614, 615, 617, 619–621, *628*

Turner, D. W., 622, *629*
Twiddy, N. D., 570, 579, 580, 591, 610, 624, *626, 627*
Tyrén, F., 231, *289*

Uman, M. A., 132, *158*
Underhill, A., 331, 332, 388, *392*, *395*
Underhill, A. B., 374, 377, *395*
Unsöld, A., 368, 369, *394*
Unzicker, A. E., 270, *291*
Utterback, N. G., 621, 622, *629*

Vainstein, L., 202, *207*
Van Citters, G. W., 374, *395*
Vanderslice, J. T., 550, *625*
Van Der Wiel, M. J., 182, 185, *207*
Van Dyck Jr., R. S., 550, 555, *624*
Van Regemorter, H., 216, 218, 232, 233, 236, 245, 246, 254, 271, *288*, 354, 359, 360, *393, 394*
Vardya, M. S., 325, 326, 329, 380, 381, *391, 396*
Vasavada, K. V., 454, 455, 459, *480*
Veldre, V., 354, *393*
Vendeneyev, V. I., 486, *539*
Victor, G. A., 269, 290, 550, *624*
Vidal, C. R., 345, 350, 365, *392*
Villarejo, D., 522, 531, *542*
Volz, D. J., 13, 62, 63, 65, 84, *89*
Von Engel, A., 592, *627*
Vorburger, T. V., 536, *544*
Voshall, R. E., 62, *90*, 130, 132, 135, *157, 158*
Voslamber, D., 345, 349, 350, *392*
Vought, R. H., 538, *544*
Vriens, L., 164, 187, 191, 192, 195, 199, 206, 448, *479*

Waddington, T. C., 525, 536, 537, *543*
Wagner, E. B., 109, 135, 136, *157*
Wahl, A. C., 537, *544*
Walker, A. B. C., 231, 242, 268, 270, 277, 286, *289–291*
Walker, D. W., 451, *479*
Walker, J. A., 522, 531, *542*
Walter, T. A., 485, 522, 537, *539*
Wannier, G. H., 99, *156*
Warmeck, P., 489, 538, *541*
Warner, B., 332, 353, 368, 369, 384, 386, 388, *392–394, 397*
Warren, R. W., 132, 148, 152, 154, *158*
Watanabe, T., 601, *628*

Watson, W. D., 383, *397*
Webb, H. W., 562, *625*
Webb, T. G., 356, 394, 441, *481*
Weidemann, V., 383, *397*
Weinberg, M., 326, *391*, 488, *541*
Weinflash, D., 486, *540*
Weiss, A. W., 231, 268, *289*, 332, 388, *392*, 452, *479*, 518, 530–532, *542*
Weiss, J., 537, *544*
Wells, D. C., 350, 374, *393*
Welsh, W. J., 381, *396*
Wentink Jr., T., 550, *625*
West, C. D., 525, *543*
Westhaus, P., 332, 333, *392*
Wetzel, W. W., 199, *207*
Wheeler, R. C., 537, 538, *544*
Whitlock, R. F., 550, *625*
Whittaker, W., 356, 388, *393*, 443, *481*
White, J. V., 132, *158*
Wickramasinghe, D. T., 383, *396*
Wickramasinghe, N. C., 381, *396*
Wiese, W. L., 231, 244, 268, *289*, 329, 331, 332, 368, 369, 388, *391, 394*
Wightman, A. S., 418, *478*
Wigner, E. P., 494, *541*
Wildt, R., 302, *390*
Wilkinson, P. G., 550, *625*
Williams, J. F., 442–444, *478*
Williams, J. M., 537, *544*
Williams, R. E., 321, 387, *390*
Williams, R. V., 286, *291*
Williams, W. L., 558, *625*
Williamson, J. H., 328, *391*, 447, *479*
Willmann, K., 167, 177, 186, 187, 199, *207*
Wilson, R., 231, 243, 255, 262, 263, 270, 283, *289–291*, 374, *395*
Wing, R. F., 380, 386, *396*
Winther, A., 363, *394*
Withbroe, G. L., 263, *290*
Wolf, K. L., 472, *480*
Wolf, M. J., 521, *542*
Wong, S. F., 536, *544*
Woo, S. B., 536, *544*
Wood, R. H., 536, *544*
Woodward, B. W., 486, 492, 496, 514, 521, 527, 532, *540*
Woosley, S., 270, *291*
Wright, A. C., 550, *625*
Wrubel, M., 368, *394*
Wu, Ta-Yon, 520, *542*
Wunderlich, R., 350, 351, *393*
Wynn, M. J., 485, *539*

Ya'akobi, B., 488, 524, 530, *540*, *543*
Yamamoto, M., 388, *393*
Yang, Y., 62, *90*
Yoshimine, M., 518, 520, 531, 532, *541*
Young, R. A., 550, *625*

Zacharias, J. R., 573, *626*
Zare, R. N., 492, *541*

Zemansky, M. W., 569, *626*
Zhirnov, V. A., 488, *540*
Zipf, E. C., 575, *627*
Zirin, H., 303, 305, *390*
Zirker, J., 372, *395*
Zollweg, R. J., 520, 527, 530–535, *542*
Zuckermann, B., 379, 381, *396*

SUBJECT INDEX

Absorption, photo, 370 ff.
Adiabatic approximation, 404
– – exact solution for hydrogen, 417–420
– – many-electron targets, 421–423
Angular correlation spectra, 190–191
Antisymmetrization, importance of, 425–426
Autoionization calculations, 459–461

Background electrons, in electron spectrometers, 178–180
Balmer line, profile, 348
Berillium like ions, energy levels, 243
– – – transition probabilities, 244
– – – collisional excitation rates, 244–249
– – – solar line intensity 253–262
Bethe–Reeh potential, 411, 424, 447
Binary encounter peak, 191
Boltzmann equation, 95–97
Branching ratios, 275

Callaway–Temkin potential, 408–410
Carbon stars, 384–387
Cascade contributions to observed intensities, 273
Characteristic energy, of a swarm, 100–103
Chemiionization, 574–613
– absolute measurements of, by afterglow technique 578–583
– – – – by a beam technique 583–588
– results of measurements, 588–592
– models of, 596–608
– detailed theoretical studies of, 608–613
Classical approximation to cross sections, 361–366
Close-coupling approximation, 352
Coincidence counting, of electrons, 174–175
Coherent scattering of radiation, 328–329
Collision strength, 216, 234, 248

Continuous spectrum, 312–316
Continuous absorption 302, 316–322
Coronal excitation equation, 213
Cross-sections, electron, momentum transfer, 103–105
– – from drift-velocity data, 130–132
– differential ionization, 161–208
– – – triple differential, 163, 189–196
– – – double differential, 164, 183–188
– – – single differential, 165
– – – experimental, in He, 185–195
– – – comparison with theory, 196–202
– excitation, theory, 354–358
– excitation, in positive ions, 232–233
– excitation in H, 443, 444
– photodetachment, table of, 488–489
– elastic scattering of electrons, 443–453
– positron–hydrogen, elastic, 468
– for ionization by He metastable, 593–596
Curve of growth analysis, 368–370
– – – – in solar atmosphere, 368
– – – – in early type stars, 373
– – – – in late type stars, 380
– – – – in stars of non-solar composition, 382

Dalgarno–Lynn functions, 412–414
Detailed balance, 217
Dielectronic recombination, 221
Diffusion equation, solution of, 139
Diffusion coefficient, 99–100
– – lateral, 137–139
– – measurement of, 140–155
Diffusion theory of recombination, 71–72
Drift-velocity, electron, 107–136
– experimental methods, 107–111
– error analysis of measurements, 111–125
– experimental data on, 127–135

SUBJECT INDEX

Einstein coefficients, 212, 301–304, 307
Electron affinity, 517–528
−− measurement of, 520–527
−− empirical fitting, 518–519
−− table of, 530–538
−− of H^-, 439
−− of S^-, 526–528
Electron swarms, theory, 95–98
−− collector, 145
Electron, energy distribution, 99
−−− in He, 104
Electron, spectrometer, 167–171
−− electronics, 172–176
−− error analysis, 176–180
−− operation 180–182
Electron temperature, 222

Fokker–Planck equation, 64–69

Geiger–Mueller counter, 108
Generalized oscillator strength, 184
Guard-electrode systems, 112

He, metastable levels, 590–596
He, elastic scattering of electrons by, 445–448
He, differential cross sections for ionization, 115–195, 198–205
He, drift velocity of electrons in, 127–153
He, line profiles 351, 376
He-like ions, energy levels, 265
−− transition probabilities, 267–268
−− excitation rates, 268–271
−− in low density plasmas, 271–274
−− in high density plasmas, 275–279
H–α line profile, 375
H-deficient stars, 383–384
H^-, electron affinity of, 439
− elastic scattering of electrons by, 453–454
− photodetachment, 457–458, 490–491

Impact approximation, in line broadening, 340
Inelastic scattering of electrons, by positive ions, 232, 244, 270
−−−− theory, 354–361
−−−− by hydrogen, 443–445
Ionization, by electron impact, theory, 196–199
− near threshold (experimental), 202–206
− from $n = 2$ levels of He-like ions, 273

Kinematics of ionizing collisions, 162

Line profiles, 333–336
Line broadening, 336–353
Lithium-like ions, energy levels, 230
−− transition probabilities, 231
−− excitation rates, 232–234
−− line intensities (laboratory), 235
−− line intensities (solar), 237–240
−− dielectronic satellite lines, 240
Lindholm–Foley theory of broadening, 338
Local thermodynamic equilibrium (LTE), 305 ff.
Ly–α profile, 349

Magnetic field deflection coefficient, 102–103
Markov processes, 66–70
Mean free path, 7
Metastable levels, in astrophysical plasma, 214
−− triplet in Ne VIII, 253
−− in collision processes, 549–624
−− (see also chemiionization)

Natanson theory of recombination, 78–80
−−−− modifications of, 80–83
Neon, elastic scattering of electrons by, 450–451
− drift velocity of electrons in, 128

Opacity, in solar photosphere, 370
− due to H_2O in cool stars, 380
− due to C^-, 381–386
Oscillator strengths, 216, 324–326

Photodetachment of negative ions, 487–491
−−−− near threshold, 492–496
−−−− table of cross sections, 488–489
−−−− double process, 491
−−−− in astrophysics, 322–328, 381
−−−− of H^-, 325, 457
−−−− of C^- and Cl^-, 327
Photoionization, 456–459
− of O II, 321
− of Si, 322
Polarizability, formula for, 324
Polarized orbital approximation, 401–477
−−− scattering from one electron targets, 433–445
−−− scattering from helium, 445–448
−−− positron scattering, 462–477

– – – electron scattering by molecules, 454–456

Quantum defect method, 313–316
Quasars, electron density in, 264

Radiative transfer, equation of, 298
– – theory, 292–299
Rate coefficient for three body recombination 3, 34–36
– – for excitation, 215
– – for electron–ion collisions, 219–222
– – for dielectronic recombination, 220
Recoil peak, in differential scattering, 195
Recombination, three-body, 1–90
– statistical analysis, 9, 13–19, 65–76
– of ions in their parent gas, 19–28
– with an arbitrary third body, 29–41
– temperature variation, 55–57
– for oxygen ions in O_2, 65, 82–84
– for NO^+, NO_2^-, 64
Recombination, dielectronic, 220

Satellite lines, 222
Scattering matrix, 336
Semi-classical approximation for cross-sections, 361
Sloan term, in theory of p-wave scattering, 435–436
Stray fields, in electron spectrometer, 179
Statistical equilibrium, for excited states, 212–214
Stellar atmosphere models, 296–312
Sternheimer's method, 421–423

Theta-pinch experiments, 227–229
Thomson theory of recombination, 85–88
Townsend energy factor, 101
Transition probabilities, 329–333
– – in Li-like ions, 231
– – in Be-like ions, 244
– – in He-like ions, 267
Transport coefficients, for electrons in gases 94–100
– – and cross sections, 103–107

SUBJECT INDEX

— electron scattering by molecules, 254, 256

Quantum defect method, 312–349
Quasarc, electron density in, 264

R-additive transfer, equation of, 235
— theory, 292–296
Rate coefficient for three body recombination, 2, 64–66
— for excitation, 215
— for electron–ion collisions, 219–222
— for dielectronic recombination, 220
Recoil peak, in differential scattering, 193
Recombination, three-body, 1–90
— cross-sections, 9, 12–17, 65–70
— of ions in their parent gas, 19–28
— with an arbitrary third body, 29–41
— temperature variation, 54–57
— for oxygen ions in O$_2$, 63, 82–84
— for NO, NO$_2^+$, 90
Recombination, dielectronic, 220

Satellite lines, 222
Scattering matrix, 336
Semi-classical approximation for cross-sections, 261
Sloan term, in theory of p-wave scattering, 415–416
Stray fields in electron spectrometer, 179
Statistical equilibrium, for excited states, 212–214
Stellar atmosphere models, 296–312
Sternheimer's method, 421–423

Townsend experiments, 217–219
Thomson theory of recombination, 85–88
Townsend energy factor, 101
Transition probabilities, 229–231
— in Li-like ions, 231
— in Be-like ions, 241
— in He-like ions, 247
Transport coefficients, for electrons in gases, 94–100
— and cross-sections, 103–107

QC
173
M224
v.2

JUN 3 1977